W9-AGN-947

PARALLELOGRAM

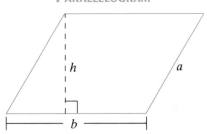

Perimeter: $P = 2a + 2b$
Area: $A = bh$

CIRCLE

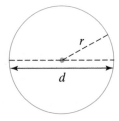

Circumference: $C = \pi d$
$C = 2\pi r$
Area: $A = \pi r^2$

RECTANGULAR SOLID

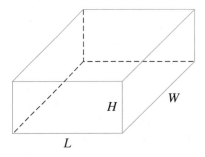

Volume: $V = LWH$
Surface Area: $A = 2HW + 2LW + 2LH$

CUBE

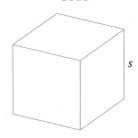

Volume: $V = s^3$

Surface Area: $A = 6s^2$

CONE

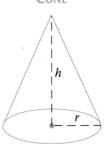

Volume: $V = \frac{1}{3}\pi r^2 h$

Surface Area: $A = \pi r\sqrt{r^2 + h^2}$

RIGHT CIRCULAR CYLINDER

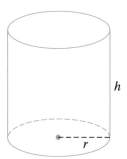

Volume: $V = \pi r^2 h$

Surface Area:
$A = 2\pi rh + 2\pi r^2$

OTHER FORMULAS

Distance: $d = rt$ (r = rate, t = time)

Temperature: $F = \frac{9}{5}C + 32$ $C = \frac{5}{9}(F - 32)$ (F = degrees Fahrenheit, C = degrees Celsius)

Simple Interest: $I = Prt$
  (P = principal, r = annual interest rate, t = time in years)

Compound Interest: $A = P\left(1 + \frac{r}{n}\right)^{nt}$

  (P = principal, r = annual interest rate, t = time in years, n = number of compoundings per year)

INTERMEDIATE ALGEBRA

SECOND EDITION

K. ELAYN MARTIN-GAY

University of New Orleans

This Annotated Instructor's Edition is exactly like your students' text, but it contains all exercise answers. Where possible the answer is displayed on the same page as the exercise. Exercise answers that require graphical solutions, and both answers and suggestions for instructional strategies for the group activities, can be found in the Instructor's Answers section in Appendix D.

The Student Edition of the text contains answers to selected exercises. Located in the back of the Student Edition are answers to odd-numbered exercises, all chapter test exercises, and all cumulative review exercises, including text references. It does not contain answers to even-numbered exercises, group activities, or instructor's notes.

Prentice Hall
Upper Saddle River
New Jersey 07458

Sponsoring Editors, Melissa Acuña and Ann Marie Jones
Editor-in-Chief, Jerome Grant
Editorial Director, Tim Bozik
Production Editor, Barbara Mack
Managing Editor, Linda Behrens
Assistant Vice President of Production and Manufacturing, David W. Riccardi
Executive Managing Editor, Kathleen Schiaparelli
Development Editor, Karen Karlin
Editor-in-Chief of Development, Ray Mullaney
Marketing Manager, Jolene Howard
Creative Director, Paula Maylahn
Art Director, Amy Rosen
Assistant to Art Director, Rod Hernandez
Art Manager, Gus Vibal
Interior Design, Geri Davis, The Davis Group, Inc.
Cover Design, Bruce Kenselaar
Photo Editor, Lori Morris-Nantz
Photo Research, Rona Tuccillo
Manufacturing Buyer, Alan Fischer
Manufacturing Manager, Trudy Pisciotti
Supplements Editor/Editorial Assistant, April Thrower
Cover Photo, Art Wolfe/Tony Stone Images

© 1997, 1993 by Prentice-Hall, Inc.
Simon & Schuster/ A Viacom Company
Upper Saddle River, New Jersey 07458

All rights reserved. No part of this book may be reproduced, in any form or by any means, without permission in writing from the publisher.

Photo credits appear on page P-1, which constitutes a continuation of the copyright page.

Printed in the United States of America

10   9   8   7   6   5   4   3   2

ISBN 0-13-258005-5

Prentice-Hall International (UK) Limited, *London*
Prentice-Hall of Australia Pty. Limited, *Sydney*
Prentice-Hall Canada Inc., *Toronto*
Prentice-Hall Hispanoamericana, S.A., *Mexico City*
Prentice-Hall of India Private Limited, *New Delhi*
Prentice-Hall of Japan, Inc., *Tokyo*
Simon & Schuster Asia Pte. Ltd., *Singapore*
Editora Prentice-Hall do Brasil, Ltda., *Rio de Janeiro*

To my husband, Clayton,
and our sons,
Eric and Bryan

CONTENTS

7

RATIONAL EXPONENTS, RADICALS, AND COMPLEX NUMBERS 413

8

QUADRATIC EQUATIONS AND FUNCTIONS 469

9

CONIC SECTIONS 527

PREFACE

ABOUT THIS BOOK

This book was written to provide students with a solid foundation in algebra as well as to help develop their problem-solving skills. Specific care has been taken to prepare students to go on to their next course in mathematics and to help students succeed in nonmathematical courses that require a grasp of algebraic fundamentals. The basic concepts of graphs and functions are introduced early. These concepts, along with problem solving, data interpretation, and geometric concepts, are emphasized and integrated throughout the book.

The many factors that contributed to the success of the first edition have been retained. In preparing this edition, I considered the comments and suggestions of colleagues throughout the country and of the many users of the first edition. The American Mathematical Association of Two-Year Colleges (AMATYC) Crossroads in Mathematics: Standards for Introductory College Mathematics before Calculus and the NCTM Standards (plus Addenda), together with advances in technology, also influenced the careful reexamination of every section of the text. All of these inputs helped me update the presentation, enhancing the content and pedagogical value.

KEY PEDAGOGICAL FEATURES OF THE SECOND EDITION

Problem-Solving Process This is formally introduced in Chapter 2, with a new six-step process that is integrated throughout the text. The six steps are UNDERSTAND, ASSIGN, ILLUSTRATE (including diagrams appropriately labeled with variables), TRANSLATE, COMPLETE, and INTERPRET. The repeated use of these steps in a variety of examples shows their wide applicability. Reinforcing the steps can increase students' comfort level and confidence in tackling problems.

Exercise Sets The exercise sets are graded in difficulty and include computational, conceptual, and applied problems. The first few exercises in each set are carefully keyed to worked examples in the text. A student can quickly gain confidence and then move on to the remaining exercises, which are not keyed to examples. Many exercises are new to the second edition. There are ample exercises throughout the book, including end-of-chapter reviews, tests, and cumulative reviews. In addition, each exercise set contains one or more of the following features.

Mental Math These problems are found at the beginning of an exercise set. They are mental warmups that reinforce concepts found in the accompanying section and increase students' confidence before they tackle an exercise set. By relying on their own mental skills, students increase not only their confidence in themselves, but also their number sense and estimation ability.

Conceptual and Writing Exercises These exercises, now found in almost every exercise set, are keyed with the icon ⬚. They call on students to demonstrate an understanding of a concept learned in the corresponding section. To do so, students are asked questions that require them to use two or more concepts together. Some exercises ask students to stop, think, and explain in their own words the concept(s) used in the exercises they have just completed. Guidelines recommended by AMATYC and other professional groups suggest that writing be incorporated into mathematics courses to reinforce concepts.

Data and Graphical Interpretation There is increased emphasis on data interpretation in exercises via tables and graphs. The ability to interpret data and read and create a variety of types of graphs is developed gradually so that students become comfortable with it.

Scientific Calculator Explorations and Exercises Scientific Calculator Explorations, although optional, contain examples and exercises to reinforce concepts or to motivate discovery learning. This feature is placed appropriately throughout the text to instruct students on the proper use of the calculator.

Additional exercises building on the skills developed in the Explorations may be found in exercise sets throughout the text and are marked with an icon ▦. The inside back cover of the text includes a quick reference to selected keys on a scientific calculator.

Graphing Calculator Explorations and Exercises For graphing calculators or computer graphing utilities, these new Explorations are integrated appropriately throughout the text. They contain optional examples and exercises to reinforce concepts, to help interpret graphs, or to motivate discovery learning.

Additional new exercises building on the skills developed in the Explorations may be found in exercise sets throughout the text and are marked with an icon ▦. The inside back cover of the text includes a quick reference to selected keys on a graphing calculator. In addition, a new appendix, "An Introduction to Using a Graphing Utility," has been added, along with exercises.

Review Exercises Formerly called Skill Review, these exercises are found at the end of each section after Chapter 1. These problems are keyed to earlier sections and review concepts that students learned earlier in the text and that are needed in the next section or in the next chapter. These exercises show the connections between earlier topics and later material.

*A **Look Ahead*** These are examples and problems similar to those found in a college algebra course. "A Look Ahead" is presented as a natural extension of the material and contains an example followed by advanced exercises. I suggest that any student who plans to take another algebra course work these problems.

Graphics The text contains numerous graphics, models, and illustrations to clarify visually and reinforce concepts and aid in problem solving. These include new bar charts, line graphs, calculator screens, application illustrations, and geometric figures. The text's inside front cover includes a quick reference to geometric figures and formulas, and the inside back cover now includes a summary of common graphs.

Applications and Connections This book contains a wealth of practical applications in worked-out examples and exercise sets. The applications, located throughout the book, help to reinforce problem-solving skills and strengthen students' understanding of mathematics in the real world. They show connections to a wide range of areas, such as biology, environmental issues, consumer applications, allied health, business, entertainment, history, art, literature, finance, sports, and music, as well as to related mathematical areas, such as geometry. Many involve interesting real-life data. Sources for data include newspapers, magazines, government publications, and reference books. Opportunities for obtaining your own real data are also included.

Group Activities Each chapter opens with a photograph and description of a real-life situation. At the close of the chapter, students can work cooperatively to apply the algebraic and critical-thinking skills they have learned to make decisions and answer the Group Activity that is related to the chapter-opening situation. These new Group Activities are multi-part, often hands-on, problems. These situations, designed for student involvement and interaction, allow for a variety of teaching and learning styles. Answers and tips for instuctional strategies for Group Activities are available in the Annotated Instructor's Edition. In addition, there are opportunities for group activities within section exercise sets.

Reminder Reminder boxes, formerly Helpful Hint boxes, contain practical advice on problem solving. Reminders appear in the context of material in the chapter and give students extra help in understanding and working problems. They are highlighted in a box for quick reference.

Chapter Highlights Found at the end of each chapter, the new Chapter Highlights contain key definitions, concepts, and examples to help students summarize and retain what they have learned.

Chapter Review and Chapter Test The end of each chapter contains a review of topics introduced in the chapter. These review problems are keyed to sections. The chapter test is not keyed to sections.

Cumulative Review Each chapter after the first contains a Cumulative Review. Each Cumulative Review problem is actually an earlier worked example in the text that is referenced in the back of the book along with the answer. Students who need to see a complete worked-out solution, with explanation, can do so by turning to the appropriate example in the text.

Readability and Connections Many reviewers of this edition as well as users of the previous edition have commented favorably on the readability and clear, organized presentation. I have tried to make the writing style as clear as possible while still retaining the mathematical integrity of the content. When a new topic is presented, I have attempted to relate the new ideas to those the students may already know. Constant reinforcement and connections within problem-solving strategies, data interpretation, geometry, patterns, graphs, and situations from everyday life can help students gradually master both new and old information.

Functional Use of Color and Design Elements of the text are highlighted with color and designed to make it easier for students to read and study.

Videotape and Software Icons At the beginning of each section, videotape and software icons are displayed. The icons help reinforce that these learning aids are available should students wish to use them in reviewing concepts and skills at their own pace. These items have direct correlation with the text and emphasize the text's methods of solution.

KEY CONTENT FEATURES OF THE SECOND EDITION

Overview In addition to the traditional topics in intermediate algebra courses, this text contains an early and intuitive introduction to graphs and functions and a strong emphasis on problem solving. Geometric concepts and reading and interpreting graphs and data are integrated throughout. The geometric concepts covered are those most important to a student's understanding of algebra, and I have included many applications and exercises devoted to this topic. Geometric figures and a review of angles, lines, and special triangles are covered in the appendices. Students are also given the opportunity to see how today's technology can help. Exercises are a critical part of student learning, and particular care was taken in writing these.

Increased Emphasis on Data Interpretation There is an increased emphasis on data interpretation via tables and graphs that begins in the first section of the book and continues throughout the text. The ability to interpret data and a variety of types of graphs, including bar, line, and circle graphs, is developed gradually so that students become comfortable with it.

Early and Intuitive Introduction to Graphs and Functions As bar and line graphs are gradually introduced in Chapters 1 and 2, an emphasis is slowly placed on the notion of paired data. This notion leads naturally to the concepts of ordered pairs, the rectangular coordinate system, and graphing equations introduced in Chapter 3.

Once students are comfortable with graphing equations, functions are introduced in Chapter 3. The concept of a function is illustrated in numerous ways to ensure student understanding: by listing ordered pairs of data, showing rectangular coordinate system graphs, visually representing set correspondences, and including numerous real-data examples. The importance of a function is continuously reinforced by not treating it as a single, stand-alone topic but by constantly integrating functions in appropriate sections of this text.

Increased Emphasis on Problem Solving Building on the strengths of the first edition, a special emphasis and strong commitment is given to contemporary and practical applications of algebra. The range of problem-solving techniques has also been expanded to include opportunities for using graphing utilities to help solve applications and for working with more real data.

Increased Opportunities to Use Technology As we noted, optional Scientific Calculator as well as Graphing Calculator Explorations are integrated appropriately throughout the text. A new appendix, "An Introduction to Using a Graphing Utility," is included, along with exercises.

New Examples Additional detailed step-by-step examples were added where needed. Many of these reflect real-life situations. Examples are used in two ways—numbered, as formal examples, and unnumbered, to introduce a topic or informally discuss the topic.

New Exercises A significant amount of time was spent on the exercise sets. New exercises and additional examples help address a wide range of student learning styles and abilities. New kinds of exercises include group activities, conceptual and writing exercises, multi-part exercises, optional graphing calculator exercises, and data analysis from tables and graphs. In addition, the mental math, drill, and word problems were refined and enhanced.

SUPPLEMENTS FOR THE INSTRUCTOR

PRINTED SUPPLEMENTS

Annotated Instructor's Edition (ISBN 0-13-258005-5)

- Answers to exercises on the same text page or in Instructor's Answers section
- Instructor's Answers section contains answers to exercises requiring graphical solutions
- Instructor's Answers section also contains answers and pedagogical suggestions for group activities
- Notes to the Instructor

Instructor's Solutions Manual (ISBN 0-13-258047-0)

- Solutions to even-numbered exercises, Chapter Tests, and Cumulative Review exercises
- Graphics computer-generated for clarity
- Answers checked for accuracy

Test Item File (ISBN 0-13-258013-6)

- Six forms (A, B, C, D, E, and F) of Chapter Tests
  —three forms contain multiple-choice items
  —three forms contain free-response items
- Two forms of Cumulative Review Tests
  —every two chapters
- Final Exams
  —four forms with free-response scrambled items
  —four forms with multiple-choice scrambled items
- Answers to all items

MEDIA SUPPLEMENTS

TestPro3 Computerized Testing (Sample Disk IBM, ISBN 0-13-058104-3; Sample Disk Mac, ISBN 0-13-258112-4; IBM, ISBN 0-13-258021-7; Mac, ISBN 0-13-258039-X)

- Comprehensive text-specific testing
- Generates test questions and drill worksheets from algorithms keyed to the text learning objectives
- Edit or add your own questions
- Compatible with Scantron or other possible scanners

USING THE INTERNET AND A WEB BROWSER

Using the Internet and a Web browser, such as Netscape, can add to your mathematical resources. The following is a list of some of the sites that may be worth your or your students' visit.

- Prentice Hall Home Page   http://www.prenhall.com
- The Mathematical Association of America   http://www.maa.org
- The American Mathematical Society   http://www.ams.org
- The National Council of Teachers of Mathematics   http://www.nctm.org
- The Census Bureau   http://www.census.gov
- Texas Instruments   http://www.ti.com/calc

INTERNET GUIDE

- This guide provides a brief history of the Internet, discusses the use of the World Wide Web, and describes how to find your way within the Internet and how to reach others on it. Contact your local Prentice Hall representative for the Internet Guide.

SUPPLEMENTS FOR THE STUDENT

PRINTED SUPPLEMENTS

Student Solutions Manual (ISBN 0-13-258096-9)

- Detailed step-by-step solutions to odd-numbered text and review exercises
- Solutions to all chapter practice tests and Cumulative Review exercises
- Solution methods reflect those emphasized in the text.
- Includes study skills and note-taking suggestions
- Ask your bookstore about ordering.

Study Guide (ISBN 0-13-258088-8)

- Additional step-by-step worked-out examples and exercises
- Practice tests and final examinations
- Solution methods reflect those emphasized in the text.
- Includes study skills and note-taking suggestions
- Ask your bookstore about ordering.

The New York Times Supplement

- A free newspaper from Prentice Hall and *The New York Times*
- Interesting and current articles on mathematics
- Invites discussion and writing about mathematics
- Created new each year

How to Study Math (ISBN 0-13-020884-1)

MEDIA SUPPLEMENTS

Videotape Series (Sample Video, ISBN 0-13-258146-9; Video Series, ISBN 0-13-258070-5)

- Specifically keyed to the text by section
- Presentation and step-by-step examples by K. Elayn Martin-Gay
- Comprehensive coverage

MathPro Tutorial Software
(IBM Network-User, ISBN 0-13-258054-3;
IBM Single-User, ISBN 0-13-268814-X;
Mac Network-User, ISBN 0-13-258062-4;
Mac Single-User, ISBN 0-13-281552-4)

- Text-specific tutorial exercises
- Interactive feedback
- Unlimited practice Warm-up Exercises
- Graded and recorded Practice Problems
- New user interface, glossary, and expressions editor for ease of use and flexibility

ACKNOWLEDGMENTS

First, as usual, I would like to thank my husband, Clayton, for his constant encouragement. I would also like to thank my children, Eric and Bryan, for continuing to eat my burnt bacon and cookies even though they now realize that they are "extra crispy."

I would also like to thank my extended family for their invaluable help and wonderful sense of humor. Their contributions are too numerous to list. They are Peter, Karen, Michael, Christopher, Matthew, and Jessica Callac; Stuart, Earline, Melissa, and Mandy Martin; Mark, Sabrina, and Madison Martin; Leo and Barbara Miller; and Jewett Gay.

A special thank you to all the users of the first edition of this text who made suggestions for improvements that were incorporated into the second edition. I would also like to thank the following reviewers of this text:

Carol Bernath *Tallahassee Community College*
Larry Blevins, *Tyler Junior College*
Douglas Cameron, *University of Akron*
Dennis Carrie, *Golden West College*
Camille Cochrane, *Shelton State Community College*
Linda F. Crabtree, *Longview Community College*
Gregory Davis, *University of Wisconsin–Green Bay*
Irene Doo, *Austin Community College*
James V. Frugale, *Middlesex Community–Technical College*
D. Michael Hamm, *Brookhaven College*
Doug Jones, *Tallahassee Community College*
Rosemary M. Karr, *Collin County Community College*
Harvey Lambert, *University of Nevada–Reno*
Peter A. Lindstrom, *North Lake College*
Marilyn Massey-Moss, *Collin County Community College*
Sofya Nayer, *Borough of Manhattan Community College*
Frank Rivera, *Glendale Community College*
Patty Sheeran Schug, *McHenry County College*
Linda Schultz, *McHenry County College*
Arthur E. Schwartz, *Mercer County Community College*

Laurence Small, *L.A. Pierce College*
John Squires, *Cleveland St. Community College*
Sharon Taylor-Riley, *Wayne County Community College*
Sally Vestal, *Southwest Missouri State University*
Najib Yazbak, *Wayne County Community College*

Cheryl Roberts, James Sellers, and Phyllis Barnidge did an excellent job of providing answers and solutions, and contributing to the overall accuracy of the book. Karen Karlin was invaluable for her many suggestions during the development of the second edition. I very much appreciated the writers and accuracy checkers of the supplements to accompany this text as well as Emily Keaton's contributions. Last, but by no means least, a special thanks to the staff at Prentice Hall for their support and assistance: Melissa Acuña, Ann Marie Jones, Barbara Mack, Linda Behrens, Alan Fischer, Amy Rosen, Gus Vibal, Paula Maylahn, April Thrower, Evan Girard, Jolene Howard, Gary June, Jerome Grant, and Tim Bozik.

K. Elayn Martin-Gay

ABOUT THE AUTHOR

K. Elayn Martin-Gay has taught mathematics at the University of New Orleans for over 16 years. She has received numerous teaching awards, including the local University Alumni Association's Award for Excellence in Teaching.

Over the years, Elayn has developed a videotaped lecture series to help her students better understand algebra. This highly successful video material is the basis of the four-book series *Prealgebra, Beginning Algebra, Intermediate Algebra,* and *Introductory and Intermediate Algebra,* a combined approach.

HOW TO USE THE TEXT: A GUIDE FOR STUDENTS

Intermediate Algebra, Second Edition, has been designed as one of several tools in a fully integrated learning package to help you develop intermediate algebra skills. Our goal is to encourage your success and mastery of the mathematical concepts introduced in this text. Take a few moments to see how this text will help you excel.

CHAPTER

4

4.1 SOLVING SYSTEMS OF LINEAR EQUATIONS IN TWO VARIABLES

4.2 SOLVING SYSTEMS OF LINEAR EQUATIONS IN THREE VARIABLES

4.3 SYSTEMS OF LINEAR EQUATIONS AND PROBLEM SOLVING

4.4 SOLVING SYSTEMS OF EQUATIONS BY MATRICES

4.5 SOLVING SYSTEMS OF EQUATIONS BY DETERMINANTS

SYSTEMS OF EQUATIONS

LOCATING LIGHTNING STRIKES

Lightning, most often produced during thunderstorms, is a rapid discharge of high-current electricity into the atmosphere. Around the world, lightning occurs at a rate of approximately 100 flashes per second. Because of lightning's potentially destructive nature, meteorologists track lightning activity by recording and plotting the positions of lightning strikes.

IN THE CHAPTER GROUP ACTIVITY ON PAGE 229, YOU WILL HAVE THE OPPORTUNITY TO PINPOINT THE LOCATION OF A LIGHTNING STRIKE.

The photo application at the opening of every chapter and **applications throughout** offer real-world scenarios that connect mathematics to your life. In addition, at the end of the chapter, a group activity or discovery-based project further shows the chapter's applicability.

Page 185

APPLY THE PROBLEM-SOLVING PROCESS

As you study, **make connections**–this text's organization can help you. There are features in this text designed to help you relate material you are learning to previously mastered material. Math topics are tied to real life as often as possible. Key learning objectives are introduced and easily identified in every section.

PROBLEM-SOLVING STEPS

1. UNDERSTAND the problem. During this step, don't work with variables, but simply become comfortable with the problem. Some ways of accomplishing this are listed next.

   - Read and reread the problem.
   - Construct a drawing to help visualize the problem.
   - Propose a solution and check. Pay careful attention as to how you check your proposed solution. This will help later when you write an equation to model the problem.

2. ASSIGN a variable to an unknown in the problem. Use this variable to represent any other unknown quantities.

3. ILLUSTRATE the problem. A diagram or chart *using the assigned variables* can often help us visualize the known facts.

4. TRANSLATE the problem into a mathematical model. This is often an equation.

5. COMPLETE the work. This often means to solve the equation.

6. INTERPRET the results. *Check* the proposed solution in the stated problem and *state* your conclusion.

Page 52

Save time by having a plan. Follow this **six-step process**, and you will find yourself successfully solving a wide range of problems.

REMINDER   Note that $f(x)$ is a special symbol in mathematics used to denote a function. The symbol $f(x)$ is read "f of x." It does **not** mean $f \cdot x$ (f times x).

"**Reminders**" contain practical advice and assist you in understanding and working problems.

VISUALIZE . . . SEE THE CONCEPTS!

Graphing is introduced early and intuitively. Fully integrated in all appropriate sections with an increased emphasis on **data interpretation** via tables and graphs, this concept is developed gradually throughout the text. Knowing how to use data and graphs is a valuable skill in the workplace as well as in other courses.

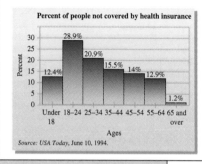

87. What percent of people aged 18 to 34 have no health insurance?

Percent of people not covered by health insurance

Source: *USA Today*, June 10, 1994.

Page 341

Real Data are integrated throughout the text, drawn from familiar sources such as magazines and newspapers.

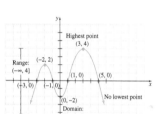

f. The y-values are greater than 0 when the graph lies above the x-axis. The x-values when this occurs are between $x = -3$ and $x = -1$ and between $x = 1$ and $x = 5$.

Page 311

Many graphics, models and illustrations provide **visual reinforcement** of concepts.

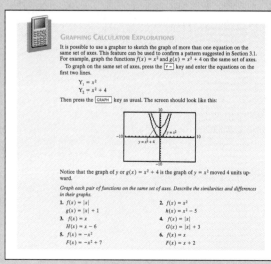

GRAPHING CALCULATOR EXPLORATIONS

It is possible to use a grapher to sketch the graph of more than one equation on the same set of axes. This feature can be used to confirm a pattern suggested in Section 3.1. For example, graph the functions $f(x) = x^2$ and $g(x) = x^2 + 4$ on the same set of axes. To graph on the same set of axes, press the $\boxed{Y=}$ key and enter the equations on the first two lines.

$$Y_1 = x^2$$
$$Y_2 = x^2 + 4$$

Then press the $\boxed{GRAPH}$ key as usual. The screen should look like this:

Notice that the graph of y or $g(x) = x^2 + 4$ is the graph of $y = x^2$ moved 4 units upward.

Graph each pair of functions on the same set of axes. Describe the similarities and differences in their graphs.

1. $f(x) = |x|$
   $g(x) = |x| + 1$
2. $f(x) = x^2$
   $h(x) = x^2 - 5$
3. $f(x) = x$
   $H(x) = x - 6$
4. $f(x) = |x|$
   $G(x) = |x| + 3$
5. $f(x) = -x^2$
   $F(x) = -x^2 + 7$
6. $f(x) = x$
   $F(x) = x + 2$

Page 131

Scientific and Graphing Calculator Explorations and exercises are woven into the appropriate sections to reinforce concepts and **motivate discovery-based learning.**

CHECK YOUR UNDERSTANDING. EXPAND IT. EXPLORE!

Good exercise sets are essential to the make-up of a solid intermediate algebra textbook. The exercises in this textbook are designed to help you understand skills and concepts as well as challenge and motivate you. Note, too, the Highlights, Test, Review, and Cumulative Review found at the end of each chapter.

MENTAL MATH

Use positive exponents to state each expression.

1. $5x^{-1}y^{-2}$ **2.** $7xy^{-4}$ **3.** $a^2b^{-1}c^{-5}$ **4.** $a^{-4}b^2c^{-6}$ **5.** $\dfrac{y^{-2}}{x^{-4}}$ **6.** $\dfrac{x^{-7}}{z^{-3}}$

Confidence-building **Mental Math** problems are in many sections.

Page 249

A Look Ahead examples and problems are similar to those found in the *next* algebra course and include more advanced exercises.

Conceptual and Writing Exercises bring together two or more concepts and often require "in your own words" written explanation.

The accompanying graph shows the daily low temperatures for one week in New Orleans, Louisiana.

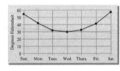

67. Which day of the week shows the greatest decrease in temperature low?

68. Which day of the week shows the greatest increase in temperature low?

69. Which day of the week had the lowest temperature?

70. Use the graph to estimate the low temperature on Thursday.

Notice that the shape of the temperature graph is similar to a parabola (see Section 5.9). In fact, this graph can be modeled by the quadratic function $f(x) = 3x^2 - 18x + 57$, where $f(x)$ is the temperature in degrees Fahrenheit and x is the number of days from Sunday. Use this function to answer Exercises 71 and 72.

71. Use the quadratic function given to approximate the temperature on Thursday. Does your answer agree with the graph above?

72. Use the function given and the quadratic formula to find when the temperature was 35°F. [*Hint:* Let $f(x) = 35$ and solve for x.] Round your answer to one decimal place and interpret your result. Does your answer agree with the graph above?

73. Use a grapher to solve Exercises 61 and 63.

74. Use a grapher to solve Exercises 62 and 64.

Review Exercises

Solve each equation. See Sections 6.7 and 7.6.

75. $\sqrt{5x - 2} = 3$ **76.** $\sqrt{y + 2} + 7 = 12$

77. $\dfrac{1}{x} + \dfrac{2}{5} = \dfrac{7}{x}$ **78.** $\dfrac{10}{z} = \dfrac{5}{z} - \dfrac{1}{3}$

Factor. See Section 5.7.

79. $x^4 + x^2 - 20$ **80.** $2y^4 + 11y^2 - 6$

81. $z^4 - 13z^2 + 36$ **82.** $x^4 - 1$

A Look Ahead

EXAMPLE

Solve $x^2 - 3\sqrt{2}x + 2 = 0$.

Solution:

In this equation, $a = 1$, $b = -3\sqrt{2}$, and $c = 2$. By the quadratic formula, we have

$$x = \frac{-b \pm \sqrt{b^2 - 4ac}}{2a}$$
$$= \frac{3\sqrt{2} \pm \sqrt{(-3\sqrt{2})^2 - 4(1)(2)}}{2(1)}$$
$$= \frac{3\sqrt{2} \pm \sqrt{18 - 8}}{2} = \frac{3\sqrt{2} \pm \sqrt{10}}{2}$$

The solution set is $\left\{ \dfrac{3\sqrt{2} + \sqrt{10}}{2}, \dfrac{3\sqrt{2} - \sqrt{10}}{2} \right\}$.

Use the quadratic formula to solve each quadratic equation. See the preceding example.

83. $3x^2 - \sqrt{12}x + 1 = 0$

84. $5x^2 + \sqrt{20}x + 1 = 0$

85. $x^2 + \sqrt{2}x + 1 = 0$

86. $x^2 - \sqrt{2}x + 1 = 0$

87. $2x^2 - \sqrt{3}x - 1 = 0$

88. $7x^2 + \sqrt{7}x - 2 = 0$

Page 487

---

21. In your own words, explain how to find x- and y-intercepts.

22. Explain why it is a good idea to use three points to graph a linear equation.

Graph each linear equation. See Examples 6 and 7.

23. $x = -1$ **24.** $y = 5$ **25.** $y = 0$

26. $x = 0$ **27.** $y + 7 = 0$ **28.** $x - 3 = 0$

Match each equation with its graph.

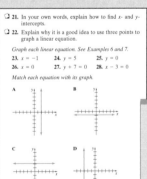

29. $y = 2$ **30.** $x = -3$

31. $x - 2 = 0$ **32.** $y + 1 = 0$

33. Discuss whether a vertical line ever has a y-intercept.

34. Discuss whether a horizontal line ever has an x-intercept.

Build your confidence with the beginning exercises; the first part of each exercise set is keyed to already worked examples. Then try the remaining exercises.

Page 143

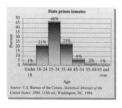

Solve each equation. Round solutions to two decimal places.

68. $\frac{1.4}{x - 2.6} = \frac{-3.5}{x + 7.1}$ **69.** $\frac{-8.5}{x + 1.9} = \frac{5.7}{x - 3.6}$

70. $\frac{10.6}{y} - 14.7 = \frac{9.92}{3.2} + 7.6$

71. $\frac{12.2}{x} + 17.3 = \frac{9.6}{x} - 14.7$

Use a grapher to verify the solution of each given exercise.

72. Exercise 20 **73.** Exercise 21
74. Exercise 30 **75.** Exercise 31

Review Exercises

Write each sentence as an equation and solve. See Section 2.2.

76. Four more than 3 times a number is 19.
77. The sum of two consecutive integers is 147.
78. The length of a rectangle is 5 inches more than the width. Its perimeter is 50 inches. Find the length and width.
79. The sum of a number and its reciprocal is $\frac{5}{2}$.

Simplify the following. See Section 1.4.

80. $-|-6 - (-5)|$ **81.** $\sqrt{49} - (10 - 6)^2$
82. $|4 - 8| + (4 - 8)$ **83.** $(-4)^2 - 5^2$

The following is from a survey of state prisons. Use this histogram to answer Exercises 84 through 88.

Page 381

Scientific and Graphing Calculator exercises are found within most exercise sets.

Review Exercises review concepts learned earlier that are needed in the next section or next chapter.

CHAPTER 5 HIGHLIGHTS

DEFINITIONS AND CONCEPTS	EXAMPLES
SECTION 5.1 EXPONENTS AND SCIENTIFIC NOTATION	
Product rule: $a^m \cdot a^n = a^{m+n}$	$x^2 \cdot x^3 = x^5$
Zero exponent: $a^0 = 1, a \neq 0$	$7^0 = 1, (-10)^0 = 1$
Quotient rule: $\frac{a^m}{a^n} = a^{m-n}$	$\frac{y^{10}}{y^4} = y^{10-4} = y^6$
Negative exponent: $a^{-n} = \frac{1}{a^n}$	$3^{-2} = \frac{1}{3^2} = \frac{1}{9}, \frac{x^2}{x^5} = x^{2-(-5)} = x^7$
A positive number is written in **scientific notation** if it is written as the product of a number a, where $1 \leq a < 10$, and an integer power of 10: $a \times 10^r$.	Numbers written in scientific notation: $568,000 = 5.68 \times 10^5$ $0.0002117 = 2.117 \times 10^{-4}$
SECTION 5.2 MORE WORK WITH EXPONENTS AND SCIENTIFIC NOTATION	
Power rules: $(a^m)^n = a^{m \cdot n}$	$(7^8)^2 = 7^{16}$

Page 321

Chapter Highlights contain key definitions, concepts, *and* examples to help you understand and recall what you have learned.

GROUP ACTIVITY

LOCATING LIGHTNING STRIKES

Weather-recording station A   Weather-recording station B

MATERIALS:
• Calculator
• Grapher (optional)

Weather-recording stations use a directional antenna to detect and measure the electro-magnetic field emitted by a lightning bolt. The

slope of the line from the station to the lightning strike as $m = -1.732$. Station B computes a slope of $m = 0.577$ from the angle it measured. Use this information to find the equations of the lines connecting each station to the position of the lightning strike.

There is an opportunity to **explore** an exercise that relates to the chapter-opening photo as a group activity or discovery-based project.

STUDENT RESOURCES

Seek out these items to match your personal learning style.

3.1 GRAPHING EQUATIONS

TAPE IA 3.1

OBJECTIVES

1. Plot ordered pairs.
2. Determine whether an ordered pair of numbers is a solution to an equation in two variables.
3. Graph linear equations.
4. Graph nonlinear equations.

When two varying quantities are measured simultaneously and repeatedly, we can record the measurement pairs and try to detect a pattern. If a pattern is recognized, we may be able to express the pattern compactly as a two-variable equation. For example, suppose that the two quantities products sold x and monthly salary y are measured simultaneously and repeatedly for an employee and recorded in the following table.

Page 108

Text-specific videos hosted by the award-winning teacher and author of *Intermediate Algebra, Second Edition,* cover each objective in every chapter section as a supplementary review.

MathPro Tutorial Software, developed around the content of *Intermediate Algebra, Second Edition,* provides interactive warm-up and graded algorithmic practice problems with step-by-step worked solutions.

ALSO AVAILABLE:

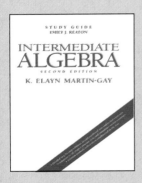

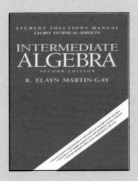

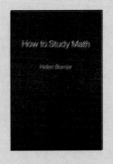

The New York Times/Themes of the Times

Newspaper-format supplement—
*ask your professor about
this exciting free supplement!*

REAL NUMBERS AND ALGEBRAIC EXPRESSIONS

ANALYZING NEWSPAPER CIRCULATION

The number of daily newspapers in business in the United States has declined steadily over the past decade, continuing the trend that started in the mid-1970s. At the turn of the century there were roughly 2300 daily newspapers operating, but by 1995 that number had dropped to only 1548. Average overall daily newspaper circulation also continues to decline, from 60,164,500 in 1992 to 59,811,600 in 1993 to 59,305,400 in 1994. (*Source: Editor & Publisher International Yearbooks.*)

IN THE CHAPTER GROUP ACTIVITY ON PAGE 33, YOU WILL HAVE THE OPPORTUNITY TO ANALYZE THE CIRCULATION OF SEVERAL NEW YORK CITY AREA NEWSPAPERS.

I n arithmetic, we add, subtract, multiply, divide, raise to powers, and take roots
   of numbers. In algebra, we add, subtract, multiply, divide, raise to powers, and
   take roots of variables. Letters, such as x, that represent numbers are called
variables. Understanding these algebraic expressions depends on your
understanding of arithmetic expressions. This chapter reviews the arithmetic
operations on real numbers and the corresponding algebraic expressions. Having
done so, we will be prepared to explore how widely and diversely useful these
algebraic expressions are for problem solving.

1.1   ALGEBRAIC EXPRESSIONS AND SETS OF NUMBERS

O B J E C T I V E S

1. Identify and evaluate algebraic expressions.
2. Identify natural numbers, whole numbers, integers, and rational and irrational real numbers.
3. Write phrases as algebraic expressions.

TAPE IA 1.1

1 Recall that letters that represent numbers are called **variables.** An **algebraic expression** is formed by numbers and variables connected by the operations of addition, subtraction, multiplication, division, raising to powers, or taking roots. For example,

$$2x + 3, \quad \frac{x + 5}{6} - \frac{z^5}{y^2}, \quad \text{and} \quad \sqrt{y} - 1.6$$

are algebraic expressions or, more simply, expressions.

   Algebraic expressions occur often during problem solving. For example, suppose that a television commercial for a watch is being filmed on the Golden Gate Bridge. A portion of this commercial consists of dropping a watch from the bridge. In order to determine the best camera angles and also whether the watch will survive the fall, it is important to know the speed of the watch at 1-second intervals. The algebraic expression

$$32t$$

gives the speed of the watch in feet per second for time t.

   To find the speed of the watch at 1 second, for example, we replace the variable t with 1 and perform the indicated multiplication. This process is called **evaluating** an expression, and the result is called the **value** of the expression for the given replacement value.

When $t = 1$ second, $32t = 3 \cdot 1 = 32$ feet per second.

When $t = 2$ seconds, $32t = 32 \cdot 2 = 64$ feet per second.

When $t = 3$ seconds, $32t = 32 \cdot 3 = 96$ feet per second.

EXAMPLE 1   The research department of a flooring company is considering a new flooring design that contains parallelograms. The area of a parallelogram with base b and height h is bh. Find the area of a parallelogram with base 10 centimeters and height 8.2 centimeters.

Solution: Replace b with 10 and h with 8.2 in the algebraic expression bh.

$$bh = 10 \cdot 8.2 = 82 \text{ square centimeters}$$

When evaluating an expression to solve a problem, we often need to think about the kind of number that is appropriate for the solution. For example, if we are asked to determine the maximum number of parking spaces for a parking lot to be constructed, an answer of $98\frac{1}{10}$ is not appropriate because $\frac{1}{10}$ of a parking space is not realistic.

2   Let's review some common sets of numbers and their graphs on a number line. To construct a number line, we draw a line and label a point 0 with which we associate the number 0. This point is called the **origin**. Choose a point to the right of 0 and label it 1. The distance from 0 to 1 is called the **unit distance** and can be used to locate more points. The **positive numbers** lie to the right of the origin, and the **negative numbers** lie to the left of the origin. The number 0 is neither positive nor negative.

Zero

Negative numbers ↓ Positive numbers

1 unit   1 unit   1 unit   1 unit   1 unit   1 unit

−3   −2   −1   0   1   2   3

A number is **graphed** on a number line by shading the point on the number line that corresponds to the number. Some common sets of numbers and their graphs include:

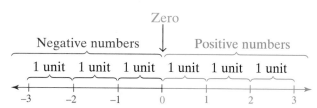

Natural numbers: $\{1, 2, 3, \ldots\}$

Whole numbers: $\{0, 1, 2, 3, \ldots\}$

Integers: $\{\ldots, -3, -2, -1, 0, 1, 2, 3, \ldots\}$

Each listing of three dots above, $\ldots$, is called an **ellipsis** and means to continue in the same pattern.

The members of a set are called its **elements**. When the elements of a set are listed, such as those displayed in the previous paragraph, the set is written in **roster** form. A set can also be written in **set builder notation**, which describes the members of a set but does not list them. The set following is written in set builder notation.

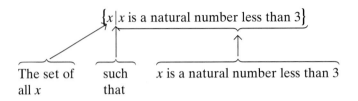

The set of
all x

such
that

x is a natural number less than 3

This same set written in roster form is {1, 2}.

A set that contains *no* elements is called the **empty set**, symbolized by { }, or the **null set**, symbolized by $\varnothing$.

$\{x \mid x$ is a month with 32 days$\}$ is $\varnothing$ or { }

because no month has 32 days. The set has no elements.

> REMINDER   Use { } to write the empty set. $\{\varnothing\}$ is **not** the empty set because it has one element: $\varnothing$.

EXAMPLE 2   List the elements in each set.

a. $\{x \mid x$ is a whole number between 1 and 6$\}$

b. $\{x \mid x$ is a natural number greater than 100$\}$

Solution: **a.** {2, 3, 4, 5}   **b.** {101, 102, 103, . . .}

The symbol $\in$ is used to denote that an element is in a particular set. The symbol $\in$ is read as "is an element of." For example, the true statement

3 is an element of {1, 2, 3, 4, 5}

can be written in symbols as

$3 \in \{1, 2, 3, 4, 5\}$

The symbol $\notin$ is read as "is not an element of." In symbols, we write the true statement "p is not an element of $\{a, 5, g, j, q\}$" as

$p \notin \{a, 5, g, j, q\}$

EXAMPLE 3   Determine whether each statement is true or false.

a. $3 \in \{x \mid x$ is a natural number$\}$ **b.** $7 \notin \{1, 2, 3\}$

Solution: **a.** True, since 3 is a natural number and therefore an element of the set.

b. True, since 7 is not an element of the set {1, 2, 3}.

We can use set builder notation to describe three other common sets of numbers.

Real numbers: $\{x \mid x$ corresponds to a point on the number line$\}$

Rational numbers: $\left\{\dfrac{a}{b} \middle| a \text{ and } b \text{ are integers and } b \neq 0\right\}$

Irrational numbers: $\{x \mid x$ is a real number and x is not a rational number$\}$

Notice that every integer is also a rational number since each integer can be written as the quotient of itself and 1:

$$3 = \frac{3}{1}, \qquad 0 = \frac{0}{1}, \qquad -8 = \frac{-8}{1}$$

Not every rational number, however, is an integer. The rational number $\frac{2}{3}$, for example, is not an integer. Some square roots are rational numbers and some are irrational numbers. For example, $\sqrt{2}$, $\sqrt{3}$, and $\sqrt{7}$ are irrational numbers while $\sqrt{25}$ is a rational number because $\sqrt{25} = 5 = \frac{5}{1}$. The number π is an irrational number. To help you make the distinction between rational and irrational numbers, here are a few examples of each.

	RATIONAL NUMBERS		**IRRATIONAL NUMBERS**
NUMBER	**EQUIVALENT QUOTIENT OF INTEGERS,** $\dfrac{a}{b}$		
$-\dfrac{2}{3}$	$\dfrac{-2}{3}$ or $\dfrac{2}{-3}$		$\sqrt{5}$
$\sqrt{36}$	$\dfrac{6}{1}$		$\dfrac{\sqrt{6}}{7}$
5	$\dfrac{5}{1}$		$-\sqrt{13}$
0	$\dfrac{0}{1}$		π
1.2	$\dfrac{12}{10}$		$\dfrac{2}{\sqrt{3}}$
$3\dfrac{7}{8}$	$\dfrac{31}{8}$		

Every rational number can be written as a decimal that either repeats or terminates. For example,

$$\frac{1}{2} = 0.5 \qquad\qquad \frac{5}{4} = 1.25$$

$$\frac{2}{3} = 0.6666666\ldots = 0.\overline{6} \qquad \frac{1}{11} = 0.090909\ldots = 0.\overline{09}$$

An irrational number written as a decimal neither terminates nor repeats. When we perform calculations with irrational numbers, we often use decimal approximations that have been rounded. For example, consider the following irrational numbers along with a 4-decimal-place approximation of each:

$$\pi \approx 3.1416 \qquad \sqrt{2} \approx 1.4142$$

Earlier we mentioned that every integer is also a rational number. In other words, all the elements of the set of integers are also elements of the set of rational numbers. When this happens, we say that the set of integers, set I, is a subset of the set of rational numbers, set Q. In symbols,

$$I \subseteq Q$$

is a subset of

The natural numbers, whole numbers, integers, rational numbers, and irrational numbers are each a subset of the set of real numbers. The relationships among these sets of numbers are shown in the following diagram.

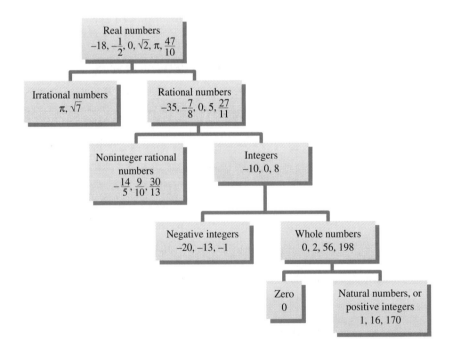

EXAMPLE 4   Determine whether the following statements are true or false.

a. 3 is a real number.   **b.** $\frac{1}{5}$ is an irrational number.

c. Every rational number is an integer.   **d.** $\{1,5\} \subseteq \{2, 3, 4, 5\}$

Solution: **a.** True. Every whole number is a real number.

b. False. The number $\frac{1}{5}$ is a rational number, since it is in the form $\frac{a}{b}$ with a and b integers and $b \neq 0$.

c. False. The number $\frac{2}{3}$, for example, is a rational number, but it is not an integer.

d. False since the element 1 in the first set is not an element of the second set.

---

3   Often, solving problems involves translating a phrase to an algebraic expression. The following is a list of key words and phrases and their translations.

ADDITION	SUBTRACTION	MULTIPLICATION	DIVISION
sum	difference of	product	quotient
plus	minus	times	divide
added to	subtracted from	multiply	into
more than	less than	twice	ratio
increased by	decreased by	of	
total	less		

EXAMPLE 5   Translate each phrase to an algebraic expression. Use the variable x to represent each unknown number.

a. Eight times a number
b. Three more than eight times a number
c. The quotient of a number and -7
d. One and six-tenths subtracted from twice a number

Solution: **a.** $8 \cdot x$ or $8x$

b. $8x + 3$

c. $x \div -7$ or $\dfrac{x}{-7}$

d. $2x - 1.6$

16. $\frac{2}{5}$ of a mile   **21.** {11, 12, 13, 14, 15, 16}   **25.** {0, 2, 4, 6, 8}   **28.**

(number line: –4 –3 –2 –1 0 1 2, points at –2, –1, 0, 1)

32.

(number line from 0 to 1, with $\frac{1}{4}$ and $\frac{1}{3}$ marked)

EXERCISE SET 1.1

29.

(number line from 0 to 2, with $\frac{1}{6}$, $\frac{1}{2}$, $\frac{2}{3}$ marked)

31.

(number line: –10 –9 –8 –7 –6 –5 –4 –3 –2 –1 0, points at –9, –6, –2)

Find the value of each algebraic expression at the given replacement values. See Example 1.

1. $5x$ when $x = 7$ 35   **2.** $3y$ when $y = 45$ 135

3. $9.8z$ when $z = 3.1$ **4.** $7.1a$ when $a = 1.5$ 10.65

5. xy when $x = 17$ and $y = 11$ 187   **3.** 30.38

6. st when $s = 6$ and $t = 1$ 6

7. ab when $a = \frac{1}{2}$ and $b = \frac{3}{4}$ $\frac{3}{8}$

8. yz when $y = \frac{2}{3}$ and $z = \frac{1}{5}$ $\frac{2}{15}$

9. $3x + y$ when $x = 6$ and $y = 4$ 22

10. $2a - b$ when $a = 12$ and $b = 7$ 17

11. $qr - s$ when $q = 1$, $r = 14$, and $s = 3$ 11

12. $xy + z$ when $x = 9$, $y = 2$, and $z = 8$ 26

13. The aircraft B737-400 flies an average speed of 400 miles per hour.

The expression

   $400t$

gives the distance traveled by the aircraft in t hours. Find the distance traveled by the B737-400 in 5 hours.   2000 miles

14. The algebraic expression $1.5x$ gives the total length of shelf space need in inches for x encyclopedias. Find the length of shelf space needed for a set of 30 encyclopedias.   45 inches

15. Employees at Wal-Mart constantly reorganize and reshelf merchandise. In doing so, they calculate floor space needed for displays. The algebraic expression $l \cdot w$ gives the floor space needed in square units for a display that measures length l units and width w units. Calculate the floor space needed for a display whose length is 5.1 feet and whose width is 4 feet.   20.4 square feet

16. The algebraic expression $\frac{x}{5}$ can be used to calculate the distance in miles that you are from a flash of lightning, where x is the number of seconds be-

tween the time you see a flash of lightning and the time you hear the thunder. Calculate the distance that you are from the flash of lightning if you hear the thunder 2 seconds after you see the lightning.

17. The B747-400 aircraft costs $7098 dollars per hour to operate. The algebraic expression

   $7098t$

gives the total cost to operate the aircraft for t hours. Find the total cost to operate the B747-400 for 5.2 hours. $36,909.60

18. Flying the SR-71A jet, Capt. Elden W. Joersz, USAF, set a record speed of 2193.16 miles per hour. At this speed, the algebraic expression $2193.16t$ gives the total distance flown in t hours. Find the distance flown by the SR-71A in 1.7 hours.   3728.372 miles

List the elements in each set. See Example 2.

19. $\{x \mid x$ is a natural number less than 6$\}$ {1, 2, 3, 4, 5}

20. $\{x \mid x$ is a natural number greater than 6$\}$ {7, 8, 9, . . .}

21. $\{x \mid x$ is a natural number between 10 and 17$\}$

22. $\{x \mid x$ is an odd natural number$\}$ {1, 3, 5, . . .}

23. $\{x \mid x$ is a whole number that is not a natural number$\}$ {0}

24. $\{x \mid x$ is a natural number less than 1$\}$ { }

25. $\{x \mid x$ is an even whole number less than 9$\}$

26. $\{x \mid x$ is an odd whole number less than 9$\}$ {1, 3, 5, 7}

Graph each set on a number line.

27. {0, 2, 4, 6}   **28.** {−1, −2, −3}   **29.** $\left\{\frac{1}{2}, \frac{2}{3}\right\}$

27. (number line: –2 –1 0 1 2 3 4 5 6 7, points at 0, 2, 4, 6)

30. {1, 3, 5, 7}   **31.** {−2, −6, −10}   **32.** $\left\{\frac{1}{4}, \frac{1}{3}\right\}$

30. (number line: 0 1 2 3 4 5 6 7, points at 1, 3, 5, 7)

33. In your own words, explain why the empty set is a subset of every set.

34. In your own words, explain why every set is a subset of itself.

List the elements of the set $\left\{3, 0, \sqrt{7}, \sqrt{36}, \frac{2}{5}, -134\right\}$ that are also elements of the given set. See Example 3.

35. Whole numbers   **36.** Integers {3, 0, $\sqrt{36}$, −134}

37. Natural numbers   **38.** Rational numbers

39. Irrational numbers   **40.** Real numbers

35. {3, 0, $\sqrt{36}$}   **37.** {3, $\sqrt{36}$}   **38.** $\left\{3, 0, \sqrt{36}, \frac{2}{5}, -134\right\}$ **39.** {$\sqrt{7}$}   **40.** $\left\{3, 0, \sqrt{7}, \sqrt{36}, \frac{2}{5}, -134\right\}$

Place ∈ or ∉ in the space provided to make each statement true. See Example 3.

41. -11 ∈ $\{x \mid x \text{ is an integer}\}$

42. -6 ∉ $\{2, 4, 6, \ldots\}$

43. 0 ∉ $\{x \mid x \text{ is a positive integer}\}$

44. 12 ∈ $\{1, 2, 3, \ldots\}$

45. 12 ∉ $\{1, 3, 5, \ldots\}$

46. $\dfrac{1}{2}$ ∉ $\{x \mid x \text{ is an irrational number}\}$

47. 0 ∉ $\{1, 2, 3, \ldots\}$

48. 0 ∉ $\{x \mid x \text{ is a natural number}\}$

Determine whether each statement is true or false. See Examples 3 and 4. Use the following sets of numbers:

N = set of natural numbers

Z = set of integers

I = set of irrational numbers

Q = set of rational numbers

$\mathbb{R}$ = set of real numbers

49. $Z \subseteq \mathbb{R}$ true   **50.** $\mathbb{R} \subseteq N$ false

51. $-1 \in Z$ true   **52.** $\dfrac{1}{2} \in Q$ true

53. $O \in N$ false   **54.** $Z \subseteq Q$ true

55. $\sqrt{5} \notin I$ false   **56.** $\pi \notin \mathbb{R}$ false

57. $N \subseteq Z$ true   **58.** $I \subseteq N$ false

59. $\mathbb{R} \subseteq Q$ false   **60.** $N \subseteq Q$ true

61. In your own words, explain why every natural number is also a rational number but not every rational number is a natural number.

62. In your own words, explain why every irrational number is a real number, but not every real number is an irrational number.

Write each phrase as an algebraic expression. Use the variable x to represent each unknown number. See Example 5.

63. Twice a number. $2x$

64. Six times a number. $6x$

65. Five more than twice a number. $2x + 5$

66. One more than six times a number. $6x + 1$

67. Ten less than a number. $x - 10$

68. A number minus seven. $x - 7$

69. The sum of a number and two. $x + 2$

70. The difference of twenty-five and a number.

71. A number divided by eleven. $\dfrac{x}{11}$

72. The quotient of twice a number and thirteen.

73. Twelve added to three times a number. $3x + 12$

74. Four subtracted from a number. $x - 4$

75. Seventeen subtracted from a number. $x - 17$

76. Four subtracted from three times a number.

77. Twice the sum of a number and three. $2(x + 3)$

78. The quotient of four and the sum of a number and one. $\dfrac{4}{x + 1}$

79. The quotient of five and the difference of four and a number. $\dfrac{5}{4 - x}$

80. Eight times the difference of a number and nine.

81. The following bar graph shows the total yearly revenue of the Walt Disney Company for the years shown. Use the graph to calculate the yearly *increase* in revenue. (Note that dollar amounts are in millions.)

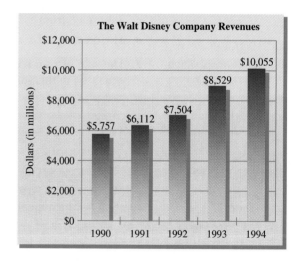

The Walt Disney Company Revenues

YEAR	YEARLY INCREASE IN REVENUE
1991	$355,000,000
1992	$1,392,000,000
1993	$1,025,000,000
1994	$1,526,000,000

70. $25 - x$ **72.** $\dfrac{2x}{13}$ **76.** $3x - 4$ **80.** $8(x - 9)$

1.2 | PROPERTIES OF REAL NUMBERS

O B J E C T I V E S

1 Use operation and order symbols to write mathematical sentences.
2 Identify identity numbers and inverses.
3 Identify and use the commutative, associative, and distributive properties.

1 In Section 1.1, we used the symbol = to mean "is equal to." All of the following key words and phrases also imply equality.

EQUALITY			
equals	is/was	represents	is the same as
gives	yields	amounts to	is equal to

EXAMPLE 1   Write each sentence using mathematical symbols.

  a. The sum of x and 5 is 20.
  b. Two times the sum of 3 and y amounts to 4.
  c. Subtract 8 from x, and the difference is the same as the product of 2 and x.
  d. The quotient of z and 9 is 3 times the difference of z and 5.

Solution: **a.** The sum of x and 5 can be written as "$x + 5$," and the word "is" means "is equal to" in this sentence, so we write $x + 5 = 20$.
  b. $2(3 + y) = 4$
  c. $x - 8 = 2x$
  d. $\dfrac{z}{9} = 3(z - 5)$

If we want to write in symbols that two numbers are not equal, we can use the symbol ≠, which means "**is not equal to.**" For example,

  $3 \neq 2$

Graphing two numbers on a number line gives us a way to compare two numbers. For two real numbers a and b, we say a **is less than** b if on the number line a lies to the left of b. Also, if b is to the right of a on the number line, then b **is greater than** a.

The symbol **>** means "**is greater than.**" Since b is greater than a, we write

  $b > a$

Also, the symbol **<** means "**is less than.**" Since a is less than b, we can write

$a < b$

Notice that if $a < b$, then also $b > a$.

The **trichotomy property** for real numbers assures us that any two real numbers a and b will either be equal or one will be greater than the other.

TRICHOTOMY PROPERTY

If a and b are real numbers, then exactly one of these statements is true:

$$a = b, \quad a < b, \quad \text{or} \quad a > b$$

EXAMPLE 2   Insert $<$, $>$, or $=$ between each pair of numbers to form a true statement.

a. $-1 \quad -2$ **b.** $\dfrac{12}{4} \quad 3$ **c.** $-5 \quad 0$ **d.** $-3.5 \quad -3.05$

Solution: **a.** $-1 > -2$ since -1 lies to the right of -2 on the number line.

b. $\dfrac{12}{4} = 3$.

c. $-5 < 0$ since -5 lies to the left of 0 on the number line.

d. $-3.5 < -3.05$ since -3.5 lies to the left of -3.05 on the number line.

R E M I N D E R   When inserting the $>$ or $<$ symbol, think of the symbols as arrowheads that "point" toward the smaller number when the statement is true.

In addition to $<$ and $>$, there are the inequality symbols $\leq$ and $\geq$. The symbol

$\leq$ means "**is less than or equal to**"

and the symbol

$\geq$ means "**is greater than or equal to**"

For example, the following are true statements:

$$10 \leq 10 \quad \text{since} \quad 10 = 10$$
$$-8 \leq 13 \quad \text{since} \quad -8 < 13$$
$$-5 \geq -5 \quad \text{since} \quad -5 = -5$$
$$-7 \geq -9 \quad \text{since} \quad -7 > -9$$

EXAMPLE 3   Write each sentence using mathematical symbols.

   a. The sum of 5 and y is greater than or equal to 7.

   b. 11 is not equal to z.

   c. 20 is less than the difference of 5 and twice x.

Solution: **a.** $5 + y \geq 7$ **b.** $11 \neq z$ **c.** $20 < 5 - 2x$

Of all the real numbers, two of them stand out as extraordinary: 0 and 1. The real number 0 is called the **additive identity** since for every real number a,

$$a + 0 = 0 + a = a$$

For example, $3 + 0 = 0 + 3 = 3$.

   The real number 1 is called the **multiplicative identity** since for every real number a,

$$a \cdot 1 = 1 \cdot a = a$$

For example, $5.2 \cdot 1 = 1 \cdot 5.2 = 5.2$.

   The uniqueness of 0 is also demonstrated by the **multiplication property of 0,** which guarantees that for every real number a,

$$a \cdot 0 = 0 \cdot a = 0$$

For example, $-7 \cdot 0 = 0 \cdot -7 = 0$.

   Two numbers whose sum is the additive identity 0 are called **opposites** or **additive inverses** of each other. If a is a real number, then the unique opposite of a is written as $-a$:

$$a + (-a) = 0$$

On the number line, we picture a real number and its opposite as being the same distance from 0 but on opposite sides of 0.

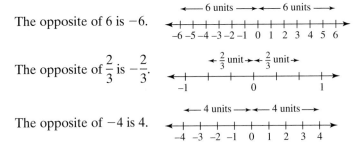

The opposite of 6 is -6.

The opposite of $\dfrac{2}{3}$ is $-\dfrac{2}{3}$.

The opposite of -4 is 4.

We stated earlier that the opposite or additive inverse of a number a is $-a$. This means that the opposite of -4 is $-(-4)$. But we stated above that the opposite of -4 is 4. This means that $-(-4) = 4$, and in general, we have the following property.

> **DOUBLE NEGATIVE PROPERTY**
> For every real number a, $-(-a) = a$.

EXAMPLE 4 Write the additive inverse, or the opposite, of each.

a. 8 **b.** $\dfrac{1}{5}$ **c.** -9.6

Solution: **a.** The opposite of 8 is -8. **b.** The opposite of $\dfrac{1}{5}$ is $-\dfrac{1}{5}$.

c. The opposite of -9.6 is $-(-9.6) = 9.6$.

Just as each real number has an opposite, each nonzero real number has a unique **reciprocal**. Two numbers whose product is 1 are called **reciprocals** or **multiplicative inverses** of each other. If a is a nonzero real number, then its reciprocal is $\dfrac{1}{a}$.

$$a \cdot \frac{1}{a} = 1$$

EXAMPLE 5 Write the multiplicative inverse, or the reciprocal, of each.

a. 11 **b.** -9 **c.** $\dfrac{7}{4}$

Solution: **a.** The reciprocal of 11 is $\dfrac{1}{11}$.

b. The reciprocal of -9 is $-\dfrac{1}{9}$.

c. The reciprocal of $\dfrac{7}{4}$ is $\dfrac{4}{7}$ because $\dfrac{7}{4} \cdot \dfrac{4}{7} = 1$.

The identities and inverses for addition and multiplication are summarized next.

	ADDITION	MULTIPLICATION
IDENTITIES	The additive identity is 0. $$a + 0 = 0 + a = a$$	The multiplicative identity is 1. $$a \cdot 1 = 1 \cdot a = a$$
INVERSES	For each number a, there is a unique number $-a$ called the **additive inverse** or **opposite** of a such that $$a + (-a) = (-a) + a = 0$$	For each nonzero a, there is a unique number $\dfrac{1}{a}$ called the **multiplicative inverse** or **reciprocal** of a such that $$a \cdot \frac{1}{a} = \frac{1}{a} \cdot a = 1$$

In addition to these special real numbers, all real numbers have certain properties that allow us to write equivalent expressions—that is, expressions that have the same value. These properties will be especially useful in Chapter 2 when we solve equations.

The **commutative properties** state that the order in which two real numbers are added or multiplied does not affect their sum or product.

COMMUTATIVE PROPERTIES

For real numbers a and b,

Addition: $a + b = b + a$

Multiplication: $a \cdot b = b \cdot a$

The **associative properties** state that regrouping numbers that are added or multiplied does not affect their sum or product.

ASSOCIATIVE PROPERTIES

For real numbers a, b, and c,

Addition: $(a + b) + c = a + (b + c)$

Multiplication: $(a \cdot b) \cdot c = a \cdot (b \cdot c)$

EXAMPLE 6   Use the commutative property of addition to write an expression equivalent to $7x + 5$.

Solution: $7x + 5 = 5 + 7x$.

EXAMPLE 7   Use the associative property of multiplication to write an expression equivalent to $4 \cdot (9y)$. Then simplify this equivalent expression.

Solution: $4 \cdot (9y) = (4 \cdot 9)y = 36y$.

The **distributive property** states that multiplication distributes over addition.

DISTRIBUTIVE PROPERTY

For real numbers a, b, and c,

$a(b + c) = ab + ac$

EXAMPLE 8   Use the distributive property to write an expression equivalent to $3(2x + y)$.

Solution: $3\,(2x + y) \;=\; 3 \cdot 2x + 3 \cdot y$ Apply the distributive property.

$\qquad\qquad = 6x + 3y$ Apply the associative property of multiplication.

EXERCISE SET 1.2

Write each statement using mathematical symbols. See Example 1.

1. The product of 4 and c is 7. $4c = 7$

2. The sum of 10 and x is -12. $10 + x = -12$

3. 3 times the sum of x and 1 amounts to 7.

4. 9 times the difference of 4 and m amounts to 1.

5. The quotient of n and 5 is 4 times n. $\dfrac{n}{5} = 4n$

6. The quotient of 8 and y is 3 more than y.

7. The difference of z and 2 is the same as the product of z and 2. $z - 2 = 2z$

8. Five added to twice q is the same as 4 more than q. $5 + 2q = q + 4$

3. $3(x + 1) = 7$ **4.** $9(4 - m) = 1$ **6.** $\dfrac{8}{y} = y + 3$

Insert $<$, $>$, or $=$ in the space provided to form a true statement. See Example 2.

9. $0 \;>\; -2$ **10.** $-5 \;<\; 0$

11. $\dfrac{12}{3} \;=\; \dfrac{8}{2}$ **12.** $\dfrac{20}{5} \;<\; \dfrac{20}{4}$

13. $-7.9 \;<\; -7.09$ **14.** $-13.07 \;>\; -13.7$

Write each statement using mathematical symbols. See Example 3.

15. The product of 7 and x is less than or equal to -21. $7x \le -21$

16. 10 subtracted from x is greater than 0. $x - 10 > 0$

17. The sum of -2 and x is not equal to 10.

18. The quotient of y and 3 is less than or equal to y.

19. Twice the difference of x and 6 is greater than the reciprocal of 11. $2(x + 6) > \dfrac{1}{11}$

20. Four times the sum of 5 and x is not equal to the opposite of 15. $4(5 + x) \ne -15$

Write the opposite (or additive inverse) of each number. See Example 4.

21. -6.2 6.2 **22.** -7.8 7.8 **23.** $\dfrac{4}{7}$ $-\dfrac{4}{7}$

24. $\dfrac{9}{5}$ $-\dfrac{9}{5}$ **25.** $-\dfrac{2}{3}$ $\dfrac{2}{3}$ **26.** $-\dfrac{14}{3}$ $\dfrac{14}{3}$

27. 0 0 **28.** 10.3 -10.3

Write the reciprocal (or multiplicative inverse) of each number if one exists. See Example 5.

29. 5 $\dfrac{1}{5}$ **30.** 9 $\dfrac{1}{9}$

31. -8 $-\dfrac{1}{8}$ **32.** -4 $-\dfrac{1}{4}$

33. $-\dfrac{1}{4}$ -4 **34.** $\dfrac{1}{9}$ 9

35. 0 undefined   **36.** $\dfrac{0}{6}$ undefined

37. $\dfrac{7}{8}$ $\dfrac{8}{7}$ **38.** $-\dfrac{23}{5}$ $-\dfrac{5}{23}$

39. Name the only real number that has no reciprocal, and explain why this is so.

40. Name the only real number that is its own opposite, and explain why this is so.   Zero.

Use a commutative property to write an equivalent expression. See Example 6.

41. $7x + y$ $y + 7x$ **42.** $3a + 2b$ $2b + 3a$

43. $z \cdot w$ $w \cdot z$ **44.** $r \cdot s$ $s \cdot r$

45. $\dfrac{1}{3} \cdot \dfrac{x}{5}$ $\dfrac{x}{5} \cdot \dfrac{1}{3}$ **46.** $\dfrac{x}{2} \cdot \dfrac{9}{10}$ $\dfrac{9}{10} \cdot \dfrac{x}{2}$

47. Is subtraction commutative? Explain why or why not.   No

48. Is division commutative? Explain why or why not.   No

17. $-2 + x \ne 10$ **18.** $\dfrac{y}{3} \le y$ **39.** Zero. For every real number x, $0 \cdot x \ne 1$, so 0 has no reciprocal. It is the only real number that has no reciprocal because if $x \ne 0$, then $x \cdot \dfrac{1}{x} = 1$ by definition.

Use an associative property to write an equivalent expression. See Example 7.

49. $5 \cdot (7x)$ $(7.5)x$ **50.** $3 \cdot (10z)$ $(3 \cdot 10)z$

51. $(x + 1.2) + y$ **52.** $5q + (2r + s)$

53. $(14z) \cdot y$ $14(z \cdot y)$ **54.** $(9.2x) \cdot y$ $9.2 (x \cdot y)$

55. Evaluate $12 - (5 - 3)$ and $(12 - 5) - 3$. Use these two expressions and discuss whether subtraction is associative.

56. Evaluate $24 \div (6 \div 3)$ and $(24 \div 6) \div 3$. Use these two expressions and discuss whether division is associative.

Use the distributive property to find the product. See Example 8.

57. $3(x + 5)$ $3x + 15$ **58.** $7(y + 2)$ $7y + 14$

59. $8(2a + b)$ $16a + 8b$ **60.** $9(c + 7d)$ $9c + 63d$

61. $2(6x + 5y + 2z)$ $12x + 10y + 4z$

62. $5(3a + b + 9c)$ $15a + 5b + 45c$

Write each sentence using mathematical symbols.

63. 7 subtracted from y is 6. $y - 7 = 6$

64. The sum of z and w is 12. $z + w = 12$

65. The product of 3 and y is less than -17. $3y < -17$

66. 5 less x is not equal to 1. $5 - x \neq 1$

67. Twice the difference of x and 6 is -27.

68. 5 times the sum of 6 and y is -35. $5(6 + y) = -35$

69. Subtract 4 from x, and the difference is greater than or equal to 3 times x. $x - 4 \geq 3x$

70. 4 times x subtracted from 10 is less than or equal to the opposite of x. $10 - 4x \leq -x$

71. 6 subtracted from twice y is the reciprocal of 8.

72. 7 subtracted from the product of 5 and n is the opposite of n. $5n - 7 = -n$

73. The sum of n and 5, divided by 2, is greater than twice n. $\dfrac{n + 5}{2} > 2n$

74. The product of 8 and x, divided by 5, is less than 3 more than x. $\dfrac{8x}{5} < x + 3$

Complete the statement to illustrate the given property.

75. $3x + 6 = \underline{\quad 6 + 3x \quad}$ Commutative property of addition

76. $8 + 0 = \underline{\quad 8 \quad}$ Additive identity property

77. $\dfrac{2}{3} + \left(-\dfrac{2}{3}\right) = \underline{\quad 0 \quad}$ Additive inverse property

78. $4(x + 3) = \underline{\quad 4x + 12 \quad}$ Distributive property

79. $7 \cdot 1 = \underline{\quad 7 \quad}$ Multiplicative identity property

80. $0 \cdot (-5.4) = \underline{\quad 0 \quad}$ Multiplication property of zero

81. $10(2y) = \underline{\quad (10 \cdot 2)y \quad}$ Associative property

82. $9y + (x + 3z) = \underline{\quad (9y + x) + 3z \quad}$ Associative property

83. To demonstrate the distributive property geometrically, represent the area of the larger rectangle in two ways: First as length a times width $b + c$, and second as the sum of the areas of the smaller rectangles. $a(b + c) = ab + ac$

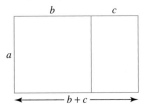

51. $x + (1.2 + y)$ **52.** $(5q + 2r) + s$ **55.** 10 and 4. Subtraction is not associative.

56. 12 and $\dfrac{4}{3}$. Division is not associative. **67.** $2(x - 6) = -27$ **71.** $2y - 6 = \dfrac{1}{8}$

1.3 | OPERATIONS ON REAL NUMBERS

O B J E C T I V E S

TAPE IA 1.3

1 Find the absolute value of a number.

2 Add and subtract real numbers.

3 Multiply and divide real numbers.

4 Simplify expressions containing exponents.

5 Find roots of numbers.

In Section 1.2, we used the number line to compare two real numbers. The number line can also be used to visualize distance, which leads to the concept of absolute value.

1   The **absolute value** of a real number a, written as $|a|$, is the distance between a and 0 on the number line. Since distance is always positive or zero, $|a|$ is always positive or zero.

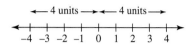

Using the number line, we see that

$$|4| = 4$$

because the distance between 4 and 0 is 4 units. Also,

$$|-4| = 4$$

because the distance between -4 and 0 is 4 units.

An equivalent definition of the absolute value of a real number a is given next.

ABSOLUTE VALUE

The absolute value of a, written as $|a|$, is

$$|a| = \begin{cases} a \text{ if } a \text{ is 0 or a positive number} \\ -a \text{ if } a \text{ is a negative number} \end{cases}$$

EXAMPLE 1   Find each absolute value.

a. $|3|$ **b.** $|-5|$ **c.** $-|2|$ **d.** $-|-8|$ **e.** $|0|$

Solution: **a.** $|3| = 3$ since 3 is located 3 units from 0 on the number line.

b. $|-5| = 5$ since -5 is 5 units from 0 on the number line.

c. $-|2| = -2$. The negative sign outside the absolute value bars means to take the opposite of the absolute value of 2.

d. $-|-8| = -8$. Since $|-8|$ is 8, $-|-8| = -8$.

e. $|0| = 0$ since 0 is located 0 units from 0 on the number line.

2   When solving problems, we often have to add real numbers. For example, if the Dallas Cowboys lose 5 yards in one play, then lose another 7 yards in the next play, their total loss may be described by $-5 + (-7)$.

The addition of two real numbers may be summarized by the following.

ADDING REAL NUMBERS

1. To add two numbers with the same sign, add their absolute values and attach their common sign.

2. To add two numbers with different signs, subtract the smaller absolute value from the larger absolute value and attach the sign of the number with the larger absolute value.

For example, to add $-5 + (-7)$, first add their absolute values.

$$|-5| = 5, |-7| = 7, \quad \text{and} \quad 5 + 7 = 12$$

Next, attach their common negative sign.

$$-5 + (-7) = -12$$

(This represents a total loss of 12 yards for the Dallas Cowboys in the example on the previous page.)

To find $(-4) + 3$, first subtract their absolute values.

$$|-4| = 4, |3| = 3, \quad \text{and} \quad 4 - 3 = 1$$

Next, attach the sign of the number with the larger absolute value.

$$(-4) + 3 = -1$$

EXAMPLE 2   Find each sum.

a. $-3 + (-11)$ **b.** $3 + (-7)$ **c.** $-10 + 15$ **d.** $-8.3 + (-1.9)$ **e.** $-\dfrac{1}{4} + \dfrac{1}{2}$

f. $-\dfrac{2}{3} + \dfrac{3}{7}$

Solution: **a.** $-3 + (-11) = -(3 + 11) = -14$ **b.** $3 + (-7) = -4$

c. $-10 + 15 = 5$ **d.** $-8.3 + (-1.9) = -10.2$ **e.** $-\dfrac{1}{4} + \dfrac{1}{2} = -\dfrac{1}{4} + \dfrac{2}{4} = \dfrac{1}{4}$

f. $-\dfrac{2}{3} + \dfrac{3}{7} = -\dfrac{14}{21} + \dfrac{9}{21} = -\dfrac{5}{21}$

Subtraction of two real numbers may be defined in terms of addition.

SUBTRACTING REAL NUMBERS

If a and b are real numbers, then the difference of a and b, written $a - b$, is defined by

$$a - b = a + (-b)$$

In other words, to subtract a second real number from a first, add the first number and the opposite of the second number.

EXAMPLE 3   Find each difference.

a. $2 - 8$ **b.** $-8 - (-1)$ **c.** $-11 - 5$ **d.** $10.7 - (-9.8)$ **e.** $\dfrac{2}{3} - \dfrac{1}{2}$

f. $1 - 0.06$ **g.** Subtract 7 from 4.

Solution: Add the opposite   Add the opposite

a. $2 - 8 = 2 + (-8) = -6$ **b.** $-8 - (-1) = -8 + (1) = -7$

c. $-11 - 5 = -11 + (-5) = -16$ **d.** $10.7 - (-9.8) = 10.7 + 9.8 = 20.5$

e. $\dfrac{2}{3} - \dfrac{1}{2} = \dfrac{2 \cdot 2}{3 \cdot 2} - \dfrac{1 \cdot 3}{2 \cdot 3} = \dfrac{4}{6} + \left(-\dfrac{3}{6}\right) = \dfrac{1}{6}$ **f.** $1 - 0.06 = 1 + (-0.06) = 0.94$

g. $4 - 7 = 4 + (-7) = -3$

To add or subtract three or more real numbers, add or subtract from left to right.

EXAMPLE 4   Simplify the following expressions.

a. $11 + 2 - 7$ **b.** $-5 - 4 + 2$

Solution: **a.** $11 + 2 - 7 = 13 - 7 = 6$ **b.** $-5 - 4 + 2 = -9 + 2 = -7$

In order to discover sign patterns when you multiply real numbers, recall that multiplication by a positive integer is the same as repeated addition. For example,

$$3(2) = 2 + 2 + 2 = 6$$
$$3(-2) = (-2) + (-2) + (-2) = -6$$

Notice here that $3(-2) = -6$. This illustrates that the product of two numbers with different signs is negative. We summarize sign patterns for multiplying any two real numbers as follows.

MULTIPLYING TWO REAL NUMBERS

The product of two numbers with the same sign is positive.
The product of two numbers with different signs is negative.

Also recall that the product of zero and any real number is zero.

$$0 \cdot a = 0$$

EXAMPLE 5 Find each product.

a. $(-8)(-1)$ **b.** $(-2)\dfrac{1}{6}$ **c.** $3(-3)$ **d.** $0(11)$ **e.** $\left(\dfrac{1}{5}\right)\left(-\dfrac{10}{11}\right)$

f. $(7)(1)(-2)(-3)$ **g.** $8(-2)(0)$

Solution: **a.** Since the signs of the two numbers are the same, the product is positive. Thus $(-8)(-1) = +8$, or 8.

b. Since the signs of the two numbers are different or unlike, the product is negative. Thus $(-2)\dfrac{1}{6} = -\dfrac{2}{6} = -\dfrac{1}{3}$.

c. $3(-3) = -9$ **d.** $0(11) = 0$ **e.** $\left(\dfrac{1}{5}\right)\left(-\dfrac{10}{11}\right) = -\dfrac{10}{55} = -\dfrac{2}{11}$

f. To multiply three or more real numbers, multiply from left to right.

$$
\begin{aligned}
(7)(1)(-2)(-3) &= 7(-2)(-3) \\
&= -14(-3) \\
&= 42
\end{aligned}
$$

g. Since zero is a factor, the product is zero.

$$(8)(-2)(0) = 0$$

REMINDER   The following sign patterns may be helpful when we are multiplying.

1. An odd number of negative factors gives a negative product.

2. An even number of negative factors gives a positive product.

Recall that $\frac{8}{4} = 2$ because $2 \cdot 4 = 8$. Likewise, $\frac{8}{-4} = -2$ because $(-2)(-4) = 8$. Also, $\frac{-8}{4} = -2$ because $(-2)4 = -8$, and $\frac{-8}{-4} = 2$ because $2(-4) = -8$. From these examples, we can see that the sign patterns for division are the same as for multiplication.

DIVIDING TWO REAL NUMBERS

The quotient of two numbers with the same signs is positive.
The quotient of two numbers with different signs is negative.

Also recall that division by a nonzero real number b is the same as multiplication by its reciprocal $\dfrac{1}{b}$. In other words,

$$a \div b = a \cdot \dfrac{1}{b}$$

Notice that *b must* be a nonzero number. We do not define division by 0. For example, $5 \div 0$, or $\frac{5}{0}$, is undefined.

EXAMPLE 6   Find each quotient.

a. $\dfrac{20}{-4}$ b. $\dfrac{-9}{-3}$ c. $-\dfrac{3}{8} \div 3$ d. $\dfrac{-40}{10}$ e. $\dfrac{-1}{10} \div \dfrac{-2}{5}$ f. $\dfrac{8}{0}$

Solution: **a.** Since the signs are different or unlike, the quotient is negative and $\dfrac{20}{-4} = -5$.

b. Since the signs are the same, the quotient is positive and $\dfrac{-9}{-3} = 3$.

c. $-\dfrac{3}{8} \div 3 = -\dfrac{3}{8} \cdot \dfrac{1}{3} = -\dfrac{1}{8}$ **d.** $\dfrac{-40}{10} = -4$

e. $\dfrac{-1}{10} \div \dfrac{-2}{5} = -\dfrac{1}{10} \cdot -\dfrac{5}{2} = \dfrac{1}{4}$ **f.** $\dfrac{8}{0}$ is undefined.

With sign rules for division, we can understand why the positioning of the negative sign in a fraction does not change the value of the fraction. For example,

$$\dfrac{-12}{3} = -4, \quad \dfrac{12}{-3} = -4, \quad \text{and} \quad -\dfrac{12}{3} = -4$$

Since all fractions equal -4, we can say that

$$\dfrac{-12}{3} = \dfrac{12}{-3} = -\dfrac{12}{3}$$

In general, the following holds true:

If a and b are real numbers and $b \neq 0$, then $\dfrac{a}{-b} = \dfrac{-a}{b} = -\dfrac{a}{b}$.

Recall that when two numbers are multiplied, they are called **factors.** For example, in $3 \cdot 5 = 15$, the 3 and 5 are called factors.

A natural number *exponent* is a shorthand notation for repeated multiplication of the same factor. This repeated factor is called the **base,** and the number of times it is used as a factor is indicated by the **exponent.** For example,

$$\overset{\text{exponent}}{\text{base} \searrow 4^3} = 4 \cdot 4 \cdot 4 = 64$$

$$\underbrace{\qquad\qquad}_{\text{4 is a factor 3 times}}$$

Also,

$$\overset{\text{exponent}}{\underset{\text{base}}{\;\;}} \;\; 2^5 = \underbrace{2 \cdot 2 \cdot 2 \cdot 2 \cdot 2}_{2 \text{ is a factor } 5 \text{ times}} = 32$$

EXPONENTS

If a is a real number and n is a natural number, then the ***n*th power of *a*,**
or ***a* raised to the *n*th power**, written as a^n, is the product of n factors,
each of which is a.

$$\overset{\text{exponent}}{\underset{\text{base}}{\;\;}} \;\; a^n = \underbrace{a \cdot a \cdot a \cdot a \cdots a}_{a \text{ is a factor } n \text{ times}}$$

It is not necessary to write an exponent of 1. For example, 3 is assumed to be 3^1.

EXAMPLE 7   Simplify each expression.

a. 3^2 **b.** $\left(\dfrac{1}{2}\right)^4$ **c.** -5^2 **d.** $(-5)^2$ **e.** -5^3 **f.** $(-5)^3$

Solution: **a.** $3^2 = 3 \cdot 3 = 9$ **b.** $\left(\dfrac{1}{2}\right)^4 = \left(\dfrac{1}{2}\right)\left(\dfrac{1}{2}\right)\left(\dfrac{1}{2}\right)\left(\dfrac{1}{2}\right) = \dfrac{1}{16}$

c. $-5^2 = -(5 \cdot 5) = -25$ **d.** $(-5)^2 = (-5)(-5) = 25$

e. $-5^3 = -(5 \cdot 5 \cdot 5) = -125$ **f.** $(-5)^3 = (-5)(-5)(-5) = -125$

REMINDER   Be very careful when simplifying expressions such as -5^2 and
$(-5)^2$.

$$-5^2 = -(5 \cdot 5) = -25 \quad \text{and} \quad (-5)^2 = (-5)(-5) = 25$$

Without parentheses, the base to square is 5, not -5.

---

5   The opposite of squaring a number is taking the **square root** of a number. For example, since the square of 4, or 4^2, is 16, we say that a square root of 16 is 4. The
notation $\sqrt{a}$ is used to denote the **positive, or principal, square root** of a nonnegative number a. We then have in symbols that

$$\sqrt{16} = 4$$

EXAMPLE 8   Find the following square roots.

a. $\sqrt{9}$ b. $\sqrt{25}$ c. $\sqrt{\dfrac{1}{4}}$

Solution: a. $\sqrt{9} = 3$ since 3 is positive and $3^2 = 9$.
b. $\sqrt{25} = 5$ since $5^2 = 25$.
c. $\sqrt{\dfrac{1}{4}} = \dfrac{1}{2}$ since $\left(\dfrac{1}{2}\right)^2 = \dfrac{1}{4}$.

We can find roots other than square roots. Since 2 cubed, written as 2^3, is 8, we say that the **cube root** of 8 is 2. This is written as

$$\sqrt[3]{8} = 2$$

Also, since $3^4 = 81$ and 3 is positive,

$$\sqrt[4]{81} = 3$$

EXAMPLE 9   Find the following roots.

a. $\sqrt[3]{27}$ b. $\sqrt[5]{1}$ c. $\sqrt[4]{16}$

Solution: a. $\sqrt[3]{27} = 3$ since $3^3 = 27$.
b. $\sqrt[5]{1} = 1$ since $1^5 = 1$.
c. $\sqrt[4]{16} = 2$ since 2 is positive and $2^4 = 16$.

Of course, as mentioned in Section 1.1, not all roots simplify to rational numbers. We study radicals further in Chapter 7.

EXERCISE SET 1.3

Find each absolute value. See Example 1.
1. $-|2|$ -2 **2.** $|8|$ 8   **3.** $|-4|$ 4
4. $|-6|$ 6   **5.** $|0|$ 0   **6.** $|-1|$ 1
7. $-|-3|$ -3 **8.** $-|-11|$ -11

🗨 **9.** Explain why $-(-2)$ and $-|-2|$ simplify to different numbers.   Answers will vary.

🗨 **10.** The boxed definition of absolute value states that $|a| = -a$ if a is a negative number. Explain why $|a|$ is always nonnegative, even though $|a| = -a$ for negative values of a.   Answers will vary.

Find each sum or difference. See Examples 2 through 4.
11. $-3 + 8$ 5   **12.** $-5 + (-9)$ **13.** $-14 + (-10)$
14. $12 + (-7)$ **15.** $-4.3 - 6.7$ **16.** $-8.2 - (-6.6)$

17. $13 - 17$ **18.** $15 - (-1)$ **19.** $\dfrac{11}{15} - \left(-\dfrac{3}{5}\right)$

20. $\dfrac{7}{10} - \dfrac{4}{5}$ **21.** $19 - 10 - 11$ **22.** $-13 - 4 + 9$

Find each product or quotient. See Examples 5 and 6.
23. $(-5)(12)$ -60 **24.** $6(-3)$ -18
25. $(-8)(-10)$ 80   **26.** $7(0)$ 0

27. $\dfrac{-12}{-4}$ 3   **28.** $\dfrac{60}{-6}$ -10

29. $\dfrac{0}{-2}$ 0   **30.** $\dfrac{-2}{0}$ undefined

31. $(-4)(-2)(-1)$ -8 **32.** $5(-3)(-2)$ 30

12. -14 **13.** -24 **14.** 5   **15.** -11 **16.** -1.6 **17.** -4 **18.** 16   **19.** $\dfrac{4}{3}$ **20.** $\dfrac{-1}{10}$ **21.** -2 **22.** -8

33. $\dfrac{-6}{7} \div 2$ $-\dfrac{3}{7}$

34. $\dfrac{-9}{13} \div (-3)$ $\dfrac{3}{13}$

35. $\left(-\dfrac{2}{7}\right)\left(-\dfrac{1}{6}\right)$ $\dfrac{1}{21}$

36. $\dfrac{5}{9}\left(-\dfrac{3}{5}\right)$ $-\dfrac{1}{3}$

Evaluate. See Example 7.

37. -7^2 -49

38. $(-7)^2$ 49

39. $(-6)^2$ 36

40. -6^2 -36

41. $(-2)^3$ -8

42. -2^3 -8

43. Explain why -3^2 and $(-3)^2$ simplify to different numbers. Answers will vary.

44. Explain why -3^3 and $(-3)^3$ simplify to the same number. Answers will vary.

Find the following roots. See Examples 8 and 9.

45. $\sqrt{49}$ 7

46. $\sqrt{81}$ 9

47. $\sqrt{\dfrac{1}{9}}$ $\dfrac{1}{3}$

48. $\sqrt{\dfrac{1}{25}}$ $\dfrac{1}{5}$

49. $\sqrt[3]{64}$ 4

50. $\sqrt[5]{32}$ 2

51. $\sqrt[4]{81}$ 3

52. $\sqrt[3]{1}$ 1

Perform the indicated operation.

53. $-4 + 7$ 3

54. $-9 + 15$ 6

55. $-9 + (-3)$ -12

56. $-17 + (-2)$ -19

57. $-4 - (-19)$ 15

58. $-5 - (-17)$ 12

59. $6.3 - 18.5$ -12.2

60. $15.9 - 21.7$ -5.8

61. $(-4)(-7)(0)$ 0

62. $(-9)(0)(-14)$ 0

63. $\left(-\dfrac{2}{3}\right)\left(\dfrac{6}{4}\right)$ -1

64. $\left(\dfrac{5}{6}\right)\left(\dfrac{-12}{15}\right)$ $-\dfrac{2}{3}$

65. $-14 - 7$ -21

66. $-6 - 31$ -37

67. $-\dfrac{4}{5} - \left(-\dfrac{3}{10}\right)$ $-\dfrac{1}{2}$

68. $-\dfrac{5}{2} - \left(-\dfrac{2}{3}\right)$ $-\dfrac{11}{6}$

69. Subtract 14 from 8 -6

70. Subtract 9 from -3 -12

71. $-\dfrac{34}{2}$ -17

72. $\dfrac{48}{-3}$ -16

73. $16 - 8 - 9$ -1

74. $-14 - 3 + 6$ -11

75. $\sqrt[3]{8}$ 2

76. $\sqrt{36}$ 6

77. $\sqrt{100}$ 10

78. $\sqrt[3]{125}$ 5

79. $-\dfrac{1}{6} \div \dfrac{9}{10}$ $-\dfrac{5}{27}$

80. $\dfrac{4}{7} \div \left(-\dfrac{1}{8}\right)$ $-\dfrac{32}{7}$

Each circle below represents a whole, or 1. Determine the unknown fractional part of each circle.

81. $\dfrac{13}{35}$

82. $\dfrac{13}{36}$

83. Most of Mount Kea, a volcano on Hawaii, lies below sea level. If this volcano begins at 5998 meters below sea level and then rises 10,203 meters, find the height of the volcano above sea level. 4205 meters

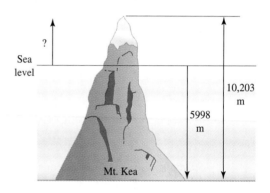

84. The highest point on land on Earth is the top of Mt. Everest, in the Himalayas, at an elevation of 29,028 feet above sea level. The lowest point on land is the Dead Sea, between Israel and Jordan, at 1319 feet below sea level. Find the difference in elevations. 30,347 feet

A fair game is one in which each team or player has the same chance of winning. Suppose that a game consists of three players taking turns spinning a spinner. If the spinner lands on yellow, player 1 gets a point. If the spinner lands on red, player 2 gets a point, and if the spinner lands on blue, player 3 gets a point. After 12 spins, the player with the most points wins.

a.   **b.**   **c.**   **d.**

85. Which spinner would lead to a fair game? b

86. If you are player 2 and want to win the game, which spinner would you choose? a

87. If you are player 1 and want to lose the game, which spinner would you choose?   d

88. Is it possible for the game to end in a three-way tie? If so, list the possible ending scores.

89. Is it possible for the game to end in a two-way tie? If so, list the possible ending scores.

88. Yes. Each player has 4 points.

Use a calculator to approximate each square root. Round to four decimal places.

 90. $\sqrt{10}$ 3.1623

 91. $\sqrt{273}$ 16.5227

92. $\sqrt{7.9}$ 2.8107

93. $\sqrt{19.6}$ 4.4272

Investment firms often advertise their gains and losses in the form of bar graphs such as the one that follows. This graph shows investment risk over time for the S&P 500 Index by showing average annual compound returns for 1 year, 5 years, 15 years, and 25 years. For example, after one year, the annual compound return in percent for an investor is anywhere from a gain of 181.5% to a loss of 64%. Use this graph to answer the questions below.

94. A person investing in the S&P 500 Index may expect at most an average annual gain of what percent after 15 years?   15.6%

95. A person investing in the S&P 500 Index may expect to lose at most an average per year of what percent after 5 years?   13.2%

96. Find the difference in percent of the highest average annual return and the lowest average annual return after 15 years.   17.7%

97. Find the difference in percent of the highest average annual return and the lowest average annual return after 25 years.   10.8%

98. Do you think that the type of investment shown in the figure is recommended for short-term investments or long-term investments? Explain your answer.   Long term. Short term is very volatile.

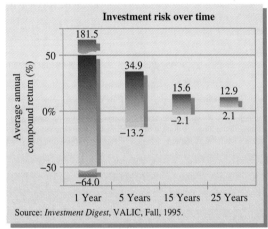

Investment risk over time

Source: *Investment Digest*, VALIC, Fall, 1995.

89. Yes. Two players have 6 points each (the third player has 0 points) or two players have 5 points each (the third has 2 points).

1.4   ORDER OF OPERATIONS AND ALGEBRAIC EXPRESSIONS

OBJECTIVES

1. Use the order of operations.

2. Identify and evaluate algebraic expressions.

3. Write word phrases as algebraic expressions.

4. Identify like terms and simplify algebraic expressions.

TAPE IA 1.4

1   The expression $3 + 2 \cdot 10$ represents the total number of disks shown.

Expressions containing more than one operation are written to follow a particular agreed-upon **order of operations.** For example, when we write $3 + 2 \cdot 10$, we mean to multiply first, and then add.

ORDER OF OPERATIONS

Simplify expressions using the following order. If grouping symbols such as parentheses are present, simplify expressions within those first, starting with the innermost set. If fraction bars are present, simplify the numerator and denominator separately.

1. Raise to powers or take roots in order from left to right.
2. Multiply or divide in order from left to right.
3. Add or subtract in order from left to right.

EXAMPLE 1   Simplify.

a. $3 + 2 \cdot 10$ **b.** $2(1 - 4)^2$ **c.** $\dfrac{|-2|^3 + 1}{-7 - \sqrt{4}}$ **d.** $\dfrac{(6 + 2) - (-4)}{2 - (-3)}$

Solution: **a.** First multiply; then add.

$$3 + 2 \cdot 10 = 3 + 20 = 23$$

b. $2(1 - 4)^2 = 2(-3)^2$ Simplify inside grouping symbols first.
$\qquad\qquad\quad\; = 2(9)$ Write $(-3)^2$ as 9.
$\qquad\qquad\quad\; = 18$ Multiply.

c. Simplify the numerator and the denominator separately; then divide.

$$\frac{|-2|^3 + 1}{-7 - \sqrt{4}} = \frac{2^3 + 1}{-7 - 2} \qquad \text{Write } |-2| \text{ as 2 and } \sqrt{4} \text{ as 2.}$$

$$= \frac{8 + 1}{-9} \qquad \text{Write } 2^3 \text{ as 8.}$$

$$= \frac{9}{-9} = -1 \qquad \text{Simplify the numerator, then divide.}$$

d. $\dfrac{(6 + 2) - (-4)}{2 - (-3)} = \dfrac{8 - (-4)}{2 - (-3)}$ Simplify inside grouping symbols first.

$$= \frac{8 + 4}{2 + 3} \qquad \text{Write subtraction as equivalent addition.}$$

$$= \frac{12}{5} \qquad \begin{array}{l}\text{Add in both the numerator and}\\ \text{denominator.}\end{array}$$

---

2   Recall from Section 1.1 that an alegbraic expression is formed by numbers and variables connected by the operations of addition, subtraction, multiplication, division, raising to powers, or taking roots. Also, if numbers are substituted for the variables in an algebraic expression and the operations performed, the result is called the value of the expression for the given replacement values. This entire process is called evaluating an expression.

EXAMPLE 2   Evaluate each algebraic expression when $x = 2$, $y = -1$, and $z = -3$.

      a. $z - y$ **b.** z^2 **c.** $\dfrac{2x + y}{z}$

Solution: **a.** $z - y = -3 - (-1) = -3 + 1 = -2$ **b.** $z^2 = (-3)^2 = 9$

      c. $\dfrac{2x + y}{z} = \dfrac{2(2) + (-1)}{-3} = \dfrac{4 + (-1)}{-3} = \dfrac{3}{-3} = -1$

Sometimes variables such as x_1 and x_2 will be used in this book. The small 1 and 2 are called **subscripts.** The variable x_1 can be read as "x sub 1," and the variable x_2 can be read as "x sub 2." The important thing to remember is that they are two different variables. For example, if $x_1 = -5$ and $x_2 = 7$, then

$$x_1 - x_2 = -5 - 7 = -12$$

EXAMPLE 3   The algebraic expression $\dfrac{5(x - 32)}{9}$ represents the equivalent temperature in degrees Celsius when x is degrees Fahrenheit. Complete the following table by evaluating this expression at given values of x.

Degrees Fahrenheit	x	-4	10	32
Degrees Celsius	$\dfrac{5(x - 32)}{9}$			

Solution: To complete the table, evaluate $\dfrac{5(x - 32)}{9}$ at each given replacement value.

When $x = -4$,

$$\frac{5(x - 32)}{9} = \frac{5(-4 - 32)}{9} = \frac{5(-36)}{9} = -20$$

When $x = 10$,

$$\frac{5(x - 32)}{9} = \frac{5(10 - 32)}{9} = \frac{5(-22)}{9} = \frac{-110}{9}$$

When $x = 32$,

$$\frac{5(x - 32)}{9} = \frac{5(32 - 32)}{9} = \frac{5 \cdot 0}{9} = 0$$

The completed table is

DEGREES FAHRENHEIT	x	-4	10	32
DEGREES CELSIUS	$\dfrac{5(x-32)}{9}$	-20	$\dfrac{-110}{9}$	0

Thus, $-4°$F is equivalent to $-20°$C, $10°$F is equivalent to $\frac{-110}{9}°$C, and $32°$F is equivalent to $0°$C.

3

As mentioned earlier, an important step in problem solving is to be able to write algebraic expressions from word phrases. Sometimes this involves a direct translation, but often an indicated operation is not directly stated but rather implied.

EXAMPLE 4 Write the following as algebraic expressions.

a. A vending machine contains x quarters. Write an expression for the *value* of the quarters.

b. The number of grams of fat in x pieces of bread if each piece of bread contains 2 grams of fat.

c. The cost of x desks if each desk costs $156.

d. Sales tax on a purchase of x dollars if the tax rate is 9%.

Solution: Each of these examples implies finding a product.

a. The value of the quarters is found by multiplying the value of a quarter (0.25 dollar) by the number of quarters.

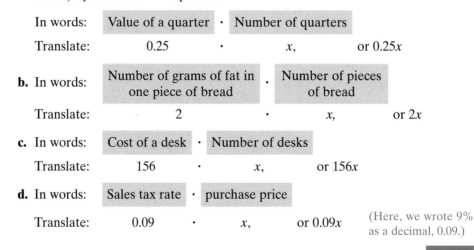

In words: Value of a quarter · Number of quarters

Translate: 0.25 · x, or $0.25x$

b. In words: Number of grams of fat in one piece of bread · Number of pieces of bread

Translate: 2 · x, or $2x$

c. In words: Cost of a desk · Number of desks

Translate: 156 · x, or $156x$

d. In words: Sales tax rate · purchase price

Translate: 0.09 · x, or $0.09x$ (Here, we wrote 9% as a decimal, 0.09.)

Two or more unknown numbers in a problem may sometimes be related. If so, try letting a variable represent one unknown number and then represent the other unknown number or numbers as expressions containing the same variable.

EXAMPLE 5   Write each as an algebraic expression.

a. Two numbers have a sum of 20. If one number is x, represent the other number as an expression in x.

b. The older sister is 8 years older than her younger sister. If the age of the younger sister is x, represent the age of the older sister as an expression in x.

c. Two angles are complementary if the sum of their measures is 90°. If the measure of one angle is x degrees, represent the measure of the other angle as an expression in x.

d. If x is the first of two consecutive integers, represent the second integer as an expression in x.

Solution: **a.** If two numbers have a sum of 20 and one number is x, the other number is "the rest of 20."

In words: | Twenty | minus | x |

Translate: 20 $-$ x

b. The older sister's age is

In words: | Eight years | added to | Younger sister's age |

Translate: 8 $+$ x

c.   In words: | Ninety | minus | x |

Translate: 90 $-$ x

d. The next consecutive integer is always one more than the previous integer.

In words: | The first integer | plus | one |

Translate: x $+$ 1

---

4   Often, an expression may be **simplified** by removing grouping symbols and combining any like terms. The **terms** of an expression are the addends of the expression. For example, in the expression $3x^2 + 4x$, the terms are $3x^2$ and $4x$.

EXPRESSION	TERMS
$-2x + y$	$-2x,\ \ y$
$3x^2 - \dfrac{y}{5} + 7$	$3x^2,\ \ -\dfrac{y}{5},\ \ 7$

Terms with the same variable(s) raised to the same power are called **like terms.** We can add or subtract like terms by using the distributive property. This process is called **combining like terms.**

EXAMPLE 6   Use the distributive property to simplify each expression.

a. $3x - 5x + 4$ **b.** $7yz + yz$ **c.** $4z + 6.1$

Solution: **a.** $3x - 5x + 4 = (3 - 5)x + 4$ Apply the distributive property.

$$= -2x + 4$$

b. $7yz + yz = (7 + 1)yz = 8yz$

c. $4z + 6.1$ cannot be simplified further since $4z$ and 6.1 are not like terms.

The distributive property can also be used to multiply. For example,

$$-2(x + 3) = -2(x) + (-2)(3) = -2x - 6$$

The associative and commutative properties may sometimes be needed when we simplify expressions in order to rearrange and group like terms.

$$-7x^2 + 5 + 3x^2 - 2 = -7x^2 + 3x^2 + 5 - 2$$
$$= (-7 + 3)x^2 + (5 - 2)$$
$$= -4x^2 + 3$$

EXAMPLE 7   Simplify each expression.

a. $3xy - 2xy + 5 - 7 + xy$ **b.** $7x^2 + 3 - 5(x^2 - 4)$

c. $(2.1x - 5.6) - (-x - 5.3)$

Solution: **a.** $3xy - 2xy + 5 - 7 + xy = 3xy - 2xy + xy + 5 - 7$ Apply the commutative property.

$$= (3 - 2 + 1)xy + (5 - 7)$$ Apply the distributive property.

$$= 2xy - 2$$ Simplify.

b. $7x^2 + 3 - 5(x^2 - 4) = 7x^2 + 3 - 5x^2 + 20$ Apply the distributive property.

$$= 2x^2 + 23$$ Simplify.

c. Think of $-(-x - 5.3)$ as $-1(-x - 5.3)$ and use the distributive property.

$$(2.1x - 5.6) - 1(-x - 5.3) = 2.1x - 5.6 + 1x + 5.3$$

$$= 3.1x - 0.3$$ Combine like terms.

EXERCISE SET 1.4

Simplify each expression. Round Exercises 19 and 20 to the nearest ten thousandth. See Example 1.

1. $3(5 - 7)^4$ 48

2. $7(3 - 8)^2$ 175

3. $-3^2 + 2^3$ -1

4. $-5^2 - 2^4$ -41

5. $\dfrac{3 - (-12)}{-5}$ -3

6. $\dfrac{-4 - (-8)}{-4}$ -1

7. $|3.6 - 7.2| + |3.6 + 7.2|$ 14.4

8. $|8.6 - 1.9| - |2.1 + 5.3|$ -0.7

9. $\dfrac{(3 - \sqrt{9}) - (-5 - 1.3)}{-3}$ -2.1

10. $\dfrac{-\sqrt{16} - (6 - 2.4)}{-2}$ -3.8

11. $\dfrac{|3 - 9| - |-5|}{-3}$ $\dfrac{-1}{3}$ **12.** $\dfrac{|-14| - |2 - 7|}{-15}$ $\dfrac{-3}{5}$

13. $(-3)^2 + 2^3$ 17 **14.** $(-15)^2 - 2^4$ 209

15. $\dfrac{3(-2 + 1)}{5} - \dfrac{-7(2 - 4)}{1 - (-2)}$ $\dfrac{-79}{15}$

16. $\dfrac{-1 - 2}{2(-3) + 10} - \dfrac{2(-5)}{-1(8) + 1}$ $\dfrac{-61}{28}$

17. $\dfrac{\left(\dfrac{-3}{10}\right)}{\left(\dfrac{42}{50}\right)}$ $\dfrac{-5}{14}$ **18.** $\dfrac{\left(\dfrac{-5}{21}\right)}{\left(\dfrac{-6}{42}\right)}$ $\dfrac{5}{3}$

19. $\dfrac{-1.682 - 17.895}{(-7.102)(-4.691)}$ -0.5876

20. $\dfrac{(-5.161)(3.222)}{7.955 - 19.676}$ 1.4187

Find the value of each expression when $x = -2$, $y = -5$, and $z = 3$. See Example 2.

21. $x^2 + z^2$ 13 **22.** $y^2 - z^2$ 16

23. $-5(-x + 3y)$ 65 **24.** $-7(-y - 4z)$ 49

25. $\dfrac{3z - y}{2x - z}$ -2 **26.** $\dfrac{5x - z}{-2y + z}$ -1

Find the value of the expression when $x_1 = 2$, $x_2 = 4$, $y_1 = -3$, $y_2 = 2$. See Example 2.

27. $\dfrac{y_2 - y_1}{x_2 - x_1}$ $\dfrac{5}{2}$

28. $\sqrt{(x_2 - x_1)^2 + (y_2 - y_1)^2}$ $\sqrt{29}$

See Example 3.

29. The algebraic expression $8 + 2y$ represents the perimeter of a rectangle with width 4 and length y.

a. Complete the table that follows by evaluating this expression at given values of y.

LENGTH	y	5	7	10	100
PERIMETER	$8 + 2y$	18	22	28	208

b. Use the results of the table in (a) to answer the following question. As the width of a rectangle remains the same and the length increases, does the perimeter increase or decrease? Explain how you arrived at your answer.   increase

30. The algebraic expression πr^2 represents the area of a circle with radius r.   **b.** increase

a. Complete the table below by evaluating this expression at given values of r. (Use 3.14 for π.)

RADIUS	r	2	3	7	10
AREA	πr^2	12.56	28.26	153.86	314

b. As the radius of a circle increases, does its area increase or decrease? Explain your answer.

31. The algebraic expression $\dfrac{100x + 5000}{x}$ represents the cost per bookshelf (in dollars) of producing x bookshelves.   **b.** decrease

a. Complete the table below.

NUMBER OF BOOKSHELVES	x	10	100	1000
COST PER BOOKSHELF	$\dfrac{100x + 5000}{x}$	600	150	105

b. As the number of bookshelves manufactured increases, does the cost per bookshelf increase or decrease? Why do you think that this is so?

32. If c is degrees Celsius, the algebraic expression $1.8c + 32$ represents the equivalent temperature in degrees Fahrenheit.

a. Complete the table below.

DEGREES CELSIUS	c	-10	0	50
DEGREES FAHRENHEIT	$1.8c + 32$	14	32	122

b. As degrees Celsius increase, do degrees Fahrenheit increase or decrease?   increase

Write each of the following as an algebraic expression. See Examples 4 and 5.

33. Write an expression for the amount of money (in dollars) in n nickels. $0.05n$

34. Write an expression for the amount of money (in dollars) in d dimes. $0.1d$

35. Write an expression for the total amount of money (in dollars) in n nickels and d dimes. $0.05n + 0.1d$

36. Write an expression for the total amount of money (in dollars) in q quarters and d dimes. $0.25q + 0.1d$

37. Two numbers have a sum of 25. If one number is x, represent the other number as an expression in x.

38. Two numbers have a sum of 112. If one number is x, represent the other number as an expression in x. $112 - x$

39. Two angles are supplementary if the sum of their measures is 180. If the measure of one angle is x degrees, represent the measure of the other angle as an expression in x. $180 - x$

40. If the measure of an angle is $5x$ degrees, represent the measure of its complement as an expression in x. $90 - 5x$

41. The cost of x compact disks if each compact disk costs $6.49. $\$6.49x$

42. The cost of y books if each book costs $35.61.

43. If x is an odd integer, represent the next odd integer as an expression in x. $x + 2$

44. If $2x$ is an even integer, represent the next even integer as an expression in x. $2x + 2$

45. Write an expression for the tax on p dollars if the tax rate is 8.25%. $\$0.0825p$

46. Write an expression for the tax on $45 if the tax rate is r%. $\$45\left(\frac{r}{100}\right)$

47. The cost for renting a car at a local agency is $31 a day and $0.29 a mile. Write an expression for the total cost for renting a car for one day and driving it x miles. $31 + 0.29x$

48. Employees at Word Processors, Inc., are paid an hourly wage of $9.25 plus $0.60 for each page typed per hour. Find the total hourly wage of an employee who types x pages per hour. $9.25 + 0.6x$

Simplify each expression. See Examples 6 and 7.

49. $-9 + 4x + 18 - 10x$ $-15x + 18$

50. $5y - 14 + 7y - 20y$ $-8y - 14$

51. $5k - (3k - 10)$ $2k + 10$

52. $-11c - (4 - 2c)$ $-9c - 4$

53. $(3x + 4) - (6x - 1)$ $-3x + 5$

54. $(8 - 5y) - (4 + 3y)$ $4 - 8y$

55. $3(x - 2) + x + 15$ $4x + 9$

56. $-4(y + 3) - 7y + 1$ $-11y - 11$

57. $-(n + 5) + (5n - 3)$ $4n - 8$

58. $-(8 - t) + (2t - 6)$ $3t - 14$

59. $4(6n - 3) - 3(8n + 4)$ -24

60. $5(2z - 6) + 10(3 - z)$ 0

61. $3x - 2(x - 5) + x$ $2x + 10$

62. $7n + 3(2n - 6) - 2$ $13n - 20$

63. $-1.2(5.7x - 3.6) + 8.75x$ $1.91x + 4.32$

64. $5.8(-9.6 - 31.2y) - 18.65$ $-180.96y - 74.33$

65. $8.1z + 7.3(z + 5.2) - 6.85$ $15.4z + 31.11$

66. $6.5y - 4.4(1.8x - 3.3) + 10.95$

67. Do figures with the same surface area always have the same volume? To see, take two $8\frac{1}{2}$ by 11 inch sheets of paper and construct two cylinders using the following figures as a guide. Working with a partner, measure the height and the radius of each resulting cylinder and use the expression $\pi r^2 h$ to approximate each volume to the nearest tenth of a cubic inch. Explain your results.   no

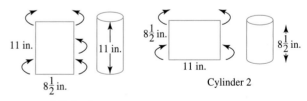

Cylinder 1
Volume = 63.2 cu. in.

Cylinder 2
Volume = 81.8 cu. in.

68. Use the same idea as in Exercise 67, work with a partner, and discover whether two rectangles with the same perimeter always have the same area. Explain your results.   no

The following graph is called a broken-line graph, or simply a line graph. This particular graph shows past, present, and future predicted population over 65. Just as with a bar graph, to find the population over 65 for a particular year, read the height of the corresponding point. To read the height, follow the point horizontally to the left until you reach the vertical axis.

69. Estimate the population over 65 in the year 1940.

70. Estimate the predicted population over 65 in the year 2030. 70 million

71. Estimate the predicted population over 65 in the year 2010. 42 million

72. Estimate the population over 65 in the year 1993.

73. Is the population over 65 increasing as time passes or decreasing? Explain how you arrived at your answer. increasing

74. The percent of Americans over 65 approximately tripled from 1900 to 1993. If this percent in 1900

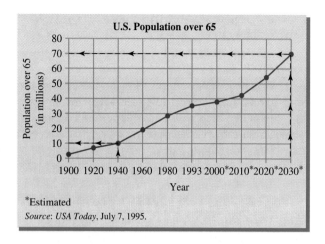

*Estimated
Source: USA Today, July 7, 1995.

was 4.1%, estimate the percent of Americans over 65 in the year 1993. 12.3%

69. 10 million **72.** 35 million

GROUP ACTIVITY

ANALYZING NEWSPAPER CIRCULATION

Use the following data to answer Questions 1 through 6 about newspaper circulation in the New York City metropolitan area.

NEWSPAPER CIRCULATION

NEWSPAPER	1993 DAILY CIRCULATION	1994 DAILY CIRCULATION	DAILY EDITION NEWSSTAND PRICE
Wall Street Journal	1,818,562	1,780,442	75¢
New York Times	1,141,366	1,114,905	60¢
New York Daily News	764,070	753,024	50¢
New York Newsday	747,890	693,556	50¢
New York Post	394,431	405,318	50¢

SOURCES: *1994 Editor & Publisher International Yearbook* and *1995 Editor & Publisher International Yearbook.*

(continued)

1. Did any newspaper gain circulation in 1994 from 1993? If so, which one(s) and by how much?

2. Construct a bar graph showing the change in circulation from 1993 to 1994 for each newspaper. Which newspaper experienced the largest change in circulation?

3. What was the total circulation of these New York-published newspapers in 1993 and in 1994? Did total circulation increase or decrease from 1993 to 1994? By how much?

4. Discuss factors that may have contributed to the overall change in daily newspaper circulation.

5. The population of New York City is approximately 7,325,000. The population of the New York City metropolitan area is approximately 16,271,000. Find the numbers of newspapers sold per person in 1994 for New York City and for the New York City metropolitan area.

Which of these figures do you think is more meaningful? Why? Why might neither of these figures be capable of describing the full circulation situation? (*Sources:* U.S. Bureau of the Census and the United Nations.)

6. Assuming that each copy was sold from a newsstand in New York City metropolitan area, use the daily edition newspaper prices given in the table to find the total amount spent each day on these New York-published newspapers in 1994. Find the total amount spent annually on these daily newspapers in 1994. (Remember: Daily editions are published only on weekdays.)

See Appendix D for Group Activity answers and suggestions.

CHAPTER 1   HIGHLIGHTS

DEFINITIONS AND CONCEPTS	EXAMPLES
SECTION 1.1 ALGEBRAIC EXPRESSIONS AND SETS OF NUMBERS	
Letters that represent numbers are called **variables.**	Examples of variables are $$x, a, m, y$$
An **algebraic expression** is formed by numbers and variables connected by the operations of addition, subtraction, multiplication, division, raising to powers, or taking roots.	Examples of algebraic expressions are $$7y, -3, \frac{x^2 - 9}{-2} + 14x, \sqrt{3} + \sqrt{m}$$
To **evaluate** an algebraic expression containing variables, substitute the given numbers for the variables and simplify. The result is called the **value** of the expression.	Evaluate $2.7x$ if $x = 3$. $$2.7x = 2.7(3)$$ $$= 8.1$$

(continued)

DEFINITIONS AND CONCEPTS	EXAMPLES

SECTION 1.1 ALGEBRAIC EXPRESSIONS AND SETS OF NUMBERS

Natural numbers: $\{1, 2, 3, ...\}$

Whole numbers: $\{0, 1, 2, 3, ...\}$

Integers: $\{..., -3, -2, -1, 0, 1, 2, 3, ...\}$

Each listing of three dots above is called an **ellipse.**

The members of a set are called its **elements.**

Set builder notation describes the elements of a set but does not list them.

Real numbers: $\{x \mid x$ corresponds to a point on the number line.$\}$

Rational numbers: $\{\frac{a}{b} \mid a$ and b are integers and $b \neq 0\}$.

Irrational numbers: $\{x \mid x$ is a real number and x is not a rational number$\}$.

If all the elements of set A are also in set B, we say that set A is a **subset** of set B, and we write $A \subseteq B$.

Given the set $\left\{-9.6, -5, -\sqrt{2}, 0, \frac{2}{5}, 101\right\}$, list the elements that belong to the set of

Natural numbers   101

Whole numbers   0, 101

Integers   −5, 0, 101

Real numbers $-9.6, -5, -\sqrt{2}, 0, \frac{2}{5}, 101$

Rational numbers $-9.6, -5, 0, \frac{2}{5}, 101$

Irrational numbers $-\sqrt{2}$

List the elements in the set $\{x \mid x$ is an integer between −2 and 5$\}$

$$\{-1, 0, 1, 2, 3, 4\}$$

$$\{1, 2, 4\} \subseteq \{1, 2, 3, 4\}.$$

SECTION 1.2 PROPERTIES OF REAL NUMBERS

Symbols: $=$ is equal to

$\neq$ is not equal to

$>$ is greater than

$<$ is less than

$\geq$ is greater than or equal to

$\leq$ is less than or equal to

$-5 = -5$

$-5 \neq -3$

$1.7 > 1.2$

$-1.7 < -1.2$

$\dfrac{5}{3} \geq \dfrac{5}{3}$

$-\dfrac{1}{2} \leq \dfrac{1}{2}$

Identity:

$a + 0 = a$ $0 + a = a$

$a \cdot 1 = a$ $1 \cdot a = a$

$3 + 0 = 3$ $0 + 3 = 3$

$-1.8 \cdot 1 = -1.8$ $1 \cdot -1.8 = -1.8$

Inverse:

$a + (-a) = 0$ $-a + a = 0$

$a \cdot \dfrac{1}{a} = 1$ $\dfrac{1}{a} \cdot a = 1$

$7 + (-7) = 0$ $-7 + 7 = 0$

$5 \cdot \dfrac{1}{5} = 1$ $\dfrac{1}{5} \cdot 5 = 1$

Commutative:

$a + b = b + a$

$a \cdot b = b \cdot a$

$x + 7 = 7 + x$

$9 \cdot y = y \cdot 9$

(continued)

DEFINITIONS AND CONCEPTS	EXAMPLES

SECTION 1.2 PROPERTIES OF REAL NUMBERS

Associative:

$(a + b) + c = a + (b + c)$

$(a \cdot b) \cdot c = a \cdot (b \cdot c)$

$(3 + 1) + 10 = 3 + (1 + 10)$

$(3 \cdot 1) \cdot 10 = 3 \cdot (1 \cdot 10)$

Distributive:

$a(b + c) = ab + ac$

$6(x + 5) = 6 \cdot x + 6 \cdot 5$

$\qquad\qquad = 6x + 30$

SECTION 1.3 OPERATIONS ON REAL NUMBERS

Absolute value:

$$|a| = \begin{cases} a \text{ if } a \text{ is } 0 \text{ or a positive number} \\ -a \text{ if } a \text{ is a negative number} \end{cases}$$

$|3| = 3, \ |0| = 0, \ |-7.2| = 7.2$

Adding real numbers:

1. To add two numbers with the same sign, add their absolute values and attach their common sign.

2. To add two numbers with different signs, subtract the smaller absolute value from the larger absolute value and attach the sign of the number with the larger absolute value.

$$\frac{2}{7} + \frac{1}{7} = \frac{3}{7}$$

$-5 + (-2.6) = -7.6$

$-18 + 6 = -12$

$20.8 + (-10.2) = 10.6$

Subtracting real numbers:

$$a - b = a + (-b)$$

$18 - 21 = 18 + (-21) = -3$

Multiplying and dividing real numbers:

The product or quotient of two numbers with the same sign is positive.

$(-8)(-4) = 32 \qquad \dfrac{-8}{-4} = 2$

$8 \cdot 4 = 32 \qquad \dfrac{8}{4} = 2$

The product or quotient of two numbers with different signs is negative.

$-17 \cdot 2 = -34 \qquad \dfrac{-14}{2} = -7$

$4(-1.6) = -6.4 \qquad \dfrac{22}{-2} = -11$

A natural number **exponent** is a shorthand notation for repeated multiplication of the same factor.

$3^4 = 3 \cdot 3 \cdot 3 \cdot 3 = 81$

The notation $\sqrt{a}$ is used to denote the **positive,** or **principal, square root** of a nonnegative number a.

$\sqrt{a} = b$ if $b^2 = a$ and b is positive.

$\sqrt{49} = 7$

Also,

$\sqrt[3]{a} = b$ if $b^3 = a$

$\sqrt[4]{a} = b$ if $b^4 = a$ and b is positive

$\sqrt[3]{64} = 4$

$\sqrt[4]{16} = 2$

(continued)

IMPORTANT CONCEPT OR SKILL	EXAMPLES

SECTION 1.4   ORDER OF OPERATIONS AND ALGEBRAIC EXPRESSIONS

Order of Operations

Simplify expressions using the order that follows. If grouping symbols such as parentheses are present, simplify expressions within those first, starting with the innermost set. If fraction bars are present, simplify the numerator and denominator separately.

1. Raise to powers or take roots in order from left to right.

2. Multiply or divide in order from left to right.

3. Add or subtract in order from left to right.

Simplify $\dfrac{42 - 2(3^2 - \sqrt{16})}{-8}$.

$$\frac{42 - 2(3^2 - \sqrt{16})}{-8} = \frac{42 - 2(9 - 4)}{-8}$$

$$= \frac{42 - 2(5)}{-8}$$

$$= \frac{42 - 10}{-8}$$

$$= \frac{32}{-8} = -4$$

CHAPTER 1 REVIEW

(1.1) *Find the value of each algebraic expression at the given replacement values.*

1. $7x$ when $x = 3$ 21

2. st when $s = 1.6$ and $t = 5$ 8

3. The humming bird has an average wing speed of 90 beats per second. The expression $90t$ gives the number of wing beats in t seconds. Calculate the number of wing beats in *1 hour* for the humming-bird.   324,000

5. {−2, 0, 2, 4, 6}
10. true   **11.** false
12. true   **13.** true
14. false   **15.** true
16. false   **17.** true
18. true   **19.** true
20. false   **21.** true
22. true   **23.** true
24. true   **25.** true
26. false   **27.** true

List the elements in each set.

4. $\{x \mid x$ is an odd integer between -2 and $4\}$ {−1, 1, 3}

5. $\{x \mid x$ is an even integer between -3 and $7\}$

6. $\{x \mid x$ is a negative whole number$\}$ { }

7. $\{x \mid x$ is a natural number that is not a rational number   { }

8. $\{x \mid x$ is a whole number greater than 5$\}$ {6, 7, 8, ...}

9. $\{x \mid x$ is an integer less than 3$\}$ {... −1, 0, 1, 2}

Determine whether each statement is true or false if $A = \{6, 10, 12\}$, $B = \{5, 9, 11\}$, $C = \{... -3, -2, -1, 0, 1, 2, 3, ...\}$, $D = \{2, 4, 6, ... 16\}$, $E = \{x \mid x$ is a rational number\}, $F = \{\ \}$, $G = \{x \mid x$ is an irrational number\}, and $H = \{x \mid x$ is a real number\}.

10. $10 \in D$ **11.** $B \in 9$ **12.** $\sqrt{169} \notin G$

13. $0 \notin F$ **14.** $\pi \in E$ **15.** $\pi \in H$

16. $\sqrt{4} \in G$ **17.** $-9 \in E$ **18.** $A \subseteq D$

19. $C \not\subseteq B$ **20.** $C \not\subseteq E$ **21.** $F \subseteq H$

22. $B \subseteq B$ **23.** $D \subseteq C$ **24.** $C \subseteq H$

25. $G \subseteq H$ **26.** $\{5\} \in B$ **27.** $\{5\} \subseteq B$

List the elements of the set $\left\{5, -\dfrac{2}{3}, \dfrac{8}{2}, \sqrt{9}, 0.3, \sqrt{7}, 1\dfrac{5}{8}, -1, \pi\right\}$ that are also elements of each given set.

28. Whole numbers   **29.** Natural numbers

30. Rational numbers   **31.** Irrational numbers

32. Real numbers   **33.** Integers

28. $\left\{5, \dfrac{8}{2}, \sqrt{9}\right\}$ **29.** $\left\{5, \dfrac{8}{2}, \sqrt{9}\right\}$ **30.** $\left\{5, -\dfrac{2}{3}, \dfrac{8}{2}, \sqrt{9}, 0.3, 1\dfrac{5}{8}, -1\right\}$ **31.** $\{\sqrt{7}, \pi\}$

32. $\left\{5, -\dfrac{2}{3}, \dfrac{8}{2}, \sqrt{9}, 0.3, \sqrt{7}, 1\dfrac{5}{8}, -1, \pi\right\}$ **33.** $\left\{5, \dfrac{8}{2}, \sqrt{9}, -1\right\}$

35. $n + 2n = -15$ **36.** $4(y + 3) = -1$ **39.** $9x - 10 = 5$ **42.** $\dfrac{2}{3} \neq 2\left(n + \dfrac{1}{4}\right)$

(1.2) *Write each statement using mathematical symbols.*

34. Twelve is the product of x and negative 4. $12 = -4x$

35. The sum of n and twice n is negative fifteen.

36. Four times the sum of y and three is -1.

37. The difference of t and five, multiplied by six is four. $6(t - 5) = 4$

38. Seven subtracted from z is six. $z - 7 = 6$

39. Ten less than the product of x and nine is five.

40. The difference of x and 5 is at least 12. $x - 5 \geq 12$

41. The opposite of four is less than the product of y and seven. $-4 < 7y$

42. Two-thirds is not equal to twice the sum of n and one-fourth.

43. The sum of t and six is not more than negative twelve. $t + 6 \leq -12$

Name the property illustrated.

44. $(M + 5) + P = M + (5 + P)$

45. $5(3x - 4) = 15x - 20$ **46.** $(-4) + 4 = 0$

47. $(3 + x) + 7 = 7 + (3 + x)$

48. $(XY)Z = (YZ)X$ **49.** $\left(-\dfrac{3}{5}\right) \cdot \left(-\dfrac{5}{3}\right) = 1$

50. $T \cdot 0 = 0$ **51.** $(ab)c = a(bc)$

52. $A + 0 = A$ **53.** $8 \cdot 1 = 8$

Find the additive inverse, or opposite.

54. $-\dfrac{3}{4}$ $\dfrac{3}{4}$ **55.** 0.6 -0.6

56. 0 0 **57.** 1 -1

Find the multiplicative inverse, or reciprocal.

58. $-\dfrac{3}{4}$ $-\dfrac{4}{3}$ **59.** 0.6 $\dfrac{1}{0.6}$

60. 0 undefined   **61.** 1 1

Complete the equation using the given property.

62. $5x - 15z = \underline{5(x - 3z)}$ Distributive property

63. $(7 + y) + (3 + x) = \underline{}$ Commutative property

64. $0 = \underline{}$ Additive inverse property

65. $1 = \underline{}$ Multiplicative inverse property

66. $[(3.4)(0.7)]5 = \underline{(3.4)[(0.7)5]}$ Associative property

67. $7 = \underline{7 + 0}$ Additive identity property

Insert $<$, $>$, or $=$ to make each statement true.

68. $-9 \;>\; -12$ **69.** $0 \;>\; -6$

70. $-3 \;<\; -1$ **71.** $7 \;=\; |-7|$

72. $-5 \;<\; -(-5)$ **73.** $-(-2) \;>\; -2$

(1.3) *Simplify.*

74. $-7 + 3$ -4 **75.** $-10 + (-25)$ -35

76. $5(-0.4)$ -2 **77.** $(-3.1)(-0.1)$ 0.31

78. $-7 - (-15)$ 8 **79.** $9 - (-4.3)$ 13.3

80. $(-6)(-4)(0)(-3)$ 0 **81.** $(-12)(0)(-1)(-5)$ 0

82. $(-24) \div 0$ undefined   **83.** $0 \div (-45)$ 0

84. $(-36) \div (-9)$ 4 **85.** $(60) \div (-12)$ -5

86. $\left(-\dfrac{4}{5}\right) - \left(-\dfrac{2}{3}\right)$ $-\dfrac{2}{15}$ **87.** $\left(\dfrac{5}{4}\right) - \left(-2\dfrac{3}{4}\right)$ 4

88. Determine the unknown fractional part.

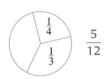

(1.4) *Simplify.*

89. $-5 + 7 - 3 - (-10)$ 9

90. $8 - (-3) + (-4) + 6$ 13

91. $3(4 - 5)^4$ 3 **92.** $6(7 - 10)^2$ 54

93. $\left(-\dfrac{8}{15}\right) \cdot \left(-\dfrac{2}{3}\right)^2$ $-\dfrac{32}{135}$ **94.** $\left(-\dfrac{3}{4}\right)^2 \cdot \left(-\dfrac{10}{21}\right)$ $-\dfrac{15}{56}$

95. $\dfrac{-\dfrac{6}{15}}{\dfrac{8}{25}}$ $-\dfrac{5}{4}$ **96.** $\dfrac{\dfrac{4}{9}}{-\dfrac{8}{25}}$ $-\dfrac{5}{2}$

97. $-\dfrac{3}{8} + 3(2) \div 6$ $\dfrac{5}{8}$

98. $5(-2) - (-3) - \dfrac{1}{6} + \dfrac{2}{3}$ $-6\dfrac{1}{2}$

99. $|2^3 - 3^2| - |5 - 7|$ -1

100. $|5^2 - 2^2| + |9 \div (-3)|$ 24

101. $(2^3 - 3^2) - (5 - 7)$ 1

102. $(5^2 - 2^4) + [9 \div (-3)]$ 6

44. associative property of addition   **45.** distributive property   **46.** additive inverse property
47. commutative property of addition   **48.** associative and commutative properties of multiplication
49. multiplicative inverse property   **50.** multiplication property of zero   **51.** associative property of multiplication
52. additive identify property   **53.** multiplicative identity property   **63.** $(3 + x) + (7 + y)$
64. $2 + (-2)$, for example   **65.** $2 \cdot 1/2$, for example

103. $\dfrac{(8 - 10)^3 - (-4)^2}{2 + 8(2) \div 4}$ -4

104. $\dfrac{(2 + 4)^2 + (-1)^5}{12 \div 2 \cdot 3 - 3}$ $\dfrac{7}{3}$ **105.** $\dfrac{(4 - 9) + 4 - 9}{10 - 12 \div 4 \cdot 8}$ $\dfrac{5}{7}$

106. $\dfrac{3 - 7 - (7 - 3)}{15 + 30 \div 6 \cdot 2}$ $-\dfrac{8}{25}$

107. $\dfrac{\sqrt{25}}{4 + 3 \cdot 7}$ $\dfrac{1}{5}$ **108.** $\dfrac{\sqrt{64}}{24 - 8 \cdot 2}$ 1

Find the value of each expression when $x = 0$, $y = 3$, and $z = -2$.

109. $x^2 - y^2 + z^2$ -5 **110.** $\dfrac{5x + z}{2y}$ $-\dfrac{1}{3}$

111. $\dfrac{-7y - 3z}{-3}$ 5 **112.** $(x - y + z)^2$ 25

CHAPTER 1 TEST

Determine whether each statement is true or false.

1. $-2.3 > 2.33$ true   **2.** $-6^2 = (-6)^2$ false

3. $-5 - 8 = -(5 - 8)$ false

4. $(-2)(-3)(0) = \dfrac{(-4)}{0}$ false

5. All natural numbers are integers.   true

6. All rational numbers are integers.   false

Simplify.

7. $5 - 12 \div 3(2)$ -3

8. $|4 - 6|^3 - (1 - 6^2)$ 43

9. $(4 - 9)^3 - |-4 - 6|^2$ -225

10. $[3|4 - 5|^5 - (-9)] \div (-6)$ -2

11. $\dfrac{6(7 - 9)^3 + (-2)}{(-2)(-5)(-5)}$ 1

Evaluate each expression when $q = 4$, $r = -2$, and $t = 1$.

12. $q^2 - r^2$ 12 **13.** $\dfrac{5t - 3q}{3r - 1}$ 1

14. The algebraic expression $5.75x$ represents the total cost for x adults to attend the theater.

   a. Complete the table that follows.

   b. As the number of adults increases does the total cost increase or decrease?   increase

16. $\dfrac{(6 - y)^2}{7} < -2$ **17.** $\dfrac{9z}{|-12|} \neq 10$ **18.** $3\left(\dfrac{n}{5}\right) = -n$ **19.** $20 = 2x - 6$ **22.** associative property of addition

113. The algebraic expression $2\pi r$ represents the circumference of (distance around) a circle of radius r.

   a. Complete the table below by evaluating the expression at given values of r. (Use 3.14 for π.)

RADIUS	r	1	10	100
CIRCUMFERENCE	$2\pi r$	6.28	62.8	628

   b. As the radius of a circle increases, does the circumference of the circle increase or decrease?

b. increase

ADULTS	x	1	3	10	20
TOTAL COST	$5.75x$	5.75	17.25	57.50	115.00

Write each statement using mathematical symbols.

15. Twice the absolute value of the sum of x and five is 30. $2|x + 5| = 30$

16. The square of the difference of six and y, divided by seven, is less than -2.

17. The product of nine and z, divided by the absolute value of -12, is not equal to 10.

18. Three times the quotient of n and five is the opposite of n.

19. Twenty is equal to 6 subtracted from twice x.

20. Negative two is equal to x divided by the sum of x and five. $-2 = \dfrac{x}{x + 5}$

Name each property illustrated.

21. $6(x - 4) = 6x - 24$ distributive property

22. $(4 + x) + z = 4 + (x + z)$

23. $(-7) + 7 = 0$ additive inverse property

24. $(-18)(0) = 0$ multiplication property of zero

25. Write an expression for the total amount of money (in dollars) in n nickels and d dimes. $0.05n + 0.1d$

EQUATIONS, INEQUALITIES, AND PROBLEM SOLVING

CERTIFICATES OF DEPOSIT
WEEK OF 10 24
$1000 MIN. DEPOSIT

	RATE	ANNUAL YIELD
3 MONTHS	2.60%	2.63%
6 MONTHS	3.20%	3.25%
1 YEAR	3.94%	4.00%
18 MONTHS	4.67%	4.75%
2 YEARS	4.67%	4.75%
30 MONTH	4.67%	4.75%

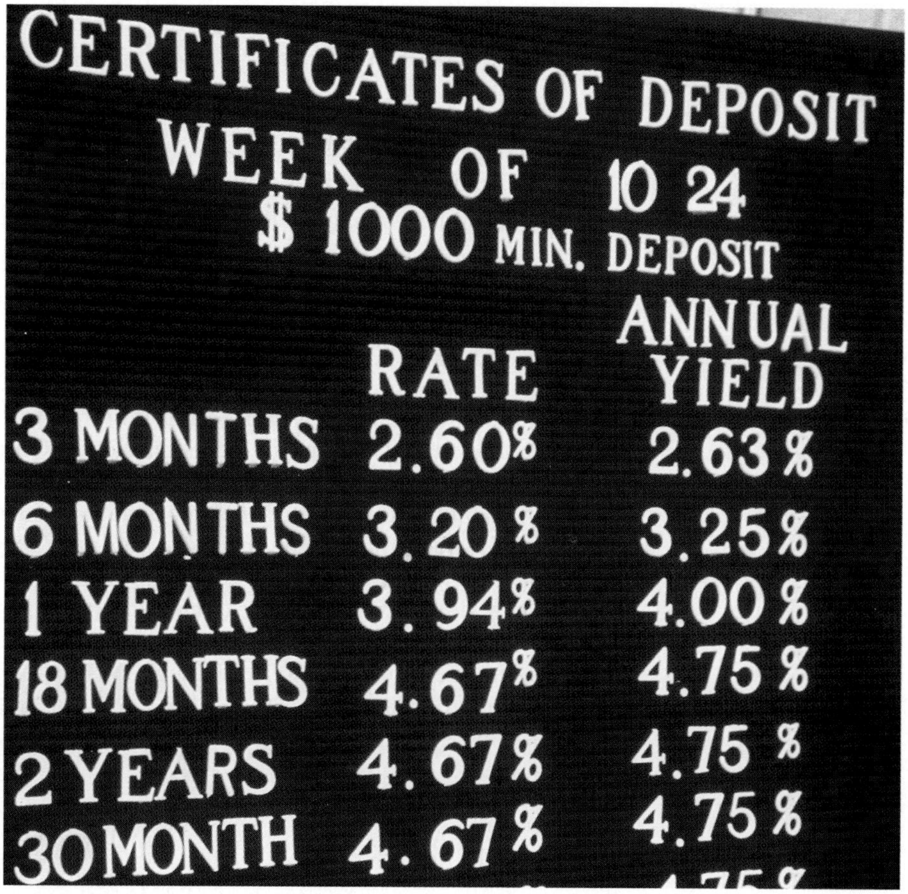

INVESTIGATING COMPOUND INTEREST

Banks offer a variety of services and products. These include savings tools such as savings accounts, money market accounts, and certificates of deposit. The interest rates that are offered for these savings tools often vary depending on how easily you may access your money, the degree of risk involved, or the length of time the money is deposited. The amount of interest you actually receive depends on the interest rate and how often the interest is compounded.

IN THE CHAPTER GROUP ACTIVITY ON PAGE 96, YOU WILL HAVE THE OPPORTUNITY TO INVESTIGATE AND COMPARE SEVERAL
DIFFERENT SAVINGS OPTIONS WITH DIFFERENT INTEREST RATES AND TYPES OF COMPOUNDING.

Mathematics is a tool for solving problems in such diverse fields as transportation, engineering, economics, medicine, business, and biology. We solve problems using mathematics by modeling real-world phenomena with mathematical equations or inequalities. Our ability to solve problems using mathematics, then, depends in part on our ability to solve equations and inequalities. In this chapter, we solve linear equations and inequalities in one variable and graph their solutions on number lines.

2.1 | LINEAR EQUATIONS IN ONE VARIABLE

O B J E C T I V E S

 Define linear equations.

 Solve linear equations using properties of equality.

3. Solve linear equations that can be simplified by combining like terms.

4. Solve linear equations containing fractions.

5. Recognize identities and equations with no solution.

6. Write algebraic expressions that can be simplified.

TAPE IA 2.1

 Suppose you are driving home from the grocery store after you have bought $3\frac{1}{2}$ pounds of steak for a special-occasion dinner. On the radio, you hear an ad from another grocery store and a special on steak, $5.60 per pound. You can't remember how much you paid for the steak that you just purchased, but you know you had $25 when you left for the store, and you've got $6 left. You wonder if the other store has a better deal.

This situation is a prime example of how we can use known number relationships to reveal unknown measurements. In this case, the unknown measurement is the cost c per pound of steak. Though we don't know the value of c, we do have known relationships involving c. We know for instance that the total cost of the meat is $3\frac{1}{2} \cdot c$ or $3.5c$, and we know that the difference of $25 and the total cost must be $6.

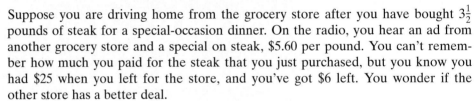

In words:	$25	minus	total cost of meat	is	$6
Translate:	25	−	3.5c	=	6

This **equation,** like every other equation, is a statement that two expressions are equal. Because we know the relationship of the measurement c to other measurements, we can eventually unravel the clues and find the value of c. In this section, we concentrate on solving equations such as this one, called **linear equations** in one variable. Linear equations are also called **first-degree equations** since the exponent on the variable is 1.

LINEAR EQUATIONS IN ONE VARIABLE

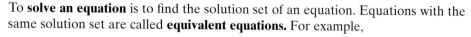

$$3x = -15 \qquad 7 - y = 3y \qquad 4n - 9n + 6 = 0 \qquad z = -2$$

LINEAR EQUATION IN ONE VARIABLE

A linear equation in one variable is an equation that can be written in the form

$$ax + b = c$$

where a, b, and c are real numbers and $a \neq 0$.

When a variable in an equation is replaced by a number and the resulting equation is true, then that number is called a **solution** of the equation. For example, 1 is a solution of the equation $3x + 4 = 7$, since $3(1) + 4 = 7$ is a true statement. But 2 is not a solution of this equation, since $3(2) + 4 = 7$ is not a true statement. The **solution set** of an equation is the set of solutions of the equation. For example, the solution set of $3x + 4 = 7$ is $\{1\}$.

To **solve an equation** is to find the solution set of an equation. Equations with the same solution set are called **equivalent equations.** For example,

$$3x + 4 = 7 \qquad 3x = 3 \qquad x = 1$$

are equivalent equations because they all have the same solution set, namely $\{1\}$. To solve an equation in x, we start with the given equation and write a series of simpler equivalent equations until we obtain an equation of the form

x = number

Two important properties are used to write equivalent equations.

THE ADDITION PROPERTY OF EQUALITY

If a, b, and c, are real numbers, then

$$a = b \quad \text{and} \quad a + c = b + c$$

are equivalent equations.

THE MULTIPLICATION PROPERTY OF EQUALITY

If $c \neq 0$, then

$$a = b \quad \text{and} \quad ac = bc$$

are equivalent equations.

The **addition property of equality** guarantees that the same number may be added to (or subtracted from) both sides of an equation, and the result is an equivalent equation. The **multiplication property of equality** guarantees that both sides of an equation may be multiplied by (or divided by) the same nonzero number, and the result is an equivalent equation.

For example, to solve $2x + 5 = 9$, use the addition and multiplication properties of equality to isolate x—that is, to write an equivalent equation of the form

$$x = \text{number}$$

EXAMPLE 1   Solve for x: $2x + 5 = 9$.

Solution: First, use the addition property of equality and subtract 5 from both sides.

$$2x + 5 = 9$$
$$2x + 5 - 5 = 9 - 5 \qquad \text{Subtract 5 from both sides.}$$
$$2x = 4 \qquad \text{Simplify.}$$

To finish solving for x, use the multiplication property of equality and divide both sides by 2.

$$\frac{2x}{2} = \frac{4}{2}$$
$$x = 2 \qquad \text{Simplify.}$$

To check, replace x in the original equation with 2.

$$2x + 5 = 9 \qquad \text{Original equation.}$$
$$2(2) + 5 = 9 \qquad \text{Let } x = 2.$$
$$4 + 5 = 9$$
$$9 = 9 \qquad \text{True.}$$

The solution set is $\{2\}$.

EXAMPLE 2   Solve for c: $25 - 3.5c = 6$.

Solution: First, use the addition property of equality and subtract 25 from both sides.

$$25 - 3.5c = 6$$
$$25 - 3.5c - 25 = 6 - 25 \qquad \text{Subtract 25 from both sides.}$$
$$-3.5c = -19 \qquad \text{Simplify.}$$
$$\frac{-3.5c}{-3.5} = \frac{-19}{-3.5} \qquad \text{Divide both sides by } -3.5.$$
$$c = \frac{38}{7} \qquad \text{Simplify } \frac{-19}{-3.5}.$$

Check to see that the solution set is $\left\{\dfrac{38}{7}\right\}$.

Recall from the beginning of this section that c in the equation given in Example 2 represents the cost of a pound of steak. Since $c = \frac{38}{7} \approx 5.43$, the steak purchased costs approximately \$5.43 per pound and is cheaper than the advertised price of \$5.60 at the other store.

3   Often, an equation can be simplified by removing any grouping symbols and combining any like terms.

EXAMPLE 3   Solve for x: $-6x - 1 + 5x = 3$.

Solution: First, the left side of this equation can be simplified by combining like terms $-6x$ and $5x$. Then use the addition property of equality and add 1 to both sides of the equation.

$$-6x - 1 + 5x = 3$$
$$-x - 1 = 3 \qquad \text{Combine like terms.}$$
$$-x - 1 + 1 = 3 + 1 \qquad \text{Add 1 to both sides of the equation.}$$
$$-x = 4 \qquad \text{Simplify.}$$

Notice that this equation is not solved for x since we have $-x$ or $-1x$, not x. To solve for x, divide both sides by -1.

$$\frac{-x}{-1} = \frac{4}{-1} \qquad \text{Divide both sides by } -1.$$
$$x = -4 \qquad \text{Simplify.}$$

Check to see that the solution set is $\{-4\}$.

If an equation contains parentheses, use the distributive property to remove them.

EXAMPLE 4   Solve for x: $2(x - 3) = 5x - 9$.

Solution: First, use the distributive property.

$$2(x - 3) = 5x - 9$$
$$2x - 6 = 5x - 9 \qquad \text{Apply the distributive property.}$$

Next, get variable terms on the same side of the equation by subtracting $5x$ from both sides.

$$2x - 6 - 5x = 5x - 9 - 5x \qquad \text{Subtract } 5x \text{ from both sides.}$$
$$-3x - 6 = -9 \qquad \text{Simplify.}$$

$$-3x - 6 + 6 = -9 + 6 \qquad \text{Add 6 to both sides.}$$

$$-3x = -3 \qquad \text{Simplify.}$$

$$\frac{-3x}{-3} = \frac{-3}{-3} \qquad \text{Divide both sides by } -3.$$

$$x = 1$$

Let $x = 1$ in the original equation to see that $\{1\}$ is the solution set.

4   If an equation contains fractions, we first clear the equation of fractions by multiplying both sides of the equation by the *least common denominator* (LCD) of all fractions in the equation.

EXAMPLE 5   Solve for y: $\dfrac{y}{3} - \dfrac{y}{4} = \dfrac{1}{6}$.

Solution: First, clear the equation of fractions by multiplying both sides of the equation by 12, the LCD of denominators 3, 4, and 6.

$$\frac{y}{3} - \frac{y}{4} = \frac{1}{6}$$

$$12\left(\frac{y}{3} - \frac{y}{4}\right) = 12\left(\frac{1}{6}\right) \qquad \text{Multiply both sides by the LCD 12.}$$

$$12\left(\frac{y}{3}\right) - 12\left(\frac{y}{4}\right) = 2 \qquad \text{Apply the distributive property.}$$

$$4y - 3y = 2 \qquad \text{Simplify.}$$

$$y = 2 \qquad \text{Simplify.}$$

To check, let $y = 2$ in the original equation.

$$\frac{y}{3} - \frac{y}{4} = \frac{1}{6} \qquad \text{Original equation.}$$

$$\frac{2}{3} - \frac{2}{4} = \frac{1}{6} \qquad \text{Let } y = 2.$$

$$\frac{8}{12} - \frac{6}{12} = \frac{1}{6} \qquad \text{Write fractions with the LCD.}$$

$$\frac{2}{12} = \frac{1}{6} \qquad \text{Subtract.}$$

$$\frac{1}{6} = \frac{1}{6} \qquad \text{Simplify.}$$

This is a true statement, so the solution set is $\{2\}$.

As a general guideline, the following steps may be used to solve a linear equation in one variable.

TO SOLVE A LINEAR EQUATION IN ONE VARIABLE

Step 1.   Clear the equation of fractions by multiplying both sides of the equation by the least common denominator (LCD) of all denominators in the equation.

Step 2.   Use the distributive property to remove grouping symbols such as parentheses.

Step 3.   Combine like terms on each side of the equation.

Step 4.   Use the addition property of equality to rewrite the equation as an equivalent equation, with variable terms on one side and numbers on the other side.

Step 5.   Use the multiplication property of equality to isolate the variable.

Step 6.   Check the proposed solution in the original equation.

EXAMPLE 6   Solve for x: $\dfrac{x+5}{2} + \dfrac{1}{2} = 2x - \dfrac{x-3}{8}$.

Solution: Multiply both sides of the equation by 8, the LCD of 2 and 8.

$$8\left(\frac{x+5}{2} + \frac{1}{2}\right) = 8\left(2x - \frac{x-3}{8}\right) \qquad \text{Multiply both sides by 8.}$$

$$4(x+5) + 4 = 16x - (x-3) \qquad \text{Apply the distributive property.}$$

$$4x + 20 + 4 = 16x - x + 3 \qquad \begin{array}{l}\text{Use the distributive property}\\ \text{to remove parentheses.}\end{array}$$

$$4x + 24 = 15x + 3 \qquad \text{Combine like terms.}$$

$$-11x + 24 = 3 \qquad \text{Subtract } 15x \text{ from both sides.}$$

$$-11x = -21 \qquad \text{Subtract 24 from both sides.}$$

$$\frac{-11x}{-11} = \frac{-21}{-11} \qquad \text{Divide both sides by } -11.$$

$$x = \frac{21}{11} \qquad \text{Simplify.}$$

To check, verify that replacing x with $\dfrac{21}{11}$ makes the original equation true. The solution set is $\left\{\dfrac{21}{11}\right\}$.

So far, each linear equation that we have solved has had a single solution. A linear equation in one variable that has exactly one solution is called a **conditional equation.** We will now look at two other types of equations: contradictions and identities.

An equation in one variable that has no solution is called a **contradiction,** and an equation in one variable that has every number (for which the equation is defined) as a solution is called an **identity.** The next examples show how to recognize contradictions and identities.

EXAMPLE 7   Solve for x: $3x + 5 = 3(x + 2)$.

Solution: First, use the distributive property and remove parentheses.

$$3x + 5 = 3(x + 2)$$
$$3x + 5 = 3x + 6 \qquad \text{Apply the distributive property.}$$
$$3x + 5 - 3x = 3x + 6 - 3x \qquad \text{Subtract } 3x \text{ from both sides.}$$
$$5 = 6$$

The equation $5 = 6$ is a false statement no matter what value the variable x might have. Thus the original equation has no solution. Its solution set is written either as $\{\ \}$ or $\varnothing$. This equation is a contradiction.

EXAMPLE 8   Solve for x: $6x - 4 = 2 + 6(x - 1)$.

Solution: First, use the distributive property and remove parentheses.

$$6x - 4 = 2 + 6(x - 1)$$
$$6x - 4 = 2 + 6x - 6 \qquad \text{Apply the distributive property.}$$
$$6x - 4 = 6x - 4 \qquad \text{Combine like terms.}$$

At this point we might notice that both sides of the equation are the same, so replacing x by any real number gives a true statement. Thus the solution set of this equation is the set of real numbers, and the equation is an identity. Continuing to "solve" $6x - 4 = 6x - 4$, we eventually arrive at the same conclusion.

$$6x - 4 + 4 = 6x - 4 + 4 \qquad \text{Add 4 to both sides.}$$
$$6x = 6x \qquad \text{Simplify.}$$
$$6x - 6x = 6x - 6x \qquad \text{Subtract } 6x \text{ from both sides.}$$
$$0 = 0 \qquad \text{Simplify.}$$

Since $0 = 0$ is a true statement for every value of x, the solution set is the set of all real numbers or $\{x | x \text{ is a real number}\}$, and the equation is called an identity.

6   In order to prepare for problem solving, we practice writing algebraic expressions that can be simplified.

EXAMPLE 9   Write the following as algebraic expressions. Then simplify.

a. The sum of two consecutive integers, if x is the first consecutive integer.

b. The perimeter of the triangle with sides of length x, $5x$, and $6x - 3$.

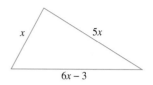

Solution: **a.** Recall that if x is the first integer, then the next consecutive integer is 1 more, or $x + 1$.

In words: | first integer | plus | next consecutive integer |

Translate: x $+$ $(x + 1)$

Then $x + (x + 1) = x + x + 1$

$\qquad\qquad\qquad = 2x + 1$ Simplify by combining like terms.

b. The perimeter of a triangle is the sum of the lengths of the sides.

In words: | side | + | side | + | side |

Translate: x $+$ $5x$ $+ (6x - 3)$.

Then $x + 5x + (6x - 3) = x + 5x + 6x - 3$

$\qquad\qquad\qquad\qquad = 12x - 3$ Simplify.

MENTAL MATH

Simplify each expression. See Example 3.

1. $3x + 5x + 6 + 15$ $8x + 21$ **2.** $8y + 3y + 7 + 11$ $11y + 18$ **3.** $5n + n + 3 - 10$ $6n - 7$

4. $m + 2m + 4 - 8$ $3m - 4$ **5.** $8x - 12x + 5 - 6$ $-4x - 1$ **6.** $4x - 10x + 13 - 16$ $-6x - 3$

EXERCISE SET 2.1

Solve for the variable. See Examples 1 and 2.

1. $x + 2.8 = 1.9$ $\{-0.9\}$ **2.** $y - 8.6 = -6.3$ $\{2.3\}$

3. $5x - 4 = 26$ $\{6\}$ **4.** $2y - 3 = 11$ $\{7\}$

5. $-4.1 - 7z = 3.6$ $\{-1.1\}$ **6.** $10.3 - 6x = -2.3$ $\{2.1\}$

7. $5y + 12 = 2y - 3$ $\{-5\}$ **8.** $4x + 14 = 6x + 8$ $\{3\}$

Solve for the variable. See Examples 3 and 4.

9. $8x - 5x + 3 = x - 7 + 10$ $\{0\}$

10. $6 + 3x + x = -x + 2 - 26$ $\{-6\}$

11. $5x + 12 = 2(2x + 7)$ {2}   **12.** $2(x + 3) = x + 5$ {−1}

13. $3(x − 6) = 5x$ {−9}   **14.** $6x = 4(5 + x)$ {10}

15. $−2(5y − 1) − y = −4(y − 3)$ $\left\{\dfrac{−10}{7}\right\}$

16. $−3(2w − 7) − 10 = 9 − 2(5w + 4)$ $\left\{\dfrac{−5}{2}\right\}$

☑ 17. a. Simplify the expression $4(x + 1) + 1$. $4x + 5$
   b. Solve the equation $4(x + 1) + 1 = −7$. {−3}
   c. Explain the difference between solving an equation for a variable and simplifying an expression.   Answers will vary.

☑ 18. Explain why the multiplication property of equality does not include multiplying both sides of an equation by 0. (Hint: Write down a false statement and then multiply both sides by 0. Is the result true or false? What does this mean?) Answers will vary.

Solve for the variable. See Examples 5 and 6.

19. $\dfrac{x}{2} + \dfrac{2}{3} = \dfrac{3}{4}$ $\left\{\dfrac{1}{6}\right\}$ **20.** $\dfrac{x}{2} + \dfrac{x}{3} = \dfrac{5}{2}$ {3}

21. $\dfrac{3t}{4} − \dfrac{t}{2} = 1$ {4}   **22.** $\dfrac{4r}{5} − 7 = \dfrac{r}{10}$ {10}

23. $\dfrac{n − 3}{4} + \dfrac{n + 5}{7} = \dfrac{5}{14}$ {1} **24.** $\dfrac{2 + h}{9} + \dfrac{h − 1}{3} = \dfrac{1}{3}$ {1}

25. $\dfrac{3x − 1}{9} + x = \dfrac{3x + 1}{3} + 4$ $\left\{\dfrac{40}{3}\right\}$

26. $\dfrac{2z + 7}{8} − 2 = z + \dfrac{z − 1}{2}$ $\left\{−\dfrac{1}{2}\right\}$

Solve the following. See Examples 7 and 8.

27. $4(n + 3) = 2(6 + 2n)$ **28.** $6(4n + 4) = 8(3 + 3n)$

29. $3(x − 1) + 5 = 3x + 2$ {$x | x$ is a real number}

30. $5x − (x + 4) = 4 + 4(x − 2)$ {$x | x$ is a real number}

☑ 31. In your own words, explain why the equation $x + 7 = x + 6$ has no solution while the solution set of the equation $x + 7 = x + 7$ contains all real numbers. Answers will vary.

☑ 32. In your own words, explain why the equation $x = −x$ has one solution, namely 0, while the solution set of the equation $x = x$ is all real numbers. Answers will vary.

Solve the following.   **33.** {4.2}

33. $−5x + 1.5 = −19.5$ **34.** $−3x − 4.7 = 11.8$ {−5.5}

35. $x − 10 = −6x + 4$ {2}   **36.** $4x − 7 = 2x − 7$ {0}

37. $3x − 4 − 5x = x + 4 + x$ {−2}

38. $13x − 15x + 8 = 4x + 2 − 24$ {5}

39. $5(y + 4) = 4(y + 5)$ {0}   **40.** $6(y − 4) = 3(y − 8)$ {0}

41. $0.6x − 10 = 1.4x − 14$ **42.** $0.3x + 2.4 = 0.1x + 4$

43. $6x − 2(x − 3) = 4(x + 1) + 4$ { }

44. $10x − 2(x + 4) = 8(x − 2) + 6$ { }

45. $\dfrac{3}{8} + \dfrac{b}{3} = \dfrac{5}{12}$ $\left\{\dfrac{1}{8}\right\}$ **46.** $\dfrac{a}{2} + \dfrac{7}{4} = 5$ $\left\{\dfrac{13}{2}\right\}$

47. $z + 3(2 + 4z) = 6(z + 1) + 5z$ {0}

48. $4(m − 6) − m = 8(m − 3) − 5m$

49. $\dfrac{3t + 1}{8} = \dfrac{5 + 2t}{7} + 2$ **50.** $4 − \dfrac{2z + 7}{9} = \dfrac{7 − z}{12}$

51. $\dfrac{m − 4}{3} − \dfrac{3m − 1}{5} = 1$ **52.** $\dfrac{n + 1}{8} − \dfrac{2 − n}{3} = \dfrac{5}{6}$

53. $\dfrac{x}{5} − \dfrac{x}{4} = \dfrac{1}{2}(x − 2)$ $\left\{\dfrac{20}{11}\right\}$ **54.** $\dfrac{y}{3} + \dfrac{y}{5} = \dfrac{1}{10}(y + 3)$

55. $5(x − 2) + 2x = 7(x + 4)$ { }

56. $3x + 2(x + 4) = 5(x + 1) + 3$

57. $y + 0.2 = 0.6(y + 3)$ {4}

58. $−(w + 0.2) = 0.3(4 − w)$ {−2}

59. $2y + 5(y − 4) = 4y − 2(y − 10)$ {8}

60. $9c − 3(6 − 5c) = c − 2(3c + 9)$ {0}

61. $2(x − 8) + x = 3(x − 6) + 2$

62. $4(x + 5) = 3(x − 4) + x$ { }

63. $\dfrac{5x − 1}{6} − 3x = \dfrac{1}{3} + \dfrac{4x + 3}{9}$ $\left\{\dfrac{−15}{47}\right\}$

64. $\dfrac{2r − 5}{3} − \dfrac{r}{5} = 4 − \dfrac{r + 8}{10}$ $\left\{\dfrac{146}{17}\right\}$

65. $−2(b − 4) − (3b − 1) = 5b + 3$ $\left\{\dfrac{3}{5}\right\}$

66. $4(t − 3) − 3(t − 2) = 2t + 8$ {−14}

67. $1.5(4 − x) = 1.3(2 − x)$ {17}

68. $2.4(2x + 3) = −0.1(2x + 3)$ {−1.5}

69. $\dfrac{1}{4}(a + 2) = \dfrac{1}{6}(5 − a)$ $\left\{\dfrac{4}{5}\right\}$

70. $\dfrac{1}{3}(8 + 2c) = \dfrac{1}{5}(3c − 5)$ {−55}

27. {$n | n$ is a real number}   **28.** {$n | n$ is a real number}   **41.** {5}   **42.** {8}   **48.** {$m | m$ is a real number}   **49.** {29}

50. {19}   **51.** {−8}   **52.** {3}   **54.** $\left\{\dfrac{9}{13}\right\}$ **56.** {$x | x$ is a real number}   **61.** {$x | x$ is a real number}

Write the following as algebraic expressions. Then simplify. See Example 9.

71. The perimeter of the square with side length y. $4y$

72. The perimeter of the rectangle with length x and width $x - 5$. $4x - 10$

73. The sum of three consecutive integers if the first integer is z. $3z + 3$

74. The sum of three consecutive odd integers if the first integer is x. $3x + 6$

75. The total amount of money (in dollars) in x nickels and $x + 3$ dimes. $0.15x + 0.3$

76. The total amount of money (in cents) in y quarters and $2y - 1$ nickels. $35y - 5$

77. A piece of land along Bayou Liberty is to be fenced and subdivided as shown so that each rectangle has the same dimensions. Express the total amount of fencing needed as an algebraic expression in x. $10x + 3$

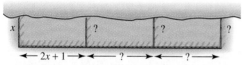

78. Write the total square feet of area of the floor plan shown as an algebraic expression in x. $8x - 14$

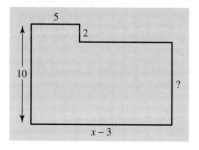

Find the value of K such that the equations are equivalent.

79. $3.2x + 4 = 5.4x - 7$
$3.2x = 5.4x + K$
$K = -11$

80. $-7.6y - 10 = -1.1y + 12$
$-7.6y = -1.1y + K$
$K = 22$

81. $\dfrac{x}{6} + 4 = \dfrac{x}{3}$
$x + K = 2x$
$K = 24$

82. $\dfrac{5x}{4} + \dfrac{1}{2} = \dfrac{x}{2}$
$5x + K = 2x$
$K = 2$

Solve and check.

83. $2.569x = -12.48534$ $\{-4.86\}$

84. $-9.112y = -47.537304$ $\{5.217\}$

85. $2.86z - 8.1258 = -3.75$ $\{1.53\}$

86. $1.25x - 20.175 = -8.15$ $\{9.62\}$

Review Exercises

87. Recall from Section 1.3 that a game is fair if each team or player has an equal chance of winning. Cut or tear a sheet of paper into 10 pieces, numbering each piece from 1 to 10. Place the pieces into a bag. Draw 2 pieces from the bag, record their sum, and then return them to the bag. If the sum is 10 or less, player 1 gets a point. If their sum is more than 10, player 2 gets a point. Is this a fair game? Try it and see.   not a fair game

Find the value of each expression when $y_1 = -3$, $y_2 = 4$, and $y_3 = 0$. See Section 1.4.

88. $(y_1)^2$ 9

89. $y_2 - y_1$ 7

90. $y_1 + y_2 + y_3$ 1

91. $\dfrac{y_3}{3y_1}$ 0

92. $\dfrac{6y_2}{y_1}$ -8

93. $y_1 - y_2$ -7

A Look Ahead

EXAMPLE

Solve for x: $5x(x - 1) + 14 = x(4x - 3) + x^2$.

Solution: $5x^2 - 5x + 14 = 4x^2 - 3x + x^2$

$5x^2 - 5x + 14 = 5x^2 - 3x$

$-5x + 14 = -3x$

$14 = 2x$

$7 = x$

Solve the following. See example.

94. $x(x - 6) + 7 = x(x + 1)$ {1}

95. $7x^2 + 2x - 3 = 6x(x + 4) + x^2$ $\left\{\dfrac{-3}{22}\right\}$

96. $3x(x + 5) - 12 = 3x^2 + 10x + 3$ {3}

97. $x(x + 1) + 16 = x(x + 5)$ {4}

2.2 | AN INTRODUCTION TO PROBLEM SOLVING

TAPE IA 2.2

OBJECTIVE

1   Apply the steps for problem solving.

Our main purpose for studying algebra is to solve problems. In previous sections, we have prepared for problem solving by writing phrases as algebraic expressions and sentences as equations. In this section, we now draw upon this experience as we write equations that describe or model a problem. The following problem-solving steps will be used throughout this text and may also be used to solve real-life problems that occur outside the mathematics classroom.

PROBLEM-SOLVING STEPS

1. UNDERSTAND the problem. During this step, don't work with variables, but simply become comfortable with the problem. Some ways of accomplishing this are listed next.
   - Read and reread the problem.
   - Construct a drawing to help visualize the problem.
   - Propose a solution and check. Pay careful attention as to how you check your proposed solution. This will help later when you write an equation to model the problem.
2. ASSIGN a variable to an unknown in the problem. Use this variable to represent any other unknown quantities.
3. ILLUSTRATE the problem. A diagram or chart *using the assigned variables* can often help us visualize the known facts.
4. TRANSLATE the problem into a mathematical model. This is often an equation.
5. COMPLETE the work. This often means to solve the equation.
6. INTERPRET the results. *Check* the proposed solution in the stated problem and *state* your conclusion.

Let's review these steps by solving a problem involving unknown numbers.

EXAMPLE 1   Find two numbers such that the second number is 3 more than twice the first number and the sum of the two numbers is 72.

Solution: **1.** UNDERSTAND the problem. Read and reread the problem and then propose a solution. For example, if the first number is 25, then the second number is 3 more than twice 25, or 53. The sum of 25 and 53 is 78, not the required sum, but we have gained some invaluable information about the problem. First, we know that the first number is less than 25 since our guess led to a sum greater than the required sum. Also, we have gained some information as to how to model the problem. Remember: The purpose of guessing a solution is not to guess correctly but to gain confidence and to help understand the problem and how to model it.

2. ASSIGN a variable. Use this variable to represent any other known quantities. If we let

the first number $= x$, then

the second number $= 2x + 3$

↑   3 more than
twice the second number

3. ILLUSTRATE the problem. No illustration is needed here.

4. TRANSLATE the problem into a mathematical model. To do so, we use the fact that the sum of the numbers is 72. First, write this relationship in words and then translate to an equation.

In words:	First number	added to	Second number	is	72
Translate:	x	$+$	$(2x + 3)$	$=$	72

5. COMPLETE the work by solving the equation.

$$x + (2x + 3) = 72$$
$$x + 2x + 3 = 72 \quad \text{Remove parentheses.}$$
$$3x + 3 = 72 \quad \text{Combine like terms.}$$
$$3x = 69 \quad \text{Subtract 3 from both sides.}$$
$$x = 23 \quad \text{Divide both sides by 3.}$$

6. INTERPRET. Here, we *check* our work and *state* the solution. Recall that if the first number $x = 23$, then the second number $2x + 3 = 2 \cdot 23 + 3 = 49$.

Check: Is the second number 3 more than twice the first number? Yes, since 3 more than twice 23 is $46 + 3$, or 49. Also, their sum, $23 + 49 = 72$, is the required sum.

State: The two numbers are 23 and 49.

Much of today's rates and statistics are given as percents. Interest rates, tax rates, nutrition labeling, and percent of households in a given category are just a few examples. Before we practice solving problems containing percents, let's take a moment and review the meaning of percent and how to find a percent of a number.

The word *percent* means *per hundred,* and the symbol % is used to denote percent. This means that 23% is 23 per hundred, or $\frac{23}{100}$. Also,

$$41\% = \frac{41}{100} = 0.41$$

To find a percent of a number, multiply. For example, to find 16% of 25, find the product of 16% (written as a decimal) and 25.

EXAMPLE 2   Find 16% of 25.

Solution:
$$16\% \cdot 25 = 0.16 \cdot 25$$
$$= 4$$

Thus, 16% of 25 is 4.

Next, we solve a problem containing percent.

EXAMPLE 3   Suppose that Service Merchandise just announced an 8% decrease in the price of their Compaq Presario computers. If one particular computer model sells for $2162 after the decrease, find the original price of this computer.

Solution: **1.** UNDERSTAND. Read and reread the problem. Recall that a percent decrease means a percent of the original price. Let's guess that the original price of the computer is $2500. The amount of decrease is then 8% of $2500, or (0.08)($2500) = $200. This means that the new price of the computer is the original price minus the decrease, or $2500 − $200 = $2300. Our guess is incorrect, but we now have an idea of how to model this problem.

2. ASSIGN a variable. Let $x =$ the original price of the computer.

3. ILLUSTRATE the problem. No illustration is needed.

4. TRANSLATE.

In words:	Original price of computer	minus	8% of original price	is	new price
Translate:	x	−	$0.08x$	=	2162

5. COMPLETE. Solve the equation.

$$x - 0.08x = 2162$$
$$0.92x = 2162 \qquad \text{Combine like terms.}$$
$$x = \frac{2162}{0.92} = 2350 \qquad \text{Divide both sides by 0.92.}$$

6. INTERPRET. *Check:* If the original price of the computer is $2350, the new price is

$$\$2350 - (0.08)(\$2350) = \$2350 - \$188$$
$$= \$2162, \text{ the given new price}$$

State: The original price of the computer is $2350.

EXAMPLE 4   A pennant in the shape of an isosceles triangle is to be constructed for the Slidell High School Athletic Club and sold as a fund raiser. The company manufacturing the pennant charges according to perimeter, and the athletic club has determined that a perimeter of 149 centimeters should make a nice profit. If each equal side of the triangle is twice the length of the third side, increased by 12 centimeters, find the lengths of the sides of the triangular pennant.

Solution: **1.** UNDERSTAND. Read and reread the problem. Recall that the perimeter of a triangle is the distance around. Let's guess that the third side of the triangular pennant is 20 centimeters. This means that each equal side is twice 20 centimeters, increased by 12 centimeters, or $2(20) + 12 = 52$ centimeters.

This gives a perimeter of $20 + 52 + 52 = 124$ centimeters. Our guess is incorrect, but we have a better understanding of how to model this problem.

2. ASSIGN a variable. Let

the third side of the triangle = x; then

the first side = twice the third side increased by 12

$$= 2 \qquad x \qquad + \qquad 12, \text{ or } 2x + 12, \text{ and}$$

the second side = $2x + 12$

3. ILLUSTRATE.

4. TRANSLATE.

In words: | First side | + | Second side | + | Third side | = | 149 |

Translate: $(2x + 12)$ + $(2x + 12)$ + x = 149

5. COMPLETE. Here, we solve the equation.

$$(2x + 12) + (2x + 12) + x = 149$$

$2x + 12 + 2x + 12 + x = 149$	Remove parentheses.
$5x + 24 = 149$	Combine like terms.
$5x = 125$	Subtract 24 from both sides.
$x = 25$	Divide both sides by 5.

6. INTERPRET. If the third side is 25 centimeters, then the first side is $2(25) + 12 = 62$ centimeters and the second side is 62 centimeters also.

Check: The first and second sides are each twice 25 centimeters increased by 12 centimeters or 62 centimeters. Also, the perimeter is $25 + 62 + 62 = 149$ centimeters, the required perimeter.

State: The dimensions of the triangle are 25 centimeters, 62 centimeters, and 62 centimeters.

EXERCISE SET 2.2

Solve. See Example 1.

1. Four times the difference of a number and 2 is the same as 6 times the number increased by 2. Find the number. -5

2. Twice the sum of a number and 3 is the same as 1 subtracted from the number. Find the number. -7

3. One number is 5 times another number. If the sum of the two numbers is 270, find the numbers.

4. One number is 6 less than another number. If the sum of the two numbers is 150, find the numbers.

Solve. See Example 2.

5. Find 30% of 260.   78   **6.** Find 70% of 180.   126

7. Find 12% of 16   1.92   **8.** Find 22% of 12.   2.64

9. The United States consists of 1943 million acres of land. Approximately 21% of this land is federally owned. Find the number of acres that are federally owned. (*Source: USA Today,* July 24, 1995.)

10. The state of Nevada contains the most federally owned acres of land in the United States. If 85% of the state's 71 million acres of land is federally owned, find the number of federally owned acres. (*Source: USA Today,* July 4, 1995.)

Nevada   60.35 million acres

11. At this writing, 47% of homes in the United States contain computers. If Charlotte, North Carolina, contains 110,000 homes, how many of these homes would you expect to have computers? (*Source: Telecommunication Research* survey.)   51,700

12. At this writing, 12% of homes in the United States contain on-line services. If Abilene, Texas, contains 40,000 homes, how many of these homes would you expect to have on-line services? (*Source: Telecommunication Research* survey.)

3. 45, 225   **4.** 72, 78   **9.** approximately 408.03 million acres   **12.** 4,800

The following graph is called a circle graph or a pie chart. The circle represents a whole or in this case 100%. This particular graph shows the kind of loans that customers get from credit unions.

Credit Union Loans

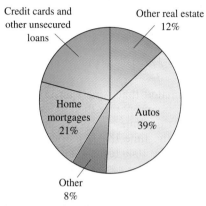

Credit cards and other unsecured loans

Other real estate 12%

Home mortgages 21%

Autos 39%

Other 8%

Source: National Credit Union Administration.

13. What percent of credit union loans are for credit cards and other unsecured loans?   20%

14. What types of loans make up most credit union loans?   autos

15. If the University of New Orleans Credit Union processed 300 loans last year, how many of these might we expect to be automobile loans?   117

16. If Homestead's Credit Union processed 537 loans last year, how many of these do you expect to be for either home mortgages or other real estate? (Round to the nearest whole.)   177

Solve. See Examples 3 and 4.

17. The B767-300 aircraft has twice as many seats as the B737-200 aircraft. If their total number of seats is 336, find the number of seats for each aircraft.

18. The governor of Delaware makes $17,000 more per year than the governor of Connecticut. If the total of these salaries is $173,000, find the salary of each governor.

19. A new FAX machine was recently purchased for an office in Hopedale for $464.40 including tax. If

the tax rate in Hopedale is 8%, find the price of the FAX machine before taxes. $430.00

20. A pre-medical student at a local university was complaining that she had just paid $86.11 for her human anatomy book, including tax. Find the price of the book before taxes if the tax rate at this university is 9%. $79.00

21. Two frames are needed with the same outside perimeter: one frame in the shape of a square and one in the shape of an equilateral triangle. Each side of the triangle is 6 centimeters longer than each side of the square. Find the dimensions of each frame.

18 cm

x

24 cm

22. The length of a rectangular sign is 2 feet less than three times its width. Find the dimensions if the perimeter is 28 feet.   width, 4 ft; length, 10 ft

Duluth, MN   153

Chicago, IL   307

x feet

23. In a blueprint of a rectangular room, the length is to be 2 centimeters greater than twice its width. Find the dimensions if the perimeter is to be 40 centimeters.   length, 14 cm; width, 6 cm

24. A plant food solution contains 5 cups of water for every 1 cup of concentrate. If the solution contains 78 cups of these two ingredients, find the number of cups of concentrate in the solution.   13

25. Manufacturers claim that a CD-ROM disk will last 20 years. Recently, statements made by the U.S. National Archives and Records Administration suggest that 20 years decreased by 75% is a more realistic lifespan because the aluminum substratum on which the data is recorded can be affected by oxidation. Find the lifespan of a CD-ROM accord-

17. B767-300: 224 seats   B737-200: 112 seats   **18.** gov. of DE $95,000; gov. of CT $78,000

ing to the U.S. National Archives and Records Administration.   5 years

CD-ROM disk

📖 **26.** In one year, 2.7% of India's forest was lost to deforestation. This percent represents 10,000 square kilometers of forest. Find the total square kilometers of forest in India before this decrease. (Round to the nearest whole square kilometer.)

27. The external tank of a NASA Space Shuttle contains the propellants used for the first 8.5 minutes after launch. Its height is 5 times the sum of its width and 1. If the sum of the height and width is 55.4 meters, find the dimensions of this tank.
width 8.4 m; height 47 m

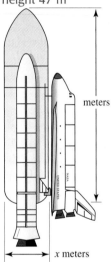

meters

x meters

28. The blue whale is the largest of whales. Its average weight is 3 times the difference of the average weight of a humpback whale and 5 tons. If the total of the average weights is 117 tons, find the average weight of each type of whale.

29. Recall that the sum of the angle measures of a triangle is 180°. Find the measures of the angles of a triangle if the measure of one angle is twice the measure of a second angle and the third angle measures 3 times the second angle decreased by 12.   64°, 32°, 84°

30. One angle is twice its complement increased by 30°. Find the measures of the two complementary angles.   20°, 70°

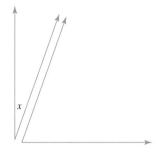

31. It is estimated that American taxpayers spent 1.5 billion hours preparing their 1995 tax form. About 44% of this time is spent on the 1040 form. Estimate the time that American taxpayers spent preparing the 1040 form. (*Source: USA Today,* July 25, 1995.)   660 million hours

32. It takes about 2.9 hours for an American taxpayer to educate himself/herself on the 1040 form. This time represents 25% of the total average time to complete the form. Find the total average time an American taxpayer requires to complete the 1040 form. (*Source: USA Today,* July 25, 1995.)

33. The zip codes of three Nevada locations—Fallon, Fernley, and Gardnerville Ranchos—are three consecutive even integers. If twice the first integer added to the third is 268,222, find each zip code.

34. During a recent year, the average SAT scores in math for the states of Alabama, Louisiana, and Michigan were 3 consecutive integers. If the sum of the first integer, second integer, and three times the third integer is 2637, find each score.

📒 **35.** Determine whether there are three consecutive integers such that their sum is three times the second integer.   any three consecutive integers

📒 **36.** Determine whether there are two consecutive odd integers such that 7 times the first exceeds 5 times the second by 54.   no solution

37. The sum of the angles in a triangle is 180°. Find the angles of a triangle whose two base angles are equal and whose third angle is 10° less than three times a base angle.   38°, 38°, 104°

---

26. 370,370 square kilometers   **28.** humpback 33 tons; blue whale 84 tons   **32.** 11.6 hours
33. Fallon's zip code is 89406; Fernley's zip code is 89408; Gardnerville's zip code is 89410
34. Alabama, 526; Louisiana, 527; Michigan, 528

38. Find an angle such that its supplement is equal to twice its complement increased by 50°.   50°

39. According to the Labor Department, the number of telephone operators will decrease to 46,000 by the year 2005. This represents a decrease of 23.3% from the number of telephone operators in 1995. Find the number of telephone operators in 1995. Round to the nearest whole.   59,974

40. The number of deaths by tornadoes from the 1940s to the 1980s has decreased by 70.86%. If the number of deaths in the 1980s was 521, find the number of deaths in the 1940s. Round to the nearest whole. (*Source:* National Weather Service.)   1788

41. In your own words, explain why you think that the need for telephone operators is decreasing. See Exercise 39.

42. In your own words, explain why you think that the number of deaths by tornadoes has decreased so much since the 1940s. See Exercise 40.

To break even in a manufacturing business, income or revenue R must equal the cost of production C. Use this information for Exercises 43 through 48.

43. The cost C to produce x number of skateboards is $C = 100 + 20x$. The skateboards are sold wholesale for $24 each, so revenue R is given by $R = 24x$. Find how many skateboards the manufacturer needs to produce and sell to break even. (*Hint:* Set the cost expression equal to the revenue expression and solve for x.)   25

44. The revenue R from selling x number of computer boards is given by $R = 60x$, and the cost C of producing them is given by $C = 50x + 5000$. Find how many boards must be sold to break even. Find how much money is needed to produce the break-even number of boards.   500 boards; $30,000

45. The cost C of producing x number of paperback books is given by $C = 4.50x + 2400$. Income R from these books is given by $R = 7.50x$. Find how many books should be produced and sold to break even.   800 books

46. Find the break-even quantity for a company that makes x number of computer monitors at a cost C given by $C = 875 + 70x$ and receives revenue R given by $R = 105x$.   25 monitors

47. In your own words, explain what happens if a company makes and sells fewer products than the break-even point.   It loses money.

48. In your own words, explain what happens if more products than the break-even point are made and sold.   The company makes a profit.

Review Exercises

Find the value of the following expressions for the given values. See Section 1.4.

49. $2a + b - c$; $a = 5$, $b = -1$, and $c = 3$ 6

50. $-3a + 2c - b$; $a = -2$; $b = 6$, and $c = -7$ -14

51. $4ab - 3bc$; $a = -5$, $b = -8$, and $c = 2$ 208

52. $ab + 6bc$; $a = 0$, $b = -1$ and $c = 9$ -54

53. $n^2 - m^2$; $n = -3$ and $m = -8$ -55

54. $2n^2 + 3m^2$; $n = -2$ and $m = 7$ 155

Use a calculator to find the value of the following expressions for the given values. See Section 1.4.

55. $P + Prt$; $P = 3,000$, $r = 0.0325$, $t = 2$ 3195

56. $\frac{1}{3}lwh$; $l = 37.8$, $w = 5.6$, $h = 7.9$ 557.424

57. $\frac{1}{3}Bh$; $B = 53.04$, $h = 6.89$ 121.8152

58. $\left(1 + \frac{r}{n}\right)^{nt}$; $r = 0.06$, $n = 3$, $t = 1$ 1.061208

41. greater use of calling cards and voice recognition technology, for example
42. better weather forecasting produces warnings to take cover, for example

2.3 | FORMULAS AND PROBLEM SOLVING

TAPE IA 2.3

O B J E C T I V E S

1. Solve a formula for a specified variable.
2. Use formulas to solve problems.

Solving problems that we encounter in the real world sometimes requires us to express relationships among measured quantities. A **formula** is an equation that describes a known relationship among measured phenomena, such as time, area, and gravity. Some examples of formulas are

FORMULA	MEANING
$I = PRT$	Interest = principal $\cdot$ rate $\cdot$ time
$A = lw$	Area of a rectangle = length $\cdot$ width
$d = rt$	Distance = rate $\cdot$ time
$C = 2\pi r$	Circumference of a circle = $2 \cdot \pi \cdot$ radius
$V = lwh$	Volume of a rectangular solid = length $\cdot$ width $\cdot$ height

Other formulas are listed in the front cover of this text. Notice that the formula for the volume of a rectangular solid $V = lwh$ is solved for V since V is by itself on one side of the equation with no V's on the other side of the equation. Suppose that the volume of a rectangular solid is known as well as its width and its length, and we wish to find its height. One way to find its height is to begin by solving the formula $V = lwh$ for h.

EXAMPLE 1   Solve $V = lwh$ for h.

Solution: To solve $V = lwh$ for h, isolate h on one side of the equation. To do so, divide both sides of the equation by lw.

$$V = lwh$$

$$\frac{V}{lw} = \frac{lwh}{lw} \qquad \text{Divide both sides by } lw.$$

$$\frac{V}{lw} = h \qquad \text{Simplify.}$$

Then to find the height of a rectangular solid, divide the volume by the product of its length and its width.

The following steps may be used to solve formulas and equations in general for a specified variable.

TO SOLVE EQUATIONS FOR A SPECIFIED VARIABLE

Step 1. Clear the equation of fractions by multiplying each side of the equation by the least common denominator.

Step 2. Use the distributive property to remove grouping symbols such as parentheses.

Step 3. Combine like terms on each side of the equation.

Step 4. Use the addition property of equality to rewrite the equation as an equivalent equation with terms containing the specified variable on one side and all other terms on the other side.

Step 5. Use the distributive property and the multiplication property of equality to isolate the specified variable.

EXAMPLE 2   Solve $3y - 2x = 7$ for y.

Solution: This is a linear equation in two variables. Often an equation such as this is solved for y in order to reveal some properties about the graph of this equation, which we will learn more about in Chapter 3. Since there are no fractions or grouping symbols, we begin with step 4 and isolate the term containing the specified variable y by adding $2x$ to both sides of the equation.

$$3y - 2x = 7$$
$$3y - 2x + 2x = 7 + 2x \qquad \text{Add } 2x \text{ to both sides.}$$
$$3y = 7 + 2x$$

To solve for y, divide both sides by 3.

$$\frac{3y}{3} = \frac{7 + 2x}{3} \qquad \text{Divide both sides by 3.}$$
$$y = \frac{2x + 7}{3} \quad \text{or} \quad y = \frac{2x}{3} + \frac{7}{3}$$

EXAMPLE 3   Solve $A = \dfrac{1}{2}(B + b)h$ for b.

Solution: Since this formula for finding the area of a trapezoid contains fractions, we begin by multiplying both sides of the equation by the LCD 2.

$$A = \frac{1}{2}(B + b)h$$

$$2 \cdot A = 2 \cdot \frac{1}{2}(B + b)h \qquad \text{Multiply both sides by 2.}$$

$$2A = (B + b)h \qquad \text{Simplify.}$$

Next, use the distributive property and remove parentheses.

$$2A = (B + b)h$$

$$2A = Bh + bh \qquad \text{Apply the distributive property.}$$

$$2A - Bh = bh \qquad \text{Isolate the term containing } b \\ \text{by subtracting } Bh \text{ from both sides.}$$

$$\frac{2A - Bh}{h} = \frac{bh}{h} \qquad \text{Divide both sides by } h.$$

$$\frac{2A - Bh}{h} = b, \quad \text{or} \quad b = \frac{2A - Bh}{h}$$

REMINDER Remember that we may isolate the specified variable on either side of the equation.

---

2 In this section, we also solve problems that can be modeled by known formulas. We use the same problem-solving steps that were introduced in the previous section. These steps have been slightly revised to include formulas.

PROBLEM-SOLVING STEPS

1. UNDERSTAND the problem. During this step don't work with variables (except for known formulas), but simply become comfortable with the problem. Some ways of accomplishing this are listed below.

   - Read and reread the problem.
   - Construct a drawing.
   - Look up an unknown formula.
   - Propose a solution and check. Pay careful attention as to how to check your proposed solution. This will help later when you write an equation to model the problem.

2. ASSIGN a variable to an unknown in the problem. Use this variable to represent any other unknown quantities.

3. ILLUSTRATE the problem. A diagram or chart using the assigned variables can often help us visualize the known facts.

4. TRANSLATE the problem into a mathematical model. This is often an equation.

5. COMPLETE the work. This often means to solve the equation.

6. INTERPRET the results: *Check* the proposed solution in the stated problem and *state* your conclusion.

Formulas are very useful in problem solving. For example, the compound interest formula

$$A = P\left(1 + \frac{r}{n}\right)^{nt}$$

is used by banks to compute the amount A in an account that pays compound interest. The variable P represents the principal, or amount invested in the account, r is the rate of interest, t is the time in years, and n is the number of times compounded per year.

EXAMPLE 4

Marial Callier just received an inheritance of $10,000 and plans to place all the money in a savings account that pays 5% compounded quarterly to help her son go to college in 3 years. How much money will be in the account in 3 years?

Solution:

1. UNDERSTAND. Read and reread the problem. The appropriate formula needed to solve this problem is the compound interest formula

$$A = P\left(1 + \frac{r}{n}\right)^{nt}$$

Make sure that you understand the meaning of all the variables in this formula.

2. ASSIGN. Since this is a direct application of the compound interest formula, the variables and their meaning are already assigned. They are for review:

   A = amount in the account after t years

   P = principal or amount invested

   t = time in years

   r = rate of interest

   n = number of times compounded per year

3. ILLUSTRATE. No illustration is needed.

4. TRANSLATE. Use the compound interest formula and let P = $10,000, r = 5% = 0.05, t = 3 years, and n = 4 since the account is compounded quarterly, or 4 times a year.

   Formula: $A = P\left(1 + \frac{r}{n}\right)^{nt}$

   Substitute: $A = 10,000\left(1 + \frac{0.05}{4}\right)^{4 \cdot 3}$

5. COMPLETE. Here, we simplify the right side of the equation.

   $A = 10,000\left(1 + \frac{0.05}{4}\right)^{4 \cdot 3}$

   $A = 10,000(1.0125)^{12}$ Simplify $\frac{1 + 0.05}{4}$ and write $4 \cdot 3$ as 12.

   $A \approx 10,000(1.160754518)$ Approximate $(1.0125)^{12}$.

   $A \approx 11,607.55$ Multiply and round to 2 decimal places.

6. INTERPRET. To *check* here, repeat your calculations to make sure that no error was made. Notice that $11,607.55 is a reasonable amount to have in the account after 3 years. *State:* In 3 years, the account contains $11,607.55. ▬▬▬▬

EXAMPLE 5

Solar cells convert the energy of sunlight into electrical energy. Suppose that 0.01 watt of electrical power is available for every square centimeter of solar cell in direct overhead sunlight. If a solar cell in the shape of a square is needed to deliver 16 watts, find the minimum side of the square.

Solution:

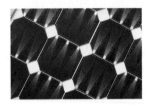

1. UNDERSTAND. Read and reread the problem. To become familiar with this problem, let's find the amount of energy available from a square with side length 20 centimeters. The area of this square is side squared, or $20^2 = 400$ square centimeters. Since 0.01 watt of power is available for every square centimeter, the power available from a square with an area of 400 square centimeters is 400(0.01) watts, or 4 watts. This is not the required 16 watts, but we now have a better understanding of the problem.

2. ASSIGN. Since we are looking for the length of a side of a square with a given area, we will use the formula for area of a square: $A = x^2$ where

A = area of the square and

x = side length of the square

3. ILLUSTRATE.

x ▢

Next, under TRANSLATE and COMPLETE, we have two parts.

First, find the square centimeters needed to generate 16 watts of power.

TRANSLATE. In words:

0.01 watts	·	area of square	=	total watts available
0.01	·	A	=	16

Translate:

COMPLETE. Solve $0.01A = 16$

$A = 1600$ Divide both sides by 0.01.

Second, find the side length of the square with an area of 1600 square centimeters.

TRANSLATE. Formula: $A = x^2$

Substitute: $1600 = x^2$

COMPLETE. Solve $1600 = x^2$

Here, x must be the number whose square is 1600, or 40.

6. INTERPRET. *Check* this solution just as we checked our proposed solution in *step 1. State:* A square with side length of 40 centimeters will generate 16 watts of power.

MENTAL MATH

Solve each equation for the specified variable. See Examples 4 and 5.

1. $2x + y = 5$; for y $y = 5 - 2x$ **2.** $7x - y = 3$; for y $y = 7x - 3$ **3.** $a - 5b = 8$; for a $a = 5b + 8$

4. $7r + s = 10$; for s $s = 10 - 7r$ **5.** $5j + k - h = 6$; for k $k = h - 5j + 6$ **6.** $w - 4y + z = 0$; for z $z = 4y - w$

4. $L = \dfrac{V}{WH}$ **5.** $y = \dfrac{9x - 16}{4}$ **6.** $y = \dfrac{17 - 2x}{3}$ **7.** $W = \dfrac{P - 2L}{2}$ **8.** $N = \dfrac{3M - A}{2}$ **9.** $A = \dfrac{J + 3}{C}$ **10.** $x = \dfrac{y - b}{m}$

11. $g = \dfrac{W}{h - 3t^2}$ **12.** $P = \dfrac{A}{rt + 1}$ **13.** $B = \dfrac{T - 2c}{AC}$ **14.** $b = \dfrac{A - 5HB}{5H}$ **15.** $r = \dfrac{C}{2\pi}$

EXERCISE SET 2.3

Solve each equation for the specified variable. See Examples 1–3.

1. $D = rt$; for t $t = \dfrac{D}{r}$ **2.** $W = gh$; for g $g = \dfrac{W}{h}$

3. $I = PRT$; for R $R = \dfrac{I}{PT}$ **4.** $V = LWH$; for L

5. $9x - 4y = 16$; for y **6.** $2x + 3y = 17$; for y

7. $P = 2L + 2W$; for W **8.** $A = 3M - 2N$; for N

9. $J = AC - 3$; for A **10.** $y = mx + b$; for x

11. $W = gh - 3gt^2$; for g **12.** $A = Prt + P$; for P

13. $T = C(2 + AB)$; for B **14.** $A = 5H(b + B)$; for b

15. $C = 2\pi r$; for r **16.** $S = 2\pi r^2 + 2\pi rh$; for h

17. $E = I(r + R)$; for r **18.** $A = P(1 + rt)$; for t

19. $s = \dfrac{n}{2}(a + L)$; for L **20.** $\dfrac{3}{4}(b - 2c) = a$; for b

21. $\dfrac{1}{u} - \dfrac{1}{v} = \dfrac{1}{w}$; for w **22.** $\dfrac{1}{r_1} + \dfrac{1}{r_2} = \dfrac{1}{R}$; for R

23. $N = 3st^4 - 5sv$; for v $v = \dfrac{3st^4 - N}{5s}$

24. $L = a + (n - 1)d$; for d

25. $S = \dfrac{a}{1 - r}$; for r **26.** $m = \dfrac{y_2 - y_1}{x_2 - x_1}$; for y_1

27. $S = 2LW + 2LH + 2WH$; for H

28. $T = 3vs - 4ws + 5vw$; for v

In this exercise set, round all dollar amounts to two decimal places.

Solve. See Example 4.

🖩 **29.** Complete the table and find the balance A if \$3500 is invested at an annual percentage rate of 3% for 10 years and compounded n times a year.

n	1	2	4	12	365
A	\$4703.71	\$4713.99	\$4719.22	\$4722.74	\$4724.45

🖩 **30.** Complete the table and find the balance A if \$5000 is invested at an annual percentage rate of 6% for 15 years and compounded n times a year.

n	1	2	4	12	365
A	\$11,982.79	\$12,136.31	\$12,216.10	\$12,270.47	\$12,297.10

31. If you are investing money in a savings account paying a rate of r, which account should you choose—an account compounded 4 times a year or 12 times a year? Explain your choice.

32. To borrow money at a rate of r, which bank should you choose—one compounding 4 times a year or 12 times a year? Explain your choice.

33. A principal of \$6000 is invested in an account paying an annual percentage rate of 4%. Find the amount

16. $h = \dfrac{S - 2\pi r^2}{2\pi r}$ **17.** $r = \dfrac{E}{I} - R$ **18.** $t = \dfrac{A - P}{Pr}$ **19.** $L = \dfrac{2s}{n} - a$ **20.** $b = \dfrac{4}{3}a + 2c$ **21.** $w = \dfrac{uv}{v - u}$

22. $R = \dfrac{r_1 r_2}{r_1 + r_2}$ **24.** $d = \dfrac{L - a}{n - 1}$ **25.** $r = \dfrac{S - a}{S}$ **26.** $y_1 = y_2 - m(x_2 - x_1)$ **27.** $H = \dfrac{S - 2LW}{2L + 2W}$ **28.** $v = \dfrac{T + 4ws}{3s + 5w}$

31. 12 times a year; you earn more interest   **32.** 4 times a year; you pay less interest

in the account after 5 years if the account is compounded.   **a.** $7313.97 **b.** $7321.14 **c.** $7325.98
(a) semiannually   **(b)** quarterly   **(c)** monthly

34. A principal of $25,000 is invested in an account paying an annual percentage rate of 5%. Find the amount in the account after 2 years if the account is compounded.
(a) semiannually   **(b)** quarterly   **(c)** monthly

Solve. See Examples 4 and 5.

35. The day's high temperature in Phoenix, Arizona, was recorded as 104°F. Write 104°F as degrees Celsius.   40°C

36. The annual low temperature in Nome, Alaska, was recorded as −15°C. Write −15°C as degrees Fahrenheit.   5°F

37. Suppose that 0.01 watt of electrical power is available for every square centimeter of solar cell in direct overhead sunlight. If a solar cell in the shape of a square is needed to deliver 25 watts, find the minimum side of the square.   50 cm

38. If a solar cell in the shape of a rectangle with width 30 centimeters is needed to deliver 15 watts, find the minimum length of the rectangle.   50 cm

39. Omaha, Nebraska, is about 90 miles from Lincoln, Nebraska. Irania must go to the law library in Lincoln to get a document for the law firm she works for. Find how long it takes her to drive **round-trip** if she averages 50 mph.

40. It took the Selby family $5\frac{1}{2}$ hours round-trip to drive from their house to their beach house 154 miles away. Find their average speed.   56 mph

41. A package of floor tiles contains 24 one-foot-square tiles. Find how many packages should be bought to cover a square ballroom floor whose side measures 64 feet.   171 packages

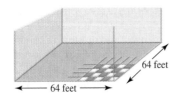

42. On-foot-square ceiling tiles are sold in packages of 50. Find how many packages must be bought for a rectangular ceiling 18 feet by 12 feet.   5 packages

43. The deepest hole in the ocean floor is beneath the Pacific Ocean and is called Hole 504B. It is located off the coast of Ecuador. Scientists are drilling it to learn more about the Earth's history. Currently, the hole is in the shape of a cylinder whose volume is approximately 3800 cubic feet and whose length is 1.3 miles. Find the radius of the hole to the nearest hundredth of a foot. (*Hint:* Make sure the same units of measurement are used.)   0.42 ft

44. The deepest manmade hole is called the Kola Superdeep Borehole. It is approximately 8 miles deep and is located near a small Russian town in the Arctic Circle. If it takes 7.5 hours to remove the drill from the bottom of the hole, find the rate that the drill can be retrieved in feet per second. Round to the nearest tenth. (*Hint:* Write 8 miles as feet, 7.5 hours as seconds, then use the formula $d = rt$.)   1.6 ft/sec

45. On April 1, 1985, *Sports Illustrated* published an April Fool's story by writer George Plimpton. He wrote that the New York Mets had discovered a man who could throw a 168-miles-per-hour fast ball. If the distance from the pitcher's mound to the plate is 60.5 feet, how long would it take for a ball thrown at that rate to travel that distance? (*Hint:* Write the rate 168 miles per hour in feet per second.   0.25 second

$$168 \text{ miles per hour} = \frac{168 \text{ miles}}{1 \text{ hour}}$$
$$= \frac{\_\_\_ \text{ feet}}{\_\_\_ \text{ seconds}}$$
$$= \frac{\_\_\_ \text{ feet}}{1 \text{ second}}$$
$$= \_\_\_ \text{ feet per second.}$$

Then use the formula $d = r \cdot t$.)

46. In 1945, Arthur C. Clarke, a scientist and science-fiction writer, predicted that an artificial satellite placed at a height of 22,248 miles directly above the equator would orbit the globe at the same speed with which the Earth was rotating. This belt along the equator is known as the Clarke belt. Use the formula for circumference of a circle and find the "length" of the Clarke belt. (*Hint:* Recall that the radius of the Earth is approximately 4000 miles. Round to the nearest whole mile.)
164,921 miles

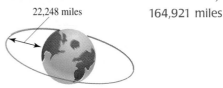

34. **a.** $27,595.32   **b.** $27,612.15   **c.** $27,623.53   39. 3.6 hr, or 3 hr and 36 min

47. An orbit such as Clarke's belt in Exercise 46 is called a geostationary orbit. In your own words, why do you think that communications satellites are placed in geostationary orbits?

48. How much do you think it costs each American to build a space shuttle? Write down your estimate. The Space Shuttle *Endeavour* was completed in 1992 and cost approximately $1.7 billion. If the population of the United States in 1992 was 250 million, find the cost per person to build the *Endeavour.* How close was your estimate?

49. Find *how much interest* $10,000 earns in 2 years in a certificate of deposit paying 8.5% interest compounded quarterly. $1831.96

50. Bryan, Eric, Mandy, and Melissa would like to go to Disneyland in 3 years. Their total cost should be $4500. If each invests $1000 in a savings account paying 5.5% interest, compounded semiannually will they have enough in 3 years?

51. A gallon of latex paint can cover 500 square feet. Find how many gallon containers of paint should be bought to paint two coats on each wall of a rectangular room whose dimensions are 14 feet by 16 feet (assume 8-foot ceilings).   2 gallons

52. A gallon of enamel paint can cover 300 square feet. Find how many gallon containers of paint should be bought to paint three coats on a wall measuring 21 feet by 8 feet.   2 gallons

53. A portion of the external tank of the Space Shuttle *Endeavour* is a liquid hydrogen tank. If the ends of the tank are hemispheres, find the volume of the tank. To do so, answer parts (a) through (c).

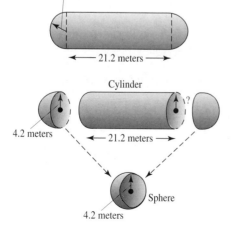

a. Find the volume of the cylinder shown. Round to 2 decimal places.   1174.86 cubic meters

b. Find the volume of the sphere shown. Round to 2 decimal places.   310.34 cubic meters

c. Add the results of parts (a) and (b). This sum is the approximate volume of the tank.

54. The space probe Pioneer 10 traveled from Mars to Jupiter, a distance of 619 million miles, in 21 months. Find the average speed of the probe in miles per hour. [*Hint:* Convert 21 months to hours (use 1 month = 30 days) and then use the formula $d = rt$.]   40,939 mph

55. Find how long it takes Mark to drive 135 miles on I-10 if he merges onto I-10 at 10 A.M. and drives nonstop with his cruise control set on 60 mph.

56. If the area of a triangular kite is 18 square feet and its base is 4 feet, find the height of the kite.   9 ft

47. to receive from and transmit to fixed locations on the Earth, for example
48. Estimates will vary; actual cost was $6.80 each.   **50.** yes, total of $4,707.07   **53. c.** 1485.20 cubic meters
55. 2.25 hr, or 2 hr and 15 min

57. The space shuttle has a cargo bay that is in the shape of a cylinder whose length is 18.3 meters and whose diameter is 4.6 meters. Find its volume.

58. Solar System distances are so great that units other than miles or kilometers are often used. For example, the astronomical unit (AU) is the average distance between the Earth and the Sun, or 92,900,000 miles. Use this information to convert each planet's distance in miles from the Sun to astronomical units. Round to three decimal places.

57. 96.807π cubic meters ≈ 304.12816 cubic meters

	MILES FROM THE SUN	AU FROM THE SUN		MILES FROM THE SUN	AU FROM THE SUN
Mercury	36 million	0.388	Saturn	886.1 million	9.538
Venus	67.2 million	0.723	Uranus	1,783 million	19.192
Earth	92.9 million	1.00	Neptune	2,793 million	30.065
Mars	141.5 million	1.523	Pluto	3,670 million	39.505
Jupiter	483.3 million	5.202			

*The measure of the chance or likelihood of an event occurring is its **probability**. A formula basic to the study of probability is the formula for the probability of an event when all the outcomes are equally likely. This formula is*

$$Probability\ of\ an\ event = \frac{number\ of\ ways\ that\ the\ event\ can\ occur}{number\ of\ possible\ outcomes}$$

For example, to find the probability that a single spin on the spinner will result in red, notice first that the spinner is divided into 8 parts, so there are 8 possible outcomes. Next, notice that there is only one sector of the spinner colored red, so the number of ways that the spinner will land on red is 1. Then this probability denoted by P(red) is

$$P(red) = \frac{1}{8}$$

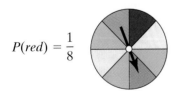

Find each probability in simplest form.

59. P(green) $\frac{1}{8}$

60. P(yellow) $\frac{1}{4}$

61. P(black) $\frac{1}{8}$

62. P(blue) $\frac{3}{8}$

63. P(green or blue) $\frac{1}{2}$

64. P(black or yellow) $\frac{3}{8}$

65. P(red, green, or black) $\frac{3}{8}$

66. P(yellow, blue, or black) $\frac{3}{4}$

67. P(white)   0

68. P(red, yellow, green, blue, or black)   1

69. From the previous probability formula, what do you think is always the probability of an event that is impossible to occur?   0

70. What do you think is always the probability of an event that is sure to occur?   1

Review Exercises

Determine which numbers in the set {−3, −2, −1, 0, 1, 2, 3 } are solutions of each inequality.

71. $x < 0$ {−3, −2, −1}

72. $x > 1$ {2, 3}

73. $x + 5 \le 6$

74. $x - 3 \ge -7$

75. In your own words, explain what real numbers are solutions of $x < 0$.   Answers will vary.

76. In your own words, explain what real numbers are solutions of $x > 1$.   Answers will vary.

73. {−3, −2, −1, 0, 1}   **74.** {−3, −2, −1, 0, 1, 2, 3}

2.4 | LINEAR INEQUALITIES AND PROBLEM SOLVING

TAPE IA 2.4

O B J E C T I V E S

1. Define a linear inequality in one variable.
2. Use interval notation.
3. Solve linear inequalities using the addition property of inequality.
4. Solve linear inequalities using the multiplication property of inequality.
5. Solve problems that can be modeled by linear inequalities.

 Relationships among measureable quantities are not always described by equations. For example, suppose that a salesperson earns a base of $600 per month plus a commission of 20% of sales. Find the minimum amount of sales needed to receive a total income of *at least* $1500 per month. Here, the phrase "at least" implies that an income of $1500 *or more* is acceptable. In symbols, we can write

$$\text{income} \ \geq \ 1500$$

This is an example of an inequality, and we will solve this problem in Example 7.

A **linear inequality** is similar to a linear equation except that the equality symbol is replaced with an inequality symbol, such as $<$, $>$, $\leq$, or $\geq$.

LINEAR INEQUALITIES IN ONE VARIABLE

$$3x + 5 \geq 4 \qquad 2y < 0 \qquad 4n \geq n - 3 \qquad 3(x - 4) < 5x \qquad \frac{x}{3} \leq 5$$

LINEAR INEQUALITY IN ONE VARIABLE

A linear inequality in one variable is an inequality that can be written in the form

$$ax + b < c$$

where a, b, and c are real numbers and $a \neq 0$.

In this section, when we make definitions, state properties, or list steps about an inequality containing the symbol $<$, we mean that the definition, property, or steps apply to an inequality containing the symbols $>$, $\leq$, and $\geq$, also.

 A **solution** of an inequality is a value of the variable that makes the inequality a true statement. The **solution set** of an inequality is the set of all solutions. Notice that the solution set of the inequality $x > 2$, for example, contains all numbers greater than 2. Its graph is an interval on the number line since an infinite number of values satisfy the variable. If we use open-closed-circle notation, the graph of $\{x \mid x > 2\}$ looks like the following.

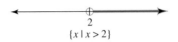

$$\{x \mid x > 2\}$$

In this text, a different graphing notation will be used to help us understand **interval notation.** Instead of an open circle, we use a parenthesis; instead of a closed circle, we use a bracket. With this new notation, the graph of $\{x \mid x > 2\}$ now looks like

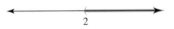

and can be represented in interval notation as $(2, \infty)$, The symbol ∞ is read "infinity" and indicates that the interval includes *all* numbers greater than 2. The left parenthesis indicates that 2 *is not* included in the interval. Using a left bracket, [, would indicate that 2 *is* included in the interval. The following table shows three equivalent ways to describe an interval: in set notation, as a graph, and in interval notation.

SET NOTATION	GRAPH	INTERVAL NOTATION
$\{x \mid x < a\}$		$(-\infty, a)$
$\{x \mid x > a\}$		(a, ∞)
$\{x \mid x \le a\}$		$(-\infty, a]$
$\{x \mid x \ge a\}$		$[a, \infty)$
$\{x \mid a < x < b\}$		(a, b)
$\{x \mid a \le x \le b\}$		$[a, b]$
$\{x \mid a < x \le b\}$		$(a, b]$
$\{x \mid a \le x < b\}$		$[a, b)$

REMINDER Notice that a parenthesis is always used to enclose ∞ and $-\infty$.

EXAMPLE 1 Graph each set on a number line and then write in interval notation.

a. $\{x \,|\, x \geq 2\}$ **b.** $\{x \,|\, x < -1\}$ **c.** $\{x \,|\, 0.5 < x \leq 3\}$

Solution: **a.**

[2, ∞)

b.

$(-\infty, -1)$

c.

(0.5, 3]

Interval notation can be used to write solutions of linear inequalities. To solve a linear inequality, we use a process similar to the one used to solve a linear equation. We use properties of inequalities to write equivalent inequalities until the variable is isolated.

ADDITION PROPERTY OF INEQUALITY

If a, b, and c are real numbers, then

$$a < b \quad \text{and} \quad a + c < b + c$$

are equivalent inequalities.

In other words, we may add the same real number to both sides of an inequality and the resulting inequality will have the same solution set. This property also allows us to subtract the same real number from both sides.

EXAMPLE 2 Solve for x: $3x + 4 \geq 2x - 6$. Graph the solution set and write it in interval notation.

Solution:

$3x + 4 \geq 2x - 6$	
$3x + 4 - 2x \geq 2x - 6 - 2x$	Subtract $2x$ from both sides.
$x + 4 \geq -6$	Combine like terms.
$x + 4 - 4 \geq -6 - 4$	Subtract 4 from both sides.
$x \geq -10$	Simplify.

The solution set is $\{x \,|\, x \geq -10\}$, which in interval notation is $[-10, \infty)$. The graph of the solution set is

$[-10, \infty)$

Thus, *every* real number greater than or equal to -10 is a solution of the linear inequality $3x + 4 \geq 2x - 6$. For example $-10, 11, 38,$ and $1,000,000$ are a few examples of solutions. To see this, replace x in $3x + 4 \geq 2x - 6$ with each of the numbers and see that the result is a true inequality.

4 Next, we introduce and use the multiplication property of inequality to solve linear inequalities. To understand this property, let's start with the true statement $-3 < 7$ and multiply both sides by 2.

$$-3 < 7$$
$$-3(2) < 7(2) \qquad \text{Multiply by 2.}$$
$$-6 < 14 \qquad \text{True.}$$

The statement remains true.

Notice what happens if both sides of $-3 < 7$ are multiplied by -2.

$$-3 < 7$$
$$-3(-2) < 7(-2)$$
$$6 < -14 \qquad \text{False.}$$

The inequality $6 < -14$ is a false statement. However, **if the direction of the inequality sign is reversed,** the result is

$$6 > -14, \text{ a true statement.}$$

These examples suggest the following property.

MULTIPLICATION PROPERTY OF INEQUALITY

If a, b, and c are real numbers and c is **positive**, then $a < b$ and $ac < bc$ are equivalent inequalities.
If a, b, and c are real numbers and c is **negative**, then $a < b$ and $ac > bc$ are equivalent inequalities.

In other words, we may multiply both sides of an inequality by the same positive real number and the result is an equivalent inequality.

We may also multiply both sides of an inequality by the same **negative number** and **reverse the direction of the inequality symbol,** and the result is an equivalent inequality. The multiplication property holds for division also, since division is defined in terms of multiplication.

R E M I N D E R   Whenever both sides of an inequality are multiplied or divided by a negative number, the direction of the inequality symbol **must be** reversed to form an equivalent inequality.

EXAMPLE 3 Solve for x.

a. $\dfrac{1}{4}x \leq \dfrac{3}{8}$

b. $-2.3x < 6.9$

Solution: **a.** $\dfrac{1}{4}x \leq \dfrac{3}{8}$

$4 \cdot \dfrac{1}{4}x \leq 4 \cdot \dfrac{3}{8}$ Multiply both sides by 4.

$x \leq \dfrac{3}{2}$ Simplify.

The solution set is $\left\{ x \mid x \leq \dfrac{3}{2} \right\}$, which in interval notation is $\left(-\infty, \dfrac{3}{2} \right]$. The graph of the solution set is

$\left(-\infty, \dfrac{3}{2} \right]$

b. $-2.3x < 6.9$

$\dfrac{-2.3x}{-2.3} > \dfrac{6.9}{-2.3}$ Divide both sides by -2.3 and reverse the inequality symbol.

$x > -3$ Simplify.

The solution set is $\{ x \mid x > -3 \}$, which is $(-3, \infty)$ in interval notation. The graph of the solution set is

$(-3, \infty)$

To solve linear inequalities in general, we follow steps similar to those for solving linear equations.

TO SOLVE A LINEAR INEQUALITY IN ONE VARIABLE

Step 1. Clear the equation of fractions by multiplying both sides of the inequality by the least common denominator (LCD) of all fractions in the inequality.

Step 2. Use the distributive property to remove grouping symbols such as parentheses.

Step 3. Combine like terms on each side of the inequality.

Step 4. Use the addition property of inequality to write the inequality as an equivalent inequality with variable terms on one side and numbers on the other side.

Step 5. Use the multiplication property of inequality to isolate the variable.

EXAMPLE 4   Solve for x: $-(x - 3) + 2 \le 3(2x - 5) + x$.

Solution:

$$-(x - 3) + 2 \le 3(2x - 5) + x$$

$-x + 3 + 2 \le 6x - 15 + x$ \qquad Apply the distributive property.

$5 - x \le 7x - 15$ \qquad Combine like terms.

$5 - x + x \le 7x - 15 + x$ \qquad Add x to both sides.

$5 \le 8x - 15$ \qquad Combine like terms.

$5 + 15 \le 8x - 15 + 15$ \qquad Add 15 to both sides.

$20 \le 8x$ \qquad Combine like terms.

$$\frac{20}{8} \le \frac{8x}{8}$$ \qquad Divide both sides by 8.

$$\frac{5}{2} \le x, \quad \text{or} \quad x \ge \frac{5}{2}$$ \qquad Simplify.

The solution set written in interval notation is $\left[\dfrac{5}{2}, \infty\right)$, and its graph is

$\left[\dfrac{5}{2}, \infty\right)$

EXAMPLE 5   Solve for x: $\dfrac{2}{5}(x - 6) \ge x - 1$.

Solution: $\dfrac{2}{5}(x - 6) \ge x - 1$

$$5\left[\frac{2}{5}(x - 6)\right] \ge 5(x - 1)$$ \qquad Multiply both sides by 5 to eliminate fractions.

$2x - 12 \ge 5x - 5$ \qquad Apply the distributive property.

$-3x - 12 \ge -5$ \qquad Subtract $5x$ from both sides.

$-3x \ge 7$ \qquad Add 12 to both sides.

$$\frac{-3x}{-3} \le \frac{7}{-3}$$ \qquad Divide both sides by -3 and reverse the inequality symbol.

$$x \le -\frac{7}{3}$$ \qquad Simplify.

The solution is graphed on a number line and written in interval notation, as $\left(-\infty, -\dfrac{7}{3}\right]$

$\left(-\infty, -\dfrac{7}{3}\right]$

EXAMPLE 6   Solve for x: $2(x + 3) > 2x + 1$.

Solution:

$$2(x + 3) > 2x + 1$$

$$2x + 6 > 2x + 1 \qquad \text{Distribute on the left side.}$$

$$2x + 6 - 2x > 2x + 1 - 2x \qquad \text{Subtract } 2x \text{ from both sides.}$$

$$6 > 1 \qquad \text{Simplify.}$$

$6 > 1$ is a true statement for all values of x, so this inequality and the original inequality are true for all numbers. The solution set is $\{x \mid x$ is a real number$\}$, or $(-\infty, \infty)$ in interval notation, and its graph is

 $(-\infty, \infty)$

5 Application problems containing words such as "at least," "at most," "between," "no more than," and "no less than" usually indicate that an inequality be solved instead of an equation. In solving applications involving linear inequalities, we use the same procedure as when we solved applications involving linear equations.

EXAMPLE 7 A salesperson earns a base of $600 per month plus a commission of 20% of sales. Find the minimum amount of sales needed to receive a total income of at least $1500 per month.

Solution: **1.** UNDERSTAND. Read and reread the problem.

2. ASSIGN. Since the unknown is the amount of sales,

let x = amount of sales.

3. ILLUSTRATE. No illustration needed.

4. TRANSLATE. As stated in the beginning of this section, we want the income to be greater than or equal to $1500. To write an inequality, notice that the salesperson's income consists of a base plus a commission (20% of sales).

In words: base + commission (20% of sales) $\geq$ 1500

Translate: 600 + 0.20x $\geq$ 1500

5. COMPLETE. Solve the inequality for x.

$$600 + 0.20x \geq 1500$$

$$600 + 0.20x - 600 \geq 1500 - 600$$

$$0.20x \geq 900$$

$$x \geq 4500$$

6. INTERPRET. *Check:* The income for sales of $4500 is

$$600 + 0.20(4500), \text{ or } 1500$$

Thus, if sales are greater than or equal to $4500, income is greater than or equal to $1500.

State: The minimum amount of sales needed for the salesperson to earn at least $1500 per month is $4500.

EXAMPLE 8 In the United States, the annual consumption of cigarettes is declining. The consumption c in billions of cigarettes per year since the year 1985 can be approximated by the formula

$$c = -14.25t + 598.69$$

where t is the number of years after 1985. Use this formula to predict the years that the consumption of cigarettes will be less than 200 billion per year.

Solution: **1.** UNDERSTAND. Read and reread the problem. To become familiar with the given formula, find the cigarette consumption after 20 years, which would be the year 1985 + 20, or 2005. To do so, substitute 20 for t in the given formula.

$$c = -14.25(20) + 598.69 = 313.69$$

Thus, in 2005, we predict cigarette consumption to be 313.69 billion.

2. ASSIGN. Variables have already been assigned from the given formula. For review, they are

c = the annual consumption of cigarettes in the United States in billions of cigarettes

t = the number of years after 1985

4. TRANSLATE. In words: We are looking for the years that the consumption of cigarettes c is less than 200. Since we are finding years t, we substitute the expression in the formula given for c, or

Translate: $-14.25t + 598.69 < 200.$

5. COMPLETE. Solve $-14.25t + 598.69 < 200$

$-14.25t < -398.69$ Subtract 598.69 from both sides.

$t > 27.98$ Divide both sides by -14.25 and round the result.

6. INTERPRET. To *check*, substitute a number greater than 27.98 and see that c is less than 200. *State:* The annual consumption of cigarettes will be less than 200 billion for the years greater than 27.98 years more than 1985, or approximately 28 + 1985 = 2013.

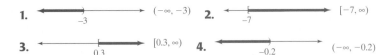

1. $(-\infty, -3)$ 2. $[-7, \infty)$ 3. $[0.3, \infty)$ 4. $(-\infty, -0.2)$

EXERCISE SET 2.4

Graph the solution set of each inequality on a number line, and write the solution set in interval notation. See Example 1.

1. $\{x | x < -3\}$ **2.** $\{x | x \geq -7\}$ **3.** $\{x | x \geq 0.3\}$

4. $\{x | x < -0.2\}$ **5.** $\{x | 5 < x\}$ **6.** $\{x | -7 \geq x\}$

7. $\{x | -2 < x < 5\}$ **8.** $\{x | -5 \leq x \leq -1\}$

9. $\{x | 5 > x > -1\}$ **10.** $\{x | -3 \geq x \geq -7\}$

11. When graphing the solution set of an inequality, explain how you know whether to use a parenthesis or a bracket.   Answers will vary.

12. Explain what is wrong with the interval notation $(-6, -\infty)$.   Answers will vary.

5. $(5, \infty)$ 6. $(-\infty, -7]$ 7. $(-2, 5)$ 8. $[-5, -1]$ 9. $(-1, 5)$ 10. $[-7, -3]$

14. $[0, \infty)$ **17.** $(-\infty, -4.7)$ **18.** $(-\infty, 2.8)$

Solve each inequality for the variable. Write the solution set in interval notation and graph the solution set. See Examples 2 and 3.   **13.** $(-1, \infty)$

13. $5x + 3 > 2 + 4x$ **14.** $7x - 1 \geq 6x - 1$

15. $8x - 7 \leq 7x - 5$ $(-\infty, 2]$

16. $12x + 14 < 11x - 2$ $(-\infty, -16)$

17. $5x < -23.5$ **18.** $-4x > -11.2$

19. $-3x \geq \dfrac{1}{2}$ **20.** $7x \leq -\dfrac{7}{3}$ $\left(-\infty, -\dfrac{1}{3}\right]$

Solve each inequality for the variable. Write the solution set in interval notation and graph the solution set. See Examples 4 through 6.   **21.** $(-\infty, 11]$

21. $15 + 2x \geq 4x - 7$ **22.** $20 + x < 6x$

23. $\dfrac{3x}{4} \geq 2$ **24.** $\dfrac{5}{6}x \geq -8$ $\left[-\dfrac{48}{5}, \infty\right)$

25. $3(x - 5) < 2(2x - 1)$ $(-13, \infty)$

26. $5(x + 4) \leq 4(2x + 3)$

27. $\dfrac{1}{2} + \dfrac{2}{3} \geq \dfrac{x}{6}$ **28.** $\dfrac{3}{4} - \dfrac{2}{3} > \dfrac{x}{6}$ $\left[\dfrac{8}{3}, \infty\right)$

29. $4(x - 1) \geq 4x - 8$ **30.** $3x + 1 < 3(x - 2)$

31. $7x < 7(x - 2)$ $\{\ \}$

32. $8(x + 3) \leq 7(x + 5) + x$ $(-\infty, \infty)$

📕 33. Explain how solving a linear inequality is similar to solving a linear equation.   Answers will vary.

📕 34. Explain how solving a linear inequality is different from solving a linear equation.   Answers will vary.

35. $(6, \infty)$

Solve each inequality. Write the solution set in interval notation and graph the solution set.

35. $7.3x > 43.8$ **36.** $9.1x < 45.5$ $5 \ (-\infty, 5)$

37. $-4x \leq \dfrac{2}{5}$ **38.** $-6x \geq \dfrac{3}{4}$ $\left(-\infty, -\dfrac{1}{8}\right]$

39. $-2x + 7 \geq 9$ **40.** $8 - 5x \leq 23$

41. $4(2x + 1) > 4$ **42.** $6(2 - x) \geq 12$

43. $\dfrac{x + 7}{5} > 1$ **44.** $\dfrac{2x - 4}{3} \leq 2$ $5 \ (-\infty, 5]$

45. $\dfrac{-5x + 11}{2} \leq 7$ **46.** $\dfrac{4x - 8}{7} < 0$ $2 \ (-\infty, 2)$

47. $8x - 16.4 \leq 10x + 2.8$ $[-9.6, \infty)$

48. $18x - 25.6 < 10x + 60.8$ $(-\infty, 10.8)$

49. $2(x - 3) > 70$ **50.** $3(5x + 6) \geq -12$

19. $-\dfrac{1}{6}$ $\left(-\infty, -\dfrac{1}{6}\right]$ **22.** $4 \ (4, \infty)$ **23.** $\dfrac{8}{3}$ $\left[\dfrac{8}{3}, \infty\right)$ **27.** $7 \ (-\infty, 7]$

28. $\dfrac{1}{2} \ \left(-\infty, \dfrac{1}{2}\right)$ **29.** $(-\infty, \infty)$ **30.** $\{\ \}$ **37.** $-\dfrac{1}{10} \ \left[-\dfrac{1}{10}, \infty\right)$

51. $-5x + 4 \leq -4(x - 1)$ $[0, \infty)$

52. $-6x + 2 < -3(x + 4)$ $\dfrac{14}{3}$ $\left(\dfrac{14}{3}, \infty\right)$

53. $\dfrac{1}{4}(x - 7) \geq x + 2$ **54.** $\dfrac{3}{5}(x + 1) \leq x + 1$

55. $\dfrac{2}{3}(x + 2) < \dfrac{1}{5}(2x + 7)$ $\dfrac{1}{4}$ $\left(-\infty, \dfrac{1}{4}\right)$

56. $\dfrac{1}{6}(3x + 10) > \dfrac{5}{12}(x - 1)$ -25 $(-25, \infty)$

57. $4(x - 6) + 2x - 4 \geq 3(x - 7) + 10x$

58. $7(2x + 3) + 4x \leq 7 + 5(3x - 4)$

59. $\dfrac{5x + 1}{7} - \dfrac{2x - 6}{4} \geq -4$ $-\dfrac{79}{3}$ $\left[-\dfrac{79}{3}, \infty\right)$

60. $\dfrac{1 - 2x}{3} + \dfrac{3x + 7}{7} > 1$ $\dfrac{7}{5}$ $\left(-\infty, \dfrac{7}{5}\right)$

61. $\dfrac{-x + 2}{2} - \dfrac{1 - 5x}{8} < -1$ -15 $(-\infty, -15)$

62. $\dfrac{3 - 4x}{6} - \dfrac{1 - 2x}{12} \leq -2$ $\dfrac{29}{6}$ $\left[\dfrac{29}{6}, \infty\right)$

63. $0.8x + 0.6x \geq 4.2$ **64.** $0.7x - x > 0.45$

65. $\dfrac{x + 5}{5} - \dfrac{3 + x}{8} \geq \dfrac{-3}{10}$ **66.** $\dfrac{x - 4}{2} - \dfrac{x - 2}{3} > \dfrac{5}{6}$

67. $\dfrac{x + 3}{12} + \dfrac{x - 5}{15} < \dfrac{2}{3}$ 5 $(-\infty, 5)$

68. $\dfrac{3x + 2}{18} - \dfrac{1 + 2x}{6} \leq -\dfrac{1}{2}$ $\dfrac{8}{3}$ $\left[\dfrac{8}{3}, \infty\right)$

Solve. See Examples 7 and 8.

69. Shureka has scores of 72, 67, 82, and 79 on her algebra tests. Use an inequality to find the minimum score she can make on the final exam to pass the course with an average of 60 or higher, given that the final exam counts as two tests.   The minimum score is 30.

70. In a Winter Olympics speed-skating event, Hans scored times of 3.52, 4.04, and 3.87 minutes on his first three trials. Use an inequality to find the maximum time he can score on his last trial so that his average time is under 4.0 minutes.   4.57 min

71. A small plane's maximum takeoff weight is 2000 pounds. Six passengers weigh an average of 160 pounds each. Use an inequality to find the maximum weight of luggage and cargo the plane can carry.   1040 lb of luggage and cargo

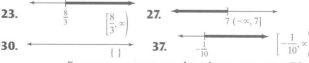

For more answer graphs, please see page 78.

39. $(-\infty, 1]$ −1   **40.** $[-3, \infty)$ −3   **41.** $(0, \infty)$ 0   **42.** $(-\infty, 0]$ 0

72. A clerk must use the elevator to move boxes of paper. The elevator's weight limit is 1500 pounds. If each box of paper weighs 66 pounds and the clerk weighs 147 pounds, use an inequality to find the maximum number of boxes she can move on the elevator at one time.   20 boxes

73. To mail an envelope first class, the U.S. Post Office charges 32 cents for the first ounce and 23 cents per ounce for each additional ounce. Use an inequality to find the maximum weight that can be mailed for $4.00.   17 oz

74. A shopping mall parking garage charges $1 for the first half-hour and 60 cents for each additional half-hour or a portion of a half-hour. Use an inequality to find how long you can park if you have only $4.00 in cash.   3 hours

75. Northeast Telephone Company offers two billing plans for local calls. Plan 1 charges $25 per month for unlimited calls, and plan 2 charges $13 per month plus 6 cents per call. Use an inequality to find the number of monthly calls for which plan 1 is more economical than plan 2. greater than 200 calls

76. A car rental company offers two subcompact rental plans. Plan A charges $32 per day for unlimited mileage, and plan B charges $24 per day plus 15 cents per mile. Use an inequality to find the number of daily miles for which plan A is more economical than plan B.   54 miles or more

77. At room temperature, glass used in windows actually has some properties of a liquid. It has a very slow, viscous flow. (Viscosity is the property of a fluid that resists internal flow. For example, lemonade flows more easily than fudge syrup. Fudge syrup has a higher viscosity than lemonade.) Glass does not become a true liquid until temperatures are greater than or equal to 500°C. Find the Fahrenheit temperatures for which glass is a liquid. (Use the formula $F = \frac{9}{5}C + 32$.) $F \geq 932°$

78. Stibnite is a silvery white mineral with a metallic luster. It is one of the few minerals that melts easily in the match flame or at temperatures of approximately 977°F or greater. Find the Celsius temperatures for which stibnite melts. [Use the formula $C = \frac{5}{9}(F - 32)$.] $C \geq 525°$

79. Although beginning salaries vary greatly according to your field of study, the equation $s = 651.2t + 27,821$ can be used to approximate and to predict average beginning salaries for candidates for bachelor's degrees. The variable s is the starting salary and t is the number of years after 1989.

  a. Approximate when beginning salaries for candidates will be greater than 30,000.   1993

  b. Determine the year you plan to graduate from college. Use this year to find the corresponding value of t and approximate your beginning salary.   Answers will vary.

80. Use the formula in Example 8 to estimate the years that the consumption of cigarettes will be less than 100 billion per year.   2020

The average consumption per year per person of whole milk w *can be approximated by the equation*

$$w = -3.26t + 87.79$$

where t *is the number of years after 1990. The average consumption of skim milk* s *per person per year can be approximated by the equation*

$$s = 1.25t + 22.75$$

where t *is the number of years after 1990. The consumption of whole milk is shown on the graph in blue and the consumption of skim milk is shown on the graph in red. Use this information for Exercises 81–89.*

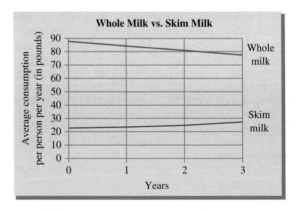

81. From the graph shown, is the consumption of whole milk increasing or decreasing over time? Explain how you arrived at your answer.   decreasing

82. From the graph shown, is the consumption of skim milk increasing or decreasing over time? Explain how you arrived at your answer.   increasing

83. Predict the consumption of whole milk in the year 2000. (*Hint:* First use the year 2000 and find the value of t.)   55.19 lb/person/yr

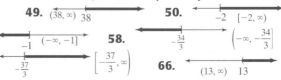

43. $(-2, \infty)$ −2   **45.** $\left[-\frac{3}{5}, \infty\right)$ $-\frac{3}{5}$ **49.** $(38, \infty)$ 38   **50.** $[-2, \infty)$ −2

53. $(-\infty, -5]$ −5   **54.** $[-1, \infty)$ −1   **57.** $(-\infty, -1)$ −1   **58.** $\left(-\infty, -\frac{34}{3}\right]$ $-\frac{34}{3}$

63. $[3, \infty)$ 3   **64.** $(-\infty, -1.5)$ −1.5   **65.** $\left[-\frac{37}{3}, \infty\right)$ $-\frac{37}{3}$ **66.** $(13, \infty)$ 13

84. Predict the consumption of skim milk in the year 2000. (*Hint:* First use the year 2000 and find the value of *t*.)   35.25 lb/person/yr

85. Determine when the consumption of whole milk will be less than 50 pounds per person per year.   2002

86. Determine when the consumption of skim milk will be greater than 50 pounds per person per year. 2012

87. For 1990 to 1993, the consumption of whole milk was greater than the consumption of skim milk. Explain how this can be determined from the graph shown.   Answers will vary.

88. How will the two lines in the graph appear when the consumption of whole milk is the same as the consumption of skim milk?   lines will intersect

89. The consumption of whole milk will be the same as the consumption of skim milk when $w = s$. Find when this will occur, by substituting the given equivalent expression for w and the given equivalent expression for s and solving for t. Round the value of t to the nearest whole and estimate the year when this will occur.   2004

Review Exercises

List or describe the integers that make both inequalities true.

90. $x < 5$ and $x > 1$

91. $x \geq 0$ and $x \leq 7$

92. $x \geq -2$ and $x \geq 2$

93. $x < 6$ and $x < -5$

Graph each set on a number line and write it in interval notation. See Section 2.4.

94. $\{x \mid 0 \leq x \leq 5\}$

95. $\{x \mid -7 < x \leq 1\}$

96. $\left\{x \mid -\dfrac{1}{2} < x < \dfrac{3}{2}\right\}$

97. $\{x \mid -2.5 \leq x < 5.3\}$

90. $2, 3, 4$ **91.** $0, 1, 2, 3, 4, 5, 6, 7$ **92.** $2, 3, 4 \ldots$ **93.** $-6, -7, -8 \ldots$ **94.** [0, 5]

95. (–7, 1]   **96.** $\left(-\frac{1}{2}, \frac{3}{2}\right)$ **97.** [–2.5, 5.3]

2.5 | COMPOUND INEQUALITIES

TAPE IA 2.5

O B J E C T I V E

1 Solve compound inequalities.

1 Two inequalities joined by the words **and** or **or** are called **compound inequalities.**

COMPOUND INEQUALITIES

$$x + 3 < 8 \quad \text{and} \quad x > 2$$
$$\frac{2x}{3} \geq 5 \quad \text{or} \quad -x + 10 < 7$$

The solution set of a compound inequality formed by the word **and** is the **intersection** of the solution sets of the two inequalities. The intersection of two sets, denoted by ∩, is the set of elements common to both sets. For example, given the sets

$$A = \{2, 4, 6, 8\}$$
$$B = \{3, 4, 5, 6\} \text{ then}$$
$$A \cap B = \{4, 6\}.$$

This is true for compound inequalities formed by the word *and*. A value of x is a solution of a compound inequality formed by the word **and** if it is a solution of **both** inequalities.

For example, the solution set of the compound inequality $x \le 5$ and $x \ge 3$ contains all values of x that make the inequality $x \le 5$ a true statement **and** the inequality $x \ge 3$ a true statement. The first graph shown next is the graph of $x \le 5$, the second graph is the graph of $x \ge 3$, and the third graph shows the intersection of the two graphs. It is the graph of $x \le 5$ **and** $x \ge 3$.

$\{x \mid x \le 5\}$ -1 0 1 2 3 4 5 6 $(-\infty, 5]$

$\{x \mid x \ge 3\}$ -1 0 1 2 3 4 5 6 $[3, \infty)$

$\{x \mid x \le 5 \text{ and } x \ge 3\}$ -1 0 1 2 3 4 5 6 $[3, 5]$

In interval notation, the set $\{x \mid x \le 5 \text{ and } x \ge 3\}$ is written as $[3, 5]$.

EXAMPLE 1   Solve for x: $x - 7 < 2$ and $2x + 1 < 9$.

Solution: First solve each inequality separately.

$$x - 7 < 2 \quad \text{and} \quad 2x + 1 < 9$$
$$x < 9 \quad \text{and} \quad 2x < 8$$
$$x < 9 \quad \text{and} \quad x < 4$$

Graph the two intervals on two number lines and find their intersection.

$\{x \mid x < 9\}$ 3 4 5 6 7 8 9 10 $(-\infty, 9)$

$\{x \mid x < 4\}$ 3 4 5 6 7 8 9 10 $(-\infty, 4)$

$\{x \mid x < 9 \text{ and } x < 4\}$ 3 4 5 6 7 8 9 10 $(-\infty, 4)$

$= \{x \mid x < 4\}$

The solution set written as an interval is $(-\infty, 4)$.

Compound inequalities containing the word **and** can be written in a more compact form. The compound inequality $2 \le x$ and $x \le 6$ can be written as

$$2 \le x \le 6$$

Recall from Section 2.4 that the graph of $2 \le x \le 6$ is all numbers between 2 and 6, including 2 and 6.

2   6

The set $\{x \mid 2 \le x \le 6\}$ written in interval notation is $[2, 6]$.

To solve a compound inequality like $2 < 4 - x < 7$, we isolate x on the "middle side." Since a compound inequality is really two inequalities in one statement, we must perform the same operation to all three "sides" of the inequality.

EXAMPLE 2 Solve $2 < 4 - x < 7$.

Solution: To isolate x, first subtract 4 from all three sides.

$$2 < 4 - x < 7$$

$$2 - 4 < 4 - x - 4 < 7 - 4 \quad \text{Subtract 4 from all three sides.}$$

$$-2 < -x < 3 \quad \text{Simplify.}$$

$$\frac{-2}{-1} > \frac{-x}{-1} > \frac{3}{-1} \quad \text{Divide all three sides by } -1 \text{ and reverse the}$$
$$\text{inequality symbols.}$$

$$2 > x > -3$$

This is equivalent to $-3 < x < 2$, and its graph is shown. The solution set in interval notation is $(-3, 2)$.

EXAMPLE 3 Solve for x: $-1 \le \dfrac{2x}{3} + 5 \le 2$.

Solution: First, clear the inequality of fractions by multiplying all three sides by the LCD of 3.

$$-1 \le \frac{2x}{3} + 5 \le 2$$

$$3(-1) \le 3\left(\frac{2x}{3} + 5\right) \le 3(2) \qquad \text{Multiply by the LCD of 3.}$$

$$-3 \le 2x + 15 \le 6 \qquad \text{Apply the distributive property and}$$
$$\text{multiply.}$$

$$-3 - 15 \le 2x + 15 - 15 \le 6 - 15 \qquad \text{Subtract 15 from all three sides.}$$

$$-18 \le 2x \le -9 \qquad \text{Simplify.}$$

$$\frac{-18}{2} \le \frac{2x}{2} \le \frac{-9}{2} \qquad \text{Divide all three sides by 2.}$$

$$-9 \le x \le -\frac{9}{2} \qquad \text{Simplify.}$$

The graph of the solution is shown.

The solution set in interval notation is $\left[-9, -\dfrac{9}{2}\right]$.

The solution set of a compound inequality formed by the word **or** is the **union** of the solution sets of the two inequalities. The union of two sets, denoted by $\cup$, is the set of elements that belong to either of the sets.

For example, given the sets

$$A = \{2, 4, 6, 8\}$$
$$B = \{3, 4, 5, 6\} \text{ then}$$
$$A \cup B = \{2, 3, 4, 5, 6, 8\}$$

This is true for compound inequalities formed by the word *or*. A value of x is a solution of a compound inequality formed by the word *or* if it is a solution of **either** inequality.

For example, the solution set of the compound inequality $x \leq 1$ or $x \geq 3$ contains all numbers that make the inequality $x \leq 1$ a true statement **or** the inequality $x \geq 3$ a true statement.

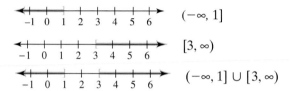

$$(-\infty, 1]$$

$$[3, \infty)$$

$$(-\infty, 1] \cup [3, \infty)$$

In interval notation, the set $\{x \mid x \leq 1 \text{ or } x \geq 3\}$ is written as $(-\infty, 1] \cup [3, \infty)$.

EXAMPLE 4   Solve $5x - 3 \leq 10$ or $x + 1 \geq 5$.

Solution: Solve each inequality separately.

$$5x - 3 \leq 10 \quad \text{or} \quad x + 1 \geq 5$$
$$5x \leq 13 \quad \text{or} \qquad x \geq 4$$
$$x \leq \frac{13}{5} \quad \text{or} \qquad x \geq 4$$

Graph each interval on a number line. Then find their union.

$$\left(-\infty, \frac{13}{5}\right]$$

$$[4, \infty)$$

$$\left(-\infty, \frac{13}{5}\right] \cup [4, \infty)$$

The solution set in interval notation is $\left(-\infty, \frac{13}{5}\right] \cup [4, \infty)$.

4. $\{x \,|\, x \text{ is an even integer or } x = 5 \text{ or } x = 7\}$ **5.** $\{\ldots -2, -1, 0, 1 \ldots\}$ **8.** $\{x \,|\, x \text{ is an odd integer or } x = 4 \text{ or } x = 6\}$

16. 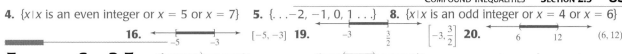 $[-5, -3]$ **19.** $\left[-3, \frac{3}{2}\right]$ **20.** $(6, 12)$

EXERCISE SET 2.5

24. $(-\infty, 1)$ **31.** $(-1, 2)$ **32.** $(-\infty, 1)$

If $A = \{x \,|\, x \text{ is an even integer}\}$, $B = \{x \,|\, x \text{ is an odd integer}\}$, $C = \{2, 3, 4, 5\}$, and $D = \{4, 5, 6, 7\}$, list the elements of each set. **1.** $\{2, 3, 4, 5, 6, 7\}$

1. $C \cup D$ **2.** $C \cap D$ $\{4, 5\}$ **3.** $A \cap D$ $\{4, 6\}$

4. $A \cup D$ **5.** $A \cup B$ **6.** $A \cap B$ $\{\}$

7. $B \cap D$ $\{5, 7\}$ **8.** $B \cup D$

Solve each compound inequality. Graph the solution set. See Example 1.

9. $x < 5$ and $x > -2$ $(-2, 5)$

10. $x \le 7$ and $x \le 1$ $(-\infty, 1]$

11. $x + 1 \ge 7$ and $3x - 1 \ge 5$ $[6, \infty)$

12. $-2x < -8$ and $x - 5 < 5$ $(4, 10)$

13. $4x + 2 \le -10$ and $2x \le 0$ $(-\infty, -3]$

14. $x + 4 > 0$ and $4x > 0$ $(0, \infty)$

Solve each compound inequality. Graph the solution set. See Examples 2 and 3. **15.** $(11, 17)$

15. $5 < x - 6 < 11$ **16.** $-2 \le x + 3 \le 0$

17. $-2 \le 3x - 5 \le 7$ $[1, 4]$

18. $1 < 4 + 2x < 7$ $\left(-\frac{3}{2}, \frac{3}{2}\right)$

19. $1 \le \frac{2}{3}x + 3 \le 4$ **20.** $-2 < \frac{1}{2}x - 5 < 1$

21. $-5 \le \frac{x + 1}{4} \le -2$ $[-21, -9]$

22. $-4 \le \frac{2x + 5}{3} \le 1$ $\left[-\frac{17}{2}, -1\right]$

Solve each compound inequality. Graph the solution set. See Example 4. **23.** $(-\infty, -1) \cup (0, \infty)$

23. $x < -1$ or $x > 0$ **24.** $x \le 1$ or $x \le -3$

25. $-2x \le -4$ or $5x - 20 \ge 5$ $[2, \infty)$

26. $x + 4 < 0$ or $6x > -12$ $(-\infty, -4) \cup (-2, \infty)$

27. $3(x - 1) < 12$ or $x + 7 > 10$ $(-\infty, \infty)$

28. $5(x - 1) \ge -5$ or $5 - x \le 11$ $[-6, \infty)$

29. Explain how solving an and-compound inequality is similar to finding the intersection of two sets. Answers will vary.

30. Explain how solving an or-compound inequality is similar to finding the union of two sets. Answers will vary.

Solve each compound inequality. Graph the solution set.

31. $x < 2$ and $x > -1$ **32.** $x < 5$ and $x < 1$

33. $x < 2$ or $x > -1$ **34.** $x < 5$ or $x < 1$

35. $x \ge -5$ and $x \ge -1$ $[-1, \infty)$

36. $x \le 0$ or $x \ge -3$ $(-\infty, \infty)$

37. $x \ge -5$ or $x \ge -1$ $[-5, \infty)$

38. $x \le 0$ and $x \ge -3$ $[-3, 0]$

39. $0 \le 2x - 3 \le 9$ **40.** $3 < 5x + 1 < 11$

41. $\frac{1}{2} < x - \frac{3}{4} < 2$ **42.** $\frac{2}{3} < x + \frac{1}{2} < 4$

43. $x + 3 \ge 3$ and $x + 3 \le 2$ $\{\}$

44. $2x - 1 \ge 3$ and $-x > 2$ $\{\}$

45. $3x \ge 5$ or $-x - 6 < 1$ $(-7, \infty)$

46. $\frac{3}{8}x + 1 \le 0$ or $-2x < -4\frac{8}{3}$ $\left(-\infty, -\frac{8}{3}\right] \cup (2, \infty)$

47. $0 < \frac{5 - 2x}{3} < 5$ $\left(-5, \frac{5}{2}\right)$

48. $-2 < \frac{-2x - 1}{3} < 2$ $\left(-\frac{7}{2}, \frac{5}{2}\right)$

49. $-6 < 3(x - 2) \le 8$ $\left(0, \frac{14}{3}\right]$

50. $-5 < 2(x + 4) < 8$ $\left(-\frac{13}{2}, 0\right)$

51. $-x + 5 > 6$ and $1 + 2x \le -5$ $(-\infty, -3]$

52. $5x \le 0$ and $-x + 5 < 8$ $(-3, 0]$

53. $3x + 2 \le 5$ or $7x > 29$ $(-\infty, 1] \cup \left(\frac{29}{7}, \infty\right)$

54. $-x < 7$ or $3x + 1 < -20$ $(-\infty, -7) \cup (-7, \infty)$

55. $5 - x > 7$ and $2x + 3 \ge 13$ $\{\}$

56. $-2x < -6$ or $1 - x > -2$ $(-\infty, 3) \cup (3, \infty)$

57. $-\frac{1}{2} \le \frac{4x - 1}{6} < \frac{5}{6}$ $\left[-\frac{1}{2}, \frac{3}{2}\right)$

58. $-\frac{1}{2} \le \frac{3x - 1}{10} < \frac{1}{2}$ $\left[-\frac{4}{3}, 2\right)$

59. $\frac{1}{15} < \frac{8 - 3x}{15} < \frac{4}{5}$ $\left(-\frac{4}{3}, \frac{7}{3}\right)$

60. $-\frac{1}{4} < \frac{6 - x}{12} < -\frac{1}{6}$ $(8, 9)$

61. $0.3 < 0.2x - 0.9 < 1.5$ $(6, 12)$

62. $-0.7 \le 0.4x + 0.8 < 0.5$ $[-3.75, -0.75)$

33. $(-\infty, \infty)$ **34.** $(-\infty, 5)$ **39.** $\left[\frac{3}{2}, 6\right]$

40. $\left(\frac{2}{5}, 2\right)$ **41.** $\left(\frac{5}{4}, \frac{11}{4}\right)$ **42.** $\left(\frac{1}{6}, \frac{7}{2}\right)$

The formula for converting Fahrenheit temperatures to Celsius temperatures is $C = \frac{5}{9}(F - 32)$. Use this formula for Exercises 63 and 64.

63. During a recent year, the temperatures in Chicago ranged from $-29°C$ to $35°C$. Use a compound inequality to convert these temperatures to Fahrenheit temperatures. $-20.2° \le F \le 95°$

64. In Oslo, the temperature ranges from $-10°$ to $18°$ Celsius. Use a compound inequality to convert these temperatures to the Fahrenheit scale.
    $14° \le F \le 64.4°$

Solve.

65. Christian D'Angelo has scores of 68, 65, 75, and 78 on his algebra tests. Use a compound inequality to find the scores he can make on his final exam to receive a C in the course. The final exam counts as two tests, and a C is received if the final course average is from 70 to 79. $67 \le$ final score ≤ 94

66. Wendy Wood has scores of 80, 90, 82, and 75 on her chemistry tests. Use a compound inequality to find the range of scores she can make on her final exam to receive a B in the course. The final exam counts as two tests, and a B is received if the final course average is from 80 to 89. $76.5 \le x \le 100$

Use the graph shown to answer Exercises 67 and 68.

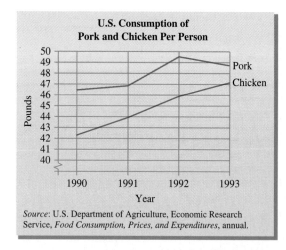

U.S. Consumption of Pork and Chicken Per Person

Source: U.S. Department of Agriculture, Economic Research Service, *Food Consumption, Prices, and Expenditures,* annual.

67. For what years was the consumption of pork greater than 45 pounds per person *and* the consumption of chicken greater than 45 pounds per person?   1992 and 1993

68. For what years was the consumption of pork less than 47 pounds per person *or* the consumption of chicken greater than 47 pounds per person?   1990, 1991, and 1993

Review Exercises

Evaluate the following. See Section 1.4.

69. $|-7| - |19|$ -12 **70.** $|-7 - 19|$ 26

71. $-(-6) - |-10|$ -4 **72.** $|-4| - (-4) + |-20|$ 28

Find by inspection all values for x that make each equation true.

73. $|x| = 7$ $-7, 7$ **74.** $|x| = 5$ $-5, 5$

75. $|x| = 0$ 0   **76.** $|x| = -2$ no solution

A Look Ahead

EXAMPLE

Solve: $x - 6 < 3x < 2x + 5$

Solution

Notice that this inequality contains a variable on the left, the right, and in the middle. To solve, rewrite the inequality using the word ***and.***

$$x - 6 < 3x \quad \text{and} \quad 3x < 2x + 5$$
$$-6 < 2x \quad \text{and} \quad x < 5$$
$$-3 < x$$
$$x > -3 \quad \text{and} \quad x < 5$$

$x > -3$

$x < 5$

$-3 < x < 5$, or $(-3, 5)$

Solve each compound inequality for x. See the example.

77. $2x - 3 < 3x + 1 < 4x - 5$ $(6, \infty)$

78. $x + 3 < 2x + 1 < 4x + 6$ $(2, \infty)$

79. $-3(x - 2) \le 3 - 2x \le 10 - 3x$ $[3, 7]$

80. $7x - 1 \le 7 + 5x \le 3(1 + 2x)$ $\{4\}$

81. $5x - 8 < 2(2 + x) < -2(1 + 2x)$ $(-\infty, -1)$

82. $1 + 2x < 3(2 + x) < 1 + 4x$ $(5, \infty)$

2.6 | ABSOLUTE VALUE EQUATIONS

TAPE IA 2.6

O B J E C T I V E

1 Solve absolute value equations.

1 In Chapter 1, we defined the absolute value of a number as its distance from 0 on a number line.

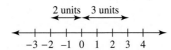

$|-2| = 2$ and $|3| = 3$

In this section, we concentrate on solving equations containing the absolute value of a variable or a variable expression. Examples of absolute value equations are

$$|x| = 3 \qquad -5 = |2y + 7| \qquad |z - 6.7| = |3z + 1.2|$$

Absolute value equations and inequalities (see Section 2.7) are extremely useful in data analysis, especially for calculating acceptable measurement error and errors that result from the way numbers are sometimes represented in computers.

Let us now consider the absolute value equation $|x| = 3$. The solution set of this equation will contain all numbers whose distance from 0 is 3 units. Two numbers are 3 units away from 0 on the number line: 3 and -3.

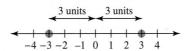

Thus the solution set of the equation $|x| = 3$ is $\{3, -3\}$. This suggests the following:

SOLVING EQUATIONS OF THE FORM $|x| = a$

If a is a positive number, then $|x| = a$ is equivalent to $x = a$ or $x = -a$.

EXAMPLE 1   Solve $|p| = 2$.

Solution: Since 2 is positive, $|p| = 2$ is equivalent to $p = 2$ or $p = -2$.

To check, let $p = 2$ and then $p = -2$ in the original equation.

$|p| = 2$	Original equation.		$|p| = 2$	Original equation.
$|2| = 2$	Let $p = 2$.		$|-2| = 2$	Let $p = -2$.
$2 = 2$	True.		$2 = 2$	True.

The solution set is $\{2, -2\}$.

If the expression inside the absolute value bars is more complicated than a single variable x, we can still apply the absolute value property.

EXAMPLE 2 Solve $|5w + 3| = 7$.

Solution: Here the expression inside the absolute value bars is $5w + 3$. If we think of the expression $5w + 3$ as x in the absolute value property, we see that $|x| = 7$ is equivalent to

$$x = 7 \quad \text{or} \quad x = -7$$

Then substitute $5w + 3$ for x, and we have

$$5w + 3 = 7 \quad \text{or} \quad 5w + 3 = -7$$

Solve these two equations for w.

$$5w + 3 = 7 \quad \text{or} \quad 5w + 3 = -7$$
$$5w = 4 \quad \text{or} \quad 5w = -10$$
$$w = \frac{4}{5} \quad \text{or} \quad w = -2$$

To check, let $w = -2$ and then $w = \frac{4}{5}$ in the original equation.

Let $w = -2$ Let $w = \frac{4}{5}$

$$|5(-2) + 3| = 7 \qquad \left|5\left(\frac{4}{5}\right) + 3\right| = 7$$
$$|-10 + 3| = 7 \qquad |4 + 3| = 7$$
$$|-7| = 7 \qquad |7| = 7$$
$$7 = 7 \quad \text{True.} \qquad 7 = 7 \quad \text{True.}$$

Both solutions check, and the solution set is $\left\{-2, \frac{4}{5}\right\}$.

EXAMPLE 3 Solve $\left|\dfrac{x}{2} - 1\right| = 11$.

Solution: $\left|\dfrac{x}{2} - 1\right| = 11$ is equivalent to

$$\frac{x}{2} - 1 = 11 \qquad \text{or} \qquad \frac{x}{2} - 1 = -11$$
$$2\left(\frac{x}{2} - 1\right) = 2(11) \qquad \text{or} \qquad 2\left(\frac{x}{2} - 1\right) = 2(-11) \qquad \text{Clear fractions.}$$
$$x - 2 = 22 \qquad \text{or} \qquad x - 2 = -22 \qquad \text{Apply the distributive}$$
$$x = 24 \qquad \text{or} \qquad x = -20 \qquad \text{property.}$$

The solution set is $\{24, -20\}$.

To apply the absolute value rule, first make sure that the absolute value expression is isolated.

REMINDER   If the equation has a single absolute value expression containing variables, isolate the absolute value expression first.

EXAMPLE 4   Solve $|2x| + 5 = 7$.

Solution: We want the absolute value expression alone on one side of the equation, so begin by subtracting 5 from both sides. Then apply the absolute value property.

$$|2x| + 5 = 7$$
$$|2x| = 2 \qquad \text{Subtract 5 from both sides.}$$
$$2x = 2 \quad \text{or} \quad 2x = -2$$
$$x = 1 \quad \text{or} \quad x = -1$$

The solution set is $\{-1, 1\}$.

EXAMPLE 5   Solve $|y| = 0$.

Solution: We are looking for all numbers whose distance from 0 is zero units. The only number is 0. The solution set is $\{0\}$.

The next two examples illustrate a special case for absolute value equations. This special case occurs when an isolated absolute value is equal to a negative number.

EXAMPLE 6   Solve $2|x| + 25 = 23$.

Solution: First, isolate the absolute value.

$$2|x| + 25 = 23$$
$$2|x| = -2 \qquad \text{Subtract 25 from both sides.}$$
$$|x| = -1 \qquad \text{Divide both sides by 2.}$$

The absolute value of a number is never negative, so this equation has no solution. The solution set is $\{ \ \}$ or $\varnothing$.

EXAMPLE 7 Solve $\left|\dfrac{3x + 1}{2}\right| = -2$.

Solution: Again, the absolute value of any expression is never negative, so no solution exists. The solution set is $\{\ \}$ or $\varnothing$.

Given two absolute value expressions, we might ask, when are the absolute values of two expressions equal? To see the answer, notice that

$$|2| = |2|, \quad |-2| = |-2|, \quad |-2| = |2|, \quad \text{and} \quad |2| = |-2|$$

  same   same   opposites   opposites

Two absolute value expressions are equal when the expressions inside the absolute value bars are equal to or are opposites of each other.

EXAMPLE 8 Solve $|3x + 2| = |5x - 8|$.

Solution: This equation is true if the expressions inside the absolute value bars are equal to or are opposites of each other.

$$3x + 2 = 5x - 8 \quad \text{or} \quad 3x + 2 = -(5x - 8)$$

Next, solve each equation.

$$3x + 2 = 5x - 8 \quad \text{or} \quad 3x + 2 = -5x + 8$$
$$-2x + 2 = -8 \quad \text{or} \quad 8x + 2 = 8$$
$$-2x = -10 \quad \text{or} \quad 8x = 6$$
$$x = 5 \quad \text{or} \quad x = \frac{3}{4}$$

The solution set is $\left\{\dfrac{3}{4}, 5\right\}$.

EXAMPLE 9 Solve $|x - 3| = |5 - x|$.

Solution:
$$x - 3 = 5 - x \quad \text{or} \quad x - 3 = -(5 - x)$$
$$2x - 3 = 5 \quad \text{or} \quad x - 3 = -5 + x$$
$$2x = 8 \quad \text{or} \quad x - 3 - x = -5 + x - x$$
$$x = 4 \quad \text{or} \quad -3 = -5 \qquad \text{False.}$$

Recall from Section 2.1 that when an equation simplifies to a false statement, the equation has no solution. Thus the only solution for the original absolute value equation is 4, and the solution set is $\{4\}$.

The following box summarizes the methods shown for solving absolute value equations.

> **ABSOLUTE VALUE EQUATIONS**
>
> $|x| = a$ $\begin{cases} \text{If } a \text{ is positive, then solve } x = a \text{ or } x = -a. \\ \text{If } a \text{ is } 0, \text{ solve } x = 0. \\ \text{If } a \text{ is negative, the equation } |x| = a \text{ has no solution.} \end{cases}$
>
> $|x| = |y|$ Solve $x = y$ or $x = -y$.

MENTAL MATH

Simplify each expression.

1. $|-7|$ 7

2. $|-8|$ 8

3. $-|5|$ -5

4. $-|10|$ -10

5. $-|-6|$ -6

6. $-|-3|$ -3

7. $|-3| + |-2| + |-7|$ 12

8. $|-1| + |-6| + |-8|$ 15

EXERCISE SET 2.6

Solve each absolute value equation. See Examples 1 through 7.

1. $|x| = 7$ $\{7, -7\}$

2. $|y| = 15$ $\{-15, 15\}$

3. $|3x| = 12.6$ $\{4.2, -4.2\}$

4. $|6n| = 12.6$ $\{2.1, -2.1\}$

5. $|2x - 5| = 9$ $\{7, -2\}$

6. $|6 + 2n| = 4$ $\{-5, -1\}$

7. $\left|\dfrac{x}{2} - 3\right| = 1$ $\{8, 4\}$

8. $\left|\dfrac{n}{3} + 2\right| = 4$ $\{-18, 6\}$

9. $|z| + 4 = 9$ $\{5, -5\}$

10. $|x| + 1 = 3$ $\{2, -2\}$

11. $|3x| + 5 = 14$ $\{3, -3\}$

12. $|2x| - 6 = 4$ $\{5, -5\}$

13. $|2x| = 0$ $\{0\}$

14. $|7z| = 0$ $\{0\}$

15. $|4n + 1| + 10 = 4$ $\{\}$

16. $|3z - 2| + 8 = 1$ $\{\}$

17. $|5x - 1| = 0$ $\left\{\dfrac{1}{5}\right\}$

18. $|3y + 2| = 0$ $\left\{-\dfrac{2}{3}\right\}$

19. Write an absolute value equation representing all numbers x whose distance from 0 is 5 units. $|x| = 5$

20. Write an absolute value equation representing all numbers x whose distance from 0 is 2 units. $|x| = 2$

Solve. See Examples 8 and 9.

21. $|5x - 7| = |3x + 11|$

22. $|9y + 1| = |6y + 4|$

23. $|z + 8| = |z - 3|$

24. $|2x - 5| = |2x + 5|$ $\{0\}$

25. Describe how solving an absolute value equation such as $|2x - 1| = 3$ is similar to solving an absolute value equation such as $|2x - 1| = |x - 5|$.

26. Describe how solving an absolute value equation such as $|2x - 1| = 3$ is different from solving an absolute value equation such as $|2x - 1| = |x - 5|$.   Answers will vary.

Solve each absolute value equation.

27. $|x| = 4$ $\{4, -4\}$

28. $|x| = 1$ $\{-1, 1\}$

29. $|y| = 0$ $\{0\}$

30. $|y| = 8$ $\{-8, 8\}$

31. $|z| = -2$ $\{\}$

32. $|y| = -9$ $\{\}$

33. $|7 - 3x| = 7$

34. $|4m + 5| = 5$

35. $|6x| - 1 = 11$ $\{2, -2\}$

36. $|7z| + 1 = 22$ $\{-3, 3\}$

37. $|4p| = -8$ $\{\}$

38. $|5m| = -10$ $\{\}$

39. $|x - 3| + 3 = 7$

40. $|x + 4| - 4 = 1$ $\{-9, 1\}$

41. $\left|\dfrac{z}{4} + 5\right| = -7$ $\{\}$

42. $\left|\dfrac{c}{5} - 1\right| = -2$ $\{\}$

43. $|9v - 3| = -8$ $\{\}$

44. $|1 - 3b| = -7$ $\{\}$

45. $|8n + 1| = 0$ $\left\{-\dfrac{1}{8}\right\}$

46. $|5x - 2| = 0$ $\left\{\dfrac{2}{5}\right\}$

47. $|1 + 6c| - 7 = -3$

48. $|2 + 3m| - 9 = -7$

21. $\left\{9, \dfrac{-1}{2}\right\}$ **22.** $\left\{-\dfrac{1}{3}, 1\right\}$ **23.** $\left\{-\dfrac{5}{2}\right\}$ **25.** Answers will vary.   **33.** $\left\{0, \dfrac{14}{3}\right\}$ **34.** $\left\{-\dfrac{5}{2}, 0\right\}$ **39.** $\{7, -1\}$

47. $\left\{\dfrac{1}{2}, \dfrac{-5}{6}\right\}$ **48.** $\left\{-\dfrac{4}{3}, 0\right\}$

49. $\left\{2, -\dfrac{12}{5}\right\}$ **50.** $\left\{\dfrac{7}{6}, \dfrac{3}{2}\right\}$ **51.** $\{3, -2\}$ **52.** $\left\{-3, -\dfrac{1}{3}\right\}$ **53.** $\left\{-8, \dfrac{2}{3}\right\}$

49. $|5x + 1| = 11$

50. $|8 - 6c| = 1$

51. $|4x - 2| = |-10|$

52. $|3x + 5| = |-4|$

53. $|5x + 1| = |4x - 7|$

54. $|3 + 6n| = |4n + 11|$

55. $|6 + 2x| = -|7|$ $\{\}$

56. $|4 - 5y| = -|-3|$ $\{\}$

57. $|2x - 6| = |10 - 2x|$

58. $|4n + 5| = |4n + 3|$ $\{-1\}$

59. $\left|\dfrac{2x - 5}{3}\right| = 7$ $\{13, -8\}$

60. $\left|\dfrac{1 + 3n}{4}\right| = 4$ $\left\{-\dfrac{17}{3}, 5\right\}$

61. $2 + |5n| = 17$ $\{3, -3\}$

62. $8 + |4m| = 24$ $\{-4, 4\}$

63. $\left|\dfrac{2x - 1}{3}\right| = |-5|$

64. $\left|\dfrac{5x + 2}{2}\right| = |-6|$

65. $|2y - 3| = |9 - 4y|$

66. $|5z - 1| = |7 - z|$ $\left\{-\dfrac{3}{2}, \dfrac{4}{3}\right\}$

67. $\left|\dfrac{3n + 2}{8}\right| = |-1|$

68. $\left|\dfrac{2r - 6}{5}\right| = |-2|$ $\{-2, 8\}$

69. $|x + 4| = |7 - x|$

70. $|8 - y| = |y + 2|$ $\{3\}$

71. $\left|\dfrac{8c - 7}{3}\right| = -|-5|$ $\{\}$

72. $\left|\dfrac{5d + 1}{6}\right| = -|-9|$ $\{\}$

73. Explain why some absolute value equations have two solutions.   Answers will vary.

74. Explain why some absolute value equations have one solution.   Answers will vary.

54. $\left\{-\dfrac{7}{5}, 4\right\}$ **57.** $\{4\}$ **63.** $\{8, -7\}$ **64.** $\left\{-\dfrac{14}{5}, 2\right\}$ **65.** $\{2, 3\}$ **67.** $\left\{2, -\dfrac{10}{3}\right\}$ **69.** $\left\{\dfrac{3}{2}\right\}$ **78.** 3, 2, 1, 0, −1 for example

79. −2, −1, 0, 1, 2, for example **80.** 0, 1, 2, 3, 4, for example

Review Exercises

The following circle graph shows the sources of Walt Disney Company's income. Use this graph to answer the questions below. See Section 2.2.

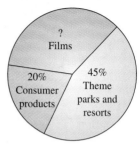

Source: USA Today, August 25, 1994.

75. What percent of Disney's income comes from films?   35%

76. A circle contains 360°. Find the number of degrees found in the 20% sector.   72°

77. If Disney's income one year is \$9.75 billion, find the income from theme parks and resorts. \$4.39 billion

List five integer solutions of each inequality.

78. $|x| \le 3$

79. $|x| \ge -2$

80. $|y| > -10$

81. $|y| < 0$ no solution

2.7   ABSOLUTE VALUE INEQUALITIES

TAPE IA 2.7

OBJECTIVES

1 Solve absolute value inequalities of the form $|x| < a$.

2 Solve absolute value inequalities of the form $|x| > a$.

1 The solution set of an absolute value inequality such as $|x| < 2$ contains all numbers whose distance from 0 is less than 2 units, as shown next.

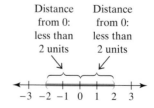

The solution set is $\{x \mid -2 < x < 2\}$, or $(-2, 2)$ in interval notation.

EXAMPLE 1   Solve $|x| \leq 3$.

Solution: The solution set of this inequality contains all numbers whose distance from 0 is less than or equal to 3. Thus 3, −3, and all numbers between 3 and −3 are in the solution set.

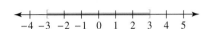

The solution set is $[-3, 3]$.

In general, we have the following:

> **SOLVING ABSOLUTE VALUE INEQUALITIES OF THE FORM $|x| < a$**
> If a is a positive number, then $|x| < a$ is equivalent to $-a < x < a$.

This property also holds true for the inequality symbol $\leq$.

EXAMPLE 2   Solve for m: $|m - 6| < 2$.

Solution: Replace x with $m - 6$ and a with 2 in the preceding property, and we see that

$$|m - 6| < 2 \quad \text{is equivalent to} \quad -2 < m - 6 < 2$$

Solve this compound inequality for m by adding 6 to all three sides.

$$-2 < m - 6 < 2$$
$$-2 + 6 < m - 6 + 6 < 2 + 6 \qquad \text{Add 6 to all three sides.}$$
$$4 < m < 8 \qquad\qquad\quad \text{Simplify.}$$

The solution set is $(4, 8)$, and its graph is shown at the left.

> R E M I N D E R   Before applying an absolute value inequality property, isolate the absolute value expression on one side of the inequality.

EXAMPLE 3   Solve for x: $|5x + 1| + 1 \leq 10$.

Solution: First, isolate the absolute value expression by subtracting 1 from both sides.

$$|5x + 1| + 1 \leq 10$$
$$|5x + 1| \leq 10 - 1 \qquad \text{Subtract 1 from both sides.}$$
$$|5x + 1| \leq 9 \qquad\quad\; \text{Simplify.}$$

Since 9 is positive, we apply the absolute value property for $|x| \leq a$.

$$-9 \leq 5x + 1 \leq 9$$

$$-9 - 1 \leq 5x + 1 - 1 \leq 9 - 1 \qquad \text{Subtract 1 from all three sides.}$$

$$-10 \leq 5x \leq 8 \qquad \text{Simplify.}$$

$$-2 \leq x \leq \frac{8}{5} \qquad \text{Divide all three sides by 5.}$$

The solution set is $\left[-2, \dfrac{8}{5}\right]$, and the graph is shown at the left.

EXAMPLE 4   Solve for x: $\left|2x - \dfrac{1}{10}\right| < -13$.

Solution: The absolute value of a number is always nonnegative and can never be less than -13. Thus this absolute value inequality has no solution. The solution set is $\{\ \}$ or $\varnothing$.

2   Let us now solve an absolute value inequality of the form $|x| > a$, such as $|x| \geq 3$. The solution set contains all numbers whose distance from 0 is 3 or more units. Thus the graph of the solution set contains 3 and all points to the right of 3 on the number line or -3 and all points to the left of -3 on the number line.

This solution set is written as $\{x \mid x \leq -3 \text{ or } x \geq 3\}$. In interval notation, the solution set is $(-\infty, -3] \cup [3, \infty)$, since "or" means "union." In general, we have the following.

> **SOLVING ABSOLUTE VALUE INEQUALITIES OF THE FORM $|x| > a$**
>
> If a is a positive number, then $|x| > a$ is equivalent to $x < -a$ or $x > a$.

This property also holds true for the inequality symbol $\geq$.

EXAMPLE 5   Solve for y: $|y - 3| > 7$.

Solution: Since 7 is positive, we apply the property for $|x| > a$.

$$|y - 3| > 7 \text{ is equivalent to } y - 3 < -7 \text{ or } y - 3 > 7$$

Next, solve the compound inequality.

$$y - 3 < -7 \qquad \text{or} \qquad y - 3 > 7$$
$$y - 3 + 3 < -7 + 3 \quad \text{or} \quad y - 3 + 3 > 7 + 3 \qquad \text{Add 3 to both sides.}$$
$$y < -4 \qquad \text{or} \qquad y > 10 \qquad \text{Simplify.}$$

The solution set is $(-\infty, -4) \cup (10, \infty)$, and its graph is shown.

Examples 6 and 8 illustrate special cases of absolute value inequalities. These special cases occur when an isolated absolute value expression is less than, less than or equal to, greater than, or greater than or equal to a negative number or 0.

EXAMPLE 6   Solve $|2x + 9| + 5 > 3$.

Solution: First isolate the absolute value expression by subtracting 5 from both sides.

$$|2x + 9| + 5 > 3$$
$$|2x + 9| + 5 - 5 > 3 - 5 \qquad \text{Subtract 5 from both sides.}$$
$$|2x + 9| > -2 \qquad \text{Simplify.}$$

The absolute value of any number is always nonnegative and thus is always greater than -2. This inequality and the original inequality are true for all values of x. The solution set is $\{x \mid x \text{ is a real number}\}$ or $(-\infty, \infty)$, and its graph is shown.

EXAMPLE 7   Solve $\left|\dfrac{x}{3} - 1\right| - 7 \geq -5$.

Solution: First, isolate the absolute value expression by adding 7 to both sides.

$$\left|\frac{x}{3} - 1\right| - 7 \geq -5$$
$$\left|\frac{x}{3} - 1\right| - 7 + 7 \geq -5 + 7 \qquad \text{Add 7 to both sides.}$$
$$\left|\frac{x}{3} - 1\right| \geq 2 \qquad \text{Simplify.}$$

Next, write the absolute value inequality as an equivalent compound inequality and solve.

$$\frac{x}{3} - 1 \leq -2 \qquad \text{or} \qquad \frac{x}{3} - 1 \geq 2$$
$$3\left(\frac{x}{3} - 1\right) \leq 3(-2) \quad \text{or} \quad 3\left(\frac{x}{3} - 1\right) \geq 3(2) \qquad \text{Clear the inequalities of fractions.}$$
$$x - 3 \leq -6 \qquad \text{or} \qquad x - 3 \geq 6 \qquad \text{Apply the distributive property.}$$
$$x \leq -3 \qquad \text{or} \qquad x \geq 9 \qquad \text{Add 3 to both sides.}$$

The solution set is $(-\infty, -3] \cup [9, \infty)$, and its graph is shown below.

EXAMPLE 8 Solve for x: $\left|\dfrac{2(x + 1)}{3}\right| \le 0$.

Solution: Recall that "$\le$" means "less than or equal to." The absolute value of any expression will never be less than 0, but it may be equal to 0. Thus, to solve $\left|\dfrac{2(x + 1)}{3}\right| \le 0$ we solve $\left|\dfrac{2(x + 1)}{3}\right| = 0$.

$$\frac{2(x + 1)}{3} = 0$$

$$3\left[\frac{2(x + 1)}{3}\right] = 3(0) \qquad \text{Clear the equation of fractions.}$$

$$2x + 2 = 0 \qquad \text{Apply the distributive property.}$$

$$2x = -2 \qquad \text{Subtract 2 from both sides.}$$

$$x = -1 \qquad \text{Divide both sides by 2.}$$

The solution set is $\{-1\}$.

The following box summarizes the types of absolute value equations and inequalities.

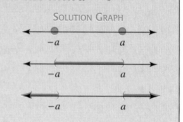

SOLVING ABSOLUTE VALUE EQUATIONS AND INEQUALITIES WITH $a > 0$

ALGEBRAIC SOLUTION	SOLUTION GRAPH
$|x| = a$ is equivalent to $x = a$ or $x = -a$.	$-a \qquad a$
$|x| < a$ is equivalent to $-a < x < a$.	$-a \qquad a$
$|x| > a$ is equivalent to $x < -a$ or $x > a$.	$-a \qquad a$

2. $\overset{-6 \qquad 6}{\longleftrightarrow}$ $(-6, 6)$ **3.** $\overset{1 \qquad 5}{\longleftrightarrow}$ $(1, 5)$ **4.** $\overset{-5 \qquad 5}{\longleftrightarrow}$ $[-5, 5]$ **5.** $\overset{-5 \qquad -1}{\longleftrightarrow}$ $(-5, -1)$

6. $\overset{-10 \qquad 2}{\longleftrightarrow}$ $(-10, 2)$ **7.** $\overset{-10 \qquad 3}{\longleftrightarrow}$ $[-10, 3]$ **8.** $\overset{-3 \qquad \frac{21}{5}}{\longleftrightarrow}$ $\left[-3, \frac{21}{5}\right]$ **9.** $\overset{-5 \qquad 5}{\longleftrightarrow}$ $[-5, 5]$

EXERCISE SET 2.7

10. $\overset{-1 \quad 1}{\longleftrightarrow}$ $[-1, 1]$ **11.** $\longleftrightarrow$ $\{\ \}$ **12.** $\longleftrightarrow$ $\{\ \}$

Solve each inequality. Then graph the solution set. See Examples 1 through 4. **1.** $\overset{-4 \qquad 4}{\longleftrightarrow}$ $[-4, 4]$

1. $|x| \le 4$ **2.** $|x| < 6$ **3.** $|x - 3| < 2$

4. $|y| \le 5$ **5.** $|x + 3| < 2$ **6.** $|x + 4| < 6$

7. $|2x + 7| \le 13$ **8.** $|5x - 3| \le 18$

9. $|x| + 7 \le 12$ **10.** $|x| + 6 \le 7$

11. $|3x - 1| < -5$ **12.** $|8x - 3| < -2$

13. $|x - 6| - 7 \le -1$ **14.** $|z + 2| - 7 < -3$

13. $\overset{0 \qquad 12}{\longleftrightarrow}$ $[0, 12]$ **14.** $\overset{-6 \qquad 2}{\longleftrightarrow}$ $(-6, 2)$ **16.** $\overset{-4 \qquad 4}{\longleftrightarrow}$ $(-\infty, -4] \cup [4, \infty)$

Solve each inequality. Graph the solution set. See Examples 5 through 7. **15.** $\overset{-3 \qquad 3}{\longleftrightarrow}$ $(-\infty, -3) \cup (3, \infty)$

15. $|x| > 3$ **16.** $|y| \ge 4$

17. $|x + 10| \ge 14$ $\overset{-24 \qquad 4}{\longleftrightarrow}$ $(-\infty, -24] \cup [4, \infty)$

18. $|x - 9| \ge 2$ $\overset{7 \qquad 11}{\longleftrightarrow}$ $(-\infty, 7] \cup [11, \infty)$

19. $|x| + 2 > 6$ **20.** $|x| - 1 > 3$

21. $|5x| > -4$ **22.** $|4x - 11| > -1$

23. $|6x - 8| + 3 > 7$ **24.** $|10 + 3x| + 1 > 2$

19. $\overset{-4 \qquad 4}{\longleftrightarrow}$ $(-\infty, -4) \cup (4, \infty)$ **20.** $\overset{-4 \qquad 4}{\longleftrightarrow}$ $(-\infty, -4) \cup (4, \infty)$ **21.** $\longleftrightarrow$ $(-\infty, \infty)$

22. $\longleftrightarrow$ $(-\infty, \infty)$ **23.** $\overset{\frac{2}{3} \qquad 2}{\longleftrightarrow}$ $\left(-\infty, \frac{2}{3}\right) \cup (2, \infty)$ **24.** $\overset{-\frac{11}{3} \quad -3}{\longleftrightarrow}$ $\left(-\infty, -\frac{11}{3}\right) \cup (-3, \infty)$

26. (graph) $(-\infty, \infty)$ **27.** (graph) $-\frac{3}{8}$ $\left(-\infty, -\frac{3}{8}\right) \cup \left(-\frac{3}{8}, \infty\right)$ **28.** (graph) { }   **33.** (graph) -2 2 $[-2, 2]$

Solve each inequality. Graph the solution set. See Example 8.   **25.** (graph) 0 $\{0\}$

25. $|x| \leq 0$ **26.** $|x| \geq 0$

27. $|8x + 3| > 0$ **28.** $|5x - 6| < 0$

29. Write an absolute value inequality representing all numbers x whose distance from 0 is less than 7 units. $|x| < 7$

30. Write an absolute value inequality representing all numbers x whose distance from 0 is greater than 4 units. $|x| > 4$

31. Write $-5 \leq x \leq 5$ as an equivalent inequality containing an absolute value. $|x| \leq 5$

32. Write $x > 1$ or $x < -1$ as an equivalent inequality containing an absolute value. $|x| > 1$

Solve each inequality. Graph the solution set.

33. $|x| \leq 2$ **34.** $|z| < 6$

35. $|y| > 1$ **36.** $|x| \geq 10$

37. $|x - 3| < 8$ **38.** $|-3 + x| \leq 10$

39. $|6x - 8| > 4$ **40.** $|1 + 0.3x| \geq 0.1$

41. $5 + |x| \leq 2$ **42.** $8 + |x| < 1$

43. $|x| > -4$ **44.** $|x| \leq -7$

45. $|2x - 7| \leq 11$ **46.** $|5x + 2| < 8$

47. $|x + 5| + 2 \geq 8$ (graph) -11 1 $(-\infty, -11] \cup [1, \infty)$

48. $|-1 + x| - 6 > 2$ (graph) -7 9 $(-\infty, -7) \cup (9, \infty)$

49. $|x| > 0$ **50.** $|x| < 0$

51. $9 + |x| > 7$ **52.** $5 + |x| \geq 4$

53. $6 + |4x - 1| \leq 9$ (graph) $-\frac{1}{2}$ 1 $\left[-\frac{1}{2}, 1\right]$

54. $-3 + |5x - 2| \leq 4$ (graph) -1 $\frac{9}{5}$ $\left[-1, \frac{9}{5}\right]$

55. $\left|\frac{2}{3}x + 1\right| > 1$ **56.** $|5x - 1| \geq 2$

57. $|5x + 3| < -6$ (graph) { }

58. $|4 + 9x| \geq -6$ (graph) $(-\infty, \infty)$

59. $|8x + 3| \geq 0$ **60.** $|5x - 6| \leq 0$

61. $|1 + 3x| + 4 < 5$

62. $|7x - 3| - 1 \leq 10$

63. $|x| - 3 \geq -3$ **64.** $|x| + 6 < 6$

65. $|8x| - 10 > -2$ (graph) -1 1 $(-\infty, -1) \cup (1, \infty)$

66. $|6x| - 13 \geq -7$ (graph) -1 1 $(-\infty, -1] \cup [1, \infty)$

67. $\left|\frac{x + 6}{3}\right| > 2$ **68.** $\left|\frac{7 + x}{2}\right| \geq 4$

69. $|2(3 + x)| > 6$ **70.** $|5(x - 3)| \geq 10$

71. $\left|\frac{5(x + 2)}{3}\right| < 7$ **72.** $\left|\frac{6(3 + x)}{5}\right| \leq 4$

73. $-15 + |2x - 7| \leq -6$ (graph) -1 8 $[-1, 8]$

74. $-9 + |3 + 4x| < -4$ (graph) -2 $\frac{1}{2}$ $\left(-2, \frac{1}{2}\right)$

75. $\left|2x + \frac{3}{4}\right| - 7 \leq -2$ (graph) $-\frac{23}{8}$ $\frac{17}{8}$ $\left[-\frac{23}{8}, \frac{17}{8}\right]$

76. $\left|\frac{3}{5} + 4x\right| - 6 < -1$ (graph) $-\frac{7}{5}$ $\frac{11}{10}$ $\left(-\frac{7}{5}, \frac{11}{10}\right)$

Solve each equation or inequality for x.

77. $|2x - 3| < 7$ $(-2, 5)$ **78.** $|2x - 3| > 7$

79. $|2x - 3| = 7$ $\{5, -2\}$ **80.** $|5 - 6x| = 29$

81. $|x - 5| \geq 12$ **82.** $|x + 4| \geq 20$

83. $|9 + 4x| = 0$ $\left\{-\frac{9}{4}\right\}$ **84.** $|9 + 4x| \geq 0$ $(-\infty, \infty)$

85. $|2x + 1| + 4 < 7$ **86.** $8 + |5x - 3| \geq 11$

87. $|3x - 5| + 4 = 5$ $\left\{2, \frac{4}{3}\right\}$ **88.** $|8x| = -5$ { }

89. $|x + 11| = -1$ { }   **90.** $|4x - 4| = -3$ { }

91. $\left|\frac{2x - 1}{3}\right| = 6$ **92.** $\left|\frac{6 - x}{4}\right| = 5$ $\{-14, 26\}$

93. $\left|\frac{3x - 5}{6}\right| > 5$ **94.** $\left|\frac{4x - 7}{5}\right| < 2$ $\left(-\frac{3}{4}, \frac{17}{4}\right)$

The expression $|x_T - x|$ is defined to be the ◌lute error in x, where x_T is the true value of a quantity and x is the measured value or value as stored in a computer.

95. If the true value of a quantity is 3.5 and the absolute error must be less than 0.05, find the acceptable measured values. $3.45 < x < 3.55$

96. If the true value of a quantity is 0.2 and the approximate value stored in a computer is $\frac{51}{256}$, find the absolute error. $\frac{1}{1280}$, or 0.00078125

Review Exercises

Recall the formula

$$\text{Probability of an event} = \frac{\substack{\text{number of ways that} \\ \text{the event can occur}}}{\substack{\text{number of possible} \\ \text{outcomes}}}$$

Find the probability of rolling each number on a single toss of a die. (Recall that a die is a cube with each of

34. (graph) -6 6 $(-6, 6)$ **35.** (graph) -1 1 $(-\infty, -1) \cup (1, \infty)$ **36.** (graph) -10 10 $(-\infty, -10] \cup [10, \infty)$

37. (graph) -5 11 $(-5, 11)$ **38.** (graph) -7 13 $[-7, 13]$ **39.** (graph) $\frac{2}{3}$ 2 $\left(-\infty, \frac{2}{3}\right) \cup (2, \infty)$

For more answer graphs, please see page 106.

its six sides containing 1, 2, 3, 4, 5, and 6 black dots, respectively.) See Section 2.3.

97. P(rolling a 2) $\frac{1}{6}$ **98.** P(rolling a 5) $\frac{1}{6}$

99. P(rolling a 7) 0   **100.** P(rolling a 0) 0

101. P(rolling a 1 or 3) $\frac{1}{3}$ **102.** P(rolling a 1, 2, 3, 4, 5, or 6) 1

Consider the equation $3x - 4y = 12$. For each value of x or y given, find the corresponding value of the other variable that makes the statement true. See Section 2.3.

103. If $x = 2$, find y. -1.5 **104.** If $y = -1$, find x. $\frac{8}{3}$

105. If $y = -3$, find x. 0 **106.** If $x = 4$, find y. 0

GROUP ACTIVITY

INVESTIGATING COMPOUND INTEREST

MATERIALS:
- Calculator
- Newspapers

Suppose you have just moved to a new city and must open new bank accounts with $5000. You would like to open a checking account, a savings account, and a certificate of deposit (CD). After doing some research you find the following four options for CDs at four local banks.

BANK	MINIMUM DEPOSIT	INTEREST RATE	NUMBER OF COMPOUNDINGS PER YEAR
A	$1000	6.55%	26
B	$1000	6.6%	6
C	$1000	6.7%	1
D	$5000	7.0%	2

1. For each CD option, use the formula for compound interest to find the values of a CD opened with the minimum deposit after an investment of 1 year, 2 years, and 10 years. Create a table of your results.

2. Which CD option would you prefer? Explain your reasoning. What other factors should you consider when choosing a savings tool such as a CD?

3. You may have noticed that it can sometimes be difficult to compare options when different interest rates and numbers of compoundings are involved. When describing investment options, banks often use *annual percentage yield* (APY) instead to help eliminate confu-

sion. The APY of an investment represents the percentage increase in the balance after one year over the original investment. Find the APY of each CD option. Do you think that the APY makes it easier to compare the options? Explain.

4. Find several bank advertisements in newspapers that outline investment opportunities, such as savings accounts or CDs. Take note of whether the investments are described in terms of APY. Prepare a brief report describing the investment opportunities you found and comparing the options. Of these options, which would you choose? Why?

See Appendix D for Group Activity answers and suggestions.

CHAPTER 2   HIGHLIGHTS

DEFINITIONS AND CONCEPTS	EXAMPLES

SECTION 2.1   LINEAR EQUATIONS IN ONE VARIABLE

An **equation** is a statement that two expressions are equal.

A **linear equation in one variable** is an equation that can be written in the form $ax + b = c$, where $a, b,$ and c are real numbers and a is not 0.

A **solution** of an equation is a value for the variable that makes the equation a true statement.

Equations:
$$5 = 5 \qquad 7x + 2 = -14 \qquad 3(x - 1)^2 = 9x^2 - 6$$

Linear equations:
$$7x + 2 = -14 \quad x = -3$$
$$5(2y - 7) = -2(8y - 1)$$

Check to see that -1 is a solution of $3(x - 1) = 4x - 2.$

$$3(-1 - 1) = 4(-1) - 2$$
$$3(-2) = -4 - 2$$
$$-6 = -6 \qquad \text{True.}$$

Thus, -1 is a solution.

Equivalent equations have the same solution.

The **addition property of equality** guarantees that the same number may be added to (or subtracted from) both sides of an equation, and the result is an equivalent equation.

The **multiplication property of equality** guarantees that both sides of an equation may be multiplied by (or divided by) the same nonzero number, and the result is an equivalent equation.

To solve linear equations in one variable:

$x - 12 = 14$ and $x = 26$ are equivalent equations.

Solve for x: $-3x - 2 = 10.$

$$-3x - 2 + 2 = 10 + 2 \qquad \text{Add 2 to both sides.}$$
$$-3x = 12$$
$$\frac{-3x}{-3} = \frac{12}{-3} \qquad \text{Divide both sides by } -3.$$
$$x = -4$$

Solve for x:
$$x - \frac{x - 2}{6} = \frac{x - 7}{3} + \frac{2}{3}$$

1. Clear the equation of fractions.

1. $6\left(x - \dfrac{x - 2}{6}\right) = 6\left(\dfrac{x - 7}{3} + \dfrac{2}{3}\right)$ Multiply both sides by 6.

$$6x - (x - 2) = 2(x - 7) + 2(2)$$

2. Remove grouping symbols such as parentheses.

3. Simplify by combining like terms.

4. Write variable terms on one side and numbers on the other side using the addition property of equality.

2. $6x - x + 2 = 2x - 14 + 4$ Remove grouping symbols.

3. $\quad 5x + 2 = 2x - 10$

4. $5x + 2 - 2 = 2x - 10 - 2$ Subtract 2.

$$5x = 2x - 12$$
$$5x - 2x = 2x - 12 - 2x \qquad \text{Subtract } 2x.$$
$$3x = -12$$

5. Isolate the variable using the multiplication property of equality.

5. $\quad \dfrac{3x}{3} = \dfrac{-12}{3}$ Divide by 3.

$$x = -4$$

(continued)

DEFINITIONS AND CONCEPTS	**EXAMPLES**

SECTION 2.1 LINEAR EQUATIONS IN ONE VARIABLE

6. Check the proposed solution in the original equation.

6. $-4 - \dfrac{-4-2}{6} = \dfrac{-4-7}{3} + \dfrac{2}{3}$ Replace x with -4 in the original equation.

$$-4 - \dfrac{-6}{6} = \dfrac{-11}{3} + \dfrac{2}{3}$$

$$-4 - (-1) = \dfrac{-9}{3}$$

$$-3 = -3 \qquad \text{True.}$$

SECTION 2.2 AN INTRODUCTION TO PROBLEM SOLVING

Colorado is shaped like a rectangle whose length is about 1.3 times its width. If the perimeter of Colorado is 2070 kilometers, find its dimensions.

1. UNDERSTAND the problem.

1. Read and reread the problem. Guess a solution and check your guess.

2. ASSIGN a variable.

2. Let x = width of Colorado in kilometers. Then $1.3x$ = length of Colorado in kilometers.

3. ILLUSTRATE the problem.

3.

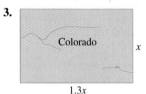

Colorado x

$1.3x$

4. TRANSLATE the problem.

4. In words: $\boxed{\text{twice the length}}$ + $\boxed{\text{twice the width}}$ = $\boxed{\text{perimeter}}$

Translate: $2(1.3x)$ + $2x$ = 2070

5. COMPLETE by solving.

5. $2.6x + 2x = 2070$

$$4.6x = 2070$$

$$x = 450$$

6. INTERPRET the results.

6. If $x = 450$ kilometers, then $1.3x = 1.3(450) = 585$ kilometers. *Check:* The perimeter of a rectangle whose width is 450 kilometers and length is 585 kilometers is $2(450) + 2(585) = 2070$ kilometers, the required perimeter. *State:* The dimensions of Colorado are approximately 450 kilometers by 585 kilometers.

SECTION 2.3 FORMULAS AND PROBLEM SOLVING

An equation that describes a known relationship among quantities is called a **formula.**

Formulas:
$$A = \pi r^2 \text{ (area of a circle)}$$
$$I = P \cdot R \cdot T \text{ (interest = principal }\cdot\text{ rate }\cdot\text{ time)}$$

(continued)

DEFINITIONS AND CONCEPTS	EXAMPLES

SECTION 2.3   FORMULAS AND PROBLEM SOLVING

To solve a formula for a specified variable, use the steps for solving an equation. Treat the specified variable as the only variable of the equation.

Solve $A = 2HW + 2LW + 2LH$ for H.

$$A - 2LW = 2HW + 2LH \qquad \text{Subtract } 2LW.$$
$$A - 2LW = H(2W + 2L) \qquad \text{Factor out } H.$$
$$\frac{A - 2LW}{2W + 2L} = \frac{H(2W + 2L)}{2W + 2L} \qquad \begin{array}{l}\text{Divide by}\\ \quad 2W + 2L.\end{array}$$
$$\frac{A - 2LW}{2W + 2L} = H \qquad \text{Simplify.}$$

SECTION 2.4   LINEAR INEQUALITIES AND PROBLEM SOLVING

A **linear inequality in one variable** is an inequality that can be written in the form $ax + b < c$, where a, b, and c are real numbers and $a \neq 0$. (The inequality symbols $\leq$, $>$, and $\geq$ also apply here.)

The **addition property of inequality** guarantees that the same number may be added to (or subtracted from) both sides of an inequality, and the resulting inequality will have the same solution set.

The **multiplication property of inequality** guarantees that both sides of an inequality may be multiplied by (or divided by) the same **positive** number, and the resulting inequality will have the same solution set. We may also multiply (or divide) both sides of an inequality by the same **negative** number and **reverse the direction of the inequality symbol,** and the result is an inequality with the same solution set.

To solve a linear inequality in one variable:

1. Clear the equation of fractions.

2. Remove grouping symbols such as parentheses.
3. Simplify by combining like terms.
4. Write variable terms on one side and numbers on the other side using the addition property of inequality.

5. Isolate the variable using the multiplication property of inequality.

Linear inequalities:

$$5x - 2 \leq -7 \qquad 3y > 1 \qquad \frac{z}{7} < -9(z - 3)$$

$$x - 9 \leq -16$$
$$x - 9 + 9 \leq -16 + 9 \qquad \text{Add 9.}$$
$$x \leq -7$$

$$6x < -66$$
$$\frac{6x}{6} < \frac{-66}{6} \qquad \begin{array}{l}\text{Divide by 6. Do not reverse}\\ \text{direction of inequality}\\ \text{symbol.}\end{array}$$
$$x < -11$$
$$-6x < -66$$
$$\frac{-6x}{-6} > \frac{-66}{-6} \qquad \begin{array}{l}\text{Divide by } -6. \text{ Reverse direc-}\\ \text{tion of inequality symbol.}\end{array}$$
$$x > 11$$

Solve for x:

$$\frac{3}{7}(x - 4) \geq x + 2$$

1. $7\left[\frac{3}{7}(x - 4)\right] \geq 7(x + 2) \qquad \text{Multiply by 7.}$

$$3(x - 4) \geq 7(x + 2)$$

2. $3x - 12 \geq 7x + 14 \qquad \begin{array}{l}\text{Apply the distributive}\\ \text{property.}\end{array}$

4. $-4x - 12 \geq 14 \qquad \text{Subtract } 7x.$

$$-4x \geq 26 \qquad \text{Add 12.}$$

$$\frac{-4x}{-4} \leq \frac{26}{-4} \qquad \begin{array}{l}\text{Divide by } -4. \text{ Reverse}\\ \text{direction of}\\ \text{inequality symbol.}\end{array}$$

$$x \leq -\frac{13}{2}$$

DEFINITIONS AND CONCEPTS	EXAMPLES

SECTION 2.5 COMPOUND INEQUALITIES

Two inequalities joined by the words *and* or *or* are called **compound inequalities.**	Compound inequalities: $$x - 7 \le 4 \quad \text{and} \quad x \ge -21$$ $$2x + 7 > x - 3 \quad \text{or} \quad 5x + 2 > -3$$

The solution set of a compound inequality formed by the word **and** is the **intersection** ∩ of the solution sets of the two inequalities.

Solve for x:

$$x < 5 \text{ and } x < 3$$

$\{x \mid x < 5\}$ [number line: −1 0 1 2 3 4 5 6] $(-\infty, 5)$

$\{x \mid x < 3\}$ [number line: −1 0 1 2 3 4 5 6] $(-\infty, 3)$

$\{x \mid x < 3$ and $x < 5\}$ [number line: −1 0 1 2 3 4 5 6] $(-\infty, 3)$

The solution set of a compound inequality formed by the word **or** is the **union** ∪ of the solution sets of the two inequalities.

Solve for x:

$$x - 2 \ge -3 \quad \text{or} \quad 2x \le -4$$
$$x \ge -1 \quad\quad \text{or} \quad\quad x \le -2$$

$\{x \mid x \ge -1\}$ [number line: −3 −2 −1 0 1] $[-1, \infty)$

$\{x \mid x \le -2\}$ [number line: −3 −2 −1 0 1] $(-\infty, -2]$

$\{x \mid x \le -2$ or $x \ge -1\}$ [number line: −3 −2 −1 0 1] $(-\infty, -2]$ $\cup [-1, \infty)$

SECTION 2.6 ABSOLUTE VALUE EQUATIONS

If a is a positive number, then $\lvert x \rvert = a$ is equivalent to $x = a$ or $x = -a$.	Solve for y: $$\lvert 5y - 1 \rvert - 7 = 4$$

$\lvert 5y - 1 \rvert = 11$ Add 7.

$5y - 1 = 11 \quad \text{or} \quad 5y - 1 = -11$

$\quad\quad 5y = 12 \quad \text{or} \quad\quad\quad 5y = -10 \quad$ Add 1.

$\quad\quad\quad y = \dfrac{12}{5} \quad \text{or} \quad\quad\quad\quad y = -2 \quad$ Divide by 5.

The solution set is $\left\{ -2, \frac{12}{5} \right\}$.

If a is negative, then $\lvert x \rvert = a$ has no solution.

Solve for x:

$$\left\lvert \frac{x}{2} - 7 \right\rvert = -1$$

The solution set is $\{ \ \}$ or $\varnothing$.

(continued)

DEFINITIONS AND CONCEPTS	EXAMPLES
SECTION 2.6 ABSOLUTE VALUE EQUATIONS	
If an absolute value equation is of the form $|x| = |y|$, solve $x = y$ or $x = -y$.	Solve for x: $$|x - 7| = |2x + 1|$$ $$x - 7 = 2x + 1 \quad \text{or} \quad x - 7 = -(2x + 1)$$ $$x = 2x + 8 \qquad\qquad x - 7 = -2x - 1$$ $$-x = 8 \qquad\qquad\qquad x = -2x + 6$$ $$x = -8 \quad \text{or} \quad 3x = 6$$ $$x = 2$$ The solution set is $\{-8, 2\}$.
SECTION 2.7 ABSOLUTE VALUE INEQUALITIES	
If a is a positive number, then $|x| < a$ is equivalent to $-a < x < a$.	Solve for y: $$|y - 5| \le 3$$ $$-3 \le y - 5 \le 3$$ $$-3 + 5 \le y - 5 + 5 \le 3 + 5 \qquad \text{Add 5.}$$ $$2 \le y \le 8$$ The solution set is $[2, 8]$.
If a is a positive number, then $|x| > a$ is equivalent to $x < -a$ or $x > a$.	Solve for x: $$\left|\frac{x}{2} - 3\right| > 7$$ $$\frac{x}{2} - 3 < -7 \quad \text{or} \quad \frac{x}{2} - 3 > 7 \qquad \text{Multiply by 2.}$$ $$x - 6 < -14 \quad \text{or} \quad x - 6 > 14$$ $$x < -8 \quad \text{or} \quad x > 20 \qquad \text{Add 6.}$$ The solution set is $(-\infty, -8) \cup (20, \infty)$.

CHAPTER 2 REVIEW

(2.1) *Solve each linear equation.*

1. $4(x - 5) = 2x - 14$ $\{3\}$

2. $x + 7 = -2(x + 8)$ $\left\{-\dfrac{23}{3}\right\}$

3. $3(2y - 1) = -8(6 + y)$ $\left\{-\dfrac{45}{14}\right\}$

4. $-(z + 12) = 5(2z - 1)$ $\left\{-\dfrac{7}{11}\right\}$

5. $n - (8 + 4n) = 2(3n - 4)$ $\{0\}$

6. $4(9v + 2) = 6(1 + 6v) - 10$ $\{\}$

7. $0.3(x - 2) = 1.2$ $\{6\}$ **8.** $1.5 = 0.2(c - 0.3)$ $\{7.8\}$

9. $\{h \mid h$ is a real number$\}$

12. $\{p \mid p$ is a real number$\}$

9. $-4(2 - 3h) = 2(3h - 4) + 6h$

10. $6(m - 1) + 3(2 - m) = 0$ $\{0\}$

11. $6 - 3(2g + 4) - 4g = 5(1 - 2g)$ $\{\}$

12. $20 - 5(p + 1) + 3p = -(2p - 15)$

13. $\dfrac{x}{3} - 4 = x - 2$ $\{-3\}$ **14.** $\dfrac{9}{4}y = \dfrac{2}{3}y$ $\{0\}$

15. $\dfrac{3n}{8} - 1 = 3 + \dfrac{n}{6}$ $\left\{\dfrac{96}{5}\right\}$ **16.** $\dfrac{z}{6} + 1 = \dfrac{z}{2} + 2$ $\{-3\}$

33. 258 miles   **34.** 250 calculators   **38.** $y = \dfrac{5x + 12}{4}$ **39.** $x = \dfrac{4y - 12}{5}$ **41.** $x = \dfrac{y - y_1 + mx_1}{m}$ **42.** $r = \dfrac{E - IR}{I}$

17. $\dfrac{y}{4} - \dfrac{y}{2} = -8$ {32}   **18.** $\dfrac{2x}{3} - \dfrac{8}{3} = x$ {−8}

19. $\dfrac{b - 2}{3} = \dfrac{b + 2}{5}$ {8}   **20.** $\dfrac{2t - 1}{3} = \dfrac{3t + 2}{15}$ {1}

21. $\dfrac{2(t + 1)}{3} = \dfrac{2(t - 1)}{3}$ {}   **22.** $\dfrac{3a - 3}{6} = \dfrac{4a + 1}{15} + 2$ {11}

23. $\dfrac{x - 2}{5} + \dfrac{x + 2}{2} = \dfrac{x + 4}{3}$ {2}

24. $\dfrac{2z - 3}{4} - \dfrac{4 - z}{2} = \dfrac{z + 1}{3}$ $\left\{\dfrac{37}{8}\right\}$

(2.2) *Solve.*

25. Twice the difference of a number and 3 is the same as 1 added to three times the number. Find the number.   −7

26. One number is 5 more than another number. If the sum of the numbers is 285, find the numbers. 140, 145

27. Find 40% of 130.   52   **28.** Find 1.5% of 8.   0.12

29. In a recent year, the average annual earnings for a worker with a college degree was $37,000. This represents an 85% increase over the average annual earnings for a high school graduate that year. Find the average annual earnings for a high school graduate that year. (*Source:* U.S. Bureau of the Census) $20,000

30. Find four consecutive integers such that twice the first subtracted from the sum of the other three integers is sixteen.   10, 11, 12, 13

31. Determine whether there are two consecutive odd integers such that 5 times the first exceeds 3 times the second by 54.   No such integers exist.

32. The length of a rectangular playing field is 5 meters less than twice its width. If 230 meters of fencing goes around the field, find the dimensions of the field.   width, 40 m; length, 75 m

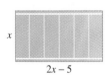

x

$2x - 5$

33. A car rental company charges $19.95 per day for a compact car plus 12 cents per mile for every mile over 100 miles driven per day. If Mr. Woo's bill for 2 days use is $46.86, find how many miles he drove.

34. The cost C of producing x number of scientific calculators is given by $C = 4.50x + 3000$, and the revenue R from selling them is given by $R = 16.50x$. Find the number of calculators that must be sold to break even. (To break even, revenue = cost.)

35. An entrepreneur can sell her musically vibrating plants for $40 each, while her cost C to produce x number of plants is given by $C = 20x + 100$. Find her break-even point. Find her revenue if she sells exactly that number of plants.   5 plants, $200

Solve each equation for the specified variable.

36. $V = LWH$; W $W = \dfrac{V}{LH}$ **37.** $C = 2\pi r$; r $r = \dfrac{C}{2\pi}$

38. $5x - 4y = -12$; y

39. $5x - 4y = -12$; x

40. $y - y_1 = m(x - x_1)$; m $m = \dfrac{y - y_1}{x - x_1}$

41. $y - y_1 = m(x - x_1)$; x

42. $E = I(R + r)$; r **43.** $S = vt + gt^2$; g

44. $T = gr + gvt$; g **45.** $I = Prt + P$; P

46. $A = \dfrac{h}{2}(B + b)$; B **47.** $V = \dfrac{1}{3}\pi r^2 h$; h $h = \dfrac{3V}{\pi r^2}$

48. $R = \dfrac{r_1 + r_2}{2}$; r_1 **49.** $\dfrac{V_1}{T_1} = \dfrac{V_2}{T_2}$; T_2 $T_2 = \dfrac{T_1 V_2}{V_1}$

50. $\dfrac{1}{a} + \dfrac{1}{b} = \dfrac{1}{c}$; b **51.** $\dfrac{2}{x} - \dfrac{3}{y} = \dfrac{1}{z}$; y

52. $R = \dfrac{R_1 R_2}{R_1 + R_2}$; R_2 **53.** $C = \dfrac{2AB}{A - B}$; A

54. $\dfrac{x - y}{5} + \dfrac{y}{4} = \dfrac{2x}{3}$; y **55.** $\dfrac{b + c}{d} - \dfrac{b}{c} = \dfrac{5}{c}$; d

Solve.

56. A principal of $3000 is invested in an account paying an annual percentage rate of 3%. Find the amount in the account after 7 years if the amount is compounded.   **(a)** $3695.27   **(b)** $3700.81

(a) semiannually   **(b)** weekly (Approximate to the nearest cent.)

57. The high temperature in Slidell, Louisiana, one day was 90° Fahrenheit. Convert this temperature to degrees Celsius. $\left(\dfrac{290}{9}\right)$°C ≈ 32.2°C

58. Angie has a photograph in which the length is 2 inches longer than the width. If she increases each dimension by 4 inches, the area is increased by 88 square inches. Find the original dimensions.

43. $g = \dfrac{s - vt}{t^2}$ **44.** $g = \dfrac{T}{r + vt}$ **45.** $P = \dfrac{I}{1 + rt}$ **46.** $B = \dfrac{2A - hb}{h}$ **48.** $r_1 = 2R - r_2$ **50.** $b = \dfrac{ac}{a - c}$

51. $y = \dfrac{3xz}{2z - x}$ **52.** $R_2 = \dfrac{RR_1}{R_1 - R}$ **53.** $A = \dfrac{CB}{C - 2B}$ **54.** $y = \dfrac{28}{3}x$ **55.** $d = \dfrac{c(b + c)}{5 + b}$ **58.** length, 10 in., width, 8 in.

60. The cylinder holds more ice cream.

74. The last judge must give her at least a 9.6 for her to win the silver medal.   **82.** $(-5, 2]$

59. One-square-foot floor tiles come 24 to a package. Find how many packages are needed to cover a rectangular floor 18 feet by 21 feet.   16 packages

60. Determine which container holds more ice cream, an 8 inch $\times$ 5 inch $\times$ 3 inch box or a cylinder with radius of 3 inches and height of 6 inches.

61. Angie left Los Angeles at 11 A.M. and drove non-stop to San Diego, 130 miles away. If she arrived at 1:15 P.M., find her average speed, rounded to the nearest mile per hour.   58 mph

(2.4) *Solve each linear inequality.*   **62.** $(3, \infty)$

62. $3(x - 5) > -(x + 3)$ **63.** $-2(x + 7) \geq 3(x + 2)$

64. $4x - (5 + 2x) < 3x - 1$ $(-4, \infty)$

65. $3(x - 8) < 7x + 2(5 - x)$ $(-17, \infty)$ **63.** $(-\infty, -4]$

66. $24 \geq 6x - 2(3x - 5) + 2x$ $(-\infty, 7]$

67. $48 + x \geq 5(2x + 4) - 2x$ $(-\infty, 4]$

68. $\dfrac{x}{3} + \dfrac{1}{2} > \dfrac{2}{3}$ $\left(\dfrac{1}{2}, \infty\right)$ **69.** $x + \dfrac{3}{4} < \dfrac{-x}{2} + \dfrac{9}{4}$

$(-\infty, 1)$

70. $\dfrac{x - 5}{2} \leq \dfrac{3}{8}(2x + 6)$ $[-19, \infty)$

71. $\dfrac{3(x - 2)}{5} > \dfrac{-5(x - 2)}{3}$ $(2, \infty)$

Solve.   **72.** It is more economical to use the housekeeper for more than 35 pounds per week.

72. George Boros can pay his housekeeper $15 per week to do his laundry, or he can have the laundromat do it at a cost of 50 cents per pound for the first 10 pounds and 40 cents for each additional pound. Use an inequality to find the weight at which it is more economical to use the housekeeper than the laundromat.

73. Ceramic firing temperatures usually range from 500° to 1000° Fahrenheit. Use a compound inequality to convert this range to the Celsius scale. Round to the nearest degree. $260° \leq C \leq 538°C$

74. In the Olympic gymnastics competition, Nana must average a score of 9.65 to win the silver medal. Seven of the eight judges have reported scores of 9.5, 9.7, 9.9, 9.7, 9.7, 9.6, and 9.5. Use an inequality to find the minimum score that the last judge can give so that Nana wins the silver medal.

75. Carol would like to pay cash for a car when she graduates from college and estimates that she can afford a car that costs between $4000 and $8000. She has saved $500 so far and plans to earn the rest of the money by working the next two summers. If Carol plans to save the same amount each summer, use a compound inequality to find the range of money she must save each summer to buy the car. $1750 to $3750

(2.5) *Solve each inequality.*   **76.** $\left[2, \dfrac{5}{2}\right]$ **77.** $\left[-2, -\dfrac{9}{5}\right)$

76. $1 \leq 4x - 7 \leq 3$ **77.** $-2 \leq 8 + 5x < -1$

78. $-3 < 4(2x - 1) < 12$ $\left(\dfrac{1}{8}, 2\right)$

79. $-6 < x - (3 - 4x) < -3$ $\left(-\dfrac{3}{5}, 0\right)$

80. $\dfrac{1}{6} < \dfrac{4x - 3}{3} \leq \dfrac{4}{5}$ $\left(\dfrac{7}{8}, \dfrac{27}{20}\right]$

81. $0 \leq \dfrac{2(3x + 4)}{5} \leq 3$ $\left[-\dfrac{4}{3}, \dfrac{7}{6}\right]$

82. $x \leq 2$ and $x > -5$ **83.** $x \leq 2$ or $x > -5$ $(-\infty, \infty)$

84. $3x - 5 > 6$ or $-x < -5$ $\left(\dfrac{11}{3}, \infty\right)$

85. $-2x \leq 6$ and $-2x + 3 < -7$ $(5, \infty)$

(2.6) *Solve each absolute value equation.*

86. $|x - 7| = 9$ $\{16, -2\}$ **87.** $|8 - x| = 3$ $\{5, 11\}$

88. $|2x + 9| = 9$ $\{0, -9\}$ **89.** $|-3x + 4| = 7$ $\left\{-1, \dfrac{11}{3}\right\}$

90. $|3x - 2| + 6 = 10$ $\left\{2, -\dfrac{2}{3}\right\}$

91. $5 + |6x + 1| = 5$ $\left\{-\dfrac{1}{6}\right\}$

92. $-5 = |4x - 3|$ $\{\}$ **93.** $|5 - 6x| + 8 = 3$ $\{\}$

94. $|7x| - 26 = -5$ $\{3, -3\}$ **95.** $-8 = |x - 3| - 10$ $\{1, 5\}$

96. $\left|\dfrac{3x - 7}{4}\right| = 2$ $\left\{5, -\dfrac{1}{3}\right\}$ **97.** $\left|\dfrac{9 - 2x}{5}\right| = -3$ $\{\}$

98. $|6x + 1| = |15 + 4x|$ $\left\{7, -\dfrac{8}{5}\right\}$

99. $|x - 3| = |7 + 2x|$ $\left\{-10, -\dfrac{4}{3}\right\}$

(2.7) *Solve each absolute value inequality. Graph the solution set and write in interval notation.*

100. $|5x - 1| < 9$ **101.** $|6 + 4x| \geq 10$

102. $|3x| - 8 > 1$ **103.** $9 + |5x| < 24$

104. $|6x - 5| \leq -1$ **105.** $|6x - 5| \geq -1$

106. $\left|3x + \dfrac{2}{5}\right| \geq 4$ **107.** $\left|\dfrac{4x - 3}{5}\right| < 1$

108. $\left|\dfrac{x}{3} + 6\right| - 8 > -5$  $(-\infty, -27) \cup (-9, \infty)$

109. $\left|\dfrac{4(x - 1)}{7}\right| + 10 < 2$ $\{\ \}$

100. $\left(-\dfrac{8}{5}, 2\right)$ **101.** $(-\infty, -4] \cup [1, \infty)$ **102.** $(-\infty, -3) \cup (3, \infty)$

103. $(-3, 3)$ **104.** $\{\ \}$ **105.** $(-\infty, \infty)$

106. $\left(-\infty, -\dfrac{22}{15}\right] \cup \left[\dfrac{6}{5}, \infty\right)$ **107.** $\left(-\dfrac{1}{2}, 2\right)$

CHAPTER 2 TEST

Solve each equation.

1. $8x + 14 = 5x + 44$ {10}

2. $3(x + 2) = 11 - 2(2 - x)$ {1}

3. $3(y - 4) + y = 2(6 + 2y)$ { }

4. $7n - 6 + n = 2(4n - 3)$ {$n|n$ is a real number}

5. $\dfrac{z}{2} + \dfrac{z}{3} = 10$ {12}   **6.** $\dfrac{7w}{4} + 5 = \dfrac{3w}{10} + 1$ $\left(-\dfrac{80}{29}\right)$

7. $|6x - 5| = 1$ $\left\{1, \dfrac{2}{3}\right\}$ **8.** $|8 - 2t| = -6$ { }

Solve each equation for the specified variable.

9. $3x - 4y = 8$; y $y = \dfrac{3x - 8}{4}$

10. $4(2n - 3m) - 3(5n - 7m) = 0$; n $n = \dfrac{9}{7}m$

11. $S = gt^2 + gvt$; g **12.** $F = \dfrac{9}{5}C + 32$; C

Solve each inequality.

13. $3(2x - 7) - 4x > -(x + 6)$ $(5, \infty)$

14. $8 - \dfrac{x}{2} \le 7$ $[2, \infty)$

15. $-3 < 2(x - 3) \le 4$ **16.** $|3x + 1| > 5$

17. $x \ge 5$ and $x \ge 4$ **18.** $x \ge 5$ or $x \ge 4$ $[4, \infty)$

19. $-x > 1$ and $3x + 3 \ge x - 3$ $[-3, -1)$

20. $6x + 1 > 5x + 4$ or $1 - x > -4$ $(-\infty, \infty)$

21. Find 12% of 80.   9.6

22. In 1995, Latter & Blum sold approximately $939.25 million in residential sales in southeast Louisiana. This represents a 121% increase over Prudential La. Properties. Find the 1995 residential sales for Prudential La. Properties. (*Source: Times/Picayune Newspaper*, March 10, 1996.) $425 million

Solve.

23. A circular dog pen has a circumference of 78.5 feet. Approximate π by 3.14 and estimate how many hunting dogs could be safely kept in the pen if each dog needs at least 60 square feet of room.

24. The company that makes Photoray sunglasses figures that the cost C to make x number of sunglasses weekly is given by $C = 3910 + 2.8x$, and the weekly revenue R is given by $R = 7.4x$. Use an inequality to find the number of sunglasses that must be made and sold to make a profit. (Revenue must exceed cost in order to make a profit.)   more than 850 sunglasses

25. Find the amount of money in an account after 10 years if a principal of $2500 is invested at 3.5% interest compounded quarterly. (Round to the nearest cent.) $3542.27

11. $g = \dfrac{S}{t^2 + vt}$ **12.** $C = \dfrac{5}{9}(F - 32)$ **15.** $\left(\dfrac{3}{2}, 5\right]$ **16.** $(-\infty, -2) \cup \left(\dfrac{4}{3}, \infty\right)$ **17.** $[5, \infty)$ **23.** approximately 8

CHAPTER 2 CUMULATIVE REVIEW

1. Find each sum.

   a. $-3 + (-11)$ -14 **b.** $3 + (-7)$ -4

   c. $-10 + 15$ 5 **d.** $-8.3 + (-1.9)$ -10.2

   e. $-\dfrac{1}{4} + \dfrac{1}{2}$ $\dfrac{1}{4}$ **f.** $-\dfrac{2}{3} + \dfrac{3}{7}$

2. Find each quotient.

   a. $\dfrac{20}{-4}$ **b.** $\dfrac{-9}{-3}$ **c.** $-\dfrac{3}{8} \div 3$

   d. $\dfrac{-40}{10}$ **e.** $\dfrac{-1}{10} \div \dfrac{-2}{5}$ **f.** $\dfrac{8}{0}$

3. Find the following square roots.

   a. $\sqrt{9}$ 3 **b.** $\sqrt{25}$ 5 **c.** $\sqrt{\dfrac{1}{4}}$ $\dfrac{1}{2}$; *Sec. 1.3, Ex. 8*

4. Simplify.   **d.** $\dfrac{12}{5}$; *Sec. 1.4, Ex. 1*

   a. $3 + 2 \cdot 10$ 23 **b.** $2(1 - 4)^2$ 18

   c. $\dfrac{|-2|^3 + 1}{-7 - \sqrt{4}}$ -1 **d.** $\dfrac{(6 + 2) - (-4)}{2 - (-3)}$

5. Write the following as algebraic expressions:

   a. A vending machine contains x quarters. Write an expression for the *value* of the quarters. $0.25x$

1f. $-\dfrac{5}{21}$; *Sec. 1.3, Ex. 2*   **2a.** -5 **b.** 3 **c.** $-\dfrac{1}{8}$ **d.** -4 **e.** $\dfrac{1}{4}$ **f.** undefined; *Sec. 1.3, Ex. 6*

b. The number of grams of fat in x pieces of bread if each piece of bread contains 2 grams of fat. $2x$

c. The cost of x desks if each desk costs $156. $156x$

d. Sales tax on a purchase of x dollars if the tax rate is 9%. $0.09x$; *Sec. 1.4, Ex. 4*

6. Simplify each expression.

a. $3xy - 2xy + 5 - 7 + xy$ $2xy - 2$

b. $7x^2 + 3 - 5(x^2 - 4)$ $2x^2 + 23$

c. $(2.1x - 5.6) - (-x - 5.3)$ $3.1x - 0.3$; *Sec. 1.4, Ex. 7*

7. Solve for x: $2x + 5 = 9$.   {2}; *Sec. 2.1, Ex. 1*

8. Solve for x: $6x - 4 = 2 + 6(x - 1)$.

9. Suppose that Service Merchandise just announced an 8% decrease in the price of their Compaq Presario computers. If one particular computer model sells for $2162 after the decrease, find the original price of this computer. $2350; *Sec. 2.2, Ex. 3*

10. Solve $V = lwh$ for h. $\dfrac{V}{lw} = h$; *Sec. 2.3, Ex. 1*

11. Solve $A = \dfrac{1}{2}(B + b)h$ for b.

11. $b = \dfrac{2A - Bh}{h}$; *Sec. 2.3, Ex. 3*

12. Solar cells convert the energy of sunlight into electrical energy. Suppose that 0.01 watt of electrical power is available for every square centimeter of solar cell in direct overhead sunlight. If a solar cell in the shape of a square is needed to deliver 16 watts, find the minimum side of the square.

13. Solve for x.   **13a.** $\left(-\infty, \dfrac{3}{2}\right]$; *Sec. 2.4, Ex. 3*

a. $\dfrac{1}{4}x \le \dfrac{3}{8}$

b. $-2.3x < 6.9$ $(-3, \infty)$

14. Solve for x: $2(x + 3) > 2x + 1$.

15. Solve for x: $x - 7 < 2$ and $2x + 1 < 9$.

16. Solve $5x - 3 \le 10$ or $x + 1 \ge 5$.

17. Solve $|p| = 2$.   {−2, 2}; *Sec. 2.6, Ex. 1*

18. Solve $|3x + 2| = |5x - 8|$. $\left\{\dfrac{3}{4}, 5\right\}$; *Sec. 2.6, Ex. 8*

19. Solve for m: $|m - 6| < 2$.   (4, 8); *Sec. 2.7, Ex. 2*

20. Solve $|2x + 9| + 5 > 3$. $(-\infty, \infty)$; *Sec. 2.7, Ex. 6*

8. {$x \mid x$ is a real number}; *Sec. 2.1, Ex. 8*   **12.** 40 cm; *Sec. 2.3, Ex. 5*   **14.** $(-\infty, \infty)$; *Sec. 2.4, Ex. 6*

15. $(-\infty, 4)$; *Sec. 2.5, Ex. 1*   **16.** $\left(-\infty, \dfrac{13}{5}\right] \cup [4, \infty)$; *Sec. 2.5, Ex. 4*

The following are more answer graphs for Exercise Set 2.7, p. 95.

40. $\left(-\infty, -\frac{11}{3}\right) \cup (-3, \infty)$ **41.** { }   **42.** { }

43. $(-\infty, \infty)$ **44.** { }   **45.** $[-2, 9]$

46. $\left(-2, \frac{6}{5}\right]$ **49.** $(-\infty, 0) \cup (0, \infty)$ **50.** { }

51. $(-\infty, \infty)$ **52.** $(-\infty, \infty)$ **55.** $(-\infty, -3) \cup (0, \infty)$

56. $\left(-\infty, -\frac{1}{5}\right] \cup \left[\frac{3}{5}, \infty\right]$ **59.** $(-\infty, \infty)$ **60.** $\left\{\frac{6}{5}\right\}$

61. $\left(-\frac{2}{3}, 0\right)$ **62.** $\left[-\frac{8}{7}, 2\right]$ **63.** $(-\infty, \infty)$

64. { }   **67.** $(-\infty, -12) \cup (0, \infty)$ **68.** $(-\infty, -15] \cup [1, \infty)$

69. $(-\infty, -6) \cup (0, \infty)$ **70.** $(-\infty, 1] \cup [5, \infty)$ **71.** $\left(-\frac{31}{5}, \frac{11}{5}\right)$

72. $\left[-\frac{19}{3}, \frac{1}{3}\right]$ **78.** $(-\infty, -2) \cup (5, \infty)$ **80.** $\left\{-4, \frac{17}{3}\right\}$ **81.** $(-\infty, -7] \cup [17, \infty)$

82. $(-\infty, -24] \cup [16, \infty)$ **85.** $(-2, 1)$ **86.** $(-\infty, 0] \cup \left[\frac{6}{5}, \infty\right)$ **91.** $\left\{\frac{19}{2}, -\frac{17}{2}\right\}$ **93.** $\left(-\infty, -\frac{25}{3}\right) \cup \left(\frac{35}{3}, \infty\right)$

GRAPHS AND FUNCTIONS

MODELING JAPANESE AUTOMOBILE IMPORTS

Annual Japanese automobile imports into the United States topped 1 million cars in 1976 and grew to an all-time high of 2.6 million cars per year in 1986. Since 1986, annual imports have steadily decreased, and a linear function adequately describes this decline.

IN THE CHAPTER GROUP ACTIVITY ON PAGE 173, YOU WILL HAVE THE OPPORTUNITY TO FIND THE LINEAR FUNCTION THAT FITS THIS PATTERN OF DECLINE IN JAPANESE IMPORTS.

The linear equations we explored in Chapter 2 are statements about a single variable. This chapter examines statements about two variables: linear equations and inequalities in two variables. We focus particularly on graphs of these equations and inequalities, which lead to the notion of relation and to the notion of function, perhaps the single most important and useful concept in all of mathematics.

3.1 GRAPHING EQUATIONS

OBJECTIVES

TAPE IA 3.1

1. Plot ordered pairs.
2. Determine whether an ordered pair of numbers is a solution to an equation in two variables.
3. Graph linear equations.
4. Graph nonlinear equations.

When two varying quantities are measured simultaneously and repeatedly, we can record the measurement pairs and try to detect a pattern. If a pattern is recognized, we may be able to express the pattern compactly as a two-variable equation. For example, suppose that the two quantities products sold x and monthly salary y are measured simultaneously and repeatedly for an employee and recorded in the following table.

Products sold	x	0	100	200	300	400	1000
Monthly salary	y	1500	1510	1520	1530	1540	1600

After studying the measurements, you may notice that there is a pattern in the quantities. The monthly salary $y = 1500 + \frac{1}{10}x$. This equation is called a linear equation in two variables. Before we discuss such equations further, let's first review the rectangular coordinate system.

In order to visualize the data in the table, the data pairs can be listed as ordered pairs of numbers. These ordered pairs of numbers can then be plotted on a **rectangular coordinate system,** which is also called a **Cartesian coordinate system** after its inventor, Rene Descartes (1596–1650).

The Cartesian coordinate system consists of two number lines that intersect at right angles at their 0 coordinates. We position these axes on paper such that one number line is horizontal and the other number line is then vertical. The horizontal number line is called the **x-axis** (or the axis of the **abscissa**), and the vertical number line is called the **y-axis** (or the axis of the **ordinate**). The point of intersection of these axes is named the **origin.**

Notice that the axes divide the plane into four regions. These regions are called **quadrants.** The top-right region is quadrant I. Quadrants II, III, and IV are numbered counterclockwise from the first quadrant as shown. The *x*-axis and the *y*-axis are not in any quadrant.

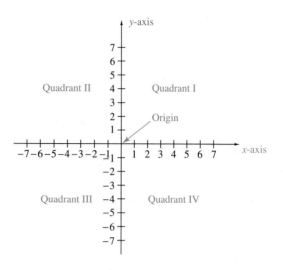

Each point in the plane can be located, or **plotted,** by describing its position in terms of distances along each axis from the origin. An **ordered pair,** represented by the notation (x, y), records these distances.

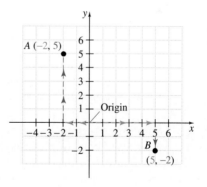

For example, the location of point *A* is described as 2 units to the left of the origin along the *x*-axis and 5 units upward parallel to the *y*-axis. Thus, we identify point *A* with the ordered pair $(-2, 5)$. Notice that the order of these numbers is critical. The *x*-value -2 is called the **x-coordinate** and is associated with the *x*-axis. The *y*-value 5 is called the **y-coordinate** and is associated with the *y*-axis. Compare the location of point *A* with the location of point *B*, which corresponds to the ordered pair (5, −2). The *x*-coordinate 5 indicates that we move 5 units to the right of the origin along the *x*-axis. The *y*-coordinate −2 indicates that we move 2 units down parallel to the *y*-axis. Point *A* is in a different position than point *B*. Two ordered pairs are considered equal and correspond to the same point if and only if their *x*-coordinates are equal and their *y*-coordinates are equal.

Keep in mind that **each ordered pair corresponds to exactly one point in the real plane and that each point in the plane corresponds to exactly one ordered pair.** Thus, we may refer to the ordered pair (*x, y*) as the point (*x, y*).

EXAMPLE 1   Plot each ordered pair on a Cartesian coordinate system and name the quadrant in which the point is located.

a. $(2, -1)$ **b.** $(0, 5)$ **c.** $(-3, 5)$ **d.** $(-2, 0)$ **e.** $\left(-\frac{1}{2}, -4\right)$ **f.** $(1.5, 1.5)$

Solution: The five points are graphed as shown:

a. $(2, -1)$ lies in quadrant IV.

b. $(0, 5)$ is not in any quadrant.

c. $(-3, 5)$ lies in quadrant II.

d. $(-2, 0)$ is not in any quadrant.

e. $\left(-\frac{1}{2}, -4\right)$ is in quadrant III.

f. $(1.5, 1.5)$ is in quadrant I.

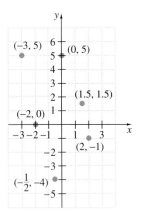

Notice that the *y*-coordinate of any point on the *x*-axis is 0. For example, the point with coordinates $(-2, 0)$ lies on the *x*-axis. Also, the *x*-coordinate of any point on the *y*-axis is 0. For example, the point with coordinates $(0, 5)$ lies on the *y*-axis. A point on an axis is called a **quadrantel** point.

2   **Solutions** of equations in two variables consist of two numbers that can be written as ordered pairs of numbers. For example, we say that the ordered pair (100, 1510) is a solution of the equation $y = 1500 + \frac{1}{10}x$ because when *x* is replaced with 100 and *y* is replaced with 1510, a true statement results.

$$y = 1500 + \frac{1}{10}x$$

$$1510 = 1500 + \frac{1}{10}(100) \qquad \text{Let } x = 100 \text{ and } y = 1510.$$

$$1510 = 1500 + 10$$

$$1510 = 1510 \qquad \text{True.}$$

EXAMPLE 2 Determine whether $(0, -12)$, $(1, 9)$, and $(2, -6)$ are solutions of the equation $3x - y = 12$.

Solution: To check each ordered pair, replace x with the x-coordinate and y with the y-coordinate and see whether a true statement results.

Let $x = 0$ and $y = -12$. Let $x = 1$ and $y = 9$. Let $x = 2$ and $y = -6$.

$$3x - y = 12 \qquad\qquad 3x - y = 12 \qquad\qquad 3x - y = 12$$
$$3(0) - (-12) = 12 \qquad 3(1) - 9 = 12 \qquad 3(2) - (-6) = 12$$
$$0 + 12 = 12 \qquad\qquad 3 - 9 = 12 \qquad\qquad 6 + 6 = 12$$
$$12 = 12 \; \text{True.} \qquad -6 = 12 \; \text{False.} \qquad 12 = 12 \; \text{True.}$$

Thus, $(1, 9)$ is not a solution but both $(2, -6)$ and $(0, -12)$ are solutions.

In fact, the equation $3x - y = 12$ has an infinite number of ordered pair solutions. Since it is impossible to list all solutions, we visualize them by graphing them.

A few more ordered pairs that satisfy $3x - y = 12$ are $(4, 0)$, $(3, -3)$, $(5, 3)$, and $(1, -9)$. These ordered pair solutions along with the ordered pair solutions from Example 2 are plotted on the following graph. The graph of $3x - y = 12$ is the single line containing these points. Every ordered pair solution of the equation corresponds to a point on this line, and every point on this line corresponds to an ordered pair solution.

x	y
4	0
1	-9
0	-12
2	-6
5	3
3	-3

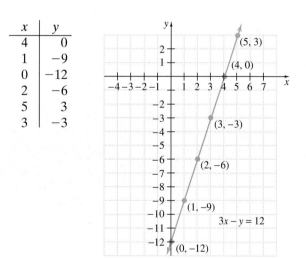

The equation $3x - y = 12$ is called a linear equation in two variables, and **the graph of every linear equation in two variables is a line.**

> **LINEAR EQUATION IN TWO VARIABLES**
>
> A linear equation in two variables is an equation that can be written in the form
>
> $$Ax + By = C$$
>
> where A, B, and C are real numbers, and A and B are not both 0.

A linear equation written in the form $Ax + By = C$ is said to be written in **standard form.**

EXAMPLES OF LINEAR EQUATIONS IN STANDARD FORM

$$3x - y = 12$$
$$-2.1x + 5.6y = 0$$

As we mentioned earlier, the equation $y = 1500 + \frac{1}{10}x$ is also a linear equation in two variables. This means that the equation $y = 1500 + \frac{1}{10}x$ can be written in standard form $Ax + By = C$ and that its graph is a line. Its solutions in ordered pair form are shown next along with a portion of its graph. Recall that x is products sold and y is monthly salary. Since we assume that the smallest amount of product sold is none, or 0, then x must be greater than or equal to 0. Therefore, only the part of the graph that lies in Quadrant I is shown. Notice that the graph gives a visual picture of the correspondence between products sold and salary.

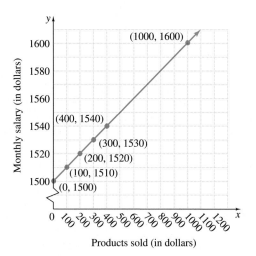

(The ⌇ symbol on the y-axis means that some values are missing along the axis.)

EXAMPLE 3   Use the graph of $y = 1500 + \dfrac{1}{10}x$ to answer the following questions:

   a. If the salesperson has $800 of products sold for a particular month, what is the salary for that month?

   b. If the salesperson wants to make more than $1600 per month, what must be the total amount of products sold?

Solution: **a.** Since x is products sold, find 800 along the x-axis and move vertically up until you reach a point on the line. From this point on the line, move horizontally to the left until you reach the y-axis. Its value on the y-axis is 1580, which means if $800 worth of products is sold, the salary for the month is $1580.

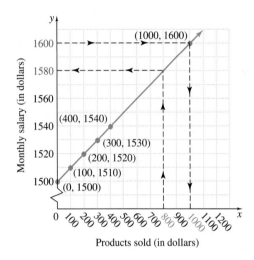

Products sold (in dollars)

   b. Since y is monthly salary, find 1600 along the y-axis and move horizontally to the right until you reach a point on the line. Either read the corresponding x-value from the labeled ordered pair, or move vertically downward until you reach the x-axis. The corresponding x-value is 1000. This means that $1000 worth of products sold gives a salary of $1600 for the month. For the salary to be greater than $1600, products sold must be greater that $1000.

   Recall from geometry that a line is determined by two points. This means that to graph a linear equation in two variables, just two solutions are needed. We will find a third solution, just to check our work. To find ordered pair solutions of linear equations in two variables, we can choose an x-value and find its corresponding y-value, or we can choose a y-value and find its corresponding x-value. The number 0 is often a convenient value to choose for x and also for y.

EXAMPLE 4   Graph the equation $5x - 2y = 10$.

Solution: Find three ordered pair solutions, and plot the ordered pairs. The line through the plotted points is the graph. Let's let x be 0, let y be 0, and then let x be 1 to find our three ordered pair solutions.

Let $x = 0$.	Let $y = 0$.	Let $x = 1$.
$5x - 2y = 10$	$5x - 2y = 10$	$5x - 2y = 10$
$5 \cdot 0 - 2y = 10$	$5x - 2 \cdot 0 = 10$	$5 \cdot 1 - 2y = 10$
$-2y = 10$ Simplify.	$5x = 10$ Simplify.	$5 - 2y = 10$ Multiply.
$y = -5$ Divide by -2.	$x = 2$ Divide by 5.	$-2y = 5$ Subtract 5.
		$y = -\dfrac{5}{2}$, or $-2\dfrac{1}{2}$

The three ordered pair solutions $(0, -5)$, $(2, 0)$, and $\left(1, -2\dfrac{1}{2}\right)$ are listed in the following table, and the graph of $5x - 2y = 10$ is shown.

x	y
0	-5
2	0
1	$-2\frac{1}{2}$

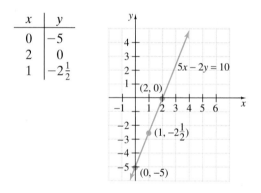

Notice that the graph crosses the x-axis at the point $(2, 0)$. This point is called the **x-intercept point,** and 2 is called the **x-intercept.** This graph also crosses the y-axis at the point $(0, -5)$. This point is called the **y-intercept point,** and -5 is called the **y-intercept.**

EXAMPLE 5   Graph the linear equation $y = \dfrac{1}{3}x$.

Solution: To graph, we find ordered pair solutions, graph the solutions, and draw a line through the solutions. We will choose x-values and substitute in the equation. To avoid fractions, we choose x-values that are multiples of 3. Recall that the equation is $y = \dfrac{1}{3}x$.

If $x = 6$, then $y = \dfrac{1}{3}(6)$, or 2.

If $x = 0$, then $y = \dfrac{1}{3}(0)$, or 0.

If $x = -3$, then $y = \dfrac{1}{3}(-3)$, or -1.

x	y
6	2
0	0
-3	-1

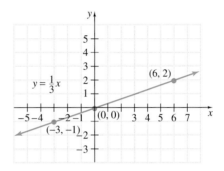

This graph crosses the x-axis at $(0, 0)$ and the y-axis at $(0, 0)$. This means that the x-intercept is 0 and that the y-intercept is 0.

4   Not all equations in two variables are linear equations, and not all graphs of equations in two variables are lines.

EXAMPLE 6   Graph $y = x^2$.

Solution: This equation is not linear, and its graph is not a line. We begin by finding ordered pair solutions. Because this graph is solved for y, we choose x-values and find corresponding y-values.

If $x = -3$, then $y = (-3)^2$, or 9.
If $x = -2$, then $y = (-2)^2$, or 4.
If $x = -1$, then $y = (-1)^2$, or 1.
If $x = 0$, then $y = 0^2$, or 0.
If $x = 1$, then $y = 1^2$, or 1.
If $x = 2$, then $y = 2^2$, or 4.
If $x = 3$, then $y = 3^2$, or 9.

x	y
-3	9
-2	4
-1	1
0	0
1	1
2	4
3	9

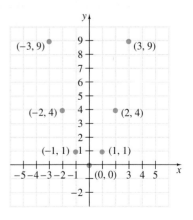

Study the table a moment and look for patterns. Notice that the ordered pair solution $(0, 0)$ contains the smallest y-value because any other x-value squared besides 0 will give a positive result. This means that the point $(0, 0)$ will be the lowest point on the graph. Also notice that all other y-values correspond to two different x-values. For example, $3^2 = 9$ and also $(-3)^2 = 9$. This means that the graph will be a mirror image of itself across the y-axis. Connect the plotted points with a smooth curve to sketch its graph as shown next.

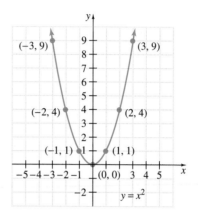

This curve is given a special name, a parabola. We will study more about parabolas in later chapters.

EXAMPLE 7   Graph the equation $y = |x|$.

Solution: This is not a linear equation, and its graph is not a line. Because we do not know the shape of this graph, we find many ordered pair solutions. We will choose x-values and substitute to find corresponding y-values.

If $x = -3$, then $y = |-3|$, or 3.
If $x = -2$, then $y = |-2|$, or 2.
If $x = -1$, then $y = |-1|$, or 1.
If $x = 0$, then $y = |0|$, or 0.
If $x = 1$, then $y = |1|$, or 1.
If $x = 2$, then $y = |2|$, or 2.
If $x = 3$, then $y = |3|$, or 3.

x	y
-3	3
-2	2
-1	1
0	0
1	1
2	2
3	3

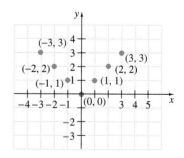

Again, study the table of values for a moment and notice any patterns.

From the plotted ordered pairs, we see that the graph of this absolute value equation is V-shaped. The completed graph is shown next.

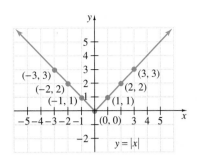

EXAMPLE 8   Graph the equation $y = |x| - 3$.

Solution: To graph $y = |x| - 3$ we choose x-values and substitute to find corresponding y-values.

x	y
-3	0
-2	-1
-1	-2
0	-3
1	-2
2	-1
3	0

If $x = -3$, then $y = |-3| - 3$, or 0.
If $x = -2$, then $y = |-2| - 3$, or -1.
If $x = -1$, then $y = |-1| - 3$, or -2.
If $x = 0$, then $y = |0| - 3$, or -3.
If $x = 1$, then $y = |1| - 3$, or -2.
If $x = 2$, then $y = |2| - 3$, or -1.
If $x = 3$, then $y = |3| - 3$, or 0.

Notice that compared with the equation $y = |x|$, the equation $y = |x| - 3$ decreases the y-values by 3 units for the same x-values. This means that the graph of $y = |x| - 3$ is the same as the graph of $y = |x|$ lowered 3 units.

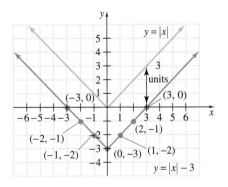

This graph shows us that it is possible to have more than one x- or y-intercept. Notice that this graph has one y-intercept point, $(0, -3)$, and two x-intercept points, $(-3, 0)$ and $(3, 0)$.

Graphing Calculator Explorations

In this section, we begin a study of graphing calculators and graphing software packages for computers. These graphers use the same point plotting technique that we introduced in this section. The advantage of this graphing technology is, of course, that graphing calculators and computers can find and plot ordered pair solutions much faster than we can. Note, however, that the features described in these boxes may not be available on all graphing calculators.

The rectangular screen where a portion of the rectangular coordinate system is displayed is called a **window.** We call it a **standard window** for graphing when both the x- and y-axes display coordinates between -10 and 10. This information is often displayed in the window menu on a graphing calculator as

$$\text{Xmin} = -10$$
$$\text{Xmax} = 10$$
$$\text{Xscl} = 1 \qquad \text{The scale on the } x\text{-axis is one unit per tick mark.}$$
$$\text{Ymin} = -10$$
$$\text{Ymax} = 10$$
$$\text{Yscl} = 1 \qquad \text{The scale on the } y\text{-axis is one unit per tick mark.}$$

To use a graphing calculator to graph the equation $y = -5x + 4$, press the $\boxed{\text{Y} =}$ key and enter the keystrokes

(Check your owner's manual to make sure the "negative" key is pressed here and not the "subtraction" key.)

The top row should now read $Y_1 = -5x + 4$. Next press the $\boxed{\text{GRAPH}}$ key, and the display should look like this:

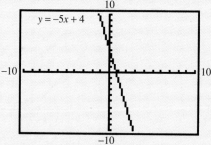

Use a standard window and graph the following equations. (Unless otherwise stated, we will use a standard window when graphing.) For Graphing Calculator Exercises 1–8, see Appendix D.

1. $y = -3.2x + 7.9$

2. $y = -x + 5.85$

3. $y = \dfrac{1}{4}x - \dfrac{2}{3}$

4. $y = \dfrac{2}{3}x - \dfrac{1}{5}$

5. $y = |x - 3| + 2$

6. $y = |x + 1| - 1$

7. $y = x^2 + 3$

8. $y = (x + 3)^2$

MENTAL MATH

Determine the coordinates of each point on the graph.

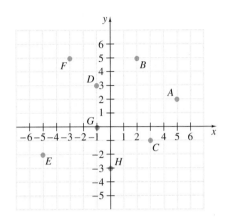

1. Point A (5, 2)   **2.** Point B (2, 5)   **3.** Point C (3, −1)   **4.** Point D (−1, 3)

5. Point E (−5, −2)   **6.** Point F (−3, 5)   **7.** Point G (−1, 0)   **8.** Point H (0, −3)

EXERCISE SET 3.1   For Exercises 1–10, see Appendix D.

Plot each point and name the quadrant or axis in which the point lies. See Example 1.

1. (3, 2)   Quadrant I

2. (2, −1)   Quadrant IV

3. (−5, 3)   Quadrant II

4. (−3, −1)   Quadrant III

5. $\left(5\frac{1}{2}, -4\right)$ Quadrant IV

6. $\left(-2, 6\frac{1}{3}\right)$ Quadrant II

7. (0, 3.5)   *y*-axis

8. (−2, 4)   Quadrant II

9. (−2, −4)   Quadrant III

10. (−4.2, 0)   *x*-axis

Given that x is a positive number and that y is a positive number, determine the quadrant or axis in which each point lies.

❏ **11.** (x, −y)

❏ **12.** (−x, y)   Quadrant II

❏ **13.** (x, 0)   *x*-axis

❏ **14.** (0, −y)   *y*-axis

❏ **15.** (−x, −y)

❏ **16.** (0, 0)   origin

11. Quadrant IV   **15.** Quadrant III

Determine whether each ordered pair is a solution of the given equation. See Example 2.

17. $y = 3x - 5$; (0, 5), (−1, −8)   no; yes

18. $y = -2x + 7$; (1, 5), (−2, 3)   yes; no

19. $-6x + 5y = -6$; (1, 0), $\left(2, \frac{6}{5}\right)$ yes; yes

20. $5x - 3y = 9$; (0, 3), $\left(\frac{12}{5}, -1\right)$ no; no

21. $y = 2x^2$; (1, 2), (3, 18)   yes; yes

22. $y = 2|x|$; (−1, 2), (0, 2)   yes; no

23. $y = x^3$; (2, 8), (3, 9)   yes; no

24. $y = x^4$; (−1, 1), (2, 16)   yes; yes

25. $y = \sqrt{x} + 2$; (1, 3), (4, 4)   yes; yes

26. $y = \sqrt[3]{x} - 4$; (1, −3), (8, 6)   yes; no

Determine whether each equation is linear or not. See Examples 3 through 8.

27. $x + y = 3$ linear   **28.** $y - x = 8$ linear

29. $y = 4x$ linear   **30.** $y = 6x$ linear

31. $y = 4x - 2$ linear   **32.** $y = 6x - 5$ linear

33. $y = |x| + 3$ not linear   **34.** $y = |x| + 2$ not linear

35. $2x - y = 5$ linear   **36.** $4x - y = 7$ linear

37. $y = 2x^2$ not linear   **38.** $y = 3x^2$ not linear

39. $y = x^2 - 3$ not linear   **40.** $y = x^2 + 3$ not linear

41. $y = -2x$ linear   **42.** $y = -3x$ linear

43. $y = -2x + 3$ linear   **44.** $y = -3x + 2$ linear

45. $y = |x + 2|$ not linear   **46.** $y = |x - 1|$ not linear

47. $y = x^3$ not linear   **48.** $y = x^3 - 2$ not linear

49. $y = -|x|$ not linear   **50.** $y = -x^2$ not linear

51. $y = -1.3x + 5.6$ linear   **52.** $y = 7.6x - 1.4$ linear

Graph each equation. If the equation is not linear, suggested x-values have been given for generating ordered pair solutions. See Examples 3 through 8.

53. $x + y = 3$ **54.** $y - x = 8$ **55.** $y = 4x$

56. $y = 6x$ **57.** $y = 4x - 2$ **58.** $y = 6x - 5$

59. $y = |x| + 3$
  Let $x = -3, -2, -1, 0, 1, 2, 3$.

60. $y = |x| + 2$
  Let $x = -3, -2, -1, 0, 1, 2, 3$.

61. $2x - y = 5$ **62.** $4x - y = 7$

63. $y = 2x^2$
  Let $x = -3, -2, -1, 0, 1, 2, 3$.

64. $y = 3x^2$
  Let $x = -3, -2, -1, 0, 1, 2, 3$.

65. $y = x^2 - 3$
  Let $x = -3, -2, -1, 0, 1, 2, 3$.

66. $y = x^2 + 3$
  Let $x = -3, -2, -1, 0, 1, 2, 3$.

67. $y = -2x$ **68.** $y = -3x$

69. $y = -2x + 3$ **70.** $y = -3x + 2$

71. $y = |x + 2|$
  Let $x = -4, -3, -2, -1, 0, 1$.

For Exercises 53–82, see Appendix D.

72. $y = |x - 1|$
  Let $x = -1, 0, 1, 2, 3, 4$.

73. $y = x^3$
  Let $x = -3, -2, -1, 0, 1, 2$.

74. $y = x^3 - 2$
  Let $x = -3, -2, -1, 0, 1, 2$.

75. $y = -|x|$
  Let $x = -3, -2, -1, 0, 1, 2, 3$.

76. $y = -x^2$
  Let $x = -3, -2, -1, 0, 1, 2, 3$.

77. $y = -1.3x + 5.6$ **78.** $y = 7.6x - 1.4$

79. Graph $y = x^2 - 4x + 7$. Let $x = 0, 1, 2, 3, 4$ to generate ordered pair solutions.

80. Graph $y = x^2 + 2x + 3$. Let $x = -3, -2, -1, 0, 1$ to generate ordered pair solutions.

81. The perimeter y of a rectangle whose width is a constant 3 inches and whose length is x inches is given by the equation
$$y = 2x + 6$$
  a. Draw a graph of this equation.
  b. Read from the graph the perimeter y of a rectangle whose length x is 4 inches.   14 inches

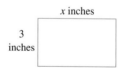

3 inches x inches

82. The distance y traveled in a train moving at a constant speed of 50 miles per hour is given by the equation
$$y = 50x$$
  where x is the time in hours traveled.
  a. Draw a graph of this equation.
  b. Read from the graph the distance y traveled after 6 hours.   300 miles

For income tax purposes, Rob Calcutta, owner of Copy Services, uses a method called **straight-line depreciation** *to show the loss in value of a copy machine he re-*

cently purchased. Rob assumes that he can use the machine for 7 years. The following graph shows the value of the machine over the years. Use this graph to answer the following questions.

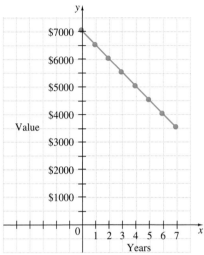

83. What was the purchase price of the copy machine?

84. What is the depreciated value of the machine in 7 years? $3500

85. What loss in value occurred during the first year?

86. What loss in value occurred during the second year? $500

87. Why do you think that this method of depreciating is called straight-line depreciation?

88. Why is the line tilted downward?

 89. On the same set of axes, graph $y = 2x$, $y = 2x - 5$, and $y = 2x + 5$. What patterns do you see in these graphs? See Appendix D. They are parallel.

90. On the same set of axes, graph $y = 2x$, $y = x$, and $y = -2x$. Describe the differences and similarities in these graphs.

91. Explain why we generally use three points to graph a line, when only two points are needed.

Write each statement as an equation in two variables. Then graph each equation.

92. The y-value is 5 more than three times the x-value.

93. The y-value is -3 decreased by twice the x-value.

94. The y-value is 2 more than the square of the x-value. $y = x^2 + 2$

95. The y-value is 5 decreased by the square of the x-value. $y = 5 - x^2$

Use a grapher to verify the graphs of the following exercises. For Exercises 96–99, see Appendix D.

96. Exercise 65   **97.** Exercise 66

98. Exercise 73   **99.** Exercise 74

Review Exercises

Solve the following equations. See Section 2.1.

100. $3(x - 2) + 5x = 6x - 16$ $\{-5\}$

101. $5 + 7(x + 1) = 12 + 10x$ $\{0\}$

102. $3x + \dfrac{2}{5} = \dfrac{1}{10}$ $\left\{-\dfrac{1}{10}\right\}$

103. $\dfrac{1}{6} + 2x = \dfrac{2}{3}$ $\left\{\dfrac{1}{4}\right\}$

Solve the following inequalities. See Section 2.4.

104. $3x \le -15$ $(-\infty, -5]$

105. $-3x > 18$ $(-\infty, -6)$

106. $2x - 5 > 4x + 3$ $(-\infty, -4)$

107. $9x + 8 \le 6x - 4$ $(-\infty, -4]$

83. $7000   **85.** $500   **87.** Depreciation is the same from year to year.   **88.** Because depreciation is a loss (the value is going down).   **90.** See Appendix D. All the lines go through the origin, but they have different slopes. **91.** Answers will vary.   **92.** $y = 3x + 5$ **93.** $y = -3 - 2x$

3.2   INTRODUCTION TO FUNCTIONS

O B J E C T I V E S

1 Define relation, domain, and range.

2 Identify functions.

3 Use the vertical line test for functions.

4 Find the domain and range of a function.

TAPE IA 3.2

Our work in Section 3.1 leads us to the notion of relation and to the notion of function, perhaps the single most important and useful notion in all of mathematics.

Equations in two variables, such as $y = 2x + 1$, describe relations between x-values and y-values. For example, if $x = 1$, then this equation describes how to find the y-value related to $x = 1$. In words, the equation $y = 2x + 1$ says that twice the x-value increased by 1 gives the corresponding y-value. The x-value of 1 corresponds to the y-value of $2(1) + 1 = 3$ for this equation, and we have the ordered pair $(1, 3)$.

There are other ways of describing relations or correspondences between two numbers or, in general, a first set (sometimes called the set of *inputs*) and a second set (sometimes called the set of *outputs*). For example,

First set: Input	Correspondence:	Second set: Output
People in a certain city	Each person's age	The set of nonnegative integers

A few examples of correspondences from this relation might be Ana, 4; Bob, 36; Trey, 21; and so on. Another example of a relation is:

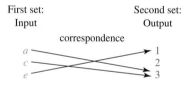

This relation shows the correspondences a, 3; c, 3; and e, 1. Notice that no element in the first set is assigned to the element 2 of the second set.

> A **relation** is a correspondence between a first set (set of inputs) and a second set (set of outputs) that assigns to each element of the first set an element of the second set. The **domain** of the relation is the first set, and the **range** of the relation is the set of elements that correspond to some element of the first set.

The domain for our relation above is $\{a, c, e\}$ and the range is $\{1, 3\}$. Notice that the range does not include the element 2 of the second set. This is because no element of the first set is assigned to this element. If a relation is defined in terms of x- and y-values, we will agree that the domain corresponds to x-values and that the range corresponds to y-values.

EXAMPLE 1   Determine the domain and range of each relation.

a. {(2, 3), (2, 4), (0, −1), (3, −1)]

b.

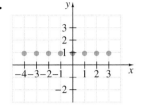

c.

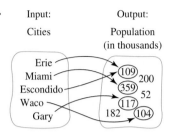

Solution: **a.** The domain is the set of all first coordinates of the ordered pairs, {2, 0, 3}. The range is the set of all second coordinates, {3, 4, −1}.

b. Ordered pairs are not listed here, but are given in graph form. The relation is {(−4, 1), (−3, 1), (−2, 1), (−1, 1), (0, 1), (1, 1), (2, 1), (3, 1)}. The domain is {−4, −3, −2, −1, 0, 1, 2, 3}.

The range is {1}.

c. The domain is the first set, {Erie, Escondido, Gary, Miami, Waco}.

The range is the numbers in the second set that correspond to elements in the first set {104, 109, 117, 359}.

 Now we consider a special kind of relation called a function.

A **function** is a relation in which each element of the first set corresponds to *exactly one element* of the second set.

REMINDER   A function is a special type of relation, so all functions are relations, but not all relations are functions.

EXAMPLE 2   Which of the following relations are also functions?

a. {(−2, 5), (2, 7), (−3, 5), (9, 9)}

b.

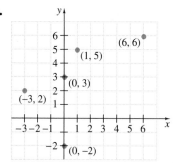

c.

INPUT:	CORRESPONDENCE:	OUTPUT
People in a certain city	Each person's age	The set of nonnegative integers

Solution: **a.** Although the ordered pairs $(-2, 5)$ and $(-3, 5)$ have the same y-value, each x-value is assigned to only one y-value, so this set of ordered pairs is a function.

b. The x-value 0 is assigned to two y-values, -2 and 3, in this graph so this relation does not define a function.

c. This relation is a function because although two different people may have the same age, each person has only one age. This means that each element in the first set is assigned to only one element in the second set.

We will call an equation such as $y = 2x + 1$ a **relation** since this equation defines a set of ordered pair solutions.

EXAMPLE 3   Is the relation $y = 2x + 1$ also a function?

Solution: The relation $y = 2x + 1$ is a function if each x-value corresponds to just one y-value. For each x-value substituted in the equation $y = 2x + 1$, the multiplication and addition performed each gives a single result, so only one y-value will be associated with each x-value. Thus, $y = 2x + 1$ is a function.

EXAMPLE 4   Is the relation $x = y^2$ also a function?

Solution: In $x = y^2$, if $y = 3$, then $x = 9$. Also, if $y = -3$, then $x = 9$. In other words, the x-value 9 corresponds to two y-values, 3 and -3. Thus, $x = y^2$ is not a function.

3   As we have seen so far, not all relations are functions. Consider the graphs of $y = 2x + 1$ and $x = y^2$ shown next. On the graph of $y = 2x + 1$, notice that each

x-value corresponds to only one y-value. Recall from Example 3 that $y = 2x + 1$ is a function.

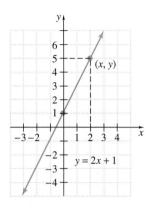
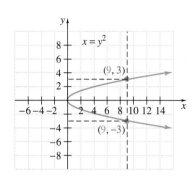

On the graph of $x = y^2$, the x-value 9, for example, corresponds to two y-values, 3 and -3, as shown by the vertical line. Recall from Example 4 that $x = y^2$ is not a function.

Graphs can be used to help determine whether a relation is also a function by the following vertical line test.

VERTICAL LINE TEST

If no vertical line can be drawn so that it intersects a graph more than once, the graph is the graph of a function.

EXAMPLE 5   Which of the following graphs are graphs of functions?

a.

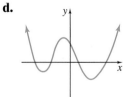

b.

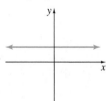

c.

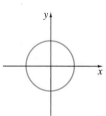

d.

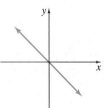

e.

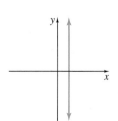

f.

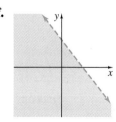

Solution: **a.** This graph is the graph of a function since no vertical line will intersect this graph more than once.

b. This graph is also the graph of a function.

c. This graph is not the graph of a function. Note that vertical lines can be drawn that intersect the graph in two points.

d. This graph is the graph of a function.

e. This graph is not the graph of a function. A vertical line can be drawn that intersects this line at every point.

f. This graph is not the graph of a function. ▬▬▬▬

Recall that the graph of a linear equation in two variables is a line, and a line that is not vertical will pass the vertical line test. Thus, **all linear equations are functions except those whose graph is a vertical line.**

4

Next, we practice finding the domain and range of a relation from its graph.

EXAMPLE 6 Find the domain and range of each relation. Determine whether the relation is also a function.

a.

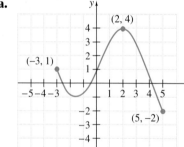

b.

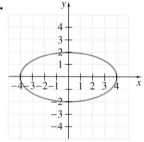

c.

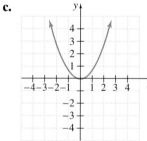

d.

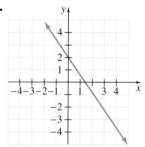

Solution: The domain is the set of values of x and the range is the set of values of y. We read these values from each graph.

a. By the vertical line test, this is the graph of a function. From the graph, x takes on values from -3 to 5 inclusive, and y takes on values from -2 to 4, inclusive. Thus, the domain is $\{x\mid -3 \le x \le 5\}$ or in interval notation is $[-3, 5]$. The range is $\{y\mid -2 \le y \le 4\}$ or in interval notation $[-2, 4]$.

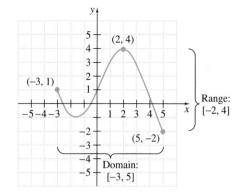

b. By the vertical line test, this is not the graph of a function.

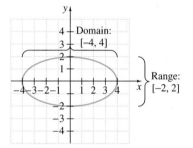

c. By the vertical line test, this is the graph of a function.

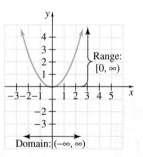

d. By the vertical line test, this is the graph of a function.

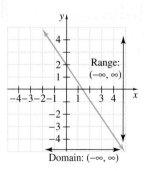

The graph of $y = 4x + 3$ for example, is a nonvertical linear equation, so it describes a function. Many times letters such as f, g, and h are used to name functions. To denote that y is a function of x in the equation $y = 4x + 3$, we can write $y = f(x)$. Then $y = 4x + 3$ can be written as $f(x) = 4x + 3$. The symbol $f(x)$ means **function of x** and is read "f of x." This notation is called **function notation.**

The notation $f(1)$ means to replace x with 1 and find the resulting y or function value. Since

$$f(x) = 4x + 3$$

then

$$f(1) = 4(1) + 3 = 7$$

This means that when $x = 1$, y or $f(x) = 7$. Here, the input is 1 and the output is $f(1)$ or 7. Now find $f(2)$, $f(0)$, and $f(-1)$.

$$f(x) = 4x + 3 \qquad\qquad f(x) = 4x + 3 \qquad\qquad f(x) = 4x + 3$$
$$f(2) = 4(2) + 3 \qquad\quad f(0) = 4(0) + 3 \qquad\quad f(-1) = 4(-1) + 3$$
$$= 8 + 3 \qquad\qquad\qquad = 0 + 3 \qquad\qquad\qquad = -4 + 3$$
$$= 11 \qquad\qquad\qquad\quad = 3 \qquad\qquad\qquad\quad = -1$$

REMINDER   Note that $f(x)$ is a special symbol in mathematics used to denote a function. The symbol $f(x)$ is read "f of x." It does **not** mean $f \cdot x$ (f times x).

EXAMPLE 7   If $f(x) = 7x^2 - 3x + 1$ and $g(x) = 3x - 2$, find the following.

a. $f(1)$ b. $g(1)$ c. $f(-2)$ d. $g(0)$

Solution: a. Substitute 1 for x in $f(x) = 7x^2 - 3x + 1$ and simplify.

$$f(x) = 7x^2 - 3x + 1$$
$$f(1) = 7(1)^2 - 3(1) + 1 = 5$$

b. $g(x) = 3x - 2$
$$g(1) = 3(1) - 2 = 1$$

c. $f(x) = 7x^2 - 3x + 1$
$$f(-2) = 7(-2)^2 - 3(-2) + 1 = 35$$

d. $g(x) = 3x - 2$
$$g(0) = 3(0) - 2 = -2$$

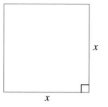

Many formulas that are familiar to you describe functions. For example, we know that the formula for the area of a square with side length x is $A = x^2$. The area of the square is actually a function of the length of the side x. If we want to write the formula using function notation, instead of

$$A = x^2$$

we may write

$$A(x) = x^2$$

Then to find the area of the square whose side measures 10 meters, we write

$$A(10) = (10 \text{ meters})^2$$
$$= 100 \text{ square meters}$$

EXAMPLE 8   The following graph shows the research and development expenditures by the Pharmaceutical Manufacturers Association as a function of time.

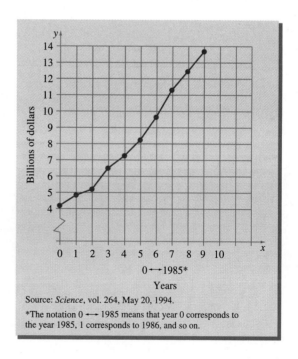

Source: *Science*, vol. 264, May 20, 1994.

*The notation 0 ◄─► 1985 means that year 0 corresponds to the year 1985, 1 corresponds to 1986, and so on.

a. Approximate the money spent on research and development in 1992.
b. In 1958, research and development expenditures were $200 million. Find the increase in expenditures from 1958 to 1994.

Solution: **a.** In 1992, approximately $11.5 billion was spent.

b. In 1994, approximately $13.8 billion, or $13,800 million was spent. The increase in spending from 1958 to 1994 is $13,800 − $200 = $13,600 million. ▬▬▬▬

Notice that the graph in Example 8 is the graph of a function since each year there is only one total amount of money spent by the Pharmaceutical Manufacturers Association on research and development. Also notice that the graph resembles the graph of a line. Often, businesses depend on equations that "closely fit" data-defined functions like this one in order to model the data and predict future trends. For example, by a method called least squares, the linear function $f(x) = 1.087x + 3.44$ approximates the data shown. Its graph and the actual data function are shown next.

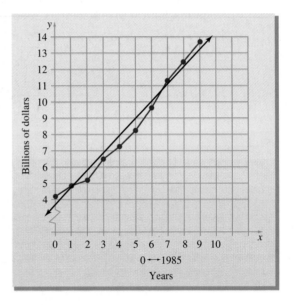

To predict the amount of money spent in the year 2000, we use $f(x) = 1.087x + 3.44$ and find $f(15)$. (Notice that year 0 on the graph corresponds to the year 1985, so year 15 corresponds to the year 2000.)

$$f(x) = 1.087x + 3.44$$
$$f(15) = 1.087(15) + 3.44$$
$$= 19.745$$

We predict that in the year 2000, $19.745 billion dollars will be spent on research and development by the Pharmaceutical Manufacturers Association.

GRAPHING CALCULATOR EXPLORATIONS

It is possible to use a grapher to sketch the graph of more than one equation on the same set of axes. This feature can be used to confirm a pattern suggested in Section 3.1. For example, graph the functions $f(x) = x^2$ and $g(x) = x^2 + 4$ on the same set of axes.

To graph on the same set of axes, press the $\boxed{Y =}$ key and enter the equations on the first two lines.

$$Y_1 = x^2$$
$$Y_2 = x^2 + 4$$

Then press the $\boxed{\text{GRAPH}}$ key as usual. The screen should look like this:

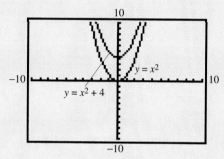

Notice that the graph of y or $g(x) = x^2 + 4$ is the graph of $y = x^2$ moved 4 units upward.

Graph each pair of functions on the same set of axes. Describe the similarities and differences in their graphs.   For Graphing Calculator Exercises 1–6, see Appendix D.

1. $f(x) = |x|$
  $g(x) = |x| + 1$

2. $f(x) = x^2$
  $h(x) = x^2 - 5$

3. $f(x) = x$
  $H(x) = x - 6$

4. $f(x) = |x|$
  $G(x) = |x| + 3$

5. $f(x) = -x^2$
  $F(x) = -x^2 + 7$

6. $f(x) = x$
  $F(x) = x + 2$

EXERCISE SET 3.2

Find the domain and the range of each relation. Also determine whether the relation is a function. See Examples 1 and 2.

1. $\{(-1, 7), (0, 6), (-2, 2), (5, 6)\}$

2. $\{(4, 9), (-4, 9), (2, 3), (10, -5)\}$

3. $\{(-2, 4), (6, 4), (-2, -3), (-7, -8)\}$

4. $\{(6, 6), (5, 6), (5, -2), (7, 6)\}$

5. $\{(1, 1), (1, 2), (1, 3), (1, 4)\}$

6. $\{(1, 1), (2, 1), (3, 1), (4, 1)\}$

1. domain: $\{-1, 0, -2, 5\}$; range: $\{7, 6, 2\}$; function   **2.** domain: $\{4, -4, 2, 10\}$; range: $\{9, 3, -5\}$; function
3. domain: $\{-2, 6, -7\}$; range: $\{4, -3, -8\}$; not a function   **4.** domain: $\{6, 5, 7\}$; range: $\{6, -2\}$; not a function
5. domain: $\{1\}$; range: $\{1, 2, 3, 4\}$; not a function   **6.** domain: $\{1, 2, 3, 4\}$; range: $\{1\}$; function

7. $\left\{\left(\dfrac{3}{2}, \dfrac{1}{2}\right), \left(1\dfrac{1}{2}, -7\right), \left(0, \dfrac{4}{5}\right)\right\}$

8. $\{(\pi, 0), (0, \pi), (-2, 4), (4, -2)\}$

9. $\{(-3, -3), (0, 0), (3, 3)\}$

10. $\left\{\left(\dfrac{1}{2}, \dfrac{1}{4}\right), \left(0\,\dfrac{7}{8}\right), (0.5, \pi)\right\}$

11.

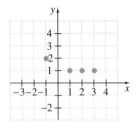

domain: {−1, 1, 2, 3}; range: {2, 1}; function

12.

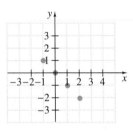

domain: {−1, 0, 1, 2}; range: {1, 0, −1, −2}; function

13.

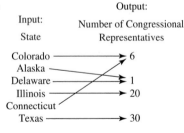

domain: {Colorado, Alaska, Delaware, Illinois, Connecticut, Texas}; range: {6, 1, 20, 30}; function

14.

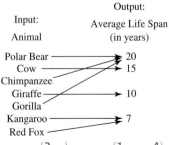

domain: {polar bear, cow, chimpanzee, giraffe, gorilla, kangaroo, red fox}; range: {20, 15, 10, 7}; function

15.

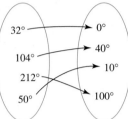

domain: {32°, 104°, 212°, 50°}; range: {0°, 40°, 10°, 100°}; function

16.

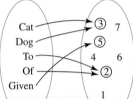

domain: {cat, dog, to, of, given}; range: {3, 5, 2}; function

17.

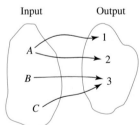

domain: {2, −1, 5, 100}; range: {0}; function

18.

domain: {A, B, C}; range: {1, 2, 3}; not a function

In Exercises 19 and 20, determine whether the relation is a function.

19.

First set: Input	Correspondence	Second set: Output
Class of algebra students	Grade average	Set of nonnegative numbers

7. domain: $\left\{\dfrac{3}{2}, 0\right\}$; range: $\left\{\dfrac{1}{2}, -7, \dfrac{4}{5}\right\}$; not a function   **8.** domain: $\{\pi, 0, -2, 4\}$; range: $\{0, \pi, 4, -2\}$; function

9. domain: $\{-3, 0, 3\}$; range: $\{-3, 0, 3\}$; function   **10.** domain: $\left\{\dfrac{1}{2}, 0\right\}$; range: $\left\{\dfrac{1}{4}, \dfrac{7}{8}, \pi\right\}$; not a function   **19.** function

20.

First set: Input	Correspondence	Second set: Output
People in New Orleans (population 500,000)	Birthdate	Days of the year

function

21. Describe a function whose domain is the set of people in your home town.   Answers will vary.

22. Describe a function whose domain is the set of people in your algebra class.   Answers will vary.

Use the vertical line test to determine whether each graph is the graph of a function. See Example 5.

23.
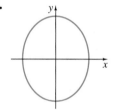

24.
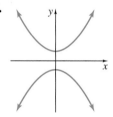

25.
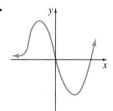

26.
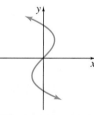

27.
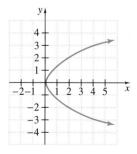

28.
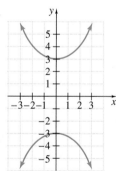

Find the domain and the range of each relation. Use the vertical line test to determine whether each graph is the graph of a function. See Example 6.

29.
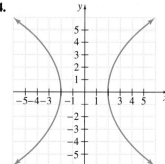
domain: $[0, \infty)$;
range: $(-\infty, \infty)$;
not a function

30.
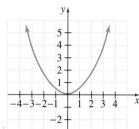
domain: $(-\infty, \infty)$;
range: $[0, \infty)$;
function

31.
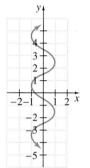
domain: $[-1, 1]$;
range: $(-\infty, \infty)$;
not a function

32.
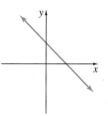
domain: $[-3, 3]$;
range: $[-3, 3]$;
not a function

33.
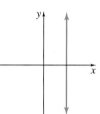
domain: $(-\infty, \infty)$;
range: $(-\infty, -3] \cup [3, \infty)$;
not a function

34.
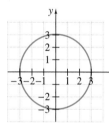
domain: $(-\infty, -2] \cup [2, \infty)$;
range: $(-\infty, \infty)$;
not a function

23. function   **24.** not a function   **25.** not a function   **26.** not a function   **27.** function   **28.** not a function

35.

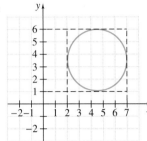

domain: [2, 7];
range: [1, 6];
not a function

36.

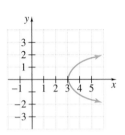

37.

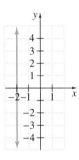

38.

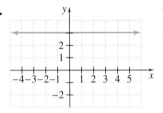

domain: $(-\infty, \infty)$;
range: {3};
function

39.

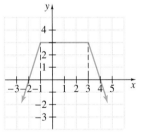

domain: $(-\infty, \infty)$;
range: $(-\infty, 3]$;
function

40.

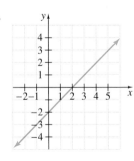

domain: $(-\infty, \infty)$;
range: $(-\infty, \infty)$;
function

41. In your own words define **(a)** function; **(b)** domain; **(c)** range.   Answers will vary.

42. Explain the vertical line test and how it is used.
Answers will vary.

Decide whether each is a function. See Examples 3 and 4.

43. $y = x + 1$ yes   **44.** $y = x - 1$ yes

45. $x = 2y^2$ no   **46.** $y = x^2$ yes

47. $y - x = 7$ yes   **48.** $2x - 3y = 9$ yes

49. $y = \dfrac{1}{x}$ yes   **50.** $y = \dfrac{1}{x - 3}$ yes

51. $y = 5x - 12$ yes   **52.** $y = \dfrac{1}{2}x + 4$ yes

53. $x = y^2$ no   **54.** $x = |y|$ no

If $f(x) = 3x + 3$, $g(x) = 4x^2 - 6x + 3$, and $h(x) = 5x^2 - 7$, find the following. See Example 7.

55. $f(4)$ 15   **56.** $f(-1)$ 0   **57.** $h(-3)$ 38 **58.** $h(0)$ -7

59. $g(2)$ 7   **60.** $g(1)$ 1   **61.** $g(0)$ 3   **62.** $h(-2)$ 13

Given the following functions, find the indicated values. See Example 7.

63. $f(x) = \dfrac{1}{2}$;

   a. $f(0)$ 0   **b.** $f(2)$ 1   **c.** $f(-2)$ -1

64. $g(x) = -\dfrac{1}{3}x$;

   a. $g(0)$ 0   **b.** $g(-1)$ $\dfrac{1}{3}$ **c.** $g(3)$ -1

65. $g(x) = 2x^2 + 4$;   **65a.** 246

   a. $g(-11)$ **b.** $g(-1)$ 6   **c.** $g\left(\dfrac{1}{2}\right)$ $\dfrac{9}{2}$

66. $h(x) = -x^2$;   **66a.** -25

   a. $h(-5)$ **b.** $h\left(-\dfrac{1}{3}\right)$ $-\dfrac{1}{9}$ **c.** $h\left(\dfrac{1}{3}\right)$ $-\dfrac{1}{9}$

67. $f(x) = -5$;

   a. $f(2)$ -5 **b.** $f(0)$ -5 **c.** $f(606)$ -5

68. $h(x) = 7$;

   a. $h(7)$ 7   **b.** $h(542)$ 7   **c.** $h\left(-\dfrac{3}{4}\right)$ 7

69. $f(x) = 1.3x^2 - 2.6x + 5.1$

   a. $f(2)$ 5.1 **b.** $f(-2)$ 15.5   **c.** $f(3.1)$ 9.533

36. domain: [3, ∞); range: $(-\infty, \infty)$; not a function   **37.** domain: {−2}; range: $(-\infty, \infty)$; not a function

70. $g(x) = 2.7x^2 + 6.8x - 10.2$

   a. $g(1)$ -0.7 **b.** $g(-5)$ 23.3   **c.** $g(7.2)$ 178.728

Use the graph in Example 8 to answer the following.

71. a. Use the graph to approximate the money spent on research and development in 1988. $6.5 billion

   b. Recall that the function $f(x) = 1.087x + 3.44$ approximates the graph of Example 8. Use this equation to approximate the money spent on research and development in 1988. [*Hint:* Find $f(3)$.] $6.701 billion

72. a. Use the graph to approximate the money spent on research and development in 1993. $12.5 billion

   b. Use the function $f(x) = 1.087x + 3.44$ to approximate the money spent on research and development in 1993. [*Hint:* Find $f(8)$.]

73. Use the function $f(x) = 1.087x + 3.44$ to predict the money spent on research and development in 2005. $25.18 billion

74. Use the function $f(x) = 1.087x + 3.44$ to predict the money spent on research and development in 2010. $30.615 billion

75. Since $y = x + 7$ describes a function, rewrite the equation using function notation. $f(x) = x + 7$

76. In your own words, explain how to find the domain of a function given its graph.

The function $A(r) = \pi r^2$ may be used to find the area of a circle if we are given its radius.

77. Find the area of a circle whose radius is 5 centimeters. (Do not approximate π.) 25π square cm

78. Find the area of a circular garden whose radius is 8 feet. (Do not approximate π.) 64π square ft

The function $V(x) = x^3$ may be used to find the volume of a cube if we are given the length x of a side.

72b. $12.136 billion   **79.** 2744 cubic in.

79. Find the volume of a cube whose side is 14 inches.

80. Find the volume of a die whose side is 1.7 centimeters.   4.913 cubic cm

Forensic scientists use the following functions to find the height of a woman if they are given the height of her femur bone f or her tibia bone t in centimeters.

$$H(f) = 2.59f + 47.24$$
$$H(t) = 2.72t + 61.28$$

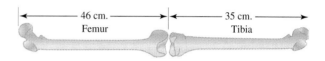

81. Find the height of a woman whose femur measures 46 centimeters.   166.38 cm

82. Find the height of a woman whose tibia measures 35 centimeters.   156.48 cm

The dosage in milligrams D of Ivermectin, a heartworm preventive, for a dog who weighs x pounds is given by $D(x) = \dfrac{136}{25}x.$

83. Find the proper dosage for a dog that weighs 30 pounds.   163.2 milligrams

84. Find the proper dosage for a dog that weighs 50 pounds.   272 milligrams

Review Exercises

Complete the given table and use the table to graph the linear equation.

85. $x - y = -5$

x	0	-5	1
y	5	0	6

86. $2x + 3y = 10$

x	0	5	2
y	$\frac{10}{3}$	0	2

87. $7x + 4y = 8$

x	0	$\frac{8}{7}$	$\frac{12}{7}$
y	2	0	-1

88. $5y - x = -15$

x	0	15	-2
y	-3	0	$-\frac{17}{5}$

89. $y = 6x$

x	0	0	-1
y	0	0	-6

90. $y = -2x$

x	0	0	-2
y	0	0	4

 91. Is it possible to find the perimeter of the following geometric figure? If so, find the perimeter.

yes, 170 meters

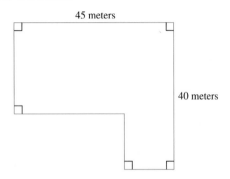

45 meters

40 meters

A Look Ahead

EXAMPLE

If $f(x) = x^2 + 2x + 1$, find the following:
a. $f(\pi)$ **b.** $f(c)$

Solution:

a. $f(x) = x^2 + 2x + 1$
$f(\pi) = \pi^2 + 2\pi + 1$

b. $f(x) = x^2 + 2x + 1$
$f(c) = (c)^2 + 2(c) + 1$
$= c^2 + 2c + 1$

Given the following functions, find the indicated values. See the previous example.

92. $f(x) = 2x + 7$; **a.** $f(2)$ 11 **b.** $f(a)$
93. $g(x) = -3x + 12$; **a.** $g(s)$ **b.** $g(r)$
94. $h(x) = x^2 + 7$; **a.** $h(3)$ 16 **b.** $h(a)$
95. $f(x) = x^2 - 12$; **a.** $f(12)$ 132 **b.** $f(a)$

92b. $2a + 7$ **93a.** $-3s + 12$ **93b.** $-3r + 12$ **94b.** $a^2 + 7$ **95b.** $a^2 - 12$

3.3 | GRAPHING LINEAR FUNCTIONS

TAPE IA 3.3

O B J E C T I V E S

1. Graph linear functions.
2. Graph linear functions by finding intercepts.
3. Graph vertical and horizontal lines.

In this section, we identify and graph linear functions. By the vertical line test, we know that all linear equations except those whose graphs are vertical lines are functions. For example, we know from Section 3.1 that $y = 2x$ is a linear equation in two variables. Its graph is shown next.

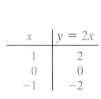

x	$y = 2x$
1	2
0	0
-1	-2

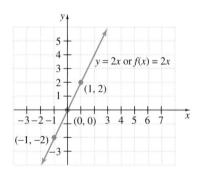

Because this graph passes the vertical line test, we know that $y = 2x$ is a function. If we want to emphasize that this equation describes a function, we may write $y = 2x$ as $f(x) = 2x$.

EXAMPLE 1   Graph $g(x) = 2x + 1$. Compare this graph with the graph of $f(x) = 2x$.

Solution: To graph $g(x) = 2x + 1$, find three ordered pair solutions.

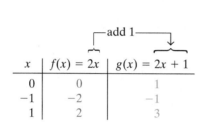

x	$f(x) = 2x$	$g(x) = 2x + 1$
0	0	1
-1	-2	-1
1	2	3

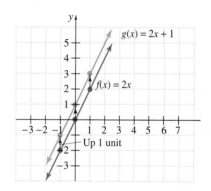

Notice that y-values for the graph of $g(x) = 2x + 1$ are obtained by adding 1 to each y-value of each corresponding point of the graph of $f(x) = 2x$. The graph of $g(x) = 2x + 1$ is the same as the graph of $f(x) = 2x$ shifted upward 1 unit.

In general, a **linear function** is a function that can be written in the form $f(x) = mx + b$. For example, $g(x) = 2x + 1$ is in this form, with $m = 2$ and $b = 1$.

EXAMPLE 2   Graph the linear functions $f(x) = -3x$ and $g(x) = -3x - 6$ on the same set of axes.

Solution: To graph $f(x)$ and $g(x)$, find ordered pair solutions.

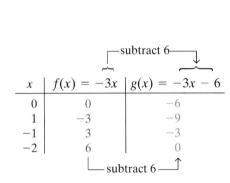

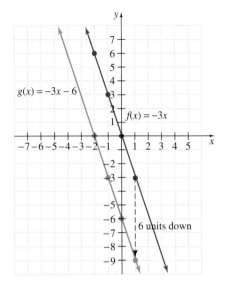

Each y-value for the graph of $g(x) = -3x - 6$ is obtained by subtracting 6 from the y-value of the corresponding point of the graph of $f(x) = -3x$. The graph of $g(x) = -3x - 6$ is the same as the graph of $f(x) = -3x$ shifted down 6 units.

In general, for any function $f(x)$, the graph of $y = f(x) + K$ is the same as the graph of $y = f(x)$ shifted $|K|$ units up if K is positive and down if K is negative.

Notice that the y-intercept of the graph of $g(x) = -3x - 6$ in the preceding figure is -6, the same as the constant in the equation. *If a linear function is written in the form $f(x) = mx + b$ or $y = mx + b$, the y-intercept point is $(0, b)$ and the y-intercept is b.* This is because if x is 0, then $f(x) = mx + b$ becomes $f(0) = m \cdot 0 + b = b$, and we have the ordered pair solution $(0, b)$. We will study this form more in the next section.

EXAMPLE 3 Find the y-intercept of the graph of each equation.

    a. $f(x) = \dfrac{1}{2}x + \dfrac{3}{7}$ **b.** $y = -2.5x - 3.2$

Solution: **a.** The y-intercept of $f(x) = \dfrac{1}{2}x + \dfrac{3}{7}$ is $\dfrac{3}{7}$. The y-intercept point is $\left(0, \dfrac{3}{7}\right)$.

        b. The y-intercept of $y = -2.5x - 3.2$ is -3.2. The y-intercept point is $(0, -3.2)$.

In general, to find the y-intercept of the graph of an equation not in the form $y = mx + b$, let $x = 0$ since any point on the y-axis has an x-coordinate of 0. To find the x-intercept of a line, let $y = 0$ or $f(x) = 0$ since any point on the x-axis has a y-coordinate of 0.

FINDING *x*- AND *y*-INTERCEPTS

To find an x-intercept, let $y = 0$ or $f(x) = 0$ and solve for x.
To find a y-intercept, let $x = 0$ and solve for y.

Intercept points are usually easy to find and plot since one coordinate is 0.

EXAMPLE 4   Graph $x - 3y = 6$ by plotting intercept points.

Solution: Let $y = 0$ to find the x-intercept and $x = 0$ to find the y-intercept.

$$
\begin{array}{ll}
\text{If } y = 0 \quad \text{then} & \text{If } x = 0 \quad \text{then} \\
x - 3(0) = 6 & 0 - 3y = 6 \\
x - 0 = 6 & -3y = 6 \\
x = 6 & y = -2
\end{array}
$$

The x-intercept is 6, and the y-intercept is -2. We find a third ordered pair solution to check our work. If we let $y = -1$, then $x = 3$. Plot the points $(6, 0)$, $(0, -2)$, and $(3, -1)$. The graph of $x - 3y = 6$ is the line drawn through these points, as shown.

x	y
6	0
0	-2
3	-1

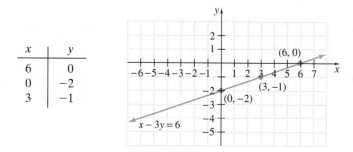

Notice that the equation $x - 3y = 6$ describes a linear function—"linear" because its graph is a line and "function" because the graph passes the vertical line test.

If we want to emphasize that the equation $x - 3y = 6$ from Example 4 describes a function, first solve the equation for y.

$$x - 3y = 6$$

$$-3y = -x + 6 \qquad \text{Subtract } x \text{ from both sides.}$$

$$\frac{-3y}{-3} = \frac{-x}{-3} + \frac{6}{-3} \qquad \text{Divide both sides by } -3.$$

$$y = \frac{1}{3}x - 2 \qquad \text{Simplify.}$$

Next, let $y = f(x)$.

$$f(x) = \frac{1}{3}x - 2$$

EXAMPLE 5   Graph $x = -2y$ by plotting intercept points.

Solution: Let $y = 0$ to find the x-intercept and $x = 0$ to find the y-intercept.

$$\begin{array}{llll}
\text{If } y = 0 & \text{then} & \text{If } x = 0 & \text{then} \\
x = -2(0) & \text{or} & 0 = -2y & \text{or} \\
x = 0 & & 0 = y &
\end{array}$$

Both the x-intercept and y-intercept are 0. In other words, when $x = 0$, then $y = 0$, which gives the ordered pair $(0, 0)$. Also, when $y = 0$, then $x = 0$, which gives the same ordered pair $(0, 0)$. This happens when the graph passes through the origin. Since two points are needed to determine a line, we must find at least one more ordered pair that satisfies $x = -2y$. Let $y = -1$ to find a second ordered pair solution and let $y = 1$ as a checkpoint.

$$\begin{array}{llll}
\text{If } y = -1 & \text{then} & \text{If } y = 1 & \text{then} \\
x = -2(-1) & \text{or} & x = -2(1) & \text{or} \\
x = 2 & & x = -2 &
\end{array}$$

The ordered pairs are $(0, 0)$, $(2, -1)$, and $(-2, 1)$. Plot these points to graph $x = -2y$.

x	y
0	0
2	-1
-2	1

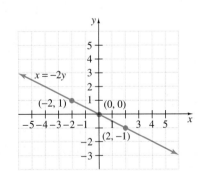

3

The equation $x = c$, where c is a real number constant, is a linear equation in two variables because it can be written in the form $x + 0y = c$. The graph of this equation is a vertical line as shown in the next example. Since a vertical line does not pass the vertical line test, the equation $x = c$ does *not* describe a function.

EXAMPLE 6   Graph $x = 2$.

Solution: The equation $x = 2$ can be written as $x + 0y = 2$. For any y-value chosen, notice that x is 2. No other value for x satisfies $x + 0y = 2$. Any ordered pair whose x-coordinate is 2 is a solution to $x + 0y = 2$ because 2 added to 0 times any value of y is $2 + 0$, or 2. We will use the ordered pairs $(2, 3)$, $(2, 0)$ and $(2, -3)$ to graph $x = 2$.

x	y
2	3
2	0
2	-3

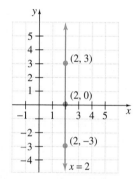

The graph is a vertical line with x-intercept 2. Notice that this graph is not the graph of a function, and it has no y-intercept because x is never 0.

VERTICAL LINES

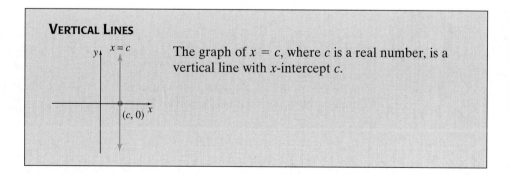

The graph of $x = c$, where c is a real number, is a vertical line with x-intercept c.

Does the equation $y = c$ describe a function? Yes, because the graph is a horizontal line, as shown next.

EXAMPLE 7 Graph $y = -3$.

Solution: The equation $y = -3$ can be written as $0x + y = -3$. For any x-value chosen, y is -3. If we choose 4, 1, and -2 as x-values, the ordered pair solutions are $(4, -3), (1, -3)$, and $(-2, -3)$. Use these ordered pairs to graph $y = -3$. The graph is a horizontal line with y-intercept -3 and no x-intercept. Recall that we may write $y = -3$ as $f(x) = -3$.

x	y
4	-3
1	-3
-2	-3

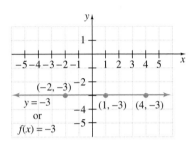

HORIZONTAL LINES

The graph of $y = c$ or $f(x) = c$, where c is a real number, is a horizontal line with y-intercept c.

GRAPHING CALCULATOR EXPLORATIONS

You may have noticed by now that to use the $\boxed{Y =}$ key on a grapher to graph an equation, the equation must be solved for y.

For Graphing Calculator Exercises 1–8, see Appendix D.

Graph each function by first solving the function for y.

1. $x = 3.5y$

2. $-2.7y = x$

3. $5.78x + 2.31y = 10.98$

4. $-7.22x + 3.89y = 12.57$

5. $y - |x| = 3.78$ $y = |x| + 3.78$

6. $3y - 5x^2 = 6x - 4$ $y = \frac{5}{3}x^2 + 2x - \frac{4}{3}$

7. $y - 5.6x^2 = 7.7x + 1.5$

8. $y + 2.6|x| = -3.2$ $y = -2.6|x| - 3.2$

---

1. $y = \dfrac{x}{3.5}$ **2.** $y = -\dfrac{x}{2.7}$ **3.** $y = -\dfrac{5.78}{2.31}x + \dfrac{10.98}{2.31}$ **4.** $y = \dfrac{7.22}{3.89}x + \dfrac{12.57}{3.89}x$ **7.** $y = 5.6x^2 + 7.7x + 1.5$

EXERCISE SET 3.3

Graph each linear function. See Examples 1 and 2.

1. $f(x) = -2x$ **2.** $f(x) = 2x$ **3.** $f(x) = -2x + 3$

4. $f(x) = 2x + 6$ **5.** $f(x) = \frac{1}{2}x$ **6.** $f(x) = \frac{1}{3}x$

7. $f(x) = \frac{1}{2}x - 4$ **8.** $f(x) = \frac{1}{3}x - 2$

The graph of $f(x) = 5x$ follows. Use this graph to match each linear function with its graph. See Examples 1 through 3.

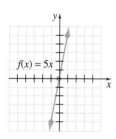

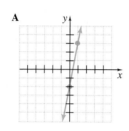

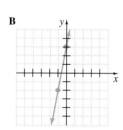

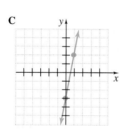

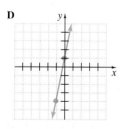

9. $f(x) = 5x - 3$ C   **10.** $f(x) = 5x - 2$ A

11. $f(x) = 5x + 1$ D   **12.** $f(x) = 5x + 3$ B

Graph each linear function by finding x- and y-intercepts. See Examples 4 and 5.

13. $x - y = 3$ **14.** $x - y = -4$ **15.** $x = 5y$

16. $2x = y$ **17.** $-x + 2y = 6$ **18.** $x - 2y = -8$

19. $2x - 4y = 8$ **20.** $2x + 3y = 6$

21. In your own words, explain how to find x- and y-intercepts.

22. Explain why it is a good idea to use three points to graph a linear equation.

Graph each linear equation. See Examples 6 and 7.

23. $x = -1$ **24.** $y = 5$ **25.** $y = 0$

26. $x = 0$ **27.** $y + 7 = 0$ **28.** $x - 3 = 0$

Match each equation with its graph.

A
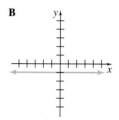

B
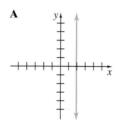

C
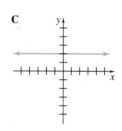

D
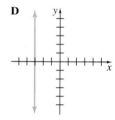

29. $y = 2$ C   **30.** $x = -3$ D

31. $x - 2 = 0$ A   **32.** $y + 1 = 0$ B

33. Discuss whether a vertical line ever has a y-intercept.   The vertical line $x = 0$ has y-intercepts.

34. Discuss whether a horizontal line ever has an x-intercept.   The horizontal line $y = 0$ has x-intercepts.

Graph each linear equation.

35. $x + 2y = 8$ **36.** $x - 3y = 3$

37. $f(x) = \frac{3}{4}x + 2$ **38.** $f(x) = \frac{4}{3}x + 2$

39. $x = -3$ **40.** $f(x) = 3$

41. $3x + 5y = 7$ **42.** $3x - 2y = 5$

43. $f(x) = x$ **44.** $f(x) = -x$

45. $x + 8y = 8$ **46.** $x - 3y = 9$

47. $5 = 6x - y$ **48.** $4 = x - 3y$

For Exercises 1−8, 13−20, 23−28, and 35−60, see Appendix D.

49. $-x + 10y = 11$ **50.** $-x + 9 = -y$

51. $y = 1$ **52.** $x = 1$

53. $f(x) = \dfrac{1}{2}x$ **54.** $f(x) = -2x$

55. $x + 3 = 0$ **56.** $y - 6 = 0$

57. $f(x) = 4x - \dfrac{1}{3}$ **58.** $f(x) = -3x + \dfrac{3}{4}$

59. $2x + 3y = 6$ **60.** $4x + y = 5$

61. Broyhill Furniture found that it takes 2 hours to manufacture each table for one of its special dining room sets. Each chair takes 3 hours to manufacture. A total of 1500 hours is available to produce tables and chairs of this style. The linear equation that models this situation is $2x + 3y = 1500$, where x represents the number of tables produced and y the number of chairs produced.

    a. Complete the ordered pair solution $(0,)$ of this equation. Describe the manufacturing situation this solution corresponds to.

    b. Complete the ordered pair solution $(, 0)$ for this equation. Describe the manufacturing situation this solution corresponds to.

    c. If 50 tables are produced, find the greatest number of chairs the company can make. 466 chairs

62. While manufacturing two different camera models, Kodak found that the basic model costs \$55 to produce, whereas the deluxe model costs \$75. The weekly budget for these two models is limited to \$33,000 in production costs. The linear equation that models this situation is $55x + 75y = 33,000$, where x represents the number of basic models and y the number of deluxe models.

    a. Complete the ordered pair solution $(0,)$ of this equation. Describe the manufacturing situation this solution corresponds to.

    b. Complete the ordered pair solution $(, 0)$ of this equation. Describe the manufacturing situation this solution corresponds to.

    c. If 350 deluxe models are produced, find the greatest number of basic models that can be made in one week. 122 basic models

63. The cost of renting a car for a day is given by the linear function $C(x) = 0.2x + 24$, where $C(x)$ is in dollars and x is the number of miles driven.

    a. Find the cost of driving the car 200 miles. \$64

    b. Graph $C(x) = 0.2x + 24$ See Appendix D.

    c. How can you tell from the graph of $C(x)$ that as the number of miles driven increases, the total cost increases also?

64. The cost of renting a piece of machinery is given by the linear function $C(x) = 4x + 10$, where $C(x)$ is in dollars and x is given in hours.

    a. Find the cost of renting the piece of machinery for 8 hours. \$42

    b. Graph $C(x) = 4x + 10$. See Appendix D.

    c. How can you tell from the graph of $C(x)$ that as the number of hours increases, the total cost increases also?

65. The yearly cost of tuition and required fees for attending a two-year college full time can be estimated by the linear function $f(t) = 91.7t + 747.8$, where t is the number of years after 1990 and $f(t)$ is the total cost.

    a. Use this function to approximate the yearly cost of attending a two-year college in the year 2010. [*Hint:* Find $f(20)$.] \$2581.80

    b. Use the given function to predict in what year will the yearly cost of tuition and required fees exceed \$2000. [*Hint:* Let $f(t) = 2000$ and solve for x.] year 2004

    c. Use this function to approximate the yearly cost of attending a two-year college in the present year. If you attend a two-year college, is this amount greater than or less than the amount that is currently charged by the college that you attend? Answers will vary.

66. The yearly cost of tuition and required fees for attending a four-year college can be estimated by the linear function $f(t) = 201.9t + 2002.2$, where t is the number of years after 1990 and $f(t)$ is the total cost in dollars.

    a. Use this function to approximate the yearly cost of attending a four-year college in the year 2010. [*Hint:* Find $f(20)$.] \$6040.20

    b. Use the given function to predict in what year will the yearly cost of tuition and required fees exceed \$4000. [*Hint:* Let $f(t) = 4000$ and solve for x.] year 2000

    c. Use this function to approximate the yearly cost of attending a four-year college in the present

61a. (0, 500); if no tables are produced, 500 chairs can be produced.
61b. (750, 0); if no chairs are produced, 750 tables can be produced.
62a. (0, 440); if no basic models are produced, 440 deluxe models can be produced.
62b. (600, 0); if no deluxe models are produced, 600 basic models can be produced.
63c. The line moves upward from left to right.   **64c.** The line moves upward from left to right.

year. If you attend a four-year college, is this amount greater than or less than the amount that is currently charged by the college that you attend?   Answers will vary.

Source: U.S. Bureau of the Census, *Statistical Abstract of the United States: 1995* (115th edition.) Washington, DC, 1995.

For Exercises 67–70, see Appendix D.

Use a grapher to verify the results of each exercise.

 67. Exercise 9   **68.** Exercise 10

 69. Exercise 17   **70.** Exercise 18

Review Exercises

Solve the following. See Sections 2.6 and 2.7.

71. $|x - 3| = 6$ $\{9, -3\}$ **72.** $|x + 2| < 4$ $(-6, 2)$

73. $|2x + 5| > 3$ **74.** $|5x| = 10$ $\{-2, 2\}$

75. $|3x - 4| \le 2$ $\left[\dfrac{2}{3}, 2\right]$ **76.** $|7x - 2| \ge 5$

Simplify.

77. $\dfrac{-6 - 3}{2 - 8}$ $\dfrac{3}{2}$ **78.** $\dfrac{4 - 5}{-1 - 0}$ 1 **79.** $\dfrac{-8 - (-2)}{-3 - (-2)}$ 6

80. $\dfrac{12 - 3}{10 - 9}$ 9 **81.** $\dfrac{0 - 6}{5 - 0}$ $-\dfrac{6}{5}$ **82.** $\dfrac{2 - 2}{3 - 5}$ 0

73. $(-\infty, -4) \cup (-1, \infty)$ **76.** $\left(-\infty, -\dfrac{3}{7}\right] \cup [1, \infty)$

3.4

THE SLOPE OF A LINE

TAPE IA 3.4

O B J E C T I V E S

1. Find the slope of a line given two points on the line.
2. Find the slope of a line given the equation of a line.
3. Compare the slopes of parallel and perpendicular lines.
4. Find the slopes of horizontal and vertical lines.

1. You may have noticed by now that different lines often tilt differently as shown next.

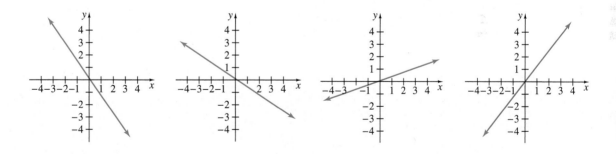

It is very important in many fields to be able to measure and compare the tilt, or steepness, of lines. For example, if the *x*-axes shown here represent years and the *y*-axes measure thousands of dollars, most business owners would prefer the graph of their profit to look like the first quadrant of the graph on the far right rather than the other graphs. In mathematics, the steepness, or tilt, of a line is also known as its **slope.** We measure the slope of a line as a ratio of **vertical change** to **horizontal change.** Slope is usually designated by the letter *m*.

Suppose that we want to measure the slope of the following line.

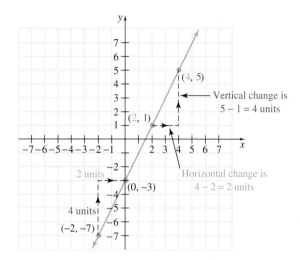

The vertical change between both pairs of points on the line is 4 units per horizontal change of 2 units. Then

$$\text{slope } m = \frac{\text{change in } y \text{ (vertical change)}}{\text{change in } x \text{ (horizontal change)}} = \frac{4}{2} = 2$$

Notice that slope is a rate of change between points. A slope of 2 or $\frac{2}{1}$ means that between pairs of points on the line, the rate of change is a vertical change of 2 units per horizontal change of 1 unit.

In general, consider the following line, which passes through the points (x_1, y_1) and (x_2, y_2). (The notation x_1 is read "x-sub-one.") The vertical change, or *rise*, between these points is the difference in the y-coordinates: $y_2 - y_1$. The horizontal change, or *run*, between the points is the difference of the x-coordinates: $x_2 - x_1$.

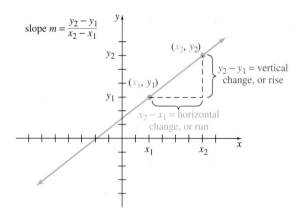

SLOPE OF A LINE

Given a line passing through points (x_1, y_1) and (x_2, y_2), the slope m of the line is

$$m = \frac{\text{rise}}{\text{run}} = \frac{y_2 - y_1}{x_2 - x_1}, \qquad \text{as long as } x_2 \neq x_1$$

EXAMPLE 1   Find the slope of the line containing the points $(0, 3)$ and $(2, 5)$. Graph the line.

Solution: Use the slope formula. It does not matter which point we call (x_1, y_1) and which point we call (x_2, y_2). Let $(x_1, y_1) = (0, 3)$ and $(x_2, y_2) = (2, 5)$.

$$m = \frac{y_2 - y_1}{x_2 - x_1}$$
$$= \frac{5 - 3}{2 - 0} = \frac{2}{2} = 1$$

Notice in this example that the slope is positive and that the graph of the line containing $(0, 3)$ and $(2, 5)$ moves upward, or increases, as we go from left to right.

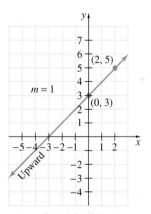

Increasing line,
positive slope

REMINDER   When we are trying to find the slope of a line through two given points, it makes no difference which given point is called (x_1, y_1) and which is called (x_2, y_2). Once an x-coordinate is called x_1, however, make sure its corresponding y-coordinate is called y_1.

EXAMPLE 2 Find the slope of the line containing the points $(5, -7)$ and $(-3, 6)$. Graph the line.

Solution: Use the slope formula. Let $(x_1, y_1) = (5, -7)$ and $(x_2, y_2) = (-3, 6)$.

$$m = \frac{y_2 - y_1}{x_2 - x_1}$$

$$= \frac{6 - (-7)}{-3 - 5} = \frac{13}{-8} = -\frac{13}{8}$$

Notice in this example that the slope is negative and that the graph of the line through $(5, -7)$ and $(-3, 6)$ moves downward, or decreases, as we go from left to right.

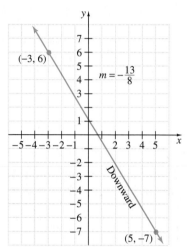

Decreasing line,
negative slope

2 As we have seen, the slope of a line is defined by two points on the line. Thus, if we know the equation of a line, we can find its slope.

EXAMPLE 3 Find the slope of the line whose equation is $f(x) = \frac{2}{3}x + 4$.

Solution: Two points are needed on the line defined by $f(x) = \frac{2}{3}x + 4$ or $y = \frac{2}{3}x + 4$ to find its slope. We will use intercepts as our two points.

If $x = 0$, then   If $y = 0$, then

$$y = \frac{2}{3} \cdot 0 + 4 \qquad\qquad 0 = \frac{2}{3}x + 4$$

$$y = 4 \qquad\qquad\qquad -4 = \frac{2}{3}x \qquad\qquad \text{Subtract 4.}$$

$$\frac{3}{2}(-4) = \frac{3}{2} \cdot \frac{2}{3}x \qquad\qquad \text{Multiply by } \frac{3}{2}.$$

$$-6 = x$$

Use the points $(0, 4)$ and $(-6, 0)$ to find the slope. Let (x_1, y_1) be $(0, 4)$ and (x_2, y_2) be $(-6, 0)$. Then

$$m = \frac{y_2 - y_1}{x_2 - x_1} = \frac{0 - 4}{-6 - 0} = \frac{-4}{-6} = \frac{2}{3}$$

Analyzing the results of Example 3, you may notice a striking pattern:

The slope of $y = \frac{2}{3}x + 4$ is $\frac{2}{3}$, the same as the coefficient of x.

Also, the y-intercept is 4, the same as the constant term, as expected.

When a linear equation is written in the form $f(x) = mx + b$ or $y = mx + b$, m is the slope of the line and b is its y-intercept. The form $y = mx + b$ is appropriately called the **slope–intercept form.**

SLOPE–INTERCEPT FORM

When a linear equation in two variables is written in slope–intercept form,

$$y = mx + b$$

then m is the slope of the line and b is the y-intercept of the line.

EXAMPLE 4 Find the slope and the y-intercept of the line whose equation is $3x - 4y = 4$.

Solution: Write the equation in slope–intercept form by solving for y.

$$3x - 4y = 4$$
$$-4y = -3x + 4 \qquad \text{Subtract } 3x \text{ from both sides.}$$
$$\frac{-4y}{-4} = \frac{-3x}{-4} + \frac{4}{-4} \qquad \text{Divide both sides by } -4.$$
$$y = \frac{3}{4}x - 1 \qquad \text{Simplify.}$$

The coefficient of x, $\frac{3}{4}$, is the slope, and the constant term -1 is the y-intercept.

---

3 Slopes of lines can help us determine whether lines are parallel. Parallel lines are distinct lines with the same steepness, so it follows that they have the same slope.

PARALLEL LINES

Nonvertical parallel lines have the same slope.

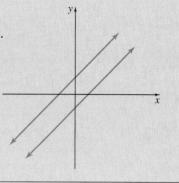

How do the slopes of perpendicular lines compare? Two lines intersecting at right angles are called **perpendicular lines.** The product of the slopes of two perpendicular lines is -1.

PERPENDICULAR LINES

If the product of the slopes of two lines is -1, then the lines are perpendicular.

(Two nonvertical lines are perpendicular if the slope of one is the negative reciprocal of the slope of the other.)

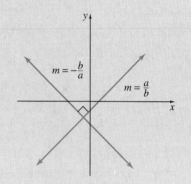

EXAMPLE 5   Are the following pairs of lines parallel, perpendicular, or neither?

a. $3x + 7y = 4$
       $6x + 14y = 7$

b. $-x + 3y = 2$
       $2x + 6y = 5$

Solution: Find the slope of each line by solving each equation for y.

a. $3x + 7y = 4$ $6x + 14y = 7$

$$7y = -3x + 4 \qquad\qquad 14y = -6x + 7$$

$$\frac{7y}{7} = \frac{-3x}{7} + \frac{4}{7} \qquad\qquad \frac{14y}{14} = \frac{-6x}{14} + \frac{7}{14}$$

$$y = -\frac{3}{7}x + \frac{4}{7} \qquad\qquad y = -\frac{3}{7}x + \frac{1}{2}$$

The slope of both lines is $-\dfrac{3}{7}$. The y-intercept of one line is $\dfrac{4}{7}$, whereas the y-intercept of the other line is $\dfrac{1}{2}$. Since these lines have the same slope and different y-intercepts, the lines are parallel.

b. $-x + 3y = 2$

$$3y = x + 2$$
$$\frac{3y}{3} = \frac{x}{3} + \frac{2}{3}$$
$$y = \frac{1}{3}x + \frac{2}{3}$$

$2x + 6y = 5$

$$6y = -2x + 5$$
$$\frac{6y}{6} = \frac{-2x}{6} + \frac{5}{6}$$
$$y = -\frac{1}{3}x + \frac{5}{6}$$

The slope of the line $-x + 3y = 2$ is $\dfrac{1}{3}$, and the slope of the line $2x + 6y = 5$ is $-\dfrac{1}{3}$. The slopes are not equal, so the lines are not parallel. The product of the slopes is $\dfrac{1}{3} \cdot -\dfrac{1}{3} = -\dfrac{1}{9}$, not -1, so the lines are not perpendicular. They are neither parallel nor perpendicular.

   Next we find the slopes of vertical and horizontal lines.

EXAMPLE 6   Find the slope of the line $x = -5$.

Solution: Recall that the graph of $x = -5$ is a vertical line with x-intercept -5.

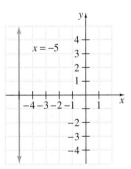

To find the slope, find two ordered pair solutions of $x = -5$. Solutions of $x = -5$ must have an x-value of -5.

Let $(x_1, y_1) = (-5, 0)$ and $(x_2, y_2) = (-5, 4)$. Then

$$m = \frac{y_2 - y_1}{x_2 - x_1} = \frac{4 - 0}{-5 - (-5)} = \frac{4}{0}$$

Since $\dfrac{4}{0}$ is undefined, we say that the slope of the vertical line $x = -5$ is undefined. Since all vertical lines are parallel, we can say that all **vertical lines have undefined slope.**

EXAMPLE 7   Find the slope of the line $y = 2$.

Solution: Find two points on the line, such as $(0, 2)$ and $(1, 2)$, and use these points to find the slope.

$$m = \frac{2 - 2}{1 - 0} = \frac{0}{1} = 0$$

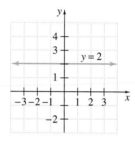

The slope of the line $y = 2$ is 0. Since all horizontal lines are parallel, we can say that all **horizontal lines have a slope of 0.**

> R E M I N D E R   Slope of 0 and undefined slope are not the same. Vertical lines have undefined slope or no slope, whereas horizontal lines have slope of 0.

The following four graphs summarize the overall appearance of lines with positive, negative, zero, or undefined slopes.

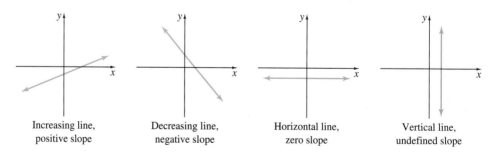

Increasing line, positive slope   Decreasing line, negative slope   Horizontal line, zero slope   Vertical line, undefined slope

The graphs of $y = \frac{1}{2}x + 1$ and $y = 5x + 1$ are shown below. Recall that the graph of $y = \frac{1}{2}x + 1$ has a slope of $\frac{1}{2}$ and that the graph of $y = 5x + 1$ has a slope of 5.

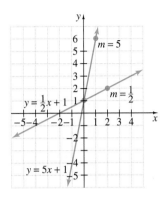

Notice that the line with the slope of 5 is steeper than the line with the slope of $\frac{1}{2}$. This is true in general for positive slopes.

For a line with positive slope m, as m increases, the line becomes steeper.

GRAPHING CALCULATOR EXPLORATIONS

Many graphing calculators have a TRACE feature. This feature allows you to trace along a graph and see the corresponding x- and y-coordinates appear on the screen. Use this feature for the following exercises.

Graph each function and then use the TRACE feature to complete each ordered pair solution. (Many times the tracer will not show an exact x- or y-value asked for. In each case, trace as closely as you can to the given x- or y-coordinate and approximate the other, unknown, coordinate to one decimal place.)

1. $y = 2.3x + 6.7$
$x = 5.1, y = ?$ 18.4

2. $y = -4.8x + 2.9$
$x = -1.8, y = ?$ 11.5

3. $y = -5.9x - 1.6$
$x = ?, y = 7.2$ −1.5

4. $y = 0.4x - 8.6$
$x = ?, y = -4.4$ 10.5

5. $y = x^2 + 5.2x - 3.3$
$x = 2.3, y = ?$
$x = ?, y = 36$ 14.0; 4.2, −9.4
(There will be two answers here.)

6. $y = 5x^2 - 6.2x - 8.3$
$x = 3.2, y = ?$
$x = ?, y = 12$ 23.1; 2.7, −1.5
(There will be two answers here.)

For Graphing Calculator Exercises 1–6, see Appendix D.

MENTAL MATH

Decide whether a line with the given slope moves upward, downward, horizontally, or vertically from left to right.

1. $m = \dfrac{7}{6}$ upward **2.** $m = -3$ downward **3.** $m = 0$ horizontally **4.** m is undefined. vertically

EXERCISE SET 3.4

Find the slope of the line that goes through the given points. See Examples 1 and 2.

1. $(3, 2), (8, 11)$ $\dfrac{9}{5}$ **2.** $(1, 6), (7, 11)$ $\dfrac{5}{6}$

3. $(3, 1), (1, 8)$ $-\dfrac{7}{2}$ **4.** $(2, 9), (6, 4)$ $-\dfrac{5}{4}$

5. $(-2, 8), (4, 3)$ $-\dfrac{5}{6}$ **6.** $(3, 7), (-2, 11)$ $-\dfrac{4}{5}$

7. $(-2, -6), (4, -4)$ $\dfrac{1}{3}$ **8.** $(-3, -4), (-1, 6)$ 5

9. $(-3, -1), (-12, 11)$ **10.** $(3, -1), (-6, 5)$ $-\dfrac{2}{3}$

11. $(-2, 5), (3, 5)$ 0 **12.** $(4, 2), (4, 0)$ undefined

13. $(-1, 1), (-1, -5)$ **14.** $(-2, -5), (3, -5)$ 0

15. $(0, 6), (-3, 0)$ 2 **16.** $(5, 2), (0, 5)$ $-\dfrac{3}{5}$

17. $(-1, 2), (-3, 4)$ -1 **18.** $(3, -2), (-1, -6)$ 1

Two lines are graphed on each set of axes. Decide whether l_1 or l_2 has the greater slope.

19.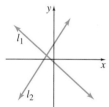

20.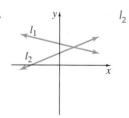

21.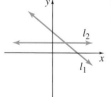

22.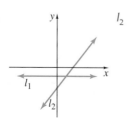

23.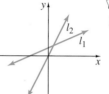

24.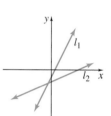

25. Each line below has negative slope.

  a. Find the slope of each line.

  b. Use the result of part (a) to fill in the blank. For lines with negative slopes, the steeper line has the _____ (greater/lesser) slope.   lesser

25a. $l_1: -2, l_2: -1, l_3: -\dfrac{2}{3}$ **26.** $m = 5, b = -2$

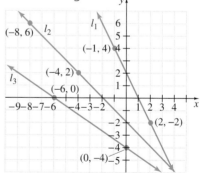

Find the slope and the y-intercept of each line. See Examples 3 and 4.

26. $f(x) = 5x - 2$ **27.** $f(x) = -2x + 6$

28. $2x + y = 7$ **29.** $-5x + y = 10$

30. $2x - 3y = 10$ **31.** $-3x - 4y = 6$

32. $f(x) = \dfrac{1}{2}x$ **33.** $f(x) = -\dfrac{1}{4}x$

9. $-\dfrac{4}{3}$ **13.** undefined **27.** $m = -2, b = 6$ **28.** $m = -2, b = 7$ **29.** $m = 5, b = 10$ **30.** $m = \dfrac{2}{3}, b = -\dfrac{10}{3}$

31. $m = -\dfrac{3}{4}, b = -\dfrac{3}{2}$ **32.** $m = \dfrac{1}{2}, 6 = 0$ **33.** $m = -\dfrac{1}{4}, b = 0$

38. undefined   **41.** undefined   **42.** undefined   **46.** $m = -1, b = 5$

Match each graph with its equation.

A

B

C

D

34. $f(x) = 2x + 3$ A   **35.** $f(x) = 2x - 3$ D
36. $f(x) = -2x + 3$ B   **37.** $f(x) = -2x - 3$ C

Find the slope of each line. See Examples 6 and 7.

38. $x = 1$ **39.** $y = -2$ 0   **40.** $y = -3$ 0

41. $x = 4$ **42.** $x + 2 = 0$ 0   **43.** $y - 7 = 0$ 0

44. Explain how merely looking at a line can tell us whether its slope is negative, positive, undefined, or zero.

45. Explain why the graph of $y = b$ is a horizontal line.

Find the slope and the y-intercept of each line.

46. $f(x) = -x + 5$ **47.** $f(x) = x + 2$ $m = 1, b = 2$

48. $-6x + 5y = 30$ **49.** $4x - 7y = 28$

50. $3x + 9 = y$ **51.** $2y - 7 = x$ $m = \frac{1}{2}, b = \frac{7}{2}$

52. $y = 4$ $m = 0, b = 4$ **53.** $x = 7$

54. $f(x) = 7x$ $m = 7, b = 0$ **55.** $f(x) = \frac{1}{7}x$ $m = \frac{1}{7}, b = 0$

56. $6 + y = 0$ **57.** $x - 7 = 0$

58. $2 - x = 3$ **59.** $2y + 4 = -7$ $m = 0, b = -\frac{11}{2}$

Determine whether the lines are parallel, perpendicular, or neither. See Example 5.

60. $f(x) = -3x + 6$ **61.** $f(x) = 5x - 6$
    $g(x) = 3x + 5$ neither $g(x) = 5x + 2$ parallel

62. $-4x + 2y = 5$ **63.** $2x - y = -10$
    $2x - y = 7$ parallel $2x + 4y = 2$

64. $-2x + 3y = 1$ **65.** $x + 4y = 7$
    $3x + 2y = 12$ $2x - 5y = 0$ neither

66. Explain whether two lines, both with positive slopes, can be perpendicular.

67. Explain why it is reasonable that nonvertical parallel lines have the same slope.

Determine the slope of each line.

68.

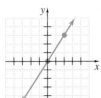

$\frac{3}{2}$ **69.**

-3

70.

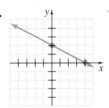

$-\frac{1}{2}$ **71.**

1

Find each slope.

72. Find the pitch, or slope, of the roof shown. $m = \frac{2}{3}$

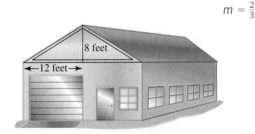

8 feet

←12 feet→

73. Upon takeoff, a Delta Airlines jet climbs to 3 miles as it passes over 25 miles of land below it. Find the slope of its climb. $\frac{3}{25}$

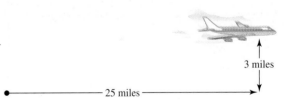

3 miles

25 miles

74. Driving down Bald Mountain in Wyoming, Bob Dean finds that he descends 1600 feet in elevation by the time he is 2.5 miles (horizontally) away from the high point on the mountain road. Find the slope of his descent (1 mile = 5280 feet).

48. $m = \frac{6}{5}, b = 6$ **49.** $m = \frac{4}{7}, b = -4$ **50.** $m = 3, b = 9$ **53.** slope is undefined, no *y*-intercept   **56.** $m = 0, b = -6$
57. slope is undefined, no *y*-intercept   **58.** slope is undefined, no *y*-intercept   **63.** perpendicular   **64.** perpendicular
74. approximately -0.12

75. Find the grade, or slope, of the road shown. $\frac{3}{20}$

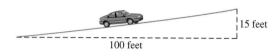

15 feet

100 feet

76. Find the slope of a line parallel to the line $f(x) = -\frac{7}{2}x - 6.$ $-\frac{7}{2}$

77. Find the slope of a line parallel to the line $f(x) = x.$ 1

78. Find the slope of a line perpendicular to the line $f(x) = -\frac{7}{2}x - 6.$ $\frac{2}{7}$

79. Find the slope of a line perpendicular to the line $f(x) = x.$ -1

80. Find the slope of a line parallel to the line $5x - 2y = 6.$ $\frac{5}{2}$

81. Find the slope of a line parallel to the line $-3x + 4y = 10.$ $\frac{3}{4}$

82. Find the slope of a line perpendicular to the line $5x - 2y = 6.$ $-\frac{2}{5}$

83. The following graph shows the altitude of a seagull in flight over a time period of 30 seconds.

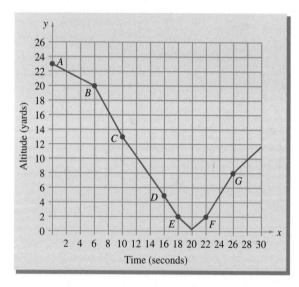

a. Find the coordinates of point B.   (6, 20)

b. Find the coordinates of point C.   (10, 13)

c. Find the rate of change of altitude between points B and C. (Recall that the rate of change between points is the slope between points. This rate of change will be in yards per second.)

84. Find the rate of change of altitude (in yards per second) between points F and G.

85. Support the result of Exercise 62 by graphing the pair of equations on a graphing calculator.

86. Support the result of Exercise 63 by graphing the pair of equations on a graphing calculator. (*Hint:* Use the window showing $[-15, 15]$ on the x-axis and $[-10, 10]$ on the y-axis.)   See Appendix D.

87. a. On a single screen, graph $y = \frac{1}{2}x + 1$, $y = x + 1$, and $y = 2x + 1$. Notice the change in slope for each graph.   See Appendix D.

b. On a single screen, graph $y = -\frac{1}{2}x + 1$, $y = -x + 1$, and $y = -2x + 1$. Notice the change in slope for each graph.   See Appendix D.

c. Determine whether the following statement is true or false for slope m of a given line. As $|m|$ becomes greater, the line becomes steeper.   true

Review Exercises

Recall the formula

$$Probability\ of\ an\ event = \frac{number\ of\ ways\ that\ the\ event\ can\ occur}{number\ of\ possible\ outcomes}$$

Suppose these cards are shuffled and one card is turned up. Find the possibility of selecting each letter.

88. $P(R)$ $\frac{1}{11}$ **89.** $P(B)$ $\frac{2}{11}$ **90.** $P(E)$ 0   **91.** $P(I$ or $T)$

92. P(selecting a letter of the alphabet)   1

93. P(vowel) $\frac{4}{11}$

Simplify and solve for y. See Section 2.3.

94. $y - 2 = 5(x + 6)$ **95.** $y - 0 = -3[x - (-10)]$

96. $y - (-1) = 2(x - 0)$ $y = 2x - 1$

97. $y - 9 = -8[x - (-4)]$ $y = -8x - 23$

83c. $-\frac{7}{4}$ or -1.75 yards per second   **84.** $\frac{3}{2}$ or 1.5 yards per second   **85.** See Appendix D.   **91.** $\frac{3}{11}$

94. $y = 5x + 32$ **95.** $y = -3x - 30$

3.5 | EQUATIONS OF LINES

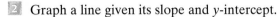

TAPE IA 3.5

O B J E C T I V E S

1. Use the slope–intercept form to find the equation of a line.
2. Graph a line given its slope and y-intercept.
3. Use the point–slope form to find the equation of a line.
4. Write equations of vertical and horizontal lines.
5. Find equations of parallel and perpendicular lines.

1 In the last section, we learned that the slope–intercept form of a linear equation is $y = mx + b$. When an equation is written in this form, the slope of the line is the same as the coefficient m of x. Also, the y-intercept of the line is the same as the constant term b. For example, the slope of the line defined by $y = 2x + 3$ is 2, and its y-intercept is 3.

We may also use the slope–intercept form to write the equation of a line given its slope and y-intercept.

EXAMPLE 1 Write an equation of the line with y-intercept -3 and slope of $\frac{1}{4}$.

Solution: We are given the slope and the y-intercept. Let $m = \frac{1}{4}$ and $b = -3$, and write the equation in slope–intercept form, $y = mx + b$.

$$y = mx + b$$

$$y = \frac{1}{4}x + (-3) \qquad \text{Let } m = \frac{1}{4} \text{ and } b = -3.$$

$$y = \frac{1}{4}x - 3 \qquad \text{Simplify.}$$

2 Given the slope and y-intercept of a line, we may graph the line as well as write its equation. Let's graph the line from Example 1. We are given that it has slope $\frac{1}{4}$ and that its y-intercept is -3. First plot the y-intercept point $(0, -3)$. To find another point on the line, recall that slope is $\dfrac{\text{rise}}{\text{run}} = \dfrac{1}{4}$. Another point may then be plotted by starting at $(0, -3)$, rising 1 unit up, and then running 4 units to the right. We are now at the point $(4, -2)$. The graph of $y = \frac{1}{4}x - 3$, which follows, is the line through points $(0, -3)$ and $(4, -2)$.

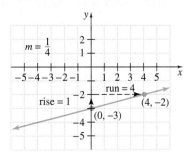

EXAMPLE 2   Graph the line through $(-1, 5)$ with slope -2.

Solution: To graph the line, we need two points. One point is $(-1, 5)$, and we will use the slope -2, which can be written as $\dfrac{-2}{1}$, to find another point.

$$m = \frac{\text{rise}}{\text{run}} = \frac{-2}{1}$$

To find another point, start at $(-1, 5)$ and move vertically 2 units down, since the numerator of the slope is -2; then move horizontally 1 unit to the right since the denominator of the slope is 1. We arrive at the point $(0, 3)$. The line through $(-1, 5)$ and $(0, 3)$ will have the required slope of -2.

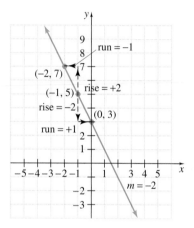

The slope -2 can also be written as $\dfrac{2}{-1}$, so to find another point we could start at $(-1, 5)$ and move 2 units up and then 1 unit left. We would stop at the point $(-2, 7)$. The line through $(-1, 5)$ and $(-2, 7)$ will have the required slope and will be the same line as shown previously through $(-1, 5)$ and $(0, 3)$.

3   When the slope of a line and a point on the line are known, the equation of the line can also be found. To do this, use the slope formula to write the slope of a line that passes through points (x, y) and (x_1, y_1). We have

$$m = \frac{y - y_1}{x - x_1}$$

Multiply both sides of this equation by $x - x_1$ to obtain

$$y - y_1 = m(x - x_1)$$

This form is called the **point–slope form** of the equation of a line.

POINT–SLOPE FORM OF THE EQUATION OF A LINE

The point–slope form of the equation of a line is $y - y_1 = m(x - x_1)$, where m is the slope of the line and (x_1, y_1) is a point on the line.

EXAMPLE 3   Find an equation of the line with slope -3 containing the point $(1, -5)$. Write the equation in standard form: $Ax + By = C$.

Solution: Because we know the slope and a point of the line, we use the point–slope form with $m = -3$ and $(x_1, y_1) = (1, -5)$.

$$y - y_1 = m(x - x_1) \qquad \text{Point–slope form.}$$
$$y - (-5) = -3(x - 1) \qquad \text{Let } m = -3 \text{ and } (x_1, y_1) = (1, -5).$$
$$y + 5 = -3x + 3 \qquad \text{Apply the distributive property.}$$
$$3x + y = -2 \qquad \text{Write in standard form.}$$

In standard form, the equation is $3x + y = -2$.

EXAMPLE 4   Find an equation of the line through points $(4, 0)$ and $(-4, -5)$. Write the equation using function notation.

Solution: First, find the slope of the line.

$$m = \frac{-5 - 0}{-4 - 4} = \frac{-5}{-8} = \frac{5}{8}$$

Next, make use of the point–slope form. Replace (x_1, y_1) by either $(4, 0)$ or $(-4, -5)$ in the point–slope equation. We will choose the point $(4, 0)$. The line through $(4, 0)$ with slope $\frac{5}{8}$ is

$$y - y_1 = m(x - x_1) \qquad \text{Point–slope form.}$$
$$y - 0 = \frac{5}{8}(x - 4) \qquad \text{Let } m = \frac{5}{8} \text{ and } (x_1, y_1) = (4, 0).$$
$$8y = 5(x - 4) \qquad \text{Multiply both sides by 8.}$$
$$8y = 5x - 20 \qquad \text{Apply the distributive property.}$$
$$-5x + 8y = -20 \qquad \text{Subtract } 5x \text{ from both sides.}$$

The standard form of the line is $-5x + 8y = -20$. To write the equation using function notation, we solve for y.

$$-5x + 8y = -20$$

$$8y = 5x - 20 \qquad \text{Add } 5x \text{ to both sides.}$$

$$y = \frac{5}{8}x - \frac{20}{8} \qquad \text{Divide both sides by 8.}$$

$$f(x) = \frac{5}{8}x - \frac{5}{2} \qquad \text{Write using function notation.}$$

> **REMINDER** Multiply both sides of the equation $-5x + 8y = -20$ by -1, and it becomes $5x - 8y = 20$. Both equations are in standard form, and their graphs are the same line.

---

4 A few special types of linear equations are linear equations whose graphs are vertical and horizontal lines.

EXAMPLE 5 Find the equation of the horizontal line containing the point $(2, 3)$.

Solution: Recall that a horizontal line has an equation of the form $y = b$. Since the line contains the point $(2, 3)$, the equation is $y = 3$.

EXAMPLE 6 Find the equation of the line containing the point $(2, 3)$ with undefined slope.

Solution: Since the line has undefined slope, the line must be vertical. A vertical line has an equation of the form $x = c$, and since the line contains the point $(2, 3)$, the equation is $x = 2$.

5 Next, we find equations of parallel and perpendicular lines.

EXAMPLE 7 Find an equation of the line containing the point $(4, 4)$ and parallel to the line $2x + 3y = -6$. Write the equation in standard form.

Solution: Because the line we want to find is *parallel* to the line $2x + 3y = -6$, the two lines must have equal slopes. Find the slope of $2x + 3y = -6$ by writing it in the form $y = mx + b$.

$$2x + 3y = -6$$

$$3y = -2x - 6 \qquad \text{Subtract } 2x \text{ from both sides.}$$

$$y = \frac{-2x}{3} - \frac{6}{3} \qquad \text{Divide by 3.}$$

$$y = -\frac{2}{3}x - 2 \qquad \text{Write in slope–intercept form.}$$

The slope of this line is $-\dfrac{2}{3}$. Thus, a line parallel to this line will also have a slope of $-\dfrac{2}{3}$. The equation we are asked to find describes a line containing the point $(4, 4)$ with a slope of $-\dfrac{2}{3}$. We use the point–slope form.

$$y - y_1 = m(x - x_1)$$

$$y - 4 = -\frac{2}{3}(x - 4) \qquad \text{Let } m = -\frac{2}{3}, x_1 = 4, \text{ and } y_1 = 4.$$

$$3(y - 4) = -2(x - 4) \qquad \text{Multiply both sides by 3.}$$

$$3y - 12 = -2x + 8 \qquad \text{Apply the distributive property.}$$

$$2x + 3y = 20 \qquad \text{Write in standard form.}$$

EXAMPLE 8 Write a function that describes the line containing the point $(4, 4)$ and is perpendicular to the line $2x + 3y = -6$.

Solution: Recall that the slope of the line $2x + 3y = -6$ is $-\dfrac{2}{3}$. A line perpendicular to this line will have a slope that is the negative reciprocal of $-\dfrac{2}{3}$, or $\dfrac{3}{2}$. From the point–slope equation, we have

$$y - y_1 = m(x - x_1)$$

$$y - 4 = \frac{3}{2}(x - 4) \qquad \text{Let } x_1 = 4, y_1 = 4, \text{ and } m = \frac{3}{2}.$$

$$2(y - 4) = 3(x - 4) \qquad \text{Multiply both sides by 2.}$$

$$2y - 8 = 3x - 12 \qquad \text{Apply the distributive property.}$$

$$2y = 3x - 4 \qquad \text{Add 8 to both sides.}$$

$$y = \frac{3}{2}x - 2 \qquad \text{Divide both sides by 2.}$$

$$f(x) = \frac{3}{2}x - 2 \qquad \text{Write using function notation.}$$

EXAMPLE 9 Southern Star Realty is an established real estate company that has enjoyed constant growth in sales since 1988. In 1990 the company sold 200 houses, and in 1995 the company sold 275 houses. Use these figures to predict the number of houses this company will sell in the year 2000.

Solution: **1.** UNDERSTAND. Read and reread this problem.

2. ASSIGN. Let

x = the number of years after 1988 and

y = the number of houses sold in x years.

The information provided then gives the ordered pairs (2, 200) and (7, 275).

3. ILLUSTRATE. To illustrate the sales of Southern Star Realty, we graph the linear equation that passes through the points (2, 200) and (7, 275).

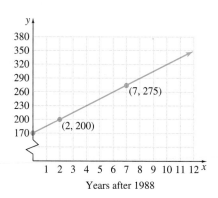

4. TRANSLATE. Write a linear equation that passes through the points (2, 200) and (7, 275). To do so, first find the slope of the line.

$$m = \frac{275 - 200}{7 - 2} = \frac{75}{5} = 15$$

Then, using the point–slope form to write the equation, we have

$$y - y_1 = m(x - x_1)$$
$$y - 200 = 15(x - 2) \qquad \text{Let } m = 15 \text{ and } (x_1, y_1) = (2, 200).$$
$$y - 200 = 15x - 30 \qquad \text{Multiply.}$$
$$y = 15x + 170 \qquad \text{Add 200 to both sides.}$$
$$f(x) = 15x + 170 \qquad \text{Write using function notation.}$$

5. COMPLETE. To predict the number of houses sold in the year 2000, we find $f(12)$ since $2000 - 1988 = 12$.

$$f(12) = 15(12) + 170$$
$$= 350$$

6. INTERPRET. *Check* to see that the point (12, 350) is a point of the line graphed in step 3. *State:* Southern Star Realty should expect to sell 350 houses in the year 2000.

FORMS OF LINEAR EQUATIONS

$Ax + By = C$	**Standard form** of a linear equation A and B are not both 0.
$y = mx + b$	**Slope–intercept form** of a linear equation The slope is m, and the y-intercept is b.
$y - y_1 = m(x - x_1)$	**Point–slope form** of a linear equation The slope is m, and (x_1, y_1) is a point on the line.
$y = c$	**Horizontal line** The slope is 0, and the y-intercept is c.
$x = c$	**Vertical line** The slope is undefined and the x-intercept is c.

PARALLEL AND PERPENDICULAR LINES

Nonvertical parallel lines have the same slope.
The product of the slopes of two nonvertical perpendicular lines is -1.

MENTAL MATH

State the slope and the y-intercept of each line with the given equation.

1. $y = -4x + 12$ $m = -4, b = 12$ **2.** $y = \dfrac{2}{3}x - \dfrac{7}{2}$ $m = \dfrac{2}{3}, b = -\dfrac{7}{2}$ **3.** $y = 5x$ $m = 5, b = 0$

4. $y = -x$ $m = -1, b = 0$ **5.** $y = \dfrac{1}{2}x + 6$ $m = \dfrac{1}{2}, b = 6$ **6.** $y = -\dfrac{2}{3}x + 5$ $m = -\dfrac{2}{3}, b = 5$

Decide whether the lines are parallel, perpendicular, or neither.

7. $y = 12x + 6$ **8.** $y = -5x + 8$ **9.** $y = -9x + 3$ **10.** $y = 2x - 12$

$\quad y = 12x - 2$ parallel $\quad y = -5x - 8$ parallel $\quad y = \dfrac{3}{2}x - 7$ neither $\quad y = \dfrac{1}{2}x - 6$ neither

EXERCISE SET 3.5

Use the slope–intercept form of the linear equation to write the equation of each line with the given slope and y-intercept. See Example 1.

1. Slope -1; y-intercept 1 $y = -x + 1$

2. Slope $\dfrac{1}{2}$; y-intercept -6 $y = \dfrac{1}{2}x - 6$

3. Slope 2; y-intercept $\dfrac{3}{4}$ $y = 2x + \dfrac{3}{4}$

4. Slope -3; y-intercept $-\dfrac{1}{5}$ $y = -3x - \dfrac{1}{5}$

5. Slope $\dfrac{2}{7}$; y-intercept 0 $y = \dfrac{2}{7}x$

6. Slope $-\dfrac{4}{5}$; y-intercept 0 $y = -\dfrac{4}{5}x$

Graph each line passing through the given point with the given slope. See Example 2.   For Exercises 7–12, see Appendix D.

7. Through $(1, 3)$ with slope $\dfrac{3}{2}$

8. Through $(-2, -4)$ with slope $\dfrac{2}{5}$

9. Through $(0, 0)$ with slope 5

10. Through $(-5, 2)$ with slope 2

11. Through $(0, 7)$ with slope -1

12. Through $(3, 0)$ with slope -3

Find an equation of the line with the given slope and containing the given point. Write the equation in standard form. See Example 3.

13. Slope 3; through $(1, 2)$ $3x - y = 1$

14. Slope 4; through $(5, 1)$ $4x - y = 19$

15. Slope -2; through $(1, -3)$ $2x + y = -1$

16. Slope -4; through $(2, -4)$ $4x + y = 4$

17. Slope $\dfrac{1}{2}$; through $(-6, 2)$ $x - 2y = -10$

18. Slope $\dfrac{2}{3}$; through $(-9, 4)$ $2x - 3y = -30$

19. Slope $-\dfrac{9}{10}$; through $(-3, 0)$ $9x + 10y = -27$

20. Slope $-\dfrac{1}{5}$; through $(4, -6)$ $x + 5y = -26$

21. $2x + y = 3$ **22.** $2x - y = 2$ **23.** $2x - 3y = -7$

Find an equation of each line graphed. Write the equation in standard form.

21.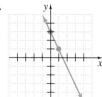

22.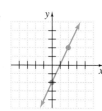

23.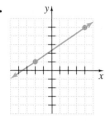

24.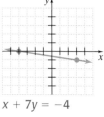

$x + 7y = -4$

Find an equation of the line passing through the given points. Use function notation to write the equation. See Example 4.   **25.** $f(x) = 3x - 6$ **26.** $f(x) = 2x - 6$

25. $(2, 0), (4, 6)$ **26.** $(3, 0), (7, 8)$

27. $(-2, 5), (-6, 13)$ **28.** $(7, -4), (2, 6)$

29. $(-2, -4), (-4, -3)$ **30.** $(-9, -2), (-3, 10)$

31. $(-3, -8), (-6, -9)$ **32.** $(8, -3), (4, -8)$

33. Describe how to check to see if the graph of $2x - 4y = 7$ passes through the points $(1.4, -1.05)$ and $(0, -1.75)$. Then follow your directions and check these points.   Answers will vary.

Use the graph of the following function $f(x)$ to find each value.

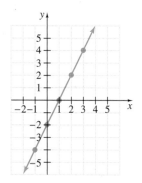

34. $f(1)$ 0   **35.** $f(0)$ -2 **36.** $f(-1)$ -4

37. $f(2)$ 2   **38.** Find x such that $f(x) = 4$.   3

39. Find x such that $f(x) = -6$. -2

Write an equation of each line. See Examples 5 and 6.

40. Vertical; through $(2, 6)$ $x = 2$

41. Slope 0; through $(-2, -4)$ $y = -4$

42. Horizontal; through $(-3, 1)$ $y = 1$

43. Vertical; through $(4, 7)$ $x = 4$

44. Undefined slope; through $(0, 5)$ $x = 0$

45. Horizontal; through $(0, 5)$ $y = 5$

46. Answer the following true or false. A vertical line is always perpendicular to a horizontal line.   true

Find an equation of each line. Write the equation using function notation. See Examples 7 and 8.

47. Through $(3, 8)$; parallel to $f(x) = 4x - 2$

48. Through $(1, 5)$; parallel to $f(x) = 3x - 4$

49. Through $(2, -5)$; perpendicular to $3y = x - 6$

27. $f(x) = -2x + 1$ **28.** $f(x) = -2x + 10$ **29.** $f(x) = -\dfrac{1}{2}x - 5$ **30.** $f(x) = 2x + 16$ **31.** $f(x) = \dfrac{1}{3}x - 7$

32. $f(x) = \dfrac{5}{4}x - 13$ **47.** $f(x) = 4x - 4$ **48.** $f(x) = 3x + 2$ **49.** $f(x) = -3x + 1$

50. $f(x) = -\dfrac{3}{2}x + 2$ **51.** $f(x) = -\dfrac{3}{2}x - 6$ **52.** $f(x) = \dfrac{2}{3}x - \dfrac{5}{3}$ **55.** $f(x) = -x + 7$

50. Through $(-4, 8)$; perpendicular to $2x - 3y = 1$

51. Through $(-2, -3)$; parallel to $3x + 2y = 5$

52. Through $(-2, -3)$; perpendicular to $3x + 2y = 5$

Find the equation of each line. Write the equation in standard form unless indicated otherwise.

53. Slope 2; through $(-2, 3)$ $2x - y = -7$

54. Slope 3; through $(-4, 2)$ $3x - y = -14$

55. Through $(1, 6)$ and $(5, 2)$; use function notation.

56. Through $(2, 9)$ and $(8, 6)$ $x + 2y = 20$

57. With slope $-\dfrac{1}{2}$; y-intercept 11 $x + 2y = 22$

58. With slope -4; y-intercept $\dfrac{2}{9}$; use function notation. $f(x) = -4x + \dfrac{2}{9}$

59. Through $(-7, -4)$ and $(0, -6)$ $2x + 7y = -42$

60. Through $(2, -8)$ and $(-4, -3)$ $5x + 6y = -38$

61. Slope $-\dfrac{4}{3}$; through $(-5, 0)$ $4x + 3y = -20$

62. Slope $-\dfrac{3}{5}$; through $(4, -1)$ $3x + 5y = 7$

63. Vertical line; through $(-2, -10)$ $x = -2$

64. Horizontal line; through $(1, 0)$ $y = 0$

65. Through $(6, -2)$; parallel to the line $2x + 4y = 9$

66. Through $(8, -3)$; parallel to the line $6x + 2y = 5$

67. Slope 0; through $(-9, 12)$ $y = 12$

68. Undefined slope; through $(10, -8)$ $x = 10$

69. Through $(6, 1)$; parallel to the line $8x - y = 9$

70. Through $(3, 5)$; perpendicular to the line $2x - y = 8$

71. Through $(5, -6)$; perpendicular to $y = 9$ $x = 5$

72. Through $(-3, -5)$; parallel to $y = 9$ $y = -5$

73. Through $(2, -8)$ and $(-6, -5)$; use function notation.

74. Through $(-4, -2)$ and $(-6, 5)$; use function notation. $f(x) = -\dfrac{7}{2}x - 16$

Solve. See Example 9.

75. A rock is dropped from the top of a 400-foot building. After 1 second, the rock is traveling 32 feet per second. After 3 seconds, the rock is traveling 96 feet per second. Let $R(x)$ be the rate of descent and x be the number of seconds since the rock was dropped.

a. Write a linear function that relates time x to rate $R(x)$. [*Hint:* Use the ordered pairs $(1, 32)$ and $(3, 96)$ and use function notation to write an equation.] $R(x) = 32x$

b. Use this function to determine the rate of the rock 4 seconds after it was dropped.
128 feet per second

76. The Whammo Company has learned that, by pricing a newly released Frisbee at $6, sales will reach 2000 per day. Raising the price to $8 will cause the sales to fall to 1500 per day. Assume that the ratio of change in price to change in daily sales is constant, and let x be the price of the Frisbee and S be number of sales.

a. Find the linear function $S(x)$ that models the price–sales relationship for this Frisbee. [*Hint:* The line must pass through $(6, 2000)$ and $(8, 1500)$.] $S(x) = -250x + 3500$

b. Predict the daily sales of Frisbees if the price is set at $7.50.   1625

77. In 1990, the median price of an existing home in the United States was $97,500. In 1994, the median price of an existing home was $109,800. Let $P(x)$ be the median price of an existing home in the year x, where $x = 0$ represents 1990. (*Source:* National Association of REALTORS®)

a. Find the linear function $P(x)$ that models the median existing home price in terms of the year x. (See the hint for Exercise 75a.)

b. Predict the median existing home price for 1999. $125,175   **77a.** $P(x) = 3075x + 97{,}500$

78. The number of births (in thousands) in the United States in 1994 was 3797. The number of births (in thousands) in the United States in 1990 was 4158. Let $B(x)$ be the number of births (in thousands) in the year x, where $x = 0$ represents 1990. (*Source:* National Center for Health Statistics)

a. Find the linear function $B(x)$ that models the number of births (in thousands) in terms of the year x. (See the hint for Exercise 76a.)

b. Predict the number of births in the United States for the year 2000.   3,255,500 births

79. Del Monte Fruit Company recently released a new applesauce. By the end of its first year, profits on this product amounted to $30,000. The anticipated profit for the end of the fourth year is $66,000. The ratio of change in time to change in profit is constant. Let x be years and P be profit.

65. $x + 2y = 2$ **66.** $3x + y = 21$ **69.** $8x - y = 47$ **70.** $x + 2y = 13$ **73.** $f(x) = -\dfrac{3}{8}x - \dfrac{29}{4}$

78a. $B(x) = -90.25x + 4158$

79a. $P(x) = 12,000x + 18,000$ **79c.** 9 years **87.**  $(-\infty, 14]$ **88.**

a. Write a linear function $P(x)$ that expresses profit as a function of time.

b. Use this function to predict the company's profit at the end of the seventh year. $102,000

c. Predict when the profit should reach $126,000.

80. The value of a computer bought in 1990 depreciates, or decreases, as time passes. Two years after the computer was bought, it was worth $2600; 5 years after it was bought, it was worth $2000.

a. If this relationship between number of years past 1990 and value of computer is linear, write an equation describing this relationship. [Use ordered pairs of the form (years past 1990, value of computer).] $y = -200x + 3000$

b. Use this equation to estimate the value of the computer in the year 2000. $1000

81. The Pool Fun Company has learned that, by pricing a newly released Fun Noodle at $3, sales will reach 10,000 Fun Noodles per day during the summer. Raising the price to $5 will cause the sales to fall to 8000 Fun Noodles per day.

a. Assume that the relationship between sales price and number of Fun Noodles sold is linear and write an equation describing this relationship. $y = -1000x + 13,000$

b. Predict the daily sales of Fun Noodles if the price is $3.50. 9500 Fun Noodles

82. The value of a building bought in 1980 appreciates, or increases, as time passes. Seven years after the building was bought, it was worth $165,000; 12 years after it was bought, it was worth $180,000.

a. If this relationship between number of years past 1980 and value of building is linear, write an equation describing this relationship. [Use ordered pairs of the form (years past 1980, value of building).] $y = 3000x + 144,000$

b. Use this equation to estimate the value of the building in the year 2000. $204,000

Use a grapher with a TRACE feature to see the results of each exercise. For Exercises 83–86, see Appendix D.

83. Exercise 55; graph the function and verify that it passes through $(1, 6)$ and $(5, 2)$.

84. Exercise 56; graph the equation and verify that it passes through $(2, 9)$ and $(8, 6)$.

85. Exercise 61; graph the equation. See that it has a negative slope and passes through $(-5, 0)$.

86. Exercise 62; graph the equation. See that it has a negative slope and passes through $(4, -1)$.

Review Exercises

Solve and graph the solution. See Section 2.4.

87. $2x - 7 \le 21$ **88.** $-3x + 1 > 0$

89. $5(x - 2) \ge 3(x - 1)$ **90.** $-2(x + 1) \le -x + 10$

91. $\dfrac{x}{2} + \dfrac{1}{4} < \dfrac{1}{8}$ **92.** $\dfrac{x}{5} - \dfrac{3}{10} \ge \dfrac{x}{2} - 1$

92. $\left(-\infty, \dfrac{7}{3}\right]$

A Look Ahead

EXAMPLE

Find an equation of the perpendicular bisector of the line segment whose endpoints are $(2, 6)$ and $(0, -2)$.

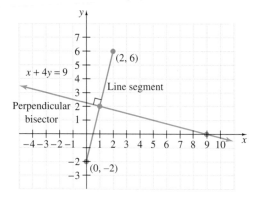

Solution:

A perpendicular bisector is a line that contains the midpoint of the given segment and is perpendicular to the segment.

Step 1. The midpoint of the segment with endpoints $(2, 6)$ and $(0, -2)$ is $(1, 2)$.

Step 2. The slope of the segment containing points $(2, 6)$ and $(0, -2)$ is 4.

Step 3. A line perpendicular to this line segment will have slope of $-\frac{1}{4}$.

Step 4. The equation of the line through the midpoint $(1, 2)$ with a slope of $-\frac{1}{4}$ will be the equation of the perpendicular bisector. This equation in standard form is $x + 4y = 9$.

89. $\left[\dfrac{7}{2}, \infty\right)$ **90.** $[-12, \infty)$ **91.** $\left(-\infty, -\dfrac{1}{4}\right)$

Find an equation of the perpendicular bisector of the line segment whose endpoints are given. See the previous example.

95. $(-2, 6); (-22, -4)$ **96.** $(5, 8); (7, 2)$

97. $(2, 3); (-4, 7)$ **98.** $(-6, 8); (-4, -2)$

93. $(3, -1); (-5, 1)$ **94.** $(-6, -3); (-8, -1)$

93. $-4x + y = 4$ **94.** $x - y = -5$ **95.** $2x + y = -23$ **96.** $x - 3y = -9$ **97.** $3x - 2y = -13$
98. $x - 5y = -20$

3.6 GRAPHING LINEAR INEQUALITIES

TAPE IA 3.6

O B J E C T I V E S

□ Graph linear inequalities.

□ Graph the intersection or union of two linear inequalities.

Recall that the graph of a linear equation in two variables is the graph of all ordered pairs that satisfy the equation, and we determined that the graph is a line. Here we graph **linear inequalities** in two variables; that is, we graph all the ordered pairs that satisfy the inequality.

If the equal sign in a linear equation in two variables is replaced with an inequality symbol, the result is a linear inequality in two variables.

EXAMPLES OF LINEAR INEQUALITIES IN TWO VARIABLES

$$3x + 5y \geq 6 \qquad 2x - 4y < -3$$
$$4x > 2 \qquad y \leq 5$$

To graph the linear inequality $x + y < 3$, for example, we first graph the related **boundary** equation $x + y = 3$. The resulting boundary line contains all ordered pairs the sum of whose coordinates is 3. This line separates the plane into two **half-planes**. All points "above" the boundary line $x + y = 3$ have coordinates that satisfy the inequality $x + y > 3$, and all points "below" the line have coordinates that satisfy the inequality $x + y < 3$.

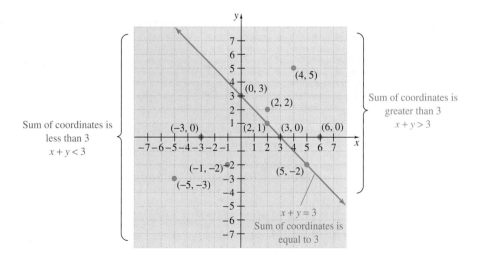

The graph, or **solution region,** for $x + y < 3$, then, is the half-plane below the boundary line and is shown shaded. The boundary line is shown dashed since it is not a part of the solution region. These ordered pairs on this line satisfy $x + y = 3$ and not $x + y < 3$.

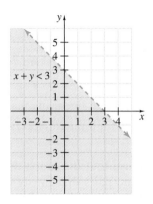

The following steps may be used to graph linear inequalities in two variables.

TO GRAPH A LINEAR INEQUALITY IN TWO VARIABLES

Step 1. Graph the boundary line found by replacing the inequality sign with an equal sign. If the inequality sign is $<$ or $>$, graph a dashed line indicating that points on the line are not solutions of the inequality. If the inequality sign is $\leq$ or $\geq$, graph a solid line indicating that points on the line are solutions of the inequality.

Step 2. Choose a **test point not on the boundary line** and substitute the coordinates of this test point into the **original inequality.**

Step 3. If a true statement is obtained in *step 2*, shade the half-plane that contains the test point. If a false statement is obtained, shade the half-plane that does not contain the test point.

EXAMPLE 1 Graph $2x - y < 6$.

Solution: First, the boundary line for this inequality is the graph of $2x - y = 6$. Graph a dashed boundary line because the inequality symbol is $<$. Next, choose a test point on either side of the boundary line. The point $(0, 0)$ is not on the boundary line, so we use this point. Replacing x with 0 and y with 0 in the *original inequality* $2x - y < 6$ leads to the following:

$$2x - y < 6$$
$$2(0) - 0 < 6 \qquad \text{Let } x = 0 \text{ and } y = 0.$$
$$0 < 6 \qquad \text{True.}$$

Because $(0, 0)$ satisfies the inequality, so does every point on the same side of the boundary line as $(0, 0)$. Shade the half-plane that contains $(0, 0)$. The half-plane graph of the inequality is shown next.

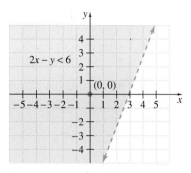

Every point in the shaded half-plane satisfies the original inequality. Notice that the inequality $2x - y < 6$ does not describe a function since its graph does not pass the vertical line test.

In general, linear inequalities of the form $Ax + By \leq C$, when A and B are not both 0, do not describe functions.

EXAMPLE 2   Graph $3x \geq y$.

Solution: First, graph the boundary line $3x = y$. Graph a solid boundary line because the inequality symbol is $\geq$. Test a point not on the boundary line to determine which half-plane contains points that satisfy the inequality. We choose $(0, 1)$ as our test point.

$$3x \geq y$$
$$3(0) \geq 1 \qquad \text{Let } x = 0 \text{ and } y = 1.$$
$$0 \geq 1 \qquad \text{False.}$$

This point does not satisfy the inequality, so the correct half-plane is on the opposite side of the boundary line from $(0, 1)$. The graph of $3x \geq y$ is the boundary line together with the shaded region shown next.

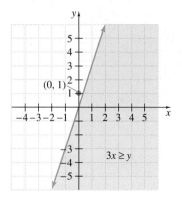

2 The intersection and the union of linear inequalities can also be graphed, as shown in the next two examples.

EXAMPLE 3 Graph the intersection of $x \geq 1$ and $y \geq 2x - 1$.

Solution: Graph each inequality. The intersection of the two graphs is all points common to both regions, as shown by the heaviest shading in the third graph.

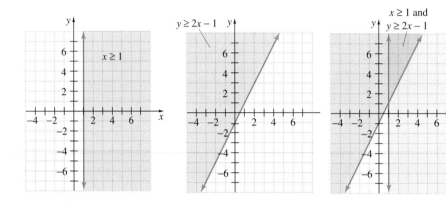

EXAMPLE 4 Graph the union of $x + \dfrac{1}{2}y \geq -4$ or $y \leq -2$.

Solution: Graph each inequality. The union of the two inequalities is both shaded regions, including the solid boundary lines shown in the third graph.

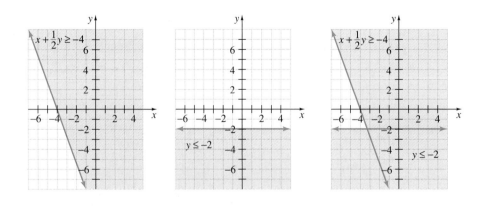

EXERCISE SET 3.6   For Exercises 1–12, see Appendix D.

Graph each inequality. See Examples 1 and 2.

1. $x < 2$ **2.** $x > -3$ **3.** $x - y \geq 7$

4. $3x + y \leq 1$ **5.** $3x + y > 6$ **6.** $2x + y > 2$

7. $y \leq -2x$ **8.** $y \leq 3x$ **9.** $2x + 4y \geq 8$

10. $2x + 6y \leq 12$ **11.** $5x + 3y > -15$

12. $2x + 5y < -20$

13. Explain when a dashed boundary line should be used in the graph of an inequality.   with < or >

14. Explain why, after the boundary line is sketched, we test a point on either side of this boundary in the original inequality.   Answers will vary.

Graph each union or intersection. See Examples 3 and 4.   For Exercises 15–46, see Appendix D.

15. The intersection of $x \geq 3$ and $y \leq -2$

16. The union of $x \geq 3$ or $y \leq -2$

17. The union of $x \leq -2$ or $y \geq 4$

18. The intersection of $x \leq -2$ and $y \geq 4$

19. The intersection of $x - y < 3$ and $x > 4$

20. The intersection of $2x > y$ and $y > x + 2$

21. The union of $x + y \leq 3$ or $x - y \geq 5$

22. The union of $x - y \leq 3$ or $x + y > -1$

Graph each inequality.

23. $y \geq -2$ **24.** $y \leq 4$ **25.** $x - 6y < 12$

26. $x - 4y < 8$ **27.** $x > 5$ **28.** $y \geq -2$

29. $-2x + y \leq 4$ **30.** $-3x + y \leq 9$

31. $x - 3y < 0$ **32.** $x + 2y > 0$

33. $3x - 2y \leq 12$ **34.** $2x - 3y \leq 9$

35. The union of $x - y \geq 2$ or $y < 5$

36. The union of $x - y < 3$ or $x > 4$

37. The intersection of $x + y \leq 1$ and $y \leq -1$

38. The intersection of $y \geq x$ and $2x - 4y \geq 6$

39. The union of $2x + y > 4$ or $x \geq 1$

40. The union of $3x + y < 9$ or $y \leq 2$

41. The intersection of $x \geq -2$ and $x \leq 1$

42. The intersection of $x \geq -4$ and $x \leq 3$

43. The union of $x + y \leq 0$ or $3x - 6y \geq 12$

44. The intersection of $x + y \leq 0$ and $3x - 6y \geq 12$

45. The intersection of $2x - y > 3$ and $x \geq 0$

46. The union of $2x - y > 3$ or $x \geq 0$

Match each inequality with its graph.

A 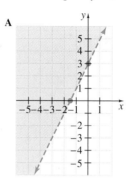   B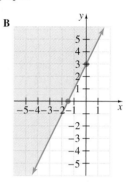

C 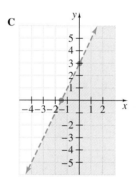   D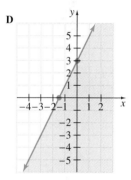

47. $y \leq 2x + 3$ D   **48.** $y < 2x + 3$ C

49. $y > 2x + 3$ A   **50.** $y \geq 2x + 3$ B

Write the inequality whose graph is given.

51. 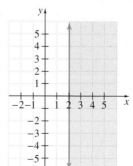 $x \geq 2$

52.

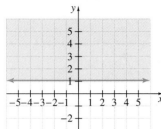

$y \geq 1$

53.

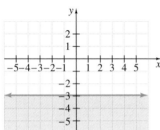

$y \leq -3$

54.

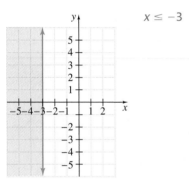

$x \leq -3$

55.

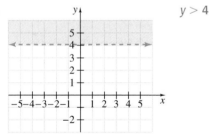

$y > 4$

56.

$y > -1$

57.

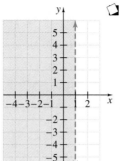

$x < 1$

58.

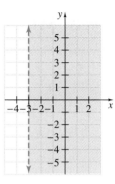

$x > -3$

Solve.

59. Rheem Abo-Zahrah decides that she will study at most 20 hours every week and that she must work at least 10 hours every week. Let x represent the hours studying and y represent the hours working. Write two inequalities that model this situation and graph their intersection.

60. The movie and TV critic for the *New York Times* spends between 2 and 6 hours daily reviewing movies and fewer than 5 hours reviewing TV shows. Let x represent the hours watching movies and y represent the time spent watching TV. Write two inequalities that model this situation and graph their intersection.

61. Chris-Craft manufactures boats out of Fiberglas and wood. Fiberglas hulls require 2 hours work, whereas wood hulls require 4 hours work. Employees work at most 40 hours a week. The following inequalities model these restrictions, where x represents the number of Fiberglas hulls produced and y represents the number of wood hulls produced.

$$\begin{cases} x \geq 0 \\ y \geq 0 \\ 2x + 4y \leq 40 \end{cases}$$

Graph the intersection of these inequalities.
See Appendix D.

Review Exercises

Evaluate each expression. See Sections 1.3 and 1.4.

62. 2^3 8   **63.** 3^2 9   **64.** -5^2 -25 **65.** $(-5)^2$ 25

66. $(-2)^4$ 16   **67.** -2^4 -16 **68.** $\left(\dfrac{3}{5}\right)^3$ $\dfrac{27}{125}$ **69.** $\left(\dfrac{2}{7}\right)^2$ $\dfrac{4}{49}$

59. $x \leq 20$ and $y \geq 10$. See Appendix D.   **60.** $2 < x < 6, 0 \leq y < 5$. See Appendix D.

Find the domain and the range of each relation. Determine whether the relation is also a function. See Section 3.2.

70.

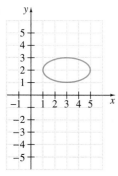

domain: [1, 5];
range: [1, 3];
no

71.

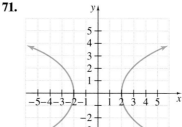

domain:
$(-\infty, 2] \cup [2, \infty)$;
range: $(-\infty, \infty)$;
no

GROUP ACTIVITY

MODELING JAPANESE AUTOMOBILE IMPORTS

OPTIONAL MATERIALS:

* Grapher with least-squares curve-fitting capabilities

The following table shows the number (in millions) of Japanese automobiles imported into the United States during the years 1986–1994.

Use the table along with your answers to the questions below to find a linear function $f(x)$ that represents the data.

JAPANESE AUTOMOBILE IMPORTS (IN MILLIONS)

YEAR	1986	1987	1988	1989	1990	1991	1992	1993	1994
x	6	7	8	9	10	11	12	13	14
IMPORTS, y	2.6	2.4	2.1	2.1	1.9	1.8	1.6	1.5	1.5

Source: U.S. Department of Commerce.

1. Plot the data given in the table as ordered pairs.

2. Use a straight edge to draw on your graph what appears to be the line that "best fits" the data you plotted.

3. Estimate the coordinates of two points that fall on your best-fitting line. Use these points to find a linear function $f(x)$ for the line.

4. Find the value of $f(19)$ and interpret its meaning in context.

(continued)

5. Compare your group's linear function with other groups' functions. Are they different? If so, explain why.

6. (Optional) Enter the data from the table into a grapher and use the linear curve-fitting feature to find a linear function for the data.

Compare this function with the one you found in question 3. How are they alike or different? Find $f(19)$, using the model you found with the grapher, and compare it with the value of $f(19)$ you found in question 4.

See Appendix D for Group Activity answers and suggestions.

CHAPTER 3   HIGHLIGHTS

DEFINITIONS AND CONCEPTS	EXAMPLES
SECTION 3.1 GRAPHING EQUATIONS	
The **rectangular coordinate system,** or **Cartesian coordinate system,** consists of a vertical and a horizontal number line on a plane intersecting at their 0 coordinate. The vertical number line is called the **y-axis,** and the horizontal number line is called the **x-axis.** The point of intersection of the axes is called the **origin.** The axes divide the plane into four regions called **quadrants.**	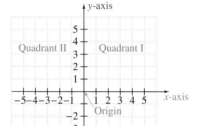
To **plot** or **graph** an ordered pair means to find its corresponding point on a rectangular coordinate system. To plot or graph the ordered pair $(-2, 5)$, start at the origin. Move 2 units to the left along the x-axis, then 5 units upward parallel to the y-axis.	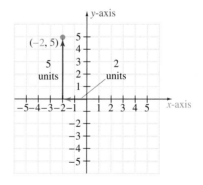
	(continued)

DEFINITIONS AND CONCEPTS	**EXAMPLES**

SECTION 3.1   GRAPHING EQUATIONS

An ordered pair is a **solution** of an equation in two variables if replacing the variables by the corresponding coordinates results in a true statement.

Determine whether $(-2, 3)$ is a solution of
$$3x + 2y = 0$$

$$3(-2) + 2(3) = 0$$
$$-6 + 6 = 0$$
$$0 = 0 \quad \text{True.}$$

$(-2, 3)$ is a solution.

A **linear equation in two variables** is an equation that can be written in the form $Ax + By = C$, where A, B, and C are real numbers and A and B are not both 0. The form $Ax + By = C$ is called **standard form.**

Linear Equations in Two Variables
$$y = -2x + 5, \quad x = 7$$

$$y - 3 = 0, \quad 6x - 4y = 10$$

$6x - 4y = 10$ is in standard form.

The graph of a linear equation in two variables is a line. To graph a linear equation in two variables, find three ordered pair solutions. Plot the solution point, and draw the line connecting the points.

Graph $3x + y = -6$.

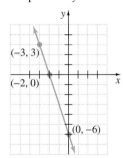

x	y
0	-6
-2	0
-3	3

To graph an equation that is not linear, find a sufficient number of ordered pair solutions so that a pattern may be discovered.

Graph $y = x^3 + 2$.

x	y
-2	-6
-1	1
0	2
1	3
2	10

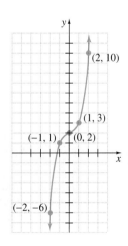

(continued)

DEFINITIONS AND CONCEPTS	EXAMPLES

SECTION 3.2 INTRODUCTION TO FUNCTIONS

A **relation** is a correspondence between a first set (set of inputs) and a second set (set of outputs) that assigns to each element of the first set an element of the second set. The **domain** of the relation is the first set, and the **range** of the relation is the set of elements that correspond to some element of the first set.

Relation

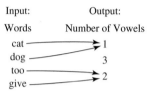

Domain: {cat, dog, too, give}
Range: {1, 2}

A **function** is a relation in which each element of the first set corresponds to exactly one element of the second set.

The previous relation is a function. Each word contains one exact number of vowels.

Vertical Line Test:

If no vertical line can be drawn so that it intersects a graph more than once, the graph is the graph of a function.

Find the domain and the range of the relation. Also determine whether the relation is a function.

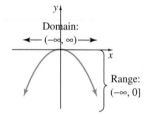

By the vertical line test, this graph is the graph of a function.

The symbol $f(x)$ means **function of x** and is called **function notation.**

If $f(x) = 2x^2 - 5$, find $f(-3)$.

$$f(-3) = 2(-3)^2 - 5 = 2(9) - 5 = 13$$

SECTION 3.3 GRAPHING LINEAR FUNCTIONS

A **linear function** is a function that can be written in the form $f(x) = mx + b$.

Linear Functions:

$$f(x) = -3, g(x) = 5x, h(x) = -\frac{1}{3}x - 7$$

To graph a linear function, find three ordered pair solutions. Graph the solutions and draw a line through the plotted points.

Graph $f(x) = -2x$.

x	y or $f(x)$
-1	2
0	0
2	-4

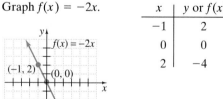

(continued)

DEFINITIONS AND CONCEPTS	EXAMPLES

SECTION 3.3   GRAPHING LINEAR FUNCTIONS

For any function $f(x)$, the graph of $y = f(x) + K$ is the same as the graph of $y = f(x)$ shifted $|K|$ units up if K is positive and $|K|$ units down if K is negative.

Graph $g(x) = -2x + 3$.

This is the same as the graph of $f(x) = -2x$ shifted 3 units up.

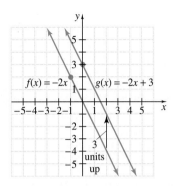

The x-coordinate of a point where a graph crosses the x-axis is called an ***x*-intercept.** The y-coordinate of a point where a graph crosses the y-axis is called a ***y*-intercept.**

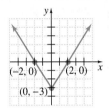

The x-intercepts of the graph are -2 and 2.
The y-intercept is -3.

To find an x-intercept, let $y = 0$ or $f(x) = 0$ and solve for x.

To find a y-intercept, let $x = 0$ and solve for y.

Graph $5x - y = -5$ by finding intercepts.

If $x = 0$, then

$$5x - y = -5$$
$$5 \cdot 0 - y = -5$$
$$-y = -5$$
$$y = 5$$

If $y = 0$, then

$$5x - y = -5$$
$$5x - 0 = -5$$
$$5x = -5$$
$$x = -1$$

Ordered pairs are $(0, 5)$ and $(-1, 0)$.

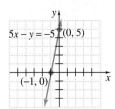

(continued)

DEFINITIONS AND CONCEPTS	EXAMPLES

SECTION 3.3   GRAPHING LINEAR FUNCTIONS

The graph of $x = c$ is a vertical line with x-intercept c.

The graph of $y = c$ is a horizontal line with y-intercept c.

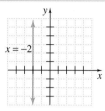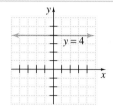

SECTION 3.4   THE SLOPE OF A LINE

The **slope** m of the line through (x_1, y_1) and (x_2, y_2) is given by

$$m = \frac{y_2 - y_1}{x_2 - x_1} \text{ as long as } x_2 \neq x_1$$

Find the slope of the line through $(-1, 7)$ and $(-2, -3)$.

$$m = \frac{y_2 - y_1}{x_2 - x_1} = \frac{-3 - 7}{-2 - (-1)} = \frac{-10}{-1} = 10$$

The **slope–intercept form** of a linear equation is $y = mx + b$, where m is the slope of the line and b is the y-intercept.

Find the slope and y-intercept of $-3x + 2y = -8$.

$$2y = 3x - 8$$
$$\frac{2y}{2} = \frac{3x}{2} - \frac{8}{2}$$
$$y = \frac{3}{2}x - 4$$

The slope of the line is $\frac{3}{2}$, and the y-intercept is -4.

Nonvertical parallel lines have the same slope.

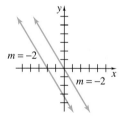

If the product of the slopes of two lines is -1, then the lines are perpendicular.

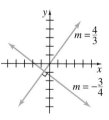

The slope of a horizontal line is 0.

The slope of a vertical line is undefined.

The slope of $y = -2$ is 0.

The slope of $x = 5$ is undefined.

(continued)

DEFINITIONS AND CONCEPTS	EXAMPLES

SECTION 3.5   EQUATIONS OF LINES

We can use the slope–intercept form to write an equation of a line given its slope and y-intercept.	Write an equation of the line with y-intercept -1 and slope $\dfrac{2}{3}$. $$y = mx + b$$ $$y = \frac{2}{3}x - 1$$
The point–slope form of the equation of a line is $y - y_1 = m(x - x_1)$, where m is the slope of the line and (x_1, y_1) is a point on the line.	Find an equation of the line with slope 2 containing the point $(1, -4)$. Write the equation in standard form: $Ax + By = C$. $$y - y_1 = m(x - x_1)$$ $$y - (-4) = 2(x - 1)$$ $$y + 4 = 2x - 2$$ $$-2x + y = -6 \qquad \text{Standard form.}$$

SECTION 3.6   GRAPHING LINEAR INEQUALITIES

If the equal sign in a linear equation in two variables is replaced with an inequality symbol, the result is a **linear inequality in two variables.**	Linear Inequalities in Two Variables: $$x \le -5y \qquad y \ge 2$$ $$3x - 2y > 7 \qquad x < -5$$
To graph a linear inequality:	Graph $2x - 4y > 4$.
1. Graph the boundary line by graphing the related equation. Draw the line solid if the inequality symbol is $\le$ or $\ge$. Draw the line dashed if the inequality symbol is $<$ or $>$.	**1.** Graph $2x - 4y = 4$. Draw a dashed line because the inequality symbol is $>$.
2. Choose a test point not on the line. Substitute its coordinates into the original inequality.	**2.** Check the test point $(0, 0)$ in the inequality $2x - 4y > 4$. $$2 \cdot 0 - 4 \cdot 0 > 4 \qquad \text{Let } x = 0 \text{ and } y = 0.$$ $$0 > 4 \qquad \text{False.}$$
3. If the resulting inequality is true, shade the **half-plane** that contains the test point. If the inequality is not true, shade the half-plane that does not contain the test point.	**3.** The inequality is false, so we shade the half-plane that does not contain $(0, 0)$. 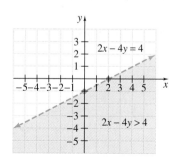

CHAPTER 3 REVIEW

(3.1) *Plot the points and name the quadrant in which each point lies.* **1.** See Appendix D. **2.** See Appendix D.

1. $A(2, -1)$, $B(-2, 1)$, $C(0, 3)$, $D(-3, -5)$

2. $A(-3, 4)$, $B(4, -3)$, $C(-2, 0)$, $D(-4, 1)$

Determine whether each ordered pair is a solution to the given equation.

3. $7x - 8y = 56$; $(0, 56)$, $(8, 0)$ no, yes

4. $-2x + 5y = 10$; $(-5, 0)$, $(1, 1)$ yes, no

5. $x = 13$; $(13, 5)$, $(13, 13)$ yes, yes

6. $y = 2$; $(7, 2)$, $(2, 7)$ yes, no

Complete the ordered pairs so that each is a solution of the given equation. **7.** $(7, 44)$

7. $-2 + y = 6x$; $(7, \ )$ **8.** $y = 3x + 5$; $(\ , -8)$ $\left(-\dfrac{13}{3}, -8\right)$

Complete the ordered pairs so that each is a solution of the given equation; then plot the ordered pairs. Use a single coordinate system for each exercise.

9. $9 = -3x + 4y$
  a. $(\ , 0)$ $(-3, 0)$
  b. $(\ , 3)$ $(1, 3)$
  c. $(9, \ )$ $(9, 9)$

10. $y = -2x$
  a. $(7, \ )$ $(7, -14)$
  b. $(-7, \ )$ $(-7, 14)$
  c. $(0, \ )$ $(0, 0)$

11. $x = 2y$
  a. $(\ , 0)$ $(0, 0)$
  b. $(\ , 5)$ $(10, 5)$
  c. $(\ , -5)$ $(-10, -5)$

Determine whether each equation is linear or not.

12. $3x - y = 4$ linear   **13.** $x - 3y = 2$ linear

14. $y = |x| + 4$ not linear   **15.** $y = x^2 + 4$ not linear

16. $y = -\dfrac{1}{2}x + 2$ linear   **17.** $y = -x + 5$ linear

Graph each equation. In some exercises, suggested x-values have been given. For Exercises 18–25, see Appendix D.

18. $3x - y = 4$ **19.** $x - 3y = 2$

20. $y = |x| + 4$ Let $x = -3, -2, -1, 0, 1, 2, 3$

21. $y = x^2 + 4$ Let $y = -3, -2, -1, 0, 1, 2, 3$

22. $y = -\dfrac{1}{2}x + 2$ **23.** $y = -x + 5$

24. $y = -1.36x$ **25.** $y = 2.1x + 5.9$

28. domain: {2, 4, 6, 8}; range: {2, 4, 5, 6}; not a function

29. domain: {triangle, square, rectangle, parallelogram}; range: {3, 4}; function

26. domain: $\left\{-\dfrac{1}{2}, 6, 0, 25\right\}$; range: $\left\{\dfrac{3}{4}, 0.75, -12, 25\right\}$; function

27. domain: $\left\{\dfrac{3}{4}, 0.75, -12, 25\right\}$; range: $\left\{-\dfrac{1}{2}, 6, 0, 25\right\}$; not a function

(3.2) *Find the domain and range of each relation. Also determine whether the relation is a function.*

26. $\left\{\left(-\dfrac{1}{2}, \dfrac{3}{4}\right), (6, 0.75), (0, -12), (25, 25)\right\}$

27. $\left\{\left(\dfrac{3}{4}, -\dfrac{1}{2}\right), (0.75, 6), (-12, 0), (25, 25)\right\}$

28.
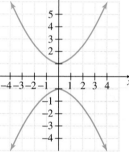

29.
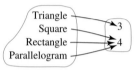

30.
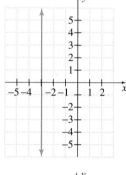
domain: $(-\infty, \infty)$; range: $(-\infty, -1] \cup [1, \infty)$; not a function

31.
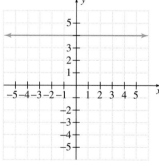
domain: $\{-3\}$; range: $(-\infty, \infty)$; not a function

32.
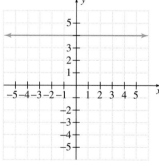
domain: $(-\infty, \infty)$; range: $\{4\}$; function

33.

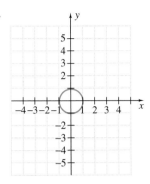

domain: $[-1, 1]$;
range: $[-1, 1]$;
not a function

C

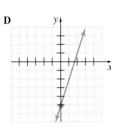

D

45. $f(x) = 3x + 1$ C   **46.** $f(x) = 3x - 2$ A
47. $f(x) = 3x + 2$ B   **48.** $f(x) = 3x - 5$ D

If $f(x) = x - 5$, $g(x) = -3x$, *and* $h(x) = 2x^2 - 6x + 1$, *find the following.*

34. $f(2)$ $f(2) = -3$ **35.** $g(0)$ $g(0) = 0$
36. $g(-6)$ $g(-6) = 18$ **37.** $h(-1)$ $h(-1) = 9$
38. $h(1)$ $h(1) = -3$ **39.** $f(5)$ $f(5) = 0$

The function $J(x) = 2.54x$ *may be used to calculate the weight of an object on Jupiter J given its weight on Earth x.*

40. If a person weighs 150 pounds on Earth, find the equivalent weight on Jupiter.   381 pounds

41. A 2000-pound probe on Earth weighs how many pounds on Jupiter?   5080 pounds

(3.3)   *Graph each linear function.*

42. $f(x) = x$ **43.** $f(x) = -\dfrac{1}{3}x$ **44.** $g(x) = 4x - 1$

The graph of $f(x) = 3x$ *is sketched next. Use this graph to match each linear function with its graph.*

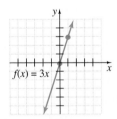

A

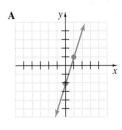

B

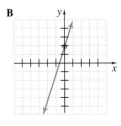

Graph each linear equation by finding intercepts if possible.   For Exercises 49–54, see Appendix D.

49. $4x + 5y = 20$ **50.** $3x - 2y = -9$
51. $4x - y = 3$ **52.** $2x + 6y = 9$
53. $y = 5$ **54.** $x = -2$

55. The cost C, in dollars, of renting a minivan for a day is given by the linear function $C(x) = 0.3x + 42$, where x is number of miles driven.

a. Find the cost of renting the minivan for a day and driving it 150 miles. $C(150) = 87$

b. Graph $C(x) = 0.3x + 42$.   See Appendix D.

(3.4)   *Find the slope of the line through each pair of points.*

56. $(2, 8)$ and $(6, -4)$ -3 **57.** $(-3, 9)$ and $(5, 13)$ $\dfrac{1}{2}$
58. $(-7, -4)$ and $(-3, 6)$ $\dfrac{5}{2}$ **59.** $(7, -2)$ and $(-5, 7)$ $-\dfrac{3}{4}$

Find the slope and y-intercept of each line.

60. $6x - 15y = 20$ **61.** $4x + 14y = 21$
$m = \dfrac{2}{5}, b = -\dfrac{4}{3}$ $m = -\dfrac{2}{7}, b = \dfrac{3}{2}$

Find the slope of each line.

62. $y - 3 = 0$ 0   **63.** $x = -5$ undefined

Two lines are graphed on each set of axes. Decide whether l_1 *or* l_2 *has the greater slope.*

64. 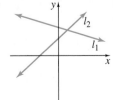 l_2 **65.** 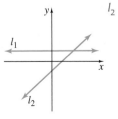 l_2

For Exercises 42–44, see Appendix D.

66.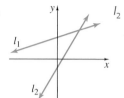

67.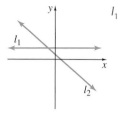

Decide whether the lines are parallel, perpendicular, or neither.

68. $f(x) = -2x + 6$
$g(x) = 2x - 1$ neither

69. $-x + 3y = 2$
$6x - 18y = 3$ parallel

(3.5) *Find an equation of the line satisfying the conditions given.*

70. Horizontal; through $(3, -1)$ $y = -1$

71. Vertical; through $(-2, -4)$ $x = -2$

72. Parallel to the line $x = 6$; through $(-4, -3)$ $x = -4$

73. Slope 0; through $(2, 5)$ $y = 5$

Find the standard-form equation of each line satisfying the conditions given.

74. Through $(-3, 5)$; slope 3 $3x - y = -14$

75. Slope 2; through $(5, -2)$ $2x - y = 12$

82. $f(x) = -2x - 2$ **83.** $f(x) = -\dfrac{3}{2}x - 8$ **84.** $f(x) = \dfrac{3}{4}x + \dfrac{7}{2}$

76. Through $(-6, -1)$ and $(-4, -2)$ $x + 2y = -8$

77. Through $(-5, 3)$ and $(-4, -8)$ $11x + y = -52$

78. Through $(-2, 3)$; perpendicular to $x = 4$ $y = 3$

79. Through $(-2, -5)$; parallel to $y = 8$ $y = -5$

Find the equation of each line satisfying the given conditions. Write each equation using function notation.

80. Slope $-\dfrac{2}{3}$; y-intercept 4 $f(x) = -\dfrac{2}{3}x + 4$

81. Slope -1; y-intercept -2 $f(x) = -x - 2$

82. Through $(2, -6)$; parallel to $6x + 3y = 5$

83. Through $(-4, -2)$; parallel to $3x + 2y = 8$

84. Through $(-6, -1)$; perpendicular to $4x + 3y = 5$

85. Through $(-4, 5)$; perpendicular to $2x - 3y = 6$
$$f(x) = -\dfrac{3}{2}x - 1$$

(3.6) *Graph each linear inequality.*

86. $3x + y > 4$ **87.** $\dfrac{1}{2}x - y < 2$ **88.** $5x - 2y \le 9$

89. $3y \ge x$ **90.** $y < 1$ **91.** $x > -2$

92. Graph the union of $y > 2x + 3$ or $x \le -3$.

93. Graph the intersection of $2x < 3y + 8$ and $y \ge -2$.

For Exercises 86–93, see Appendix D.

CHAPTER 3 TEST

1. Plot the points, and name the quadrant in which each is located: $A(6, -2), B(4, 0), C(-1, 6)$.

2. Complete the ordered pair solution $(-6, \ )$ of the equation $2y - 3x = 12$. $(-6, -3)$

Graph each line.   For Exercises 3–6, see Appendix D.

3. $2x - 3y = -6$ **4.** $4x + 6y = 7$

5. $f(x) = \dfrac{2}{3}x$ **6.** $y = -3$

7. Find the slope of the line that passes through $(5, -8)$ and $(-7, 10)$.

8. Find the slope and the y-intercept of the line $3x + 12y = 8$. $m = -\dfrac{1}{4}, b = \dfrac{2}{3}$

Graph each function. Suggested x-values have been given for ordered pair solutions.

9. $f(x) = (x - 1)^2$ For Exercises 9–10, see Appendix D.
Let $x = -2, -1, 0, 1, 2, 3, 4$

10. $g(x) = |x| + 2$
Let $x = -3, -2, -1, 0, 1, 2, 3$

Find an equation of each line satisfying the conditions given. Write Exercises 11–15 in standard form. Write Exercises 16–18 using function notation.

11. Horizontal; through $(2, -8)$ $y = -8$

12. Vertical; through $(-4, -3)$ $x = -4$

13. Perpendicular to $x = 5$; through $(3, -2)$ $y = -2$

14. Through $(4, -1)$; slope -3 $3x + y = 11$

15. Through $(0, -2)$; slope 5 $5x - y = 2$

16. Through $(4, -2)$ and $(6, -3)$ $f(x) = -\dfrac{1}{2}x$

17. Through $(-1, 2)$; perpendicular to $3x - y = 4$

18. Parallel to $2y + x = 3$; through $(3, -2)$

19. Line L_1 has the equation $2x - 5y = 8$. Line L_2 passes through the points $(1, 4)$ and $(-1, -1)$.

1. See Appendix D.   **7.** $-\dfrac{3}{2}$ **17.** $f(x) = -\dfrac{1}{3}x + \dfrac{5}{3}$ **18.** $f(x) = -\dfrac{1}{2}x - \dfrac{1}{2}$

Determine whether these lines are parallel lines, perpendicular lines, or neither.   neither

Graph each inequality.

20. $x \leq -4$ **21.** $y > -2$ **22.** $2x - y > 5$

23. The intersection of $2x + 4y < 6$ and $y \leq -4$

For Exercises 20–23, see Appendix D.

Find the domain and range of each relation. Also determine whether the relation is a function.

24.

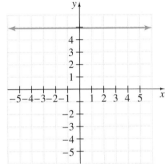

domain: $(-\infty, \infty)$;
range: {5};
function

25.

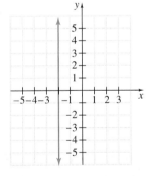

domain: {−2};
range: $(-\infty, \infty)$;
not a function

26.

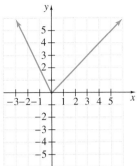

domain: $(-\infty, \infty)$;
range: $[0, \infty)$;
function

27.

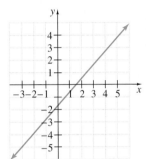

domain: $(-\infty, \infty)$;
range: $(-\infty, \infty)$;
function

28. The average yearly earnings for high school graduates age 18 and older is given by the linear function

$$f(x) = 708x + 13{,}570$$

where x is the number of years since 1985 that a person graduated.

a. Find the average earnings in 1990 for high school graduates. $17,110

b. Predict the average earnings for high school graduates in the year 2000. $24,190

c. Predict the first year that the average earnings for high school graduates will be greater than $30,000.   2009

CHAPTER 3   CUMULATIVE REVIEW

1. Simplify the following expressions.

  a. $11 + 2 - 7$ 6   **b.** $-5 - 4 + 2$

2. Solve for x: $-6x - 1 + 5x = 3$.   {−4}; *Sec. 2.1, Ex. 3*

3. Find two numbers such that the second number is twice the first number increased by 3 and the sum of the two numbers is 72.   23, 49; *Sec. 2.2, Ex. 1*

4. Solve $3y - 2x = 7$ for y. $y = \dfrac{2x}{3} + \dfrac{7}{3}$; *Sec. 2.3, Ex. 2*

1b. −7; *Sec. 1.3, Ex. 4*   **6.** $\left[-9, -\dfrac{9}{2}\right]$; *Sec. 2.5, Ex. 3*

5. A salesperson earns a base of $600 per month plus a commission of 20% of sales. Find the minimum amount of sales needed to receive a total income of at least $1500 per month. $4500; *Sec. 2.4, Ex. 7*

6. Solve for x: $-1 \leq \dfrac{2x}{3} + 5 \leq 2$.

7. Solve $\left|\dfrac{x}{2} - 1\right| = 11$.   {−20, 24}; *Sec. 2.6, Ex. 3*

10. The five points are graphed as shown in Appendix D; *Sec. 3.1, Ex. 1* **a.** quadrant IV

8. Solve for *y*: $|y - 3| > 7$. $(-\infty, -4) \cup (10, \infty)$; *Sec. 2.7, Ex. 5*

9. Solve for *x*: $\left|\dfrac{2(x + 1)}{3}\right| \leq 0$. $\{-1\}$; *Sec. 2.7, Ex. 8*

10. Plot each ordered pair on a Cartesian coordinate system and name the quadrant in which the point is located.

   a. $(2, -1)$ **b.** $(0, 5)$ **c.** $(-3, 5)$

   d. $(-2, 0)$ **e.** $\left(-\dfrac{1}{2}, -4\right)$ **f.** $(1.5, 1.5)$

11. Graph the equation $y = |x|$.

12. Determine the domain and range of each relation.

   a. $\{(2, 3), (2, 4), (0, -1), (3, -1)\}$

   b.

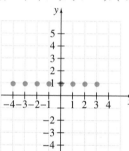

12a. domain: $\{2, 0, 3\}$; range: $\{3, 4, -1\}$ *(Sec. 3.2, Ex. 1)*

domain: $\{-4, -3, -2, -1, 0, 1, 2, 3\}$; range: $\{1\}$

   c.

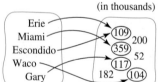

Cities   Population (in thousands)

domain: {Erie, Escondido, Gary, Miami, Waco}; range: {104, 109, 117, 359}

13. Which of the following graphs are graphs of functions?

   a.

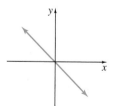

yes

   b.

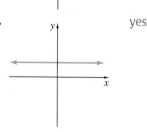

yes

c.

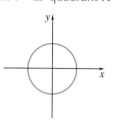

no

d.

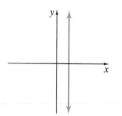

yes

e.

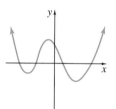

no

f.

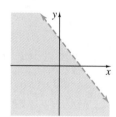

no; *Sec. 3.2, Ex. 5*

14. See Appendix D; *Sec. 3.3, Ex. 4*

14. Graph $x - 3y = 6$ by plotting intercept points.

15. Graph $y = -3$.   See Appendix D; *Sec. 3.3, Ex. 7*

16. Find the slope of the line containing the points $(5, -7)$ and $(-3, 6)$. $-\dfrac{13}{8}$; *Sec. 3.4, Ex. 2*

17. Find the slope and the *y*-intercept of the line whose equation is $3x - 4y = 4$.

18. Find an equation of the line with slope -3 containing the point $(1, -5)$. Write the equation in standard form: $Ax + By = C$.

19. Find the equation of the horizontal line containing the point $(2, 3)$. $y = 3$; *Sec. 3.5, Ex. 5*

20. Graph $3x \geq y$.   See Appendix D; *Sec. 3.6, Ex. 2*

21. Graph the union of $x + \dfrac{1}{2}y \geq -4$ or $y \leq -2$.

10b. not in any quadrant. A point on an axis is called a quadrantal point.   **c.** quadrant II
d. not in any quadrant.   **e.** quadrant III   **f.** quadrant I   **11.** See Appendix D; *Sec. 3.1, Ex. 7*

17. $m = \dfrac{3}{4}$; *y*-intercept -1; *Sec. 3.4, Ex. 4*   **18.** $3x + y = -2$; *Sec. 3.5, Ex. 3*   **21.** See Appendix D; *Sec. 3.6, Ex. 4*

SYSTEMS OF EQUATIONS

LOCATING LIGHTNING STRIKES

Lightning, most often produced during thunderstorms, is a rapid discharge of high-current electricity into the atmosphere. Around the world, lightning occurs at a rate of approximately 100 flashes per second. Because of lightning's potentially destructive nature, meteorologists track lightning activity by recording and plotting the positions of lightning strikes.

IN THE CHAPTER GROUP ACTIVITY ON PAGE 229, YOU WILL HAVE THE OPPORTUNITY TO PINPOINT THE LOCATION OF A LIGHTNING STRIKE.

I n this chapter, two or more equations in two or more variables are solved simultaneously. Such a collection of equations is called a **system of equations.** Systems of equations are good mathematical models for many real-world problems because these problems may involve several related patterns.

4.1 | SOLVING SYSTEMS OF LINEAR EQUATIONS IN TWO VARIABLES

O B J E C T I V E S

TAPE IA 4.1

1. Solve a system by graphing.
2. Solve a system by substitution.
3. Solve a system by elimination.

An important problem that often occurs in the fields of business and economics concerns the concepts of revenue and cost. For example, suppose that a small manufacturing company begins to manufacture and sell compact disc storage units. The revenue of a company is the company's income from selling these units, and the cost is the amount of money that a company spends to manufacture these units. The following coordinate system shows the graph of revenue and cost for the storage units.

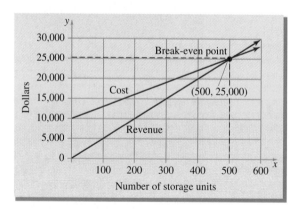

These lines intersect at the point (500, 25,000). This means that when 500 storage units are manufactured and sold, both cost and revenue are $25,000. In business, this point of intersection is called the *break-even point.* Notice that for *x*-values (units sold) less than 500, the cost graph is above the revenue graph, meaning that cost of manufacturing is greater than revenue, and so the company is losing money. For *x*-values (units sold) greater than 500, the revenue graph is above the cost graph, meaning that revenue is greater than cost, and so the company is making money.

Recall from Chapter 3 that each line is a graph of some linear equation in two variables. Both equations together form a **system of equations.** The common point of intersection is called the **solution of the system.** Some examples of systems of linear equations in two variables are

SYSTEMS OF LINEAR EQUATIONS IN TWO VARIABLES

$$\begin{cases} x - 2y = -7 \\ 3x + y = 0 \end{cases} \qquad \begin{cases} x = 5 \\ x + \dfrac{y}{2} = 9 \end{cases} \qquad \begin{cases} x - 3 = 2y + 6 \\ y = 1 \end{cases}$$

Recall that a solution of an equation in two variables is an ordered pair (x, y) that makes the equation true. A **solution of a system** of two equations in two variables is an ordered pair (x, y) that makes both equations true.

We can estimate the solutions of a system by graphing each equation on the same coordinate system and estimating the coordinates of any point of intersection.

EXAMPLE 1 Solve each system by graphing. If the system has just one solution, estimate the solution.

a. $\begin{cases} x + y = 2 \\ 3x - y = -2 \end{cases}$

b. $\begin{cases} x - 2y = 4 \\ x = 2y \end{cases}$

c. $\begin{cases} 2x + 4y = 10 \\ x + 2y = 5 \end{cases}$

Solution: Since the graph of a linear equation in two variables is a line, graphing two such equations yields two lines in a plane.

a. $\begin{cases} x + y = 2 \\ 3x - y = -2 \end{cases}$

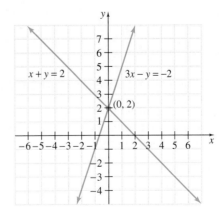

These lines intersect at one point as shown. The coordinates of the point of intersection appear to be $(0, 2)$. Check this estimated solution by replacing x with 0 and y with 2 in **both** equations.

$x + y = 2$	First equation.	$3x - y = -2$	Second equation.
$0 + 2 = 2$	Let $x = 0$ and $y = 2$.	$3 \cdot 0 - 2 = -2$	Let $x = 0$ and $y = 2$.
$2 = 2$	True.	$-2 = -2$	True.

The ordered pair $(0, 2)$ does satisfy both equations. We conclude therefore that $(0, 2)$ is the solution of the system. A system that has at least one solution, such as this one, is said to be **consistent.**

b. $\begin{cases} x - 2y = 4 \\ x = 2y \end{cases}$

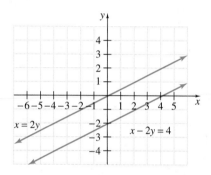

The lines appear to be parallel. To be sure, write each equation in point-slope form, $y = mx + b$.

$x - 2y = 4$	First equation.	$x = 2y$	Second equation.
$-2y = -x + 4$	Subtract x from both sides.	$\dfrac{1}{2}x = y$	Divide both sides by 2.
$y = \dfrac{1}{2}x - 2$	Divide both sides by -2.	$y = \dfrac{1}{2}x$	

The graphs of these equations have the same slope, $\dfrac{1}{2}$, but different y-intercepts, so we have confirmed that these lines are parallel. Therefore, the system has no solution since the equations have no common solution (there are no intersection points). A system that has no solution is said to be **inconsistent.**

c. $\begin{cases} 2x + 4y = 10 \\ x + 2y = 5 \end{cases}$

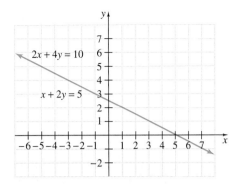

The graph of each equation is the same line. To confirm this, notice that if both sides of the second equation are multiplied by 2, the result is the first equation. This means that the equations have identical solutions. Any ordered pair solution of one equation satisfies the other equation also. Thus, these equations are said to be **dependent equations.** The solution set of the system is $\{(x, y)\,|\,x + 2y = 5\}$ or, equivalently, $\{(x, y)\,|\,2x + 4y = 10\}$ since the lines describe identical ordered pairs. Written this way, the solution set is read "the set of all ordered pairs (x, y), such that $2x + 4y = 10$." There are therefore an infinite number of solutions to this system.

We can summarize the information discovered in Example 1 as follows.

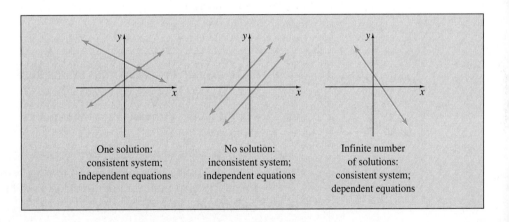

| One solution: consistent system; independent equations | No solution: inconsistent system; independent equations | Infinite number of solutions: consistent system; dependent equations |

Graphing the equations of a system by hand is often a good method of finding approximate solutions of a system, but it is not a reliable method of finding exact solutions of a system. We turn instead to two algebraic methods of solving systems. We use the first method, the **substitution method,** to solve the system

$$\begin{cases} y = x + 5 & \text{First equation.} \\ 3x = 2y - 9 & \text{Second equation.} \end{cases}$$

Remember that we are looking for an ordered pair, if there is one, that satisfies both equations. Satisfying the first equation, $y = x + 5$, means that y must be $x + 5$. **Substituting** the expression $x + 5$ for y in the second equation yields an equation in one variable, which we can solve for x.

$$3x = 2y - 9 \qquad \text{Second equation.}$$

$$3x = 2(x + 5) - 9 \qquad \text{Replace } y \text{ with } x + 5 \text{ in the second equation.}$$
$$3x = 2x + 10 - 9$$
$$x = 1$$

The x-coordinate of the solution of the system is 1. The y-coordinate is the y-value corresponding to the x-value 1. Choose either equation of the system and solve for y when x is 1.

$$y = x + 5 \qquad \text{First equation.}$$
$$= 1 + 5 \qquad \text{Let } x = 1.$$
$$= 6$$

The y-coordinate is 6, so the solution of the system is $(1, 6)$. This means that, when both equations are graphed, the one point of intersection occurs at the point with coordinates $(1, 6)$. We can check this solution by substituting 1 for x and 6 for y in both equations of the system.

TO SOLVE A SYSTEM OF TWO EQUATIONS BY THE SUBSTITUTION METHOD

Step 1. Solve one of the equations for one of its variables.

Step 2. Substitute the expression for the variable found in *step 1* into the other equation.

Step 3. Find the value of one variable by solving the equation from *step 2*.

Step 4. Find the value of the other variable by substituting the value found in *step 3* in any equation of the system.

Step 5. Check the ordered pair solution in **both** of the original equations.

EXAMPLE 2 Use the substitution method to solve the system

$$\begin{cases} 2x + 4y = -6 & \text{First equation.} \\ x - 2y = -5 & \text{Second equation.} \end{cases}$$

Solution: We begin by solving the second equation for x since the coefficient 1 on the x-term keeps us from introducing tedious fractions. The equation $x - 2y = -5$ solved for x is $x = 2y - 5$. Substitute $2y - 5$ for x in the first equation.

$$2x + 4y = -6 \qquad \text{First equation.}$$

$$2(2y - 5) + 4y = -6 \qquad \text{Substitute } 2y - 5 \text{ for } x.$$

$$4y - 10 + 4y = -6$$

$$8y = 4$$

$$y = \frac{4}{8} = \frac{1}{2} \qquad \text{Solve for } y.$$

The y-coordinate of the solution is $\frac{1}{2}$. To find the x-coordinate, replace y with $\frac{1}{2}$ in the equation $x = 2y - 5$.

$$x = 2y - 5$$

$$x = 2\left(\frac{1}{2}\right) - 5 = 1 - 5 = -4$$

The solution set is $\left\{\left(-4, \frac{1}{2}\right)\right\}$. To check, see that $\left(-4, \frac{1}{2}\right)$ satisfies both equations of the system.

EXAMPLE 3   Use substitution to solve the system

$$\begin{cases} -\dfrac{x}{6} + \dfrac{y}{2} = \dfrac{1}{2} \\ \dfrac{x}{3} - \dfrac{y}{6} = -\dfrac{3}{4} \end{cases}$$

Solution: First, multiply each equation by its least common denominator in order to write this system as an equivalent system without fractions. We multiply the first equation by 6 and the second equation by 12. (Just as for equations, equivalent systems are systems that have the same solution.)

$$\begin{cases} 6\left(-\dfrac{x}{6} + \dfrac{y}{2}\right) = 6\left(\dfrac{1}{2}\right) \\ 12\left(\dfrac{x}{3} - \dfrac{y}{6}\right) = 12\left(-\dfrac{3}{4}\right) \end{cases}$$

simplifies to

$$\begin{cases} -x + 3y = 3 & \text{First equation.} \\ 4x - 2y = -9 & \text{Second equation.} \end{cases}$$

We now solve the first equation for x.

$$-x + 3y = 3 \qquad \text{First equation.}$$
$$3y - 3 = x \qquad \text{Solve for } x.$$

Next, replace x with $3y - 3$ in the second equation.

$$4x - 2y = -9 \qquad \text{Second equation.}$$
$$4(3y - 3) - 2y = -9$$
$$12y - 12 - 2y = -9$$
$$10y = 3$$
$$y = \frac{3}{10} \qquad \text{Solve for } y.$$

The y-coordinate is $\frac{3}{10}$. To find the x-coordinate, replace y with $\frac{3}{10}$ in the equation $x = 3y - 3$. Then

$$x = 3\left(\frac{3}{10}\right) - 3 = \frac{9}{10} - 3 = \frac{9}{10} - \frac{30}{10} = -\frac{21}{10}$$

The solution set is $\left\{\left(-\frac{21}{10}, \frac{3}{10}\right)\right\}$.

3   The **elimination method,** or **addition method,** is a second algebraic technique for solving systems of equations. For this method, we rely on a version of the addition property of equality, which states that "equals added to equals are equal." In symbols,

$$\text{if } A = B \text{ and } C = D \text{ then } A + C = B + D.$$

EXAMPLE 4   Use the elimination method to solve the system

$$\begin{cases} x - 5y = -12 & \text{First equation.} \\ -x + y = 4 & \text{Second equation.} \end{cases}$$

Solution: Since the left side of each equation is equal to the right side, we add equal quantities by adding the left sides of the equations and the right sides of the equations. This sum gives us an equation in one variable, y, which we can solve for y.

$$\begin{array}{ll} x - 5y = -12 & \text{First equation.} \\ \underline{-x + y = 4} & \text{Second equation.} \\ -4y = -8 & \text{Add.} \\ y = 2 & \text{Solve for } y. \end{array}$$

The y-coordinate of the solution is 2. To find the corresponding x-coordinate, replace y with 2 in either original equation of the system.

$$-x + y = 4 \qquad \text{Second equation.}$$
$$-x + 2 = 4 \qquad \text{Let } y = 2.$$
$$-x = 2$$
$$x = -2$$

The solution set is $\{(-2, 2)\}$. Check to see that $(-2, 2)$ satisfies both equations of the system. ▬▬▬▬

TO SOLVE A SYSTEM OF TWO LINEAR EQUATIONS BY THE ELIMINATION METHOD

Step 1.   Rewrite each equation in standard form, $Ax + By = C$.

Step 2.   If necessary, multiply one or both equations by some nonzero number so that the coefficient of one variable in one equation is the opposite of its coefficient in the other equation.

Step 3.   Add the equations.

Step 4.   Find the value of one variable by solving the equation from *step 3*.

Step 5.   Find the value of the second variable by substituting the value found in *step 4* into either of the original equations.

Step 6.   Check the proposed ordered pair solution in both of the original equations.

EXAMPLE 5   Use the elimination method to solve the system

$$\begin{cases} 3x + \dfrac{y}{2} = 2 \\ 6x + y = 5 \end{cases}$$

Solution: If we add the two equations, the sum will still be an equation in two variables. Notice that if we multiply both sides of the first equation by -2, the coefficients of x in the two equations will be opposites. Then

$$\begin{cases} -2\left(3x + \dfrac{y}{2}\right) = -2(2) \\ 6x + y = 5 \end{cases} \quad \text{simplifies to} \quad \begin{cases} -6x - y = -4 \\ 6x + y = 5 \end{cases}$$

Next, add the left sides and add the right sides.

$$\begin{array}{r} -6x - y = -4 \\ 6x + y = 5 \\ \hline 0 = 1 \qquad \text{False.} \end{array}$$

The resulting equation, $0 = 1$, is false for all values of y or x. Thus, the system has no solution. The solution set is $\{ \}$ or $\varnothing$.

This system is inconsistent, and the graphs of the equations are parallel lines.

EXAMPLE 6   Use the elimination method to solve the system

$$\begin{cases} 3x - 2y = 10 \\ 4x - 3y = 15 \end{cases}$$

Solution: To eliminate y when the equations are added, multiply both sides of the first equation by 3 and both sides of the second equation by -2. Then

$$\begin{cases} 3(3x - 2y) = 3(10) \\ -2(4x - 3y) = -2(15) \end{cases} \quad \text{simplifies to} \quad \begin{cases} 9x - 6y = 30 \\ -8x + 6y = -30 \end{cases}$$

Next, add the left sides and add the right sides.

$$\begin{array}{r} 9x - 6y = 30 \\ -8x + 6y = -30 \\ \hline x \quad\quad = 0 \end{array}$$

To find y, let $x = 0$ in either equation of the system.

$$\begin{aligned} 3x - 2y &= 10 \\ 3(0) - 2y &= 10 \qquad \text{Let } x = 0. \\ -2y &= 10 \\ y &= -5 \end{aligned}$$

The solution set is $\{(0, -5)\}$. Check to see that $(0, -5)$ satisfies both equations.

EXAMPLE 7   Use the elimination method to solve

$$\begin{cases} -5x - 3y = 9 \\ 10x + 6y = -18 \end{cases}$$

Solution: To eliminate x when the equations are added, multiply both sides of the first equation by 2. Then

$$\begin{cases} 2(-5x - 3y) = 2(9) \\ 10x + 6y = -18 \end{cases} \quad \text{simplifies to} \quad \begin{cases} -10x - 6y = 18 \\ 10x + 6y = -18 \end{cases}$$

Next, add the equations.

$$-10x - 6y = 18$$
$$\underline{10x + 6y = -18}$$
$$0 = 0$$

The resulting equation, $0 = 0$, is true for all possible values of y or x. Notice in the original system that if both sides of the first equation are multiplied by -2, the result is the second equation. This means that the two equations are equivalent and that they have the same solution set. Thus, the equations of this system are dependent, and the solution set of the system is

$$\{(x, y) \mid -5x - 3y = 9\} \text{ or equivalently } \{(x, y) \mid 10x + 6y = -18\} \quad \blacksquare$$

EXAMPLE 8   A small manufacturing company manufactures and sells compact disc storage units. The revenue equation for these units is

$$y = 50x$$

where x is the number of units sold and y is the revenue, or income, in dollars for selling x units. The cost equation for these units is

$$y = 30x + 10,000$$

where x is the number of units manufactured and y is the total cost in dollars for manufacturing x units. Use these equations to find the number of units to be sold for the company to break even.

Solution: The break-even point is found by solving the system

$$\begin{cases} y = 50x & \text{First equation.} \\ y = 30x + 10,000 & \text{Second equation.} \end{cases}$$

Since both equations are solved for a variable, use the substitution method and substitute $50x$ for y in the second equation.

$y = 30x + 10,000$	Second equation.
$50x = 30x + 10,000$	Substitute $50x$ for y.
$20x = 10,000$	Subtract $30x$.
$x = 500$	Divide by 20.

The x-value of the ordered pair solution is 500. To find the corresponding y-value, we substitute 500 for x in the equation $y = 50x$. Then $y = 50(500) = 25,000$. The ordered pair solution is $(500, 25,000)$. This means that the business must sell 500 compact disc storage units to break even. The graph of the equations in this system can be found at the beginning of this section.   ▬▬▬▬

GRAPHING CALCULATOR EXPLORATIONS

A grapher may be used to approximate solutions of systems of equations by graphing each equation on the same set of axes and approximating any points of intersection. For example, approximate the solution of the system

$$\begin{cases} y = -2.6x + 5.6 \\ y = 4.3x - 4.9 \end{cases}$$

First use a standard window and graph both equations on a single screen.

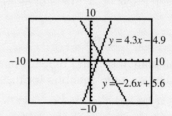

The two lines intersect. To approximate the point of intersection, trace to the point of intersection and use an Intersect feature of the grapher, a Zoom In feature of the grapher, or redefine the window to [0, 3] by [0, 3]. If we redefine the window to [0, 3] by [0, 3], the screen should look like the following:

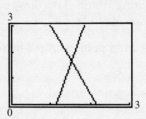

By tracing along the curves, we can see that the point of intersection has an x-value between 1.5 and 1.532. Continue to zoom and trace or redefine the window until the coordinates of the point of intersection can be determined to the nearest hundredth. The approximate point of intersection is (1.52, 1.64).

Solve each system of equations. Approximate the solutions to two decimal places.

1. $y = -1.65x + 3.65$ {(2.11, 0.17)}
$y = 4.56x - 9.44$

2. $y = 7.61x + 3.48$ {(−1.12, −5.02)}
$y = -1.26x - 6.43$

3. $2.33x - 4.72y = 10.61$ {(0.57, −1.97)}
$5.86x + 6.22y = -8.89$

4. $-7.89x - 5.68y = 3.26$ {(−1.38, 1.35)}
$- 3.65x + 4.98y = 11.77$

EXERCISE SET 4.1

For Problems 1–6, see Appendix D. **1.** $\{(2, -1)\}$ **2.** $\{(5, 2)\}$ **3.** $\{(1, 2)\}$
4. $\{(2, 2)\}$ **5.** $\{\}$ **6.** $\{\}$

Solve each system by graphing. See Example 1.

1. $\begin{cases} x + y = 1 \\ x - 2y = 4 \end{cases}$ **2.** $\begin{cases} 2x - y = 8 \\ x + 3y = 11 \end{cases}$ **3.** $\begin{cases} 2y - 4 = 0 \\ x + 2y = 5 \end{cases}$

4. $\begin{cases} 4x - y = 6 \\ x - y = 0 \end{cases}$ **5.** $\begin{cases} 3x - y = 4 \\ 6x - 2y = 4 \end{cases}$ **6.** $\begin{cases} -x + 3y = 6 \\ 3x - 9y = 9 \end{cases}$

7. Can a system consisting of two linear equations have exactly two solutions? Explain why or why not. No

8. Suppose the graph of the equations in a system of two equations in two variables consists of a circle and a line. Discuss the possible number of solutions for this system. 0, 1, or 2

Solve each system of equations by the substitution method. See Examples 2 and 3.

9. $\begin{cases} x + y = 10 \\ y = 4x \end{cases}$ $\{(2, 8)\}$ **10.** $\begin{cases} 5x + 2y = -17 \\ x = 3y \end{cases}$ $\{(-3, -1)\}$

11. $\begin{cases} 4x - y = 9 \\ 2x + 3y = -27 \end{cases}$ $\{(0, -9)\}$ **12.** $\begin{cases} 3x - y = 6 \\ -4x + 2y = -8 \end{cases}$ $\{(2, 0)\}$

13. $\begin{cases} \dfrac{1}{2}x + \dfrac{3}{4}y = -\dfrac{1}{4} \\ \dfrac{3}{4}x - \dfrac{1}{4}y = 1 \end{cases}$ $\{(1, -1)\}$ **14.** $\begin{cases} \dfrac{2}{5}x + \dfrac{1}{5}y = -1 \\ x + \dfrac{2}{5}y = -\dfrac{8}{5} \end{cases}$ $\{(2, -9)\}$

15. $\begin{cases} \dfrac{x}{3} + y = \dfrac{4}{3} \\ -x + 2y = 11 \end{cases}$ $\{(-5, 3)\}$ **16.** $\begin{cases} \dfrac{x}{8} - \dfrac{y}{2} = 1 \\ \dfrac{x}{3} - y = 2 \end{cases}$ $\{(0, -2)\}$

Solve each system of equations by the elimination method. See Examples 4–7.

17. $\begin{cases} 2x - 4y = 0 \\ x + 2y = 5 \end{cases}$ $\left\{\left(\dfrac{5}{2}, \dfrac{5}{4}\right)\right\}$ **18.** $\begin{cases} 2x - 3y = 0 \\ 2x + 6y = 3 \end{cases}$ $\left\{\left(\dfrac{1}{2}, \dfrac{1}{3}\right)\right\}$

19. $\begin{cases} 5x + 2y = 1 \\ x - 3y = 7 \end{cases}$ $\{(1, -2)\}$ **20.** $\begin{cases} 6x - y = -5 \\ 4x - 2y = 6 \end{cases}$ $\{(-2, -7)\}$

21. $\begin{cases} 5x - 2y = 27 \\ -3x + 5y = 18 \end{cases}$ $\{(9, 9)\}$ **22.** $\begin{cases} 3x + 4y = 2 \\ 2x + 5y = -1 \end{cases}$ $\{(2, -1)\}$

23. $\begin{cases} 3x - 5y = 11 \\ 2x - 6y = 2 \end{cases}$ $\{(7, 2)\}$ **24.** $\begin{cases} 6x - 3y = -3 \\ 4x + 5y = -9 \end{cases}$ $\{(-1, -1)\}$

25. $\begin{cases} x - 2y = 4 \\ 2x - 4y = 4 \end{cases}$ $\{\}$ **26.** $\begin{cases} -x + 3y = 6 \\ 3x - 9y = 9 \end{cases}$ $\{\}$

27. $\begin{cases} 3x + y = 1 \\ 2y = 2 - 6x \end{cases}$ **28.** $\begin{cases} y = 2x - 5 \\ 8x - 4y = 20 \end{cases}$

29. Write a system of two linear equations in x and y that has the ordered pair solution (2, 5).

30. Which method would you use to solve the system

$$\begin{cases} 5x - 2y = 6 \\ 2x + 3y = 5 \end{cases}$$

Explain your choice. Answers will vary.

Solve each system of equations.

31. $\begin{cases} 2x + 5y = 8 \\ 6x + y = 10 \end{cases}$ $\left\{\left(\dfrac{3}{2}, 1\right)\right\}$ **32.** $\begin{cases} x - 4y = -5 \\ -3x - 8y = 0 \end{cases}$

33. $\begin{cases} x + y = 1 \\ x - 2y = 4 \end{cases}$ $\{(2, -1)\}$ **34.** $\begin{cases} 2x - y = 8 \\ x + 3y = 11 \end{cases}$ $\{(5, 2)\}$

35. $\begin{cases} \dfrac{1}{3}x + y = \dfrac{4}{3} \\ -\dfrac{1}{4}x - \dfrac{1}{2}y = -\dfrac{1}{4} \end{cases}$ **36.** $\begin{cases} \dfrac{3}{4}x - \dfrac{1}{2}y = -\dfrac{1}{2} \\ x + y = -\dfrac{3}{2} \end{cases}$

37. $\begin{cases} 2x + 6y = 8 \\ 3x + 9y = 12 \end{cases}$ **38.** $\begin{cases} x = 3y - 1 \\ 2x - 6y = -2 \end{cases}$

39. $\begin{cases} 4x + 2y = 5 \\ 2x + y = -1 \end{cases}$ $\{\}$ **40.** $\begin{cases} 3x + 6y = 15 \\ 2x + 4y = 3 \end{cases}$ $\{\}$

41. $\begin{cases} 10y - 2x = 1 \\ 5y = 4 - 6x \end{cases}$ $\left\{\left(\dfrac{1}{2}, \dfrac{1}{5}\right)\right\}$ **42.** $\begin{cases} 3x + 4y = 0 \\ 7x = 3y \end{cases}$ $\{(0, 0)\}$

43. $\begin{cases} \dfrac{3}{4}x + \dfrac{5}{2}y = 11 \\ \dfrac{1}{16}x - \dfrac{3}{4}y = -1 \end{cases}$ $\{(8, 2)\}$ **44.** $\begin{cases} \dfrac{2}{3}x + \dfrac{1}{4}y = -\dfrac{3}{2} \\ \dfrac{1}{2}x - \dfrac{1}{4}y = -2 \end{cases}$ $\{(-3, 2)\}$

45. $\begin{cases} x = 3y + 2 \\ 5x - 15y = 10 \end{cases}$ **46.** $\begin{cases} y = \dfrac{1}{7}x + 3 \\ x - 7y = -21 \end{cases}$

47. $\begin{cases} 2x - y = -1 \\ y = -2x \end{cases}$ **48.** $\begin{cases} x = \dfrac{1}{5}y \\ x - y = -4 \end{cases}$ $\{(1, 5)\}$

49. $\begin{cases} 2x = 6 \\ y = 5 - x \end{cases}$ $\{(3, 2)\}$ **50.** $\begin{cases} x = 3y + 4 \\ -y = 5 \end{cases}$ $\{(-11, -5)\}$

51. $\begin{cases} \dfrac{x + 5}{2} = \dfrac{6 - 4y}{3} \\ \dfrac{3x}{5} = \dfrac{21 - 7y}{10} \end{cases}$ **52.** $\begin{cases} \dfrac{y}{5} = \dfrac{8 - x}{2} \\ x = \dfrac{2y - 8}{3} \end{cases}$ $\{(4, 10)\}$

27. $\{(x, y) | 3x + y = 1\}$ **28.** $\{(x, y) | y = 2x - 5\}$ **29.** Answers will vary. One possibility: $\begin{cases} -2x + y = 1 \\ x - 2y = -8 \end{cases}$

32. $\left\{\left(-2, \dfrac{3}{4}\right)\right\}$ **35.** $\{(-5\ 3)\}$ **36.** $\left\{\left(-1, -\dfrac{1}{2}\right)\right\}$ **37.** $\{(x, y) | 3x + 9y = 12\}$ **38.** $\{(x, y) | x = 3y - 1\}$

45. $\{(x, y) | x = 3y + 2\}$ **46.** $\{(x, y) | y = \dfrac{1}{7}x + 3\}$ **47.** $\left\{\left(-\dfrac{1}{4}, \dfrac{1}{2}\right)\right\}$ **51.** $\{(7, -3)\}$

53. $\begin{cases} 4x - 7y = 7 \\ 12x - 21y = 24 \end{cases}$ {} **54.** $\begin{cases} 2x - 5y = 12 \\ -4x + 10y = 20 \end{cases}$ {}

55. $\begin{cases} \dfrac{2}{3}x - \dfrac{3}{4}y = -1 \\ -\dfrac{1}{6}x + \dfrac{3}{8}y = 1 \end{cases}$ **56.** $\begin{cases} \dfrac{1}{2}x - \dfrac{1}{3}y = -3 \\ \dfrac{1}{8}x + \dfrac{1}{6}y = 0 \end{cases}$ {(−4, 3)}

57. $\begin{cases} 0.7x - 0.2y = -1.6 \\ 0.2x - y = -1.4 \end{cases}$ **58.** $\begin{cases} -0.7x + 0.6y = 1.3 \\ 0.5x - 0.3y = -0.8 \end{cases}$

59. $\begin{cases} 4x - 1.5y = 10.2 \\ 2x + 7.8y = -25.68 \end{cases}$ **60.** $\begin{cases} x - 3y = -5.3 \\ 6.3x + 6y = 3.96 \end{cases}$

The concept of supply and demand is used often in business. In general, as the unit price of a commodity increases, the demand for that commodity decreases. Also, as a commodity's unit price increases, the manufacturer normally increases the supply. The point where supply is equal to demand is called the equilibrium point. The following graph shows the graph of a demand equation and a supply equation for ties. The x-axis represents number of ties in thousands, and the y-axis represents the cost of a tie. Use this graph for Exercises 61–64.

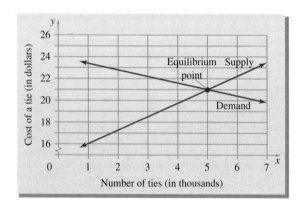

61. Find the number of ties and the price per tie when supply equals demand.   5000 ties; $21

62. When *x* is between 3 and 4, is supply greater than demand or is demand greater than supply?

63. When *x* is greater than 7, is supply greater than demand or is demand greater than supply?

64. For what *x*-values are the *y*-values corresponding to the supply equation greater than the *y*-values corresponding to the demand equation?
for *x*-values greater than 5

The revenue equation for a certain brand of toothpaste is $y = 2.5x$, where x is the number of tubes of toothpaste sold and y is the total income for selling x tubes. The cost equation is $y = 0.9x + 3000$, where x is the number of tubes of toothpaste manufactured and y is the cost of producing x tubes. The following set of axes shows the graph of the cost and revenue equations. Use this graph for Exercises 65–70. See Example 8.

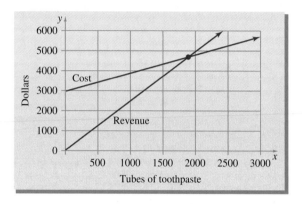

65. Find the coordinates of the point of intersection by solving the system

$$\begin{cases} y = 2.5x \\ y = 0.9x + 3000 \end{cases}$$ (1875, 4687.5)

66. Explain the meaning of the ordered pair point of intersection. 1875 tubes must be sold to break even.

67. If the company sells 2000 tubes of toothpaste, does the company make money or lose money?

68. If the company sells 1000 tubes of toothpaste, does the company make money or lose money?

69. For what *x*-values will the company make a profit? (*Hint:* For what *x*-values is the revenue graph "higher" than the cost graph?)

70. For what *x*-values will the company lose money? (*Hint:* For what *x*-values is the revenue graph "lower" than the cost graph?)

55. {(3, 4)}   **57.** {(−2, 1)}   **58.** {(−1, 1)}   **59.** {(1.2, −3.6)}   **60.** {(−0.8, 1.5)}
62. Demand is greater than supply.   **63.** Supply is greater than demand.   **67.** makes money
68. loses money   **69.** for *x*-values greater than 1875   **70.** for *x*-values less than 1875

71. The amount of U.S. federal government income y (in billions of dollars) for the fiscal year x, from 1991 through 1994 ($x = 0$ represents 1990), can be modeled by the linear equation $y = 67x + 971$. The amount of U.S. federal government expenditures (in billions of dollars) for the same period can be modeled by the linear equation $y = 44x + 1285$. If these patterns in income and expenditures continue into the future, will income ever equal expenditures? If so, in what year? (*Source*: U.S. Department of the Treasury)   2003

72. The number of milk cows y (in thousands) on farms in the United States for the year x, from 1970 through 1995 ($x = 0$ represents 1980), can be modeled by the linear equation $107x + y = 11{,}096$. The number of sheep (in thousands) on farms in the United States for the same period can be modeled by the linear equation $y = -399x + 15{,}149$. In which year were there the same number of milk cows as sheep? (*Source*: National Agricultural Statistics Service)   1993

Review Exercises

Determine whether the given replacement values make each equation true or false. See Section 2.1.

73. $3x - 4y + 2z = 5; x = 1, y = 2,$ and $z = 5$ true

74. $x + 2y - z = 7; x = 2, y = -3,$ and $z = 3$ false

75. $-x - 5y + 3z = 15; x = 0, y = -1,$ and $z = 5$ false

76. $-4x + y - 8z = 4; x = 1, y = 0,$ and $z = -1$ true

Add the equations. See Section 4.1.

77. $3x + 2y - 5z = 10$
$-3x + 4y + \ z = 15$

78. $x + 4y - 5z = 20$
$2x - 4y - 2z = -17$

79. $10x + 5y + 6z = 14$
$-9x + 5y - 6z = -12$
$x + 10y = 2$

80. $-9x - 8y - z = 31$
$9x + 4y - z = 12$
$-4y - 2z = 33$

A Look Ahead

EXAMPLE   Solve the system $\begin{cases} -\dfrac{4}{x} - \dfrac{4}{y} = -11 \\ \dfrac{1}{x} + \dfrac{1}{y} = 1 \end{cases}$

Solution: First, make the following substitution.

77. $6y - 4z = 25$ **78.** $3x - 7z = 3$

Let $a = \dfrac{1}{x}$ and $b = \dfrac{1}{y}$ in both equations.

Then

$$\begin{cases} -4\left(\dfrac{1}{x}\right) - 4\left(\dfrac{1}{y}\right) = -11 \\ \dfrac{1}{x} + \dfrac{1}{y} = 1 \end{cases}$$

is equivalent to $\begin{cases} -4a - 4b = -11 \\ a + b = \ \ 1 \end{cases}$

We solve by the elimination method. Multiplying both sides of the second equation by 4 and adding the left sides and the right sides of the equations,

$$\begin{cases} -4a - 4b = -11 \\ {}^4(a + b) = \ \ {}^4(1) \end{cases}$$

simplifies to $\begin{cases} -4a - 4b = -11 \\ 4a + 4b = \ \ 4 \end{cases}$

$$0 = -7 \quad \text{False.}$$

The equation $0 = -7$ is false for all values of a and hence for all values of $\dfrac{1}{x}$ and all values of x. This system has no solution.

Solve each system. See the preceding example.

81. $\begin{cases} \dfrac{1}{x} + y = 12 \\ \dfrac{3}{x} - y = 4 \end{cases}$ $\left\{\left(\dfrac{1}{4}, 8\right)\right\}$

82. $\begin{cases} x + \dfrac{2}{y} = 7 \\ 3x + \dfrac{3}{y} = 6 \end{cases}$ $\left\{\left(-3, \dfrac{1}{5}\right)\right\}$

83. $\begin{cases} \dfrac{1}{x} + \dfrac{1}{y} = 5 \\ \dfrac{1}{x} - \dfrac{1}{y} = 1 \end{cases}$ $\left\{\left(\dfrac{1}{3}, \dfrac{1}{2}\right)\right\}$

84. $\begin{cases} \dfrac{2}{x} + \dfrac{3}{y} = 5 \\ \dfrac{5}{x} - \dfrac{3}{y} = 2 \end{cases}$ $\{(1, 1)\}$

85. $\begin{cases} \dfrac{2}{x} + \dfrac{3}{y} = -1 \\ \dfrac{3}{x} - \dfrac{2}{y} = 18 \end{cases}$ $\left\{\left(\dfrac{1}{4}, -\dfrac{1}{3}\right)\right\}$

86. $\begin{cases} \dfrac{3}{x} - \dfrac{2}{y} = -18 \\ \dfrac{2}{x} + \dfrac{3}{y} = 1 \end{cases}$ $\left\{\left(-\dfrac{1}{4}, \dfrac{1}{3}\right)\right\}$

87. $\begin{cases} \dfrac{2}{x} - \dfrac{4}{y} = 5 \\ \dfrac{1}{x} - \dfrac{2}{y} = \dfrac{3}{2} \end{cases}$ $\{\ \}$

88. $\begin{cases} \dfrac{5}{x} + \dfrac{7}{y} = 1 \\ -\dfrac{10}{x} - \dfrac{14}{y} = 0 \end{cases}$ $\{\ \}$

4.2 | SOLVING SYSTEMS OF LINEAR EQUATIONS IN THREE VARIABLES

TAPE IA 4.2

OBJECTIVES

1. Recognize a linear equation in three variables.
2. Solve a system of three linear equations in three variables.

In this section, the algebraic methods of solving systems of two linear equations in two variables are extended to systems of three linear equations in three variables. We call the equation $3x - y + z = -15$, for example, a **linear equation in three variables** since there are three variables and each variable is raised only to the power 1. A solution of this equation is an **ordered triple (x, y, z)** that makes the equation a true statement. For example, the ordered triple $(2, 0, -21)$ is a solution of $3x - y + z = -15$ since replacing x with 2, y with 0, and z with -21 yields the true statement $3(2) - 0 + (-21) = -15$. The graph of this equation is a plane in three-dimensional space, just as the graph of a linear equation in two variables is a line in two-dimensional space.

Although we will not discuss the techniques for graphing equations in three variables, visualizing the possible patterns of intersecting planes gives us insight into the possible patterns of solutions of a system of three three-variable linear equations. There are four possible patterns.

1. Three planes have a single point in common. This point represents the single solution of the system. The system is **consistent.**

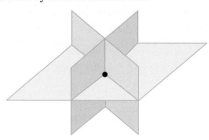

2. Three planes intersect at no point common to all three. This system has no solution. A few ways that this can occur are shown. This system is **inconsistent.**

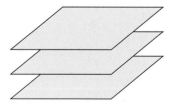

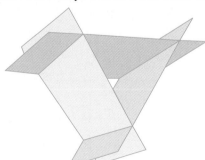

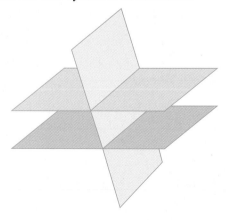

3. Three planes intersect at all the points of a single line. The system has infinitely many solutions. This system is **consistent.**

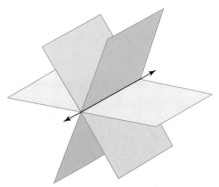

4. Three planes coincide at all points on the plane. The system is consistent, and the equations are **dependent.**

2 Using the elimination method to solve a system in three variables, we eliminate a variable and obtain a system in two variables.

EXAMPLE 1   Solve the system:

$$\begin{cases} 3x - y + z = -15 & \text{Equation (1)} \\ x + 2y - z = 1 & \text{Equation (2)} \\ 2x + 3y - 2z = 0 & \text{Equation (3)} \end{cases}$$

Solution: Add equations (1) and (2) to eliminate z.

$$\begin{array}{r} 3x - y + z = -15 \\ x + 2y - z = 1 \\ \hline 4x + y = -14 \quad \text{Equation (4)} \end{array}$$

Next, add two *other* equations and *eliminate z again*. To do so, multiply both sides of equation (1) by 2 and add this resulting equation to equation (3). Then

$$\begin{cases} 2(3x - y + z) = 2(-15) \\ 2x + 3y - 2z = 0 \end{cases} \quad \begin{array}{l} \text{simplifies} \\ \text{to} \end{array} \quad \begin{array}{r} 6x - 2y + 2z = -30 \\ 2x + 3y - 2z = 0 \\ \hline 8x + y = -30 \quad \text{Equation (5)} \end{array}$$

Now solve equations (4) and (5) for x and y. To solve by elimination, multiply both sides of equation (4) by -1 and add this resulting equation to equation (5). Then

$$\begin{cases} -1(4x + y) = -1(-14) \\ \quad\;\; 8x + y = -30 \end{cases} \quad \begin{matrix} \text{simplifies} \\ \text{to} \end{matrix} \quad \begin{cases} -4x - y = 14 \\ \;\; 8x + y = -30 \end{cases}$$

$$\begin{aligned} 4x \quad\;\;\; &= -16 \qquad \text{Add the equations.} \\ x &= -4 \qquad \text{Solve for } x. \end{aligned}$$

Replace x with -4 in equation (4) or (5).

$$\begin{aligned} 4x + y &= -14 \qquad \text{Equation (4)} \\ 4(-4) + y &= -14 \qquad \text{Let } x = -4. \\ y &= 2 \qquad \text{Solve for } y. \end{aligned}$$

Finally, replace x with -4 and y with 2 in equation (1), (2), or (3).

$$\begin{aligned} x + 2y - z &= 1 \qquad \text{Equation (2)} \\ -4 + 2(2) - z &= 1 \qquad \text{Let } x = -4 \text{ and } y = 2. \\ -4 + 4 - z &= 1 \\ -z &= 1 \\ z &= -1 \end{aligned}$$

The solution is $(-4, 2, -1)$. To check, let $x = -4$, $y = 2$, and $z = -1$ in all three original equations of the system.

Equation (1)	Equation (2)	Equation (3)
$3x - y + z = -15$	$x + 2y - z = 1$	$2x + 3y - 2z = 0$
$3(-4) - 2 + (-1) = -15$	$-4 + 2(2) - (-1) = 1$	$2(-4) + 3(2) - 2(-1) = 0$
$-12 - 2 - 1 = -15$	$-4 + 4 + 1 = 1$	$-8 + 6 + 2 = 0$
$-15 = -15$	$1 = 1$	$0 = 0$
True.	True.	True.

All three statements are true, so the solution set is $\{(-4, 2, -1)\}$.

EXAMPLE 2 Solve the system:

$$\begin{cases} 2x - 4y + 8z = 2 & (1) \\ -x - 3y + \;\; z = 11 & (2) \\ \;\; x - 2y + 4z = 0 & (3) \end{cases}$$

Solution: Add equations (2) and (3) to eliminate x, and the new equation is

$$-5y + 5z = 11 \quad (4)$$

To eliminate x again, multiply both sides of equation (2) by 2, and add the resulting equation to equation (1). Then

$$\begin{cases} 2x - 4y + 8z = 2 \\ 2(-x - 3y + z) = 2(11) \end{cases} \quad \begin{matrix} \text{simplifies} \\ \text{to} \end{matrix} \quad \begin{cases} 2x - 4y + 8z = 2 \\ -2x - 6y + 2z = 22 \end{cases}$$

$$-10y + 10z = 24 \quad (5)$$

Next, solve for y and z using equations (4) and (5). Multiply both sides of equation (4) by -2, and add the resulting equation to equation (5).

$$\begin{cases} -2(-5y + 5z) = -2(11) \\ -10y + 10z = 24 \end{cases} \quad \begin{matrix} \text{simplifies} \\ \text{to} \end{matrix} \quad \begin{cases} 10y - 10z = -22 \\ \underline{-10y + 10z = 24} \\ \qquad\qquad 0 = 2 \quad \text{False.} \end{cases}$$

Since the statement is false, this system is inconsistent and has no solution. The solution set is the empty set { } or $\varnothing$. ▬▬▬

The elimination method is summarized next.

To Solve a System of Three Linear Equations by the Elimination Method

Step 1.   Write each equation in standard form $Ax + By + Cz = D$.

Step 2.   Choose a pair of equations and use the equations to eliminate a variable.

Step 3.   Choose any other pair of equations and eliminate the **same variable** as in *step 2*.

Step 4.   Two equations in two variables should be obtained from *step 2* and *step 3*. Use methods from Section 4.1 to solve this system for both variables.

Step 5.   To solve for the third variable, substitute the values of the variables found in *step 4* into any of the original equations containing the third variable.

EXAMPLE 3   Solve the system:

$$\begin{cases} 2x + 4y \qquad\;\; = \;\;\; 1 & (1) \\ 4x \qquad\; - 4z = -1 & (2) \\ \qquad\;\; y - 4z = -3 & (3) \end{cases}$$

Solution: Notice that equation (2) has no term containing the variable y. Let us eliminate y using equations (1) and (3). Multiply both sides of equation (3) by -4, and add the resulting equation to equation (1). Then

$$\begin{cases} 2x + 4y \qquad\; = 1 \\ -4(\; y - 4z) = -4(-3) \end{cases} \quad \begin{matrix} \text{simplifies} \\ \text{to} \end{matrix} \quad \begin{cases} 2x + 4y \qquad\quad = \;\;\; 1 \\ \underline{\quad - 4y + 16z = 12} \\ 2x \qquad\; + 16z = 13 \quad (4) \end{cases}$$

Next, solve for z using equations (4) and (2). Multiply both sides of equation (4) by -2 and add the resulting equation to equation (2).

$$\begin{cases} -2(2x + 16z) = -2(13) \\ 4x - 4z = -1 \end{cases} \begin{array}{l} \text{simplifies} \\ \text{to} \end{array} \begin{cases} -4x - 32z = -26 \\ 4x - 4z = -1 \end{cases}$$

$$-36z = -27$$

$$z = \frac{3}{4}$$

Replace z with $\frac{3}{4}$ in equation (3) and solve for y.

$$y - 4\left(\frac{3}{4}\right) = -3 \qquad \text{Let } z = \frac{3}{4} \text{ in equation (3).}$$

$$y - 3 = -3$$

$$y = 0$$

Replace y with 0 in equation (1) and solve for x.

$$2x + 4(0) = 1$$

$$2x = 1$$

$$x = \frac{1}{2}$$

The solution set is $\left\{\left(\frac{1}{2}, 0, \frac{3}{4}\right)\right\}$. Check to see that this solution satisfies all three equations of the system.

EXAMPLE 4   Solve the system:

$$\begin{cases} x - 5y - 2z = 6 & (1) \\ -2x + 10y + 4z = -12 & (2) \\ \frac{1}{2}x - \frac{5}{2}y - z = 3 & (3) \end{cases}$$

Solution: Multiply both sides of equation (3) by 2 to eliminate fractions, and multiply both sides of equation (2) by $-\frac{1}{2}$ so that the coefficient of x is 1. The resulting system is then

$$\begin{cases} x - 5y - 2z = 6 & (1) \\ x - 5y - 2z = 6 & \text{Multiply (2) by } -\frac{1}{2}. \\ x - 5y - 2z = 6 & \text{Multiply (3) by 2.} \end{cases}$$

All three equations are identical, and therefore equations (1), (2), and (3) are all equivalent. There are infinitely many solutions of this system. The equations are dependent. The solution set can be written as $\{(x, y, z) \mid x - 5y - 2z = 6\}$.

EXERCISE SET 4.2

9. Answers will vary. One possibility is $\begin{cases} 3x && = -3 \\ 2x + 4y && = 6 \\ x - 3y + z = -11 \end{cases}$

10. Answers will vary. One possibility is $\begin{cases} -x + y - 2z = -11 \\ 3x && + z = 11 \\ && 2z = 10 \end{cases}$

Solve each system. See Examples 1 and 3.

1. $\begin{cases} x + y && = 3 \\ 2y && = 10 \\ 3x + 2y - 3z = 1 \end{cases}$
2. $\begin{cases} 5x && = 5 \\ 2x + y && = 4 \\ 3x + y - 4z = -15 \end{cases}$

3. $\begin{cases} 2x + 2y + z = 1 \\ -x + y + 2z = 3 \\ x + 2y + 4z = 0 \end{cases}$
4. $\begin{cases} 2x - 3y + z = 5 \\ x + y + z = 0 \\ 4x + 2y + 4z = 4 \end{cases}$

1. $\{(-2, 5, 1)\}$ **2.** $\{(1, 2, 5)\}$ **3.** $\{(-2, 3, -1)\}$
4. $\{(-3, -2, 5)\}$

Solve each system. See Examples 2 and 4.

5. $\begin{cases} x - 2y + z = -5 \\ -3x + 6y - 3z = 15 \\ 2x - 4y + 2z = -10 \end{cases}$ $\{(x, y, z) \mid x - 2y + z = -5\}$

6. $\begin{cases} 3x + y - 2z = 2 \\ -6x - 2y + 4z = -2 \\ 9x + 3y - 6z = 6 \end{cases}$ $\{\}$

7. $\begin{cases} 4x - y + 2z = 5 \\ 2y + z = 4 \\ 4x + y + 3z = 10 \end{cases}$ $\{\}$ **8.** $\begin{cases} 5y - 7z = 14 \\ 2x + y + 4z = 10 \\ 2x + 6y - 3z = 30 \end{cases}$ $\{\}$

📃 **9.** Write a linear equation in three variables that has $(-1, 2\ -4)$ as a solution. (There are many possibilities.)

📃 **10.** Write a system of three linear equations in three variables that has $(2, 1, 5)$ as a solution. (There are many possibilities.)

Solve each system.

11. $\begin{cases} x + 5z = 0 \\ 5x + y = 0 \\ y - 3z = 0 \end{cases}$
12. $\begin{cases} x - 5y = 0 \\ x - z = 0 \\ -x + 5z = 0 \end{cases}$

13. $\begin{cases} 6x - 5z = 17 \\ 5x - y + 3z = -1 \\ 2x + y = -41 \end{cases}$
14. $\begin{cases} x + 2y = 6 \\ 7x + 3y + z = -33 \\ x - z = 16 \end{cases}$

15. $\begin{cases} x + y + z = 8 \\ 2x - y - z = 10 \\ x - 2y - 3z = 22 \end{cases}$
16. $\begin{cases} 5x + y + 3z = 1 \\ x - y + 3z = -7 \\ -x + y = 1 \end{cases}$

17. $\begin{cases} x + 2y - z = 5 \\ 6x + y + z = 7 \\ 2x + 4y - 2z = 5 \end{cases}$ $\{\}$ **18.** $\begin{cases} 4x - y + 3z = 10 \\ x + y - z = 5 \\ 8x - 2y + 6z = 10 \end{cases}$ $\{\}$

19. $\begin{cases} 2x - 3y + z = 2 \\ x - 5y + 5z = 3 \\ 3x + y - 3z = 5 \end{cases}$
20. $\begin{cases} 4x + y - z = 8 \\ x - y + 2z = 3 \\ 3x - y + z = 6 \end{cases}$

21. $\begin{cases} -2x - 4y + 6z = -8 \\ x + 2y - 3z = 4 \\ 4x + 8y - 12z = 16 \end{cases}$ $\{(x, y, z) \mid x + 2y - 3z = 4\}$

22. $\begin{cases} -6x + 12y + 3z = -6 \\ 2x - 4y - z = 2 \\ -x + 2y + \dfrac{z}{2} = -1 \end{cases}$

23. $\begin{cases} 2x + 2y - 3z = 1 \\ y + 2z = -14 \\ 3x - 2y = -1 \end{cases}$
24. $\begin{cases} 7x + 4y = 10 \\ x - 4y + 2z = 6 \\ y - 2z = -1 \end{cases}$

25. $\begin{cases} \dfrac{3}{4}x - \dfrac{1}{3}y + \dfrac{1}{2}z = 9 \\ \dfrac{1}{6}x + \dfrac{1}{3}y - \dfrac{1}{2}z = 2 \\ \dfrac{1}{2}x - y + \dfrac{1}{2}z = 2 \end{cases}$
26. $\begin{cases} \dfrac{1}{3}x - \dfrac{1}{4}y + z = -9 \\ \dfrac{1}{2}x - \dfrac{1}{3}y - \dfrac{1}{4}z = -6 \\ x - \dfrac{1}{2}y - z = -8 \end{cases}$

📃 **27.** The fraction $\frac{1}{24}$ can be written as the following sum:

$$\frac{1}{24} = \frac{x}{8} + \frac{y}{4} + \frac{z}{3}$$

where the numbers x, y, and z are solutions of

$$\begin{cases} x + y + z = 1 \\ 2x - y + z = 0 \\ -x + 2y + 2z = -1 \end{cases}$$

Solve the system and see that the sum of the fractions is $\frac{1}{24}$. $\{(1, 1, -1)\}$

📃 **28.** The fraction $\frac{1}{18}$ can be written as the following sum.

$$\frac{1}{18} = \frac{x}{2} + \frac{y}{3} + \frac{z}{9}$$

where the numbers x, y, and z are solutions of

$$\begin{cases} x + 3y + z = -3 \\ -x + y + 2z = -14 \\ 3x + 2y - z = 12 \end{cases}$$

Solve the system and see that the sum of the fractions is $\frac{1}{18}$. $\{(1, 1, -7)\}$

11. $\{(0, 0, 0)\}$ **12.** $\{(0, 0, 0)\}$ **13.** $\{(-3, -35, -7)\}$ **14.** $\{(-4, 5, -20)\}$ **15.** $\{(6, 22, -20)\}$ **16.** $\{(1, 2, -2)\}$
19. $\{(3, 2, 2)\}$ **20.** $\{(2, 1, 1)\}$ **22.** $\{(x, y, z) \mid -6x + 12y + 3z = -6\}$ **23.** $\{(-3, -4, -5)\}$ **24.** $\{(2, -1, 0)\}$
25. $\{(12, 6, 4)\}$ **26.** $\{(-6, 12, -4)\}$

Review Exercises

Solve. See Section 2.2.

29. The sum of two numbers is 45 and one number is twice the other. Find the numbers.   15 and 30

30. The difference between two numbers is 5. Twice the smaller number added to five times the larger number is 53. Find the numbers.   9 and 4

Solve. See Section 2.1.

31. $2(x - 1) - 3x = x - 12$ {5}

32. $7(2x - 1) + 4 = 11(3x - 2)$ {1}

33. $-y - 5(y + 5) = 3y - 10$ $\left\{-\dfrac{15}{9}\right\}$

34. $z - 3(z + 7) = 6(2z + 1)$ $\left\{-\dfrac{27}{14}\right\}$

37. {(1, −1, 2, 3)}   **38.** {(1, 2, 3, 4)}

A Look Ahead

Solve each system.

35. $\begin{cases} x + y \qquad - w = \ \ 0 \\ \qquad y + 2z + w = \ \ 3 \\ x \qquad - z \qquad = \ \ 1 \\ 2x - y \qquad - w = -1 \end{cases}$ {(1, 1, 0, 2)}

36. $\begin{cases} 5x + 4y \qquad\quad = \ \ 29 \\ \qquad y + z - w = -2 \\ 5x \qquad + z \qquad = \ \ 23 \\ \qquad y - z + w = \ \ 4 \end{cases}$ {(5, 1, −2, 1)}

37. $\begin{cases} x + y + z + w = 5 \\ 2x + y + z + w = 6 \\ x + y + z \qquad = 2 \\ x + y \qquad\quad = 0 \end{cases}$

38. $\begin{cases} 2x \qquad - z \qquad = -1 \\ \qquad y + z + \ w = \ \ 9 \\ \qquad y \qquad - 2w = -6 \\ x + y \qquad\quad = \ \ 3 \end{cases}$

4.3 | SYSTEMS OF LINEAR EQUATIONS AND PROBLEM SOLVING

TAPE IA 4.3

OBJECTIVE

1 Solve problems that can be modeled by a system of linear equations.

Thus far, we have solved problems by writing one-variable equations and solving for the variable. Some of these problems can be solved, perhaps more easily, by writing a system of equations, as illustrated in this section. We begin with a problem about numbers.

EXAMPLE 1   A first number is 4 less than a second number. Four times the first number is 6 more than twice the second. Find the numbers.

Solution: **1. UNDERSTAND.**   Read and reread the problem and guess a solution. If one number is 10 and this is 4 less than a second number, the second number is 14. Four times the first number is 4(10), or 40. This is not equal to 6 more than twice the second number, which is 2(14) + 6 or 34. Although we guessed incorrectly, we now have a better understanding of the problem.

2. ASSIGN.   Since we are looking for two numbers, we will let

$x = $ first number

$y = $ second number

4. TRANSLATE.   Since we have assigned two variables to this problem, we will translate the given facts into two equations.

In words:	the first number	is	4 less than the second number
Translate:	x	$=$	$y - 4$

Next we translate the second statement into an equation.

In words:	four times the first number	is	6 more than twice the second number
Translate:	$4x$	$=$	$2y + 6$

5. **COMPLETE.**   Here we solve the system

$$\begin{cases} x = y - 4 \\ 4x = 2y + 6 \end{cases}$$

Since the first equation expresses x in terms of y, use substitution.

Substitute $y - 4$ for x in the second equation and solve for y.

$$4x = 2y + 6$$
$$4(y - 4) = 2y + 6 \qquad \text{Let } x = y - 4.$$
$$4y - 16 = 2y + 6$$
$$2y = 22$$
$$y = 11$$

Now replace y with 11 in the equation $x = y - 4$ and solve for x. Then $x = y - 4$ becomes $x = 11 - 4 = 7$. The solution of the system is $(7, 11)$.

6. **INTERPRET.**   Since the solution of the system is $(7, 11)$, then the first number we are looking for is 7 and the second number is 11. *Checking,* notice that 7 *is* 4 less than 11, and 4 times 7 *is* 6 more than twice 11. The proposed numbers, 7 and 11, are correct.

State: The numbers are 7 and 11.

EXAMPLE 2   Two cars leave Indianapolis, one traveling east and the other west. After 3 hours they are 297 miles apart. If one car is traveling 5 mph faster than the other, what is the speed of each?

Solution: 1. **UNDERSTAND.**   Read and reread the problem. Let's guess a solution and use the formula $d = r \cdot t$ to check. Suppose that one car is traveling at a rate of 55 miles per hour. This means that the other car is traveling at a rate of 50 miles per hour since we are told that one car is traveling 5 mph faster than the other. To find the distance apart after 3 hours, we will first find the distance traveled by each car. One car's distance is rate $\cdot$ time $= 55(3) = 165$ miles. The other car's distance is rate $\cdot$ time $= 50(3) = 150$ miles. Since one car is traveling east and the other west, their distance apart is the sum of their distances, or 165 miles $+$ 150 miles $= 315$ miles. This distance apart is not the required distance of 297 miles.

50(3) = 150 miles   55(3) = 165 miles

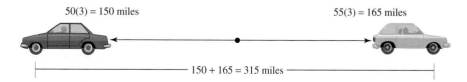

150 + 165 = 315 miles

Now that we have a better understanding of the problem, let's model it with a system of equations.

2. **ASSIGN.** Let

x = speed of one car

y = speed of the other car

3. **ILLUSTRATE.** We summarize the information on the following chart. Both cars have traveled 3 hours. Since distance = rate · time, their distances are $3x$ and $3y$ miles, respectively.

	RATE ·	**TIME** =	**DISTANCE**
ONE CAR	x	3	$3x$
OTHER CAR	y	3	$3y$

4. **TRANSLATE.** We translate into two equations.

In words:	one car's distance	added to	the other car's distance	is	297
Translate:	$3x$	$+$	$3y$	$=$	297

In words:	one car	is	5 mph faster than the other
Translate:	x	$=$	$y + 5$

5. **COMPLETE.** Here we solve the system.

$$\begin{cases} 3x + 3y = 297 \\ x = y + 5 \end{cases}$$

Again, the substitution method is appropriate. Replace x with $y + 5$ in the first equation and solve for y.

$$3x + 3y = 297$$
$$3(y + 5) + 3y = 297 \qquad \text{Let } x = y + 5.$$
$$3y + 15 + 3y = 297$$
$$6y = 282$$
$$y = 47$$

To find x, replace y with 47 in the equation $x = y + 5$. Then $x = 47 + 5 = 52$. The solution of the system is $(52, 47)$.

6. **INTERPRET.** The solution $(52, 47)$ means that the cars are traveling at 52 mph and 47 mph, respectively. To *check,* notice that one car is traveling 5 mph faster than the other. Also, if one car travels 52 mph for 3 hours, the distance is

3(52) = 156 miles. The other car traveling for 3 hours at 47 mph travels a distance of 3(47) = 141 miles. The sum of the distances 156 + 141 is 297 miles, the required distance.

State: The cars are traveling at 52 mph and 47 mph.

EXAMPLE 3   Lynn Pike, a pharmacist, needs 70 liters of 50% alcohol solution. She has available a 30% alcohol solution and an 80% alcohol solution. How many liters of each solution should she mix to obtain 70 liters of a 50% alcohol solution?

1. UNDERSTAND.   Read and reread the problem. Next, guess the solution. Suppose that we need 20 liters of the 30% solution. Then we need 70 − 20 = 50 liters of the 80% solution. To see if this gives us 70 liters of a 50% alcohol solution, let's find the amount of pure alcohol in each solution.

number of liters	×	alcohol strength	=	amount of pure alcohol
20 liters	×	0.30	=	6 liters
50 liters	×	0.80	=	40 liters
70 liters	×	0.50	=	35 liters

Since 6 liters + 40 liters = 46 liters and not 35 liters, our guess is incorrect, but we have gained some insight as to how to model and check this problem.

2. ASSIGN.   Let

x = amount of 30% solution, in liters

y = amount of 80% solution, in liters

3. ILLUSTRATE.   Use a table to organize the given data.

	NUMBER OF LITERS	ALCOHOL STRENGTH	AMOUNT OF PURE ALCOHOL
30% SOLUTION	x	30%	$0.30x$
80% SOLUTION	y	80%	$0.80y$
50% SOLUTION NEEDED	70	50%	$(0.50)(70)$

4. TRANSLATE.   We translate into two equations.

In words: amount of 30% solution + amount of 80% solution = 70

Translate: x + y = 70

In words: amount of pure alcohol in 30% solution + amount of pure alcohol in 80% solution = amount of pure alcohol in 50% solution

Translate: $0.30x$ + $0.80y$ = $(0.50)(70)$

5. COMPLETE. Here we solve the system

$$\begin{cases} x + y = 70 \\ 0.30x + 0.80y = (0.50)(70) \end{cases}$$

To solve this system, use the elimination method. Multiply both sides of the first equation by -3 and both sides of the second equation by 10. Then

$$\begin{cases} -3(x + y) = -3(70) \\ 10(0.30x + 0.80y) = 10(0.50)(70) \end{cases} \quad \text{simplifies to} \quad \begin{cases} -3x - 3y = -210 \\ 3x + 8y = 350 \end{cases}$$
$$\begin{array}{rcl} 5y &=& 140 \\ y &=& 28 \end{array}$$

Replace y with 28 in the equation $x + y = 70$ and find that $x + 28 = 70$, or $x = 42$.

The solution of the system is $(42, 28)$.

6. INTERPRET. To *check*, recall how we checked our guess. *State:* The pharmacist needs to mix 42 liters of 30% solution and 28 liters of 80% solution to obtain 70 liters of 50% solution. ▬▬▬▬▬▬▬

Recall that businesses are often computing cost and revenue functions or equations in order to predict needed sales, to determine whether prices need to be adjusted, and also to see whether the company is making money or losing money. Recall also that the value at which revenue equals cost is called the break-even point. When revenue is less than cost, the company is losing money, and when revenue is greater than cost, the company is making money.

▬▬▬▬▬

EXAMPLE 4 A manufacturing company recently purchased $3000 worth of new equipment in order to offer new personalized stationery to its customers. The cost of producing a package of personalized stationery is $3.00, and it is sold for $5.50. Find the number of packages that must be sold to break even.

Solution: **1. UNDERSTAND.** Read and reread the problem. Notice that the cost to the company will include a one-time cost of $3000 for the equipment and then $3.00 per package produced. The revenue will be $5.50 per package sold.

2. ASSIGN.

Let x = number of packages of personalized stationery

Let $C(x)$ = total cost for producing x packages of stationery

Let $R(x)$ = total revenue for selling x packages of stationery

4. TRANSLATE. The revenue equation is

In words:	revenue for selling x packages of stationery	=	price per package	·	number of packages

Translate:	$R(x)$	=	5.5	·	x

The cost equation is

In words:

cost for producing x packages of stationery	=	cost per package	·	number of packages	+	cost for equipment

Translate: $C(x)$ =   3   · x +   3000

Since the break-even point is when $R(x) = C(x)$, we solve the equation

$$5.5x = 3x + 3000$$

5. COMPLETE.

$$5.5x = 3x + 3000$$
$$2.5x = 3000 \qquad \text{Subtract } 3x \text{ from both sides.}$$
$$x = 1200 \qquad \text{Divide both sides by 2.5.}$$

6. INTERPRET.   *Check:* To see whether the break-even point occurs when 1200 packages are produced and sold, see if revenue equals cost when $x = 1200$. When $x = 1200$, $R(x) = 5.5x = 5.5(1200) = 6600$ and $C(x) = 3x + 3000 = 3(1200) + 3000 = 6600$. Since $R(x) = C(x) = 6600$, the break-even point is 1200.

State: The company must sell 1200 packages of stationery in order to break even. The graph of this system follows.

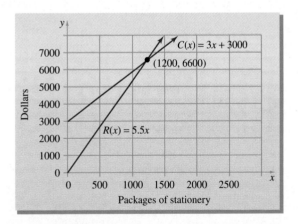

In order to introduce problem solving by writing a system of three linear equations in three unknowns, we solve a problem about triangles.

EXAMPLE 5   The measure of the largest angle of a triangle is 80° more than the measure of the smallest angle, and the measure of the remaining angle is 10° more than the measure of the smallest angle. Find the measure of each angle.

Solution: **1. UNDERSTAND.**   Read and reread the problem. Recall that the sum of the measures of the angles of a triangle is 180°. Then guess a solution. If the smallest angle measures 20°, the measure of the largest angle is 80° more, or

$20° + 80° = 100°$. The measure of the remaining angle is $10°$ more than the measure of the smallest angle, or $20° + 10° = 30°$. The sum of these three angles is $20° + 100° + 30° = 150°$, not the required $180°$. We now know that the measure of the smallest angle is greater than $20°$.

2. **ASSIGN.** Let

   x = degree measure of the smallest angle

   y = degree measure of the largest angle

   z = degree measure of the remaining angle

3. **ILLUSTRATE.**

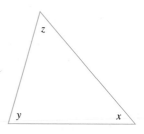

4. **TRANSLATE.**   We translate into three equations.

   In words: | the sum of the angles | = | 180

   Translate: $x + y + z = 180$

   In words: | the largest angle | is | 80 more than the smallest angle

   Translate: $y = x + 80$

   In words: | the remaining angle | is | 10 more than the smallest angle

   Translate: $z = x + 10$

5. **COMPLETE.**   We solve the system

   $$\begin{cases} x + y + z = 180 \\ y = x + 80 \\ z = x + 10 \end{cases}$$

   Since y and z are both expressed in terms of x, we will solve using the substitution method.

Substitute $y = x + 80$ and $z = x + 10$ in the first equation. Then

becomes
$$x + y + z = 180$$
$$x + (x + 80) + (x + 10) = 180$$
$$3x + 90 = 180$$
$$3x = 90$$
$$x = 30$$

Then $y = x + 80 = 30 + 80 = 110$, and $z = x + 10 = 30 + 10 = 40$. The angles measure 30°, 40°, and 110°.

6. INTERPRET.   To *check*, notice that $30° + 40° + 110° = 180°$. Also, the measure of the largest angle, 110°, is 80° more than the measure of the smallest angle, 30°. The measure of the remaining angle, 40°, is 10° more than the measure of the smallest angle, 30°.

State: The angles measure 30°, 40°, and 110°.

EXERCISE SET 4.3

Solve. See Examples 1–3.

1. One number is two more than a second number. Twice the first is 4 less than 3 times the second. Find the numbers.   10 and 8

2. Three times one number minus a second is 8, and the sum of the numbers is 12. Find the numbers.

3. A Delta 727 traveled 560 mph with the wind and 480 mph against the wind. Find the speed of the plane in still air and the speed of the wind.

4. Terry Watkins can row about 10.6 kilometers in 1 hour downstream and 6.8 kilometers upstream in 1 hour. Find how fast he can row in still water, and find the speed of the current.

5. Find how many quarts of 4% butterfat milk and 1% butterfat milk should be mixed to yield 60 quarts of 2% butterfat milk.

6. A pharmacist needs 500 milliliters of a 20% phenobarbital solution but has only 5% and 25% phenobarbital solutions available. Find how many milliliters of each he should mix to get the desired solution.   125 ml of 5%; 375 ml of 25%

7. Karen Karlin bought some large frames for $15 each and some small frames for $8 each at a closeout sale. If she bought 22 frames for $239, find how many of each type she bought.

8. Hilton University Drama Club sold 311 tickets for a play. Student tickets cost 50 cents each; nonstudent tickets cost $1.50. If total receipts were $385.50, find how many tickets of each type were sold.

9. One number is two less than a second number. Twice the first is 4 more than 3 times the second. Find the numbers.   −10 and −8

10. Twice one number plus a second number is 42, and the one number minus the second number is −6. Find the numbers.   12 and 18

11. An office supply store in San Diego sells seven tablets and 4 pens for $6.40. Also, two tablets and 19 pens cost $5.40. Find the price of each.

12. A Candy Barrel shop manager mixes M&M's worth $2.00 per pound with trail mix worth $1.50 per pound. Find how many pounds of each she should use to get 50 pounds of a party mix worth $1.80 per pound.   30 lb of M&M's; 20 lb of trail mix

2. 5 and 7   **3.** plane, 520 mph; wind, 40 mph   **4.** still water, 8.7 km/hr; current, 1.9 km/hr   **5.** 20 quarts of 4%; 40 quarts of 1%   **7.** 9 large frames; 13 small frames   **8.** 81 student tickets, 230 nonstudent tickets   **11.** tablets, $0.80; pens, $0.20

13. An airplane takes 3 hours to travel a distance of 2160 miles with the wind. The return trip takes 4 hours against the wind. Find the speed of the plane in still air and the speed of the wind.

14. Two cyclists start at the same point and travel in opposite directions. One travels 4 mph faster than the other. In 4 hours they are 112 miles apart. Find how fast each is traveling.   12 mph and 16 mph

15. The perimeter of a quadrilateral (four-sided polygon) is 29 inches. The longest side is twice as long as the shortest side. The other two sides are equally long and are 2 inches longer than the shortest side. Find the length of all four sides.

16. The perimeter of a triangle is 93 centimeters. If two sides are equally long and the third side is 9 centimeters longer than the others, find the lengths of the three sides.   28 cm, 28 cm, 37 cm

17. The sum of three numbers is 40. One number is five more than a second and twice the third. Find the numbers.   18, 13, and 9

18. The sum of the digits of a three-digit number is 15. The tens-place digit is twice the hundreds-place digit, and the ones-place digit is 1 less than the hundreds-place digit. Find the three-digit number.   483

19. Jack Reinholt, a car salesman, has a choice of two pay arrangements: a weekly salary of $200 plus 5% commission on sales, or a straight 15% commission. Find the amount of sales for which Jack's earnings are the same regardless of the pay arrangement.

20. Hertz car rental agency charges $25 daily plus 10 cents per mile. Budget charges $20 daily plus 25 cents per mile. Find the daily mileage for which the Budget charge for the day is twice that of the Hertz charge for the day.   600 miles

21. Carroll Blakemore, a drafting student, bought 3 templates and a pencil one day for $6.45. Another day he bought 2 pads of paper and 4 pencils for $7.50. If the price of a pad of paper is three times the price of a pencil, find the price of each type of item.

Given the cost function $C(x)$ and the revenue function $R(x)$, find the number of units x that must be sold to break even. See Example 4.

22. $C(x) = 30x + 10,000$
$R(x) = 46x$ 625 units

23. $C(x) = 12x + 15,000$
$R(x) = 32x$ 750 units

24. $C(x) = 1.2x + 1500$
$R(x) = 1.7x$ 3000 units

25. $C(x) = 0.8x + 900$
$R(x) = 2x$ 750 units

26. $C(x) = 75x + 160,000$
$R(x) = 200x$ 1280 units

27. $C(x) = 105x + 70,000$
$R(x) = 245x$ 500 units

28. The planning department of Abstract Office Supplies has been asked to determine whether the company should introduce a new computer desk next year. The department estimates that $6000 of new equipment will need to be purchased and that the cost of manufacturing each desk will be $200. The department also estimates that the revenue from each desk will be $450.

   a. Determine the revenue function $R(x)$ from the sale of x desks. $R(x) = 450x$

   b. Determine the cost function $C(x)$ for manufacturing x desks. $C(x) = 200x + 6000$

   c. Find the break-even point.   24 desks

29. Baskets, Inc., is planning to introduce a new woven basket. The company estimates that $500 worth of new equipment will be needed to manufacture this new type of basket and that it will cost $15 per basket to manufacture. The company also estimates that the revenue from each basket will be $31.

   a. Determine the revenue function $R(x)$ from the sale of x baskets. $R(x) = 31x$

   b. Determine the cost function $C(x)$ for manufacturing x baskets. $C(x) = 15x + 500$

   c. Find the break-even point.   31.25, or 32 baskets

30. Line l and line m are parallel lines cut by transversal t. Find the values of x and y.

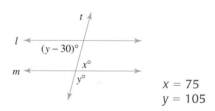

$x = 75$
$y = 105$

31. Find the values of x and y in the following isosceles triangle.

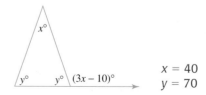

$x = 40$
$y = 70$

13. plane, 630 mph; wind, 90 mph   **15.** 5 in., 7 in., 7 in., and 10 in.   **19.** $2000 in sales
21. $1.90 for a template; $0.75 for a pencil; $2.25 for a pad of paper

Solve. See Example 5.

32. Rabbits in a lab are to be kept on a strict daily diet to include 30 grams of protein, 16 grams of fat, and 24 grams of carbohydrates. The scientist has only three food mixes available with the following grams of nutrients per unit. two units of Mix A, 3 units of Mix B, and 1 unit of Mix C

	PROTEIN	FAT	CARBOHYDRATE
MIX A	4	6	3
MIX B	6	1	2
MIX C	4	1	12

Find how many units of each mix are needed daily to meet each rabbit's dietary needs.

33. Gerry Gundersen mixes different solutions with concentrations of 25%, 40%, and 50% to get 200 liters of a 32% solution. If he uses twice as much of the 25% solution as of the 40% solution, find how many liters of each kind he uses.

34. Find the values of a, b, and c such that the equation $y = ax^2 + bx + c$ has ordered pair solutions $(1, 6)$, $(-1, -2)$, and $(0, -1)$. To do so, substitute each ordered pair solution into the equation. Each time, the result is an equation in three unknowns: a, b, and c. Then solve the resulting system of three linear equations in three unknowns, a, b, and c. $a = 3, b = 4, c = -1$

35. Find the values of a, b, and c such that the equation $y = ax^2 + bx + c$ has ordered pair solutions $(1, 2)$, $(2, 3)$ and $(-1, 6)$. (See Exercise 34.)

36. Find the values of x, y, and z in the following triangle.

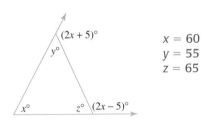

$x = 60$
$y = 55$
$z = 65$

37. The sum of the measures of the angles of a quadrilateral is $360°$. Find the value of x, y, and z in the following quadrilateral.

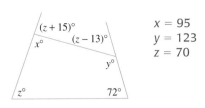

$x = 95$
$y = 123$
$z = 70$

38. Data (x, y) for the total number y (in thousands) of college-bound students who took the ACT assessment in the year x are $(-1, 855)$, $(1, 796)$, $(4, 892)$, where $x = 0$ represents 1990. Find the values of a, b, and c such that the equation $y = ax^2 + bx + c$ models this data. According to your model, how many students will take the ACT assessment in 1999? (*Source*: The American College Testing Program)

39. Monthly normal rainfall data (x, y) for Portland, Oregon, are $(4, 2.47)$, $(7, 0.6)$, $(8, 1.1)$, where x represents time in months (with $x = 1$ representing January) and y represents rainfall in inches. Find the values of a, b, and c rounded to 2 decimal places such that the equation $y = ax^2 + bx + c$ models this data. According to your model, how much rain should Portland expect during September? (*Source*: National Climatic Data Center)

Review Exercises

Multiply both sides of equation (1) by 2, and add the resulting equation to equation (2). See Section 4.2.

40. $3x - y + z = 2$ (1)   **41.** $2x + y + 3z = 7$ (1)
  $-x + 2y + 3z = 6$ (2) $-4x + y + 2z = 4$ (2)

Multiply both sides of equation (1) by -3, and add the resulting equation to equation (2). See Section 4.2.

42. $x + 2y - z = 0$ (1)   **43.** $2x - 3y + 2z = 5$ (1)
  $3x + y - z = 2$ (2) $x - 9y + z = -1$ (2)

Given the spinner below, find the probability of the spinner landing on the indicated color in one spin. See Section 2.3.

44. $P(\text{red})$ $\dfrac{3}{8}$ **45.** $P(\text{green})$ $\dfrac{3}{8}$

46. $P(\text{white})$ 0 **47.** $P(\text{red or blue})$ $\dfrac{5}{8}$

33. 120 liters of 25%, 60 liters of 40%, 20 liters of 50%   **35.** $a = 1, b = -2, c = 3$
38. $a = 12.3, b = -29.5, c = 813.2$; 1544 students in 1999   **39.** $a = 0.28, b = -3.70, c = 12.78$; 2.16 inches in Sept.
40. $5x + 5z = 10$ **41.** $3y + 8z = 18$ **42.** $-5y + 2z = 2$ **43.** $-5x - 5z = -16$

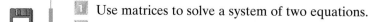

4.4 | SOLVING SYSTEMS OF EQUATIONS BY MATRICES

O B J E C T I V E S

1. Use matrices to solve a system of two equations.
2. Use matrices to solve a system of three equations.

TAPE IA 4.4

By now, you have seen that the solution of a system of equations depends on the coefficients of the equations in the system and not on the variables. In this section, we introduce solving a system of equations by a **matrix.**

A matrix (plural: **matrices**) is a rectangular array of numbers. The following are examples of matrices.

$$\begin{bmatrix} 1 & 0 \\ 0 & 1 \end{bmatrix} \qquad \begin{bmatrix} 2 & 1 & 3 & -1 \\ 0 & -1 & 4 & 5 \\ -6 & 2 & 1 & 0 \end{bmatrix} \qquad \begin{bmatrix} a & b & c \\ d & e & f \end{bmatrix}$$

2×2 matrix 3×4 matrix 2×3 matrix
2 rows, 2 columns   3 rows, 4 columns   2 rows, 3 columns

To see the relationship between systems of equations and matrices, consider this system of equations, written in standard form.

$$\begin{cases} 2x - 3y = 6 \\ x + y = 0 \end{cases}$$

A corresponding matrix associated with this system is

$$\begin{bmatrix} 2 & -3 & \vdots & 6 \\ 1 & 1 & \vdots & 0 \end{bmatrix}$$

The coefficients of each variable are placed to the left of a vertical dashed line. The constants are placed to the right. This 2×3 matrix is called the **augmented matrix of the system.** Observe that the rows of this augmented matrix correspond to the equations in the system. The first equation corresponds to the first row; the second equation corresponds to the second row. Each number in the matrix is called an **element.**

The method of solving systems by matrices is to write the augmented matrix as an equivalent matrix from which we easily identify the solution. Two matrices are equivalent if they represent systems that have the same solution set. The following **row operations** can be performed on matrices, and the result is an equivalent matrix.

ELEMENTARY ROW OPERATIONS

1. Any two rows in a matrix may be interchanged.

2. The elements of any row may be multiplied (or divided) by the same nonzero number.

3. The elements of any row may be multiplied (or divided) by a nonzero number and added to its corresponding elements in any other row.

EXAMPLE 1   Solve the system using matrices.

$$\begin{cases} x + 3y = 5 \\ 2x - y = -4 \end{cases}$$

Solution: The augmented matrix is $\begin{bmatrix} 1 & 3 & \vdots & 5 \\ 2 & -1 & \vdots & -4 \end{bmatrix}$. Use elementary row operations to write an equivalent matrix that has 1's along the main diagonal and 0's below each 1 in the main diagonal. The main diagonal of a matrix is the left-to-right diagonal starting with row 1, column 1. For the matrix given, the element in the first row, first column is already 1, as desired. Next we write an equivalent matrix with a 0 below the 1. To do this, multiply row 1 by -2 and add to row 2. *We will change only row 2.*

$$\begin{bmatrix} 1 & 3 & \vdots & 5 \\ -2(1) + 2 & -2(3) + (-1) & \vdots & -2(5) + (-4) \end{bmatrix} \text{simplifies to} \begin{bmatrix} 1 & 3 & \vdots & 5 \\ 0 & -7 & \vdots & -14 \end{bmatrix}$$

$$\uparrow \quad \uparrow \qquad \uparrow \quad \uparrow \qquad \uparrow \quad \uparrow$$

| row 1 | row 2 | row 1 | row 2 | row 1 | row 2 |
| element | element | element | element | element | element |

Now continue down the main diagonal and change the -7 to a 1 by use of an elementary row operation. Divide row 2 by -7. Then

$$\begin{bmatrix} 1 & 3 & \vdots & 5 \\ \dfrac{0}{-7} & \dfrac{-7}{-7} & \vdots & \dfrac{-14}{-7} \end{bmatrix} \text{simplifies to} \begin{bmatrix} 1 & 3 & \vdots & 5 \\ 0 & 1 & \vdots & 2 \end{bmatrix}$$

This last matrix corresponds to the system

$$\begin{cases} 1x + 3y = 5 \\ 0x + 1y = 2 \end{cases} \text{or} \begin{cases} x + 3y = 5 \\ y = 2 \end{cases}$$

To find x, let $y = 2$ in the first equation, $x + 3y = 5$.

$$x + 3y = 5 \qquad \text{\small First equation.}$$
$$x + 3(2) = 5 \qquad \text{\small Let } y = 2.$$
$$x = -1$$

The solution set is $\{(-1, 2)\}$. Check to see that this ordered pair satisfies both equations.

---

2   Solving a system of three equations in three variables using matrices means writing the corresponding matrix and finding an equivalent matrix that has 1's along the main diagonal and 0's below the 1's.

EXAMPLE 2   Solve the system using matrices.

$$\begin{cases} x + 2y + z = 2 \\ -2x - y + 2z = 5 \\ x + 3y - 2z = -8 \end{cases}$$

Solution: The corresponding matrix is $\begin{bmatrix} 1 & 2 & 1 & \vdots & 2 \\ -2 & -1 & 2 & \vdots & 5 \\ 1 & 3 & -2 & \vdots & -8 \end{bmatrix}$. Our goal is to write an equiva-

lent matrix with 1's on the main diagonal and 0's below the 1's. The element in row 1, column 1 is already 1. Next we must get 0's for each element in the rest of column 1. To do this, first we multiply the elements of row 1 by 2 and add the new elements to row 2. Also, we multiply the elements of row 1 by -1 and add the new elements to the elements of row 3. We *do not change row 1*. Then

$$\begin{bmatrix} 1 & 2 & 1 & \vdots & 2 \\ 2(1)-2 & 2(2)-1 & 2(1)+2 & \vdots & 2(2)+5 \\ -1(1)+1 & -1(2)+3 & -1(1)-2 & \vdots & -1(2)-8 \end{bmatrix} \text{ simplifies to } \begin{bmatrix} 1 & 2 & 1 & \vdots & 2 \\ 0 & 3 & 4 & \vdots & 9 \\ 0 & 1 & -3 & \vdots & -10 \end{bmatrix}$$

We continue down the diagonal and use elementary row operations to get 1 where the element 3 is now. To do this, interchange rows 2 and 3.

$$\begin{bmatrix} 1 & 2 & 1 & \vdots & 2 \\ 0 & 3 & 4 & \vdots & 9 \\ 0 & 1 & -3 & \vdots & -10 \end{bmatrix} \text{ is equivalent to } \begin{bmatrix} 1 & 2 & 1 & \vdots & 2 \\ 0 & 1 & -3 & \vdots & -10 \\ 0 & 3 & 4 & \vdots & 9 \end{bmatrix}$$

Next we want the new row 3, column 2 element to be 0. We multiply the elements of row 2 by -3 and add the result to the elements of row 3.

$$\begin{bmatrix} 1 & 2 & 1 & \vdots & 2 \\ 0 & 1 & -3 & \vdots & -10 \\ -3(0)+0 & -3(1)+3 & -3(-3)+4 & \vdots & -3(-10)+9 \end{bmatrix} \begin{matrix} \text{simplifies} \\ \text{to} \end{matrix} \begin{bmatrix} 1 & 2 & 1 & \vdots & 2 \\ 0 & 1 & -3 & \vdots & -10 \\ 0 & 0 & 13 & \vdots & 39 \end{bmatrix}$$

Finally, we divide the elements of row 3 by 13 so that the final main diagonal element is 1.

$$\begin{bmatrix} 1 & 2 & 1 & \vdots & 2 \\ 0 & 1 & -3 & \vdots & -10 \\ \frac{0}{13} & \frac{0}{13} & \frac{13}{13} & \vdots & \frac{39}{13} \end{bmatrix} \text{ simplifies to } \begin{bmatrix} 1 & 2 & 1 & \vdots & 2 \\ 0 & 1 & -3 & \vdots & -10 \\ 0 & 0 & 1 & \vdots & 3 \end{bmatrix}$$

This matrix corresponds to the system

$$\begin{cases} x + 2y + z = 2 \\ y - 3z = -10 \\ z = 3 \end{cases}$$

We identify the z-coordinate of the solution as 3. Next, we replace z with 3 in the second equation and solve for y.

$$\begin{aligned} y - 3z &= -10 && \text{Second equation.} \\ y - 3(3) &= -10 && \text{Let } z = 3. \\ y &= -1 \end{aligned}$$

To find x, we let $z = 3$ and $y = -1$ in the first equation.

$$x + 2y + z = 2 \qquad \text{First equation.}$$
$$x + 2(-1) + 3 = 2 \qquad \text{Let } z = 3 \text{ and } y = -1.$$
$$x = 1$$

The solution set is $\{(1, -1, 3)\}$. Check to see that it satisfies the original system.

EXAMPLE 3   Solve the system using matrices.

$$\begin{cases} 2x - y = 3 \\ 4x - 2y = 5 \end{cases}$$

Solution: The corresponding augmented matrix is $\begin{bmatrix} 2 & -1 & | & 3 \\ 4 & -2 & | & 5 \end{bmatrix}$. To get 1 in the row 1, column 1 position, divide the elements of row 1 by 2.

$$\begin{bmatrix} \dfrac{2}{2} & -\dfrac{1}{2} & | & \dfrac{3}{2} \\ 4 & -2 & | & 5 \end{bmatrix} \text{ simplifies to } \begin{bmatrix} 1 & -\dfrac{1}{2} & | & \dfrac{3}{2} \\ 4 & -2 & | & 5 \end{bmatrix}$$

To get 0 under the 1, multiply the elements of row 1 by -4 and add the new elements to the elements of row 2.

$$\begin{bmatrix} 1 & -\dfrac{1}{2} & | & \dfrac{3}{2} \\ -4(1) + 4 & -4\left(-\dfrac{1}{2}\right) - 2 & | & -4\left(\dfrac{3}{2}\right) + 5 \end{bmatrix} \text{ simplifies to } \begin{bmatrix} 1 & -\dfrac{1}{2} & | & \dfrac{3}{2} \\ 0 & 0 & | & -1 \end{bmatrix}$$

The corresponding system is $\begin{cases} x - \dfrac{1}{2}y = \dfrac{3}{2} \\ 0 = -1 \end{cases}$. The equation $0 = -1$ is false for all y or x values; hence the system is inconsistent and has no solution.

EXERCISE SET 4.4

Solve each system of linear equations using matrices. See Example 1.

1. $\begin{cases} x + y = 1 \\ x - 2y = 4 \end{cases}$ $\{(2, -1)\}$ **2.** $\begin{cases} 2x - y = 8 \\ x + 3y = 11 \end{cases}$ $\{(5, 2)\}$

3. $\begin{cases} x + 3y = 2 \\ x + 2y = 0 \end{cases}$ $\{(-4, 2)\}$ **4.** $\begin{cases} 4x - y = 5 \\ 3x - 3 = 0 \end{cases}$ $\{(1, -1)\}$

Solve each system of linear equations using matrices. See Example 2.

5. $\begin{cases} x + y = 3 \\ 2y = 10 \\ 3x + 2y - 4z = 12 \end{cases}$ **6.** $\begin{cases} 5x = 5 \\ 2x + y = 4 \\ 3x + y - 5z = -15 \end{cases}$

$\{(-2, 5, -2)\}$ $\qquad\qquad$ $\{(1, 2, 4)\}$

7. $\begin{cases} 2y - z = -7 \\ x + 4y + z = -4 \\ 5x - y + 2z = 13 \end{cases}$ **8.** $\begin{cases} 4y + 3z = -2 \\ 5x - 4y = 1 \\ -5x + 4y + z = -3 \end{cases}$

$\{(1, -2, 3)\}$ $\qquad\qquad$ $\{(1, 1, -2)\}$

Solve each system of linear equations using matrices. See Example 3.

9. $\begin{cases} x - 2y = 4 \\ 2x - 4y = 4 \end{cases}$ $\{\ \}$ **10.** $\begin{cases} -x + 3y = 6 \\ 3x - 9y = 9 \end{cases}$ $\{\ \}$

11. $\begin{cases} 3x - 3y = 9 \\ 2x - 2y = 6 \end{cases}$ **12.** $\begin{cases} 9x - 3y = 6 \\ -18x + 6y = -12 \end{cases}$

$\{(x, y) | x - y = 3\}$ $\qquad\qquad$ $\{(x, y) | 9x - 3y = 6\}$

Solve each system of linear equations using matrices.

13. $\begin{cases} x - 4 = 0 \\ x + y = 1 \end{cases}$ $\{(4, -3)\}$

14. $\begin{cases} 3y = 6 \\ x + y = 7 \end{cases}$ $\{(5, 2)\}$

15. $\begin{cases} x + y + z = 2 \\ 2x \quad\;\; - z = 5 \\ \quad\;\; 3y + z = 2 \end{cases}$ $\{(2, 1, -1)\}$

16. $\begin{cases} x + 2y + z = 5 \\ x - y - z = 3 \\ \quad\;\; y + z = 2 \end{cases}$ $\{(5, -2, 4)\}$

17. $\begin{cases} 5x - 2y = 27 \\ -3x + 5y = 18 \end{cases}$ $\{(9, 9)\}$

18. $\begin{cases} 4x - y = 9 \\ 2x + 3y = -27 \end{cases}$ $\{(0, -9)\}$

19. $\begin{cases} 4x - 7y = 7 \\ 12x - 21y = 24 \end{cases}$ $\{\ \}$

20. $\begin{cases} 2x - 5y = 12 \\ -4x + 10y = 20 \end{cases}$ $\{\ \}$

21. $\begin{cases} 4x - y + 2z = 5 \\ \quad\;\; 2y + z = 4 \\ 4x + y + 3z = 10 \end{cases}$ $\{\ \}$

22. $\begin{cases} \quad\;\; 5y - 7z = 14 \\ 2x + y + 4z = 10 \\ 2x + 6y - 3z = 30 \end{cases}$ $\{\ \}$

23. $\begin{cases} 4x + y + z = 3 \\ -x + y - 2z = -11 \\ x + 2y + 2z = -1 \end{cases}$ $\{(1, -4, 3)\}$

24. $\begin{cases} x + y + z = 9 \\ 3x - y + z = -1 \\ -2x + 2y - 3z = -2 \end{cases}$ $\{(0, 5, 4)\}$

Review Exercises

Determine whether each graph is the graph of a function. See Section 3.2.

25.
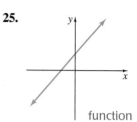
function

26.
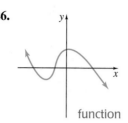
function

27.
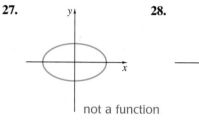
not a function

28.
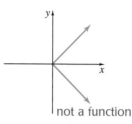
not a function

Evaluate. See Section 1.3. **31.** -36 **32.** -33 **33.** 0

29. $(-1)(-5) - (6)(3)$ -13 **30.** $(2)(-8) - (-4)(1)$ -12

31. $(4)(-10) - (2)(-2)$ **32.** $(-7)(3) - (-2)(-6)$

33. $(-3)(-3) - (-1)(-9)$ **34.** $(5)(6) - (10)(10)$ -70

4.5 SOLVING SYSTEMS OF EQUATIONS BY DETERMINANTS

TAPE IA 4.5

OBJECTIVES

1. Define and evaluate a 2×2 determinant.
2. Use Cramer's rule to solve a system of two linear equations in two variables.
3. Define and evaluate a 3×3 determinant.
4. Use Cramer's rule to solve a linear system of three equations in three variables.

1. We have solved systems of two linear equations in two variables in four different ways: graphically, by substitution, by elimination, and by matrices. Now we analyze another method called *Cramer's rule*.

Recall that a matrix is a rectangular array of numbers. If a matrix has the same number of rows and columns, it is called a **square matrix.** Examples of square matrices are

$$\begin{bmatrix} 1 & 6 \\ 5 & 2 \end{bmatrix} \qquad \begin{bmatrix} 2 & 4 & 1 \\ 0 & 5 & 2 \\ 3 & 6 & 9 \end{bmatrix}$$

A **determinant** is a real number associated with a square matrix. The determinant of a square matrix is denoted by placing vertical bars about the array of numbers. Thus,

The determinant of the square matrix $\begin{bmatrix} 1 & 6 \\ 5 & 2 \end{bmatrix}$ is $\begin{vmatrix} 1 & 6 \\ 5 & 2 \end{vmatrix}$.

The determinant of the square matrix $\begin{bmatrix} 2 & 4 & 1 \\ 0 & 5 & 2 \\ 3 & 6 & 9 \end{bmatrix}$ is $\begin{vmatrix} 2 & 4 & 1 \\ 0 & 5 & 2 \\ 3 & 6 & 9 \end{vmatrix}$.

We define the determinant of a 2×2 matrix first.

DETERMINANT OF A 2×2 MATRIX

$$\begin{vmatrix} a & b \\ c & d \end{vmatrix} = ad - bc$$

EXAMPLE 1   Evaluate.

a. $\begin{vmatrix} -1 & 2 \\ 3 & -4 \end{vmatrix}$ 　　　　　　　　　　　　　**b.** $\begin{vmatrix} 2 & 0 \\ 7 & -5 \end{vmatrix}$

Solution: First, identify the values of a, b, c, and d.

a. Here $a = -1$, $b = 2$, $c = 3$, and $d = -4$.

$$\begin{vmatrix} -1 & 2 \\ 3 & -4 \end{vmatrix} = ad - bc = (-1)(-4) - (2)(3) = -2$$

b. In this example, $a = 2$, $b = 0$, $c = 7$, and $d = -5$.

$$\begin{vmatrix} 2 & 0 \\ 7 & -5 \end{vmatrix} = ad - bc = 2(-5) - (0)(7) = -10$$

2   To develop Cramer's rule, we solve by elimination the system $\begin{cases} ax + by = h \\ cx + dy = k \end{cases}$.

First, eliminate y by multiplying both sides of the first equation by d and both sides of the second equation by $-b$ so that the coefficients of y are opposites. The result is the following system.

$$\begin{cases} d(ax + by) = & d \cdot h \\ -b(cx + dy) = -b \cdot k \end{cases} \quad \text{simplifies to} \quad \begin{cases} adx + bdy = & hd \\ -bcx - bdy = -kb \end{cases}$$

We now add the two equations and solve for x.

$$adx + bdy = hd$$

$$\underline{- bcx - bdy = -kb}$$

$$adx - bcx \quad\quad = hd - kb \quad\quad \text{Add the equations.}$$

$$(ad - bc)x \quad\quad = hd - kb$$

$$x \quad\quad = \frac{hd - kb}{ad - bc} \quad\quad \text{Solve for } x.$$

When we replace x with $\dfrac{hd - kb}{ad - bc}$ in the equation $ax + by = h$ and solve for y, we

find that $y = \dfrac{ak - ch}{ad - bc}$.

Notice that the numerator of the value of x is the determinant of

$$\begin{vmatrix} h & b \\ k & d \end{vmatrix} = hd - kb$$

Also, the numerator of the value of y is the determinant of

$$\begin{vmatrix} a & h \\ c & k \end{vmatrix} = ak - hc$$

Finally, the denominator of the values of x and y is the same and is the determinant of

$$\begin{vmatrix} a & b \\ c & d \end{vmatrix} = ad - bc$$

This means that the values of x and y can be written in determinant notation:

$$x = \frac{\begin{vmatrix} h & b \\ k & d \end{vmatrix}}{\begin{vmatrix} a & b \\ c & d \end{vmatrix}} \quad \text{and} \quad y = \frac{\begin{vmatrix} a & h \\ c & k \end{vmatrix}}{\begin{vmatrix} a & b \\ c & d \end{vmatrix}}$$

For convenience, we label the determinants D, D_x, and D_y.

x-coefficients

y-coefficients

$$\begin{vmatrix} a & b \\ c & d \end{vmatrix} = D \quad\quad \begin{vmatrix} h & b \\ k & d \end{vmatrix} = D_x \quad\quad \begin{vmatrix} a & h \\ c & k \end{vmatrix} = D_y$$

x-column replaced by constants

y-column replaced by constants

These determinant formulas for the coordinates of the solution of a system are known as **Cramer's rule.**

CRAMER'S RULE FOR TWO LINEAR EQUATIONS IN TWO VARIABLES

The solution of the system $\begin{cases} ax + by = h \\ cx + dy = k \end{cases}$ is given by

$$x = \frac{\begin{vmatrix} h & b \\ k & d \end{vmatrix}}{\begin{vmatrix} a & b \\ c & d \end{vmatrix}} = \frac{D_x}{D}; \qquad y = \frac{\begin{vmatrix} a & h \\ c & k \end{vmatrix}}{\begin{vmatrix} a & b \\ c & d \end{vmatrix}} = \frac{D_y}{D}$$

as long as $D = ad - bc$ is not 0.

When $D = 0$, the system is either inconsistent or the equations are dependent. When this happens, use another method to see which is the case.

EXAMPLE 2   Use Cramer's rule to solve each system.

a. $\begin{cases} 3x + 4y = -7 \\ x - 2y = -9 \end{cases}$ **b.** $\begin{cases} 5x + y = 5 \\ -7x - 2y = -7 \end{cases}$

Solution: **a.** Find D, D_x, and D_y.

$$\begin{array}{ccc} a & b & h \\ \downarrow & \downarrow & \downarrow \end{array}$$
$$\begin{cases} 3x + 4y = -7 \\ x - 2y = -9 \end{cases}$$
$$\begin{array}{ccc} \uparrow & \uparrow & \uparrow \\ c & d & k \end{array}$$

$$D = \begin{vmatrix} a & b \\ c & d \end{vmatrix} = \begin{vmatrix} 3 & 4 \\ 1 & -2 \end{vmatrix} = 3(-2) - 4(1) = -10$$

$$D_x = \begin{vmatrix} h & b \\ k & d \end{vmatrix} = \begin{vmatrix} -7 & 4 \\ -9 & -2 \end{vmatrix} = (-7)(-2) - 4(-9) = 50$$

$$D_y = \begin{vmatrix} a & h \\ c & k \end{vmatrix} = \begin{vmatrix} 3 & -7 \\ 1 & -9 \end{vmatrix} = 3(-9) - (-7)(1) = -20$$

Then $x = \dfrac{D_x}{D} = \dfrac{50}{-10} = -5$ and $y = \dfrac{D_y}{D} = \dfrac{-20}{-10} = 2$. The solution set is $\{(-5, 2)\}$.

As always, check the solution in both original equations.

b. $\begin{cases} 5x + y = 5 \\ -7x - 2y = -7 \end{cases}$ Find D, D_x, and D_y.

$$D = \begin{vmatrix} 5 & 1 \\ -7 & -2 \end{vmatrix} = 5(-2) - (-7)(1) = -3$$

$$D_x = \begin{vmatrix} 5 & 1 \\ -7 & -2 \end{vmatrix} = 5(-2) - (-7)(1) = -3$$

$$D_y = \begin{vmatrix} 5 & 5 \\ -7 & -7 \end{vmatrix} = 5(-7) - 5(-7) = 0$$

$$x = \frac{D_x}{D} = \frac{-3}{-3} = 1, \qquad y = \frac{D_y}{D} = \frac{0}{-3} = 0$$

The solution set is $\{(1, 0)\}$.

3 Three-by-three determinants can be used to solve systems of three equations in three variables. The determinant of a 3×3 matrix, however, is considerably more complex than the 2×2 case.

DETERMINANT OF A 3×3 MATRIX

$$\begin{vmatrix} a_1 & b_1 & c_1 \\ a_2 & b_2 & c_2 \\ a_3 & b_3 & c_3 \end{vmatrix} = a_1 \cdot \begin{vmatrix} b_2 & c_2 \\ b_3 & c_3 \end{vmatrix} - a_2 \cdot \begin{vmatrix} b_1 & c_1 \\ b_3 & c_3 \end{vmatrix} + a_3 \cdot \begin{vmatrix} b_1 & c_1 \\ b_2 & c_2 \end{vmatrix}$$

The determinant of a 3×3 matrix, then, is related to the determinants of three 2×2 matrices. Each determinant of these 2×2 matrices is called a **minor,** and every element of a 3×3 matrix has a minor associated with it. For example, the minor of c_2 is the determinant of the 2×2 matrix found by deleting the row and column containing c_2.

$$\begin{array}{ccc} a_1 & b_1 & c_1 \\ a_2 & b_2 & c_2 \\ a_3 & b_3 & c_3 \end{array}$$

The minor of c_2 is

$$\begin{vmatrix} a_1 & b_1 \\ a_3 & b_3 \end{vmatrix}$$

Also, the minor of element a_1 is the determinant of the 2×2 matrix that has no row or column containing a_1.

$$\begin{array}{ccc} a_1 & b_1 & c_1 \\ a_2 & b_2 & c_2 \\ a_3 & b_3 & c_3 \end{array}$$

The minor of a_1 is

$$\begin{vmatrix} b_2 & c_2 \\ b_3 & c_3 \end{vmatrix}$$

So the determinant of a 3×3 matrix can be written as

$$a_1 \cdot (\text{minor of } a_1) - a_2(\text{minor of } a_2) + a_3(\text{minor of } a_3)$$

Finding the determinant by using minors of elements in the first row is called **expanding** by the minors of the first row. *The value of a determinant can be found by expanding by the minors of any row or column.* The following **array of signs** is helpful in determining whether to add or subtract the product of an element and its minor.

$$\begin{matrix} + & - & + \\ - & + & - \\ + & - & + \end{matrix}$$

If an element is in a position marked $+$, we add. If marked $-$, we subtract.

EXAMPLE 3   Evaluate by expanding by the minors of the given row or column.

$$\begin{vmatrix} 0 & 5 & 1 \\ 1 & 3 & -1 \\ -2 & 2 & 4 \end{vmatrix}$$

a. First column   **b.** Second row

Solution: **a.** The elements of the first column are $0, 1$, and -2. The first column of the array of signs is $+, -, +$.

$$\begin{vmatrix} 0 & 5 & 1 \\ 1 & 3 & -1 \\ -2 & 2 & 4 \end{vmatrix} = 0 \cdot \begin{vmatrix} 3 & -1 \\ 2 & 4 \end{vmatrix} - 1 \cdot \begin{vmatrix} 5 & 1 \\ 2 & 4 \end{vmatrix} + (-2) \cdot \begin{vmatrix} 5 & 1 \\ 3 & -1 \end{vmatrix}$$

$$= 0(12 + 2) - 1(20 - 2) + (-2)(-5 - 3)$$

$$= 0 - 18 + 16 = -2$$

b. The elements of the second row are $1, 3$, and -1. This time, the signs begin with $-$ and again alternate.

$$\begin{vmatrix} 0 & 5 & 1 \\ 1 & 3 & -1 \\ -2 & 2 & 4 \end{vmatrix} = -1 \cdot \begin{vmatrix} 5 & 1 \\ 2 & 4 \end{vmatrix} + 3 \cdot \begin{vmatrix} 0 & 1 \\ -2 & 4 \end{vmatrix} - (-1) \cdot \begin{vmatrix} 0 & 5 \\ -2 & 2 \end{vmatrix}$$

$$= -1(20 - 2) + 3(0 - (-2)) - (-1)(0 - (-10))$$

$$= -18 + 6 + 10 = -2$$

Notice that the determinant of the 3×3 matrix is the same regardless of the row or column you select to expand by.

4   A system of three equations in three variables may be solved with Cramer's rule also. Using the elimination process to solve a system with unknown constants as coefficients leads to the following.

CRAMER'S RULE FOR THREE EQUATIONS IN THREE VARIABLES

The solution of the system $\begin{cases} a_1x + b_1y + c_1z = k_1 \\ a_2x + b_2y + c_2z = k_2 \\ a_3x + b_3y + c_3z = k_3 \end{cases}$ is given by

$$x = \frac{D_x}{D}, \; y = \frac{D_y}{D}, \text{ and } z = \frac{D_z}{D},$$

where $D = \begin{vmatrix} a_1 & b_1 & c_1 \\ a_2 & b_2 & c_2 \\ a_3 & b_3 & c_3 \end{vmatrix}$ $D_x = \begin{vmatrix} k_1 & b_1 & c_1 \\ k_2 & b_2 & c_2 \\ k_3 & b_3 & c_3 \end{vmatrix}$

$D_y = \begin{vmatrix} a_1 & k_1 & c_1 \\ a_2 & k_2 & c_2 \\ a_3 & k_3 & c_3 \end{vmatrix}$ $D_z = \begin{vmatrix} a_1 & b_1 & k_1 \\ a_2 & b_2 & k_2 \\ a_3 & b_3 & k_3 \end{vmatrix}$

as long as D is not 0.

EXAMPLE 4 Use Cramer's rule to solve the system

$$\begin{cases} x - 2y + z = 4 \\ 3x + y - 2z = 3 \\ 5x + 5y + 3z = -8 \end{cases}$$

Solution: First, find D, D_x, D_y, and D_z. Beginning with D, we expand by the minors of the first column.

$$D = \begin{vmatrix} 1 & -2 & 1 \\ 3 & 1 & -2 \\ 5 & 5 & 3 \end{vmatrix} = 1 \cdot \begin{vmatrix} 1 & -2 \\ 5 & 3 \end{vmatrix} - 3 \cdot \begin{vmatrix} -2 & 1 \\ 5 & 3 \end{vmatrix} + 5 \cdot \begin{vmatrix} -2 & 1 \\ 1 & -2 \end{vmatrix}$$

$$= 1(3 - (-10)) - 3(-6 - 5) + 5(4 - 1)$$

$$= 13 + 33 + 15 = 61$$

$$D_x = \begin{vmatrix} 4 & -2 & 1 \\ 3 & 1 & -2 \\ -8 & 5 & 3 \end{vmatrix} = 4 \cdot \begin{vmatrix} 1 & -2 \\ 5 & 3 \end{vmatrix} - 3 \cdot \begin{vmatrix} -2 & 1 \\ 5 & 3 \end{vmatrix} + (-8) \cdot \begin{vmatrix} -2 & 1 \\ 1 & -2 \end{vmatrix}$$

$$= 4(3 - (-10)) - 3(-6 - 5) + (-8)(4 - 1)$$

$$= 52 + 33 - 24 = 61$$

$$D_y = \begin{vmatrix} 1 & 4 & 1 \\ 3 & 3 & -2 \\ 5 & -8 & 3 \end{vmatrix} = 1 \cdot \begin{vmatrix} 3 & -2 \\ -8 & 3 \end{vmatrix} - 3 \cdot \begin{vmatrix} 4 & 1 \\ -8 & 3 \end{vmatrix} + 5 \cdot \begin{vmatrix} 4 & 1 \\ 3 & -2 \end{vmatrix}$$

$$= 1(9 - 16) - 3(12 + 8) + 5(-8 - 3)$$

$$= -7 - 60 - 55 = -122$$

$$D_z = \begin{vmatrix} 1 & -2 & 4 \\ 3 & 1 & 3 \\ 5 & 5 & -8 \end{vmatrix} = 1 \cdot \begin{vmatrix} 1 & 3 \\ 5 & -8 \end{vmatrix} - 3 \cdot \begin{vmatrix} -2 & 4 \\ 5 & -8 \end{vmatrix} + 5 \cdot \begin{vmatrix} -2 & 4 \\ 1 & 3 \end{vmatrix}$$

$$= 1(-8 - 15) - 3(16 - 20) + 5(-6 - 4)$$

$$= -23 + 12 - 50 = -61$$

From these determinants, we calculate the solution:

$$x = \frac{D_x}{D} = \frac{61}{61} = 1, \quad y = \frac{D_y}{D} = \frac{-122}{61} = -2, \quad z = \frac{D_z}{D} = \frac{-61}{61} = -1$$

The solution set of the system is $\{(1, -2, -1)\}$. Check this solution by verifying that it satisfies each equation of the system.

EXERCISE SET 4.5

Evaluate. See Example 1.

1. $\begin{vmatrix} 3 & 5 \\ -1 & 7 \end{vmatrix}$ 26

2. $\begin{vmatrix} -5 & 1 \\ 0 & -4 \end{vmatrix}$ 20

3. $\begin{vmatrix} 9 & -2 \\ 4 & -3 \end{vmatrix}$ -19

4. $\begin{vmatrix} 4 & 0 \\ 9 & 8 \end{vmatrix}$ 32

5. $\begin{vmatrix} -2 & 9 \\ 4 & -18 \end{vmatrix}$ 0

6. $\begin{vmatrix} -40 & 8 \\ 70 & -14 \end{vmatrix}$ 0

Use Cramer's rule, if possible, to solve each system of linear equations. See Example 2.

7. $\begin{cases} 2y - 4 = 0 \\ x + 2y = 5 \end{cases}$ $\{(1, 2)\}$

8. $\begin{cases} 4x - y = 5 \\ 3x - 3 = 0 \end{cases}$ $\{(1, -1)\}$

9. $\begin{cases} 3x + y = 1 \\ 2y = 2 - 6x \end{cases}$

10. $\begin{cases} y = 2x - 5 \\ 8x - 4y = 20 \end{cases}$

11. $\begin{cases} 5x - 2y = 27 \\ -3x + 5y = 18 \end{cases}$ $\{(9, 9)\}$

12. $\begin{cases} 4x - y = 9 \\ 2x + 3y = -27 \end{cases}$ $\{(0, -9)\}$

Evaluate. See Example 3.

13. $\begin{vmatrix} 2 & 1 & 0 \\ 0 & 5 & -3 \\ 4 & 0 & 2 \end{vmatrix}$ 8

14. $\begin{vmatrix} -6 & 4 & 2 \\ 1 & 0 & 5 \\ 0 & 3 & 1 \end{vmatrix}$ 92

15. $\begin{vmatrix} 4 & -6 & 0 \\ -2 & 3 & 0 \\ 4 & -6 & 1 \end{vmatrix}$ 0

16. $\begin{vmatrix} 5 & 2 & 1 \\ 3 & -6 & 0 \\ -2 & 8 & 0 \end{vmatrix}$ 12

17. $\begin{vmatrix} 3 & 6 & -3 \\ -1 & -2 & 3 \\ 4 & -1 & 6 \end{vmatrix}$ 54

18. $\begin{vmatrix} 2 & -2 & 1 \\ 4 & 1 & 3 \\ 3 & 1 & 2 \end{vmatrix}$ -3

Use Cramer's rule, if possible, to solve each system of linear equations. See Example 4.

19. $\begin{cases} 3x + z = -1 \\ -x - 3y + z = 7 \\ 3y + z = 5 \end{cases}$

20. $\begin{cases} 4y - 3z = -2 \\ 8x - 4y = 4 \\ -8x + 4y + z = -2 \end{cases}$

21. $\begin{cases} x + y + z = 8 \\ 2x - y - z = 10 \\ x - 2y + 3z = 22 \end{cases}$ $\{(6, -2, 4)\}$

22. $\begin{cases} 5x + y + 3z = 1 \\ x - y - 3z = -7 \\ -x + y = 1 \end{cases}$ $\{(-1, 0, 2)\}$

Evaluate.

23. $\begin{vmatrix} 10 & -1 \\ -4 & 2 \end{vmatrix}$ 16

24. $\begin{vmatrix} -6 & 2 \\ 5 & -1 \end{vmatrix}$ -4

25. $\begin{vmatrix} 1 & 0 & 4 \\ 1 & -1 & 2 \\ 3 & 2 & 1 \end{vmatrix}$ 15

26. $\begin{vmatrix} 0 & 1 & 2 \\ 3 & -1 & 2 \\ 3 & 2 & -2 \end{vmatrix}$ 30

27. $\begin{vmatrix} \frac{3}{4} & \frac{5}{2} \\ -\frac{1}{6} & \frac{7}{3} \end{vmatrix}$ $\frac{13}{6}$

28. $\begin{vmatrix} \frac{5}{7} & \frac{1}{3} \\ \frac{6}{7} & \frac{2}{3} \end{vmatrix}$ $\frac{4}{21}$

9. $\{(x, y) \mid 3x + y = 1\}$ **10.** $\{(x, y) \mid y = 2x - 5\}$ **19.** $\{(-2, 0, 5)\}$ **20.** $\{(1, 1, 2)\}$

29. $\begin{vmatrix} 4 & -2 & 2 \\ 6 & -1 & 3 \\ 2 & 1 & 1 \end{vmatrix}$ 0   **30.** $\begin{vmatrix} 1 & 5 & 0 \\ 7 & 9 & -4 \\ 3 & 2 & -2 \end{vmatrix}$ 0

31. $\begin{vmatrix} -2 & 5 & 4 \\ 5 & -1 & 3 \\ 4 & 1 & 2 \end{vmatrix}$ 56   **32.** $\begin{vmatrix} 5 & -2 & 4 \\ -1 & 5 & 3 \\ 1 & 4 & 2 \end{vmatrix}$ −56

33. If all the elements in a single row of a determinant are zero, to what does the determinant evaluate? Explain your answer.   0

34. If all the elements in a single column of a determinant are 0, to what does the determinant evaluate? Explain your answer.   0

Find the value of x such that each is a true statement.

35. $\begin{vmatrix} 1 & x \\ 2 & 7 \end{vmatrix} = -3$ 5   **36.** $\begin{vmatrix} 6 & 1 \\ -2 & x \end{vmatrix} = 26$ 4

Use Cramer's rule, if possible, to solve each system of linear equations.

37. $\begin{cases} 2x - 5y = 4 \\ x + 2y = -7 \end{cases}$ **38.** $\begin{cases} 3x - y = 2 \\ -5x + 2y = 0 \end{cases}$ {(4, 10)}

39. $\begin{cases} 4x + 2y = 5 \\ 2x + y = -1 \end{cases}$ { }   **40.** $\begin{cases} 3x + 6y = 15 \\ 2x + 4y = 3 \end{cases}$ { }

41. $\begin{cases} 2x + 2y + z = 1 \\ -x + y + 2z = 3 \\ x + 2y + 4z = 0 \end{cases}$ **42.** $\begin{cases} 2x - 3y + z = 5 \\ x + y + z = 0 \\ 4x + 2y + 4z = 4 \end{cases}$

43. $\begin{cases} \dfrac{2}{3}x - \dfrac{3}{4}y = -1 \\ -\dfrac{1}{6}x + \dfrac{3}{4}y = \dfrac{5}{2} \end{cases}$ {(3, 4)}   **44.** $\begin{cases} \dfrac{1}{2}x - \dfrac{1}{3}y = -3 \\ \dfrac{1}{8}x + \dfrac{1}{6}y = 0 \end{cases}$ {(−4, 3)}

45. $\begin{cases} 0.7x - 0.2y = -1.6 \\ 0.2x - y = -1.4 \end{cases}$ **46.** $\begin{cases} -0.7x + 0.6y = 1.3 \\ 0.5x - 0.3y = -0.8 \end{cases}$

47. $\begin{cases} -2x + 4y - 2z = 6 \\ x - 2y + z = -3 \\ 3x - 6y + 3z = -9 \end{cases}$ **48.** $\begin{cases} -x - y + 3z = 2 \\ 4x + 4y - 12z = -8 \\ -3x - 3y + 9z = 6 \end{cases}$

49. $\begin{cases} x - 2y + z = -5 \\ 3y + 2z = 4 \\ 3x - y = -2 \end{cases}$ **50.** $\begin{cases} 4x + 5y = 10 \\ 3y + 2z = -6 \\ x + y + z = 3 \end{cases}$
{(0, 2, −1)}   {(5, −2, 0)}

Review Exercises

Simplify each expression. See Section 1.4.

51. $5x - 6 + x - 12$ **52.** $4y + 3 - 15y - 1$

53. $2(3x - 6) + 3(x - 1)$ $9x - 15$

54. $-3(2y - 7) - 1(11 + 12y)$ $-18y + 10$

Graph each function. See Section 3.3.

55. $f(x) = 5x - 6$ **56.** $g(x) = -x + 1$

57. $h(x) = 3$ **58.** $f(x) = -3$

For Exercises 55–58, see Appendix D.

A Look Ahead

EXAMPLE   Evaluate the determinant.

$$\begin{vmatrix} 2 & 0 & -1 & 3 \\ 0 & 5 & -2 & -1 \\ 3 & 1 & 0 & 1 \\ 4 & 2 & -2 & 0 \end{vmatrix}$$

Solution:

To evaluate a 4×4 determinant, select any row or column and expand by the minors. The array of signs for a 4×4 determinant is the same as for a 3×3 determinant except expanded. We expand using the fourth row.

$$\begin{vmatrix} 2 & 0 & -1 & 3 \\ 0 & 5 & -2 & -1 \\ 3 & 1 & 0 & 1 \\ \rightarrow 4 & 2 & -2 & 0 \end{vmatrix}$$

$$= -4 \cdot \begin{vmatrix} 0 & -1 & 3 \\ 5 & -2 & -1 \\ 1 & 0 & 1 \end{vmatrix} + 2 \cdot \begin{vmatrix} 2 & -1 & 3 \\ 0 & -2 & -1 \\ 3 & 0 & 1 \end{vmatrix}$$

$$- (-2) \cdot \begin{vmatrix} 2 & 0 & 3 \\ 0 & 5 & -1 \\ 3 & 1 & 1 \end{vmatrix} + 0 \cdot \begin{vmatrix} 2 & 0 & -1 \\ 0 & 5 & -2 \\ 3 & 1 & 0 \end{vmatrix}$$

Now find the value of each 3×3 determinant. The value of the 4×4 determinant is

$$-4(12) + 2(17) + 2(-33) + 0 = -80$$

Find the value of each determinant. See the preceding example.

59. $\begin{vmatrix} 5 & 0 & 0 & 0 \\ 0 & 4 & 2 & -1 \\ 1 & 3 & -2 & 0 \\ 0 & -3 & 1 & 2 \end{vmatrix}$ −125   **60.** $\begin{vmatrix} 1 & 7 & 0 & -1 \\ 1 & 3 & -2 & 0 \\ 1 & 0 & -1 & 2 \\ 0 & -6 & 2 & 4 \end{vmatrix}$ 0

61. $\begin{vmatrix} 4 & 0 & 2 & 5 \\ 0 & 3 & -1 & 1 \\ 0 & 0 & 2 & 0 \\ 0 & 0 & 0 & 1 \end{vmatrix}$ 24   **62.** $\begin{vmatrix} 2 & 0 & -1 & 4 \\ 6 & 0 & 4 & 1 \\ 2 & 4 & 3 & -1 \\ 4 & 0 & 5 & -4 \end{vmatrix}$ 56

37. {(−3, −2)}   **41.** {(−2, 3, −1)}   **42.** {(−3, −2, 5)}   **45.** {(−2, 1)}   **46.** {(−1, 1)}

47. {(x, y, z) | x − 2y + z = −3}   **48.** {(x, y, z) | −x − y + 3z = 2}   **51.** 6x − 18   **52.** −11y + 2

GROUP ACTIVITY

LOCATING LIGHTNING STRIKES

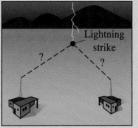

Weather-recording station *A*   Weather-recording station *B*

MATERIALS:
- Calculator
- Grapher (optional)

Weather-recording stations use a directional antenna to detect and measure the electromagnetic field emitted by a lightning bolt. The antenna can determine the angle between a fixed point and the position of the lightning strike (see figure) but cannot determine the distance to the lightning strike. However, the angle measured by the antenna can be used to find the slope of the line connecting the positions of the weather station and the lightning strike. From there, the equation of the line connecting the positions of the weather station and the lightning strike may be found. If two such lines may be found—that is, if another weather station's antenna detects the same lightning flash—the coordinates of the lightning strike's position may be pinpointed.

1. A weather-recording station *A* is located at the coordinates (35, 28), and a second station *B* is at (52, 12). Plot the positions of the two weather-recording stations.

2. A lightning strike is detected by both stations. Station *A* uses a measured angle to find the slope of the line from the station to the lightning strike as $m = -1.732$. Station *B* computes a slope of $m = 0.577$ from the angle it measured. Use this information to find the equations of the lines connecting each station to the position of the lightning strike.

3. Have each group member solve the resulting system of equations in one of the following five ways:
   (a) Graphing—graph the two equations on your plot of the positions of the two weather-recording stations. Estimate the coordinates of their point of intersection.
   (b) Using either the method of substitution or of elimination (whichever you prefer)
   (c) Using matrices
   (d) Using Cramer's rule
   (e) (Optional) Using a grapher to graph the lines and an intersect feature to estimate the coordinates of their point of intersection

Compare your results. What are the coordinates of the lightning strike?

See Appendix D for Group Activity answers and suggestions.

CHAPTER 4 HIGHLIGHTS

DEFINITIONS AND CONCEPTS	EXAMPLES

SECTION 4.1 SOLVING SYSTEMS OF LINEAR EQUATIONS IN TWO VARIABLES

A **system of linear equations** consists of two or more linear equations.

Systems of Linear Equations:

$$\begin{cases} x - 3y = 6 \\ y = \dfrac{1}{2}x \end{cases} \qquad \begin{cases} x + 2y - z = 1 \\ 3x - y + 4z = 0 \\ 5y + z = 6 \end{cases}$$

A **solution** of a system of two equations in two variables is an ordered pair (x, y) that makes both equations true.

Determine whether $(2, -5)$ is a solution of the system.

$$\begin{cases} x + y = -3 \\ 2x - 3y = 19 \end{cases}$$

Replace x with 2 and y with -5 in both equations.

$$x + y = -3 \qquad\qquad 2x - 3y = 19$$
$$2 + (-5) = -3 \qquad\quad 2(2) - 3(-5) = 19$$
$$-3 = -3 \quad \text{True.} \qquad\quad 4 + 15 = 19$$
$$19 = 19 \quad \text{True.}$$

$(2, -5)$ is a solution of the system.

Geometrically, a solution of a system in two variables is a point common to the graphs of the equations.

Solve by graphing $\begin{cases} y = 2x - 1 \\ x + 2y = 13 \end{cases}$

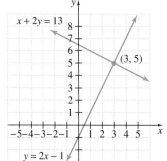

A system of equations with at least one solution is a **consistent system.** A system that has no solution is an **inconsistent system.**

If the graphs of two linear equations are identical, the equations are **dependent.**

If their graphs are different, the equations are **independent.**

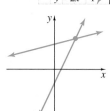

One solution;
consistent and
independent

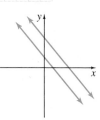

No solution;
inconsistent and
independent

(continued)

DEFINITIONS AND CONCEPTS	EXAMPLES

SECTION 4.1 SOLVING SYSTEMS OF LINEAR EQUATIONS IN TWO VARIABLES

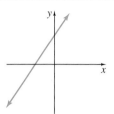

Infinite number
of solutions; consistent
and dependent

To solve a system of linear equations by the **substitution method:**

Step 1.   Solve one equation for a variable.

Step 2.   Substitute the expression for the variable into the other equation.

Step 3.   Solve the equation from *step 2* to find the value of one variable.

Step 4.   Substitute the value from *step 3* in either original equation to find the value of the other variable.

Step 5.   Check the solution in both equations.

Solve by substitution:

$$\begin{cases} y = x + 2 \\ 3x - 2y = -5 \end{cases}$$

Substitute $x + 2$ for y in the second equation.

$$3x - 2y = -5$$
$$3x - 2(x + 2) = -5$$
$$3x - 2x - 4 = -5$$
$$x - 4 = -5 \qquad \text{Simplify.}$$
$$x = -1 \qquad \text{Add 4.}$$

To find y, let $x = -1$ in $y = x + 2$, so $y = -1 + 2 = 1$. The solution $(-1, 1)$ checks.

To solve a system of linear equations by the **elimination method:**

Step 1.   Rewrite each equation in standard form $Ax + By = C$.

Step 2.   Multiply one or both equations by a nonzero number so that the coefficients of a variable are opposites.

Step 3.   Add the equations.

Step 4.   Find the value of one variable by solving the resulting equation.

Step 5.   Substitute the value from *step 4* into either original equation to find the value of the other variable.

Step 6.   Check the solution in both equations.

Solve by elimination:

$$\begin{cases} x - 3y = -3 \\ -2x + y = 6 \end{cases}$$

Multiply both sides of the first equation by 2.

$$2x - 6y = -6$$
$$\underline{-2x + y = 6}$$
$$-5y = 0 \qquad \text{Add.}$$
$$y = 0 \qquad \text{Divide by } -5.$$

To find x, let $y = 0$ in an original equation.

$$x - 3y = -3$$
$$x - 3 \cdot 0 = -3$$
$$x = -3$$

The solution $(-3, 0)$ checks.   *(continued)*

DEFINITIONS AND CONCEPTS	**EXAMPLES**

SECTION 4.2   SOLVING SYSTEMS OF LINEAR EQUATIONS IN THREE VARIABLES

A **solution** of an equation in three variables x, y, and z is an **ordered triple** (x, y, z) that makes the equation a true statement.

Verify that $(-2, 1, 3)$ is a solution of
$$2x + 3y - 2z = -7.$$
Replace x with -2, y with 1, and z with 3.
$$2(-2) + 3(1) - 2(3) = -7$$
$$-4 + 3 - 6 = -7$$
$$-7 = -7 \qquad \text{True.}$$
$(-2, 1, 3)$ is a solution.

To solve a system of three linear equations by the elimination method:

Step 1.   Write each equation in standard form, $Ax + By + Cz = D$.

Step 2.   Choose a pair of equations and use the equations to eliminate a variable.

Step 3.   Choose any other pair of equations and eliminate the same variable.

Step 4.   Solve the system of two equations in two variables from *steps 1* and *2*.

Step 5.   Solve for the third variable by substituting the values of the variables from *step 4* into any of the original equations.

Solve:
$$\begin{cases} 2x + y - z = 0 \ (1) \\ x - y - 2z = -6 \ (2) \\ -3x - 2y + 3z = -22 \ (3) \end{cases}$$

1. Each equation is written in standard form.

2.
$$\begin{array}{r} 2x + y - z = 0 \ (1) \\ x - y - 2z = -6 \ (2) \\ \hline 3x \qquad - 3z = -6 \ (4) \qquad \text{Add.} \end{array}$$

3. Eliminate y from equations (1) and (3) also.
$$\begin{array}{r} 4x + 2y - 2z = 0 \qquad \text{Multiply equation} \\ -3x - 2y + 3z = -22 \ (3) \qquad (1) \text{ by 2.} \\ \hline x \qquad + z = -22 \ (5) \qquad \text{Add.} \end{array}$$

4. Solve:
$$\begin{cases} 3x - 3z = -6 \ (4) \\ x + z = -22 \ (5) \end{cases}$$
$$\begin{array}{r} x - z = -2 \qquad \text{Divide equation (4) by 3.} \\ x + z = -22 \ (5) \\ \hline 2x \qquad = -24 \\ x \qquad = -12 \end{array}$$
To find z, use equation (5).
$$x + z = -22$$
$$-12 + z = -22$$
$$z = -10$$

5. To find y, use equation (1).
$$2x + y - z = 0$$
$$2(-12) + y - (-10) = 0$$
$$-24 + y + 10 = 0$$
$$y = 14$$
The solution is $(-12, 14, -10)$.   *(continued)*

DEFINITIONS AND CONCEPTS	EXAMPLES

SECTION 4.3   SYSTEMS OF EQUATIONS AND PROBLEM SOLVING

1. UNDERSTAND the problem.	Two numbers have a sum of 11. Twice one number is 3 less than 3 times the other. Find the numbers. **1.** Read and reread.
2. ASSIGN.	**2.** x = one number y = other number
3. ILLUSTRATE.	**3.** No illustration needed.

4. TRANSLATE.	**4.** In words: sum of numbers is 11
	
	    Translate: $x + y$ = 11
	
	    In words: twice one number is 3 less than 3 times the other number
	
	    Translate: $2x$ = $3y - 3$

5. COMPLETE.	**5.** Solve the system $\begin{cases} x + y = 11 \\ 2x = 3y - 3 \end{cases}$. In the first equation $x = 11 - y$. Substitute into the other equation. $$2x = 3y - 3$$ $$2(11 - y) = 3y - 3$$ $$22 - 2y = 3y - 3$$ $$-5y = -25$$ $$y = 5$$ Replace y with 5 in the equation $x = 11 - y$. Then $x = 11 - 5 = 6$. The solution is $(6, 5)$.
6. INTERPRET.	**6.** To *check*, see that $6 + 5 = 11$, the required sum, and that twice 6 is 3 times 5 less 3. *State:* The numbers are 6 and 5.

SECTION 4.4   SOLVING SYSTEMS OF EQUATIONS BY MATRICES

A **matrix** is a rectangular array of numbers.	Matrices: $$\begin{bmatrix} -7 & 0 & 3 \\ 1 & 2 & 4 \end{bmatrix} \quad \begin{bmatrix} a & b & c \\ d & e & f \\ g & h & i \end{bmatrix}$$
The **augmented matrix of the system** is obtained by writing a matrix composed of the coefficients of the variables and the constants of the system.	The augmented matrix of the system $\begin{cases} x - y = 1 \\ 2x + y = 11 \end{cases}$ is $\begin{bmatrix} 1 & -1 & \vdots & 1 \\ 2 & 1 & \vdots & 11 \end{bmatrix}$.

(continued)

DEFINITIONS AND CONCEPTS	EXAMPLES

SECTION 4.4 SOLVING SYSTEMS OF EQUATIONS BY MATRICES

The following **row operations** can be performed on matrices, and the result is an equivalent matrix.

Elementary row operations:

1. Interchange any two rows.

2. Multiply (or divide) the elements of one row by the same nonzero number.

3. Multiply (or divide) the elements of one row by the same nonzero number and add to its corresponding elements in any other row.

Solve $\begin{cases} x - y = 1 \\ 2x + y = 11 \end{cases}$ using matrices.

The augmented matrix is

$$\left[\begin{array}{cc|c} 1 & -1 & 1 \\ 2 & 1 & 11 \end{array}\right]$$

Use row operations to write an equivalent matrix with 1's along the main diagonal and 0's below each 1 in the main diagonal. Multiply row 1 by -2 and add to row 2. Change row 2 only.

$$\left[\begin{array}{cc|c} 1 & -1 & 1 \\ -2(1) + 2 & -2(-1) + 1 & -2(1) + 11 \end{array}\right]$$

simplifies to $\left[\begin{array}{cc|c} 1 & -1 & 1 \\ 0 & 3 & 9 \end{array}\right]$

Divide row 2 by 3.

$$\left[\begin{array}{cc|c} 1 & -1 & 1 \\ \dfrac{0}{3} & \dfrac{3}{3} & \dfrac{9}{3} \end{array}\right] \text{ simplifies to } \left[\begin{array}{cc|c} 1 & -1 & 1 \\ 0 & 1 & 3 \end{array}\right]$$

This matrix corresponds to the system

$$\begin{cases} x - y = 1 \\ y = 3 \end{cases}$$

Let $y = 3$ in the first equation.

$$x - 3 = 1$$
$$x = 4$$

The solution set is $\{4, 3\}$.

SECTION 4.5 SOLVING SYSTEMS OF EQUATIONS BY DETERMINANTS

A **square matrix** is a matrix with the same number of rows and columns.

Square Matrices:

$$\begin{bmatrix} -2 & 1 \\ 6 & 8 \end{bmatrix} \qquad \begin{bmatrix} 4 & -1 & 6 \\ 0 & 2 & 5 \\ 1 & 1 & 2 \end{bmatrix}$$

A **determinant** is a real number associated with a square matrix. To denote the determinant, place vertical bars about the array of numbers.

The determinant of $\begin{bmatrix} -2 & 1 \\ 6 & 8 \end{bmatrix}$ is $\begin{vmatrix} -2 & 1 \\ 6 & 8 \end{vmatrix}$.

(continued)

DEFINITIONS AND CONCEPTS	EXAMPLES

SECTION 4.5 SOLVING SYSTEMS OF EQUATIONS BY DETERMINANTS

The determinant of a 2×2 matrix is

$$\begin{vmatrix} a & b \\ c & d \end{vmatrix} = ad - bc$$

$$\begin{vmatrix} -2 & 1 \\ 6 & 8 \end{vmatrix} = -2 \cdot 8 - 1 \cdot 6 = -22$$

Cramer's rule for two linear equations in two variables:

The solution of the system $\begin{cases} ax + by = h \\ cx + dy = k \end{cases}$ is given by

$$x = \frac{\begin{vmatrix} h & b \\ k & d \end{vmatrix}}{\begin{vmatrix} a & b \\ c & d \end{vmatrix}} = \frac{D_x}{D}; \qquad y = \frac{\begin{vmatrix} a & h \\ c & k \end{vmatrix}}{\begin{vmatrix} a & b \\ c & d \end{vmatrix}} = \frac{D_y}{D}$$

as long as $D = ad - bc$ is not 0.

Use Cramer's rule to solve

$$\begin{cases} 3x + 2y = 8 \\ 2x - y = -11 \end{cases}$$

$$D = \begin{vmatrix} 3 & 2 \\ 2 & -1 \end{vmatrix} = 3(-1) - 2(2) = -7$$

$$D_x = \begin{vmatrix} 8 & 2 \\ -11 & -1 \end{vmatrix} = 8(-1) - 2(-11) = 14$$

$$D_y = \begin{vmatrix} 3 & 8 \\ 2 & -11 \end{vmatrix} = 3(-11) - 8(2) = -49$$

$$x = \frac{D_x}{D} = \frac{14}{-7} = -2, \; y = \frac{D_y}{D} = \frac{-49}{-7} = 7$$

The solution set is $\{(-2, 7)\}$.

Determinant of a 3×3 matrix:

$$\begin{vmatrix} a_1 & b_1 & c_1 \\ a_2 & b_2 & c_2 \\ a_3 & b_3 & c_3 \end{vmatrix} = a_1 \cdot \begin{vmatrix} b_2 & c_2 \\ b_3 & c_3 \end{vmatrix} - a_2 \cdot \begin{vmatrix} b_1 & c_1 \\ b_3 & c_3 \end{vmatrix} + a_3 \cdot \begin{vmatrix} b_1 & c_1 \\ b_2 & c_2 \end{vmatrix}$$

Each 2×2 matrix above is called a **minor.**

$$\begin{vmatrix} 0 & 2 & -1 \\ 5 & 3 & 0 \\ 2 & -2 & 4 \end{vmatrix} = 0 \begin{vmatrix} 3 & 0 \\ -2 & 4 \end{vmatrix} - 2 \begin{vmatrix} 5 & 0 \\ 2 & 4 \end{vmatrix} + (-1) \begin{vmatrix} 5 & 3 \\ 2 & -2 \end{vmatrix}$$

$$= 0(12 - 0) - 2(20 - 0) - 1(-10 - 6)$$

$$= 0 - 40 + 16 = -24$$

Cramer's rule for three equations in three variables:

The solution of the system $\begin{cases} a_1x + b_1y + c_1z = k_1 \\ a_2x + b_2y + c_2z = k_2 \\ a_3x + b_3y + c_3z = k_3 \end{cases}$

is given by

$$x = \frac{D_x}{D}, \; y = \frac{D_y}{D}, \text{ and } z = \frac{D_z}{D}$$

where

$$D = \begin{vmatrix} a_1 & b_1 & c_1 \\ a_2 & b_2 & c_2 \\ a_3 & b_3 & c_3 \end{vmatrix} \qquad D_x = \begin{vmatrix} k_1 & b_1 & c_1 \\ k_2 & b_2 & c_2 \\ k_3 & b_3 & c_3 \end{vmatrix}$$

$$D_y = \begin{vmatrix} a_1 & k_1 & c_1 \\ a_2 & k_2 & c_2 \\ a_3 & k_3 & c_3 \end{vmatrix} \qquad D_z = \begin{vmatrix} a_1 & b_1 & k_1 \\ a_2 & b_2 & k_2 \\ a_3 & b_3 & k_3 \end{vmatrix}$$

as long as D is not 0.

Use Cramer's rule to solve

$$\begin{cases} 3y + 2z = 8 \\ x + y + z = 3 \\ 2x - y + z = 2 \end{cases}$$

$$D = \begin{vmatrix} 0 & 3 & 2 \\ 1 & 1 & 1 \\ 2 & -1 & 1 \end{vmatrix} = -3$$

$$D_x = \begin{vmatrix} 8 & 3 & 2 \\ 3 & 1 & 1 \\ 2 & -1 & 1 \end{vmatrix} = 3$$

$$D_y = \begin{vmatrix} 0 & 8 & 2 \\ 1 & 3 & 1 \\ 2 & 2 & 1 \end{vmatrix} = 0$$

$$D_z = \begin{vmatrix} 0 & 3 & 8 \\ 1 & 1 & 3 \\ 2 & -1 & 2 \end{vmatrix} = -12$$

(continued)

DEFINITIONS AND CONCEPTS	EXAMPLES
SECTION 4.5 SOLVING SYSTEMS OF EQUATIONS BY DETERMINANTS	
	$$x = \frac{D_x}{D} = \frac{3}{-3} = -1, \ y = \frac{D_y}{D} = \frac{0}{-3} = 0$$ $$z = \frac{D_z}{D} = \frac{-12}{-3} = 4$$ The solution is $(-1, 0, 4)$.

CHAPTER 4 REVIEW

For Exercises 1–5, see Appendix D.

(4.1) *Solve each system of equations in two variables by each of three methods: (1) graphing, (2) substitution, and (3) elimination.*

1. $\begin{cases} 3x + 10y = 1 \\ x + 2y = -1 \end{cases}$ $\{(-3, 1)\}$

2. $\begin{cases} y = \frac{1}{2}x + \frac{2}{3} \\ 4x + 6y = 4 \end{cases}$ $\left\{ \left(0, \frac{2}{3}\right) \right\}$

3. $\begin{cases} 2x - 4y = 22 \\ 5x - 10y = 16 \end{cases}$ $\{ \ \}$

4. $\begin{cases} 3x - 6y = 12 \\ 2y = x - 4 \end{cases}$ $\{(x, y) \,|\, 3x - 6y = 12\}$

5. $\begin{cases} \frac{1}{2}x - \frac{3}{4}y = -\frac{1}{2} \\ \frac{1}{8}x + \frac{3}{4}y = \frac{19}{8} \end{cases}$ $\left\{ \left(3, \frac{8}{3}\right) \right\}$

6. The revenue equation for a certain style of backpack is

$$y = 32x$$

where x is the number of backpacks sold and y is the income in dollars for selling x backpacks. The cost equation for these units is

$$y = 15x + 25{,}500$$

where x is the number of backpacks manufactured and y is the cost in dollars for manufacturing x backpacks. Find the number of units to be sold for the company to break even.   1500 backpacks

(4.2) *Solve each system of equations in three variables.*

7. $\begin{cases} x + z = 4 \\ 2x - y = 4 \\ x + y - z = 0 \end{cases}$ $\{(2, 0, 2)\}$

8. $\begin{cases} 2x + 5y = 4 \\ x - 5y + z = -1 \\ 4x - z = 11 \end{cases}$ $\{(2, 0, -3)\}$

9. $\left\{ \left(-\frac{1}{2}, \frac{3}{4}, 1\right) \right\}$ **10.** $\{(-1, 2, 0)\}$ **12.** $\{(5, 3, 0)\}$

9. $\begin{cases} 4y + 2z = 5 \\ 2x + 8y = 5 \\ 6x + 4z = 1 \end{cases}$

10. $\begin{cases} 5x \quad 7y = 9 \\ + 14y - z = 28 \\ 4x + 2z = -4 \end{cases}$

11. $\begin{cases} 3x - 2y + 2z = 5 \\ -x + 6y + z = 4 \\ 3x + 14y + 7z = 20 \end{cases}$

12. $\begin{cases} x + 2y + 3z = 11 \\ y + 2z = 3 \\ 2x + 2z = 10 \end{cases}$ $\{ \ \}$

13. $\begin{cases} 7x - 3y + 2z = 0 \\ 4x - 4y - z = 2 \\ 5x + 2y + 3z = 1 \end{cases}$ $\{(1, 1, -2)\}$

14. $\begin{cases} x - 3y - 5z = -5 \\ 4x - 2y + 3z = 13 \\ 5x + 3y + 4z = 22 \end{cases}$ $\{(3, 1, 1)\}$

(4.3) *Use systems of equations to solve the following applications.*

15. The sum of three numbers is 98. The sum of the first and second is two more than the third number, and the second is four times the first. Find the numbers.   10, 40, and 48

16. Alice's coin purse has 95 coins in it—all dimes, nickels, and pennies—worth $4.03 total. There are twice as many pennies as dimes and one fewer nickel than dimes. Find how many of each type of coin the purse contains.
24 dimes, 23 nickels, and 48 pennies

17. One number is 3 times a second number, and twice the sum of the numbers is 168. Find the numbers.   63 and 21

18. Sue is 16 years older than Pat. In 15 years Sue will be twice as old as Pat is then. How old is each now?   Sue is 17 years old, and Pat is 1 year old.

19. Two cars leave Chicago, one traveling east and the other west. After 4 hours they are 492 miles apart. If one car is traveling 7 mph faster than the other, find the speed of each.   58 mph, 65 mph

20. The foundation for a rectangular Hardware Warehouse has a length three times the width and is 296 feet around. Find the dimensions of the building.   width, 37 ft; length, 111 ft

21. James has available a 10% alcohol solution and a 60% alcohol solution. Find how many liters of each solution he should mix to make 50 liters of 40% alcohol solution.

22. An employee at See's Candy Store needs a special mixture of candy. She has creme-filled chocolates that sell for $3.00 per pound, chocolate-covered nuts that sell for $2.70 per pound, and chocolate-covered raisins that sell for $2.25 per pound. She wants to have twice as many raisins as nuts in the mixture. Find how many pounds of each she should use to make 45 pounds worth $2.80 per pound.

23. Chris has $2.77 in his coin jar—all in pennies, nickels, and dimes. If he has 53 coins in all and four more nickels than dimes, find how many of each type of coin he has.

24. If $10,000 and $4000 are invested such that $1250 is earned in one year, and if the rate of interest on the larger investment is 2% more than that of the smaller investment, find the rates of interest.

25. The perimeter of an isosceles (two sides equal) triangle is 73 centimeters. If two sides are of equal length and the third side is 7 centimeters longer than the others, find the lengths of the three sides.

26. The sum of three numbers is 295. One number is five more than a second and twice the third. Find the numbers.   120, 115, and 60

(4.4) *Use matrices to solve each system.*

27. $\begin{cases} 3x + 10y = 1 \\ x + 2y = -1 \end{cases}$

28. $\begin{cases} 3x - 6y = 12 \\ 2y = x - 4 \end{cases}$

29. $\begin{cases} 3x - 2y = -8 \\ 6x + 5y = 11 \end{cases}$

30. $\begin{cases} 6x - 6y = -5 \\ 10x - 2y = 1 \end{cases}$ $\left\{\left(\frac{1}{3}, \frac{7}{6}\right)\right\}$

31. $\begin{cases} 3x - 6y = 0 \\ 2x + 4y = 5 \end{cases}$ $\left\{\left(\frac{5}{4}, \frac{5}{8}\right)\right\}$

32. $\begin{cases} 5x - 3y = 10 \\ -2x + y = -1 \end{cases}$ $\{(-7, -15)\}$

33. $\begin{cases} 0.2x - 0.3y = -0.7 \\ 0.5x + 0.3y = 1.4 \end{cases}$ $\{(1, 3)\}$

34. $\begin{cases} 3x + 2y = 8 \\ 3x - y = 5 \end{cases}$ $\{(2, 1)\}$

35. $\begin{cases} x + z = 4 \\ 2x - y = 0 \\ x + y - z = 0 \end{cases}$ $\{(1, 2, 3)\}$

36. $\begin{cases} 2x + 5y = 4 \\ x - 5y + z = -1 \\ 4x - z = 11 \end{cases}$ $\{(2, 0, -3)\}$

37. $\begin{cases} 3x - y = 11 \\ x + 2z = 13 \\ y - z = -7 \end{cases}$ $\{(3, -2, 5)\}$

38. $\begin{cases} 5x + 7y + 3z = 9 \\ 14y - z = 28 \\ 4x + 2z = -4 \end{cases}$ $\{(-1, 2, 0)\}$

39. $\begin{cases} 7x - 3y + 2z = 0 \\ 4x - 4y - z = 2 \\ 5x + 2y + 3z = 1 \end{cases}$ $\{(1, 1, -2)\}$

40. $\begin{cases} x + 2y + 3z = 14 \\ y + 2z = 3 \\ 2x - 2z = 10 \end{cases}$ $\{\ \}$

(4.5) *Evaluate.*

41. $\begin{vmatrix} -1 & 3 \\ 5 & 2 \end{vmatrix}$ -17

42. $\begin{vmatrix} 3 & -1 \\ 2 & 5 \end{vmatrix}$ 17

43. $\begin{vmatrix} 2 & -1 & -3 \\ 1 & 2 & 0 \\ 3 & -2 & 2 \end{vmatrix}$ 34

44. $\begin{vmatrix} -2 & 3 & 1 \\ 4 & 4 & 0 \\ 1 & -2 & 3 \end{vmatrix}$ -72

Use Cramer's rule to solve each system of equations.

45. $\begin{cases} 3x - 2y = -8 \\ 6x + 5y = 11 \end{cases}$

46. $\begin{cases} 6x - 6y = -5 \\ 10x - 2y = 1 \end{cases}$ $\left\{\left(\frac{1}{3}, \frac{7}{6}\right)\right\}$

47. $\begin{cases} 3x + 10y = 1 \\ x + 2y = -1 \end{cases}$ $\{(-3, 1)\}$

48. $\begin{cases} y = \frac{1}{2}x + \frac{2}{3} \\ 4x + 6y = 4 \end{cases}$ $\left\{\left(0, \frac{2}{3}\right)\right\}$

49. $\begin{cases} 2x - 4y = 22 \\ 5x - 10y = 16 \end{cases}$ $\{\ \}$

50. $\begin{cases} 3x - 6y = 12 \\ 2y = x - 4 \end{cases}$ $\{(x, y) | x - 2y = 4\}$

51. $\begin{cases} x + z = 4 \\ 2x - y = 0 \\ x + y - z = 0 \end{cases}$ $\{(1, 2, 3)\}$

52. $\begin{cases} 2x + 5y = 4 \\ x - 5y + z = -1 \\ 4x - z = 11 \end{cases}$ $\{(2, 0, -3)\}$

53. $\begin{cases} x + 3y - z = 5 \\ 2x - y - 2z = 3 \\ x + 2y + 3z = 4 \end{cases}$ $\{(2, 1, 0)\}$

54. $\begin{cases} 2x - z = 1 \\ 3x - y + 2z = 3 \\ x + y + 3z = -2 \end{cases}$

55. $\begin{cases} x + 2y + 3z = 14 \\ y + 2z = 3 \\ 2x - 2z = 10 \end{cases}$ $\{\ \}$

56. $\begin{cases} 5x + 7y = 9 \\ 14y - z = 28 \\ 4x + 2z = -4 \end{cases}$

21. 20 liters of 10% solution, 30 liters of 60% solution   **22.** 30 lb of creme-filled; 5 lb of chocolate-covered nuts; 10 lb of chocolate-covered raisins   **23.** 17 pennies, 20 nickels, and 16 dimes   **24.** larger investment, 9.5%; smaller investment, 7.5%   **25.** Two sides are 22 cm each; third side is 29 cm.   **27.** $\{(-3, 1)\}$ **28.** $\{(x, y) | x - 2y = 4\}$

29. $\left\{\left(-\frac{2}{3}, 3\right)\right\}$ **45.** $\left\{\left(-\frac{2}{3}, 3\right)\right\}$ **54.** $\left\{\left(\frac{3}{7}, -2, -\frac{1}{7}\right)\right\}$ **56.** $\{(-1, 2, 0)\}$

CHAPTER 4 TEST

Evaluate each determinant.

1. $\begin{vmatrix} 4 & -7 \\ 2 & 5 \end{vmatrix}$ 34

2. $\begin{vmatrix} 4 & 0 & 2 \\ 1 & -3 & 5 \\ 0 & -1 & 2 \end{vmatrix}$ -6

Solve each system of equations graphically and then solve by the elimination method or the substitution method. For Exercises 3–4, see Appendix D.

3. $\begin{cases} 2x - y = -1 \\ 5x + 4y = 17 \end{cases}$ $\{(1, 3)\}$

4. $\begin{cases} 7x - 14y = 5 \\ x = 2y \end{cases}$ $\{\ \}$

Solve each system. **5.** $\{(2, -3)\}$

5. $\begin{cases} 4x - 7y = 29 \\ 2x + 5y = -11 \end{cases}$

6. $\begin{cases} 15x + 6y = 15 \\ 10x + 4y = 10 \end{cases}$

7. $\begin{cases} 2x - 3y = 4 \\ 3y + 2z = 2 \\ x - z = -5 \end{cases}$

8. $\begin{cases} 3x - 2y - z = -1 \\ 2x - 2y = 4 \\ 2x - 2z = -12 \end{cases}$ $\{\ \}$

9. $\begin{cases} \dfrac{x}{2} + \dfrac{y}{4} = -\dfrac{3}{4} \\ x + \dfrac{3}{4}y = -4 \end{cases}$ $\left(\dfrac{7}{2}, -10\right)$

Use Cramer's rule to solve each system.

10. $\begin{cases} 3x - y = 7 \\ 2x + 5y = -1 \end{cases}$ $\{(2, -1)\}$

11. $\begin{cases} 4x - 3y = -6 \\ -2x + y = 0 \end{cases}$ $\{3, 6\}$

12. $\begin{cases} x + y + z = 4 \\ 2x + 5y = 1 \\ x - y - 2z = 0 \end{cases}$

13. $\begin{cases} 3x + 2y + 3z = 3 \\ x - z = 9 \\ 4y + z = -4 \end{cases}$

Use matrices to solve each system.

14. $\begin{cases} x - y = -2 \\ 3x - 3y = -6 \end{cases}$

15. $\begin{cases} x + 2y = -1 \\ 2x + 5y = -5 \end{cases}$

16. $\begin{cases} x - y - z = 0 \\ 3x - y - 5z = -2 \\ 2x + 3y = -5 \end{cases}$

17. $\begin{cases} 2x - y + 3z = 4 \\ 3x - 3z = -2 \\ -5x + y = 0 \end{cases}$

18. Frame Masters, Inc., recently purchased $5500 worth of new equipment in order to offer a new style of eyeglass frame. The marketing department of Frame Masters estimates that the cost of producing this new frame is $18 and that the frame will be sold to stores for $38. Find the number of frames that must be sold in order to break even.

19. A motel in New Orleans charges $90 per day for double occupancy and $80 per day for single occupancy. If 80 rooms are occupied for a total of $6930, how many rooms of each kind are there?

20. The research department of a company that manufactures children's fruit drinks is experimenting with a new flavor. A 17.5% fructose solution is needed, but only 10% and 20% solutions are available. How many gallons of a 10% fructose solution should be mixed with a 20% fructose solution in order to obtain 20 gallons of a 17.5% fructose solution? 5 gal. of 10%, 15 gal. of 20%

6. $\{(x, y) \mid 10x + 4y = 10\}$ **7.** $\{(-1, -2, 4)\}$ **12.** $\{(3, -1, 2)\}$ **13.** $\{(5, 0, -4)\}$ **14.** $\{(x, y) \mid x - y = -2\}$ **15.** $\{(5, -3)\}$ **16.** $\{(-1, -1, 0)\}$ **17.** $\{\ \}$ **18.** 275 frames **19.** 53 double rooms and 27 single rooms

CHAPTER 4 CUMULATIVE REVIEW

1. Write each sentence with mathematical symbols.

   a. The sum of 5 and y is greater than or equal to 7.

   b. 11 is not equal to z. $11 \neq z$

   c. 20 is less than the difference of 5 and twice x.

2. Find each difference.

   a. $2 - 8$ -6 **b.** $-8 - (-1)$ -7 **c.** $-11 - 5$ -16

   d. $10.7 - (-9.8)$ 20.5 **e.** $\dfrac{2}{3} - 1\dfrac{1}{6}$ **f.** $1 - 0.06$ 0.94

   g. Subtract 7 from 4. -3; *Sec. 1.3, Ex. 3*

3. Solve for y: $\dfrac{y}{3} - \dfrac{y}{4} = \dfrac{1}{6}$. $\{2\}$; *Sec. 2.1, Ex. 5*

4. Find 16% of 25. 4; *Sec. 2.2, Ex. 2*

5. Solve for x: $3x + 4 \geq 2x - 6$. Graph the solution set and write it in interval notation. *Sec. 2.4, Ex. 2*

6. Solve $2 < 4 - x < 7$. $(-3, 2)$; *Sec. 2.5, Ex. 2*

7. Solve $|5w + 3| = 7$. $\left\{-2, \dfrac{4}{5}\right\}$; *Sec. 2.6, Ex. 2*

8. Graph the equation $5x - 2y = 10$.

9. Find the domain and range of each relation. Determine whether the relation is also a function.

1a. $5 + y \geq 7$ **1c.** $20 < 5 - 2x$; *Sec. 1.2, Ex. 3* **5.** $\xleftarrow{\qquad}_{-10}$ $[-10, \infty)$ **8.** See Appendix D; *Sec. 3.1, Ex. 4*

a.

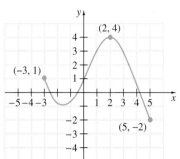

domain: $[-3, 5]$; range: $[-2, 4]$; function

b.

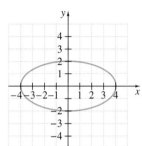

domain: $[-4, 4]$; range: $[-2, 2]$; not a function

c.

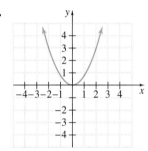

domain: $(-\infty, \infty)$; range: $[0, \infty)$; function

d.

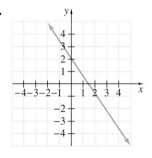

domain: $(-\infty, \infty)$; range: $[-\infty, \infty)$; function; *Sec. 3.2, Ex. 6*

10. Graph $x = -2y$ by plotting intercept points.

11. Find the slope of the line containing the points $(0, 3)$ and $(2, 5)$.   *1; Sec. 3.4, Ex. 1*

12. Find the slope of the line $x = -5$.

13. Find an equation of the line through points $(4, 0)$ and $(-4, -5)$. Write the equation using function notation. $f(x) = \frac{5}{8}x - \frac{5}{2}$; *Sec. 3.5, Ex. 4*

14. Graph $2x - y < 6$.   See Appendix D; *Sec. 3.6, Ex. 1*

15. Use the substitution method to solve the system

$$\begin{cases} 2x + 4y = -6 \\ x - 2y = -5 \end{cases} \quad \left\{\left(-4, \frac{1}{2}\right)\right\}; \text{ Sec. 4.1, Ex. 2}$$

16. Use the elimination method to solve the system

$$\begin{cases} 3x + \dfrac{y}{2} = 2 \\ 6x + y = 5 \end{cases} \quad \{\ \}; \text{ Sec. 4.1, Ex. 5}$$

17. Solve the system

$$\begin{cases} 3x - y + z = -15 \\ x + 2y - z = 1 \\ 2x + 3y - 2z = 0 \end{cases} \{(-4, 2, -1)\}; \text{ Sec. 4.2, Ex. 1}$$

18. Two cars leave Indianapolis, one traveling east and the other west. After 3 hours they are 297 miles apart. If one car is traveling 5 mph faster than the other, what is the speed of each?

19. Use matrices to solve the system:

$$\begin{cases} x + 2y + z = 2 \\ -2x - y + 2z = 5 \\ x + 3y - 2z = -8 \end{cases}$$

20. Evaluate by expanding by the minors of the given row or column.

$$\begin{vmatrix} 0 & 5 & 1 \\ 1 & 3 & -1 \\ -2 & 2 & 4 \end{vmatrix}$$

  a. First column **b.** Second row -2; *Sec. 4.5, Ex. 3*

21. Use Cramer's rule to solve the system:

$$\begin{cases} x - 2y + z = 4 \\ 3x + y - 2z = 3 \\ 5x + 5y + 3z = -8 \end{cases} \quad \{(1, -2, -1)\}; \text{ Sec. 4.5, Ex. 4}$$

10. See Appendix D; *Sec. 3.3, Ex. 5*   **12.** undefined; *Sec. 3.4, Ex. 6*   **18.** 47 mph and 52 mph; *Sec. 4.3, Ex. 2*
19. $\{(1, -1, 3)\}$; *Sec. 4.4, Ex. 2*

EXPONENTS, POLYNOMIALS, AND POLYNOMIAL FUNCTIONS

FINDING THE LARGEST AREA

Picture framers must often find either the optimal or the most pleasing dimensions of a frame or mat to fit the area of the item to be framed. If a standard-sized frame won't fit, the framer will need to build a custom frame. Sometimes custom framing requires working with a limited amount of framing or matting material.

IN THE CHAPTER GROUP ACTIVITY ON PAGE 320, YOU WILL HAVE THE OPPORTUNITY TO FIND THE DIMENSIONS OF A PICTURE FRAME MADE FROM A FIXED AMOUNT OF MATERIAL THAT WILL ALLOW THE LARGEST AMOUNT OF INTERIOR DISPLAY AREA.

5.1 | EXPONENTS AND SCIENTIFIC NOTATION

O B J E C T I V E S

TAPE IA 5.1

1. Use the product rule for exponents.
2. Evaluate a raised to the 0 power.
3. Use the quotient rule for exponents.
4. Define a raised to the negative nth power.
5. Write numbers in scientific notation.
6. Convert numbers from scientific notation to standard notation.

1

Recall that exponents may be used to write repeated factors in a more compact form. As we have seen in the previous chapters, exponents can be used when the repeated factor is a number or a variable. For example,

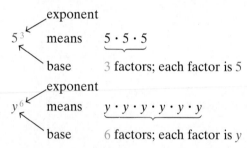

Expressions such as 5^3 and y^6 that contain exponents are called exponential expressions.

Exponential expressions can be multiplied, divided, added, subtracted, and themselves raised to powers. In this section, we review operations on exponential expressions.

We review multiplication first. To multiply x^2 by x^3, use the definition of a^n:

$$x^2 \cdot x^3 = \underbrace{(x \cdot x)(x \cdot x \cdot x)}_{x \text{ is a factor 5 times}}$$

$$= x^5$$

Notice that the result is exactly the same if we add the exponents.

$$x^2 \cdot x^3 = x^{2+3} = x^5$$

This suggests the following.

PRODUCT RULE FOR EXPONENTS

If m and n are positive integers and a is a real number, then

$$a^m \cdot a^n = a^{m+n}$$

In other words, the *product* of exponential expressions with a common base is the common base raised to a power equal to the *sum* of the exponents of the factors.

EXAMPLE 1   Use the product rule to simplify.

      a. $2^2 \cdot 2^5$ **b.** $x^7 x^3$ **c.** $y \cdot y^2 \cdot y^4$

Solution: **a.** $2^2 \cdot 2^5 = 2^{2+5} = 2^7$

      b. $x^7 x^3 = x^{7+3} = x^{10}$

      c. $y \cdot y^2 \cdot y^4 = (y^1 \cdot y^2) \cdot y^4$
                             $= y^3 \cdot y^4$
                             $= y^7$

EXAMPLE 2   Use the product rule to multiply.

      a. $(3x^6)(5x)$ **b.** $(-2x^3 p^2)(4xp^{10})$

Solution: Here, we use properties of multiplication to group together like bases.

      a. $(3x^6)(5x) = 3(5)x^6 x^1 = 15x^7$

      b. $(-2x^3 p^2)(4xp^{10}) = -2(4)x^3 x^1 p^2 p^{10} = -8x^4 p^{12}$

    2 The definition of a^n does not include the possibility that n might be 0. But if it did, then, by the product rule,

$$a^0 \cdot a^n = a^{0+n} = a^n = 1 \cdot a^n.$$

From this, we reasonably define that $a^0 = 1$, as long as a does not equal 0.

> **ZERO EXPONENT**
>
> If a does not equal 0, then $a^0 = 1$.

EXAMPLE 3   Evaluate the following.

      a. 7^0 **b.** -7^0 **c.** $(2x + 5)^0$ **d.** $2x^0$

Solution: **a.** $7^0 = 1$

      b. Without parentheses, only 7 is raised to the 0 power.

$$-7^0 = -(7^0) = -(1) = -1$$

      c. $(2x + 5)^0 = 1$

      d. $2x^0 = 2(1) = 2$

To find quotients of exponential expressions, we again begin with the definition of a^n to simplify $\dfrac{x^9}{x^2}$. For example,

$$\frac{x^9}{x^2} = \frac{\boxed{x} \cdot \boxed{x} \cdot x \cdot x \cdot x \cdot x \cdot x \cdot x \cdot x}{\boxed{x} \cdot \boxed{x}} = x^7$$

(Assume that denominators containing variables are not 0.)

Notice that the result is exactly the same if we subtract the exponents.

$$\frac{x^9}{x^2} = x^{9-2} = x^7$$

This suggests the following.

QUOTIENT RULE FOR EXPONENTS

If a is a nonzero real number and n and m are integers, then

$$\frac{a^m}{a^n} = a^{m-n}$$

In other words, the *quotient* of exponential expressions with a common base is the common base raised to a power equal to the *difference* of the exponents.

EXAMPLE 4 Use the quotient rule to simplify.

a. $\dfrac{x^7}{x^4}$ **b.** $\dfrac{5^8}{5^2}$ **c.** $\dfrac{20x^6}{4x^5}$ **d.** $\dfrac{12y^{10}z^7}{14y^8z^7}$

Solution: **a.** $\dfrac{x^7}{x^4} = x^{7-4} = x^3$

b. $\dfrac{5^8}{5^2} = 5^{8-2} = 5^6$

c. $\dfrac{20x^6}{4x^5} = 5x^{6-5} = 5x^1,$ or $5x$

d. $\dfrac{12y^{10}z^7}{14y^8z^7} = \dfrac{6}{7}y^{10-8} \cdot z^{7-7} = \dfrac{6}{7}y^2z^0 = \dfrac{6}{7}y^2,$ or $\dfrac{6y^2}{7}$

When the exponent of the denominator is larger than the exponent of the numerator, applying the quotient rule yields a negative exponent. For example,

$$\frac{x^3}{x^5} = x^{3-5} = x^{-2}$$

Using the definition of a^n, though, gives us

$$\frac{x^3}{x^5} = \frac{\boxed{x} \cdot \boxed{x} \cdot \boxed{x}}{\boxed{x} \cdot \boxed{x} \cdot \boxed{x} \cdot x \cdot x} = \frac{1}{x^2}$$

From this, we reasonably define $x^{-2} = \dfrac{1}{x^2}$ or, in general, $a^{-n} = \dfrac{1}{a^n}$.

NEGATIVE EXPONENTS

If a is a real number other than 0 and n is a positive integer, then

$$a^{-n} = \frac{1}{a^n}$$

EXAMPLE 5   Use only positive exponents to write the following. Simplify if possible.

a. 5^{-2} **b.** $2x^{-3}$ **c.** $(3x)^{-1}$ **d.** $\dfrac{m^5}{m^{15}}$ **e.** $\dfrac{3^3}{3^6}$ **f.** $2^{-1} + 3^{-2}$ **g.** $\dfrac{1}{t^{-5}}$

Solution: **a.** $5^{-2} = \dfrac{1}{5^2} = \dfrac{1}{25}$

b. Without parentheses, only x is raised to the -3 power.

$$2x^{-3} = 2 \cdot \frac{1}{x^3} = \frac{2}{x^3}$$

c. With parentheses, both 3 and x are raised to the -1 power.

$$(3x)^{-1} = \frac{1}{(3x)^1} = \frac{1}{3x}$$

d. $\dfrac{m^5}{m^{15}} = m^{5-15} = m^{-10} = \dfrac{1}{m^{10}}$

e. $\dfrac{3^3}{3^6} = 3^{3-6} = 3^{-3} = \dfrac{1}{3^3} = \dfrac{1}{27}$

f. $2^{-1} + 3^{-2} = \dfrac{1}{2^1} + \dfrac{1}{3^2} = \dfrac{1}{2} + \dfrac{1}{9} = \dfrac{9}{18} + \dfrac{2}{18} = \dfrac{11}{18}$

g. $\dfrac{1}{t^{-5}} = \dfrac{1}{\dfrac{1}{t^5}} = 1 \div \dfrac{1}{t^5} = 1 \cdot \dfrac{t^5}{1} = t^5$

REMINDER   Notice that when a factor containing an exponent is moved from the numerator to the denominator or from the denominator to the numerator, the sign of its exponent changes.

$$x^{-3} = \frac{1}{x^3}, \qquad 5^{-2} = \frac{1}{5^2} = \frac{1}{25}$$

$$\frac{1}{y^{-4}} = y^4, \qquad \frac{1}{2^{-3}} = 2^3 = 8$$

EXAMPLE 6   Simplify each expression. Use positive exponents to write answers.

a. $\dfrac{x^{-9}}{x^2}$ **b.** $\dfrac{p^4}{p^{-3}}$ **c.** $\dfrac{2^{-3}}{2^{-1}}$ **d.** $\dfrac{2x^{-7}y^2}{10xy^{-5}}$ **e.** $\dfrac{(3x^{-3})(x^2)}{x^6}$

Solution: **a.** $\dfrac{x^{-9}}{x^2} = x^{-9-2} = x^{-11} = \dfrac{1}{x^{11}}$

b. $\dfrac{p^4}{p^{-3}} = p^{4-(-3)} = p^7$

c. $\dfrac{2^{-3}}{2^{-1}} = 2^{-3-(-1)} = 2^{-2} = \dfrac{1}{2^2} = \dfrac{1}{4}$

d. $\dfrac{2x^{-7}y^2}{10xy^{-5}} = \dfrac{x^{-7-1} \cdot y^{2-(-5)}}{5} = \dfrac{x^{-8}y^7}{5} = \dfrac{y^7}{5x^8}$

e. Simplify the numerator first.

$$\dfrac{(3x^{-3})(x^2)}{x^6} = \dfrac{3x^{-3+2}}{x^6} = \dfrac{3x^{-1}}{x^6} = 3x^{-1-6} = 3x^{-7} = \dfrac{3}{x^7}$$

EXAMPLE 7   Simplify. Assume that a and t are nonzero integers and that x is not 0.

a. $x^{2a} \cdot x^3$ **b.** $\dfrac{x^{2t-1}}{x^{t-5}}$

Solution: **a.** $x^{2a} \cdot x^3 = x^{2a+3}$ Use the product rule.

b. $\dfrac{x^{2t-1}}{x^{t-5}} = x^{(2t-1)-(t-5)}$ Use the quotient rule.

$\qquad = x^{2t-1-t+5} = x^{t+4}$

Very large and very small numbers occur frequently in nature. For example, the distance between the Earth and the Sun is approximately 150,000,000 kilometers. A helium atom has a diameter of 0.000 000 022 centimeters. It can be tedious to write these very large and very small numbers in standard notation like this. **Scientific notation** is a convenient shorthand notation for writing very large and very small numbers.

> **SCIENTIFIC NOTATION**
>
> A positive number is written in **scientific notation** if it is written as the product of a number a, where $1 \le a < 10$, and an integer power r of 10: $a \times 10^r$.

The following are examples of numbers written in scientific notation:

$$2.03 \times 10^2, \quad 7.362 \times 10^7, \quad 8.1 \times 10^{-5}$$

To write the approximate distance between the Earth and the Sun in scientific notation, move the decimal point to the left until the number is between 1 and 10.

150,000,000.

The decimal point was moved 8 places to the left, so

$$150,000,000 = 1.5 \times 10^8$$

Next, to write the diameter of a helium atom in scientific notation, again move the decimal point until we have a number between 1 and 10.

0. 000 000 022

The decimal point was moved 8 places to the right, so

$$0.\,000\,000\,022 = 2.2 \times 10^{-8}$$

TO WRITE A NUMBER IN SCIENTIFIC NOTATION

Step 1.   Move the decimal point in the original number until the new number has a value between 1 and 10.

Step 2.   Count the number of decimal places the decimal point was moved in *step 1*. If the decimal point was moved to the left, the count is positive. If the decimal point was moved to the right, the count is negative.

Step 3.   Multiply the new number in *step 1* by 10 raised to an exponent equal to the count found in *step 2*.

EXAMPLE 8   Write each number in scientific notation.

a. 730,000   **b.** 0.00000104

Solution: **a.** *Step 1.*   Move the decimal point until the number is between 1 and 10.

730,000.

Step 2.   The decimal point is moved 5 places to the left, so the count is positive 5.

Step 3. $730,000 = 7.3 \times 10^5$.

b. *Step 1.*   Move the decimal point until the number is between 1 and 10.

0.00000104

Step 2.   The decimal point is moved 6 places to the right, so the count is -6.

Step 3. $0.00000104 = 1.04 \times 10^{-6}$.

6 To write a scientific notation number in standard form, we reverse the preceding steps.

TO WRITE A SCIENTIFIC NOTATION NUMBER IN STANDARD NOTATION

Move the decimal point in the number the same number of places as the exponent on 10. If the exponent is positive, move the decimal point to the right. If the exponent is negative, move the decimal point to the left.

EXAMPLE 9 Write each number in standard notation.

a. 7.7×10^8 **b.** 1.025×10^{-3}

Solution: **a.** Since the exponent is positive, move the decimal point 8 places to the right. Add zeros as needed.

$$7.7 \times 10^8 = 770,000,000$$

b. Since the exponent is negative, move the decimal point 3 places to the left. Add zeros as needed.

$$1.025 \times 10^{-3} = 0.001025$$

SCIENTIFIC CALCULATOR EXPLORATIONS

Multiply 5,000,000 by 700,000 on your calculator. The display should read $\boxed{3.5 \qquad 12}$ or $\boxed{3.5\ E\ 12}$, which is the product written in scientific notation. Both these notations mean 3.5×10^{12}.

To enter a number written in scientific notation on a calculator, find the key marked $\boxed{EE}$. (On some calculators, this key may be marked $\boxed{EXP}$.)

To enter 7.26×10^{13}, press the keys

$$\boxed{7.26}\ \boxed{EE}\ \boxed{13}$$

The display will read $\boxed{7.26 \qquad 13}$ or $\boxed{7.26\ E\ 13}$.

Use your calculator to perform each operation indicated.

1. Multiply 3×10^{11} and 2×10^{32}. 6×10^{43}

2. Divide 6×10^{14} by 3×10^9. 2×10^5

3. Multiply 5.2×10^{23} and 7.3×10^4. 3.796×10^{28}

4. Divide 4.38×10^{41} by 3×10^{17}. 1.46×10^{24}

MENTAL MATH

Use positive exponents to state each expression.

1. $5x^{-1}y^{-2}$ $\dfrac{5}{xy^2}$ **2.** $7xy^{-4}$ $\dfrac{7x}{y^4}$ **3.** $a^2b^{-1}c^{-5}$ $\dfrac{a^2}{bc^5}$ **4.** $a^{-4}b^2c^{-6}$ $\dfrac{b^2}{a^4c^6}$ **5.** $\dfrac{y^{-2}}{x^{-4}}\dfrac{x^4}{y^2}$ **6.** $\dfrac{x^{-7}}{z^{-3}}\dfrac{z^3}{x^7}$

EXERCISE SET 5.1   **5.** $-140x^{12}$ **6.** $27y^5$ **7.** $-20x^2y$ **9.** $-16x^6y^3p^2$ **28.** $\dfrac{1}{2^3}=\dfrac{1}{8}$ **32.** $\dfrac{10}{b}$

Use the product rule to simplify each expression. See Examples 1 and 2.

1. $4^2 \cdot 4^3 \; 4^5$ **2.** $3^3 \cdot 3^5 \; 3^8$ **3.** $x^5 \cdot x^3 \; x^8$ **4.** $a^2 \cdot a^9 \; a^{11}$

5. $-7x^3 \cdot 20x^9$ **6.** $-3y \cdot -9y^4$ **7.** $(4xy)(-5x)$

8. $(7xy)(7aby) \; 49abxy^2$ **9.** $(-4x^3p^2)(4y^3x^3)$

10. $(-6a^2b^3)(-3ab^3) \; 18a^3b^6$

Evaluate the following. See Example 3.

11. $-8^0 \; -1$ **12.** $(-9)^0 \; 1$ **13.** $(4x+5)^0 \; 1$

14. $8x^0 + 1 \; 9$ **15.** $(5x)^0 + 5x^0 \; 6$ **16.** $4y^0 - (4y)^0 \; 3$

17. Explain why $(-5)^0$ simplifies to 1 but -5^0 simplifies to -1.

18. Explain why both $4x^0 - 3y^0$ and $(4x - 3y)^0$ simplify to 1.

Find each quotient. See Example 4.

19. $\dfrac{a^5}{a^2} \; a^3$ **20.** $\dfrac{x^9}{x^4} \; x^5$ **21.** $\dfrac{x^9y^6}{x^8y^6} \; x$ **22.** $\dfrac{a^{12}b^2}{a^9b} \; a^3b$

23. $-\dfrac{26z^{11}}{2z^7} \; -13z^4$ **24.** $\dfrac{16x^5}{8x} \; 2x^4$

25. $\dfrac{-36a^5b^7c^{10}}{6ab^3c^4} \; -6a^4b^4c^6$ **26.** $\dfrac{49a^3bc^{14}}{-7abc^8} \; -7a^2c^6$

Simplify each expression. Write answers with positive exponents. See Examples 5 and 6.

27. $4^{-2} \; \dfrac{1}{16}$ **28.** 2^{-3} **29.** $\dfrac{x^7}{x^{15}} \; \dfrac{1}{x^8}$ **30.** $\dfrac{z}{z^3} \; \dfrac{1}{z^2}$

31. $5a^{-4} \; \dfrac{5}{a^4}$ **32.** $10b^{-1}$ **33.** $\dfrac{x^{-2}}{x^5} \; \dfrac{1}{x^7}$ **34.** $\dfrac{y^{-6}}{y^{-9}} \; y^3$

35. $\dfrac{8r^4}{2r^{-4}} \; 4r^8$ **36.** $\dfrac{3s^3}{15s^{-3}} \; \dfrac{s^6}{5}$ **37.** $\dfrac{x^{-9}x^4}{x^{-5}} \; 1$ **38.** $\dfrac{y^{-7}y}{y^8} \; \dfrac{1}{y^{14}}$

Simplify. Assume that variables in the exponent represent nonzero integers and that x, y, and z are not 0. See Example 7.

39. $x^5 \cdot x^{7a} \; x^{7a+5}$ **40.** $y^{2p} \cdot y^{9p} \; y^{11p}$ **41.** $\dfrac{x^{3t-1}}{x^t} \; x^{2t-1}$

42. $\dfrac{y^{4p-2}}{y^{3p}} \; y^{p-2}$ **43.** $x^{4a} \cdot x^7 \; x^{4a+7}$ **44.** $x^{9y} \cdot x^{-7y} \; x^{2y}$

45. $\dfrac{z^{6x}}{z^7} \; z^{6x-7}$ **46.** $\dfrac{y^6}{y^{4z}} \; y^{6-4z}$ **47.** $\dfrac{x^{3t} \cdot x^{4t-1}}{x^t} \; x^{6t-1}$

48. $\dfrac{z^{5x} \cdot z^{x-7}}{z^x} \; z^{5x-7}$

Simplify the following. Write answers with positive exponents.

49. $4^{-1} + 3^{-2} \; \dfrac{13}{36}$ **50.** $1^{-3} - 4^{-2} \; \dfrac{15}{16}$ **51.** $4x^0 + 5 \; 9$

52. $-5x^0 \; -5$ **53.** $x^7 \cdot x^8 \; x^{15}$ **54.** $y^6 \cdot y \; y^7$

55. $2x^3 \cdot 5x^7$ **56.** $-3z^4 \cdot 10z^7$ **57.** $\dfrac{z^{12}}{z^{15}} \; \dfrac{1}{z^3}$

58. $\dfrac{x^{11}}{x^{20}} \; \dfrac{1}{x^9}$ **59.** $\dfrac{y^{-3}}{y^{-7}} \; y^4$ **60.** $\dfrac{z^{-12}}{z^{10}} \; \dfrac{1}{z^{22}}$ **61.** $3x^{-1} \; \dfrac{3}{x}$

62. $(4x)^{-1} \; \dfrac{1}{4x}$ **63.** $3^0 - 3t^0 \; -2$ **64.** $4^0 + 4x^0 \; 5$ **65.** $\dfrac{r^4}{r^{-4}} \; r^8$

66. $\dfrac{x^{-5}}{x^3} \; \dfrac{1}{x^8}$ **67.** $\dfrac{x^{-7}y^{-2}}{x^2y^2} \; \dfrac{1}{x^9y^4}$ **68.** $\dfrac{a^{-5}b^7}{a^{-2}b^{-3}} \; \dfrac{b^{10}}{a^3}$ **69.** $\dfrac{2a^{-6}b^2}{18ab^{-5}} \; \dfrac{b^7}{9a^7}$

70. $\dfrac{18ab^{-6}}{3a^{-3}b^6} \; \dfrac{6a^4}{b^{12}}$ **71.** $\dfrac{(24x^8)(x)}{20x^{-7}} \; \dfrac{6x^{16}}{5}$ **72.** $\dfrac{(30z^2)(z^5)}{55z^{-4}} \; \dfrac{6z^{11}}{11}$

Write each number in scientific notation. See Example 8.

73. $31{,}250{,}000$ **74.** $678{,}000$ **75.** 0.016

76. 0.007613 **77.** $67{,}413$ **78.** $36{,}800{,}000$

55. $10x^{10}$ **56.** $-30z^{11}$ **62.** $\dfrac{1}{4x}$ **63.** -2 **64.** 5 **67.** $\dfrac{1}{x^9y^4}$ **68.** $\dfrac{b^{10}}{a^3}$ **69.** $\dfrac{b^7}{9a^7}$ **71.** $\dfrac{6x^{16}}{5}$ **72.** $\dfrac{6z^{11}}{11}$

73. 3.125×10^7 **74.** 6.78×10^5 **75.** 1.6×10^{-2} **76.** 7.613×10^{-3} **77.** 6.7413×10^4 **78.** 3.68×10^7

79. 0.0125   **80.** 0.00084   **81.** 0.000053

82. 98,700,000,000 9.87×10^{10}

Write each number in standard notation, without exponents. See Example 9.

83. 3.6×10^{-9} **84.** 2.7×10^{-5} **85.** 9.3×10^{7}

86. 6.378×10^{8} **87.** 1.278×10^{6} **88.** 7.6×10^{4}

89. 7.35×10^{12} **90.** 1.66×10^{-5} **91.** 4.03×10^{-7}

92. 8.007×10^{8} 800,700,000

 93. Explain how to convert a number from standard notation to scientific notation.

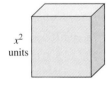 **94.** Explain how to convert a number from scientific notation to standard notation.

Express each number in scientific notation.

95. The approximate distance between Jupiter and Earth is 918,000,000 kilometers. 9.18×10^{8} km

96. The number of millimeters in 50 kilometers is 50,000,000. 5×10^{7} mm

97. A computer can perform 48,000,000,000 arithmetic operations in one minute. 4.8×10^{10} operations/min.

98. The center of the Sun is about 27,000,000°F. 2.7×10^{7} °F

99. A pulsar is a rotating neutron star that gives off sharp, regular pulses of radio waves. For one par-

ticular pulsar, the rate of pulses is every 0.001 second. 1.0×10^{-3}

100. To convert from cubic inches to cubic meters, multiply by 0.0000164. 1.64×10^{-5}

Review Exercises

Evaluate. See Section 1.3.

101. $(5 \cdot 2)^2$ 100   **102.** $5^2 \cdot 2^2$ 100   **103.** $\left(\dfrac{3}{4}\right)^3$ $\dfrac{27}{64}$

104. $\dfrac{3^3}{4^3}$ $\dfrac{27}{64}$ **105.** $(2^3)^2$ 64   **106.** $(2^2)^3$ 64

107. $(2^{-1})^4$ $\dfrac{1}{16}$ **108.** $(2^4)^{-1}$ $\dfrac{1}{16}$

79. 1.25×10^{-2} **80.** 8.4×10^{-4} **81.** 5.3×10^{-5} **83.** 0.0000000036   **84.** 0.000027   **85.** 93,000,000
86. 637,800,000   **87.** 1,278,000   **88.** 76,000   **89.** 7,350,000,000,000   **90.** 0.0000166   **91.** 0.000000403

5.2 | MORE WORK WITH EXPONENTS AND SCIENTIFIC NOTATION

O B J E C T I V E S

TAPE IA 5.2

1 Use the power rules for exponents.

2 Use exponent rules and definitions to simplify exponential expressions.

3 Compute, using scientific notation.

1

The volume of the cube shown whose side measures x^2 units is $(x^2)^3$ cubic units. To simplify an expression such as $(x^2)^3$, we use the definition of a^n. Then

$$(x^2)^3 = \underbrace{(x^2)(x^2)(x^2)}_{x^2 \text{ is a factor 3 times}} = x^{2+2+2} = x^6$$

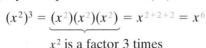

x^2 units

Notice that the result is exactly the same if the exponents are multiplied:

$$(x^2)^3 = x^{2 \cdot 3} = x^6$$

This suggests that the power of an exponential expression raised to a power is the product of the exponents. Two additional power rules for exponents are given in the following box.

POWER RULES FOR EXPONENTS

If a and b are real numbers and m and n are integers, then

$$(a^m)^n = a^{m \cdot n} \qquad \text{Power of a power.}$$

$$(ab)^m = a^m b^m \qquad \text{Power of a product.}$$

$$\left(\frac{a}{b}\right)^n = \frac{a^n}{b^n} \ (b \neq 0) \qquad \text{Power of a quotient.}$$

EXAMPLE 1   Use the power of a power rule to simplify the following expressions. Use positive exponents to write all results.

a. $(x^5)^7$ **b.** $(2^2)^3$ **c.** $(5^{-1})^2$ **d.** $(y^{-3})^{-4}$

Solution: **a.** $(x^5)^7 = x^{5 \cdot 7} = x^{35}$

b. $(2^2)^3 = 2^{2 \cdot 3} = 2^6 = 64$

c. $(5^{-1})^2 = 5^{-1 \cdot 2} = 5^{-2} = \dfrac{1}{5^2} = \dfrac{1}{25}$

d. $(y^{-3})^{-4} = y^{-3(-4)} = y^{12}$

EXAMPLE 2   Use the power rules to simplify the following. Use positive exponents to write all results.

a. $(5x^2)^3$ **b.** $\left(\dfrac{2}{3}\right)^3$ **c.** $\left(\dfrac{3p^4}{q^5}\right)^2$ **d.** $\left(\dfrac{2^{-3}}{y}\right)^{-2}$ **e.** $(x^{-5}y^2z^{-1})^7$

Solution: **a.** $(5x^2)^3 = 5^3 \cdot (x^2)^3 = 5^3 \cdot x^{2 \cdot 3} = 125x^6$

b. $\left(\dfrac{2}{3}\right)^3 = \dfrac{2^3}{3^3} = \dfrac{8}{27}$

c. $\left(\dfrac{3p^4}{q^5}\right)^2 = \dfrac{(3p^4)^2}{(q^5)^2} = \dfrac{3^2 \cdot (p^4)^2}{(q^5)^2} = \dfrac{9p^8}{q^{10}}$

d. $\left(\dfrac{2^{-3}}{y}\right)^{-2} = \dfrac{(2^{-3})^{-2}}{y^{-2}}$

$$= \dfrac{2^6}{y^{-2}} = 64y^2 \qquad \text{Use the negative exponent rule.}$$

e. $(x^{-5}y^2z^{-1})^7 = (x^{-5})^7 \cdot (y^2)^7 \cdot (z^{-1})^7$

$$= x^{-35}y^{14}z^{-7} = \dfrac{y^{14}}{x^{35}z^7}$$

In the next few examples, we practice the use of several of the rules and definitions for exponents. The following is a summary of these rules and definitions.

SUMMARY OF RULES FOR EXPONENTS

If a and b are real numbers and m and n are integers, then

Product rule for exponents	$a^m \cdot a^n = a^{m+n}$	
Zero exponent	$a^0 = 1$	$(a \neq 0)$
Negative exponent	$a^{-n} = \dfrac{1}{a^n}$	$(a \neq 0)$
Quotient rule	$\dfrac{a^m}{a^n} = a^{m-n}$	$(a \neq 0)$
Power rules	$(a^m)^n = a^{m \cdot n}$	
	$(ab)^m = a^m \cdot b^m$	
	$\left(\dfrac{a}{b}\right)^m = \dfrac{a^m}{b^m}$	$(b \neq 0)$

EXAMPLE 3   Simplify each expression. Use positive exponents to write answers.

a. $(2x^0y^{-3})^{-2}$ **b.** $\left(\dfrac{x^{-5}}{x^{-2}}\right)^{-3}$ **c.** $\left(\dfrac{2}{7}\right)^{-2}$ **d.** $\dfrac{5^{-2}x^{-3}y^{11}}{x^2y^{-5}}$

Solution: **a.** $(2x^0y^{-3})^{-2} = 2^{-2}(x^0)^{-2}(y^{-3})^{-2}$

$= 2^{-2}x^0y^6$

$= \dfrac{1(y^6)}{2^2}$ Write x^0 as 1.

$= \dfrac{y^6}{4}$

b. $\left(\dfrac{x^{-5}}{x^{-2}}\right)^{-3} = \dfrac{(x^{-5})^{-3}}{(x^{-2})^{-3}} = \dfrac{x^{15}}{x^6} = x^{15-6} = x^9$

c. $\left(\dfrac{2}{7}\right)^{-2} = \dfrac{2^{-2}}{7^{-2}} = \dfrac{7^2}{2^2} = \dfrac{49}{4}$

d. $\dfrac{5^{-2}x^{-3}y^{11}}{x^2y^{-5}} = \left(5^{-2}\right)\left(\dfrac{x^{-3}}{x^2}\right)\left(\dfrac{y^{11}}{y^{-5}}\right) = 5^{-2}x^{-3-2}y^{11-(-5)} = 5^{-2}x^{-5}y^{16}$

$= \dfrac{y^{16}}{5^2x^5} = \dfrac{y^{16}}{25x^5}$

EXAMPLE 4   Simplify each expression. Use positive exponents to write answers.

a. $\left(\dfrac{3x^2y}{y^{-9}z}\right)^{-2}$ b. $\left(\dfrac{3a^2}{2x^{-1}}\right)^3\left(\dfrac{x^{-3}}{4a^{-2}}\right)^{-1}$

Solution: There is often more than one way to simplify exponential expressions. Here, we will simplify inside parentheses if possible before we apply power rules for exponents.

a. $\left(\dfrac{3x^2y}{y^{-9}z}\right)^{-2} = \left(\dfrac{3x^2y^{10}}{z}\right)^{-2} = \dfrac{3^{-2}x^{-4}y^{-20}}{z^{-2}} = \dfrac{z^2}{3^2x^4y^{20}} = \dfrac{z^2}{9x^4y^{20}}$

b. $\left(\dfrac{3a^2}{2x^{-1}}\right)^3\left(\dfrac{x^{-3}}{4a^{-2}}\right)^{-1} = \dfrac{27a^6}{8x^{-3}} \cdot \dfrac{x^3}{4^{-1}a^2}$

$= \dfrac{27 \cdot 4 \cdot a^6x^3x^3}{8 \cdot a^2} = \dfrac{27a^4x^6}{2}$

EXAMPLE 5   Simplify the expression. Assume that a and b are integers and that x and y are not 0.

a. $x^{-b}(2x^b)^2$ b. $\dfrac{(y^{3a})^2}{y^{a-6}}$

Solution: a. $x^{-b}(2x^b)^2 = x^{-b}2^2x^{2b} = 4x^{-b+2b} = 4x^b$

b. $\dfrac{(y^{3a})^2}{y^{a-6}} = \dfrac{y^{2(3a)}}{y^{a-6}} = \dfrac{y^{6a}}{y^{a-6}} = y^{6a-(a-6)} = y^{6a-a+6} = y^{5a+6}$

3   To perform operations on numbers written in scientific notation, we use properties of exponents.

EXAMPLE 6   Perform the indicated operations. Write each result in scientific notation.

a. $(8.1 \times 10^5)(5 \times 10^{-7})$ b. $\dfrac{1.2 \times 10^4}{3 \times 10^{-2}}$

Solution: a. $(8.1 \times 10^5)(5 \times 10^{-7}) = 8.1 \times 5 \times 10^5 \times 10^{-7}$

$= 40.5 \times 10^{-2}$

$= (4.05 \times 10^1) \times 10^{-2}$

$= 4.05 \times 10^{-1}$

b. $\dfrac{1.2 \times 10^4}{3 \times 10^{-2}} = \left(\dfrac{1.2}{3}\right)\left(\dfrac{10^4}{10^{-2}}\right) = 0.4 \times 10^{4-(-2)}$

$= 0.4 \times 10^6 = (4 \times 10^{-1}) \times 10^6 = 4 \times 10^5$

EXAMPLE 7   Use scientific notation to simplify $\dfrac{2000 \times 0.000021}{700}$.

Solution: $\dfrac{2000 \times 0.000021}{700} = \dfrac{(2 \times 10^3)(2.1 \times 10^{-5})}{7 \times 10^2} = \dfrac{2(2.1)}{7} \cdot \dfrac{10^3 \cdot 10^{-5}}{10^2}$

$$= 0.6 \times 10^{-4}$$
$$= (6 \times 10^{-1}) \times 10^{-4}$$
$$= 6 \times 10^{-5}$$

MENTAL MATH

Simplify. See Example 1.

1. $(x^4)^5 \; x^{20}$ **2.** $(5^6)^2 \; 5^{12}$ **3.** $x^4 \cdot x^5 \; x^9$ **4.** $x^7 \cdot x^8 \; x^{15}$ **5.** $(y^6)^7 \; y^{42}$

6. $(x^3)^4 \; x^{12}$ **7.** $(z^4)^5 \; z^{20}$ **8.** $(z^3)^7 \; z^{21}$ **9.** $(z^{-6})^{-3} \; z^{18}$ **10.** $(y^{-4})^{-2} \; y^8$

3. $\dfrac{1}{x^{36}}$ **5.** $\dfrac{1}{y^5}$ **6.** $\dfrac{1}{z^{10}}$ **7.** $9x^4y^6$ **8.** $16x^6y^2z^2$ **9.** $16x^{20}y^{12}$ **12.** $\dfrac{x^{12}}{36y^{14}}$ **13.** $\dfrac{y^{15}}{x^{35}z^{20}}$

18. $\dfrac{8}{x^{12}y^{18}}$ **23.** $\dfrac{1}{125}$ **24.** $\dfrac{1}{64}$ **29.** $16x^4$ **32.** $4^{24}y^{24}$

EXERCISE SET 5.2

Simplify. Use positive exponents to write each answer.
See Examples 1 and 2.

1. $(3^{-1})^2 \; \dfrac{1}{9}$ **2.** $(2^{-2})^2 \; \dfrac{1}{16}$ **3.** $(x^4)^{-9}$ **4.** $(y^7)^{-3} \; \dfrac{1}{y^{21}}$

5. $(y)^{-5}$ **6.** $(z^{-1})^{10}$ **7.** $(3x^2y^3)^2$ **8.** $(4x^3yz)^2$

9. $\left(\dfrac{2x^5}{y^{-3}}\right)^4$ **10.** $\left(\dfrac{3a^{-4}}{b^7}\right)^3 \; \dfrac{27}{a^{12}b^{21}}$ **11.** $(a^2bc^{-3})^{-6} \; \dfrac{c^{18}}{a^{12}b^6}$

12. $(6x^{-6}y^7z^0)^{-2}$ **13.** $\left(\dfrac{x^7y^{-3}}{z^{-4}}\right)^{-5}$ **14.** $\left(\dfrac{a^{-2}b^{-5}}{c^{-11}}\right)^{-6} \; \dfrac{a^{12}b^{30}}{c^{66}}$

Simplify. Use positive exponents to write each answer.
See Examples 3 and 4.

15. $\left(\dfrac{a^{-4}}{a^{-5}}\right)^{-2} \; \dfrac{1}{a^2}$ **16.** $\left(\dfrac{x^{-9}}{x^{-4}}\right)^{-3} \; x^{15}$ **17.** $\left(\dfrac{2a^{-2}b^5}{4a^2b^7}\right)^{-2} \; 4a^8b^4$

18. $\left(\dfrac{5x^7y^4}{10x^3y^{-2}}\right)^{-3}$ **19.** $\dfrac{4^{-1}x^2yz}{x^{-2}yz^3} \; \dfrac{x^4}{4z^2}$ **20.** $\dfrac{8^{-2}x^{-3}y^{11}}{x^2y^{-5}} \; \dfrac{y^{16}}{64x^5}$

📌 **21.** Is there a number a such that $a^{-1} = a^1$? If so, give the value of a.   yes; $a = \pm 1$

📌 **22.** Is there a number a such that a^{-2} is a negative number? If so, give the value of a.   no

Simplify. Use positive exponents to write each answer.

23. $(5^{-1})^3$ **24.** $(8^2)^{-1}$ **25.** $(x^7)^{-9} \; \dfrac{1}{x^{63}}$ **26.** $(y^{-4})^5 \; \dfrac{1}{y^{20}}$

27. $\left(\dfrac{7}{8}\right)^3 \; \dfrac{343}{512}$ **28.** $\left(\dfrac{4}{3}\right)^2 \; \dfrac{16}{9}$ **29.** $(4x^2)^2$ **30.** $(-8x^3)^2 \; 64x^6$

31. $(-2^{-2}y)^3 \; -\dfrac{y^3}{64}$ **32.** $(-4^{-6}y^{-6})^{-4}$ **33.** $\left(\dfrac{4^{-4}}{y^3x}\right)^{-2} \; 4^8x^2y^6$

37. $-\dfrac{a^6}{512x^3y^9}$ **39.** $\dfrac{x^{14}y^{14}}{a^{21}}$ **40.** $\dfrac{x^5y^{10}}{5^{15}}$ **42.** 8^6y^4 **50.** $125x^3y^3z^6$ **51.** $\dfrac{1}{x^{30}b^6c^6}$ **55.** $\dfrac{2}{x^4y^{10}}$ **56.** $\dfrac{y^2}{z^7}$

57. x^{9a+18} **58.** x^{4b+14} **59.** x^{12a+2} **60.** x^{-3y+1} **61.** b^{10x^2-4x} **62.** c^{6a+9} **63.** y^{15a+3} **64.** y^{26a+1}

34. $\left(\dfrac{7^{-3}}{ab^2}\right)^{-2} \; 7^6a^2b^4$ **35.** $\left(\dfrac{6p^6}{p^{12}}\right)^2 \; \dfrac{36}{p^{12}}$ **36.** $\left(\dfrac{4p^6}{p^9}\right)^3 \; \dfrac{64}{p^9}$

37. $(-8y^3xa^{-2})^{-3}$ **38.** $(-xy^0x^2a^3)^{-3} \; -\dfrac{1}{x^9a^9}$

39. $\left(\dfrac{x^{-2}y^{-2}}{a^{-3}}\right)^{-7}$ **40.** $\left(\dfrac{x^{-1}y^{-2}}{5^{-3}}\right)^{-5}$ **41.** $\left(\dfrac{3x^5}{6x^4}\right)^4 \; \dfrac{x^4}{16}$

42. $\left(\dfrac{8^{-3}}{y^2}\right)^{-2}$ **43.** $\left(\dfrac{1}{4}\right)^{-3} \; 64$ **44.** $\left(\dfrac{1}{8}\right)^{-2} \; 64$ **45.** $\dfrac{(y^3)^{-4}}{y^3} \; \dfrac{1}{y^{15}}$

46. $\dfrac{2(y^3)^{-3}}{y^{-3}} \; \dfrac{2}{y^6}$ **47.** $\dfrac{8p^7}{4p^9} \; \dfrac{2}{p^2}$ **48.** $\left(\dfrac{2x^4}{x^2}\right)^3 \; 8x^6$

49. $(4x^6y^5)^{-2} (6x^4y^3) \; \dfrac{3}{8x^8y^7}$ **50.** $(5xy)^3(z^{-2})^{-3}$

51. $x^6(x^6bc)^{-6}$ **52.** $2(y^2b)^{-4} \; \dfrac{2}{y^8b^4}$ **53.** $\dfrac{2^{-3}x^2y^{-5}}{5^{-2}x^7y^{-1}} \; \dfrac{25}{8x^5y^4}$

54. $\dfrac{7^{-1}a^{-3}b^5}{a^2b^{-2}} \; \dfrac{b^7}{7a^5}$ **55.** $\left(\dfrac{2x^2}{y^4}\right)^3 \cdot \left(\dfrac{2x^5}{y}\right)^{-2}$ **56.** $\left(\dfrac{3z^{-2}}{y}\right)^2 \cdot \left(\dfrac{9y^{-4}}{z^{-3}}\right)^{-1}$

Simplify the following. Assume that variables in the exponents represent nonzero integers and that all other variables are not 0. See Example 5.

57. $(x^{3a+6})^3$ **58.** $(x^{2b+7})^2$ **59.** $\dfrac{x^{4a}(x^{4a})^3}{x^{4a-2}}$ **60.** $\dfrac{x^{-5y+2}x^{2y}}{x}$

61. $(b^{5x-2})^{2x}$ **62.** $(c^{2a+3})^3$ **63.** $\dfrac{(y^{2a})^8}{y^{a-3}}$ **64.** $\dfrac{(y^{4a})^7}{y^{2a-1}}$

65. $\left(\dfrac{2x^{3t}}{x^{2t-1}}\right)^4 \; 16x^{4t+4}$ **66.** $\left(\dfrac{3y^{5a}}{y^{-a+1}}\right)^2 \; 9y^{12a-2}$

Perform indicated operations. Write each result in scientific notation. See Examples 6 and 7.

67. $(5 \times 10^{11})(2.9 \times 10^{-3})$ **68.** $(3.6 \times 10^{-12})(6 \times 10^9)$

69. $(2 \times 10^5)^3$ **70.** $(3 \times 10^{-7})^3$ **71.** $\dfrac{3.6 \times 10^{-4}}{9 \times 10^2}$ 4×10^{-7}

72. $\dfrac{1.2 \times 10^9}{2 \times 10^{-5}}$ **73.** $\dfrac{0.0069}{0.023}$ **74.** $\dfrac{0.00048}{0.0016}$ 3.0×10^{-1}

75. $\dfrac{18,200 \times 100}{91,000}$ 2.0×10^1 **76.** $\dfrac{0.0003 \times 0.0024}{0.0006 \times 20}$ 6.0×10^{-5}

77. $\dfrac{6000 \times 0.006}{0.009 \times 400}$ 1.0×10^1 **78.** $\dfrac{0.00016 \times 300}{0.064 \times 100}$ 7.5×10^{-3}

79. $\dfrac{0.00064 \times 2000}{16,000}$ **80.** $\dfrac{0.00072 \times 0.003}{0.00024}$ 9.0×10^{-3}

81. $\dfrac{66,000 \times 0.001}{0.002 \times 0.003}$ **82.** $\dfrac{0.0007 \times 11,000}{0.001 \times 0.0001}$ 7.7×10^7

83. $\dfrac{1.25 \times 10^{15}}{(2.2 \times 10^{-2})(6.4 \times 10^{-5})}$ $8.877840909 \times 10^{20}$

84. $\dfrac{(2.6 \times 10^{-3})(4.8 \times 10^{-4})}{1.3 \times 10^{-12}}$ 9.6×10^5

Solve.

85. A fast computer can add two numbers in about 10^{-8} second. Express in scientific notation how long it would take this computer to do this task 200,000 times. $0.002 = 2 \times 10^{-3}$ sec

86. The density D of an object is equivalent to the quotient of its mass M and volume V. Thus $D = \dfrac{M}{V}$. Express in scientific notation the density of an object whose mass is 500,000 pounds and whose volume is 250 cubic feet.

87. The density of ordinary water is 3.12×10^{-2} tons per cubic foot. The volume of water in the largest of the Great Lakes, Lake Superior, is 4.269×10^{14} cubic feet. Use the formula $D = \dfrac{M}{V}$ (see Exercise 86) to find the mass (in tons) of the water in Lake Superior. Express your answer in scientific notation. (*Source:* National Ocean Service)

88. The population of the United States in 1994 was 2.603×10^8. The land area of the United States is 3.536×10^6 square miles. Find the population density (number of people per square mile) for the United States in 1994. Round to the nearest whole. (*Source:* U.S. Bureau of the Census)

89. Each side of the cube shown is $\dfrac{2x^{-2}}{y}$ meters. Find its volume. $\dfrac{8}{x^6 y^3}$ cubic meters

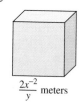

$\dfrac{2x^{-2}}{y}$ meters

90. The lot shown is in the shape of a parallelogram with base $\dfrac{3x^{-1}}{y^{-3}}$ feet and height $5x^{-7}$ feet. Find its area.

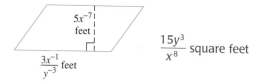

$5x^{-7}$ feet

$\dfrac{3x^{-1}}{y^{-3}}$ feet

$\dfrac{15y^3}{x^8}$ square feet

91. To convert from square inches to square meters, multiply by 6.452×10^{-4}. The area of the following square is 4×10^{-2} square inches. Convert this area to square meters. 2.5808×10^{-5} square meters

4×10^{-2}
square inches

92. To convert from cubic inches to cubic meters, multiply by 1.64×10^{-5}. A grain of salt is in the shape of a cube. If an average size of a grain of salt is 3.8×10^{-6} cubic inches, convert this volume to cubic meters. 6.232×10^{-11} cubic meters

93. Explain whether 0.4×10^{-5} is written in scientific notation.   Answers will vary.

Review Exercises

Simplify each expression. See Section 1.4.

94. $-5y + 4y - 18 - y$ $-2y - 18$

95. $12m - 14 - 15m - 1$ $-3m - 15$

96. $-3x - (4x - 2)$ $-7x + 2$

97. $-9y - (5 - 6y)$ $-3y - 5$

98. $3(z - 4) - 2(3z + 1)$ $-3z - 14$

99. $5(x - 3) - 4(2x - 5)$ $-3x + 5$

67. 1.45×10^9 **68.** 2.16×10^{-2} **69.** 8.0×10^{15} **70.** 2.7×10^{-20} **72.** 6.0×10^{13} **73.** 3.0×10^{-1}
79. 8.0×10^{-5} **81.** 1.1×10^7 **86.** 2.0×10^3 pounds/cubic feet   **87.** 1.331928×10^{13} tons
88. 74 people per square mile

5.3 | POLYNOMIALS AND POLYNOMIAL FUNCTIONS

TAPE IA 5.3

OBJECTIVES

1. Define *term, constant, polynomial, monomial, binomial,* and *trinomial.*
2. Identify the degree of a term and of a polynomial.
3. Define polynomial functions.
4. Add polynomials.
5. Subtract polynomials.

A **term** is a number or the product of a number and one or more variables raised to powers. The **numerical coefficient,** or simply the **coefficient,** is the numerical factor of a term.

TERM	NUMERICAL COEFFICIENT
$-12x^5$	-12
x^3y	1
$-z$	-1
2	2

If a term contains only a number, it is called a **constant term,** or simply a **constant.**

A **polynomial** is a finite sum of terms in which all variables are raised to non-negative integer powers and no variables appear in the denominator.

POLYNOMIALS	NOT POLYNOMIALS	
$4x^5y + 7xz$	$5x^{-3} + 2x$	(negative integer exponent)
$-5x^3 + 2x + \dfrac{2}{3}$	$\dfrac{6}{x^2} - 5x + 1$	(variable in denominator)

A polynomial that contains only one variable is called a **polynomial in one variable.** For example, $3x^2 - 2x + 7$ is a **polynomial in x.** This polynomial in x is written in *descending order* since the terms are listed in descending order of the variable's exponents. (The term 7 can be thought of as $7x^0$.) The following examples are polynomials in one variable written in **descending order.**

$$4x^3 - 7x^2 + 5, \qquad y^2 - 4, \qquad 8a^4 - 7a^2 + 4a$$

A **monomial** is a polynomial consisting of one term. A **binomial** is a polynomial consisting of two terms. A **trinomial** is a polynomial consisting of three terms.

MONOMIALS	BINOMIALS	TRINOMIALS
ax^2	$x + y$	$x^2 + 4xy + y^2$
$-3x$	$6y^2 - 2$	$-x^4 + 3x^3 + 1$
4	$\dfrac{5}{7}z^3 - 2z$	$8y^2 - 2y - 10$

By definition, all monomials, binomials, and trinomials are also polynomials.

2 Each term of a polynomial has a **degree**.

> **DEGREE OF A TERM**
>
> The **degree of a term** is the sum of the exponents on the *variables* contained in the term.

EXAMPLE 1   Find the degree of each term.

a. $3x^2$ **b.** -2^3x^5 **c.** y **d.** $12x^2yz^3$ **e.** 5

Solution: **a.** The exponent on x is 2, so the degree of the term is 2.

b. The exponent on x is 5, so the degree of the term is 5. (Recall that the degree is the sum of the exponents on only the *variables*.)

c. The degree of y, or y^1, is 1.

d. The degree is the sum of the exponents on the variables, or $2 + 1 + 3 = 6$.

c. The degree of 5, which can be written as $5x^0$, is 0.

From the preceding example, we can say that the degree of a constant is 0. Also, the term 0 has no degree.

Each polynomial also has a degree.

> **DEGREE OF A POLYNOMIAL**
>
> The **degree of a polynomial** is the largest degree of all its terms.

EXAMPLE 2   Find the degree of each polynomial and indicate whether the polynomial is a monomial, binomial, trinomial, or none of these.

a. $7x^3 - 3x + 2$ **b.** $-xyz$ **c.** $x^2 - 4$ **d.** $2xy + x^2y^2 - 5x^2 - 6$

Solution: **a.** The degree of the trinomial $7x^3 - 3x + 2$ is 3, the largest degree of any term.

b. The degree of the monomial $-xyz$, or $-x^1y^1z^1$, is 3.

c. The degree of the binomial $x^2 - 4$ is 2.

d. The degree of each term of the polynomial $2xy + x^2y^2 - 5x^2 - 6$ is

TERM	DEGREE
$2xy$, or $2x^1y^1$	$1 + 1 = 2$
x^2y^2	$2 + 2 = 4$
$-5x^2$	2
-6	0

The largest degree of any term is 4, so the degree of this polynomial is 4.

3   At times, it is convenient to use function notation to represent polynomials. For example, we may write $P(x)$ to represent the polynomial $3x^2 - 2x - 5$. In symbols, this is

$$P(x) = 3x^2 - 2x - 5$$

This function is called a **polynomial function** because the expression $3x^2 - 2x - 5$ is a polynomial.

> R E M I N D E R   Recall that the symbol $P(x)$ **does not mean** P times x. It is a special symbol used to denote a function.

EXAMPLE 3   If $P(x) = 3x^2 - 2x - 5$, find the following:

a. $P(1)$ **b.** $P(-2)$

Solution: **a.** Substitute 1 for x in $P(x) = 3x^2 - 2x - 5$ and simplify.

$$P(x) = 3x^2 - 2x - 5$$
$$P(1) = 3(1)^2 - 2(1) - 5 = -4$$

b. $P(x) = 3x^2 - 2x - 5$
$$P(-2) = 3(-2)^2 - 2(-2) - 5 = 11$$

Many real-world phenomena are modeled by polynomial functions. If the polynomial function model is given, we can often find the solution of a problem by evaluating the function at a certain value.

EXAMPLE 4   The Royal Gorge suspension bridge, the world's highest bridge, in Colorado is 1053 feet above the Arkansas River. An object is dropped from the top of this bridge. Neglecting air resistance, the height of the object at time t seconds is given by the polynomial function $P(t) = -16t^2 + 1053$. Find the height of the object when $t = 1$ second and when $t = 8$ seconds.

Solution: To find the height of the object at 1 second, we find $P(1)$.

$$P(t) = -16t^2 + 1053$$
$$P(1) = -16(1)^2 + 1053$$
$$P(1) = 1037$$

When $t = 1$ second, the height of the object is 1037 feet.

To find the height of the object at 8 seconds, we find $P(8)$.

$$P(t) = -16t^2 + 1037$$
$$P(8) = -16(8)^2 + 1037$$
$$P(8) = -1024 + 1037$$
$$P(8) = 13$$

When $t = 8$ seconds, the height of the object is 13 feet.

Notice that when time t increases, the height of the object decreases.

4 Before we add polynomials, recall that terms are considered to be **like terms** if they contain exactly the same variables raised to exactly the same powers.

LIKE TERMS	UNLIKE TERMS
$-5x^2, -x^2$	$4x^2, 3x$
$7xy^3z, -2xzy^3$	$12x^2y^3, -2xy^3$

To simplify a polynomial, **combine like terms** by using the distributive property. For example, by the distributive property,

$$5x + 7x = (5 + 7)x = 12x$$

EXAMPLE 5   Simplify by combining like terms.

    a. $-12x^2 + 7x^2 - 6x$ **b.** $3xy - 2x + 5xy - x$

Solution: By the distributive property,

    a. $-12x^2 + 7x^2 - 6x = (-12 + 7)x^2 - 6x = -5x^2 - 6x$

b. Use the associative and commutative properties to group together like terms; then combine:

$$3xy - 2x + 5xy - x = 3xy + 5xy - 2x - x$$
$$= (3 + 5)xy + (-2 - 1)x$$
$$= 8xy - 3x$$

Now we have reviewed the necessary skills to add polynomials.

To Add Polynomials

Combine all like terms.

EXAMPLE 6   Add.

a. $(7x^3y - xy^3 + 11) + (6x^3y - 4)$ **b.** $(3a^3 - b + 2a - 5) + (a + b + 5)$

Solution: **a.** To add, remove the parentheses and group like terms.

$$(7x^3y - xy^3 + 11) + (6x^3y - 4)$$
$$= 7x^3y - xy^3 + 11 + 6x^3y - 4$$
$$= 7x^3y + 6x^3y - xy^3 + 11 - 4 \qquad \text{Group like terms.}$$
$$= 13x^3y - xy^3 + 7 \qquad \text{Combine like terms.}$$

b. $(3a^3 - b + 2a - 5) + (a + b + 5)$
$$= 3a^3 - b + 2a - 5 + a + b + 5$$
$$= 3a^3 - b + b + 2a + a - 5 + 5 \qquad \text{Group like terms.}$$
$$= 3a^3 + 3a \qquad \text{Combine like terms.}$$

EXAMPLE 7   Add $11x^3 - 12x^2 + x - 3$ and $x^3 - 10x + 5$.

Solution: $(11x^3 - 12x^2 + x - 3) + (x^3 - 10x + 5)$
$$= 11x^3 + x^3 - 12x^2 + x - 10x - 3 + 5 \qquad \text{Group like terms.}$$
$$= 12x^3 - 12x^2 - 9x + 2 \qquad \text{Combine like terms.}$$

Sometimes it is more convenient to add polynomials vertically. To do this, line up like terms beneath one another and add like terms.

EXAMPLE 8 Add $11x^3 - 12x^2 + x - 3$ and $x^3 - 10x + 5$ vertically.

Solution:

$$11x^3 - 12x^2 + \quad x - 3$$
$$\underline{\quad x^3 \qquad\qquad - 10x + 5} \qquad \text{Line up like terms.}$$
$$12x^3 - 12x^2 - \quad 9x + 2 \qquad \text{Combine like terms.}$$

This example is the same as Example 7, only here we added vertically.

5 The definition of subtraction of real numbers can be extended to apply to polynomials. To subtract a number, we add its opposite:

$$a - b = a + (-b)$$

To subtract a polynomial, we add its opposite. In other words, if P and Q are polynomials, then

$$P - Q = P + (-Q)$$

The polynomial $-Q$ is the **opposite,** or **additive inverse,** of the polynomial Q. We can find $-Q$ by changing the sign of each term of Q.

> **TO SUBTRACT POLYNOMIALS**
>
> To subtract polynomials, change the signs of the terms of the polynomial being subtracted; then add.

For example,

To subtract, change the signs; then

Add

$$(3x^2 + 4x - 7) - \overbrace{(3x^2 - 2x - 5)} = (3x^2 + 4x - 7) + \overbrace{(-3x^2 + 2x + 5)}$$
$$= 3x^2 + 4x - 7 - 3x^2 + 2x + 5$$
$$= 6x - 2 \qquad \text{Combine like terms.}$$

EXAMPLE 9 Subtract: $(12z^5 - 12z^3 + z) - (-3z^4 + z^3 + 12z)$

Solution: To subtract, change the sign of each term of the second polynomial and add the result to the first polynomial.

$$(12z^5 - 12z^3 + z) - (-3z^4 + z^3 + 12z)$$
$$= 12z^5 - 12z^3 + z + 3z^4 - z^3 - 12z \qquad \text{Change signs and add.}$$
$$= 12z^5 + 3z^4 - 12z^3 - z^3 + z - 12z \qquad \text{Group like terms.}$$
$$= 12z^5 + 3z^4 - 13z^3 - 11z \qquad \text{Combine like terms.}$$

EXAMPLE 10   Subtract $4x^3y^2 - 3x^2y^2 + 2y^2$ from $10x^3y^2 - 7x^2y^2$.

Solution: If we subtract 2 from 8, the difference is $8 - 2 = 6$. Notice the order of the numbers, and then write "Subtract $4x^3y^2 - 3x^2y^2 + 2y^2$ from $10x^3y^2 - 7x^2y^2$" as a mathematical expression.

$$(10x^3y^2 - 7x^2y^2) - (4x^3y^2 - 3x^2y^2 + 2y^2)$$
$$= 10x^3y^2 - 7x^2y^2 - 4x^3y^2 + 3x^2y^2 - 2y^2 \qquad \text{Remove parentheses.}$$
$$= 6x^3y^2 - 4x^2y^2 - 2y^2 \qquad \text{Combine like terms.}$$

EXAMPLE 11   Perform the subtraction $(10x^3y^2 - 7x^2y^2) - (4x^3y^2 - 3x^2y^2 + 2y^2)$ vertically.

Solution: Add the opposite of the second polynomial.

$$\begin{array}{r} 10x^3y^2 - 7x^2y^2 \\ - (4x^3y^2 - 3x^2y^2 + 2y^2) \\ \hline \end{array} \quad \text{is equivalent to} \quad \begin{array}{r} 10x^3y^2 - 7x^2y^2 \\ -4x^3y^2 + 3x^2y^2 - 2y^2 \\ \hline 6x^3y^2 - 4x^2y^2 - 2y^2 \end{array}$$

Polynomial functions, like polynomials, can be added, subtracted, multiplied, and divided. For example, if

$$P(x) = x^2 + x + 1$$

then

$$2P(x) = 2(x^2 + x + 1) = 2x^2 + 2x + 2$$

Also, if $Q(x) = 5x^2 - 1$, then $P(x) + Q(x) = (x^2 + x + 1) + (5x^2 - 1) = 6x^2 + x$.

A useful business and economics application of subtracting polynomial functions is finding the profit function $P(x)$ when given a revenue function $R(x)$ and a cost function $C(x)$. In business, it is true that

profit = revenue − cost, or
$$P(x) = R(x) - C(x)$$

For example, if the revenue function is $R(x) = 7x$ and the cost function is $C(x) = 2x + 5000$, then the profit function is

$$P(x) = R(x) - C(x)$$

or

$$P(x) = 7x - (2x + 5000) \qquad \text{Substitute } R(x) = 7x \text{ and } C(x) = 2x + 5000.$$
$$P(x) = 5x - 5000$$

Problem-solving exercises involving profit are in the exercise set.

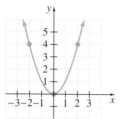

In this section, we reviewed how to find the degree of a polynomial. Knowing the degree of a polynomial can help us recognize the graph of the related polynomial function. For example, we know from Section 3.1 that the graph of the polynomial function $f(x) = x^2$ is shown at left.

The polynomial x^2 has degree 2. The graphs of all polynomial functions of degree 2 will have this same general shape—opening upward, as shown, or downward. Graphs of polynomial functions of degree 2 or 3 will, in general, resemble one of the graphs shown next.

Degree 2

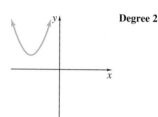

Coefficient of x^2
is a positive number.

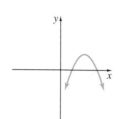

Coefficient of x^2
is a negative number.

Degree 3

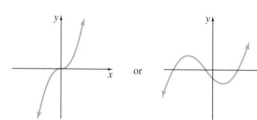   or

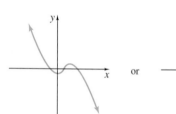   or

Coefficient of x^3
is a positive number.

Coefficient of x^3
is a negative number.

General Shapes of Graphs of Polynomial Functions

EXAMPLE 12   Determine which of the following graphs is the graph of
$f(x) = 5x^3 - 6x^2 + 2x + 3$.

A 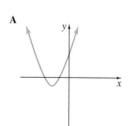   B 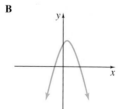   C 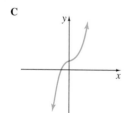   D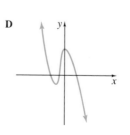

Solution: The degree of $f(x)$ is 3, which means that its graph has the shape of C or D. The coefficient of x^3 is 5, a positive number, so the graph has the shape of C. ▬▬▬

GRAPHING CALCULATOR EXPLORATIONS

A grapher may be used to visualize addition and subtraction of polynomials in one variable. For example, to visualize the following polynomial subtraction statement:

$$(3x^2 - 6x + 9) - (x^2 - 5x + 6) = 2x^2 - x + 3$$

graph both

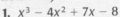
LEFT SIDE OF EQUATION

$$Y_1 = (3x^2 - 6x + 9) - (x^2 - 5x + 6)$$

and

RIGHT SIDE OF EQUATION

$$Y_2 = 2x^2 - x + 3$$

on the same screen and see that their graphs coincide. (*Note:* If the graphs do not coincide, we can be sure that a mistake has been made in combining polynomials or in calculator keystrokes. If the graphs appear to coincide, we cannot be sure that our work is correct. This is because it is possible for the graphs to differ so slightly that we do not notice it.)

1. $x^3 - 4x^2 + 7x - 8$
2. $-15x^3 + 3x^2 + 3x + 2$
3. $-2.1x^2 - 3.2x - 1.7$
4. $-7.9x^2 + 20.3x - 7.8$
5. $7.69x^2 - 1.26x + 5.3$
6. $-0.98x^2 + 2.8x + 1.86$

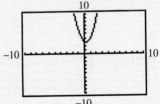

The graphs of Y_1 and Y_2 are shown. The graphs appear to coincide so the subtraction statement

$$(3x^2 - 6x + 9) - (x^2 - 5x + 6) = 2x^2 - x + 3$$

appears to be correct.

Perform the indicated operations. Then visualize by using the procedure described above.

1. $(2x^2 + 7x + 6) + (x^3 - 6x^2 - 14)$ **2.** $(-14x^3 - x + 2) + (-x^3 + 3x^2 + 4x)$

3. $(1.8x^2 - 6.8x - 1.7) - (3.9x^2 - 3.6x)$ **4.** $(-4.8x^2 + 12.5x - 7.8) - (3.1x^2 - 7.8x)$

5. $(1.29x - 5.68) + (7.69x^2 - 2.55x + 10.98)$ **6.** $(-0.98x^2 - 1.56x + 5.57) + (4.36x - 3.71)$

EXERCISE SET 5.3

Find the degree of each term. See Example 1.

1. 4   0

2. 7   0

3. $5x^2$ 2

4. $-z^3$ 3

5. $-3xy^2$ 3

6. $12x^3z$ 3

Find the degree of each polynomial and indicate whether the polynomial is a monomial, binomial, trinomial, or none of these. See Example 2.

7. $6x + 3$

8. $7x - 8$ binomial of degree 1

9. $3x^2 - 2x + 5$

10. $5x^2 - 3x^2y - 2x^3$

11. $-xyz$

12. -9 monomial of degree 0

13. $x^2y - 4xy^2 + 5x + y$ degree 3; none of these

14. $-2x^2y - 3y^2 + 4x + y^5$ degree 5; none of these

15. In your own words, describe how to find the degree of a term.

16. In your own words, describe how to find the degree of a polynomial.

If $P(x) = x^2 + x + 1$ and $Q(x) = 5x^2 - 1$, find the following. See Example 3.

17. $P(7)$ 57

18. $Q(4)$ 79

19. $Q(-10)$ 499

20. $P(-4)$ 13

21. $P(0)$ 1

22. $Q(0)$ -1

Refer to Example 4 for Exercises 23 through 26.

23. Find the height of the object at $t = 2$ seconds. 989 feet

24. Find the height of the object at $t = 4$ seconds. 797 feet

25. Find the height of the object at $t = 6$ seconds. 477 feet

26. Approximate (to the nearest second) how long it takes before the object hits the ground. (*Hint:* The object hits the ground when $P(x) = 0$.)   8 sec

Simplify by combining like terms. See Example 5.

27. $5y + y$ 6y

28. $-x + 3x$ 2x

29. $4x + 7x - 3$ $11x - 3$

30. $-8y + 9y + 4y^2$ $4y^2 + y$

31. $4xy + 2x - 3xy - 1$ $xy + 2x - 1$

32. $-8xy^2 + 4x - x + 2xy^2$ $-6xy^2 + 3x$

Perform indicated operations. See Examples 6 through 11.

33. $(9y^2 - 8) + (9y^2 - 9)$ $18y^2 - 17$

34. $(x^2 + 4x - 7) + (8x^2 + 9x - 7)$ $9x^2 + 13x - 14$

35. $(x^2 + xy - y^2)$ and $(2x^2 - 4xy + 7y^2)$ $3x^2 - 3xy + 6y^2$

36. $(4x^3 - 6x^2 + 5x + 7)$ and $(2x^2 + 6x - 3)$

37.
$$\begin{array}{r} x^2 - 6x + 3 \\ +\ \underline{(2x + 5)} \\ x^2 - 4x + 8 \end{array}$$

38.
$$\begin{array}{r} -2x^2 + 3x - 9 \\ +\ \underline{(2x - 3)} \\ -2x^2 + 5x - 12 \end{array}$$

39. $(9y^2 - 7y + 5) - (8y^2 - 7y + 2)$ $y^2 + 3$

40. $(2x^2 + 3x + 12) - (5x - 7)$ $2x^2 - 2x + 19$

41. $(6x^2 - 3x)$ from $(4x^2 + 2x)$ $-2x^2 + 5x$

42. $(xy + x - y)$ from $(xy + x - 3)$ $y - 3$

43.
$$\begin{array}{r} 3x^2 - 4x + 8 \\ -\ \underline{(5x^2 - 7)} \\ -2x^2 - 4x + 15 \end{array}$$

44.
$$\begin{array}{r} -3x^2 - 4x + 8 \\ -\ \underline{(5x + 12)} \\ -3x^2 - 9x - 4 \end{array}$$

45. $(5x - 11) + (-x - 2)$ $4x - 13$

46. $(3x^2 - 2x) + (5x^2 - 9x)$ $8x^2 - 11x$

47. $(7x^2 + x + 1) - (6x^2 + x - 1)$ $x^2 + 2$

48. $(4x - 4) - (-x - 4)$ 5x

49. $(7x^3 - 4x + 8) + (5x^3 + 4x + 8x)$ $12x^3 + 8x + 8$

50. $(9xyz + 4x - y) + (-9xyz - 3x + y + 2)$ $x + 2$

51. $(9x^3 - 2x^2 + 4x - 7) - (2x^3 - 6x^2 - 4x + 3)$

52. $(3x^2 + 6xy + 3y^2) - (8x^2 - 6xy - y^2)$

53. Add $(y^2 + 4yx + 7)$ and $(-19y^2 + 7yx + 7)$.

54. Subtract $(x - 4)$ from $(3x^2 - 4x + 5)$.

55. $(3x^3 - b + 2a - 6) + (-4x^3 + b + 6a - 6)$

56. $(5x^2 - 6) + (2x^2 - 4x + 8)$ $7x^2 - 4x + 2$

57. $(4x^2 - 6x + 2) - (-x^2 + 3x + 5)$ $5x^2 - 9x - 3$

58. $(5x^2 + x + 9) - (2x^2 - 9)$ $3x^2 + x + 18$

59. $(-3x + 8) + (-3x^2 + 3x - 5)$ $-3x^2 + 3$

60. $(5y^2 - 2y + 4) + (3y + 7)$ $5y^2 + y + 11$

61. $(-3 + 4x^2 + 7xy^2) + (2x^3 - x^2 + xy^2)$

62. $(-3x^2y + 4) - (-7x^2y - 8y)$ $4x^2y + 8y + 4$

63.
$$\begin{array}{r} 6y^2 - 6y + 4 \\ -\underline{(-y^2 - 6y + 7)} \\ 7y^2 - 3 \end{array}$$

64.
$$\begin{array}{r} -4x^3 + 4x^2 - 4x \\ -\ \underline{(2x^3 - 2x^2 + 3x)} \\ -6x^3 + 6x^2 - 7x \end{array}$$

65.
$$\begin{array}{r} 3x^2 + 15x + 8 \\ +\ \underline{(2x^2 + 7x + 8)} \\ 5x^2 + 22x + 16 \end{array}$$

66.
$$\begin{array}{r} 9x^2 + 9x - 4 \\ +\ \underline{(7x^2 - 3x - 4)} \\ 16x^2 + 6x - 8 \end{array}$$

67. Find the sum of $(5q^4 - 2q^2 - 3q)$ and $(-6q^4 + 3q^2 + 5)$. $-q^4 + q^2 - 3q + 5$

7. binomial of degree 1   **9.** trinomial of degree 2   **10.** trinomial of degree 3   **11.** monomial of degree 3

36. $4x^3 - 4x^2 + 11x + 4$ **51.** $7x^3 + 4x^2 + 8x - 10$ **52.** $-5x^2 + 12xy + 4y^2$

53. $-18y^2 + 11yx + 14$ **54.** $3x^2 - 5x + 9$ **55.** $-x^3 + 8a - 12$ **61.** $8xy^2 + 2x^3 + 3x^2 - 3$

68. Find the sum of $(5y^4 - 7y^2 + x^2 - 3)$ and $(-3y^4 + 2y^2 + 4)$. $2y^4 - 5y^2 + x^2 + 1$

69. Subtract $(3x + 7)$ from the sum of $(7x^2 + 4x + 9)$ and $(8x^2 + 7x - 8)$. $15x^2 + 8x - 6$

70. Subtract $(9x + 8)$ from the sum of $(3x^2 - 2x - x^3 + 2)$ and $(5x^2 - 8x - x^3 + 4)$.

71. Find the sum of $(4x^4 - 7x^2 + 3)$ and $(2 - 3x^4)$.

72. Find the sum of $(8x^4 - 14x^2 + 6)$ and $(-12x^6 - 21x^4 - 9x^2)$. $-12x^6 - 13x^4 - 23x^2 + 6$

73. $(8x^{2y} - 7x^y + 3) + (-4x^{2y} + 9x^y - 14)$ $4x^{2y} + 2x^y - 11$

74. $(14z^{5x} + 3z^{2x} + z) - (2z^{5x} - 10z^{2x} + 3z)$

75. The polynomial function $P(x) = 45x - 100,000$ models the relationship between the number of lamps x that Sherry's Lamp Shop sells and the profit the shop makes, $P(x)$. Find $P(4000)$, the profit from selling 4000 lamps. $80,000

76. The polynomial $P(t) = -32t + 500$ models the relationship between the length of time t in seconds a particle flies through space, beginning at a velocity of 500 feet per second, and its accrued velocity, $P(t)$. Find $P(3)$, the accrued velocity after 3 seconds. 404 feet per second

77. The function $f(x) = 4.31x^2 + 29.17x + 343.44$ can be used to approximate spending for health care in the United States, where x is the number of years since 1970 and $f(x)$ is the amount of money spent per capita. (Sources: Office of National Health Statistics and U.S. Bureau of the Census)

  a. Approximate the amount of money spent on health care per capita in the year 1980. $1066.14

  b. Approximate the amount of money spent on health care per capita in the year 1990. $2650.84

  c. Use the given function to predict the amount of money spent on health care per capita in the year 2000. $5097.54

  d. From parts a, b, and c, is the amount of money spent rising at a steady rate? Why or why not?

78. An object is thrown upward with an initial velocity of 50 feet per second from the top of the 350-foot-high City Hall in Milwaukee, Wisconsin. The height of the object at any time t can be described by the polynomial function $P(t) = -16t^2 + 50t + 350$. Find the height of the projectile when $t = 1$ second, $t = 2$ seconds, and $t = 3$ seconds. (Source: World Almanac) 384 feet; 386 feet; 356 feet

79. The total cost (in dollars) for MCD, Inc., Manufacturing Company to produce x blank audiocassette tapes per week is given by the polynomial function $C(x) = 0.8x + 10,000$. Find the total cost in producing 20,000 tapes per week. $26,000

80. The total revenues (in dollars) for MCD, Inc., Manufacturing Company to sell x blank audiocassette tapes per week is given by the polynomial function $R(x) = 2x$. Find the total revenue in selling 20,000 tapes per week. $40,000

A projectile is fired upward from the ground with an initial velocity of 300 feet per second. Neglecting air resistance, the height of the projectile at any time t can be described by the polynomial function

$$P(t) = -16t^2 + 300t$$

81. Find the height of the projectile at the given times.

  a. $t = 1$ second 284 ft **b.** $t = 2$ seconds 536 ft

  c. $t = 3$ seconds 756 ft **d.** $t = 4$ seconds 944 ft

82. Explain why the height increases and then decreases as time passes.

83. Approximate (to the nearest second) how long before the object hits the ground. 19 seconds

If $P(x) = 3x + 3$, $Q(x) = 4x^2 - 6x + 3$, and $R(x) = 5x^2 - 7$, find the following.

84. $P(x) + Q(x)$ **85.** $R(x) + P(x)$

86. $Q(x) - R(x)$ **87.** $P(x) - Q(x)$

88. $2[Q(x)] - R(x)$ **89.** $-5[P(x)] - Q(x)$

90. $3[R(x)] + 4[P(x)]$ **91.** $2[Q(x)] + 7[R(x)]$

92. If the revenue function of a certain company is given by $R(x) = 5.5x$, where x is the number of packages of personalized stationery sold and the cost function is given by $C(x) = 3x + 3000$,

  a. Find the profit function. (Recall that revenue − cost = profit.) $P(x) = 2.5x - 3000$

  b. Find the profit when 2000 packages of stationery are sold. $2000

Match each equation with its graph found on the next page. See Example 12.

93. $f(x) = 3x^2 - 2$ A   **94.** $h(x) = 5x^3 - 6x + 2$ B

95. $g(x) = -2x^3 - 3x^2 + 3x - 2$ D

---

70. $-2x^3 + 8x^2 - 19x - 2$ **71.** $x^4 - 7x^2 + 5$ **74.** $12z^{5x} + 13z^{2x} - 2z$ **77d.** No, $f(x)$ is not linear.
84. $4x^2 - 3x + 6$ **85.** $5x^2 + 3x - 4$ **86.** $-x^2 - 6x + 10$ **87.** $-4x^2 + 9x$ **88.** $3x^2 - 12x + 13$
89. $-4x^2 - 9x - 18$ **90.** $15x^2 + 12x - 9$ **91.** $43x^2 - 12x - 43$

96. $g(x) = -2x^2 - 6x + 2$ C

A
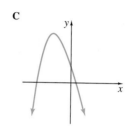

B
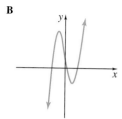

C

D
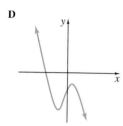

Find each perimeter.

97.

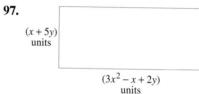

$(x + 5y)$ units

$(3x^2 - x + 2y)$ units

$(6x^2 + 14y)$ units

98.

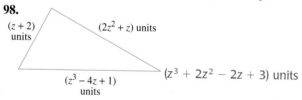

$(z + 2)$ units

$(2z^2 + z)$ units

$(z^3 - 4z + 1)$ units

$(z^3 + 2z^2 - 2z + 3)$ units

Review Exercises

Multiply. See Section 1.2.

99. $5(3x - 2)$ $15x - 10$

100. $-7(2z - 6y)$ $-14z + 42y$

101. $-2(x^2 - 5x + 6)$
$-2x^2 + 10x - 12$

102. $5(-3y^2 - 2y + 7)$
$-15y^2 - 10y + 35$

A Look Ahead

EXAMPLE

If $P(x) = -3x + 5$, find the following:

a. $P(a)$ **b.** $P(-x)$ **c.** $P(x + h)$

Solution: **a.** $P(x) = -3x + 5$
$$P(a) = -3a + 5$$
b. $P(x) = -3x + 5$
$$P(-x) = -3(-x) + 5$$
$$= 3x + 5$$
c. $P(x) = -3x + 5$
$$P(x + h) = -3(x + h) + 5$$
$$= -3x - 3h + 5$$

103a. $2a - 3$ **b.** $-2x - 3$ **c.** $2x + 2h - 3$

If $P(x)$ is the polynomial given, find $P(a)$, $P(-x)$, and $P(x + h)$. See the preceding example.

103. $P(x) = 2x - 3$ **104.** $P(x) = 8x + 3$

105. $P(x) = 4x$ **106.** $P(x) = -4x$

107. $P(x) = 4x - 1$ **108.** $P(x) = 3x - 2$

104a. $8a + 3$ **b.** $-8x + 3$ **c.** $8x + 8h + 3$ **105a.** $4a$ **b.** $-4x$ **c.** $4x + 4h$ **106a.** $-4a$ **b.** $4x$
c. $-4x - 4h$ **107a.** $4a - 1$ **b.** $-4x - 1$ **c.** $4x + 4h - 1$ **108a.** $3a - 2$ **b.** $-3x - 2$ **c.** $3x + 3h - 2$

5.4   MULTIPLYING POLYNOMIALS

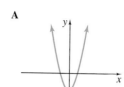

TAPE IA 5.4

OBJECTIVES

1. Multiply two polynomials.
2. Multiply binomials.
3. Square binomials.
4. Multiply the sum and difference of two terms.

1   Properties of real numbers and exponents are used continually in the process of multiplying polynomials. To multiply monomials, for example, we apply the commutative and associative properties of real numbers and the product rule for *exponents*.

EXAMPLE 1 Multiply.

a. $(2x^3)(5x^6)$ **b.** $(7y^4z^4)(-xy^{11}z^5)$

Solution: Group like bases and apply the product rule for exponents.

a. $(2x^3)(5x^6) = 2(5)(x^3)(x^6) = 10x^9$

b. $(7y^4z^4)(-xy^{11}z^5) = 7(-1)x(y^4y^{11})(z^4z^5) = -7xy^{15}z^9$

To multiply a monomial by a polynomial other than a monomial, we use an expanded form of the distributive property:

$$a(b + c + d + \cdots + z) = ab + ac + ad + \cdots + az$$

Notice that the monomial a is multiplied by each term of the polynomial.

EXAMPLE 2 Multiply.

a. $2x(5x - 4)$ **b.** $-3x^2(4x^2 - 6x + 1)$ **c.** $-xy(7x^2y + 3xy - 11)$

Solution: Apply the distributive property.

a. $2x(5x - 4) = 2x(5x) + 2x(-4)$

$$= 10x^2 - 8x$$

b. $-3x^2(4x^2 - 6x + 1) = -3x^2(4x^2) + (-3x^2)(-6x) + (-3x^2)(1)$

$$= -12x^4 + 18x^3 - 3x^2$$

c. $-xy(7x^2y + 3xy - 11) = -xy(7x^2y) + (-xy)(3xy) + (-xy)(-11)$

$$= -7x^3y^2 - 3x^2y^2 + 11xy$$

In general, we have the following:

> To multiply any two polynomials, use the distributive property, and multiply each term of one polynomial by each term of the other polynomial. Then combine any like terms.

EXAMPLE 3 Multiply and simplify the product if possible.

a. $(x + 3)(2x + 5)$ **b.** $(2x^3 - 3)(5x^2 - 6x + 7)$

Solution: **a.** Multiply each term of $(x + 3)$ by $(2x + 5)$.

$(x + 3)(2x + 5) = x(2x + 5) + 3(2x + 5)$ Apply the distributive property.

$= 2x^2 + 5x + 6x + 15$ Apply the distributive property again.

$= 2x^2 + 11x + 15$ Combine like terms.

b. Multiply each term of $(2x^3 - 3)$ by each term of $(5x^2 - 6x + 7)$.

$$(2x^3 - 3)(5x^2 - 6x + 7) = 2x^3(5x^2 - 6x + 7) + (-3)(5x^2 - 6x + 7)$$
$$= 10x^5 - 12x^4 + 14x^3 - 15x^2 + 18x - 21$$

Sometimes polynomials are easier to multiply vertically, in the same way we multiply real numbers. When multiplying vertically, line up like terms in the **partial products** vertically. This makes combining like terms easier.

EXAMPLE 4   Find the product of $(4x^2 + 7)$ and $(x^2 + 2x + 8)$.

Solution:
$$
\begin{array}{r}
x^2 + 2x + 8 \\
4x^2 + 7 \\
\hline
7x^2 + 14x + 56 \\
4x^4 + 8x^3 + 32x^2 \\
\hline
4x^4 + 8x^3 + 39x^2 + 14x + 56
\end{array}
$$

$7(x^2 + 2x + 8)$

$4x^2(x^2 + 2x + 8)$

Combine like terms.

2   When multiplying a binomial by a binomial, we can use a special order of multiplying terms, called the **FOIL** order. The letters of FOIL stand for "First–Outer–Inner–Last." To illustrate this method, multiply $(2x - 3)$ by $(3x + 1)$.

Multiply the **First** terms of each binomial. $(2x - 3)(3x + 1)$ **F** $2x(3x) = 6x^2$

Multiply the **Outer** terms of each binomial. $(2x - 3)(3x + 1)$ **O** $2x(1) = 2x$

Multiply the **Inner** terms of each binomial. $(2x - 3)(3x + 1)$ **I** $-3(3x) = -9x$

Multiply the **Last** terms of each binomial. $(2x - 3)(3x + 1)$ **L** $-3(1) = -3$

Combine like terms.

$$6x^2 + 2x - 9x - 3 = 6x^2 - 7x - 3$$

EXAMPLE 5   Multiply $(x - 1)(x + 2)$. Use the FOIL order.

Solution:

First   Outer   Inner   Last

$$(x - 1)(x + 2) = x \cdot x + 2 \cdot x + (-1)x + (-1)(2)$$
$$= x^2 + 2x - x - 2$$
$$= x^2 + x - 2 \qquad \text{Combine like terms.}$$

EXAMPLE 6 Multiply $(2x - 7)(3x - 4)$.

Solution:

$$\begin{array}{cccc} \text{First} & \text{Outer} & \text{Inner} & \text{Last} \\ \downarrow & \downarrow & \downarrow & \downarrow \end{array}$$

$$(2x - 7)(3x - 4) = 2x(3x) + 2x(-4) + (-7)(3x) + (-7)(-4)$$
$$= 6x^2 - 8x - 21x + 28$$
$$= 6x^2 - 29x + 28$$

3 Certain products and powers of polynomials create striking patterns. For example, the **square of a binomial** is a special case of the product of two binomials. Find $(a + b)^2$ by the FOIL order for multiplying two binomials.

$$(a + b)^2 = (a + b)(a + b)$$
$$\quad\quad\quad\quad \text{F} \quad \text{O} \quad \text{I} \quad \text{L}$$
$$= a^2 + ab + ba + b^2$$
$$= a^2 + 2ab + b^2$$

This product can be visualized geometrically by analyzing areas. The area of the square shown next can be written as $(a + b)^2$ or as the sum of the areas of the smaller rectangles.

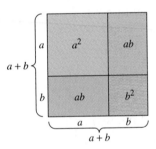

The sum of the areas of the smaller rectangles is $a^2 + ab + ab + b^2$, or $a^2 + 2ab + b^2$. Thus,

$$(a + b)^2 = a^2 + 2ab + b^2$$

We can use this pattern as the basis of a quick method for squaring a binomial. Two forms are shown for squaring a binomial: the sum of terms and the difference of terms. We call these **special products.**

SQUARE OF A BINOMIAL

$$(a + b)^2 = a^2 + 2ab + b^2, \quad (a - b)^2 = a^2 - 2ab + b^2$$

A binomial squared is the sum of the first term squared, twice the product of both terms, and the second term squared.

EXAMPLE 7   Multiply.

   a. $(x + 5)^2$ **b.** $(x - 9)^2$ **c.** $(3x + 2z)^2$ **d.** $(4m^2 - 3n)^2$

Solution: **a.** $(x + 5)^2 = x^2 + 2 \cdot x \cdot 5 + 5^2 = x^2 + 10x + 25$
   b. $(x - 9)^2 = x^2 - 2 \cdot x \cdot 9 + 9^2 = x^2 - 18x + 81$
   c. $(3x + 2z)^2 = (3x)^2 + 2(3x)(2z) + (2z)^2 = 9x^2 + 12xz + 4z^2$
   d. $(4m^2 - 3n)^2 = (4m^2)^2 - 2(4m^2)(3n) + (3n)^2 = 16m^4 - 24m^2n + 9n^2$

R E M I N D E R   Note that $(a + b)^2 = a^2 + 2ab + b^2$, **not** $a^2 + b^2$. Also, $(a - b)^2 = a^2 - 2ab + b^2$, **not** $a^2 - b^2$.

---

4   Another special product applies to the sum and difference of the same two terms. Multiply $(a + b)(a - b)$ to see a pattern.

$$(a + b)(a - b) = a^2 - ab + ba - b^2$$
$$= a^2 - b^2$$

PRODUCT OF THE SUM AND DIFFERENCE OF TWO TERMS

$$(a + b)(a - b) = a^2 - b^2$$

The product of the sum and difference of the same two terms is the difference of the first term squared and the second term squared.

EXAMPLE 8   Multiply.

   a. $(x - 3)(x + 3)$ **b.** $(4y + 1)(4y - 1)$ **c.** $(x^2 + 2y)(x^2 - 2y)$

Solution: **a.** $(x - 3)(x + 3) = x^2 - 3^2 = x^2 - 9$
   b. $(4y + 1)(4y - 1) = (4y)^2 - 1^2 = 16y^2 - 1$
   c. $(x^2 + 2y)(x^2 - 2y) = (x^2)^2 - (2y)^2 = x^4 - 4y^2$

EXAMPLE 9   Multiply: $[3 + (2a + b)]^2$

Solution: Think of 3 as the first term and $(2a + b)$ as the second term, and apply the method for squaring a binomial.

$$[3 + (2a + b)]^2 = \quad (3)^2 \quad + 2(3)(2a + b) + (2a + b)^2$$

$$\begin{array}{ccc} \text{First term} & \text{Twice the} & \text{Last term} \\ \text{squared} & \text{product of} & \text{squared} \\ & \text{both terms} & \end{array}$$

$$= 9 + 6(2a + b) + (2a + b)^2$$
$$= 9 + 12a + 6b + (2a)^2 + 2(2a)(b) + b^2 \quad \text{Square } (2a + b).$$
$$= 9 + 12a + 6b + 4a^2 + 4ab + b^2$$

EXAMPLE 10   Multiply: $[(5x - 2y) - 1][(5x - 2y) + 1]$

Solution: Think of $(5x - 2y)$ as the first term and 1 as the second term, and apply the method for the product of the sum and difference of two terms.

$$[(5x - 2y) - 1][(5x - 2y) + 1] = (5x - 2y)^2 - 1^2$$

$$\begin{array}{cc} \text{First term} & \text{Second term} \\ \text{squared} & \text{squared} \end{array}$$

$$= (5x - 2y)^2 - 1$$
$$= (5x)^2 - 2(5x)(2y) + (2y)^2 - 1 \quad \text{Square } (5x - 2y).$$
$$= 25x^2 - 20xy + 4y^2 - 1$$

Our work in multiplying polynomials is often useful in evaluating polynomial functions.

EXAMPLE 11   If $f(x) = x^2 + 5x - 2$, find $f(a + 1)$.

Solution: To find $f(a + 1)$, replace x with the expression $a + 1$ in the polynomial function $f(x)$.

$$f(x) = x^2 + 5x - 2$$
$$f(a + 1) = (a + 1)^2 + 5(a + 1) - 2$$
$$= a^2 + 2a + 1 + 5a + 5 - 2$$
$$= a^2 + 7a + 4$$

GRAPHING CALCULATOR EXPLORATIONS

In the previous section, we used a grapher to visualize addition and subtraction of polynomials in one variable. In this section, the same method is used to visualize multiplication of polynomials in one variable. For example, to see that

$$(x - 2)(x + 1) = x^2 - x - 2,$$

graph both $Y_1 = (x - 2)(x + 1)$ and $Y_2 = x^2 - x - 2$ on the same screen and see whether their graphs coincide.

1. $x^2 - 16$
2. $x^2 + 6x + 9$
3. $9x^2 - 42x + 49$
4. $25x^2 - 20x + 4$
5. $5x^3 - 14x^2 - 13x - 2$
6. $14x^3 + 29x^2 - 23x - 20$

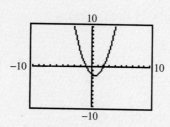

By tracing along both graphs, we see that the graphs of Y_1 and Y_2 appear to coincide, and thus $(x - 2)(x + 1) = x^2 - x - 2$ appears to be correct.

Multiply. Then use a grapher to see the results.

1. $(x + 4)(x - 4)$ **2.** $(x + 3)(x + 3)$ **3.** $(3x - 7)^2$
4. $(5x - 2)^2$ **5.** $(5x + 1)(x^2 - 3x - 2)$ **6.** $(7x + 4)(2x^2 + 3x - 5)$

---

1. $-12x^5$ **2.** $-24a^2$ **3.** $12x^2 + 21x$ **4.** $30x^2 - 20x$ **5.** $-24x^2y - 6xy^2$
6. $-48xy^2 - 32xy$ **9.** $2x^2 - 2x - 12$ **10.** $3y^2 + 13y - 10$
11. $2x^4 + 3x^3 - 2x^2 + x + 6$ **12.** $3a^3 + 5a^2 + 3a + 10$

EXERCISE SET 5.4

Multiply. See Examples 1 through 4.

1. $(-4x^3)(3x^2)$ **2.** $(-6a)(4a)$ **3.** $3x(4x + 7)$

4. $5x(6x - 4)$ **5.** $-6xy(4x + y)$ **6.** $-8y(6xy + 4x)$

7. $-4ab(xa^2 + ya^2 - 3)$ $-4a^3bx - 4a^3by + 12ab$

8. $-6b^2z(z^2a + baz - 3b)$ $-6b^2z^3a - 6b^3az^2 + 18b^3z$

9. $(x - 3)(2x + 4)$ **10.** $(y + 5)(3y - 2)$

11. $(2x + 3)(x^3 - x + 2)$ **12.** $(a + 2)(3a^2 - a + 5)$

13. $\begin{array}{r} 3x - 2 \\ 5x + 1 \end{array}$ $15x^2 - 7x - 2$ **14.** $\begin{array}{r} 2z - 4 \\ 6z - 2 \end{array}$ $12z^2 - 28z + 8$

15. $\begin{array}{r} 3m^2 + 2m - 1 \\ 5m + 2 \end{array}$ **16.** $\begin{array}{r} 2x^2 - 3x - 4 \\ x + 5 \end{array}$

17. Explain how to multiply a polynomial by a polynomial.

18. Explain why $(3x + 2)^2$ does not equal $9x^2 + 4$.

Multiply the binomials. See Examples 5 and 6.

19. $(x - 3)(x + 4)$ **20.** $(c - 3)(c + 1)$
21. $(5x + 8y)(2x - y)$ **22.** $(2n - 9m)(n - 7m)$
23. $(3x - 1)(x + 3)$ **24.** $(5d - 3)(d + 6)$
25. $\left(3x + \dfrac{1}{2}\right)\left(3x - \dfrac{1}{2}\right)$ **26.** $\left(2x - \dfrac{1}{3}\right)\left(2x + \dfrac{1}{3}\right)$

Multiply, using special product methods. See Examples 7 through 9. **29.** $36y^2 - 1$

27. $(x + 4)^2$ $x^2 + 8x + 16$ **28.** $(x - 5)^2$ $x^2 - 10x + 25$
29. $(6y - 1)(6y + 1)$ **30.** $(x - 9)(x + 9)$ $x^2 - 81$
31. $(3x - y)^2$ **32.** $(4x - z)^2$ $16x^2 - 8xz + z^2$

15. $15m^3 + 16m^2 - m - 2$ **16.** $2x^3 + 7x^2 - 19x - 20$ **19.** $x^2 + x - 12$ **20.** $c^2 - 2c - 3$

21. $10x^2 + 11xy - 8y^2$ **22.** $2n^2 - 23mn + 63m^2$ **23.** $3x^2 + 8x - 3$ **24.** $5d^2 + 27d - 18$

25. $9x^2 - \dfrac{1}{4}$ **26.** $4x^2 - \dfrac{1}{9}$ **31.** $9x^2 - 6xy + y^2$

35. $16b^2 + 32b + 16$ **36.** $9b^2 - 48b + 64$ **47.** $49x^2 - 9$ **49.** $9x^3 + 30x^2 + 12x - 24$ **51.** $16x^2 - \dfrac{2}{3}x - \dfrac{1}{6}$

33. $(3b - 6y)(3b + 6y)$ **34.** $(2x - 4y)(2x + 4y)$
$9b^2 - 36y^2$ $4x^2 - 16y^2$

Multiply, using special product methods. See Example 10.

35. $[3 + (4b + 1)]^2$ **36.** $[5 - (3b - 3)]^2$

37. $[(2s - 3) - 1][(2s - 3) + 1]$ $4s^2 - 12s + 8$

38. $[(2y + 5) + 6][(2y + 5) - 6]$ $4y^2 + 20y - 11$

39. $[(xy + 4) - 6]^2$ $x^2y^2 - 4xy + 4$

40. $[(2a^2 + 4a) + 1]^2$ $4a^4 + 16a^3 + 20a^2 + 8a + 1$

41. Explain when the FOIL method can be used to multiply polynomials.

42. Explain why the product of $(a + b)$ and $(a - b)$ is not a trinomial.

43. $9x^2 + 18x + 5$ **44.** $20x^2 - x - 30$

Multiply.   **45.** $10x^5 + 8x^4 + 2x^3 + 25x^2 + 20x + 5$

43. $(3x + 1)(3x + 5)$ **44.** $(4x - 5)(5x + 6)$

45. $(2x^3 + 5)(5x^2 + 4x + 1)$

46. $(3y^3 - 1)(3y^3 - 6y + 1)$ $9y^6 - 18y^4 + 6y - 1$

47. $(7x - 3)(7x + 3)$ **48.** $(4x + 1)(4x - 1)$ $16x^2 - 1$

49. $3x^2 + 4x - 4$ **50.** $6x^2 + 2x - 1$ $18x^3 - 30x^2 - 15x + 6$
 $\underline{\quad 3x + 6 \quad}$ $\underline{\quad 3x - 6 \quad}$

51. $\left(4x + \dfrac{1}{3}\right)\left(4x - \dfrac{1}{2}\right)$ **52.** $\left(4y - \dfrac{1}{3}\right)\left(3y - \dfrac{1}{8}\right)$ $12y^2 - \dfrac{3}{2}y + \dfrac{1}{24}$

53. $(6x + 1)^2$ **54.** $(4x + 7)^2$ $16x^2 + 56x + 49$

55. $(x^2 + 2y)(x^2 - 2y)$ $x^4 - 4y^2$

56. $(3x + 2y)(3x - 2y)$ $9x^2 - 4y^2$

57. $-6a^2b^2(5a^2b^2 - 6a - 6b)$ $-30a^4b^4 + 36a^3b^2 + 36a^2b^3$

58. $7x^2y^3(-3ax - 4xy + z)$ $-21ax^3y^3 - 28x^3y^4 + 7x^2y^3z$

59. $(a - 4)(2a - 4)$ **60.** $(2x - 3)(x + 1)$ $2x^2 - x - 3$

61. $(7ab + 3c)(7ab - 3c)$ $49a^2b^2 - 9c^2$

62. $(3xy - 2b)(3xy + 2b)$ $9x^2y^2 - 4b^2$

63. $(m - 4)^2$ $m^2 - 8m + 16$ **64.** $(x + 2)^2$ $x^2 + 4x + 4$

65. $(3x + 1)^2$ $9x^2 + 6x + 1$ **66.** $(4x + 6)^2$ $16x^2 + 48x + 36$

67. $(y - 4)(y - 3)$ **68.** $(c - 8)(c + 2)$ $c^2 - 6c - 16$

69. $(x + y)(2x - 1)(x + 1)$ $2x^3 + 2x^2y + x^2 + xy - x - y$

70. $(z + 2)(z - 3)(2z + 1)$ $2z^3 - z^2 - 13z - 6$

71. $(3x^2 + 2x - 1)^2$ **72.** $(4x^2 + 4x - 4)^2$

73. $(3x + 1)(4x^2 - 2x + 5)$ $12x^3 - 2x^2 + 13x + 5$

74. $(2x - 1)(5x^2 - x - 2)$ $10x^3 - 7x^2 - 3x + 2$

If $R(x) = x + 5$, $Q(x) = x^2 - 2$, and $P(x) = 5x$, find the following.

75. $P(x) \cdot R(x)$ $5x^2 + 25x$ **76.** $P(x) \cdot Q(x)$ $5x^3 - 10x$

77. $[Q(x)]^2$ $x^4 - 4x^2 + 4$ **78.** $[R(x)]^2$ $x^2 + 10x + 25$

79. $R(x) \cdot Q(x)$ **80.** $P(x) \cdot R(x) \cdot Q(x)$

81. Perform the indicated operations. Explain the difference between the two problems.

 a. $(3x + 5) + (3x + 7)$ **b.** $(3x + 5)(3x + 7)$

82. Find the area of the circle. Do not approximate π.

$\pi(25x^2 - 20x + 4)$ sq. km

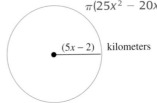

$(5x - 2)$ kilometers

83. Find the volume of the cylinder. Do not approximate π. $\pi(7y^3 - 42y^2 + 63y)$ cu. cm

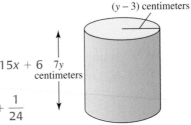

$(y - 3)$ centimeters

$7y$ centimeters

If $f(x) = x^2 - 3x$, find the following. See Example 11.

84. $f(a)$ $a^2 - 3a$ **85.** $f(c)$ $c^2 - 3c$

86. $f(a + h)$ **87.** $f(a + 5)$ $a^2 + 7a + 10$

88. $f(b - 2)$ $b^2 - 7b + 10$ **89.** $f(a - b)$

90. $F(x) = x^2 + 3x + 2$, find (a) $F(a + h)$, (b) $F(a)$, (c) $F(a + h) - F(a)$.

91. If $g(x) = x^2 + 2x + 1$, find (a) $g(a + h)$, (b) $g(a)$, (c) $g(a + h) - g(a)$. **a.** $a^2 + 2ah + h^2 + 2a + 2h + 1$
 b. $a^2 + 2a + 1$ **c.** $2ah + h^2 + 2h$

Multiply. Assume that variables represent positive integers.   **92.** $30x^2y^{2n+1} - 10x^2y^n$ **93.** $-6y^4z^{3n} + 3yz^n$

92. $5x^2y^n(6y^{n+1} - 2)$ **93.** $-3yz^n(2y^3z^{2n} - 1)$

94. $(x^a + 5)(x^{2a} - 3)$ **95.** $(x^a + y^{2b})(x^a - y^{2b})$
 $x^{3a} + 5x^{2a} - 3x^a - 15$ $x^{2a} - y^{4b}$

Review Exercises

Use the slope–intercept form of a line, $y = mx + b$, to find the slope of each line. See Section 3.4.

96. $y = -2x + 7$ -2 **97.** $y = \dfrac{3}{2}x - 1$ $\dfrac{3}{2}$

98. $3x - 5y = 14$ $\dfrac{3}{5}$ **99.** $x + 7y = 2$ $-\dfrac{1}{7}$

53. $36x^2 + 12x + 1$ **59.** $2a^2 - 12a + 16$ **67.** $y^2 - 7y + 12$ **71.** $9x^4 + 12x^3 - 2x^2 - 4x + 1$
72. $16x^4 + 32x^3 - 16x^2 - 32x + 16$ **79.** $x^3 + 5x^2 - 2x - 10$ **80.** $5x^4 + 25x^3 - 10x^2 - 50x$ **81a.** $6x + 12$
81b. $9x^2 + 36x + 35$ **86.** $a^2 + 2ah + h^2 - 3a - 3h$ **89.** $a^2 - 2ab + b^2 - 3a + 3b$
90a. $a^2 + 2ah + h^2 + 3a + 3h + 2$ **90b.** $a^2 + 3a + 2$ **90c.** $2ah + h^2 + 3h$

Use the vertical line test to determine which of the following are graphs of functions. See Section 3.2.

100.

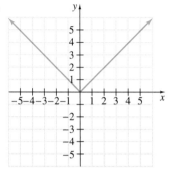

function

101.

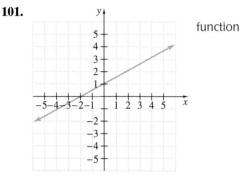

function

5.5 | THE GREATEST COMMON FACTOR AND FACTORING BY GROUPING

TAPE IA 5.5

O B J E C T I V E S

1. Identify the GCF.
2. Factor out the GCF of a polynomial's terms.
3. Factor polynomials by grouping.

Factoring is the reverse process of multiplying. It is the process of writing a polynomial as a product.

$$6x^2 + 13x - 5 = (3x - 1)(2x + 5)$$

factoring → ← multiplying

In the next few sections, we review techniques for factoring polynomials. These techniques are used at the end of this chapter to solve polynomial equations and to graph polynomial functions.

To factor a polynomial, we first factor out the greatest common factor (GCF) of its terms, using the distributive property. The GCF of a list of terms or monomials is the product of the GCF of the numerical coefficients and each GCF of the powers of a common variable.

TO FIND THE GCF OF A LIST OF MONOMIALS

Step 1.   Find the GCF of the numerical coefficients.

Step 2.   Find the GCF of the variable factors.

Step 3.   The product of the factors found in *steps 1* and *2* is the GCF of the monomials.

EXAMPLE 1 Find the GCF of $20x^3y$, $10x^2y^2$, and $35x^3$.

Solution: The GCF of the numerical coefficients 20, 10, and 35 is 5, the largest integer that is a factor of each integer. The GCF of the variable factors x^3, x^2, and x^3 is x^2 because x^2 is the largest factor common to all three powers of x. The variable y is not a common factor because it does not appear in all three monomials. The GCF is thus

$$5 \cdot x^2, \quad \text{or} \quad 5x^2$$

A first step in factoring polynomials is to use the distributive property and write the polynomial as a product of the GCF of its monomial terms and a simpler polynomial. This is called **factoring out** the GCF.

EXAMPLE 2 Factor.

a. $8x + 4$ **b.** $5y - 2z$ **c.** $6x^2 - 3x^3$

Solution: **a.** The GCF of terms $8x$ and 4 is 4.

$$8x + 4 = 4(2x) + 4(1) \qquad \text{Factor out 4 from each term.}$$
$$= 4(2x + 1) \qquad \text{Apply the distributive property.}$$

The factored form of $8x + 4$ is $4(2x + 1)$. To check, multiply $4(2x + 1)$ to see that the product is $8x + 4$.

b. There is no common factor of the terms $5y$ and $-2z$ other than 1 (or -1).

c. The greatest common factor of $6x^2$ and $-3x^3$ is $3x^2$. Thus,

$$6x^2 - 3x^3 = 3x^2(2) - 3x^2(x)$$
$$= 3x^2(2 - x)$$

> **REMINDER** To verify that the GCF has been factored out correctly, multiply the factors together and see that their product is the original polynomial.

EXAMPLE 3 Factor $17x^3y^2 - 34x^4y^2$.

Solution: The GCF of the two terms is $17x^3y^2$, which we factor out of each term.

$$17x^3y^2 - 34x^4y^2 = 17x^3y^2(1) - 17x^3y^2(2x)$$
$$= 17x^3y^2(1 - 2x)$$

> **REMINDER** If the GCF happens to be one of the terms in the polynomial, a factor of 1 will remain for this term when the GCF is factored out. For example, in the polynomial $21x^2 + 7x$, the GCF of $21x^2$ and $7x$ is $7x$, so
>
> $$21x^2 + 7x = 7x(3x) + 7x(1) = 7x(3x + 1)$$

EXAMPLE 4   Factor $-3x^3y + 2x^2y - 5xy$.

Solution: Two possibilities are shown for factoring this polynomial. First, the common factor xy is factored out.

$$-3x^3y + 2x^2y - 5xy = xy(-3x^2 + 2x - 5)$$

Also, the common factor $-xy$ can be factored out as shown.

$$-3x^3y + 2x^2y - 5xy = -xy(3x^2) + (-xy)(-2x) + (-xy)(5)$$
$$= -xy(3x^2 - 2x + 5)$$

Both of these alternatives are correct.

EXAMPLE 5   Factor $2(x - 5) + 3a(x - 5)$.

Solution: The GCF is the binomial factor $(x - 5)$. Thus,

$$2(x - 5) + 3a(x - 5) = (x - 5)(2 + 3a)$$

EXAMPLE 6   Factor $7x(x^2 + 5y) - (x^2 + 5y)$.

Solution: The GCF is the expression $(x^2 + 5y)$. Factor this from each term.

$$7x(x^2 + 5y) - (x^2 + 5y) = 7x(x^2 + 5y) - 1(x^2 + 5y) = (x^2 + 5y)(7x - 1)$$

Notice that we write $-(x^2 + 5y)$ as $-1(x^2 + 5y)$ to aid in factoring.

3   Sometimes it is possible to factor a polynomial by grouping the terms of the polynomial and looking for common factors in each group. This method of factoring is called **factoring by grouping.**

EXAMPLE 7   Factor $ab - 6a + 2b - 12$.

Solution: First look for the GCF of all four terms. The GCF of all four terms is 1. Next group the first two terms and the last two terms and factor out common factors from each group.

$$ab - 6a + 2b - 12 = (ab - 6a) + (2b - 12)$$

Factor a from the first group and 2 from the second group.

$$= a(b - 6) + 2(b - 6)$$

Now we see a GCF of $(b - 6)$. Factor out $(b - 6)$ to get

$$a(b - 6) + 2(b - 6) = (b - 6)(a + 2)$$

This factorization can be checked by multiplying $(b - 6)$ by $(a + 2)$ to verify that the product is the original polynomial.

> REMINDER   Notice that the polynomial $a(b - 6) + 2(b - 6)$ is **not** in factored form. It is a **sum,** not a **product.** The factored form is $(b - 6)(a + 2)$.

EXAMPLE 8   Factor $m^2n^2 + m^2 - 2n^2 - 2$.

Solution: Once again, the GCF of all four terms is 1. Try grouping the first two terms together and the last two terms together.

$$m^2n^2 + m^2 - 2n^2 - 2 = (m^2n^2 + m^2) + (-2n^2 - 2)$$

Factor m^2 from the first group and 2 from the second group.

$$= m^2(n^2 + 1) + 2(-n^2 - 1)$$

Notice that the polynomial is not in factored form since it is still written as a sum and not a product.

There is no common factor in this resulting polynomial, but notice that $(n^2 + 1)$ and $(-n^2 - 1)$ are opposites. Try grouping the terms differently, as follows:

$$m^2n^2 + m^2 - 2n^2 - 2 = (m^2n^2 + m^2) - (2n^2 + 2) \qquad \text{Watch the signs!}$$
$$= m^2(n^2 + 1) - 2(n^2 + 1) \qquad \text{Factor common factors}$$
$$\qquad\qquad\qquad\qquad\qquad\qquad \text{from the groups of terms.}$$
$$= (n^2 + 1)(m^2 - 2) \qquad \text{Factor out a GCF of}$$
$$\qquad\qquad\qquad\qquad\qquad (n^2 + 1).$$

MENTAL MATH

Find the GCF of each list of monomials.

1. 6, 12   6
2. 9, 27   9
3. $15x$, 10   5
4. $9x$, 12   3
5. $13x$, $2x$ x
6. $4y$, $5y$ y
7. $7x$, $14x$ $7x$
8. $8z$, $4z$ $4z$

EXERCISE SET 5.5

Find the GCF of each list of monomials. See Example 1.

1. a^8, a^5, a^3 a^3
2. b^9, b^2, b^5 b^2
3. $x^2y^3z^3, y^2z^3, xy^2z^2$ y^2z^2
4. $xy^2z^3, x^2y^2z^2, x^2y^3$ xy^2
5. $6x^3y, 9x^2y^2, 12x^2y$ $3x^2y$
6. $4xy^2, 16xy^3, 8x^2y^2$ $4xy^2$
7. $10x^3yz^3, 20x^2z^5, 45xz^3$
8. $12y^2z^4, 9xy^3z^4, 15x^2y^2z^3$

Factor out the GCF in each polynomial. See Examples 2 through 6.

9. $18x - 12$ $6(3x - 2)$
10. $21x + 14$ $7(3x + 2)$
11. $4y^2 - 16xy^3$
12. $3z - 21xz^4$
13. $6x^5 - 8x^4 + 2x^3$
14. $9x + 3x^2 - 6x^3$

7. $5xz^3$
8. $3y^2z^3$
11. $4y^2(1 - 4xy)$
12. $3z(1 - 7xz^3)$
13. $2x^3(3x^2 - 4x + 1)$
14. $3x(3 + x - 2x^2)$

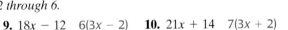

17. $(x + 3)(6 + 5a)$ **18.** $(x - 4)(2 + 3y)$ **19.** $(z + 7)(2x + 1)$ **27.** $(a - 2)(c + 4)$ **28.** $(b - 3)(c + 8)$

15. $8a^3b^3 - 4a^2b^2 + 4ab + 16ab^2 \; 4ab(2a^2b^2 - ab + 1 + 4b)$

16. $12a^3b - 6ab + 18ab^2 - 18a^2b \; 6ab(2a^2 - 1 + 3b - 3a)$

17. $6(x + 3) + 5a(x + 3)$ **18.** $2(x - 4) + 3y(x - 4)$

19. $2x(z + 7) + (z + 7)$ **20.** $x(y - 2) + (y - 2)$

21. $3x(x^2 + 5) - 2(x^2 + 5)$ $(x^2 + 5)(3x - 2)$ $(y - 2)(x + 1)$

22. $4x(2y + 3) - 5(2y + 3)$ $(2y + 3)(4x - 5)$

23. When $3x^2 - 9x + 3$ is factored, the result is $3(x^2 - 3x + 1)$. Explain why it is necessary to include the term 1 in this factored form.

24. Construct a trinomial whose GCF is $5x^2y^3$.
Answers will vary.

Factor each polynomial by grouping. See Examples 7 and 8. **25.** $(a + 2)(b + 3)$ **26.** $(a + 5)(b + 2)$

25. $ab + 3a + 2b + 6$ **26.** $ab + 2a + 5b + 10$

27. $ac + 4a - 2c - 8$ **28.** $bc + 8b - 3c - 24$

29. $2xy - 3x - 4y + 6$ **30.** $12xy - 18x - 10y + 15$

31. $12xy - 8x - 3y + 2$ **32.** $20xy - 15 - 4y + 3$

33. The material needed to manufacture a tin can is given by the polynomial

$$2\pi r^2 + 2\pi rh$$

where the radius is r and height is h. Factor this expression. $2\pi r(r + h)$

34. The amount E of current in an electrical circuit is given by the formula

$$IR_1 + IR_2 = E$$

Write an equivalent equation by factoring the expression $IR_1 + IR_2$. $I(R_1 + R_2) = E$

35. At the end of T years, the amount of money A in a savings account earning simple interest from an initial investment of P dollars at rate R is given by the formula

$$A = P + PRT$$

Write an equivalent equation by factoring the expression $P + PRT$. $A = P(1 + RT)$

36. An open-topped box has a square base and a height of 10 inches. If each of the bottom edges of the box has length x inches, find the amount of

material needed to construct the box. Write the answer in factored form. $2x(x + 20)$ sq. in.

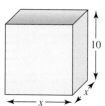

37. An object is thrown upward from the ground with an initial velocity of 64 feet per second. The height $h(t)$ of the object after t seconds is given by the polynomial function **37a.** $h(t) = -16t(t - 4)$

$$h(t) = -16t^2 + 64t$$

a. Write an equivalent factored expression for the function $h(t)$ by factoring $-16t^2 + 64t$.

b. Find $h(1)$ by using $h(t) = -16t^2 + 64t$ and then by using the factored form of $h(t)$. 48 feet

c. Explain why the values found in part (b) are the same.

38. An object is dropped from the gondola of a hot-air balloon at a height of 224 feet. The height $h(t)$ of the object after t seconds is given by the polynomial function

$$h(t) = -16t^2 + 224$$

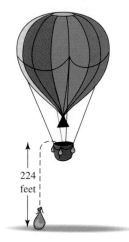

a. $h(t) = -16(t^2 - 14)$
b. 160 feet

a. Write an equivalent factored expression for the function $h(t)$ by factoring $-16t^2 + 224$.

b. Find $h(2)$ by using $h(t) = -16t^2 + 224$ and then by using the factored form of the function.

c. Explain why the values found in part (b) are the same.

29. $(x - 2)(2y - 3)$ **30.** $(6x - 5)(2y - 3)$ **31.** $(4x - 1)(3y - 2)$ **32.** $4(5xy - y - 3)$

45. $-4xy(5x - 4y^2)$ **46.** $9xy(-2y^2 + 3x^3)$ **55.** $(x + 3)(y - 5)$ **56.** $(x + 4)(y - 3)$ **57.** $(2a - 3)(3b - 1)$

Factor each polynomial.

39. $6x^3 + 9$ $3(2x^3 + 3)$ **40.** $6x^2 - 8$ $2(3x^2 - 4)$

41. $x^3 + 3x^2$ $x^2(x + 3)$ **42.** $x^4 - 4x^3$ $x^3(x - 4)$

43. $8a^3 - 4a$ $4a(2a^2 - 1)$ **44.** $12b^4 + 3b^2$ $3b^2(4b^2 + 1)$

45. $-20x^2y + 16xy^3$ **46.** $-18xy^3 + 27x^4y$

47. $10a^2b^3 + 5ab^2 - 15ab^3$ $5ab^2(2ab + 1 - 3b)$

48. $10ef - 20e^2f^3 + 30e^3f$ $10ef(1 - 2ef^2 + 3e^2)$

49. $9abc^2 + 6a^2bc - 6ab + 3bc$ $3b(3ac^2 + 2a^2c - 2a + c)$

50. $4a^2b^2c - 6ab^2c - 4ac + 8a$ $2a(2ab^2c - 3b^2c - 2c + 4)$

51. $4x(y - 2) - 3(y - 2)$ $(y - 2)(4x - 3)$

52. $8y(z + 8) - 3(z + 8)$ $(z + 8)(8y - 3)$

53. $6xy + 10x + 9y + 15$ $(2x + 3)(3y + 5)$

54. $15xy + 20x + 6y + 8$ $(3y + 4)(5x + 2)$

55. $xy + 3y - 5x - 15$ **56.** $xy + 4y - 3x - 12$

57. $6ab - 2a - 9b + 3$ **58.** $16ab - 8a - 6b + 3$

59. $12xy + 18x + 2y + 3$ **60.** $20xy + 8x + 5y + 2$

61. $2m(n - 8) - (n - 8)$ **62.** $3a(b - 4) - (b - 4)$

63. $15x^3y^2 - 18x^2y^2$ **64.** $12x^4y^2 - 16x^3y^3$

65. $2x^2 + 3xy + 4x + 6y$ **66.** $3x^2 + 12x + 4xy + 16y$

67. $5x^2 + 5xy - 3x - 3y$ **68.** $4x^2 + 2xy - 10x - 5y$

69. $x^3 + 3x^2 + 4x + 12$ **70.** $x^3 + 4x^2 + 3x + 12$

71. $x^3 - x^2 - 2x + 2$ **72.** $x^3 - 2x^2 - 3x + 6$

 73. A factored polynomial can be in many forms. For example, a factored form of $xy - 3x - 2y + 6$ is $(x - 2)(y - 3)$. Which of the following is not a factored form of $xy - 3x - 2y + 6$?   none

a. $(2 - x)(3 - y)$ **b.** $(-2 + x)(-3 + y)$

c. $(x - 2)(y - 3)$ **d.** $(-x + 2)(-y + 3)$

 74. Consider the following sequence of algebraic steps:

$$x^3 - 6x^2 + 2x - 10 = (x^3 - 6x^2) + (2x - 10)$$
$$= x^2(x - 6) + 2(x - 5)$$

Explain whether the final result is the factored form of the original polynomial.

Review Exercises

Simplify the following. See Section 5.1.

75. $(5x^2)(11x^5)$ $55x^7$ **76.** $(7y)(-2y^3)$ $-14y^4$

77. $(5x^2)^3$ $125x^6$ **78.** $(-2y^3)^4$ $16y^{12}$

Find each product by using the FOIL order of multiplying binomials. See Section 5.4. **79.** $x^2 - 3x - 10$

79. $(x + 2)(x - 5)$ **80.** $(x - 7)(x - 1)$

81. $(x + 3)(x + 2)$ **82.** $(x - 4)(x + 2)$

83. $(y - 3)(y - 1)$ **84.** $(s + 8)(s + 10)$
   $y^2 - 4y + 3$ $s^2 + 18s + 80$

A Look Ahead

EXAMPLE

Factor $x^{5a} - x^{3a} + x^{7a}$.

Solution:

The variable x is common to all three terms, and the power $3a$ is the smallest of the exponents. So factor out the common factor x^{3a}.

$$x^{5a} - x^{3a} + x^{7a} = x^{3a}(x^{2a}) - x^{3a}(1) + x^{3a}(x^{4a})$$
$$= x^{3a}(x^{2a} - 1 + x^{4a})$$

Factor. Assume that variables used as exponents represent positive integers. **85.** $x^n(x^{2n} - 2x^n + 5)$

85. $x^{3n} - 2x^{2n} + 5x^n$ **86.** $3y^n + 3y^{2n} + 5y^{8n}$

87. $6x^{8a} - 2x^{5a} - 4x^{3a}$ **88.** $3x^{5a} - 6x^{3a} + 9x^{2a}$

58. $(2b - 1)(8a - 3)$ **59.** $(6x + 1)(2y + 3)$ **60.** $(5y + 2)(4x + 1)$ **61.** $(n - 8)(2m - 1)$ **62.** $(b - 4)(3a - 1)$
63. $3x^2y^2(5x - 6)$ **64.** $4x^3y^2(3x - 4y)$ **65.** $(2x + 3y)(x + 2)$ **66.** $(x + 4)(3x + 4y)$ **67.** $(5x - 3)(x + y)$
68. $(2x + y)(2x - 5)$ **69.** $(x^2 + 4)(x + 3)$ **70.** $(x + 4)(x^2 + 3)$ **71.** $(x^2 - 2)(x - 1)$ **72.** $(x - 2)(x^2 - 3)$
80. $x^2 - 8x + 7$ **81.** $x^2 + 5x + 6$ **82.** $x^2 - 2x - 8$ **86.** $y^n(3 + 3y^n + 5y^{7n})$ **87.** $2x^{3a}(3x^{5a} - x^{2a} - 2)$
88. $3x^{2a}(x^{3a} - 2x^a + 3)$

5.6 | FACTORING TRINOMIALS

OBJECTIVES

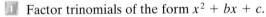

1. Factor trinomials of the form $x^2 + bx + c$.
2. Factor trinomials of the form $ax^2 + bx + c$.
3. Factor by substitution.
4. Factor trinomials by the AC method.

TAPE IA 5.6

1

In the previous section, we used factoring by grouping to factor four-term polynomials. In this section, we present techniques for factoring trinomials. Since $(x - 2)(x + 5) = x^2 + 3x - 10$, we say that $(x - 2)(x + 5)$ is a factored form of $x^2 + 3x - 10$. Taking a close look at how $(x - 2)$ and $(x + 5)$ are multiplied suggests a pattern for factoring trinomials of the form

$$x^2 + bx + c.$$

The pattern for factoring is summarized next.

TO FACTOR A TRINOMIAL OF THE FORM $x^2 + bx + c$, find two numbers whose product is c and whose sum is b. The factored form of $x^2 + bx + c$ is

$(x + $ one number$)(x + $ other number$)$

EXAMPLE 1   Factor $x^2 + 10x + 16$.

Solution: Look for two integers whose product is 16 and whose sum is 10. Since our integers must have a positive product and a positive sum, we look at only positive factors of 16.

POSITIVE FACTORS OF 16	SUM OF FACTORS
1, 16	$1 + 16 = 17$
4, 4	$4 + 4 = 8$
2, 8	$2 + 8 = 10$

The correct pair of numbers is 2 and 8 because their product is 16 and their sum is 10. Thus,

$$x^2 + 10x + 16 = (x + \text{one number})(x + \text{other number})$$

or

$$x^2 + 10x + 16 = (x + 2)(x + 8)$$

To check, see that $(x + 2)(x + 8) = x^2 + 10x + 16$.

EXAMPLE 2   Factor $x^2 - 12x + 35$.

Solution: Find two integers whose product is 35 and whose sum is -12. Since our integers must have a positive product and a negative sum, we consider only negative factors of 35. The numbers are -5 and -7.

$$x^2 - 12x + 35 = [x + (-5)][x + (-7)]$$
$$= (x - 5)(x - 7)$$

To check, see that $(x - 5)(x - 7) = x^2 - 12x + 35$.

EXAMPLE 3   Factor $5x^3 - 30x^2 - 35x$.

Solution: First, factor out a GCF of $5x$.

$$5x^3 - 30x^2 - 35x = 5x(x^2 - 6x - 7)$$

Next, factor $x^2 - 6x - 7$ by finding two numbers whose product is -7 and whose sum is -6. The numbers are 1 and -7.

$$5x^3 - 30x^2 - 35x = 5x(x^2 - 6x - 7)$$
$$= 5x(x + 1)(x - 7)$$

R E M I N D E R   If the polynomial to be factored contains a common factor that is factored out, don't forget to include that common factor in the final factored form of the original polynomial.

EXAMPLE 4   Factor $2n^2 - 38n + 80$.

Solution: The terms of this polynomial have a GCF of 2, which we factor out first.

$$2n^2 - 38n + 80 = 2(n^2 - 19n + 40)$$

Next, factor $n^2 - 19n + 40$ by finding two numbers whose product is 40 and whose sum is -19. Both numbers must be negative since their sum is -19. Possibilities are

$$-1 \text{ and } -40, \qquad -2 \text{ and } -20, \qquad -4 \text{ and } -10, \qquad -5 \text{ and } -8$$

None of the pairs has a sum of -19, so no further factoring with integers is possible. The factored form of $2n^2 - 38n + 80$ is

$$2n^2 - 38n + 80 = 2(n^2 - 19n + 40)$$

We call a polynomial such as $n^2 - 19n + 40$ that cannot be factored with integers a **prime polynomial.**

2

Next, we factor trinomials of the form $ax^2 + bx + c$, where the coefficient a of x^2 is not 1. Don't forget that the first step in factoring any polynomial is to factor out the GCF of its terms.

EXAMPLE 5   Factor $2x^2 + 11x + 15$.

Solution: Factors of $2x^2$ are $2x$ and x. Try these factors as first terms of the binomials.

$$2x^2 + 11x + 15 = (2x + \;)(x + \;)$$

Next, try combinations of factors of 15 until the correct middle term, $11x$, is obtained. We will try only positive factors of 15 since the coefficient of the middle term, 11, is positive. Positive factors of 15 are 1 and 15 and 3 and 5.

$(2x + 1)(x + 15)$ $\qquad\qquad$ $(2x + 15)(x + 1)$

$\underbrace{\qquad 1x \qquad}$ $\qquad\qquad\qquad$ $\underbrace{\qquad 15x \qquad}$

$\dfrac{30x}{31x}$, incorrect middle term $\qquad$ $\dfrac{2x}{17x}$, incorrect middle term

$(2x + 3)(x + 5)$ $\qquad\qquad$ $(2x + 5)(x + 3)$

$\underbrace{\qquad 3x \qquad}$ $\qquad\qquad\qquad$ $\underbrace{\qquad 5x \qquad}$

$\dfrac{10x}{13x}$, incorrect middle term $\qquad$ $\dfrac{6x}{11x}$, *correct* middle term

Thus, the factored form of $2x^2 + 11x + 15$ is $(2x + 5)(x + 3)$.

TO FACTOR A TRINOMIAL OF THE FORM $ax^2 + bx + c$

Step 1.   Write all pairs of factors of ax^2.

Step 2.   Write all pairs of factors of c, the constant term.

Step 3.   Try various combinations of these factors until the correct middle term bx is found.

Step 4.   If no combination exists, the polynomial is **prime.**

EXAMPLE 6   Factor $3x^2 - x - 4$.

Solution: Factors of $3x^2$: $3x \cdot x$.

Factors of -4: $-4 = -1 \cdot 4$, $-4 = 1 \cdot -4$, $-4 = -2 \cdot 2$, $-4 = 2 \cdot -2$

Try possible combinations of these factors.

$$(3x - 1)(x + 4) \qquad\qquad (3x + 4)(x - 1)$$

$$-1x \qquad\qquad\qquad\qquad 4x$$

$$12x \qquad\qquad\qquad\qquad -3x$$
$$11x, \text{ incorrect middle term} \qquad 1x, \text{ incorrect middle term}$$

$$(3x - 4)(x + 1)$$

$$-4x$$

$$3x$$
$$-1x, \text{\textit{correct} middle term}$$

Thus, $3x^2 - x - 4 = (3x - 4)(x + 1)$.

EXAMPLE 7 Factor $12x^3y - 22x^2y + 8xy$.

Solution: First factor out the GCF of the terms of this trinomial, $2xy$.

$$12x^3y - 22x^2y + 8xy = 2xy(6x^2 - 11x + 4)$$

Now try to factor the trinomial $6x^2 - 11x + 4$.

Factors of $6x^2$: $6x^2 = 2x \cdot 3x, \qquad 6x^2 = 6x \cdot x$

Try $2x$ and $3x$.

$$2xy(6x^2 - 11x + 4) = 2xy(2x + \ )(3x + \ )$$

The constant term 4, is positive and the coefficient of the middle term -11 is negative, so factor 4 into negative factors only.

Negative factors of 4: $4 = -4(-1), \qquad 4 = -2(-2)$

Try -4 and -1.

$$2xy(2x - 4)(3x - 1)$$

$$-12x$$

$$-2x$$
$$-14x, \text{ incorrect middle term}$$

This combination cannot be correct, because one of the factors $(2x - 4)$ has a common factor of 2. This cannot happen if the polynomial $6x^2 - 11x + 4$ has no common factors. Try -1 and -4.

$$2xy(2x - 1)(3x - 4)$$

$$-3x$$

$$-8x$$
$$-11x, \text{\textit{correct} middle term}$$

If this combination had not worked, we would try -2 and -2 and then $6x$ and x as factors of $6x^2$.

$$12x^3y - 22x^2y + 8xy = 2xy(2x - 1)(3x - 4)$$

R E M I N D E R If a trinomial has no common factor (other than 1), then none of its binomial factors will contain a common factor (other than 1).

EXAMPLE 8 Factor $16x^2 + 24xy + 9y^2$.

Solution: No GCF can be factored out of this trinomial. Factors of $16x^2$ are

$$16x^2 = 16x \cdot x, \qquad 16x^2 = 8x \cdot 2x, \qquad 16x^2 = 4x \cdot 4x$$

Factors of $9y^2$ are

$$9y^2 = y \cdot 9y, \qquad 9y^2 = 3y \cdot 3y$$

Try possible combinations until the correct factorization is found.

$$16x^2 + 24xy + 9y^2 = (4x + 3y)(4x + 3y), \quad \text{or} \quad (4x + 3y)^2$$

The trinomial $16x^2 + 24xy + 9y^2$ in Example 8 is an example of a **perfect square trinomial** since its factors are two identical binomials. In the next section, we examine a special method for factoring perfect square trinomials.

3 A complicated-looking polynomial may be a simpler trinomial "in disguise." Revealing the simpler trinomial is possible by substitution.

EXAMPLE 9 Factor $2(a + 3)^2 - 5(a + 3) - 7$.

Solution: The quantity $(a + 3)$ is in two of the terms of this polynomial. **Substitute** x for $(a + 3)$, and the result is the following simpler trinomial:

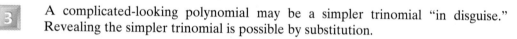

$$2(a + 3)^2 - 5(a + 3) - 7 \qquad \text{Original trinomial.}$$
$$\downarrow \qquad\qquad \downarrow$$
$$= \quad 2(x)^2 \quad - \quad 5(x) \quad - 7 \qquad \text{Substitute } x \text{ for } (a + 3).$$

Now factor $2x^2 - 5x - 7$.

$$2x^2 - 5x - 7 = (2x - 7)(x + 1)$$

But the quantity in the original polynomial was $(a + 3)$, not x. Thus we need to reverse the substitution and replace x with $(a + 3)$.

$$(2x - 7)(x + 1) \qquad\qquad \text{Factored expression.}$$
$$= [2(a + 3) - 7][(a + 3) + 1] \qquad \text{Substitute } (a + 3) \text{ for } x.$$
$$= (2a + 6 - 7)(a + 3 + 1) \qquad \text{Remove inside parentheses.}$$
$$= (2a - 1)(a + 4) \qquad\qquad \text{Simplify.}$$

Thus, $2(a + 3)^2 - 5(a + 3) - 7 = (2a - 1)(a + 4)$.

EXAMPLE 10 Factor $5x^4 + 29x^2 - 42$.

Solution: Again, substitution may help us factor this polynomial more easily. Let $y = x^2$, so $y^2 = (x^2)^2$, or x^4. Then

$$5x^4 + 29x^2 - 42$$

becomes

$$5y^2 + 29y - 42$$

which factors as

$$5y^2 + 29y - 42 = (5y - 6)(y + 7)$$

Next, replace y with x^2 to get

$$(5x^2 - 6)(x^2 + 7)$$

---

4 There is another method we can use when factoring trinomials of the form $ax^2 + bx + c$: Write the trinomial as a four-term polynomial, and then factor by grouping. The method is called the AC method.

TO FACTOR A TRINOMIAL OF THE FORM $ax^2 + bx + c$ BY THE AC METHOD

Step 1. Find two numbers whose product is $a \cdot c$ and whose sum is b.

Step 2. Write the term bx as a sum by using the factors found in *step 1*.

Step 3. Factor by grouping.

EXAMPLE 11 Factor $6x^2 + 13x + 6$.

Solution: In this trinomial, $a = 6, b = 13$, and $c = 6$.

Step 1. Find two numbers whose product is $a \cdot c$, or $6 \cdot 6 = 36$, and whose sum is b, 13.

The two numbers are 4 and 9.

Step 2. Write the middle term $13x$ as the sum $4x + 9x$.

$$6x^2 + 13x + 6 = 6x^2 + 4x + 9x + 6$$

Step 3. Factor $6x^2 + 4x + 9x + 6$ by grouping.

$$(6x^2 + 4x) + (9x + 6) = 2x(3x + 2) + 3(3x + 2)$$
$$= (3x + 2)(2x + 3)$$

MENTAL MATH

1. Find two numbers whose product is 10 and whose sum is 7. 5 and 2
2. Find two numbers whose product is 12 and whose sum is 8. 2 and 6
3. Find two numbers whose product is 24 and whose sum is 11. 8 and 3
4. Find two numbers whose product is 30 and whose sum is 13. 10 and 3

1. $(x + 3)(x + 6)$ **3.** $(x - 8)(x - 4)$ **5.** $(x + 12)(x - 2)$ **7.** $(x - 6)(x + 4)$ **9.** $3(x - 2)(x - 4)$
10. $y^2(x + 1)(x + 3)$ **11.** $4z(x + 2)(x + 5)$ **12.** $5(x - 2)(x - 7)$ **13.** $2(x + 18)(x - 3)$ **14.** $3(x - 4)(x + 8)$

EXERCISE SET 5.6

17. $(5x + 1)(x + 3)$ **19.** $(2x - 3)(x - 4)$ **21.** prime polynomial **25.** $2(3x - 5)(2x + 5)$

Factor each trinomial. See Examples 1 through 4.

1. $x^2 + 9x + 18$
2. $x^2 + 9x + 20$ $(x + 4)(x + 5)$
3. $x^2 - 12x + 32$
4. $x^2 - 12x + 27$ $(x - 3)(x - 9)$
5. $x^2 + 10x - 24$
6. $x^2 + 3x - 54$ $(x - 6)(x + 9)$
7. $x^2 - 2x - 24$
8. $x^2 - 9x - 36$ $(x + 3)(x - 12)$
9. $3x^2 - 18x + 24$
10. $x^2y^2 + 4xy^2 + 3y^2$
11. $4x^2z + 28xz + 40z$
12. $5x^2 - 45x + 70$
13. $2x^2 + 30x - 108$
14. $3x^2 + 12x - 96$

15. Find all positive and negative integers b such that $x^2 + bx + 6$ factors. ±5, ±7

16. Find all positive and negative integers b such that $x^2 + bx - 10$ factors. ±3, ±9

Factor each trinomial. See Examples 5 through 8 and 11.

17. $5x^2 + 16x + 3$
18. $3x^2 + 8x + 4$ $(3x + 2)(x + 2)$
19. $2x^2 - 11x + 12$
20. $3x^2 - 19x + 20$ $(3x - 4)(x - 5)$
21. $2x^2 + 25x - 20$
22. $6x^2 - 13x - 8$ $(3x - 8)(2x + 1)$
23. $4x^2 - 12x + 9$ $(2x - 3)^2$
24. $25x^2 - 30x + 9$ $(5x - 3)^2$
25. $12x^2 + 10x - 50$
26. $12y^2 - 48y + 45$
27. $3y^4 - y^3 - 10y^2$
28. $2x^2z + 5xz - 12z$
29. $6x^3 + 8x^2 + 24x$
30. $18y^3 + 12y^2 + 2y$ $2y(3y + 1)^2$
31. $x^2 + 8xz + 7z^2$
32. $a^2 - 2ab - 15b^2$
33. $2x^2 - 5xy - 3y^2$
34. $6x^2 + 11xy + 4y^2$
35. $x^2 - x - 12$
36. $x^2 + 4x - 5$ $(x - 1)(x + 5)$
37. $28y^2 + 22y + 4$
38. $24y^3 - 2y^2 - y$
39. $2x^2 + 15x - 27$
40. $3x^2 + 14x + 15$

41. Find all positive and negative integers b such that $3x^2 + bx + 5$ factors. ±8, ±16

42. Find all positive and negative integers b such that $2x^2 + bx + 7$ factors. ±9, ±15

Use substitution to factor each polynomial completely. See Examples 9 and 10.

43. $x^4 + x^2 - 6$
44. $x^4 - x^2 - 20$
45. $(5x + 1)^2 + 8(5x + 1) + 7$ $(5x + 8)(5x + 2)$
46. $(3x - 1)^2 + 5(3x - 1) + 6$ $(3x + 1)(3x + 2)$
47. $x^6 - 7x^3 + 12$
48. $x^6 - 4x^3 - 12$
49. $(a + 5)^2 - 5(a + 5) - 24$ $(a - 3)(a + 8)$
50. $(3c + 6)^2 + 12(3c + 6) - 28$ $(3c + 4)(3c + 20)$

51. The volume $V(x)$ of a box in terms of its height x is given by the function $V(x) = 3x^3 - 2x^2 - 8x$. Factor this expression for $V(x)$. $x(3x + 4)(x - 2)$

52. Based on your results from Exercise 51, find the length and width of the box if the height is 5 inches and the dimensions of the box are whole numbers. width, 3 in., height, 19 in.

Factor each polynomial completely. **53.** $(x - 27)(x + 3)$
53. $x^2 - 24x - 81$
54. $x^2 - 48x - 100$
55. $x^2 - 15x - 54$
56. $x^2 - 15x + 54$
57. $3x^2 - 6x + 3$ $3(x - 1)^2$ **58.** $8x^2 - 8x + 2$ $2(2x - 1)^2$

26. $3(2y - 3)(2y - 5)$ **27.** $y^2(3y + 5)(y - 2)$ **28.** $z(2x - 3)(x + 4)$ **29.** $2x(3x^2 + 4x + 12)$ **31.** $(x + 7z)(x + z)$
32. $(a + 3b)(a - 5b)$ **33.** $(2x + y)(x - 3y)$ **34.** $(3x + 4y)(2x + y)$ **35.** $(x - 4)(x + 3)$ **37.** $2(7y + 2)(2y + 1)$
38. $y(6y + 1)(4y - 1)$ **39.** $(2x - 3)(x + 9)$ **40.** $(3x + 5)(x + 3)$ **43.** $(x^2 + 3)(x^2 - 2)$ **44.** $(x^2 - 5)(x^2 + 4)$
47. $(x^3 - 4)(x^3 - 3)$ **48.** $(x^3 + 2)(x^3 - 6)$ **54.** $(x + 2)(x - 50)$ **55.** $(x - 18)(x + 3)$ **56.** $(x - 6)(x - 9)$

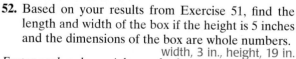

59. $(3x + 1)(x - 2)$ **60.** $(5x + 1)(x - 3)$ **61.** $(4x - 3)(2x - 5)$ **62.** $(4x - 3)(3x - 2)$ **63.** $3x^2(2x + 1)(3x + 2)$

59. $3x^2 - 5x - 2$ **60.** $5x^2 - 14x - 3$

61. $8x^2 - 26x + 15$ **62.** $12x^2 - 17x + 6$

63. $18x^4 + 21x^3 + 6x^2$ **64.** $20x^5 + 54x^4 + 10x^3$

65. $3a^2 + 12ab + 12b^2$ **66.** $2x^2 + 16xy + 32y^2$

67. $x^2 + 4x + 5$ **68.** $x^2 + 6x + 8$ $(x + 2)(x + 4)$

69. $2(x + 4)^2 + 3(x + 4) - 5$ $(2x + 13)(x + 3)$

70. $3(x + 3)^2 + 2(x + 3) - 5$ $(3x + 14)(x + 2)$

71. $6x^2 - 49x + 30$ **72.** $4x^2 - 39x + 27$

73. $x^4 - 5x^2 - 6$ **74.** $x^4 - 5x^2 + 6$

75. $6x^3 - x^2 - x$ **76.** $12x^3 + x^2 - x$

77. $12a^2 - 29ab + 15b^2$ **78.** $16y^2 + 6yx - 27x^2$

79. $9x^2 + 30x + 25$ **80.** $4x^2 + 6x + 9$

81. $3x^2y - 11xy + 8y$ **82.** $5xy^2 - 9xy + 4x$

83. $2x^2 + 2x - 12$ **84.** $3x^2 + 6x - 45$

85. $(x - 4)^2 + 3(x - 4) - 18$ $(x + 2)(x - 7)$

86. $(x - 3)^2 - 2(x - 3) - 8$ $(x - 1)(x - 7)$

87. $2x^6 + 3x^3 - 9$ **88.** $3x^6 - 14x^3 + 8$

89. $72xy^4 - 24xy^2z + 2xz^2$ $2x(6y^2 - z)^2$

90. $36xy^2 - 48xyz^2 + 16xz^4$ $4x(3y - 2z^2)^2$

Recall that a grapher may be used to visualize addition, subtraction, and multiplication of polynomials. In the same manner, a grapher may be used to visualize factoring of polynomials in one variable. For example, to see that

$$2x^3 - 9x^2 - 5x = x(2x + 1)(x - 5)$$

graph $Y_1 = 2x^3 - 9x^2 - 5x$ *and* $Y_2 = x(2x + 1)(x - 5)$. *Then trace along both graphs to see that they coincide.*

64. $2x^3(2x + 5)(5x + 1)$ **65.** $3(a + 2b)^2$ **66.** $2(x + 4y)^2$ **67.** prime polynomial   **71.** $(3x - 2)(2x - 15)$
72. $(4x - 3)(x - 9)$ **73.** $(x^2 - 6)(x^2 + 1)$ **74.** $(x^2 - 2)(x^2 - 3)$ **75.** $x(3x + 1)(2x - 1)$ **76.** $x(4x - 1)(3x + 1)$
77. $(4a - 3b)(3a - 5b)$ **78.** $(8y - 9x)(2y + 3x)$ **79.** $(3x + 5)^2$ **80.** prime polynomial   **81.** $y(3x - 8)(x - 1)$
82. $x(5y - 4)(y - 1)$ **83.** $2(x + 3)(x - 2)$ **84.** $3(x - 3)(x + 5)$ **87.** $(2x^3 - 3)(x^3 + 3)$ **88.** $(3x^3 - 2)(x^3 - 4)$

Factor the following and use this method to check your results.   **91.** $x^2(x + 5)(x + 1)$ **92.** $x(x + 2)(x + 4)$

 91. $x^4 + 6x^3 + 5x^2$ **92.** $x^3 + 6x^2 + 8x$

 93. $30x^3 + 9x^2 - 3x$ **94.** $-6x^4 + 10x^3 - 4x^2$
$3x(5x - 1)(2x + 1)$ $-2x^2(3x - 2)(x - 1)$

Review Exercises   **95.** $x^3 - 8$ **96.** $y^3 + 1$

Multiply the following. See Section 5.4.

95. $(x - 2)(x^2 + 2x + 4)$ **96.** $(y + 1)(y^2 - y + 1)$

If $P(x) = 3x^2 + 2x - 9$, *find the following. See Section 5.3.*   **99.** -8 **100.** -1

97. $P(0)$ -9 **98.** $P(1)$ -4 **99.** $P(-1)$ **100.** $P(-2)$

A Look Ahead

EXAMPLE

Factor $x^{2n} + 7x^n + 12$.

Solution:

Factors of x^{2n} are x^n and x^n, so $x^{2n} + 7x^n + 12 = (x^n + \text{one number})(x^n + \text{other number})$. Factors of 12 whose sum is 7 are 3 and 4. Thus

$$x^{2n} + 7x^n + 12 = (x^n + 4)(x^n + 3)$$

Factor. Assume that variables used as exponents represent positive integers. See the preceding example.

101. $(x^n + 8)(x^n + 2)$
102. $(x^n - 4)(x^n - 3)$
103. $(x^n - 6)(x^n + 3)$
104. $(x^n - 2)(x^n + 9)$
105. $(2x^n + 1)(x^n + 5)$
106. $(3x^n - 2)(x^n - 2)$ **107.** $(2x^n - 3)^2$ **108.** $(3x^n + 4)^2$

101. $x^{2n} + 10x^n + 16$ **102.** $x^{2n} - 7x^n + 12$

103. $x^{2n} - 3x^n - 18$ **104.** $x^{2n} + 7x^n - 18$

105. $2x^{2n} + 11x^n + 5$ **106.** $3x^{2n} - 8x^n + 4$

107. $4x^{2n} - 12x^n + 9$ **108.** $9x^{2n} + 24x^n + 16$

5.7 | FACTORING BY SPECIAL PRODUCTS AND FACTORING STRATEGIES

O B J E C T I V E S

1. Factor a perfect square trinomial.
2. Factor the difference of two squares.
3. Factor the sum or difference of two cubes.
4. Practice techniques for factoring polynomials.

TAPE IA 5.7

In the previous section, we considered a variety of ways to factor trinomials of the form $ax^2 + bx + c$. In one particular example, we factored $16x^2 + 24xy + 9y^2$ as

$$16x^2 + 24xy + 9y^2 = (4x + 3y)^2$$

Recall that we called $16x^2 + 24xy + 9y^2$ a perfect square trinomial because its factors are two identical binomials. A perfect square trinomial can be factored quickly if you recognize the trinomial as a perfect square.

A trinomial is a perfect square trinomial if it can be written so that its first term is the square of some quantity a, its last term is the square of some quantity b, and its middle term is twice the product of the quantities a and b. The following special formulas can be used to factor perfect square trinomials.

PERFECT SQUARE TRINOMIALS

$$a^2 + 2ab + b^2 = (a + b)^2$$
$$a^2 - 2ab + b^2 = (a - b)^2$$

Notice that these equations are the same special products from Section 5.4 for the square of a binomial.

From $a^2 + 2ab + b^2 = (a + b)^2$, we see that

$$16x^2 + 24xy + 9y^2 = (4x)^2 + 2(4x)(3y) + (3y)^2 = (4x + 3y)^2$$

EXAMPLE 1   Factor $m^2 + 10m + 25$.

Solution: Notice that the first term is a square: $m^2 = (m)^2$,
the last term is a square: $25 = 5^2$,
and $10m = 2 \cdot 5 \cdot m$.
Thus,

$$m^2 + 10m + 25 = m^2 + 2(m)(5) + 5^2 = (m + 5)^2$$

EXAMPLE 2   Factor $3a^2x - 12abx + 12b^2x$.

Solution: The terms of this trinomial have a GCF of $3x$, which we factor out first.

$$3a^2x - 12abx + 12b^2x = 3x(a^2 - 4ab + 4b^2)$$

Now, the polynomial $a^2 - 4ab + 2b^2$ is a perfect square trinomial. Notice that
the first term is a square: $a^2 = (a)^2$,
the last term is a square: $4b^2 = (2b)^2$,
and $4ab = 2(a)(2b)$.
The factoring can now be completed as

$$3x(a^2 - 4ab + 4b^2) = 3x(a - 2b)^2$$

R E M I N D E R   If you recognize a trinomial as a perfect square trinomial, use the special formulas to factor. However, methods for factoring trinomials in general from Section 5.6 will also result in the correct factored form.

We now factor special types of binomials, beginning with the **difference of two squares.** The special product pattern presented in Section 5.4 for the product of a sum and a difference of two terms is used again here. However, the emphasis is now on factoring rather than on multiplying.

DIFFERENCE OF TWO SQUARES

$$a^2 - b^2 = (a + b)(a - b)$$

Notice that a binomial is a difference of two squares when it is the difference of the square of some quantity a and the square of some quantity b.

EXAMPLE 3   Factor the following:

a. $x^2 - 9$ b. $16y^2 - 9$ c. $50 - 8y^2$

Solution: a. $x^2 - 9 = x^2 - 3^2$ b. $16y^2 - 9 = (4y)^2 - 3^2$

$= (x + 3)(x - 3)$ $= (4y + 3)(4y - 3)$

c. First factor out the common factor of 2.

$50 - 8y^2 = 2(25 - 4y^2)$

$= 2(5 + 2y)(5 - 2y)$

The binomial $x^2 + 9$ is a **sum of two squares** and cannot be factored by using real numbers. **In general, except for factoring out a GCF, the sum of two squares usually cannot be factored by using real numbers.**

R E M I N D E R   The sum of two squares whose GCF is 1 usually cannot be factored by using real numbers.

EXAMPLE 4   Factor the following:

a. $p^4 - 16$ b. $(x + 3)^2 - 36$

Solution: a. $p^4 - 16 = (p^2)^2 - 4^2$

$= (p^2 + 4)(p^2 - 4)$

The binomial factor $p^2 + 4$ cannot be factored by using real numbers, but the binomial factor $p^2 - 4$ is a difference of squares.

$$(p^2 + 4)(p^2 - 4) = (p^2 + 4)(p + 2)(p - 2)$$

b. Factor $(x + 3)^2 - 36$ as the difference of squares.

$$(x + 3)^2 - 36 = (x + 3)^2 - 6^2$$

$$= [(x + 3) + 6][(x + 3) - 6] \quad \text{Factor.}$$

$$= [x + 3 + 6][x + 3 - 6] \quad \text{Remove parentheses.}$$

$$= (x + 9)(x - 3) \quad \text{Simplify.}$$

3 Although the sum of two squares usually cannot be factored, the sum of two cubes, as well as the difference of two cubes, can be factored as follows.

SUM AND DIFFERENCE OF TWO CUBES

$$a^3 + b^3 = (a + b)(a^2 - ab + b^2)$$
$$a^3 - b^3 = (a - b)(a^2 + ab + b^2)$$

To check the first pattern, find the product of $(a + b)$ and $(a^2 - ab + b^2)$.

$$
\begin{array}{r}
a^2 \;- ab \;+ b^2 \\
a \;+ b \\
\hline
a^2b - ab^2 + b^3 \\
a^3 - a^2b + ab^2 \\
\hline
a^3 \qquad\qquad\quad + b^3
\end{array}
$$

Then $a^3 + b^3 = (a + b)(a^2 - ab + b^2)$.

EXAMPLE 5   Factor $x^3 + 8$.

Solution: First, write the binomial in the form $a^3 + b^3$. Then, use the formula

$$a^3 + b^3 = (a + b)(a^2 - a \cdot b + b^2), \text{ where } a \text{ is } x \text{ and } b \text{ is } 2.$$

$$\downarrow \quad \downarrow \quad \downarrow \quad \downarrow \ \downarrow \quad \downarrow \ \downarrow \quad \downarrow$$

$$x^3 + 8 = x^3 + 2^3 = (x + 2)(x^2 - x \cdot 2 + 2^2)$$

Thus, $x^3 + 8 = (x + 2)(x^2 - 2x + 4)$.

EXAMPLE 6   Factor $p^3 + 27q^3$.

Solution: $p^3 + 27q^3 = p^3 + (3q)^3$

$$= (p + 3q)[p^2 - (p)(3q) + (3q)^2]$$

$$= (p + 3q)(p^2 - 3pq + 9q^2)$$

EXAMPLE 7   Factor $y^3 - 64$.

Solution: This is a difference of cubes since $y^3 - 64 = y^3 - 4^3$.

From $a^3 - b^3 = (a - b)(a^2 + a \cdot b + b^2)$ we have that

$$\downarrow \quad \downarrow \quad \downarrow \quad \downarrow \quad \downarrow \quad \downarrow \quad \downarrow \quad \downarrow$$

$$y^3 - 4^3 = (y - 4)(y^2 + y \cdot 4 + 4^2)$$

$$= (y - 4)(y^2 + 4y + 16)$$

REMINDER   When factoring sums or differences of cubes, be sure to notice the sign patterns.

same sign

$$x^3 + y^3 = (x + y)(x^2 - xy + y^2)$$

opposite sign   always positive

same sign

$$x^3 - y^3 = (x - y)(x^2 + xy + y^2)$$

opposite sign   always positive

EXAMPLE 8   Factor $125q^2 - n^3q^2$.

Solution: First, factor out a common factor of q^2.

$$125q^2 - n^3q^2 = q^2(125 - n^3)$$

$$= q^2(5^3 - n^3)$$

opposite sign   positive

$$= q^2(5 - n)[5^2 + (5)(n) + (n^2)]$$

$$= q^2(5 - n)(25 + 5n + n^2)$$

Thus $125q^2 - n^3q^2 = q^2(5 - n)(25 + 5n + n^2)$. The trinomial $25 + 5n + n^2$ cannot be factored further.

EXAMPLE 9   Factor $x^2 + 4x + 4 - y^2$.

Solution: Factoring by grouping comes to mind since the sum of the first three terms of this polynomial is a perfect square trinomial.

$$x^2 + 4x + 4 - y^2 = (x^2 + 4x + 4) - y^2 \quad \text{Group the first three terms.}$$

$$= (x + 2)^2 - y^2 \qquad \text{Factor the perfect square trinomial.}$$

This is not factored yet since we have a *difference,* not a *product.* Since $(x + 2)^2 - y^2$ is a difference of squares, we have

$$(x + 2)^2 - y^2 = [(x + 2) + y][(x + 2) - y]$$
$$= (x + 2 + y)(x + 2 - y)$$

4   The key to proficiency in factoring polynomials is to practice until you are comfortable with each technique. A strategy for factoring polynomials is given next.

TO FACTOR A POLYNOMIAL

Step 1.   Are there any common factors? If so, factor out the GCF.

Step 2.   How many terms are in the polynomial?

    a. If there are **two** terms, decide if one of the following formulas may be applied:

        i. Difference of two squares: $a^2 - b^2 = (a - b)(a + b)$.

        ii. Difference of two cubes: $a^3 - b^3 = (a - b)(a^2 + ab + b^2)$.

        iii. Sum of two cubes: $a^3 + b^3 = (a + b)(a^2 - ab + b^2)$.

    b. If there are **three** terms, try one of the following:

        i. Perfect square trinomial: $a^2 + 2ab + b^2 = (a + b)^2$.

                      $a^2 - 2ab + b^2 = (a - b)^2$

        ii. If not a perfect square trinomial, factor by using the methods presented in Section 5.6.

    c. If there are **four** or more terms, try factoring by grouping.

Step 3.   See if any factors in the factored polynomial can be factored further.

EXAMPLE 10   Factor each polynomial completely.

    a. $8a^2b - 4ab$ **b.** $36x^2 - 9$ **c.** $2x^2 - 5x - 7$

Solution: **a.** *Step 1.* The terms have a common factor of $4ab$, which we factor out.

$$8a^2b - 4ab = 4ab(2a - 1)$$

Step 2. There are two terms, but the binomial $2a - 1$ is not the difference of two squares or the sum or difference of two cubes.

Step 3. The factor $2a - 1$ cannot be factored further.

b. *Step 1.* Factor out a common factor of 9.

$$36x^2 - 9 = 9(4x^2 - 1)$$

Step 2. The factor $4x^2 - 1$ has two terms, and it is the difference of two squares.

$$9(4x^2 - 1) = 9(2x + 1)(2x - 1)$$

Step 3. No factor with more than one term can be factored further.

c. *Step 1.* The terms of $2x^2 - 5x - 7$ contain no common factor other than 1 or -1.

Step 2. There are three terms. The trinomial is not a perfect square, so we factor by methods from Section 5.6.

$$2x^2 - 5x - 7 = (2x - 7)(x + 1)$$

Step 3. No factor with more than one term can be factored further. ▬▬▬▬

EXAMPLE 11

Factor each polynomial completely.

     a. $5p^2 + 5 + qp^2 + q$ **b.** $9x^2 + 24x + 16$ **c.** $y^2 + 25$

Solution:

a. *Step 1.* There is no common factor of all terms of $5p^2 + 5 + qp^2 + q$.

Step 2. The polynomial has four terms, so try factoring by grouping.

$$
\begin{aligned}
5p^2 + 5 + qp^2 + q &= (5p^2 + 5) + (qp^2 + q) &&\text{Group the terms.}\\
&= 5(p^2 + 1) + q(p^2 + 1)\\
&= (p^2 + 1)(5 + q)
\end{aligned}
$$

Step 3. No factor can be factored further.

b. *Step 1.* The terms of $9x^2 + 24x + 16$ contain no common factor other than 1 or -1.

Step 2. The trinomial $9x^2 + 24x + 16$ is a perfect square trinomial, and $9x^2 + 24x + 16 = (3x + 4)^2$.

Step 3. No factor can be factored further.

c. *Step 1.* There is no common factor of $y^2 + 25$ other than 1.

Step 2. This binomial is the sum of two squares and is prime.

Step 3. The binomial $y^2 + 25$ cannot be factored further. ▬▬▬▬

EXAMPLE 12

Factor each completely.

     a. $27a^3 - b^3$ **b.** $3n^2m^4 - 48m^6$ **c.** $2x^2 - 12x + 18 - 2z^2$

     d. $8x^4y^2 + 125xy^2$ **e.** $(x - 5)^2 - 49y^2$

Solution:

a. This binomial is a difference of two cubes.

$$
\begin{aligned}
27a^3 - b^3 &= (3a)^3 - b^3\\
&= (3a - b)[(3a)^2 + (3a)(b) + b^2]\\
&= (3a - b)(9a^2 + 3ab + b^2)
\end{aligned}
$$

b. $3n^2m^4 - 48m^6 = 3m^4(n^2 - 16m^2)$ Factor out the GCF, $3m^4$.

$ = 3m^4(n + 4m)(n - 4m)$ Factor the difference of squares.

c. $2x^2 - 12x + 18 - 2z^2 = 2(x^2 - 6x + 9 - z^2)$ The GCF is 2.

$= 2[(x^2 - 6x + 9) - z^2]$ Group the first three terms together.

$= 2[(x - 3)^2 - z^2]$ Factor the perfect square trinomial.

$= 2[(x - 3) + z][(x - 3) - z]$ Factor the difference of squares.

$= 2(x - 3 + z)(x - 3 - z)$

d. $8x^4y^2 + 125xy^2 = xy^2(8x^3 + 125)$ The GCF is xy^2.

$= xy^2[(2x)^3 + 5^3]$

$= xy^2(2x + 5)[(2x)^2 - (2x)(5) + 5^2]$ Factor the sum of cubes.

$= xy^2(2x + 5)(4x^2 - 10x + 25)$

e. This binomial is a difference of squares.

$(x - 5)^2 - 49y^2 = (x - 5)^2 - (7y)^2$

$= [(x - 5) + 7y][(x - 5) - 7y]$

$= (x - 5 + 7y)(x - 5 - 7y)$

13. $(y + 9)(y - 5)$ **14.** $(x - 1 + z)(x - 1 - z)$ **17.** $(x + 3)(x^2 - 3x + 9)$ **19.** $(z - 1)(z^2 + z + 1)$
21. $(m + n)(m^2 - mn + n^2)$ **22.** $(r + 5)(r^2 - 5r + 25)$ **23.** $y^2(x - 3)(x^2 + 3x + 9)$ **24.** $(4 - p)(16 + 4p + p^2)$
25. $b(a + 2b)(a^2 - 2ab + 4b^2)$ **26.** $a(2b + 3a)(4b^2 - 6ab + 9a^2)$ **27.** $(5y - 2x)(25y^2 + 10yx + 4x^2)$

EXERCISE SET 5.7

28. $2(3y - 4)(9y^2 + 12y + 16)$ **29.** $(x + 3 + y)(x + 3 - y)$
30. $(x + 6 + y)(x + 6 - y)$ **31.** $(x - 5 + y)(x - 5 - y)$ **32.** $(x - 9 + y)(x - 9 - y)$

Factor the following. See Examples 1 and 2.

1. $x^2 + 6x + 9$ $(x + 3)^2$ **2.** $x^2 - 10x + 25$ $(x - 5)^2$

3. $4x^2 - 12x + 9$ $(2x - 3)^2$ **4.** $25x^2 + 10x + 1$ $(5x + 1)^2$

5. $3x^2 - 24x + 48$ $3(x - 4)^2$ **6.** $x^3 + 14x^2 + 49x$ $x(x + 7)^2$

7. $9y^2x^2 + 12yx^2 + 4x^2$ **8.** $32x^2 - 16xy + 2y^2$
$x^2(3y + 2)^2$ $2(4x - y)^2$

Factor the following. See Examples 3 and 4.

9. $x^2 - 25$ $(x + 5)(x - 5)$ **10.** $y^2 - 100$ $(y + 10)(y - 10)$

11. $9 - 4z^2$ $(3 + 2z)(3 - 2z)$ **12.** $16x^2 - y^2$ $(4x + y)(4x - y)$

13. $(y + 2)^2 - 49$ **14.** $(x - 1)^2 - z^2$

15. $64x^2 - 100$ **16.** $4x^2 - 36$ $4(x + 3)(x - 3)$
$4(4x + 5)(4x - 5)$

Factor the following. See Examples 5 through 8.

17. $x^3 + 27$ **18.** $y^3 + 1$ $(y + 1)(y^2 - y + 1)$

19. $z^3 - 1$ **20.** $x^3 - 8$ $(x - 2)(x^2 + 2x + 4)$

21. $m^3 + n^3$ **22.** $r^3 + 125$

23. $x^3y^2 - 27y^2$ **24.** $64 - p^3$

25. $a^3b + 8b^4$ **26.** $8ab^3 + 27a^4$

27. $125y^3 - 8x^3$ **28.** $54y^3 - 128$

Factor the following. See Example 9.

29. $x^2 + 6x + 9 - y^2$ **30.** $x^2 + 12x + 36 - y^2$

31. $x^2 - 10x + 25 - y^2$ **32.** $x^2 - 18x + 81 - y^2$

33. $4x^2 + 4x + 1 - z^2$ **34.** $9y^2 + 12y + 4 - x^2$
$(2x + 1 + z)(2x + 1 - z)$ $(3y + 2 - x)(3y + 2 + x)$
Factor each polynomial completely.   **35.** $(3x + 7)(3x - 7)$

35. $9x^2 - 49$ **36.** $25x^2 - 4$

37. $x^4 - 81$ **38.** $x^4 - 256$

39. $x^2 + 8x + 16 - 4y^2$ $(x + 4 + 2y)(x + 4 - 2y)$

40. $x^2 + 14x + 49 - 9y^2$ $(x + 7 + 3y)(x + 7 - 3y)$

41. $(x + 2y)^2 - 9$ **42.** $(3x + y)^2 - 25$

43. $x^3 - 1$ **44.** $x^3 - 8$

45. $x^3 + 125$ **46.** $x^3 + 216$

47. $4x^2 + 25$ **48.** $16x^2 + 25$

49. $4a^2 + 12a + 9$ **50.** $9a^2 - 30a + 25$

51. $18x^2y - 2y$ **52.** $12xy^2 - 108x$

53. $x^6 - y^3$ **54.** $x^3 - y^6$

55. $x^2 + 16x + 64 - x^4$ $(x + 8 + x^2)(x + 8 - x^2)$

36. $(5x + 2)(5x - 2)$ **37.** $(x^2 + 9)(x + 3)(x - 3)$ **38.** $(x^2 + 16)(x + 4)(x - 4)$ **41.** $(x + 2y + 3)(x + 2y - 3)$
42. $(3x + y + 5)(3x + y - 5)$ **43.** $(x - 1)(x^2 + x + 1)$ **44.** $(x - 2)(x^2 + 2x + 4)$ **45.** $(x + 5)(x^2 - 5x + 25)$
46. $(x + 6)(x^2 - 6x + 36)$ **47.** prime polynomial   **48.** prime polynomial   **49.** $(2a + 3)^2$ **50.** $(3a - 5)^2$
51. $2y(3x + 1)(3x - 1)$ **52.** $12x(y + 3)(y - 3)$ **53.** $(x^2 - y)(x^4 + x^2y + y^2)$ **54.** $(x - y^2)(x^2 + xy^2 + y^4)$

57. $3y^2(x^2 + 3)(x^4 - 3x^2 + 9)$ **58.** $x^2y^3(y^2 + 1)(y^4 - y^2 + 1)$ **59.** $(x + y + 5)(x^2 + 2xy + y^2 - 5x - 5y + 25)$

56. $x^2 + 20x + 100 - x^4$ $(x + 10 + x^2)(x + 10 - x^2)$

57. $3x^6y^2 + 81y^2$ **58.** $x^2y^9 + x^2y^3$

59. $(x + y)^3 + 125$ **60.** $(x + y)^3 + 27$

61. $(2x + 3)^3 - 64$ **62.** $(4x + 2)^3 - 125$

63. The manufacturer of Antonio's Metal Washers needs to determine the cross-sectional area of each washer. If the outer radius of the washer is R and the radius of the hole is r, express the area of the washer as a polynomial. Factor this polynomial completely. $\pi R^2 - \pi r^2 = \pi(R + r)(R - r)$

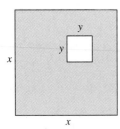

64. Express the area of the shaded region as a polynomial. Factor the polynomial completely.
$$A = x^2 - y^2 = (x + y)(x - y)$$

65. The manufacturer of Tootsie Roll Pops plans to change the size of its candy. To compute the new cost, the company needs a formula for the volume of the candy coating without the Tootsie Roll center. Given the diagram, express the volume as a polynomial. Factor this polynomial completely.

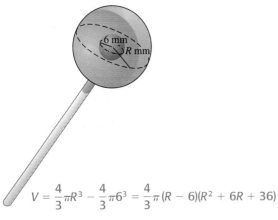

$$V = \frac{4}{3}\pi R^3 - \frac{4}{3}\pi 6^3 = \frac{4}{3}\pi(R - 6)(R^2 + 6R + 36)$$

66. Express the area of the shaded region as a polynomial. Factor the polynomial completely.
$$A = 9 - 4x^2 = (3 + 2x)(3 - 2x)$$

71. $2xy(7x - 1)$
72. $6ab(4b - 1)$
73. $4(x + 2)(x - 2)$
74. $9(x - 3)(x + 3)$
75. $(3x - 11)(x + 1)$
76. $(5x + 3)(x - 1)$
77. $4(x + 3)(x - 1)$
78. $6(x + 1)(x - 2)$
79. $(2x + 9)^2$
80. $(5x + 4)^2$

Factor completely. See Examples 10 through 12.

67. $x^2 - 8x + 16 - y^2$ $(x - 4 + y)(x - 4 - y)$

68. $12x^2 - 22x - 20$ $2(3x + 2)(2x - 5)$

69. $x^4 - x$ $x(x - 1)(x^2 + x + 1)$

70. $(2x + 1)^2 - 3(2x + 1) + 2$ $2x(2x - 1)$

71. $14x^2y - 2xy$ **72.** $24ab^2 - 6ab$

73. $4x^2 - 16$ **74.** $9x^2 - 81$

75. $3x^2 - 8x - 11$ **76.** $5x^2 - 2x - 3$

77. $4x^2 + 8x - 12$ **78.** $6x^2 - 6x - 12$

79. $4x^2 + 36x + 81$ **80.** $25x^2 + 40x + 16$

81. $8x^3 + 27y^3$ **82.** $125x^3 + 8y^3$

83. $64x^2y^3 - 8x^2$ **84.** $27x^5y^4 - 216x^2y$

85. $(x + 5)^3 + y^3$ **86.** $(y - 1)^3 + 27x^3$

87. $(5a - 3)^2 - 6(5a - 3) + 9$ $(5a - 6)^2$

88. $(4r + 1)^2 + 8(4r + 1) + 16$ $(4r + 5)^2$

Find a value of c that makes each trinomial a perfect square trinomial.   **89.** $c = 9$ **91.** $c = 49$ **93.** $c = \pm 8$

89. $x^2 + 6x + c$ **90.** $y^2 + 10y + c$ $c = 25$

91. $m^2 - 14m + c$ **92.** $n^2 - 2n + c$ $c = 1$

93. $x^2 + cx + 16$ **94.** $x^2 + cx + 36$ $c = \pm 12$

95. Factor $x^6 - 1$ completely, using the following methods from this chapter.

  a. Factor the expression by treating it as the difference of two squares $(x^3)^2 - 1^2$.

  b. Factor the expression treating it as the difference of two cubes, $(x^2)^3 - 1^3$.

  c. Are the answers to parts (a) and (b) the same? Why or why not?   Answers will vary.

60. $(x + y + 3)(x^2 + 2xy + y^2 - 3x - 3y + 9)$ **61.** $(2x - 1)(4x^2 + 20x + 37)$ **62.** $(4x - 3)(16x^2 + 36x + 39)$
81. $(2x + 3y)(4x^2 - 6xy + 9y^2)$ **82.** $(5x + 2y)(25x^2 - 10xy + 4y^2)$ **83.** $8x^2(2y - 1)(4y^2 + 2y + 1)$
84. $27x^2y(xy - 2)(x^2y^2 + 2xy + 4)$ **85.** $(x + 5 + y)(x^2 + 10x - xy - 5y + y^2 + 25)$
86. $(y - 1 + 3x)(y^2 - 2y + 1 - 3xy + 3x + 9x^2)$ **95a.** $(x + 1)(x^2 - x + 1)(x - 1)(x^2 + x + 1)$
95b. $(x + 1)(x - 1)(x^4 + x^2 + 1)$

Review Exercises

Solve the following equations. See Section 2.1.

96. $x - 5 = 0$ $\{5\}$

97. $x + 7 = 0$ $\{-7\}$

98. $3x + 1 = 0$ $\left\{-\dfrac{1}{3}\right\}$

99. $5x - 15 = 0$ $\{3\}$

100. $-2x = 0$ $\{0\}$

101. $3x = 0$ $\{0\}$

102. $-5x + 25 = 0$ $\{5\}$

103. $-4x - 16 = 0$ $\{-4\}$

A Look Ahead

EXAMPLE

Factor $x^{2n} - 100$.

Solution:

This binomial is a difference of squares.

$$x^{2n} - 100 = (x^n)^2 - 10^2$$
$$= (x^n + 10)(x^n - 10)$$

Factor each expression. Assume that variables used as exponents represent positive integers. See the preceding example.

104. $x^{2n} - 25$

105. $x^{2n} - 36$

106. $36x^{2n} - 49$

107. $25x^{2n} - 81$

108. $x^{4n} - 16$

109. $x^{4n} - 625$

104. $(x^n - 5)(x^n + 5)$ **105.** $(x^n - 6)(x^n + 6)$ **106.** $(6x^n - 7)(6x^n + 7)$ **107.** $(5x^n + 9)(5x^n - 9)$
108. $(x^n - 2)(x^n + 2)(x^{2n} + 4)$ **109.** $(x^n - 5)(x^n + 5)(x^{2n} + 25)$

5.8   SOLVING EQUATIONS BY FACTORING AND PROBLEM SOLVING

TAPE IA 5.8

O B J E C T I V E S

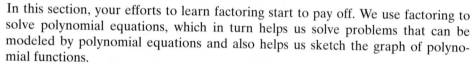

    1   Solve polynomial equations by factoring.
    2   Solve problems that can be modeled by quadratic equations.

    1   In this section, your efforts to learn factoring start to pay off. We use factoring to solve polynomial equations, which in turn helps us solve problems that can be modeled by polynomial equations and also helps us sketch the graph of polynomial functions.

A *polynomial equation* is the result of setting two polynomials equal to each other. Examples of polynomial equations are

$$3x^3 - 2x^2 = x^2 + 2x - 1 \quad 2.6x + 7 = -1.3 \quad -5x^2 - 5 = -9x^2 - 2x + 1$$

A polynomial equation is in *standard form* if one side of the equation is 0. In standard form the polynomial equations above are

$$3x^3 - 3x^2 - 2x + 1 = 0 \quad 2.6x + 8.3 = 0 \quad 4x^2 + 2x - 6 = 0$$

The degree of a simplified polynomial equation in standard form is the same as the highest degree of any of its terms. A polynomial equation of degree 2 is also called a **quadratic equation.**

A solution of a polynomial equation in one variable is a value of the variable that makes the equation true. The method presented in this section for solving polynomial equations is called the **factoring method.** This method is based on the **zero-factor property.**

> **ZERO-FACTOR PROPERTY**
>
> If a and b are real numbers and $a \cdot b = 0$, then $a = 0$ or $b = 0$.
> This property is true for three or more factors also.

In other words, if the product of two or more real numbers is zero, then at least one number must be zero.

EXAMPLE 1   Solve $(x + 2)(x - 6) = 0$.

Solution: By the zero-factor property, $(x + 2)(x - 6) = 0$ only if $x + 2 = 0$ or $x - 6 = 0$.

$$x + 2 = 0 \quad \text{or} \quad x - 6 = 0 \qquad \text{Apply the zero-factor property.}$$
$$x = -2 \quad \text{or} \qquad x = 6 \qquad \text{Solve each linear equation.}$$

To check, let $x = -2$ and then let $x = 6$ in the original equation.

Let $x = -2$.	Let $x = 6$.
Then $(x + 2)(x - 6) = 0$	Then $(x + 2)(x - 6) = 0$
becomes $(-2 + 2)(-2 - 6) = 0$	becomes $(6 + 2)(6 - 6) = 0$
$(0)(-8) = 0$	$(8)(0) = 0$
$0 = 0$ True.	$0 = 0$ True.

Both -2 and 6 check, and the solution set is $\{-2, 6\}$.

EXAMPLE 2   Solve $2x^2 + 9x - 5 = 0$.

Solution: To use the zero-factor property, one side of the equation must be 0, and the other side must be in factored form.

$$2x^2 + 9x - 5 = 0$$
$$(2x - 1)(x + 5) = 0 \qquad \text{Factor.}$$
$$2x - 1 = 0 \quad \text{or} \quad x + 5 = 0 \qquad \text{Set each factor equal to zero.}$$
$$2x = 1$$
$$x = \frac{1}{2} \quad \text{or} \qquad x = -5 \qquad \text{Solve each linear equation.}$$

The solution set is $\left\{-5, \dfrac{1}{2}\right\}$. To check, let $x = \dfrac{1}{2}$ in the original equation; then let $x = -5$ in the original equation.

TO SOLVE POLYNOMIAL EQUATIONS BY FACTORING

Step 1.   Write the equation in standard form so that one side of the equation is 0.

Step 2.   Factor the polynomial completely.

Step 3.   Set each factor containing a variable equal to 0.

Step 4.   Solve the resulting equations.

Step 5.   Check each solution in the original equation.

Since it is not always possible to factor a polynomial, not all polynomial equations can be solved by factoring. Other methods of solving polynomial equations are presented in Chapter 8.

EXAMPLE 3   Solve $x(2x - 7) = 4$.

Solution: First write the equation in standard form; then factor.

$$x(2x - 7) = 4$$
$$2x^2 - 7x = 4 \qquad \text{Multiply.}$$
$$2x^2 - 7x - 4 = 0 \qquad \text{Write in standard form.}$$
$$(2x + 1)(x - 4) = 0 \qquad \text{Factor.}$$
$$2x + 1 = 0 \quad \text{or} \quad x - 4 = 0 \qquad \text{Set each factor equal to zero.}$$
$$2x = -1 \quad \text{or} \qquad x = 4 \qquad \text{Solve.}$$
$$x = -\frac{1}{2} \quad \text{or} \qquad x = 4$$

The solution set is $\left\{-\dfrac{1}{2}, 4\right\}$. Check both solutions in the original equation.

R E M I N D E R   To apply the zero-factor property, one side of the equation must be 0, and the other side of the equation must be factored. To solve the equation $x(2x - 7) = 4$, for example, you may **not** set each factor equal to 4.

EXAMPLE 4   Solve $3(x^2 + 4) + 5 = -6(x^2 + 2x) + 13$.

Solution: Rewrite the equation so that one side is 0.

$$3(x^2 + 4) + 5 = -6(x^2 + 2x) + 13$$

$$3x^2 + 12 + 5 = -6x^2 - 12x + 13 \qquad \text{Apply the distributive property.}$$

$$9x^2 + 12x + 4 = 0 \qquad \text{Rewrite the equation so that one side is 0.}$$

$$(3x + 2)(3x + 2) = 0 \qquad \text{Factor.}$$

$$3x + 2 = 0 \quad \text{or} \quad 3x + 2 = 0 \qquad \text{Set each factor equal to 0.}$$

$$3x = -2 \quad \text{or} \quad 3x = -2$$

$$x = -\frac{2}{3} \quad \text{or} \quad x = -\frac{2}{3} \qquad \text{Solve each equation.}$$

The solution set is $\left\{-\frac{2}{3}\right\}$. Check by substituting $-\frac{2}{3}$ into the original equation.

If the equation contains fractions, we clear the equation of fractions as a first step.

EXAMPLE 5   Solve $2x^2 = \frac{17}{3}x + 1$.

Solution:

$$2x^2 = \frac{17}{3}x + 1$$

$$3(2x^2) = 3\left(\frac{17}{3}x + 1\right) \qquad \text{Clear the equation of fractions.}$$

$$6x^2 = 17x + 3 \qquad \text{Apply the distributive property.}$$

$$6x^2 - 17x - 3 = 0 \qquad \text{Rewrite the equation in standard form.}$$

$$(6x + 1)(x - 3) = 0 \qquad \text{Factor.}$$

$$6x + 1 = 0 \quad \text{or} \quad x - 3 = 0 \qquad \text{Set each factor equal to zero.}$$

$$6x = -1$$

$$x = -\frac{1}{6} \quad \text{or} \quad x = 3 \qquad \text{Solve each equation.}$$

The solution set is $\left\{-\frac{1}{6}, 3\right\}$.

EXAMPLE 6   Solve $x^3 = 4x$.

Solution:

$$x^3 = 4x$$

$$x^3 - 4x = 0$$ Rewrite the equation so that one side is 0.

$$x(x^2 - 4) = 0$$ Factor out the GCF, x.

$$x(x + 2)(x - 2) = 0$$ Factor the difference of squares.

$x = 0$ or $x + 2 = 0$ or $x - 2 = 0$ Set each factor equal to 0.

$x = 0$ or $x = -2$ or $x = 2$ Solve each equation.

The solution set is $\{-2, 0, 2\}$. Check by substituting into the original equation. ▬▬▬

Notice that the *third*-degree equation of Example 6 yielded *three* solutions.

EXAMPLE 7   Solve $x^3 + 5x^2 = x + 5$.

Solution: First write the equation so that one side is 0.

$$x^3 + 5x^2 - x - 5 = 0$$

$$(x^3 - x) + (5x^2 - 5) = 0$$ Factor by grouping.

$$x(x^2 - 1) + 5(x^2 - 1) = 0$$

$$(x^2 - 1)(x + 5) = 0$$

$$(x + 1)(x - 1)(x + 5) = 0$$ Factor the difference of squares.

$x + 1 = 0$ or $x - 1 = 0$ or $x + 5 = 0$ Set each factor equal to 0.

$x = -1$ or $x = 1$ or $x = -5$ Solve each equation.

The solution set is $\{-5, -1, 1\}$. Check in the original equation.   ▬▬▬

2   Some problems may be modeled by polynomial equations. To solve these problems, we use the same problem-solving steps that were introduced in Section 2.2. When solving these problems, keep in mind that a solution of an equation that models a problem is not always a solution to the problem. For example, a person's weight or the length of a side of a geometric figure is always a positive number. Discard solutions that do not make sense as solutions of the problem.

EXAMPLE 8   An Alpha III model rocket is launched from the ground with an A8–3 engine. Without a parachute the height of the rocket $h(t)$ at time t seconds is approximated by the function

$$h(t) = -16t^2 + 144t$$

Find how long it takes the rocket to return to the ground.

Solution:

1. UNDERSTAND. Read and reread the problem. The function $h(t) = -16t^2 + 144t$ models the height of the rocket. Familiarize yourself with this function by finding a few function values.

> When $t = 1$ second, the height of the rocket is
> $h(1) = -16(1)^2 + 144(1) = 128$ feet.

> When $t = 2$ seconds, the height of the rocket is
> $h(2) = -16(2)^2 + 144(2) = 224$ feet.

Since we have been given the needed function, we proceed to *step 4.*

4. TRANSLATE. To find how long it takes the rocket to hit the ground, we want to know for what value of t is the height $h(t)$ equal to 0. That is, we want to solve $h(t) = 0$.

$$-16t^2 + 144t = 0$$

5. COMPLETE. Solve the quadratic equation by factoring.

$$-16t^2 + 144t = 0$$
$$-16t(t - 9) = 0$$
$$-16t = 0 \text{ or } t - 9 = 0$$
$$t = 0 \quad \text{or} \quad t = 9$$

6. INTERPRET. The height $h(t)$ is 0 feet at time 0 seconds (when the rocket is launched) and at time 9 seconds. To *check*, see that the height of the rocket at 9 seconds, $h(9)$, equals 0.

$$h(9) = -16(9)^2 + 144(9) = -1296 + 1296 = 0$$

State: The rocket returns to the ground 9 seconds after it is launched.   ▬▬▬

Some of the exercises at the end of this section make use of the **Pythagorean theorem.** Before we review this theorem, recall that a **right triangle** is a triangle that contains a 90° angle, or right angle. The **hypotenuse** of a right triangle is the side opposite the right angle and is the longest side of the triangle. The **legs** of a right triangle are the other sides of the triangle.

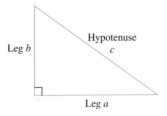

> **PYTHAGOREAN THEOREM**
>
> In a right triangle, the sum of the squares of the lengths of the two legs is equal to the square of the length of the hypotenuse.
>
> $$(\text{leg})^2 + (\text{leg})^2 = (\text{hypotenuse})^2 \quad \text{or} \quad a^2 + b^2 = c^2$$

EXAMPLE 9   When framing a new house, carpenters frequently make use of the Pythagorean theorem in order to determine whether a wall is "square"—that is, whether the wall forms a right angle with the floor. Often, they use a triangle whose sides are three consecutive integers. Find a right triangle whose sides are three consecutive integers.

Solution: **1.** UNDERSTAND.   Read and reread the problem.

2. ASSIGN.   Let $x, x + 1$, and $x + 2$ be three consecutive integers. Since these integers represent lengths of the sides of a right triangle, we have

$$x = \text{one leg}$$
$$x + 1 = \text{other leg}$$
$$x + 2 = \text{hypotenuse (longest side)}$$

3. ILLUSTRATE.   An illustration is to the left.

4. TRANSLATE.   By the Pythagorean theorem, we have

In words: $(\text{leg})^2 + (\text{leg})^2 = (\text{hypotenuse})^2$

Translate: $(x)^2 + (x + 1)^2 = (x + 2)^2$

5. COMPLETE.   Solve the equation.

$$x^2 + (x + 1)^2 = (x + 2)^2$$
$$x^2 + x^2 + 2x + 1 = x^2 + 4x + 4 \qquad \text{Multiply.}$$
$$2x^2 + 2x + 1 = x^2 + 4x + 4$$
$$x^2 - 2x - 3 = 0 \qquad \text{Write in standard form.}$$
$$(x - 3)(x + 1) = 0$$
$$x - 3 = 0 \quad \text{or} \quad x + 1 = 0$$
$$x = 3 \quad \text{or} \qquad x = -1$$

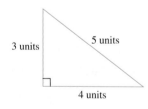

6. INTERPRET.   Discard $x = -1$ since length cannot be negative. If $x = 3$, then $x + 1 = 4$ and $x + 2 = 5$.

Check: To check, see that $(\text{leg})^2 + (\text{leg})^2 = (\text{hypotenuse})^2$

$$3^2 + 4^2 = 5^2$$

or

$$9 + 16 = 25 \qquad \text{True.}$$

State: The lengths of the sides of the right triangle are 3, 4, and 5 units.

A carpenter uses this information, for example, by marking off lengths of 3 and 4 feet on the floor and framing respectively. If the diagonal length between these marks is 5 feet, the wall is "square." If not, adjustments must be made.

Recall that to find the x-intercepts of the graph of a function, let $f(x) = 0$, or $y = 0$, and solve for x. This fact gives us a visual interpretation of the results of this section.

From Example 1, we know that the solutions of the equation $(x + 2)(x - 6) = 0$ are -2 and 6. These solutions give us important information about the related polynomial function $p(x) = (x + 2)(x - 6)$. We know that when x is -2 or when x is 6, the value of $p(x)$ is 0.

$$p(x) = (x + 2)(x - 6)$$
$$p(-2) = (-2 + 2)(-2 - 6) = (0)(-8) = 0$$
$$p(6) = (6 + 2)(6 - 6) = (8)(0) = 0$$

Thus, we know that $(-2, 0)$ and $(6, 0)$ are the x-intercept points of the graph of $p(x)$.

We also know that the graph of $p(x)$ does not cross the x-axis at any other point. For this reason, and the fact that $p(x) = (x + 2)(x - 6) = x^2 - 4x - 12$ has degree 2, we conclude that the graph of p must look something like one of these two graphs:

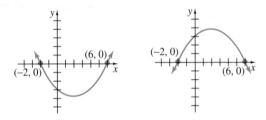

In the following section and in a later chapter, we explore these graphs more fully. For the moment, know that the solutions of a polynomial equation are the x-intercepts of the graph of the related function and that the x-intercepts of the graph of a polynomial function are the solutions of the related polynomial equation. These values are also called **roots,** or **zeros,** of a polynomial function.

EXAMPLE 10   Match each function with its graph:

$$f(x) = (x - 3)(x + 2), \quad g(x) = x(x + 2)(x - 2), \quad h(x) = (x - 2)(x + 2)(x - 1)$$

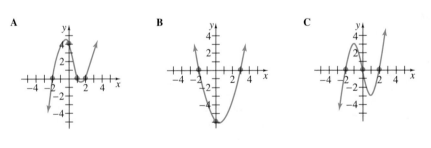

Solution: The graph of the function $f(x) = (x - 3)(x + 2)$ has two x-intercept points, $(3, 0)$ and $(-2, 0)$, because the equation $0 = (x - 3)(x + 2)$ has two solutions, 3 and -2. The graph of $f(x)$ is graph B.

The graph of the function $g(x) = x(x + 2)(x - 2)$ has three x-intercept points, $(0, 0)$, $(-2, 0)$, and $(2, 0)$, because the equation $0 = x(x + 2)(x - 2)$ has three solutions, 0, -2, and 2. The graph of $g(x)$ is graph C.

The graph of the function $h(x) = (x - 2)(x + 2)(x - 1)$ has three intercept points, $(-2, 0)$, $(1, 0)$, and $(2, 0)$, because the equation $0 = (x - 2)(x + 2)(x - 1)$ has three solutions, -2, 1, and 2. The graph of $h(x)$ is graph A.

GRAPHING CALCULATOR EXPLORATIONS

We can use a grapher to approximate real number solutions of any quadratic equation in standard form, whether the associated polynomial is factorable or not. For example, let's solve the quadratic equation $x^2 - 2x - 4 = 0$. The solutions of this equation will be the x-intercepts of the graph of the function $f(x) = x^2 - 2x - 4$. (Recall that to find x-intercepts, we let $f(x) = 0$, or $y = 0$.) When we use a standard window, the graph of this function looks like this:

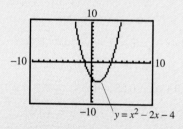

The graph appears to have one x-intercept between -2 and -1 and one between 3 and 4. To find the x-intercept between 3 and 4 to the nearest hundredth, we can use a Root feature, a Zoom feature, which magnifies a portion of the graph around the cursor, or we can redefine our window. If we redefine our window to

$$\text{Xmin} = 2 \qquad \text{Ymin} = -1$$
$$\text{Xmax} = 5 \qquad \text{Ymax} = 1$$
$$\text{Xscl} = 1 \qquad \text{Yscl} = 1$$

(continued)

the resulting screen is

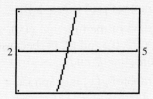

By using the Trace feature, we can now see that one of the intercepts is between 3.21 and 3.25. To approximate to the nearest hundredth, Zoom again or redefine the window to

Xmin = 3.2   Ymin = −0.1
Xmax = 3.3   Ymax = 0.1
Xscl = 1   Yscl = 1

If we use the Trace feature again, we see that, to the nearest hundredth, the x-intercept is 3.23. By repeating this process, we can approximate the other x-intercept to be −1.23.

To check, find $f(3.23)$ and $f(-1.23)$. Both of these values should be close to 0. (They will not be exactly 0 since we approximated these solutions.)

$$f(3.23) = -0.027 \quad \text{and} \quad f(-1.23) = -0.0271$$

Solve each of these quadratic equations by graphing a related function and approximating the x-intercepts.

1. $x^2 + 3x - 2 = 0$ $\{-3.562, 0.562\}$

2. $5x^2 - 7x + 1 = 0$ $\{0.161, 1.239\}$

3. $2.3x^2 - 4.4x - 5.6 = 0$ $\{-0.874, 2.787\}$

4. $0.2x^2 + 6.2x + 2.1 = 0$ $\{-30.658, -0.342\}$

5. $0.09x^2 - 0.13x - 0.08 = 0$ $\{-0.465, 1.910\}$

6. $x^2 + 0.08x - 0.01 = 0$ $\{-0.148, 0.068\}$

MENTAL MATH

Solve each equation for the variable. See Example 1.

1. $(x - 3)(x + 5) = 0$ $\{3, -5\}$

2. $(y + 5)(y + 3) = 0$ $\{-5, -3\}$

3. $(z - 3)(z + 7) = 0$ $\{3, -7\}$

4. $(c - 2)(c - 4) = 0$ $\{2, 4\}$

5. $x(x - 9) = 0$ $\{0, 9\}$

6. $w(w + 7) = 0$ $\{0, -7\}$

EXERCISE SET 5.8

1. $\left\{-3, \frac{4}{3}\right\}$ **2.** $\left\{-\frac{1}{5}, 2\right\}$ **3.** $\left\{\frac{5}{2}, -\frac{3}{4}\right\}$ **4.** $\left\{\frac{4}{3}, \frac{7}{2}\right\}$ **7.** $\left\{\frac{1}{4}, -\frac{2}{3}\right\}$ **8.** $\left\{-2, \frac{7}{3}\right\}$ **11.** $\left\{\frac{3}{5}, -1\right\}$

12. $\left\{-\frac{1}{2}, 2\right\}$ **25.** $\{2, 1, -1\}$ **33.** $\{-3, 5\}$ **35.** $\left\{-\frac{1}{2}, \frac{1}{3}\right\}$ **36.** $\left\{-1, -\frac{5}{8}\right\}$ **39.** $\left\{\frac{4}{5}\right\}$

Solve each equation. See Example 1.

1. $(x + 3)(3x - 4) = 0$ **2.** $(5x + 1)(x - 2) = 0$

3. $3(2x - 5)(4x + 3) = 0$ **4.** $8(3x - 4)(2x - 7) = 0$

Solve each equation. See Examples 2 through 5. **5.** $\{-3, -8\}$

5. $x^2 + 11x + 24 = 0$ **6.** $y^2 - 10y + 24 = 0$ $\{4, 6\}$

7. $12x^2 + 5x - 2 = 0$ **8.** $3y^2 - y - 14 = 0$

9. $z^2 + 9 = 10z$ $\{1, 9\}$ **10.** $n^2 + n = 72$ $\{-9, 8\}$

11. $x(5x + 2) = 3$ **12.** $n(2n - 3) = 2$

13. $x^2 - 6x = x(8 + x)$ $\{0\}$ **14.** $n(3 + n) = n^2 + 4n$ $\{0\}$

15. $\frac{z^2}{6} - \frac{z}{2} - 3 = 0$ $\{6, -3\}$ **16.** $\frac{c^2}{20} - \frac{c}{4} + \frac{1}{5} = 0$ $\{1, 4\}$

17. $\frac{x^2}{2} + \frac{x}{20} = \frac{1}{10}$ $\left\{\frac{2}{5}, -\frac{1}{2}\right\}$ **18.** $\frac{y^2}{30} = \frac{y}{15} + \frac{1}{2}$ $\{-3, 5\}$

19. $\frac{4t^2}{5} = \frac{t}{5} + \frac{3}{10}$ $\left\{\frac{3}{4}, -\frac{1}{2}\right\}$ **20.** $\frac{5x^2}{6} - \frac{7x}{2} + \frac{2}{3} = 0$ $\left\{\frac{1}{5}, 4\right\}$

Solve each equation. See Examples 6 and 7.

21. $(x + 2)(x - 7)(3x - 8) = 0$ $\left\{-2, 7, \frac{8}{3}\right\}$

22. $(4x + 9)(x - 4)(x + 1) = 0$ **22.** $\left\{-\frac{9}{4}, -1, 4\right\}$

23. $y^3 = 9y$ $\{0, 3, -3\}$ **24.** $n^3 = 16n$ $\{-4, 0, 4\}$

25. $x^3 - x = 2x^2 - 2$ **26.** $m^3 = m^2 + 12m$ $\{-3, 0, 4\}$

27. Explain how solving $2(x - 3)(x - 1) = 0$ differs from solving $2x(x - 3)(x - 1) = 0$.

28. Explain why the zero-factor property works for more than two numbers whose product is 0.

Solve each equation. **29.** $\left\{-\frac{7}{2}, 10\right\}$ **30.** $\left\{-4, \frac{1}{5}\right\}$

29. $(2x + 7)(x - 10) = 0$ **30.** $(x + 4)(5x - 1) = 0$

31. $3x(x - 5) = 0$ $\{0, 5\}$ **32.** $4x(2x + 3) = 0$ $\left\{-\frac{3}{2}, 0\right\}$

33. $x^2 - 2x - 15 = 0$ **34.** $x^2 + 6x - 7 = 0$ $\{-7, 1\}$

35. $12x^2 + 2x - 2 = 0$ **36.** $8x^2 + 13x + 5 = 0$

37. $w^2 - 5w = 36$ $\{-4, 9\}$ **38.** $x^2 + 32 = 12x$ $\{4, 8\}$

39. $25x^2 - 40x + 16 = 0$ **40.** $9n^2 + 30n + 25 = 0$ $\left\{-\frac{5}{3}\right\}$

41. $2r^3 + 6r^2 = 20r$ **42.** $-2t^3 = 108t - 30t^2$ $\{0, 6, 9\}$

43. $z(5z - 4)(z + 3) = 0$ **44.** $2r(r + 3)(5r - 4) = 0$

45. $2z(z + 6) = 2z^2 + 12z - 8$ $\{\}$

46. $3c^2 - 8c + 2 = c(3c - 8)$ $\{\}$

41. $\{-5, 0, 2\}$ **43.** $\left\{-3, 0, \frac{4}{5}\right\}$ **44.** $\left\{-3, 0, \frac{4}{5}\right\}$ **47.** $\{-7, 4\}$ **48.** $\left\{-1, -\frac{1}{2}\right\}$ **50.** $\{-6, -3\}$ **57.** $\{-6, 5\}$

58. $\{-5, 3\}$ **59.** $\left\{-\frac{1}{3}, 0, 1\right\}$ **60.** $\left\{-\frac{13}{5}, 0, 2\right\}$ **61.** $\left\{-\frac{1}{3}, 0\right\}$

47. $(x - 1)(x + 4) = 24$ **48.** $(2x - 1)(x + 2) = -3$

49. $\frac{x^2}{4} - \frac{5}{2}x + 6 = 0$ $\{4, 6\}$ **50.** $\frac{x^2}{18} + \frac{x}{2} + 1 = 0$

51. $y^2 + \frac{1}{4} = -y$ $\left\{-\frac{1}{2}\right\}$ **52.** $\frac{x^2}{10} + \frac{5}{2} = x$ $\{5\}$

53. $y^3 + 4y^2 = 9y + 36$ $\{-4, -3, 3\}$

54. $x^3 + 5x^2 = x + 5$ $\{-5, -1, 1\}$

55. $2x^3 = 50x$ $\{-5, 0, 5\}$ **56.** $m^5 = 36m^3$ $\{-6, 0, 6\}$

57. $x^2 + (x + 1)^2 = 61$ **58.** $y^2 + (y + 2)^2 = 34$

59. $m^2(3m - 2) = m$ **60.** $x^2(5x + 3) = 26x$

61. $3x^2 = -x$ **62.** $y^2 = -5y$ $\{-5, 0\}$

63. $x(x - 3) = x^2 + 5x + 7$ $\left\{-\frac{7}{8}\right\}$

64. $z^2 - 4z + 10 = z(z - 5)$ $\{-10\}$

65. $3(t - 8) + 2t = 7 + t$ $\left\{\frac{31}{4}\right\}$

66. $7c - 2(3c + 1) = 5(4 - 2c)$ $\{2\}$

67. $-3(x - 4) + x = 5(3 - x)$ $\{1\}$

68. $-4(a + 1) - 3a = -7(2a - 3)$ $\left\{\frac{25}{7}\right\}$

69. Describe two ways a linear equation differs from a quadratic equation.   Answers will vary.

70. Is the following step correct? Why or why not?

$$x(x - 3) = 5$$
$$x = 5 \text{ or } x - 3 = 5 \quad \text{no}$$

Solve. See Examples 8 and 9. **71.** -11 and -6 or 6 and 11

71. One number exceeds another by five, and their product is 66. Find the numbers.

72. If the sum of two numbers is 4 and their product is $\frac{15}{4}$, find the numbers. $\frac{3}{2}$ and $\frac{5}{2}$

73. An electrician needs to run a cable from the top of a 60-foot tower to a transmitter box located 45 feet away from the base of the tower. Find how long he should cut the cable.   75 ft

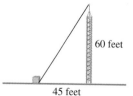

60 feet

45 feet

74. A stereo-system installer needs to run speaker wire along the two diagonals of a rectangular room whose dimensions are 40 feet by 75 feet. Find how much speaker wire she needs.   170 ft

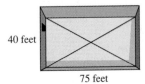

40 feet

75 feet

75. If the cost, $C(x)$, for manufacturing x units of a certain product is given by $C(x) = x^2 - 15x + 50$, find the number of units manufactured at a cost of $9500.   105 units

76. Determine whether any three consecutive integers represent the lengths of the sides of a right triangle.   yes; 3, 4, 5

77. The shorter leg of a right triangle is 3 centimeters less than the other leg. Find the length of the two legs if the hypotenuse is 15 centimeters.

78. Marie Mulroney has a rectangular board 12 inches by 16 inches around which she wants to put a uniform border of shells. If she has enough shells for a border whose area is 128 square inches, determine the width of the border.   2 in.

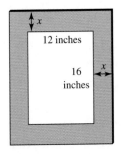

x

12 inches

16 inches

x

79. A gardener has a rose garden that measures 30 feet by 20 feet. He wants to put a uniform border of pine bark around the outside of the garden. Find how wide the border should be if he has enough pine bark to cover 336 square feet.   3 ft

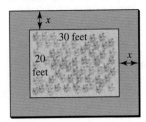

x

30 feet

20 feet

x

77. 12 cm and 9 cm

80. While hovering near the top of Ribbon Falls in Yosemite National Park at 1600 feet, a helicopter pilot accidentally drops his sunglasses. The height $h(t)$ of the sunglasses after t seconds is given by the polynomial function
$$h(t) = -16t^2 + 1600$$
When will the sunglasses hit the ground?   10 sec.

81. After t seconds, the height $h(t)$ of a model rocket launched from the ground into the air is given by the function
$$h(t) = -16t^2 + 80t$$
Find how long it takes the rocket to reach a height of 96 feet.   ascending, 2 sec.; descending, 3 sec.

Match each polynomial function (A–F) with its graph. See Example 10.

82. $f(x) = (x - 2)(x + 5)$ E

83. $g(x) = (x + 1)(x - 6)$ D

84. $h(x) = x(x + 3)(x - 3)$ F

85. $F(x) = (x + 1)(x - 2)(x + 5)$ A

86. $G(x) = 2x^2 + 9x + 4$ B

87. $H(x) = 2x^2 - 7x - 4$ C

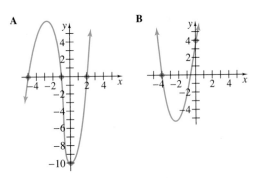

A

B

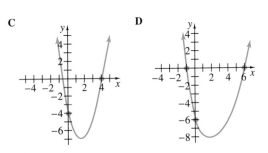

C

D

E

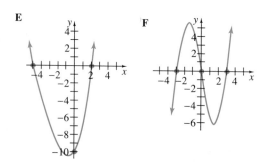

F

92.

93.

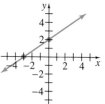

Write a quadratic equation that has the given numbers as solutions.

88. 5, 3   **89.** 6, 7   **90.** $-1, 2$ **91.** $4, -3$

88. Answers will vary. Ex.: $f(x) = x^2 - 8x + 15$
89. Answers will vary. Ex.: $f(x) = x^2 - 13x + 42$

Review Exercises

Write the x- and y-intercept points for each graph and determine whether the graph is the graph of a function. See Sections 3.1 and 3.2.

94.

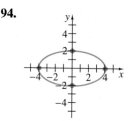

95.

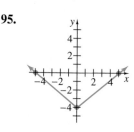

96. Draw a function with intercept points $(-3, 0)$, $(5, 0)$, and $(0, 4)$.   Answers will vary.

97. Draw a function with intercept points $(-7, 0)$,

$\left(-\dfrac{1}{2}, 0\right)$, $(4, 0)$, and $(0, -1)$.   Answers will vary.

90. Answers will vary. Ex.: $f(x) = x^2 - x - 2$ **91.** Answers will vary. Ex.: $f(x) = x^2 - x - 12$
92. $(-3, 0), (0, 2)$; function   **93.** $(-4, 0), (0, 0), (3, 0)$; function   **94.** $(-4, 0), (0, 2), (4, 0), (0, -2)$; not a function
95. $(-5, 0), (5, 0), (0, -4)$; function

5.9 | AN INTRODUCTION TO GRAPHING POLYNOMIAL FUNCTIONS

TAPE IA 5.9

OBJECTIVES

1. Analyze the graph of a polynomial function.
2. Graph quadratic functions.
3. Find the vertex of a parabola by using the vertex formula.
4. Graph cubic functions.

We discussed linear functions of the form $f(x) = mx + b$ in Chapter 3. In this chapter, we have thus far briefly discussed polynomial functions. In this section, we further discuss polynomial functions. As mentioned earlier, some polynomial functions are given special names according to their degree. For example,

$f(x) = 2x - 6$ is called a **linear function; its degree is one.**

$f(x) = 5x^2 - x + 3$ is called a **quadratic function; its degree is two.**

$f(x) = 7x^3 + 3x^2 - 1$ is called a **cubic function; its degree is three.**

$f(x) = -8x^4 - 3x^3 + 2x^2 + 20$ is called a **quartic function; its degree is four.**

All the above functions are also polynomial functions.

Before we practice graphing polynomial functions, let's analyze the graph of a polynomial function.

EXAMPLE 1 Given the graph of the function $g(x)$:

   a. Find the domain and the range of the function.
   b. List the x- and y-intercept points.
   c. Find the coordinates of the point with the greatest y-value.
   d. Find the coordinates of the point with the least y-value.
   e. List the x-values whose y-values are equal to 0.
   f. List the x-values whose y-values are greater than 0.
   g. Find the solutions of $g(x) = 0$.

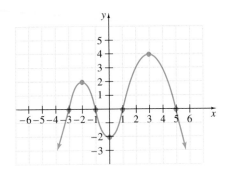

Solution: **a.** The domain is the set of all real numbers, or in interval notation, $(-\infty, \infty)$. The range is $(-\infty, 4]$.

   b. The x-intercept points are $(-3, 0)$, $(-1, 0)$, $(1, 0)$, and $(5, 0)$. The y-intercept point is $(0, -2)$.

   c. The point with the greatest y-value corresponds to the "highest" point. This is the point with coordinates $(3, 4)$. (This means that for all real number values for x, the greatest y-value, or $f(x)$ value, is 4.)

   d. The point with the least y-value corresponds to the "lowest" point. This graph contains no "lowest" point, so there is no point with the least y-value.

   e. The y-values are equal to 0 when the graph lies on the x-axis. The x-values when this occurs are the x-intercepts $-3, -1, 1$, and 5. Notice that this tells us that $g(-3) = 0, g(-1) = 0, g(1) = 0$, and $g(5) = 0$.

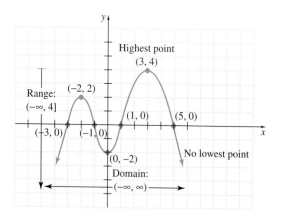

f. The y-values are greater than 0 when the graph lies above the x-axis. The x-values when this occurs are between $x = -3$ and $x = -1$ and between $x = 1$ and $x = 5$.

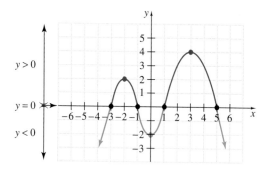

g. The solutions of $g(x) = 0$ are the x-intercepts of the graph. The x-intercept points are $(-3, 0), (-1, 0), (1, 0)$, and $(5, 0)$. This means that when $x = -3, -1, 1$, or 5, y or $g(x) = 0$. The solutions are $-3, -1, 1$, and 5.

The graph of any polynomial function (linear, quadratic, cubic, and so on) can be sketched by plotting a sufficient number of ordered pairs that satisfy the function and connecting them to form a smooth curve. The graph of all polynomial functions will pass the vertical line test since they are graphs of functions. To graph a linear function defined by $f(x) = mx + b$, recall that two ordered pair solutions will suffice since its graph is a line. To graph other polynomial functions, we need to find and plot more ordered pair solutions to ensure a reasonable picture of its graph.

Since we know how to graph linear functions (see Chapter 3), we will now graph quadratic functions and discuss special characteristics of their graphs.

> **QUADRATIC FUNCTION**
>
> A quadratic function is a function that can be written in the form
>
> $$f(x) = ax^2 + bx + c$$
>
> where a, b, and c are real numbers and $a \neq 0$.

We know that an equation of the form $f(x) = ax^2 + bx + c$ may be written as $y = ax^2 + bx + c$. Thus, both $f(x) = ax^2 + bx + c$ and $y = ax^2 + bx + c$ define quadratic functions as long as a is not 0.

Recall the graph of the quadratic function defined by $f(x) = x^2$ by plotting points. Choose $-3, -2, -1, 0, 1, 2,$ and 3 as x-values, and find corresponding $f(x)$ or y-values.

x	$y = f(x)$
-3	9
-2	4
-1	1
0	0
1	1
2	4
3	9

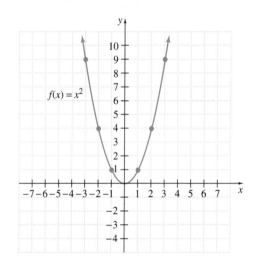

Notice that the graph passes the vertical line test, as it should since it is a function. Recall that this curve is called a **parabola.** The highest point on a parabola that opens downward or the lowest point on a parabola that opens upward is called the **vertex** of the parabola. The vertex of this parabola is $(0, 0)$, the lowest point on the graph. If we fold the graph along the y-axis, we can see that the two sides of the graph coincide. This means that this curve is symmetric about the y-axis, and the y-axis, or the line $x = 0$, is called the **axis of symmetry.** The graph of every quadratic function is a parabola and has an axis of symmetry: the vertical line that passes through the vertex of the parabola.

EXAMPLE 2   Graph the quadratic function $f(x) = -x^2 + 2x - 3$ by plotting points.

Solution: To graph, choose values for x and find corresponding $f(x)$ or y-values.

x	$y = f(x)$
-2	-11
-1	-6
0	-3
1	-2
2	-3
3	-6

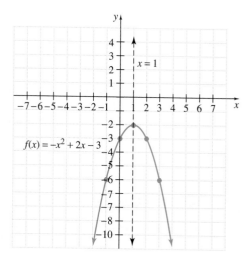

$f(x) = -x^2 + 2x - 3$

The vertex of this parabola is $(1, -2)$, the highest point on the graph. The vertical line $x = 1$ is the axis of symmetry. Recall that to find the x-intercepts of a graph, let $f(x)$ or $y = 0$. Since this graph has no x-intercepts, it means that $0 = -x^2 + 2x - 3$ has no real number solutions.

Notice that the parabola $f(x) = -x^2 + 2x - 3$ opens downward, whereas $f(x) = x^2$ opens upward. When the equation of a quadratic function is written in the form $f(x) = ax^2 + bx + c$, recall that the coefficient of the squared variable a, determines whether the parabola opens downward or upward. If $a > 0$, the parabola opens upward, and if $a < 0$, the parabola opens downward.

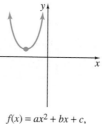

$f(x) = ax^2 + bx + c,$
$a > 0$, opens upward

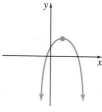

$f(x) = ax^2 + bx + c,$
$a < 0$, opens downward

In both $f(x) = x^2$ and $f(x) = -x^2 + 2x - 3$, the vertex happens to be one of the points we chose to plot. Since this is not always the case, and since plotting the vertex allows us to draw the graph quickly, we need a consistent method for finding the vertex. One method is to use the following formula, which we shall derive in Chapter 8.

> **VERTEX FORMULA**
>
> The graph of $f(x) = ax^2 + bx + c, a \neq 0$, is a parabola with vertex
> $$\left(\frac{-b}{2a}, f\left(\frac{-b}{2a} \right) \right)$$

We can also find the x- and y-intercepts of a parabola to aid in graphing. Recall that x-intercepts of the graph of any equation may be found by letting $y = 0$ in the equation and solving for x. Also, y-intercepts may be found by letting $x = 0$ in the equation and solving for y or $f(x)$.

EXAMPLE 3　Graph $f(x) = x^2 + 2x - 3$. Find the vertex and any intercepts.

Solution:　To find the vertex, use the vertex formula. For the function $f(x) = x^2 + 2x - 3$, $a = 1$ and $b = 2$. Thus,

$$x = \frac{-b}{2a} = \frac{-2}{2(1)} = -1$$

Next find $f(-1)$.

$$f(-1) = (-1)^2 + 2(-1) - 3$$
$$= 1 - 2 - 3$$
$$= -4$$

The vertex is $(-1, -4)$, and since $a = 1$ is greater than 0, this parabola opens upward. This parabola will have two x-intercepts because its vertex lies below the x-axis and it opens upward. To find the x-intercepts, let y or $f(x) = 0$ and solve for x.

$$f(x) = x^2 + 2x - 3$$
$$0 = x^2 + 2x - 3 \qquad \text{Let } f(x) = 0.$$
$$0 = (x + 3)(x - 1) \qquad \text{Factor.}$$
$$x + 3 = 0 \quad \text{or} \quad x - 1 = 0 \qquad \text{Set each factor equal to 0.}$$
$$x = -3 \quad \text{or} \qquad x = 1 \qquad \text{Solve.}$$

The x-intercepts are -3 and 1.

To find the y-intercept, let $x = 0$.

$$f(x) = x^2 + 2x - 3$$
$$f(0) = 0^2 + 2(0) - 3$$
$$f(0) = -3$$

The y-intercept is -3.

Now plot these points and connect them with a smooth curve.

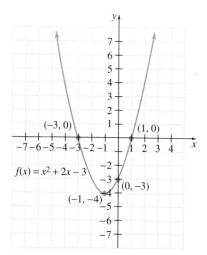

$f(x) = x^2 + 2x - 3$

REMINDER   Not all graphs of parabolas have x-intercepts. To see this, first plot the vertex of the parabola and decide whether the parabola opens upward or downward. Then use this information to decide whether the graph of the parabola has x-intercepts.

EXAMPLE 4   Graph $f(x) = 3x^2 - 12x + 13$. Find the vertex and any intercepts.

Solution: To find the vertex, use the vertex formula. For the function $y = 3x^2 - 12x + 13$, $a = 3$ and $b = -12$. Thus,

$$x = \frac{-b}{2a} = \frac{-(-12)}{2(3)} = \frac{12}{6} = 2$$

Next find $f(2)$.

$$f(2) = 3(2)^2 - 12(2) + 13$$
$$= 3(4) - 24 + 13$$
$$= 1$$

The vertex is $(2, 1)$. Also, this parabola opens upward, since $a = 3$, which is greater than 0. Notice that this parabola has no x-intercepts: Its vertex lies above the x-axis, and it opens upward.

To find the y-intercept, let $x = 0$.

$$f(0) = 3(0)^2 - 12(0) + 13$$
$$= 0 - 0 + 13$$
$$= 13$$

The y-intercept is 13. Use this information along with symmetry of a parabola to sketch the graph of $f(x) = 3x^2 - 12x + 13$.

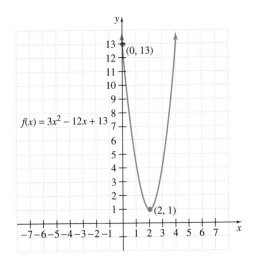

$f(x) = 3x^2 - 12x + 13$

In Section 8.4, we study the graphing of quadratic functions further.

4 To sketch the graph of a cubic function, we again plot points and then connect the points with a smooth curve.

When graphing cubic functions, keep in mind their general shape from Section 5.3.

**Graph of a Polynomial Function
of Degree 3**

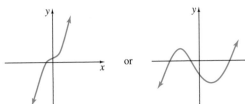

Coefficient of x^3
is a positive number.

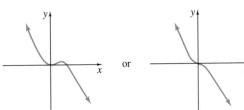

Coefficient of x^3
is a negative number.

EXAMPLE 5 Graph $f(x) = x^3 - 4x$. Find any intercepts.

Solution: To find x-intercepts, let y or $f(x) = 0$ and solve for x.

$$f(x) = x^3 - 4x$$
$$0 = x^3 - 4x \qquad \text{Let } f(x) = 0$$
$$0 = x(x^2 - 4)$$
$$0 = x(x + 2)(x - 2) \qquad \text{Factor.}$$
$$x = 0 \quad \text{or} \quad x + 2 = 0 \quad \text{or} \quad x - 2 = 0 \qquad \text{Set each factor equal to 0.}$$
$$x = 0 \quad \text{or} \qquad x = -2 \quad \text{or} \qquad x = 2 \qquad \text{Solve.}$$

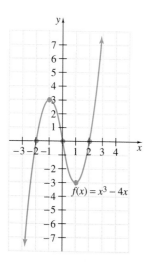

This graph has three x-intercepts. They are 0, -2 and 2.
To find the y-intercept, let $x = 0$.

$$f(0) = 0^3 - 4(0) = 0$$

Next select some x-values and find their corresponding $f(x)$ or y-values.

$$f(x) = x^3 - 4x$$
$$f(-3) = (-3)^3 - 4(-3) = -27 + 12 = -15$$
$$f(-1) = (-1)^3 - 4(-1) = -1 + 4 = 3$$
$$f(1) = 1^3 - 4(1) = 1 - 4 = -3$$
$$f(3) = 3^3 - 4(3) = 27 - 12 = 15$$

x	$f(x)$
-3	-15
-1	3
1	-3
3	15

Plot the intercepts and points and connect them with a smooth curve.

REMINDER   When a graph has an x-intercept of 0, notice that the y-intercept will also be 0.

REMINDER   If unsure about the graph of a function, plot more points.

EXAMPLE 6   Graph $f(x) = -x^3$. Find any intercepts.

Solution: To find x-intercepts, let y or $f(x) = 0$ and solve for x.

$$f(x) = -x^3$$
$$0 = -x^3$$
$$0 = x$$

The only x-intercept is 0. This means that the y-intercept is 0 also.
Next choose some x-values and find corresponding y-values.

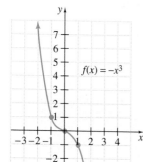

$$f(x) = -x^3$$
$$f(-2) = -(-2)^3 = 8$$
$$f(-1) = -(-1)^3 = 1$$
$$f(1) = -(1)^3 = -1$$
$$f(2) = -2^3 = -8$$

x	$f(x)$
-2	8
-1	1
1	-1
2	-8

Plot the points and sketch the graph of $f(x) = -x^3$.

MENTAL MATH

State whether the graph of each quadratic function, a parabola, opens upward or downward.

1. $f(x) = 2x^2 + 7x + 10$ upward

2. $f(x) = -3x^2 - 5x$ downward

3. $f(x) = -x^2 + 5$ downward

4. $f(x) = x^2 + 3x + 7$ upward

3a. domain, $(-\infty, \infty)$; range, $[-4, \infty)$ **b.** x-intercept points, $(-3, 0)$, $(1, 0)$; y-intercept point, $(0, -3)$ **c.** There is no such point. **d.** $(-1, -4)$ **e.** $-3, 1$ **f.** $x < -3$ or $x > 1$ **g.** $\{-3, 1\}$

EXERCISE SET 5.9

For the graph of each function $f(x)$, answer the following. See Example 1.

a. *Find the domain and the range of the function.*

b. *List the x- and y-intercept points.*

c. *Find the coordinates of the point with the greatest y-value.*

d. *Find the coordinates of the point with the least y-value.*

e. *List the x-values whose y-values are equal to 0.*

f. *List the x-values whose y-values are greater than 0.*

g. *Find the solution set of $f(x) = 0$.*

2.

a. domain, (∞, ∞); range, $[1, \infty)$
b. no x-intercept points; y-intercept point, $(0, 4)$
c. There is no such point.
d. $(-3, 1)$
e. None
f. all x-values in the domain
g. $\{\}$

3. The graph in Example 3 of this section.

4. The graph in Example 4 of this section.

5. The graph in Example 5 of this section.

6. The graph in Example 6 of this section.

Graph each quadratic function by plotting points. See Example 2. For Exercises 7–12, see Appendix D.

7. $f(x) = 2x^2$

8. $f(x) = -3x^2$

9. $f(x) = x^2 + 1$

10. $f(x) = x^2 - 2$

11. $f(x) = -x^2$

12. $f(x) = \dfrac{1}{2}x^2$

1.

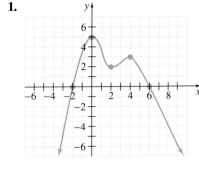

a. domain, $(-\infty, \infty)$; range, $(-\infty, 5]$
b. x-intercept points, $(-2, 0)$, $(6, 0)$; y-intercept point, $(0, 5)$ **c.** $(0, 5)$ **d.** There is no such point.
e. $-2, 6$
f. between $x = -2$ and $x = 6$
g. $\{-2, 6\}$

4a. domain, $(-\infty, \infty)$; range, $[1, \infty)$ **b.** no x-intercept points; y-intercept point, $(0, 13)$ **c.** There is no such point. **d.** $(2, 1)$ **e.** none **f.** all x-values in the domain **g.** $\{\}$ **5a.** domain, $(-\infty, \infty)$; range, $(-\infty, \infty)$ **b.** x-intercept points, $(-2, 0)$, $(0, 0)$, $(2, 0)$; y-intercept point, $(0, 0)$ **c.** There is no such point. **d.** There is no such point. **e.** $-2, 0, 2$ **f.** between $x = -2$ and 0; $x > 2$ **g.** $\{-2, 0, 2\}$ **6a.** domain $(-\infty, \infty)$; range, $(-\infty, \infty)$ **b.** x-intercept point, $(0, 0)$; y-intercept point, $(0, 0)$ **c.** There is no such point. **d.** There is no such point. **e.** 0 **f.** $x < 0$ **g.** $\{0\}$

Find the vertex of the graph of each function. See Examples 3 and 4.

13. $f(x) = x^2 + 8x + 7$ $(-4, -9)$

14. $f(x) = x^2 + 6x + 5$ $(-3, -4)$

15. $f(x) = 3x^2 + 6x + 4$ $(-1, 1)$

16. $f(x) = -2x^2 + 2x + 1$ $\left(\dfrac{1}{2}, \dfrac{3}{2}\right)$

17. $f(x) = -x^2 + 10x + 5$ $(5, 30)$

18. $f(x) = -x^2 - 8x + 2$ $(-4, 18)$

19. If the vertex of a parabola lies below the x-axis and the parabola opens upward, how many x-intercepts will the graph have?   2

20. If the vertex of a parabola lies below the x-axis and the parabola opens downward, how many x-intercepts will the graph have?   0

21. If the vertex of a parabola lies above the x-axis and the parabola opens upward, how many x-intercepts will the graph have?   0

22. If the vertex of a parabola lies above the x-axis and the parabola opens downward, how many x-intercepts will the graph have?   2

23. If the vertex of a parabola is the origin, how many x-intercepts and how many y-intercepts will the graph have?   1 x-intercept; 1 y-intercept

For Exercises 24–33, see Appendix D.
Graph each quadratic function. Find and label the vertex and intercepts. See Examples 3 and 4.

24. $f(x) = x^2 + 8x + 7$

25. $f(x) = x^2 + 6x + 5$

26. $f(x) = x^2 - 2x - 24$

27. $f(x) = x^2 - 12x + 35$

28. $f(x) = 2x^2 - 6x$

29. $f(x) = -3x^2 + 6x$

Graph each cubic function. Find any intercepts. See Examples 5 and 6.

30. $f(x) = 4x^3 - 9x$

31. $f(x) = 2x^3 - 5x^2 - 3x$

32. $f(x) = x^3 + 3x^2 - x - 3$

33. $f(x) = x^3 + x^2 - 4x - 4$

34. Can the graph of a function ever have more than one y-intercept point? Why?   No. Answers will vary.

35. In general, is there a limit to the number of x-intercepts for the graph of a function?   no

For Exercises 36–67, see Appendix D.
Graph each function. Find intercepts. If the function is a quadratic function, find the vertex.

36. $f(x) = x^2 + 4x - 5$

37. $f(x) = x^2 + 2x - 3$

38. $f(x) = (x - 2)(x + 2)(x + 1)$

39. $f(x) = x^3 - 4x^2 + 3x$

40. $f(x) = x^2 + 1$

41. $f(x) = x^2 + 4$ **42.** $f(x) = -5x^2 + 5x$

43. $f(x) = 3x^2 - 12x$ **44.** $f(x) = x^3 - 9x$

45. $f(x) = x^3 + x^2 - 12x$

46. $f(x) = -x^3 - x^2 + 2x$

47. $f(x) = x^3 + x^2 - 9x - 9$

48. $f(x) = x^2 - 4x + 4$ **49.** $f(x) = x^2 - 2x + 1$

50. $f(x) = -x^3 + x$ **51.** $f(x) = x^2 + 6x$

52. $f(x) = 2x^2 - x - 3$

53. $f(x) = (x + 2)(x - 2)$

54. $f(x) = -x^3 + 3x^2 + x - 3$

55. $f(x) = -x^3 + 25x$ **56.** $f(x) = x^2 - 10x + 26$

57. $f(x) = x^2 + 2x + 4$

58. $f(x) = x(x - 4)(x + 2)$

59. $f(x) = 3x(x - 3)(x + 5)$

60. $g(x) = x(x - 2)(x + 3)(x + 5)$

61. $h(x) = (x - 4)(x - 2)(2x + 1)(x + 3)$

Use a grapher to verify the graph in each exercise.

62. Exercise 36   **63.** Exercise 37

64. Exercise 54   **65.** Exercise 55

Use a grapher to approximate all x-intercepts to the nearest tenth.

66. $F(x) = -x^4 + 2.1x^2 + 5.6$ $x = -1.9, x = 1.9$

67. $G(x) = x^4 - 6.2x^2 - 6.2$ $x = -2.7, x = 2.7$

Review Exercises   **72.** $\dfrac{m^3}{2n^{10}}$

Simplify each fraction. See Sections 5.1 and 5.2.

68. $-\dfrac{8}{10}$ $-\dfrac{4}{5}$ **69.** $-\dfrac{45}{100}$ $-\dfrac{9}{20}$ **70.** $\dfrac{x^7y^{10}}{x^3y^{15}}$ $\dfrac{x^4}{y^5}$

71. $\dfrac{a^{14}b^2}{ab^4}$ $\dfrac{a^{13}}{b^2}$ **72.** $\dfrac{7n^{-9}m^{-2}}{14nm^{-5}}$ **73.** $\dfrac{20x^{-3}y^5}{25y^{-2}x}$ $\dfrac{4y^7}{5x^4}$

Group Activity

Finding the Largest Area

Materials:
• Calculator, graphing calculator (optional)

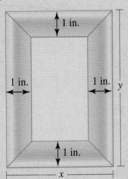

A picture framer has a piece of wood that measures 1 inch wide × 50 inches long with which she would like to make a picture frame with the largest possible interior area. Complete the following activity to help her determine the dimensions of the frame that she should use to achieve her goal.

1. Use the situation given in the figure to write an equation in x and y for the *outer* perimeter of the frame. (Remember that the perimeter will equal 50 inches.)

2. Complete the following table, letting $A(x)$ be the interior area. From the table, what appears to be the largest interior area? Which exterior dimensions of the frame provide this area?*

3. Use the table and write $A(x)$ as a function of x alone. (*Hint*: Use the equation from Question 1.)

4. Graph the function $A(x)$. Locate and label the point from the table that represents the maximum interior area. Describe the location of the point in relation to the rest of the graph. See Appendix D for Group Activity answers and suggestions.

| | | Frame's Interior Dimensions | | |
| | | Interior Width | Interior Height | $A(x)$ Interior Area |
x	y			
2.0				
2.5				
3.0				
3.5				
4.0				
4.5				
5.0				
5.5				
6.0				
6.5				
7.0				
7.5				
8.0				
8.5				
9.0				
9.5				
10.0				
10.5				
11.0				
11.5				
12.0				
12.5				
13.0				
13.5				
14.0				
14.5				
15.0				

* If using a graphing calculator, find the first few entries in the table to write $A(x)$. Then use the features Table Set and Δ Table to complete the table.

CHAPTER 5   HIGHLIGHTS

DEFINITIONS AND CONCEPTS	EXAMPLES
SECTION 5.1 EXPONENTS AND SCIENTIFIC NOTATION	

Product rule: $a^m \cdot a^n = a^{m+n}$

Zero exponent: $a^0 = 1, a \neq 0$

Quotient rule: $\dfrac{a^m}{a^n} = a^{m-n}$

Negative exponent: $a^{-n} = \dfrac{1}{a^n}$

$x^2 \cdot x^3 = x^5$

$7^0 = 1, \ (-10)^0 = 1$

$\dfrac{y^{10}}{y^4} = y^{10-4} = y^6$

$3^{-2} = \dfrac{1}{3^2} = \dfrac{1}{9}, \ \dfrac{x^{-5}}{x^{-7}} = x^{-5-(-7)} = x^2$

A positive number is written in **scientific notation** if it is written as the product of a number a, where $1 \leq a < 10$, and an integer power of 10: $a \times 10^r$.

Numbers written in scientific notation:
$568,000 = 5.68 \times 10^5$
$0.0002117 = 2.117 \times 10^{-4}$

| **SECTION 5.2   MORE WORK WITH EXPONENTS AND SCIENTIFIC NOTATION** | |

Power rules:
$$(a^m)^n = a^{m \cdot n}$$
$$(ab)^m = a^m b^m$$
$$\left(\frac{a}{b}\right)^n = \frac{a^n}{b^n}$$

$$(7^8)^2 = 7^{16}$$
$$(2y)^3 = 2^3 y^3 = 8y^3$$
$$\left(\frac{5x^{-3}}{x^2}\right)^{-2} = \frac{5^{-2}x^6}{x^{-4}}$$
$$= 5^{-2} \cdot x^{6-(-4)}$$
$$= \frac{x^{10}}{5^2}, \ \text{or} \ \frac{x^{10}}{25}$$

| **SECTION 5.3   POLYNOMIALS AND POLYNOMIAL FUNCTIONS** | |

A **polynomial** is a finite sum of terms in which all variables have exponents raised to nonnegative integer powers and no variables appear in the denominator.

Polynomials

$1.3x^2$ (monomial)

$-\dfrac{1}{3}y + 5$ (binomial)

$6z^2 - 5z + 7$ (trinomial)

A function P is a **polynomial function** if $P(x)$ is a polynomial.

For the polynomial function
$P(x) = -x^2 + 6x - 12$, find $P(-2)$.
$P(-2) = -(-2)^2 + 6(-2) - 12 = -28$.

To add polynomials, combine all like terms.

Add:
$$(3y^2x - 2yx + 11) + (-5y^2x - 7)$$
$$= -2y^2x - 2yx + 4$$

To subtract polynomials, change the signs of the terms of the polynomial being subtracted, then add.

Subtract:
$$(-2z^3 - z + 1) - (3z^3 + z - 6)$$
$$= -2z^3 - z + 1 - 3z^3 - z + 6$$
$$= -5z^3 - 2z + 7$$

DEFINITIONS AND CONCEPTS	EXAMPLES

SECTION 5.4 MULTIPLYING POLYNOMIALS

To multiply two polynomials, use the distributive property and multiply each term of one polynomial by each term of the other polynomial; then combine like terms.

Multiply:

$$(x^2 - 2x)(3x^2 - 5x + 1)$$
$$= 3x^4 - 5x^3 + x^2 - 6x^3 + 10x^2 - 2x$$
$$= 3x^4 - 11x^3 + 11x^2 - 2x$$

Special products:

$$(a + b)^2 = a^2 + 2ab + b^2$$
$$(a - b)^2 = a^2 - 2ab + b^2$$
$$(a + b)(a - b) = a^2 - b^2$$

$$(3m + 2n)^2 = 9m^2 + 12mn + 4n^2$$
$$(z^2 - 5)^2 = z^4 - 10z^2 + 25$$
$$(7y + 1)(7y - 1) = 49y^2 - 1$$

The FOIL method may be used when multiplying two binomials.

Multiply:

$$(x^2 + 5)(2x^2 - 9)$$
$$\quad \text{F} \qquad \text{O} \qquad \text{I} \qquad \text{L}$$
$$= x^2(2x^2) + x^2(-9) + 5(2x^2) + 5(-9)$$
$$= 2x^4 - 9x^2 + 10x^2 - 45$$

SECTION 5.5 THE GREATEST COMMON FACTOR AND FACTORING BY GROUPING

The greatest common factor (GCF) of the terms of a polynomial is the product of the GCF of the numerical coefficients and the GCF of the variable factors.

Factor $14xy^3 - 2xy^2 = 2 \cdot 7 \cdot x \cdot y^3 - 2 \cdot x \cdot y^2$.
The GCF is $2 \cdot x \cdot y^2$, or $2xy^2$.

$$14xy^3 - 2xy^2 = 2xy^2(7y - 1)$$

To factor a polynomial by grouping, group the terms so that each group has a common factor. Factor out these common factors. Then see if the new groups have a common factor.

Factor: $x^4y - 5x^3 + 2xy - 10$.

$$= x^3(xy - 5) + 2(xy - 5)$$
$$= (xy - 5)(x^3 + 2)$$

SECTION 5.6 FACTORING TRINOMIALS

To factor $ax^2 + bx + c$,

Step 1. Write all pairs of factors of ax^2.

Step 2. Write all pairs of factors of c.

Step 3. Try combinations of these factors until the middle term bx is found.

Factor $28x^2 - 27x - 10$.

Factors of $28x^2$: $28x$ and x, $2x$ and $14x$, $4x$ and $7x$.

Factors of -10: -2 and 5, 2 and -5, -10 and 1, 10 and -1.

$$28x^2 - 27x - 10 = (7x - 2)(4x + 5)$$

SECTION 5.7 FACTORING BY SPECIAL PRODUCTS AND FACTORING STRATEGIES

Perfect square trinomial:

$$a^2 + 2ab + b^2 = (a + b)^2$$
$$a^2 - 2ab + b^2 = (a - b)^2$$

Factor:

$$25x^2 + 30x + 9 = (5x + 3)^2$$
$$49z^2 - 28z + 4 = (7z - 2)^2$$

(continued)

DEFINITIONS AND CONCEPTS	EXAMPLES

SECTION 5.7 FACTORING BY SPECIAL PRODUCTS AND FACTORING STRATEGIES

Difference of two squares

$$a^2 - b^2 = (a + b)(a - b)$$

$$36x^2 - y^2 = (6x + y)(6x - y)$$

Sum and difference of two cubes

$$a^3 + b^3 = (a + b)(a^2 - ab + b^2)$$
$$a^3 - b^3 = (a - b)(a^2 + ab + b^2)$$

$$8y^3 + 1 = (2y + 1)(4y^2 - 2y + 1)$$
$$27p^3 - 64q^3 = (3p - 4q)(9p^2 + 12pq + 16q^2)$$

To factor a polynomial,

Step 1. Factor out the GCF.

Step 2. If the polynomial is a binomial, see if it is a difference of two squares or a sum or difference of two cubes. If it is a trinomial, see if it is a perfect square trinomial. If not, try factoring by methods of Section 5.6. If it is a polynomial with 4 or more terms, try factoring by grouping.

Step 3. See if any factors can be factored further.

Factor: $10x^4y + 5x^2y - 15y$.

$$= 5y(2x^4 + x^2 - 3)$$
$$= 5y(2x^2 + 3)(x^2 - 1)$$
$$= 5y(2x^2 + 3)(x + 1)(x - 1)$$

SECTION 5.8 SOLVING POLYNOMIAL EQUATIONS BY FACTORING AND PROBLEM SOLVING

To solve polynomial equations by factoring:

Step 1. Write the equation so that one side is 0.

Step 2. Factor the polynomial completely.

Step 3. Set each factor equal to 0.

Step 4. Solve the resulting equations.

Step 5. Check each solution.

Solve:

$$2x^3 - 5x^2 = 3x$$
$$2x^3 - 5x^2 - 3x = 0$$
$$x(2x + 1)(x - 3) = 0$$
$$x = 0 \quad \text{or} \quad 2x + 1 = 0 \quad \text{or} \quad x - 3 = 0$$
$$x = 0 \quad \text{or} \quad x = -\frac{1}{2} \quad \text{or} \quad x = 3$$

SECTION 5.9 AN INTRODUCTION TO GRAPHING POLYNOMIAL FUNCTIONS

To graph a polynomial function, find and plot x- and y-intercepts and a sufficient number of ordered pair solutions. Then connect the plotted points with a smooth curve.

Graph $f(x) = x^3 + 2x^2 - 3x$.

$$0 = x^3 + 2x^2 - 3x$$
$$0 = x(x - 1)(x + 3)$$
$$x = 0 \text{ or } x = 1 \text{ or } x = -3$$

The x-intercept points are $(0, 0)$, $(1, 0)$, and $(-3, 0)$.

$$f(0) = 0^3 + 2 \cdot 0^2 - 3 \cdot 0 = 0.$$

The y-intercept point is $(0, 0)$.

(continued)

DEFINITIONS AND CONCEPTS	EXAMPLES

SECTION 5.9 AN INTRODUCTION TO GRAPHING POLYNOMIAL FUNCTIONS

x	$f(x)$
-4	-20
-2	6
-1	4
$\dfrac{1}{2}$	$-\dfrac{7}{8}$
2	10

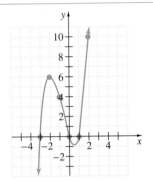

A quadratic function is a function that can be written in the form

$$f(x) = ax^2 + bx + c, a \neq 0$$

The graph of this quadratic function is a parabola with vertex $\left(\dfrac{-b}{2a}, f\left(\dfrac{-b}{2a}\right)\right)$.

Find the vertex of the graph of the quadratic function

$$f(x) = 2x^2 - 8x + 1$$

Here $a = 2$ and $b = -8$.

$$\frac{-b}{2a} = \frac{-(-8)}{2 \cdot 2} = 2$$

$$f(2) = 2 \cdot 2^2 - 8 \cdot (2) + 1 = -7$$

The vertex has coordinates $(2, -7)$.

CHAPTER 5 REVIEW

(5.1) *Evaluate.*

1. $(-2)^2$ 4   **2.** $(-3)^4$ 81   **3.** -2^2 -4

4. -3^4 -81 **5.** 8^0 1   **6.** -9^0 -1

7. -4^{-2} $-\dfrac{1}{16}$ **8.** $(-4)^{-2}$ $\dfrac{1}{16}$

Simplify each expression. Use only positive exponents.

9. $-xy^2 \cdot y^3 \cdot xy^2z$ $-x^2y^7z$

10. $(-4xy)(-3xy^2b)$ $12x^2y^3b$

11. $a^{-14} \cdot a^5$ $\dfrac{1}{a^9}$ **12.** $\dfrac{a^{16}}{a^{17}}$ $\dfrac{1}{a}$ **13.** $\dfrac{x^{-7}}{x^4}$ $\dfrac{1}{x^{11}}$

14. $\dfrac{9a(a^{-3})}{18a^{15}}$ $\dfrac{1}{2a^{17}}$ **15.** $\dfrac{y^{6p-3}}{y^{6p+2}}$ $\dfrac{1}{y^5}$

Write in scientific notation.

16. $36{,}890{,}000$ 3.689×10^7 **17.** -0.000362 -3.62×10^{-4}

18. 0.000001678

Write each number without exponents.

18. 1.678×10^{-6} **19.** 4.1×10^5 $410{,}000$

(5.2) *Simplify. Use only positive exponents.*

20. $(8^5)^3$ 8^{15} **21.** $\left(\dfrac{a}{4}\right)^2$ $\dfrac{a^2}{16}$ **22.** $(3x)^3$ **23.** $(-4x)^{-2}$

24. $\left(\dfrac{6x}{5}\right)^2$ **25.** $(8^6)^{-3}$ **26.** $\left(\dfrac{4}{3}\right)^{-2}$ $\dfrac{9}{16}$ **27.** $(-2x^3)^{-3}$

28. $\left(\dfrac{8p^6}{4p^4}\right)^{-2}$ $\dfrac{1}{4p^4}$ **29.** $(-3x^{-2}y^2)^3$ **30.** $\left(\dfrac{x^{-5}y^{-3}}{z^3}\right)^{-5}$

31. $\dfrac{4^{-1}x^3yz}{x^{-2}yx^4}$ $\dfrac{xz}{4}$ **32.** $(5xyz)^{-4}(x^{-2})^{-3}$ **33.** $\dfrac{2(3yz)^{-3}}{y^{-3}}$

Simplify each expression.

34. $x^{4a}(3x^{5a})^3$ $27x^{19a}$ **35.** $\dfrac{4y^{3x-3}}{2y^{2x+4}}$ $2y^{x-7}$

22. $27x^3$ **23.** $\dfrac{1}{16x^2}$ **24.** $\dfrac{36x^2}{25}$ **25.** $\dfrac{1}{8^{18}}$ **27.** $-\dfrac{1}{8x^9}$ **29.** $\dfrac{-27y^6}{x^6}$ **30.** $x^{25}y^{15}z^{15}$

32. $\dfrac{x^2}{625y^4z^4}$ **33.** $\dfrac{2}{27z^3}$

Use scientific notation to find the quotient. Express each quotient in scientific notation.

36. $\dfrac{(0.00012)(144,000)}{0.0003}$ **37.** $\dfrac{(-0.00017)(0.00039)}{3000}$

5.76×10^4 -2.21×10^{-11}

Simplify. Use only positive exponents.

38. $\dfrac{27x^{-5}y^5}{18x^{-6}y^2} \cdot \dfrac{x^4y^{-2}}{x^{-2}y^3}$ $\dfrac{3x^7}{2y^2}$ **39.** $\dfrac{3x^5}{y^{-4}} \cdot \dfrac{(3xy^{-3})^{-2}}{(z^{-3})^{-4}}$ $\dfrac{x^3y^{10}}{3z^{12}}$

40. $\dfrac{(x^w)^2}{(x^{w-4})^{-2}}$ x^{4w-8}

(5.3)   *Find the degree of each polynomial.*

41. $x^2y - 3xy^3z + 5x + 7y$ 5

42. $3x + 2$ 1

Simplify by combining like terms.

43. $4x + 8x - 6x^2 - 6x^2y$ $12x - 6x^2 - 6x^2y$

44. $-8xy^3 + 4xy^3 - 3x^3y$ $-4xy^3 - 3x^3y$

Add or subtract as indicated.

45. $(3x + 7y) + (4x^2 - 3x + 7) + (y - 1)$ $4x^2 + 8y + 6$

46. $(4x^2 - 6xy + 9y^2) - (8x^2 - 6xy - y^2)$ $-4x^2 + 10y^2$

47. $(3x^2 - 4b + 28) + (9x^2 - 30) - (4x^2 - 6b + 20)$

48. Add $(9xy + 4x^2 + 18)$ and $(7xy - 4x^3 - 9x)$.

49. Subtract $(x - 7)$ from the sum of $(3x^2y - 7xy - 4)$ and $(9x^2y + x)$. $12x^2y - 7xy + 3$

50. $\begin{array}{r} x^2 - 5x + 7 \\ -\ \ (x + 4) \\ \hline x^2 - 6x + 3 \end{array}$ **51.** $\begin{array}{r} x^3 \quad\quad + 2xy^2 - y \\ +\ (x - 4xy^2 \quad - 7) \\ \hline x^3 + x - 2xy^2 - y - 7 \end{array}$

If $P(x) = 9x^2 - 7x + 8$, find the following.

52. $P(6)$ 290   **53.** $P(-2)$ 58   **54.** $P(-3)$ 110

If $P(x) = 2x - 1$ and $Q(x) = x^2 + 2x - 5$, find the following.   **55.** $x^2 + 4x - 6$ **56.** $-x^2 + 2x + 3$

55. $P(x) + Q(x)$ **56.** $2[P(x)] - Q(x)$

57. Find the perimeter of the rectangle.
$(6x^2y - 12x + 12)$ cm

$x^2y + 5$
cm

$2x^2y - 6x + 1$
cm

(5.4)   *Multiply.*

58. $-6x(4x^2 - 6x + 1)$ **59.** $-4ab^2(3ab^3 + 7ab + 1)$

60. $(x - 4)(2x + 9)$ **61.** $(-3xa + 4b)^2$

62. $(9x^2 + 4x + 1)(4x - 3)$ $36x^3 - 11x^2 - 8x - 3$

63. $(5x - 9y)(3x + 9y)$ **64.** $\left(x - \dfrac{1}{3}\right)\left(x + \dfrac{2}{3}\right)$

65. $(x^2 + 9x + 1)^2$ $x^4 + 18x^3 + 83x^2 + 18x + 1$

Multiply, using special products.

66. $(3x - y)^2$ **67.** $(4x + 9)^2$

68. $(x + 3y)(x - 3y)$ $x^2 - 9y^2$

69. $[4 + (3a - b)][4 - (3a - b)]$
$-9a^2 + 6ab - b^2 + 16$

70. If $P(x) = 2x - 1$ and $Q(x) = x^2 + 2x - 5$, find $P(x) \cdot Q(x)$. $2x^3 + 3x^2 - 12x + 5$

71. Find the area of the rectangle. $(9y^2 - 49z^2)$ sq. units

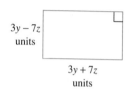

$3y - 7z$
units

$3y + 7z$
units

Multiply. Assume that all variable exponents represent integers. **72.** $12a^{2b+2} - 28a^b$ **73.** $16x^2y^{2z} - 8xy^zb + b^2$

72. $4a^b(3a^{b+2} - 7)$ **73.** $(4xy^z - b)^2$

74. $(3x^a - 4)(3x^a + 4)$ $9x^{2a} - 16$

(5.5)   *Factor out the greatest common factor.*
75. $8x^2(2x - 3)$ **76.** $12y(3 - 2y)$
75. $16x^3 - 24x^2$ **76.** $36y - 24y^2$

77. $6ab^2 + 8ab - 4a^2b^2$ $2ab(3b + 4 - 2ab)$

78. $14a^2b^2 - 21ab^2 + 7ab$ $7ab(2ab - 3b + 1)$

79. $6a(a + 3b) - 5(a + 3b)$ $(a + 3b)(6a - 5)$

80. $4x(x - 2y) - 5(x - 2y)$ $(x - 2y)(4x - 5)$

81. $xy - 6y + 3x - 18$ **82.** $ab - 8b + 4a - 32$

83. $pq - 3p - 5q + 15$ **84.** $x^3 - x^2 - 2x + 2$

85. A smaller square is cut from a larger rectangle. Write the area of the shaded region as a factored polynomial. $x(2y - x)$

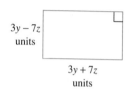

x

y x

$2x$

47. $8x^2 + 2b - 22$ **48.** $-4x^3 + 4x^2 + 16xy - 9x + 18$ **58.** $-24x^3 + 36x^2 - 6x$ **59.** $-12a^2b^5 - 28a^2b^3 - 4ab^2$

60. $2x^2 + x - 36$ **61.** $9x^2a^2 - 24xab + 16b^2$ **63.** $15x^2 + 18xy - 81y^2$ **64.** $x^2 + \dfrac{1}{3}x - \dfrac{2}{9}$ **66.** $9x^2 - 6xy + y^2$

67. $16x^2 + 72x + 81$ **81.** $(x - 6)(y + 3)$ **82.** $(a - 8)(b + 4)$ **83.** $(p - 5)(q - 3)$ **84.** $(x^2 - 2)(x - 1)$

86. $(x - 18)(x + 4)$ **87.** $(x - 4)(x + 20)$ **92.** $(6x + 5)(x + 2)$ **93.** $(15x - 1)(x - 6)$ **94.** $2(2x - 3)(x + 2)$

(5.6)   *Completely factor each polynomial.*

86. $x^2 - 14x - 72$ **87.** $x^2 + 16x - 80$

88. $2x^2 - 18x + 28$ $2(x - 2)(x - 7)$

89. $3x^2 + 33x + 54$ $3(x + 2)(x + 9)$

90. $2x^3 - 7x^2 - 9x$ $x(2x - 9)(x + 1)$

91. $3x^2 + 2x - 16$ $(3x + 8)(x - 2)$

92. $6x^2 + 17x + 10$ **93.** $15x^2 - 91x + 6$

94. $4x^2 + 2x - 12$ **95.** $9x^2 - 12x - 12$

96. $y^2(x + 6)^2 - 2y(x + 6)^2 - 3(x + 6)^2$

97. $(x + 5)^2 + 6(x + 5) + 8$ $(x + 7)(x + 9)$

98. $x^4 - 6x^2 - 16$ **99.** $x^4 + 8x^2 - 20$

(5.7)   *Factor each polynomial completely.*

100. $x^2 - 100$ **101.** $x^2 - 81$ $(x - 9)(x + 9)$

102. $2x^2 - 32$ **103.** $6x^2 - 54$ $6(x - 3)(x + 3)$

104. $81 - x^4$ **105.** $16 - y^4$ $(4 + y^2)(2 - y)(2 + y)$

106. $(y + 2)^2 - 25$ **107.** $(x - 3)^2 - 16$ $(x - 7)(x + 1)$

108. $x^3 + 216$ **109.** $y^3 + 512$ $(y + 8)(y^2 - 8y + 64)$

110. $8 - 27y^3$ **111.** $1 - 64y^3$ $(1 - 4y)(1 + 4y + 16y^2)$

112. $6x^4y + 48xy$ **113.** $2x^5 + 16x^2y^3$

114. $x^2 - 2x + 1 - y^2$ $(x - 1 + y)(x - 1 - y)$

115. $x^2 - 6x + 9 - 4y^2$ $(x - 3 - 2y)(x - 3 + 2y)$

116. $4x^2 + 12x + 9$ $(2x + 3)^2$

117. $16a^2 - 40ab + 25b^2$ $(4a - 5b)^2$

118. The volume of the cylindrical shell is $\pi R^2 h - \pi r^2 h$ cubic units. Write this volume as a factored expression. $\pi h(R + r)(R - r)$ cu. units

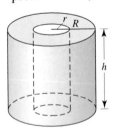

(5.8)   *Solve each polynomial equation for the variable.*

119. $(3x - 1)(x + 7) = 0$ $\left\{\frac{1}{3}, -7\right\}$

120. $3(x + 5)(8x - 3) = 0$ $\left\{-5, \frac{3}{8}\right\}$

121. $5x(x - 4)(2x - 9) = 0$ $\left\{0, 4, \frac{9}{2}\right\}$

122. $6(x + 3)(x - 4)(5x + 1) = 0$ $\left\{-3, -\frac{1}{5}, 4\right\}$

123. $2x^2 = 12x$ $\{0, 6\}$

124. $4x^3 - 36x = 0$ $\{-3, 0, 3\}$

125. $(1 - x)(3x + 2) = -4x$ $\left\{-\frac{1}{3}, 2\right\}$

126. $2x(x - 12) = -40$ $\{2, 10\}$

127. $3x^2 + 2x = 12 - 7x$ $\{-4, 1\}$

128. $2x^2 + 3x = 35$ $\left\{\frac{7}{2}, -5\right\}$

129. $x^3 - 18x = 3x^2$ $\{0, 6, -3\}$

130. $19x^2 - 42x = -x^3$ $\{-21, 0, 2\}$

131. $12x = 6x^3 + 6x^2$ $\{0, -2, 1\}$

132. $8x^3 + 10x^2 = 3x$ $\left\{-\frac{3}{2}, 0, \frac{1}{4}\right\}$

133. The sum of a number and twice its square is 105. Find the number.

134. The length of a rectangular piece of carpet is 2 meters less than 5 times its width. Find the dimensions of the carpet if its area is 16 square meters.   width, 2 m; length, 8 m

135. A scene from an adventure film calls for a stunt dummy to be dropped from above the second-story platform of the Eiffel Tower, a distance of 400 feet. Its height $h(t)$ at the time t seconds is given by

$$h(t) = -16t^2 + 400$$

Determine how long before the stunt dummy reaches the ground.   5 sec.

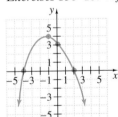

400 feet

(5.9)   *Exercises 136–139 refer to the following graph.*

136. domain, $(-\infty, \infty)$; range, $(-\infty, 4]$

136. Find the domain and the range of the function.

95. $3(x - 2)(3x + 2)$ **96.** $(x + 6)^2(y - 3)(y + 1)$ **98.** $(x^2 - 8)(x^2 + 2)$ **99.** $(x^2 - 2)(x^2 + 10)$ **100.** $(x + 10)(x - 10)$
102. $2(x + 4)(x - 4)$ **104.** $(9 + x^2)(3 + x)(3 - x)$ **106.** $(y + 7)(y - 3)$ **108.** $(x + 6)(x^2 - 6x + 36)$
110. $(2 - 3y)(4 + 6y + 9y^2)$ **112.** $6xy(x + 2)(x^2 - 2x + 4)$ **113.** $2x^2(x + 2y)(x^2 - 2xy + 4y^2)$ **133.** $-\frac{15}{2}, 7$

137. *x*-intercepts, $(-4, 0)$, $(2, 0)$; *y*-intercept, $(0, 3)$

137. List the *x*- and *y*-intercept points.

138. Find the coordinates of the point with the greatest *y*-value. $(-1, 4)$

139. List the *x*-values for which the *y*-values are greater than 0.   between $x = -4$ and $x = 2$

Graph each polynomial function defined by the equation. Find all intercepts. If the function is a quadratic function, find the vertex. For Exercises 140–147, see Appendix D.

140. $f(x) = x^2 + 6x + 9$

141. $f(x) = x^2 - 5x + 4$

142. $f(x) = (x - 1)(x^2 - 2x - 3)$

143. $f(x) = (x + 3)(x^2 - 4x + 3)$

144. $f(x) = 2x^2 - 4x + 5$

145. $f(x) = x^2 - 2x + 3$

146. $f(x) = x^3 - 16x$

147. $f(x) = x^3 + 5x^2 + 6x$

CHAPTER 5 TEST

Simplify. Use positive exponents to write answers.

1. $(-9x)^{-2}$ $\dfrac{1}{81x^2}$

2. $-3xy^{-2}(4xy^2)z$ $-12x^2z$

3. $\dfrac{6^{-1}a^2b^{-3}}{3^{-2}a^{-5}b^2}$ $\dfrac{3a^7}{2b^5}$

4. $\left(\dfrac{-xy^{-5}z}{xy^3}\right)^{-5}$ $-\dfrac{y^{40}}{z^5}$

Write in scientific notation.

5. 630,000,000 6.3×10^8

6. 0.01200 1.2×10^{-2}

7. Write 5.0×10^{-6} without exponents.   0.000005

8. Use scientific notation to find the quotient.

$$\frac{(0.0024)(0.00012)}{0.00032}$$ 0.0009

Perform the indicated operations.

9. $(4x^3 - 3x - 4) - (9x^3 + 8x + 5)$ $-5x^3 - 11x - 9$

10. $-3xy(4x + y)$

11. $(3x + 4)(4x - 7)$

12. $(5a - 2b)(5a + 2b)$ $25a^2 - 4b^2$

13. $(6m + n)^2$ $36m^2 + 12mn + n^2$

14. $(2x - 1)(x^2 - 6x + 4)$ $2x^3 - 13x^2 + 14x - 4$

Factor each polynomial completely.

15. $16x^3y - 12x^2y^4$ $4x^2y(4x - 3y^3)$

16. $x^2 - 13x - 30$ $(x - 15)(x + 2)$

17. $4y^2 + 20y + 25$ $(2y + 5)^2$

18. $6x^2 - 15x - 9$ $3(2x + 1)(x - 3)$

19. $4x^2 - 25$

20. $x^3 + 64$

21. $3x^2y - 27y^3$

22. $6x^2 + 24$ $6(x^2 + 4)$

23. $x^2y - 9y - 3x^2 + 27$ $(x + 3)(x - 3)(y - 3)$

Solve the equation for the variable.

24. $3(n - 4)(7n + 8) = 0$ $\left\{4, -\dfrac{8}{7}\right\}$

25. $(x + 2)(x - 2) = 5(x + 4)$ $\{-3, 8\}$

26. $2x^3 + 5x^2 - 8x - 20 = 0$ $\left\{-\dfrac{5}{2}, -2, 2\right\}$

27. Write the area of the shaded region as a factored polynomial. $(x + 2y)(x - 2y)$

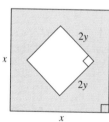

28a. 960 feet   **28b.** 953.44 feet

28. A pebble is hurled upward from the top of the Canada Trust Tower, which is 880 feet tall, with an initial velocity of 96 feet per second. Neglecting air resistance, the height $h(t)$ of the pebble after t seconds is given by the polynomial function

$$h(t) = -16t^2 + 96t + 880$$

a. Find the height of the pebble when $t = 1$.

b. Find the height of the pebble when $t = 5.1$.

c. When will the pebble hit the ground?   11 sec.

Graph. Find and label x- and y-intercepts. If the graph is a parabola, find its vertex.

29. $f(x) = x^2 - 4x - 5$ See Appendix D.

30. $f(x) = x^3 - 1$ See Appendix D.

10. $-12x^2y - 3xy^2$ **11.** $12x^2 - 5x - 28$ **19.** $(2x + 5)(2x - 5)$ **20.** $(x + 4)(x^2 - 4x + 16)$
21. $3y(x + 3y)(x - 3y)$

CHAPTER 5 CUMULATIVE REVIEW

1. Translate each phrase to an algebraic expression. Use the variable x to represent each unknown number.

  a. Eight times a number $8x$

  b. Three more than eight times a number $8x + 3$

  c. The quotient of a number and -7 $\dfrac{x}{-7}$

  d. One and six-tenths subtracted from twice a number $2x - 1.6$; *Sec. 1.1, Ex. 5*

2. Find each absolute value.

  a. $|3|$ 3   **b.** $|-5|$ 5   **c.** $-|2|$ -2

  d. $-|-8|$ -8 **e.** $|0|$ 0; *Sec 1.3, Ex. 1*

3. Evaluate each algebraic expression when $x = 2$, $y = -1$, and $z = -3$. **3c.** -1; *Sec 1.4, Ex. 2*

  a. $z - y$ -2 **b.** z^2 9   **c.** $\dfrac{2x + y}{z}$

4. Solve for x: $2(x - 3) = 5x - 9$. $\{1\}$; *Sec 2.1, Ex. 4*

5. Solve for x: $-(x - 3) + 2 \le 3(2x - 5) + x$.

6. Plot each ordered pair on a Cartesian coordinate system and name the quadrant in which the point is located.   See Appendix D.

  a. $(2, -1)$ **b.** $(0, 5)$ **c.** $(-3, 5)$

  d. $(-2, 0)$ **e.** $\left(-\dfrac{1}{2}, -4\right)$ **f.** $(1.5, 1.5)$

7. Determine the domain and range of each relation.

  a. $\{(2, 3), (2, 4), (0, -1), (3, -1)\}$

  b.

  a. domain: $\{2, 0, 3\}$; range: $\{3, 4, -1\}$

  b. domain: $\{-4, -3, -2, -1, 0, 1, 2, 3\}$; range: $\{1\}$

  c. domain: {Erie, Escondido, Gary, Miami, Waco}; range: {104, 109, 117, 359}; *Sec. 3.2, Ex. 1*

c.   Input: Output:

    Cities   Population (in thousands)

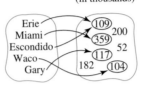

8. Graph $x - 3y = 6$ by plotting intercept points. See Appendix D; *Sec. 3.3, Ex. 4*

9. Find the slope of the line whose equation is $f(x) = \frac{2}{3}x + 4$. $m = \frac{2}{3}$; *Sec. 3.4, Ex. 3*

10. Graph the intersection of $x \ge 1$ and $y \ge 2x - 1$. See Appendix D; *Sec. 3.6, Ex. 3*

11. Use the elimination method to solve the system.

$$\begin{cases} x - 5y = -12 & \text{First equation.} \\ -x + y = 4 & \text{Second equation.} \end{cases}$$

  $\{(-2, 2)\}$; *Sec. 4.1, Ex. 4*

12. Use matrices to solve the system.

$$\begin{cases} 2x - y = 3 \\ 4x - 2y = 5 \end{cases} \quad \begin{array}{l} \text{inconsistent system,} \\ \text{no solution; } \textit{Sec. 4.4, Ex. 3} \end{array}$$

13. Evaluate.

  a. $\begin{vmatrix} -1 & 2 \\ 3 & -4 \end{vmatrix}$ -2

  b. $\begin{vmatrix} 2 & 0 \\ 7 & -5 \end{vmatrix}$ -10; *Sec. 4.5, Ex. 1*

14. Use the product rule to simplify.

  a. $2^2 \cdot 2^5$ 2^7 **b.** $x^7 x^3$ x^{10} **c.** $y \cdot y^2 \cdot y^4$ y^7; *Sec. 5.1, Ex. 1*

15. Use the power of a power rule to simplify the following expressions. Use positive exponents to write all results.   **d.** y^{12}; *Sec. 5.2, Ex. 1*

  a. $(x^5)^7$ x^{35} **b.** $(2^2)^3$ 64

  c. $(5^{-1})^2$ $\dfrac{1}{25}$ **d.** $(y^{-3})^{-4}$

5. $\left[\dfrac{5}{2}, \infty\right)$; *Sec. 2.4, Ex. 4*   **6a.** quadrant IV   **6b.** not in any quadrant   **6c.** quadrant II   **6d.** not in any quadrant

6e. quadrant III   **6f.** quadrant I *Sec. 3.1, Ex. 1*

16. Find the degree of each polynomial and indicate whether the polynomial is a monomial, binomial, trinomial, or none of these.

   a. $7x^3 - 3x + 2$ degree 3, trinomial

   b. $-xyz$ degree 3, monomial

   c. $x^2 - 4$ degree 2, binomial

   d. $2xy + x^2y^2 - 5x^2 - 6$

           degree 4, polynomial; *Sec. 5.3, Ex. 2*

17. Multiply.

   a. $2x(5x - 4)$ $10x^2 - 8x$

   b. $-3x^2(4x^2 - 6x + 1)$ $-12x^4 + 18x^3 - 3x^2$

   c. $-xy(7x^2y + 3xy - 11)$ $-7x^3y^2 - 3x^2y^2 + 11xy$;

                                Sec. 5.4, Ex. 2

18. Factor $2(x - 5) + 3a(x - 5)$.

18. $(x - 5)(2 + 3a)$; *Sec. 5.5, Ex. 5*   **19.** $2xy(2x - 1)(3x - 4)$; *Sec. 5.6, Ex. 7*

19. Factor $12x^3y - 22x^2y + 8xy$.

20. Factor $x^3 + 8$. $(x + 2)(x^2 - 2x + 4)$; *Sec. 5.7, Ex. 5*

21. Solve $2x^2 = \frac{17}{3}x + 1$. $\left\{-\frac{1}{6}, 3\right\}$; *Sec. 5.8, Ex. 5*

22. An Alpha III model rocket is launched from the ground with an A8-3 engine. Without a parachute the height of the rocket $h(t)$ at time t seconds is approximated by the function

$$h(t) = -16t^2 + 144t$$

Find how long it takes the rocket to return to the ground.   9 sec.; *Sec. 5.8, Ex. 8*

23. Graph $f(x) = x^2 + 2x - 3$. Find the vertex and any intercepts.   See Appendix D; *Sec. 5.9, Ex. 3*

RATIONAL EXPRESSIONS

MODELING ELECTRICITY PRODUCTION

In 1994, energy produced by renewable sources, including hydroelectric, geothermal, wind, and solar powers, accounted for over 8% of the United States' total production of energy (*Source*: U.S. Department of Energy). Wind energy can be harnessed by windmills to produce electricity, but it is the least utilized of these renewable energy sources. However, progressive communities are experimenting with fields of windmills for communal electricity needs, examining exactly how wind speed affects the amount of electricity produced.

IN THE CHAPTER GROUP ACTIVITY ON PAGE 399, YOU WILL HAVE THE OPPORTUNITY TO MODEL AND INVESTIGATE THE RELATIONSHIP BETWEEN WIND SPEED AND THE AMOUNT OF ELECTRICITY PRODUCED HOURLY BY A WINDMILL.

Polynomials are to algebra what integers are to arithmetic. We have added, subtracted, multiplied, and raised polynomials to powers, each operation yielding another polynomial, just as these operations on integers yield another integer. But when we divide one integer by another, the result may or may not be an integer. Likewise, when we divide one polynomial by another, we may or may not get a polynomial in return. The quotient $x \div (x + 1)$ is not a polynomial; it is a *rational expression* that can be written as $\dfrac{x}{x + 1}$.

In this chapter, we study these new algebraic forms known as rational expressions and the *rational functions* they generate.

6.1 RATIONAL FUNCTIONS AND SIMPLIFYING RATIONAL EXPRESSIONS

OBJECTIVES

TAPE IA 6.1

1 Define a rational expression and a rational function.

2 Find the domain of a rational function.

3 Write a rational expression in lowest terms.

4 Write a rational expression equivalent to a rational expression with a different denominator.

Recall that a *rational number*, or *fraction*, is a number that can be written as the quotient $\dfrac{p}{q}$ of two integers p and q as long as q is not 0. A **rational expression** is an expression that can be written as the quotient $\dfrac{P}{Q}$ of two polynomials P and Q as long as Q is not 0.

EXAMPLES OF RATIONAL EXPRESSIONS

$$\frac{3x + 7}{2} \qquad \frac{5x^2 - 3}{x - 1} \qquad \frac{7x - 2}{2x^2 + 7x + 6}$$

Rational expressions are sometimes used to describe functions. For example, we call the function $f(x) = \dfrac{x^2 + 2}{x - 3}$ a **rational function** since $\dfrac{x^2 + 2}{x - 3}$ is a rational expression.

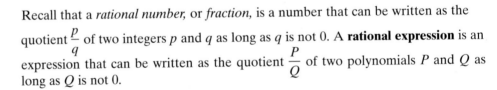

EXAMPLE 1 For the ICL Production Company, the rational function $C(x) = \dfrac{2.6x + 10,000}{x}$ describes the company's cost per disc of pressing x compact discs. Find the cost per disc for pressing:

a. 100 compact discs

b. 1000 compact discs

Solution: **a.** $C(100) = \dfrac{2.6(100) + 10{,}000}{100} = \dfrac{10{,}260}{100} = 102.6$

The cost per disc for pressing 100 compact discs is $102.60.

b. $C(1000) = \dfrac{2.6(1000) + 10{,}000}{1000} = \dfrac{12{,}600}{1000} = 12.6.$

The cost per disc for pressing 1000 compact discs is $12.60. Notice that as more compact discs are produced, the cost per disc decreases.

2   As with fractions, a rational expression is **undefined** if the denominator is 0. If a variable in a rational expression is replaced with a number that makes the denominator 0, we say that the rational expression is **undefined** for this value of the variable. For example, the rational expression $\dfrac{x^2 + 2}{x - 3}$ is undefined when x is 3, because replacing x with 3 results in a denominator of 0. For this reason, we must exclude 3 from the domain of the function defined by $f(x) = \dfrac{x^2 + 2}{x - 3}.$
The domain of f is then

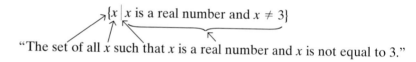

$$\{x \mid x \text{ is a real number and } x \neq 3\}$$

"The set of all x such that x is a real number and x is not equal to 3."

Unless told otherwise, we assume that the domain of a function described by an equation is the set of all real numbers for which the equation is defined.

EXAMPLE 2   Find the domain of each rational function.

a. $f(x) = \dfrac{8x^3 + 7x^2 + 20}{2}$ **b.** $g(x) = \dfrac{5x^2 - 3}{x - 1}$ **c.** $f(x) = \dfrac{7x - 2}{2x^2 + 7x + 6}$

Solution: The domain of each function will contain all real numbers except those values that make the denominator 0.

a. No matter what the value of x, the denominator of $f(x) = \dfrac{8x^3 + 7x^2 + 20}{2}$ is never 0, so the domain of f is $\{x \mid x \text{ is a real number}\}$.

b. To find the values of x that make the denominator of $g(x)$ equal to 0, we solve the equation "denominator = 0":

$$x - 1 = 0, \text{ or } x = 1$$

The domain of $g(x)$ must exclude 1 since the rational expression is undefined when x is 1. The domain of g is $\{x \mid x \text{ is a real number and } x \neq 1\}$.

c. We find the domain by setting the denominator equal to 0.

$$2x^2 + 7x + 6 = 0 \qquad \text{Set the denominator equal to 0.}$$
$$(2x + 3)(x + 2) = 0 \qquad \text{Factor.}$$
$$2x + 3 = 0 \quad \text{or} \quad x + 2 = 0 \qquad \text{Set each factor equal to 0.}$$
$$x = -\frac{3}{2} \quad \text{or} \quad x = -2 \qquad \text{Solve.}$$

The domain of f is $\{x \mid x$ is a real number and $x \neq -\frac{3}{2}$ and $x \neq -2\}$. ▬▬▬

Recall that a fraction is in lowest terms or simplest form if the numerator and denominator have no common factors other than 1 (or -1). For example, $\frac{3}{13}$ is in lowest terms since 3 and 13 have no common factors other than 1 (or -1).

To **simplify** a rational expression, or to write it in lowest terms, we use the fundamental principle of rational expressions.

FUNDAMENTAL PRINCIPLE OF RATIONAL EXPRESSIONS

For any rational expression $\dfrac{P}{Q}$ and any polynomial R, $R \neq 0$,

$$\frac{P \cdot R}{Q \cdot R} = \frac{P}{Q}$$

Thus, the fundamental principle says that multiplying or dividing the numerator and denominator of a rational expression by the same nonzero polynomial yields an equivalent rational expression.

To simplify a rational expression such as $\dfrac{(x + 2)^2}{x^2 - 4}$, factor the numerator and the denominator and then use the fundamental principle of rational expressions to divide out common factors.

$$\frac{(x + 2)^2}{x^2 - 4} = \frac{(x + 2)(x + 2)}{(x + 2)(x - 2)}$$
$$= \frac{x + 2}{x - 2}$$

This means that the rational expression $\dfrac{(x + 2)^2}{x^2 - 4}$ has the same value as the rational expression $\dfrac{x + 2}{x - 2}$ for all values of x except 2 and -2. (Remember that when x is 2, the denominator of both rational expressions is 0 and that when x is -2, the original rational expression has a denominator of 0.) As we simplify rational expressions, we will assume that the simplified rational expression is equivalent to the original rational expression for all real numbers except those for which either denominator is 0.

In general, the following steps may be used to simplify rational expressions or to write a rational expression in lowest terms.

> ## To Simplify, or Write a Rational Expression in Lowest Terms
>
> *Step 1.*   Completely factor the numerator and denominator of the rational expression.
>
> *Step 2.*   Use the fundamental principle of rational expressions to divide both the numerator and denominator by the GCF.

For now, we assume that variables in a rational expression do not represent values that make the denominator 0.

EXAMPLE 3   Write each rational expression in lowest terms.

a. $\dfrac{24x^6y^5}{8x^7y}$ **b.** $\dfrac{2x^2}{10x^3 - 2x^2}$

Solution: **a.** The GCF of the numerator and denominator is $8x^6y$.

$$\frac{24x^6y^5}{8x^7y} = \frac{(8x^6y)3y^4}{(8x^6y)x}$$ Factor the numerator and denominator.

$$= \frac{3y^4}{x}$$ Apply the fundamental principle and divide out common factors.

b. Factor out $2x^2$ from the denominator. Then divide numerator and denominator by their GCF, $2x^2$.

$$\frac{2x^2}{10x^3 - 2x^2} = \frac{2x^2 \cdot 1}{2x^2(5x - 1)} = \frac{1}{5x - 1}$$

EXAMPLE 4   Write each rational expression in lowest terms.

a. $\dfrac{2 + x}{x + 2}$ **b.** $\dfrac{2 - x}{x - 2}$ **c.** $\dfrac{18 - 2x^2}{x^2 - 2x - 3}$

Solution: **a.** By the commutative property of addition, $2 + x = x + 2$, so

$$\frac{2 + x}{x + 2} = \frac{x + 2}{x + 2} = 1$$

b. The terms in the numerator of $\dfrac{2 - x}{x - 2}$ differ by sign from the terms of the denominator, so the polynomials are opposites of each other and the expression simplifies to -1. To see this, factor out -1 from the numerator or the denominator. If -1 is factored from the numerator, then

$$\frac{2 - x}{x - 2} = \frac{-1(-2 + x)}{x - 2} = \frac{-1(x - 2)}{x - 2} = -1$$

REMINDER   When the numerator and the denominator of a rational expression are opposites of each other, the expression simplifies to -1.

c. $\dfrac{18 - 2x^2}{x^2 - 2x - 3} = \dfrac{2(9 - x^2)}{(x + 1)(x - 3)}$ Factor.

$\qquad\qquad = \dfrac{2(3 + x)(3 - x)}{(x + 1)(x - 3)}$ Factor completely.

Notice the opposites $3 - x$ and $x - 3$. We write $3 - x$ as $-1(x - 3)$ and simplify.

$$\dfrac{2(3 + x)(3 - x)}{(x + 1)(x - 3)} = \dfrac{2(3 + x) \cdot -1(x - 3)}{(x + 1)(x - 3)} = -\dfrac{2(3 + x)}{x + 1}$$

REMINDER   Recall that for a fraction $\dfrac{a}{b}$,

$$\dfrac{a}{-b} = \dfrac{-a}{b} = -\dfrac{a}{b}$$

For example,

$$\dfrac{-(x + 1)}{(x + 2)} = \dfrac{(x + 1)}{-(x + 2)} = -\dfrac{x + 1}{x + 2}$$

EXAMPLE 5   Write each rational expression in lowest terms.

a. $\dfrac{x^3 + 8}{2 + x}$

b. $\dfrac{2y^2 + 2}{y^3 - 5y^2 + y - 5}$

Solution: **a.** $\dfrac{x^3 + 8}{2 + x} = \dfrac{(x + 2)(x^2 - 2x + 4)}{x + 2}$ Factor the sum of the two cubes.

$\qquad\qquad = x^2 - 2x + 4$ Divide out common factors.

b. First factor the denominator by grouping.

$$\begin{aligned} y^3 - 5y^2 + y - 5 &= (y^3 - 5y^2) + (y - 5) \\ &= y^2(y - 5) + 1(y - 5) \\ &= (y - 5)(y^2 + 1) \end{aligned}$$

Then

$$\frac{2y^2 + 2}{y^3 - 5y^2 + y - 5} = \frac{2(y^2 + 1)}{(y - 5)(y^2 + 1)} = \frac{2}{y - 5}$$

4

The fundamental principle of rational expressions also allows us to write a rational expression as an equivalent rational expression with a different denominator. It is often necessary to do so in adding and subtracting rational expressions.

EXAMPLE 6   Write each rational expression as an equivalent rational expression with the given denominator.

a. $\dfrac{3x}{2y}$, denominator $10xy^3$

b. $\dfrac{3x + 1}{x - 5}$, denominator $2x^2 - 11x + 5$

Solution: **a.** $\dfrac{3x}{2y} = \dfrac{?}{10xy^3}$

If the denominator $2y$ is multiplied by $5xy^2$, the result is the given denominator $10xy^3$.

$$2y(5xy^2) = 10xy^3$$

original   given
denominator   denominator

Use the fundamental principle of rational expressions and multiply the numerator and the denominator of the original rational expression by $5xy^2$. Then

$$\frac{3x}{2y} = \frac{3x(5xy^2)}{2y(5xy^2)} = \frac{15x^2y^2}{10xy^3}$$

b. The factored form of the given denominator, $2x^2 - 11x + 5$, is $(x - 5)(2x - 1)$.

$$\frac{3x + 1}{x - 5} = \frac{?}{2x^2 - 11x + 5} = \frac{?}{(x - 5)(2x - 1)}$$

Use the fundamental principle of rational expressions and multiply the numerator and denominator of the original rational expression by $2x - 1$.

$$\frac{3x + 1}{x - 5} = \frac{(3x + 1)(2x - 1)}{(x - 5)(2x - 1)} = \frac{6x^2 - x - 1}{(x - 5)(2x - 1)}$$

To prepare for adding and subtracting rational expressions, we multiply the binomials in the numerator but leave the denominator in factored form.

GRAPHING CALCULATOR EXPLORATIONS

A grapher may be used to graph and to find or confirm the domain of a rational function. For example, the domain of $f(x) = \dfrac{7x - 2}{(x - 2)(x + 5)}$ is all real numbers except 2 and -5. This means that the graph of $f(x)$ should not cross the vertical lines $x = 2$ and $x = -5$. The graph of $f(x)$ in *connected* mode follows. In connected mode the grapher tries to connect all dots of the graph so that the result is a smooth curve. This is what has happened in the graph. Notice that the graph appears to contain vertical lines at $x = 2$ and at $x = -5$. We know that this cannot happen because the graph is not defined at $x = 2$ and at $x = -5$. We also know that this cannot happen because the graph of this function would not pass the vertical line test.

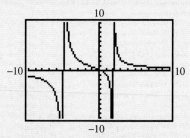

If we graph $f(x)$ in *dot* mode, the graph appears as follows. In dot mode the grapher will not connect dots with a smooth curve. Notice that the vertical lines have disappeared, and we have a better picture of the graph. The graph, however, actually appears more like the hand-drawn graph to its right. By using a Table feature, a Calculate Value feature, or by tracing, we can see that the function is not defined at $x = 2$ and at $x = -5$.

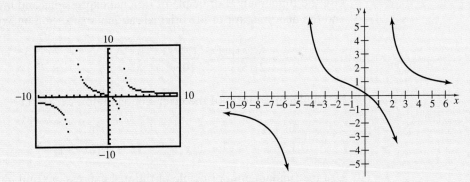

Find the domain of each rational function. Then graph each rational function and use the graph to confirm the domain.

1. $f(x) = \dfrac{x + 1}{x^2 - 4}$

2. $g(x) = \dfrac{5x}{x^2 - 9}$

3. $h(x) = \dfrac{x^2}{2x^2 + 7x - 4}$

4. $f(x) = \dfrac{3x + 2}{4x^2 - 19x - 5}$

1. $\{x\mid x$ is a real number and $x \neq -2, x \neq 2\}$ **2.** $\{x\mid x$ is a real number and $x \neq 3, x \neq -3\}$

3. $\{x\mid x$ is a real number and $x \neq -4, x \neq \frac{1}{2}\}$ **4.** $\{x\mid x$ is a real number and $x \neq -\frac{1}{4}, x \neq 5\}$

5. $\{x \mid x$ is a real number$\}$ **6.** $\{x \mid x$ is a real number$\}$ **7.** $\{t \mid t$ is a real number and $t \neq 0\}$
8. $\{t \mid t$ is a real number and $t \neq 0\}$ **9.** $\{x \mid x$ is a real number and $x \neq 7\}$ **10.** $\{x \mid x$ is a real number and $x \neq 2\}$

EXERCISE SET 6.1

11. $\{x \mid x$ is a real number and $x \neq 0, x \neq -3\}$ **12.** $\{x \mid x$ is a real number and $x \neq 0, x \neq 2\}$
13. $\{x \mid x$ is a real number and $x \neq -2, x \neq 0, x \neq 1\}$
14. $\{x \mid x$ is a real number and $x \neq 5, x \neq 2\}$

Find each function value. See Example 1.

1. $f(x) = \dfrac{x + 8}{2x - 1}; f(2), f(0), f(-1)$ $\dfrac{10}{3}, -8, -\dfrac{7}{3}$

2. $f(y) = \dfrac{y - 2}{-5 + y}; f(-5), f(0), f(10)$ $\dfrac{7}{10}, \dfrac{2}{5}, \dfrac{8}{5}$

3. $g(x) = \dfrac{x^2 + 8}{x^3 - 25x}; g(3), g(-2), g(1)$ $-\dfrac{17}{48}, \dfrac{2}{7}, -\dfrac{3}{8}$

4. $s(t) = \dfrac{t^3 + 1}{t^2 + 1}; s(-1), s(1), s(2)$ $0, 1, \dfrac{9}{5}$

Find the domain of each rational function. See Example 2.

5. $f(x) = \dfrac{5x - 7}{4}$

6. $g(x) = \dfrac{4 - 3x}{2}$

7. $s(t) = \dfrac{t^2 + 1}{2t}$

8. $v(t) = -\dfrac{5t + t^2}{3t}$

9. $f(x) = \dfrac{3x}{7 - x}$

10. $f(x) = \dfrac{-4x}{-2 + x}$

11. $g(x) = \dfrac{2 - 3x^2}{3x^2 + 9x}$

12. $C(x) = \dfrac{5 - 4x^2}{4x - 2x^2}$

13. $R(x) = \dfrac{3 + 2x}{x^3 + x^2 - 2x}$

14. $h(x) = \dfrac{5 - 3x}{2x^2 - 14x + 20}$

15. $C(x) = \dfrac{x + 3}{x^2 - 4}$

16. $R(x) = \dfrac{5}{x^2 - 7x}$

17. In your own words, explain how to find the domain of a rational function.

18. In your own words, explain how to simplify a rational expression or to write it in lowest terms.

Write each rational expression in lowest terms. See Examples 3–5.

19. $\dfrac{10x^3}{18x}$ $\dfrac{5x^2}{9}$ **20.** $-\dfrac{48a^7}{16a^{10}}$ $-\dfrac{3}{a^3}$ **21.** $\dfrac{9x^6y^3}{18x^2y^5}$ $\dfrac{x^4}{2y^2}$

22. $\dfrac{10ab^5}{15a^3b^5}$ $\dfrac{2}{3a^2}$ **23.** $\dfrac{8q^2}{16q^3 - 16q^2}$ $\dfrac{1}{2(q - 1)}$

24. $\dfrac{3y}{6y^2 - 30y}$ $\dfrac{1}{2(y - 5)}$ **25.** $\dfrac{x + 5}{5 + x}$ 1

26. $\dfrac{x - 5}{5 - x}$ -1 **27.** $\dfrac{x - 1}{1 - x^2}$ $\dfrac{-1}{x + 1}$

28. $\dfrac{10 + 5x}{x^2 + 2x}$ $\dfrac{5}{x}$ **29.** $\dfrac{7 - x}{x^2 - 14x + 49}$ $\dfrac{-1}{x - 7}$

30. $\dfrac{x^2 - 9}{2x^2 - 5x - 3}$ $\dfrac{x + 3}{2x + 1}$ **31.** $\dfrac{4x - 8}{3x - 6}$ $\dfrac{4}{3}$

32. $\dfrac{12 - 6x}{30 - 15x}$ $\dfrac{2}{5}$ **33.** $\dfrac{2x - 14}{7 - x}$ -2

34. $\dfrac{9 - x}{5x - 45}$ $-\dfrac{1}{5}$ **35.** $\dfrac{x^2 - 2x - 3}{x^2 - 6x + 9}$ $\dfrac{x + 1}{x - 3}$

36. $\dfrac{x^2 + 10x + 25}{x^2 + 8x + 15}$ $\dfrac{x + 5}{x + 3}$ **37.** $\dfrac{2x^2 + 12x + 18}{x^2 - 9}$

38. $\dfrac{x^2 - 4}{2x^2 + 8x + 8}$ $\dfrac{x - 2}{2(x + 2)}$ **39.** $\dfrac{3x + 6}{x^2 + 2x}$ $\dfrac{3}{x}$

40. $\dfrac{3x + 4}{9x^2 + 4}$ $\dfrac{3x + 4}{9x^2 + 4}$ **41.** $\dfrac{x + 2}{x^2 - 4}$ $\dfrac{1}{x - 2}$

42. $\dfrac{x^2 - 9}{x - 3}$ $x + 3$ **43.** $\dfrac{2x^2 - x - 3}{2x^3 - 3x^2 + 2x - 3}$

44. $\dfrac{3x^2 - 5x - 2}{6x^3 + 2x^2 + 3x + 1}$ $\dfrac{x - 2}{2x^2 + 1}$

45. $\dfrac{x^4 - 16}{x^2 + 4}$ $x^2 - 4$

46. $\dfrac{x^2 + y^2}{x^4 - y^4}$ $\dfrac{1}{(x + y)(x - y)}$ **47.** $\dfrac{x^2 + 6x - 40}{10 + x}$ $x - 4$

48. $\dfrac{x^2 - 8x + 16}{4 - x}$ $4 - x$ **49.** $\dfrac{2x^2 - 7x - 4}{x^2 - 5x + 4}$ $\dfrac{2x + 1}{x - 1}$

50. $\dfrac{3x^2 - 11x + 10}{x^2 - 7x + 10}$ $\dfrac{3x - 5}{x - 5}$ **51.** $\dfrac{x^3 - 125}{5 - x}$

52. $\dfrac{4x + 4}{2x^3 + 2}$ $\dfrac{2}{x^2 - x + 1}$ **53.** $\dfrac{8x^3 - 27}{4x - 6}$ $\dfrac{4x^2 + 6x + 9}{2}$

54. $\dfrac{9x^2 - 15x + 25}{27x^3 + 125}$ **55.** $\dfrac{x + 5}{x^2 + 5}$ $\dfrac{x + 5}{x^2 + 5}$

56. $\dfrac{5x}{5x^2 + 5x}$ $\dfrac{1}{x + 1}$

Write each rational expression as an equivalent rational expression with the given denominator. See Example 6.

57. $\dfrac{5}{2y}, 4y^3z$ $\dfrac{10y^2z}{4y^3z}$ **58.** $\dfrac{1}{z}, 5z^5$ $\dfrac{5z^4}{5z^5}$

59. $\dfrac{3x}{2x - 1}, 2x^2 + 9x - 5$ $\dfrac{3x^2 + 15x}{2x^2 + 9x - 5}$

60. $\dfrac{5}{3x + 2}, 3x^2 - 13x - 10$ $\dfrac{5x - 25}{3x^2 - 13x - 10}$

61. $\dfrac{x - 2}{1}, x + 2$ $\dfrac{x^2 - 4}{x + 2}$ **62.** $\dfrac{x - 5}{1}, x + 1$

63. $\dfrac{5}{m}, 6m^3$ $\dfrac{30m^2}{6m^3}$ **64.** $\dfrac{3}{x}, 3x$ $\dfrac{9}{3x}$

15. $\{x \mid x$ is a real number and $x \neq 2, x \neq -2\}$ **16.** $\{x \mid x$ is a real number and $x \neq 0, x \neq 7\}$

37. $\dfrac{2(x + 3)}{x - 3}$ **43.** $\dfrac{x + 1}{x^2 + 1}$ **51.** $-x^2 - 5x - 25$ **54.** $\dfrac{1}{3x + 5}$ **62.** $\dfrac{x^2 - 4x - 5}{x + 1}$

65. $\dfrac{7}{m-2}$, $5m - 10$

66. $\dfrac{-2}{x+1}$, $10x + 10$ $\dfrac{-20}{10x+10}$

67. $\dfrac{y+4}{y-4}$, $y^2 - 16$

68. $\dfrac{5}{x-1}$, $x^2 - 1$ $\dfrac{5x+5}{x^2-1}$

69. $\dfrac{12x}{x+2}$, $x^2 + 4x + 4$

70. $\dfrac{x+6}{x-4}$, $x^2 - 8x + 16$

71. $\dfrac{1}{x+2}$, $x^3 + 8$

72. $\dfrac{x}{3x-1}$, $27x^3 - 1$

73. $\dfrac{a}{a+2}$, $ab - 3a + 2b - 6$ $\dfrac{ab-3a}{ab-3a+2b-6}$

74. $\dfrac{5}{x-y}$, $2x^2 + 5x - 2xy - 5y$ $\dfrac{10x+25}{2x^2+5x-2xy-5y}$

75. Graph a portion of the function $f(x) = \dfrac{20x}{100-x}$.

To do so, complete the given table, plot the points, and then connect the plotted points with a smooth curve.   See Appendix D.

x	0	10	30	50	70	90	95	99
y or $f(x)$	0	$\frac{20}{9}$	$\frac{60}{7}$	20	$\frac{140}{3}$	180	380	1980

76. The domain of the function $f(x) = \dfrac{1}{x}$ is all real numbers except 0. This means that the graph of this function will be in two pieces: one piece corresponding to x values less than 0 and one piece corresponding to x values greater than 0. Graph the function by completing the following tables, separately plotting the points and connecting each set of plotted points with a smooth curve.

x	$\frac{1}{4}$	$\frac{1}{2}$	1	2	4
y or $f(x)$	4	2	1	$\frac{1}{2}$	$\frac{1}{4}$

x	-4	-2	-1	$-\frac{1}{2}$	$-\frac{1}{4}$
y or $f(x)$	$-\frac{1}{4}$	$-\frac{1}{2}$	-1	-2	-4

See Appendix D.

77. The function $f(x) = \dfrac{100{,}000x}{100-x}$ models the cost in dollars for removing x percent of the pollutants

from a bayou in which a nearby company dumped creosol.

a. What is the domain of $f(x)$? $\{x \mid 0 \le x < 100\}$

b. Find the cost of removing 30% of the pollutants from the bayou. (*Hint:* Find $f(30)$.) $42{,}857.14

c. Find the cost of removing 60% of the pollutants and then 80% of the pollutants.

d. Find $f(90)$, then $f(95)$, and then $f(99)$. What happens to the cost as x approaches 100%?

78. The total revenue from the sale of a popular book is approximated by the rational function $R(x) = \dfrac{1000x^2}{x^2+4}$ where x is the number of years since publication and $R(x)$ is the total revenue in millions of dollars.

a. Find the total revenue at the end of the first year. $200 million

b. Find the total revenue at the end of the second year. $500 million

c. Find the revenue during the second year only.
   $300 million

Review Exercises

Perform the indicated operations. See Section 1.3.

79. $\dfrac{6}{35} \cdot \dfrac{28}{9}$ $\dfrac{8}{15}$

80. $\dfrac{3}{8} \cdot \dfrac{4}{27}$ $\dfrac{1}{18}$

81. $\dfrac{8}{35} \div \dfrac{4}{5}$ $\dfrac{2}{7}$

82. $\dfrac{6}{11} \div \dfrac{2}{11}$ 3

83. $\left(\dfrac{1}{2} \cdot \dfrac{1}{4}\right) \div \dfrac{3}{8}$ $\dfrac{1}{3}$

84. $\dfrac{3}{7} \cdot \left(\dfrac{1}{5} \div \dfrac{3}{10}\right)$ $\dfrac{2}{7}$

*Recall that a **histogram** is a special type of bar graph in which the width of the bar stands for a single number or a range of numbers and the height of each bar represents the number of occurrences. For example, the width of each bar in the following histogram represents a range of ages, and the height represents the number of occurrences, or percents. Use this graph to answer Exercises 85 through 87. See Section 1.1.*

85. What percent of people aged 45 and over have no health insurance?   28.1%

65. $\dfrac{35}{5m-10}$ **67.** $\dfrac{y^2+8y+16}{y^2-16}$ **69.** $\dfrac{12x^2+24x}{x^2+4x+4}$ **70.** $\dfrac{x^2+2x-24}{x^2-8x+16}$ **71.** $\dfrac{x^2-2x+4}{x^3+8}$ **72.** $\dfrac{9x^3+3x^2+x}{27x^3-1}$

77c. $150,000; $400,000   **77d.** 900,000; $1,900,000; $9,900,000

86. Find the age group of people who have the lowest percent of no health insurance.   65 and over

87. What percent of people aged 18 to 34 have no health insurance?   49.8%

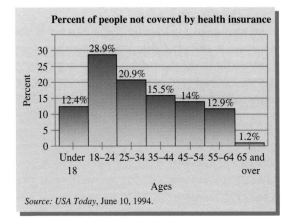

Percent of people not covered by health insurance

Source: *USA Today*, June 10, 1994.

A Look Ahead

EXAMPLE

Write $\dfrac{x^{2n} + 2x^n y^n + y^{2n}}{x^n + y^n}$ in lowest terms.

Solution: Factor the numerator.

$$\frac{x^{2n} + 2x^n y^{2n} + y^{2n}}{x^n + y^n} = \frac{(x^n + y^n)(x^n + y^n)}{(x^n + y^n)} = x^n + y^n$$

Simplify or write each rational expression in lowest terms. Assume that no denominator is 0.

88. $\dfrac{p^x - 4}{4 - p^x}$ -1

89. $\dfrac{3 + q^n}{q^n + 3}$ 1

90. $\dfrac{x^n + 4}{x^{2n} - 16}$ $\dfrac{1}{x^n - 4}$

91. $\dfrac{x^{2k} - 9}{3 + x^k}$ $x^k - 3$

92. $\dfrac{x^{2k} - 4x^k + 16}{x^{3k} + 64}$ $\dfrac{1}{x^k + 4}$

93. $\dfrac{4x^k - 12}{x^{2k} + 4}$ $\dfrac{4x^k - 12}{x^{2k} + 4}$

6.2 MULTIPLYING AND DIVIDING RATIONAL EXPRESSIONS

TAPE IA 6.2

OBJECTIVES

 Multiply rational expressions.

2 Divide by a rational expression.

Arithmetic operations on rational expressions are performed in the same way as they are on rational numbers.

1

> **MULTIPLYING RATIONAL EXPRESSIONS**
> Let P, Q, R, and S be polynomials. Then
> $$\frac{P}{Q} \cdot \frac{R}{S} = \frac{PR}{QS}$$
> as long as $Q \neq 0$ and $S \neq 0$.

In other words, to multiply rational expressions, the product of their numerators is the numerator of the product and the product of their denominators is the denominator of the product.

EXAMPLE 1   Multiply.

a. $\dfrac{2x^3}{9y} \cdot \dfrac{y^2}{4x^3}$

b. $\dfrac{1 + 3n}{2n} \cdot \dfrac{2n - 4}{3n^2 - 2n - 1}$

Solution: **a.** $\dfrac{2x^3}{9y} \cdot \dfrac{y^2}{4x^3} = \dfrac{2x^3 y^2}{36x^3 y}$

To simplify, divide the numerator and the denominator by the common factor, $2x^3 y$.

$$\dfrac{2x^3 y^2}{36x^3 y} = \dfrac{y\,(2x^3 y)}{18\,(2x^3 y)} = \dfrac{y}{18}$$

b. $\dfrac{1 + 3n}{2n} \cdot \dfrac{2n - 4}{3n^2 - 2n - 1} = \dfrac{1 + 3n}{2n} \cdot \dfrac{2(n - 2)}{(3n + 1)(n - 1)}$ Factor.

$$= \dfrac{(1 + 3n) \cdot 2\,(n - 2)}{2n\,(3n + 1)\,(n - 1)}$$ Multiply.

$$= \dfrac{n - 2}{n(n - 1)}$$ Divide out common factors.

When we multiply rational expressions, it is usually best to factor each numerator and denominator first. This will help when we apply the fundamental principle to write the product in lowest terms. The following steps may be used to multiply rational expressions.

TO MULTIPLY RATIONAL EXPRESSIONS

Step 1.   Completely factor the numerators and denominators.

Step 2.   Multiply the numerators and multiply the denominators.

Step 3.   Write the product in lowest terms by applying the fundamental principle of rational expressions and dividing both the numerator and the denominator by their GCF.

EXAMPLE 2   Multiply.

a. $\dfrac{2x^2 + 3x - 2}{-4x - 8} \cdot \dfrac{16x^2}{4x^2 - 1}$

b. $(ac - ad + bc - bd) \cdot \dfrac{a + b}{d - c}$

Solution: **a.** $\dfrac{2x^2 + 3x - 2}{-4x - 8} \cdot \dfrac{16x^2}{4x^2 - 1} = \dfrac{(2x - 1)(x + 2)}{-4(x + 2)} \cdot \dfrac{16x^2}{(2x + 1)(2x - 1)}$ Factor.

$$= \dfrac{4 \cdot 4x^2 \, (2x - 1)(x + 2)}{-1 \cdot \, 4(x + 2) \, (2x + 1) \, (2x - 1)}$$ Multiply.

$$= -\dfrac{4x^2}{2x + 1}$$ Divide out common factors.

b. First factor $ac - ad + bc - bd$ by grouping.

$$ac - ad + bc - bd = (ac - ad) + (bc - bd) \qquad \text{Group terms.}$$
$$= a(c - d) + b(c - d)$$
$$= (c - d)(a + b)$$

To multiply, write $(c - d)(a + b)$ as a fraction whose denominator is 1.

$$(ac - ad + bc - bd) \cdot \dfrac{a + b}{d - c} = \dfrac{(c - d)(a + b)}{1} \cdot \dfrac{a + b}{d - c} \qquad \text{Factor.}$$

$$= \dfrac{(c - d)(a + b)(a + b)}{d - c} \qquad \text{Multiply.}$$

Write $(c - d)$ as $-1(d - c)$ and simplify.

$$\dfrac{-1\,(d - c)\,(a + b)(a + b)}{d - c} = -(a + b)^2$$

To divide by a rational expression, multiply by its reciprocal. Recall that two numbers are reciprocals of each other if their product is 1. Similarly, if $\dfrac{P}{Q}$ is a rational expression, then $\dfrac{Q}{P}$ is its **reciprocal**, since

$$\dfrac{P}{Q} \cdot \dfrac{Q}{P} = \dfrac{P \cdot Q}{Q \cdot P} = 1$$

The following are examples of expressions and their reciprocals.

Expression	Reciprocal
$\dfrac{3}{x}$	$\dfrac{x}{3}$
$\dfrac{2 + x^2}{4x - 3}$	$\dfrac{4x - 3}{2 + x^2}$
x^3	$\dfrac{1}{x^3}$
0	no reciprocal

Division of rational expressions is defined as follows.

DIVIDING RATIONAL EXPRESSIONS
Let P, Q, R, and S be polynomials. Then

$$\frac{P}{Q} \div \frac{R}{S} = \frac{P}{Q} \cdot \frac{S}{R} = \frac{PS}{QR}$$

as long as $Q \neq 0$, $S \neq 0$, and $R \neq 0$.

Notice that division of rational expressions is the same as for rational numbers.

EXAMPLE 3   Divide.

a. $\dfrac{3x}{5y} \div \dfrac{9y}{x^5}$

b. $\dfrac{8m^2}{3m^2 - 12} \div \dfrac{40}{2 - m}$

Solution: **a.** $\dfrac{3x}{5y} \div \dfrac{9y}{x^5} = \dfrac{3x}{5y} \cdot \dfrac{x^5}{9y}$ Multiply by the reciprocal of the divisor.

$= \dfrac{x^6}{15y^2}$ Simplify.

b. $\dfrac{8m^2}{3m^2 - 12} \div \dfrac{40}{2 - m} = \dfrac{8m^2}{3m^2 - 12} \cdot \dfrac{2 - m}{40}$ Multiply by the reciprocal of the divisor.

$= \dfrac{8m^2(2 - m)}{3(m + 2)(m - 2) \cdot 40}$ Factor and multiply.

$= \dfrac{8\,m^2 \cdot -1(m - 2)}{3(m + 2)(m - 2) \cdot 8 \cdot 5}$ Write $(2 - m)$ as $-1(m - 2)$.

$= -\dfrac{m^2}{15(m + 2)}$ Simplify.   ▬▬▬

R E M I N D E R   When dividing rational expressions, do not divide out common factors until the division problem is rewritten as a multiplication problem.

EXAMPLE 4   Divide: $\dfrac{8x^3 + 125}{x^4 + 5x^2 + 4} \div \dfrac{2x + 5}{2x^2 + 8}$

Solution: $\dfrac{8x^3 + 125}{x^4 + 5x^2 + 4} \div \dfrac{2x + 5}{2x^2 + 8} = \dfrac{8x^3 + 125}{x^4 + 5x^2 + 4} \cdot \dfrac{2x^2 + 8}{2x + 5}$

$$= \frac{(2x + 5)(4x^2 - 10x + 25) \cdot 2(x^2 + 4)}{(x^2 + 1)(x^2 + 4) \cdot (2x + 5)}$$

$$= \frac{2(4x^2 - 10x + 25)}{x^2 + 1}$$

As we have seen in earlier chapters, it is possible to add, subtract, and multiply functions. It is also possible to divide functions. Although we have not stated it as such, the sums, differences, products, and quotients of functions are themselves functions. For example, if $f(x) = \dfrac{3}{x}$ and $g(x) = \dfrac{x + 1}{5}$, their product, $f(x) \cdot g(x) = \dfrac{3}{x} \cdot \dfrac{x + 1}{5} = \dfrac{3(x + 1)}{5x}$, is a new function as long as the denominator is not 0. We can use the notation $(f \cdot g)(x)$ to denote this new function. Finding the sum, difference, product, and quotient of functions to generate new functions is called the **algebra of functions**.

This *algebra* of functions is defined next.

ALGEBRA OF FUNCTIONS

Let f and g be functions. Their **sum**, written $f + g$, is defined by

$$(f + g)(x) = f(x) + g(x)$$

Their **difference**, written as $f - g$, is defined by

$$(f - g)(x) = f(x) - g(x)$$

Their **product**, written as $f \cdot g$, is defined by

$$(f \cdot g)(x) = f(x) \cdot g(x)$$

Their **quotient**, written as $\dfrac{f}{g}$, is defined by

$$\left(\frac{f}{g}\right)(x) = \frac{f(x)}{g(x)}, \qquad g(x) \neq 0$$

EXAMPLE 5   If $f(x) = x - 1$ and $g(x) = 2x - 3$, find the following.

   a. $(f + g)(x)$ **b.** $(f - g)(x)$ **c.** $(f \cdot g)(x)$ **d.** $\left(\dfrac{f}{g}\right)(x)$

Solution: Replace $f(x)$ by $x - 1$ and $g(x)$ by $2x - 3$. Then simplify.

   a. $(f + g)(x) = f(x) + g(x)$

   $= (x - 1) + (2x - 3)$

   $= 3x - 4$

   b. $(f - g)(x) = f(x) - g(x)$

   $= (x - 1) - (2x - 3)$

   $= x - 1 - 2x + 3$

   $= -x + 2$

   c. $(f \cdot g)(x) = f(x) \cdot g(x)$

   $= (x - 1)(2x - 3)$

   $= 2x^2 - 5x + 3$

   d. $\left(\dfrac{f}{g}\right)(x) = \dfrac{f(x)}{g(x)} = \dfrac{x - 1}{2x - 3}, \quad x \neq \dfrac{3}{2}$

There is an interesting but not surprising relationship between the graphs of functions and the graph of their sum, difference, product, and quotient. For example, the graph of $(f + g)(x)$ can be found by adding the graph of $f(x)$ to the graph of $g(x)$. We add two graphs by adding corresponding y-values.

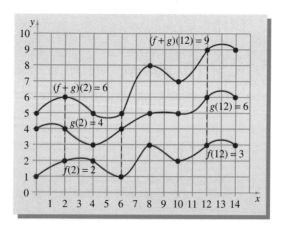

GRAPHING CALCULATOR EXPLORATIONS

If $f(x) = \dfrac{1}{2}x + 2$ and $g(x) = \dfrac{1}{3}x^2 + 4$, then $(f + g)(x) = f(x) + g(x)$

$$= \left(\dfrac{1}{2}x + 2\right) + \left(\dfrac{1}{3}x^2 + 4\right)$$

$$= \dfrac{1}{3}x^2 + \dfrac{1}{2}x + 6.$$

To visualize this addition of functions with a grapher, graph

$$Y_1 = \dfrac{1}{2}x + 2, \quad Y_2 = \dfrac{1}{3}x^2 + 4, \quad \text{and} \quad Y_3 = \dfrac{1}{3}x^2 + \dfrac{1}{2}x + 6$$

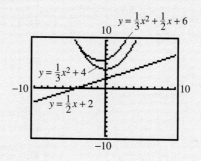

Use a Table feature to verify that for a given x value, $Y_1 + Y_2 = Y_3$.

EXERCISE SET 6.2

Multiply as indicated. Write all answers in lowest terms.
See Examples 1 and 2.

1. $\dfrac{4}{x} \cdot \dfrac{x^2}{8}$ $\dfrac{x}{2}$

2. $\dfrac{x}{3} \cdot \dfrac{9}{x^3}$ $\dfrac{3}{x^2}$

3. $\dfrac{2a^2b}{6ac} \cdot \dfrac{3c^2}{4ab}$ $\dfrac{c}{4}$

4. $\dfrac{5ab^4}{6abc} \cdot \dfrac{2bc^2}{10ab^2}$ $\dfrac{b^2c}{6a}$

5. $\dfrac{2x}{5} \cdot \dfrac{5x + 10}{6(x + 2)}$ $\dfrac{x}{3}$

6. $\dfrac{3x}{7} \cdot \dfrac{14 - 7x}{9(2 - x)}$ $\dfrac{x}{3}$

7. $\dfrac{2x - 4}{15} \cdot \dfrac{6}{2 - x}$ $-\dfrac{4}{5}$

8. $\dfrac{10 - 2x}{7} \cdot \dfrac{14}{5x - 25}$ $-\dfrac{4}{5}$

9. $\dfrac{18a - 12a^2}{4a^2 + 4a + 1} \cdot \dfrac{4a^2 + 8a + 3}{4a^2 - 9}$ $-\dfrac{6a}{2a + 1}$

10. $\dfrac{a - 5b}{a^2 + ab} \cdot \dfrac{b^2 - a^2}{10b - 2a}$ $\dfrac{a - b}{2a}$

32. $(g \cdot f)(x) = 5x^3 + 5x$ **33.** $(f - g)(x) = x^2 - 5x + 1$ **36.** $(g - f)(x) = -x^2 + 5x - 1$

11. $\dfrac{x^2 - 6x - 16}{2x^2 - 128} \cdot \dfrac{x^2 + 16x + 64}{3x^2 + 30x + 48}$ $\dfrac{1}{6}$

12. $\dfrac{2x^2 + 12x - 32}{x^2 + 16x + 64} \cdot \dfrac{x^2 + 10x + 16}{x^2 - 3x - 10}$ $\dfrac{2(x - 2)}{x - 5}$

13. $\dfrac{4x + 8}{x + 1} \cdot \dfrac{2 - x}{3x - 15} \cdot \dfrac{2x^2 - 8x - 10}{x^2 - 4}$ $-\dfrac{8}{3}$

14. $\dfrac{3x - 15}{2 - x} \cdot \dfrac{x + 1}{4x + 8} \cdot \dfrac{x^2 - 4}{2x^2 - 8x - 10}$ $-\dfrac{3}{8}$

Solve.

15. Find the area of the rectangle.

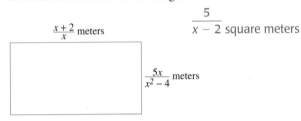

$\dfrac{x + 2}{x}$ meters $\dfrac{5}{x - 2}$ square meters

$\dfrac{5x}{x^2 - 4}$ meters

16. Find the area of the triangle.

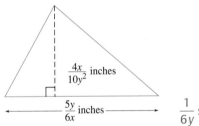

$\dfrac{4x}{10y^2}$ inches

$\dfrac{5y}{6x}$ inches $\dfrac{1}{6y}$ square inches

Divide as indicated. Write all answers in lowest terms. See Examples 3 and 4.

17. $\dfrac{4}{x} \div \dfrac{8}{x^2}$ $\dfrac{x}{2}$

18. $\dfrac{x}{3} \div \dfrac{x^3}{9}$ $\dfrac{3}{x^2}$

19. $\dfrac{4ab}{3c^2} \div \dfrac{2a^2b}{6ac}$ $\dfrac{4}{c}$

20. $\dfrac{6abc}{5ab^4} \div \dfrac{2bc^2}{10ab^2}$ $\dfrac{6a}{b^2c}$

21. $\dfrac{2x}{5} \div \dfrac{6x + 12}{5x + 10}$ $\dfrac{x}{3}$

22. $\dfrac{7}{3x} \div \dfrac{14 - 7x}{18 - 9x}$ $\dfrac{3}{x}$

23. $\dfrac{2(x + y)}{5} \div \dfrac{6(x + y)}{25}$ $\dfrac{5}{3}$

24. $\dfrac{4}{x - 2y} \div \dfrac{9}{3x - 6y}$ $\dfrac{4}{3}$

25. $\dfrac{x^2 - 6x + 9}{x^2 - x - 6} \div \dfrac{x^2 - 9}{4}$ $\dfrac{4}{(x + 2)(x + 3)}$

41. $\dfrac{1}{4a(a - b)}$ **42.** $\dfrac{1}{2a(a + 2)}$

26. $\dfrac{x^2 - 4}{3x + 6} \div \dfrac{2x^2 - 8x + 8}{x^2 + 4x + 4}$ $\dfrac{(x + 2)^2}{6(x - 2)}$

27. $\dfrac{x^2 - 6x - 16}{2x^2 - 128} \div \dfrac{x^2 + 10x + 16}{x^2 + 16x + 64}$ $\dfrac{1}{2}$

28. $\dfrac{a^2 - a - 6}{a^2 - 81} \div \dfrac{a^2 - 7a - 18}{4a + 36}$ $\dfrac{4(a - 3)}{(a - 9)^2}$

29. $\dfrac{14x^4}{y^5} \div \dfrac{2x^2}{y^7} \div \dfrac{2x^4}{7y^5}$ $\dfrac{49y^7}{2x^2}$

30. $\dfrac{x^2}{7y^2} \div \left(\dfrac{2x^2}{y^7} \div \dfrac{2x^2}{7y^5}\right)$ $\dfrac{x^2}{49}$

If $f(x) = x^2 + 1$ and $g(x) = 5x$, find the following. See Example 5.   **31.** $(f + g)(x) = x^2 + 5x + 1$

31. $(f + g)(x)$ **32.** $(g \cdot f)(x)$

33. $(f - g)(x)$ **34.** $\left(\dfrac{f}{g}\right)(x)$ $\left(\dfrac{f}{g}\right)(x) = \dfrac{x^2 + 1}{5x}$

35. $\left(\dfrac{g}{f}\right)(x)$ $\left(\dfrac{g}{f}\right)(x) = \dfrac{5x}{x^2 + 1}$ **36.** $(g - f)(x)$

Perform the indicated operation. Write all answers in lowest terms.

37. $\dfrac{3xy^3}{4x^3y^2} \cdot \dfrac{-8x^3y^4}{9x^4y^7}$ $-\dfrac{2}{3x^3y^2}$ **38.** $-\dfrac{2xyz^3}{5x^2z^2} \cdot \dfrac{10xy}{x^3}$ $-\dfrac{4y^2z}{x^3}$

39. $\dfrac{8a}{3a^4b^2} \div \dfrac{4b^5}{6a^2b}$ $\dfrac{4}{ab^6}$ **40.** $\dfrac{3y^3}{14x^4} \div \dfrac{8y^3}{7x}$ $\dfrac{3}{16x^3}$

41. $\dfrac{a^2b}{a^2 - b^2} \cdot \dfrac{a + b}{4a^3b}$ **42.** $\dfrac{3ab^2}{a^2 - 4} \cdot \dfrac{a - 2}{6a^2b^2}$

43. $\dfrac{x^2 - 9}{4} \div \dfrac{x^2 - 6x + 9}{x^2 - x - 6}$ $\dfrac{x^2 + 5x + 6}{4}$

44. $\dfrac{a - 5b}{a^2 + ab} \div \dfrac{15b - 3a}{b^2 - a^2}$ $\dfrac{a - b}{3a}$

45. $\dfrac{9x + 9}{4x + 8} \cdot \dfrac{2x + 4}{3x^2 - 3}$ $\dfrac{3}{2(x - 1)}$

46. $\dfrac{x^2 - 1}{10x + 30} \cdot \dfrac{12x + 36}{3x - 3}$ $\dfrac{2(x + 1)}{5}$

47. $\dfrac{a + b}{ab} \div \dfrac{a^2 - b^2}{4a^3b}$ $\dfrac{4a^2}{a - b}$

48. $\dfrac{6a^2b^2}{a^2 - 4} \div \dfrac{3ab^2}{a - 2}$ $\dfrac{2a}{a + 2}$

49. $\dfrac{2x^2 - 4x - 30}{5x^2 - 40x - 75} \div \dfrac{x^2 - 8x + 15}{x^2 - 6x + 9}$ $\dfrac{2x^2 - 18}{5(x^2 - 8x - 15)}$

53. $\dfrac{3b}{a - b}$ **54.** $\dfrac{4(x + 1)}{x - 5}$ **67.** $\dfrac{(y + 5)(2x - 1)}{(y + 2)(5x + 1)}$

50. $\dfrac{4a + 36}{a^2 - 7a - 18} \div \dfrac{a^2 - a - 6}{a^2 - 81}$ $\dfrac{4(a + 9)^2}{(a + 2)^2(a - 3)}$

51. $\dfrac{2x^3 - 16}{6x^2 + 6x - 36} \cdot \dfrac{9x + 18}{3x^2 + 6x + 12}$ $\dfrac{x + 2}{x + 3}$

52. $\dfrac{x^2 - 3x + 9}{5x^2 - 20x - 105} \cdot \dfrac{x^2 - 49}{x^3 + 27}$ $\dfrac{x + 7}{5(x + 3)^2}$

53. $\dfrac{15b - 3a}{b^2 - a^2} \div \dfrac{a - 5b}{ab + b^2}$ **54.** $\dfrac{4x + 4}{x - 1} \div \dfrac{x^2 - 4x - 5}{x^2 - 1}$

55. $\dfrac{a^3 + a^2b + a + b}{a^3 + a} \cdot \dfrac{6a^2}{2a^2 - 2b^2}$ $\dfrac{3a}{a - b}$

56. $\dfrac{a^2 - 2a}{ab - 2b + 3a - 6} \cdot \dfrac{8b + 24}{3a + 6}$ $\dfrac{8a}{3(a + 2)}$

57. $\dfrac{5a}{12} \cdot \dfrac{2}{25a^2} \cdot \dfrac{15a}{2}$ $\dfrac{1}{4}$ **58.** $\dfrac{4a}{7} \div \dfrac{a^2}{14} \cdot \dfrac{3}{a}$ $\dfrac{24}{a^2}$

59. $\dfrac{3x - x^2}{x^3 - 27} \div \dfrac{x}{x^2 + 3x + 9}$ -1

60. $\dfrac{x^2 - 3x}{x^3 - 27} \div \dfrac{2x}{2x^2 + 6x + 18}$ 1

61. $\dfrac{4a}{7} \div \left(\dfrac{a^2}{14} \cdot \dfrac{3}{a}\right)$ $\dfrac{8}{3}$ **62.** $\dfrac{a^2}{14} \cdot \dfrac{3}{a} \div \dfrac{4a}{7}$ $\dfrac{3}{8}$

63. $\dfrac{8b + 24}{3a + 6} \div \dfrac{ab - 2b + 3a - 6}{a^2 - 4a + 4}$ $\dfrac{8(a - 2)}{3(a + 2)}$

64. $\dfrac{2a^2 - 2b^2}{a^3 + a^2b + a + b} \div \dfrac{6a^2}{a^3 + a}$ $\dfrac{a - b}{3a}$

65. $\dfrac{4}{x} \div \dfrac{3xy}{x^2} \cdot \dfrac{6x^2}{x^4}$ $\dfrac{8}{x^2 y}$ **66.** $\dfrac{4}{x} \cdot \dfrac{3xy}{x^2} \div \dfrac{6x^2}{x^4}$ $2y$

67. $\dfrac{3x^2 - 5x - 2}{y^2 + y - 2} \cdot \dfrac{y^2 + 4y - 5}{12x^2 + 7x + 1} \div \dfrac{5x^2 - 9x - 2}{8x^2 - 2x - 1}$

68. $\dfrac{x^2 + x - 2}{3y^2 - 5y - 2} \cdot \dfrac{12y^2 + y - 1}{x^2 + 4x - 5} \div \dfrac{8y^2 - 6y + 1}{5y^2 - 9y - 2}$

69. $\dfrac{5a^2 - 20}{3a^2 - 12a} \div \dfrac{a^3 + 2a^2}{2a^2 - 8a} \cdot \dfrac{9a^3 + 6a^2}{2a^2 - 4a}$ $\dfrac{15a + 10}{a}$

70. $\dfrac{5a^2 - 20}{3a^2 - 12a} \div \left(\dfrac{a^3 + 2a^2}{2a^2 - 8a} \cdot \dfrac{9a^3 + 6a^2}{2a^2 - 4a}\right)$ $\dfrac{20(a - 2)^2}{9a^3(3a + 2)}$

71. $\dfrac{5x^4 + 3x^2 - 2}{x - 1} \cdot \dfrac{x + 1}{x^4 - 1}$ $\dfrac{5x^2 - 2}{(x - 1)^2}$

72. $\dfrac{3x^4 - 10x^2 - 8}{x - 2} \cdot \dfrac{3x + 6}{15x^2 + 10}$ $\dfrac{3(x + 2)^2}{5}$

68. $\dfrac{(x + 2)(5y + 1)}{(x + 5)(2y - 1)}$ **77.** $(g - h)(x) = x^2 - 4x - 1$ **78.** $(h \cdot g)(x) = 4x^3 + 3x^2 + 8x + 6$ **79.** $(h + f)(x) = 2x + 3$

80. $(h - g)(x) = -x^2 + 4x + 1$ **81.** $f(a + b) = -2a - 2b$ **82.** $g(a + b) = a^2 + 2ab + b^2 + 2$ **83.** $\left(\dfrac{f}{h}\right)(x) = -\dfrac{2x}{4x + 3}$

If $f(x) = -2x$, $g(x) = x^2 + 2$, and $h(x) = 4x + 3$, find the following.   **73.** $(f + g)(x) = x^2 - 2x + 2$

73. $(f + g)(x)$ **74.** $(g - f)(x)$
                               $(g - f)(x) = x^2 + 2x + 2$

75. $\left(\dfrac{f}{g}\right)(x)$ $\left(\dfrac{f}{g}\right)(x) = -\dfrac{2x}{x^2 + 2}$ **76.** $\left(\dfrac{g}{f}\right)(x)$ $\left(\dfrac{g}{f}\right)(x) = -\dfrac{x^2 + 2}{2x}$

77. $(g - h)(x)$ **78.** $(h \cdot g)(x)$

79. $(h + f)(x)$ **80.** $(h - g)(x)$

81. $f(a + b)$ **82.** $g(a + b)$

83. $\left(\dfrac{f}{h}\right)(x)$ **84.** $\left(\dfrac{h}{g}\right)(x)$ $\left(\dfrac{h}{g}\right)(x) = \dfrac{4x + 3}{x^2 + 2}$

🔲 **85.** In our definition of division for

$$\dfrac{P}{Q} \div \dfrac{R}{S}$$

we stated that $Q \neq 0$, $S \neq 0$, and $R \neq 0$. Explain why R cannot equal 0.

🔲 **86.** Find the polynomial in the second numerator such that the following statement is true. $(x - 5)(2x + 7)$

$$\dfrac{x^2 - 4}{x^2 - 7x + 10} \cdot \dfrac{?}{2x^2 + 11x + 14} = 1$$

🔲 **87.** A parallelogram has area $\dfrac{x^2 + x - 2}{x^3}$ square feet and height $\dfrac{x^2}{x - 1}$ feet. Express the length of its base as a rational expression in x. (*Hint:* Since $A = b \cdot h$, then $b = \dfrac{A}{h}$ or $b = A \div h$.)

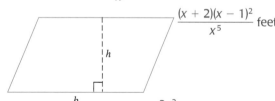

$\dfrac{(x + 2)(x - 1)^2}{x^5}$ feet

88. $\dfrac{3x^2}{y^2}$ dollars per person

🔲 **88.** A lottery prize of $\dfrac{15x^3}{y^2}$ dollars is to be divided among $5x$ people. Express the amount of money each person is to receive as a rational expression in x and y.

89. Business people are concerned with cost functions, revenue functions, and profit functions. Recall that the profit $P(x)$ obtained from x units of a product is equal to the revenue $R(x)$ from selling the x units minus the cost $C(x)$ of manufacturing the x units.

Write an equation expressing this relationship among $C(x)$, $R(x)$, and $P(x)$. $P(x) = R(x) - C(x)$

90. Suppose the revenue $R(x)$ for x units of a product can be described by $R(x) = 25x$, and the cost $C(x)$ can be described by $C(x) = 50 + x^2 + 4x$. Find the profit $P(x)$ for x units. $P(x) = 21x - x^2 - 50$

Review Exercises

Perform the indicated operation. See Section 1.3.

91. $\dfrac{4}{5} + \dfrac{3}{5}$ $\dfrac{7}{5}$ **92.** $\dfrac{4}{10} - \dfrac{7}{10}$ $-\dfrac{3}{10}$ **93.** $\dfrac{5}{28} - \dfrac{2}{21}$ $\dfrac{1}{12}$

94. $\dfrac{5}{13} + \dfrac{2}{7}$ $\dfrac{61}{91}$ **95.** $\dfrac{3}{8} + \dfrac{1}{2} - \dfrac{3}{16}$ $\dfrac{11}{16}$ **96.** $\dfrac{2}{9} - \dfrac{1}{6} + \dfrac{2}{3}$ $\dfrac{13}{18}$

A Look Ahead

EXAMPLE

Perform the following operation.

$$\frac{x^{2n} - 3x^n - 18}{x^{2n} - 9} \cdot \frac{3x^n + 9}{x^{2n}}$$

Solution: $\dfrac{x^{2n} - 3x^n - 18}{x^{2n} - 9} \cdot \dfrac{3x^n + 9}{x^{2n}}$

$$= \frac{(x^n + 3)(x^n - 6) \cdot 3(x^n + 3)}{(x^n + 3)(x^n - 3) \cdot x^{2n}}$$

$$= \frac{3(x^n - 6)(x^n + 3)}{x^{2n}(x^n - 3)}$$

Perform the indicated operation. Write all answers in lowest terms. See the preceding example.

97. $\dfrac{x^{2n} - 4}{7x} \cdot \dfrac{14x^3}{x^n - 2}$ $2x^2(x^n + 2)$

98. $\dfrac{x^{2n} + 4x^n + 4}{4x - 3} \cdot \dfrac{8x^2 - 6x}{x^n + 2}$ $2x(x^n + 2)$

99. $\dfrac{y^{2n} + 9}{10y} \cdot \dfrac{y^n - 3}{y^{4n} - 81}$ **100.** $\dfrac{y^{4n} - 16}{y^{2n} + 4} \cdot \dfrac{6y}{y^n + 2}$

101. $\dfrac{y^{2n} - y^n - 2}{2y^n - 4} \div \dfrac{y^{2n} - 1}{1 + y^n}$ $\dfrac{y^n + 1}{2(y^n - 1)}$

102. $\dfrac{y^{2n} + 7y^n + 10}{10} \div \dfrac{y^{2n} + 4y^n + 4}{5y^n + 25}$ $\dfrac{(y^n + 5)^2}{2(y^n + 2)}$

99. $\dfrac{1}{10y(y^n + 3)}$ **100.** $6y(y^n - 2)$

6.3 | ADDING AND SUBTRACTING RATIONAL EXPRESSIONS

O B J E C T I V E S

1 Add or subtract rational expressions with common denominators.

2 Identify the least common denominator of two or more rational expressions.

3 Add and subtract rational expressions with unlike denominators.

1 Rational expressions, like rational numbers, can be added or subtracted. We define the sum or difference of rational expressions in the same way that we defined the sum or difference of rational numbers.

ADDING OR SUBTRACTING RATIONAL EXPRESSIONS WITH COMMON DENOMINATORS

If $\dfrac{P}{Q}$ and $\dfrac{R}{Q}$ are rational expressions, then

$$\frac{P}{Q} + \frac{R}{Q} = \frac{P + R}{Q} \quad \text{and} \quad \frac{P}{Q} - \frac{R}{Q} = \frac{P - R}{Q}$$

To add or subtract rational expressions with common denominators, add or subtract the numerators and write the sum or difference over the common denominator.

EXAMPLE 1   Add or subtract.

a. $\dfrac{5}{7} + \dfrac{x}{7}$

b. $\dfrac{x}{4} + \dfrac{5x}{4}$

c. $\dfrac{x^2}{x + 7} - \dfrac{49}{x + 7}$

d. $\dfrac{x}{3y^2} - \dfrac{x + 1}{3y^2}$

Solution: The rational expressions have common denominators, so add or subtract their numerators and place the sum or difference over their common denominator.

a. $\dfrac{5}{7} + \dfrac{x}{7} = \dfrac{5 + x}{7}$

b. $\dfrac{x}{4} + \dfrac{5x}{4} = \dfrac{x + 5x}{4} = \dfrac{6x}{4} = \dfrac{3x}{2}$

c. $\dfrac{x^2}{x + 7} - \dfrac{49}{x + 7} = \dfrac{x^2 - 49}{x + 7}$

Next write this rational expression in lowest terms.

$$\dfrac{x^2 - 49}{x + 7} = \dfrac{(x + 7)(x - 7)}{x + 7} = x - 7$$

d. $\dfrac{x}{3y^2} - \dfrac{x + 1}{3y^2} = \dfrac{x - (x + 1)}{3y^2}$ Subtract numerators.

$$= \dfrac{x - x - 1}{3y^2}$$ Apply the distributive property.

$$= -\dfrac{1}{3y^2}$$ Simplify.

2   To add or subtract rational expressions with unlike denominators, first write the rational expressions as equivalent rational expressions with common denominators.

The **least common denominator (LCD)** is usually the easiest common denominator to work with. The LCD of a list of rational expressions is a polynomial of least degree whose factors include the denominator factors in the list.

Use the following steps to find the LCD.

TO FIND THE LEAST COMMON DENOMINATOR (LCD)

Step 1. Factor each denominator completely.

Step 2. The LCD is the product of all unique factors formed in *step 1*, each raised to a power equal to the greatest number of times that the factor appears in any one factored denominator.

EXAMPLE 2 Find the LCD of rational expressions in each list.

a. $\dfrac{2}{15x^5y^2}, \dfrac{3z}{5xy^3}$

b. $\dfrac{7}{z+1}, \dfrac{z}{z-1}$

c. $\dfrac{m-1}{m^2-25}, \dfrac{2m}{2m^2-9m-5}, \dfrac{7}{m^2-10m+25}$

Solution: **a.** Factor each denominator.

$$15x^5y^2 = 3 \cdot 5 \cdot x^5 \cdot y^2$$
$$5xy^3 = 5 \cdot x \cdot y^3$$

The unique factors are 3, 5, x, and y.

The greatest number of times that 3 appears in one denominator is 1.
The greatest number of times that 5 appears in one denominator is 1.
The greatest number of times that x appears in one denominator is 5.
The greatest number of times that y appears in one denominator is 3.
The LCD is the product of $3^1 \cdot 5^1 \cdot x^5 \cdot y^3$, or $15x^5y^3$.

b. The denominators $z + 1$ and $z - 1$ do not factor further. Each factor appears once so the

$$\text{LCD} = (z + 1)(z - 1)$$

c. First factor each denominator.

$$m^2 - 25 = (m + 5)(m - 5)$$
$$2m^2 - 9m - 5 = (2m + 1)(m - 5)$$
$$m^2 - 10m + 25 = (m - 5)(m - 5)$$

The LCD $= (m + 5)(2m + 1)(m - 5)^2$, which is the product of each unique factor raised to a power equal to the greatest number of times it appears in any one factored denominator.

To add or subtract rational expressions with unlike denominators, we write each rational expression as an equivalent rational expression so that their denominators are alike.

TO ADD OR SUBTRACT RATIONAL EXPRESSIONS WITH UNLIKE DENOMINATORS

Step 1.   Find the LCD of the rational expressions.

Step 2.   Write each rational expression as an equivalent rational expression whose denominator is the LCD found in *step 1*.

Step 3.   Add or subtract numerators, and write the sum or difference over the common denominator.

Step 4.   Simplify, or write the resulting rational expression in lowest terms.

EXAMPLE 3   Perform the indicated operation.

a. $\dfrac{2}{x^2 y} + \dfrac{5}{3x^3 y}$ **b.** $\dfrac{3x}{x + 2} + \dfrac{2x}{x - 2}$ **c.** $\dfrac{5k}{k^2 - 4} - \dfrac{2}{k^2 + k - 2}$

Solution: **a.** The LCD is $3x^3 y$. Write each fraction as an equivalent fraction with denominator $3x^3 y$.

$$\frac{2}{x^2 y} + \frac{5}{3x^3 y} = \frac{2 \cdot 3x}{x^2 y \cdot 3x} + \frac{5}{3x^3 y}$$

$$= \frac{6x}{3x^3 y} + \frac{5}{3x^3 y}$$

$$= \frac{6x + 5}{3x^3 y}   \qquad \text{Add the numerators.}$$

b. The LCD is the product of the two denominators: $(x + 2)(x - 2)$.

$$\frac{3x}{x + 2} + \frac{2x}{x - 2} = \frac{3x \cdot (x - 2)}{(x + 2) \cdot (x - 2)} + \frac{2x \cdot (x + 2)}{(x - 2) \cdot (x + 2)}   \qquad \begin{array}{l}\text{Write equivalent}\\\text{rational expressions.}\end{array}$$

$$= \frac{3x(x - 2) + 2x(x + 2)}{(x + 2)(x - 2)}   \qquad \text{Add the numerators.}$$

$$= \frac{3x^2 - 6x + 2x^2 + 4x}{(x + 2)(x - 2)}   \qquad \begin{array}{l}\text{Apply the distributive}\\\text{property.}\end{array}$$

$$= \frac{5x^2 - 2x}{(x + 2)(x - 2)}   \qquad \text{Simplify the numerator.}$$

c. $\dfrac{5k}{k^2 - 4} - \dfrac{2}{k^2 + k - 2}$

$$= \frac{5k}{(k + 2)(k - 2)} - \frac{2}{(k + 2)(k - 1)}   \qquad \begin{array}{l}\text{Factor each}\\\text{denominator to}\\\text{find the LCD.}\end{array}$$

The LCD is $(k + 2)(k - 2)(k - 1)$. Write equivalent rational expressions with the LCD as denominators.

$$\frac{5k}{(k + 2)(k - 2)} - \frac{2}{(k + 2)(k - 1)} = \frac{5k(k - 1)}{(k + 2)(k - 2)(k - 1)} - \frac{2(k - 2)}{(k + 2)(k - 1)(k - 2)}$$

$$= \frac{5k(k - 1) - 2(k - 2)}{(k + 2)(k - 2)(k - 1)} \quad \text{Subtract the numerators.}$$

$$= \frac{5k^2 - 5k - 2k + 4}{(k + 2)(k - 2)(k - 1)}$$

$$= \frac{5k^2 - 7k + 4}{(k + 2)(k - 2)(k - 1)} \quad \text{Simplify the numerator.}$$

Since the numerator polynomial is prime, the numerator and denominator have no common factors and this rational expression is in lowest terms. ▬▬▬▬

EXAMPLE 4 Perform the indicated operation.

a. $\dfrac{x}{x - 3} - 5$ **b.** $\dfrac{7}{x - y} + \dfrac{3}{y - x}$

Solution: **a.** $\dfrac{x}{x - 3} - 5 = \dfrac{x}{x - 3} - \dfrac{5}{1}$ Write 5 as $\dfrac{5}{1}$.

The LCD is $x - 3$.

$$= \frac{x}{x - 3} - \frac{5 \cdot (x - 3)}{1 \cdot (x - 3)}$$

$$= \frac{x - 5(x - 3)}{x - 3} \quad \text{Subtract.}$$

$$= \frac{x - 5x + 15}{x - 3} \quad \text{Apply the distributive property.}$$

$$= \frac{-4x + 15}{x - 3} \quad \text{Simplify.}$$

b. Notice that the denominators $x - y$ and $y - x$ are opposites of one another. To write equivalent rational expressions with the LCD, write one denominator such as $y - x$ as $-1(x - y)$:

$$\frac{7}{x - y} + \frac{3}{y - x} = \frac{7}{x - y} + \frac{3}{-1(x - y)}$$

$$= \frac{7}{x - y} + \frac{-3}{x - y}$$

$$= \frac{4}{x - y} \quad \text{Add the numerators.} \quad ▬▬▬▬$$

EXAMPLE 5   Add:

$$\frac{2x - 1}{2x^2 - 9x - 5} + \frac{x + 3}{6x^2 - x - 2}$$

Solution:

$$\frac{2x - 1}{2x^2 - 9x - 5} + \frac{x + 3}{6x^2 - x - 2} = \frac{2x - 1}{(2x + 1)(x - 5)} + \frac{x + 3}{(2x + 1)(3x - 2)} \qquad \text{Factor the} \atop \text{denominators.}$$

The LCD is $(2x + 1)(x - 5)(3x - 2)$.

$$= \frac{(2x - 1) \cdot (3x - 2)}{(2x + 1)(x - 5) \cdot (3x - 2)} + \frac{(x + 3) \cdot (x - 5)}{(2x + 1)(3x - 2) \cdot (x - 5)}$$

$$= \frac{6x^2 - 7x + 2}{(2x + 1)(x - 5)(3x - 2)} + \frac{x^2 - 2x - 15}{(2x + 1)(x - 5)(3x - 2)}$$

$$= \frac{6x^2 - 7x + 2 + x^2 - 2x - 15}{(2x + 1)(x - 5)(3x - 2)} \qquad \text{Add.}$$

$$= \frac{7x^2 - 9x - 13}{(2x + 1)(x - 5)(3x - 2)} \qquad \text{Simplify the numerator.}$$

The rational expression is in lowest terms.

EXAMPLE 6   Perform the indicated operations.

$$\frac{7}{x - 1} + \frac{2(x + 1)}{(x - 1)^2} - \frac{2}{x^2}$$

Solution: The LCD is $x^2(x - 1)^2$.

$$\frac{7}{x - 1} + \frac{2(x + 1)}{(x - 1)^2} - \frac{2}{x^2} = \frac{7 \cdot x^2(x - 1)}{(x - 1) \cdot x^2(x - 1)} + \frac{2(x + 1) \cdot x^2}{(x - 1)^2 \cdot x^2} - \frac{2 \cdot (x - 1)^2}{x^2 \cdot (x - 1)^2}$$

$$= \frac{7x^3 - 7x^2}{x^2(x - 1)^2} + \frac{2x^3 + 2x^2}{x^2(x - 1)^2} - \frac{2x^2 - 4x + 2}{x^2(x - 1)^2}$$

$$= \frac{7x^3 - 7x^2 + 2x^3 + 2x^2 - 2x^2 + 4x - 2}{x^2(x - 1)^2} \qquad \text{Watch your} \atop \text{signs!}$$

$$= \frac{9x^3 - 7x^2 + 4x - 2}{x^2(x - 1)^2} \qquad \text{Add or subtract like} \atop \text{terms in the numerator.}$$

Since the numerator and denominator have no common factors, this rational expression is in lowest terms.

GRAPHING CALCULATOR EXPLORATIONS

A grapher can be used to support the results of operations on rational expressions. For example, to verify the result of Example 3b, graph

$$Y_1 = \frac{3x}{x+2} + \frac{2x}{x-2} \text{ and } Y_2 = \frac{5x^2 - 2x}{(x+2)(x-2)}$$

on the same set of axes. The graphs should be the same. Use a Table feature or a Trace feature to see that this is true.

EXERCISE SET 6.3

Perform the indicated operation. Write each answer in lowest terms. See Example 1.

1. $\dfrac{2}{x} - \dfrac{5}{x} - \dfrac{3}{x}$

2. $\dfrac{4}{x^2} + \dfrac{2}{x^2} \dfrac{6}{x^2}$

3. $\dfrac{2}{x-2} + \dfrac{x}{x-2} \dfrac{x+2}{x-2}$

4. $\dfrac{x}{5-x} + \dfrac{2}{5-x} \dfrac{x+2}{5-x}$

5. $\dfrac{x^2}{x+2} - \dfrac{4}{x+2} \quad x-2$

6. $\dfrac{4}{x-2} - \dfrac{x^2}{x-2} \quad -2-x$

7. $\dfrac{2x-6}{x^2+x-6} + \dfrac{3-3x}{x^2+x-6} \dfrac{1}{2-x}$

8. $\dfrac{5x+2}{x^2+2x-8} + \dfrac{2-4x}{x^2+2x-8} \dfrac{1}{x-2}$

9. Find the perimeter and the area of the square.

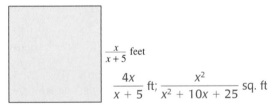

$\dfrac{x}{x+5}$ feet

$\dfrac{4x}{x+5}$ ft; $\dfrac{x^2}{x^2+10x+25}$ sq. ft

10. Find the perimeter of the quadrilateral.

$(2x+3)$ cm

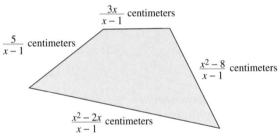

$\dfrac{3x}{x-1}$ centimeters

$\dfrac{5}{x-1}$ centimeters

$\dfrac{x^2-8}{x-1}$ centimeters

$\dfrac{x^2-2x}{x-1}$ centimeters

Find the LCD of the rational expressions in the list. See Example 2.

11. $\dfrac{2}{7}, \dfrac{3}{5x} \quad 35x$

12. $\dfrac{4}{5y}, \dfrac{3}{4y^2} \quad 20y^2$

13. $\dfrac{3}{x}, \dfrac{2}{x+1} \quad x(x+1)$

14. $\dfrac{5}{2x}, \dfrac{7}{2+x} \quad 2x(2+x)$

15. $\dfrac{12}{x+7}, \dfrac{8}{x-7}$

16. $\dfrac{1}{2x-1}, \dfrac{x}{2x+1} \quad (2x-1)(2x+1)$

17. $\dfrac{5}{3x+6}, \dfrac{2x}{2x-4}$

18. $\dfrac{2}{3a+9}, \dfrac{5}{5a-15}$

19. $\dfrac{2a}{a^2-b^2}, \dfrac{1}{a^2-2ab+b^2} \quad (a+b)(a-b)^2$

20. $\dfrac{2a}{a^2+8a+16}, \dfrac{7a}{a^2+a-12} \quad (a+4)^2(a-3)$

15. $(x+7)(x-7)$ **17.** $6(x+2)(x-2)$ **18.** $15(a+3)(a-3)$

21. When is the LCD of two rational expressions equal to the product of their denominators? (*Hint:* What is the LCD of $\dfrac{1}{x}$ and $\dfrac{7}{x+5}$?) Answers will vary.

22. When is the LCD of two rational expressions with different denominators equal to one of the denominators? (*Hint:* What is the LCD of $\dfrac{3x}{x+2}$ and $\dfrac{7x+1}{(x+2)^3}$?) Answers will vary.

Perform the indicated operation. Write each answer in the lowest terms. See Example 3.

23. $\dfrac{4}{3x} + \dfrac{3}{2x}$ $\dfrac{17}{6x}$

24. $\dfrac{10}{7x} - \dfrac{5}{2x}$ $-\dfrac{15}{14x}$

25. $\dfrac{5}{2y^2} - \dfrac{2}{7y}$ $\dfrac{35 - 4y}{14y^2}$

26. $\dfrac{4}{11x^4y} - \dfrac{1}{4x^2y^3}$ $\dfrac{16y^2 - 11x^2}{44x^4y^3}$

27. $\dfrac{x-3}{x+4} - \dfrac{x+2}{x-4}$

28. $\dfrac{x-1}{x-5} - \dfrac{x+2}{x+5}$

29. $\dfrac{1}{x-5} + \dfrac{x}{x^2-x-20}$ $\dfrac{2x+4}{(x-5)(x+4)}$

30. $\dfrac{x+1}{x^2-x-20} - \dfrac{2}{x+4}$ $\dfrac{-x+11}{(x+4)(x-5)}$

Perform the indicated operation. Write each answer in lowest terms. See Example 4.

31. $\dfrac{1}{a-b} + \dfrac{1}{b-a}$ 0

32. $\dfrac{1}{a-3} - \dfrac{1}{3-a}$ $\dfrac{2}{a-3}$

33. $x + 1 + \dfrac{1}{x-1}$ $\dfrac{x^2}{x-1}$

34. $5 - \dfrac{1}{x-1}$ $\dfrac{5x-6}{x-1}$

35. $\dfrac{5}{x-2} + \dfrac{x+4}{2-x}$ $\dfrac{-x+1}{x-2}$

36. $\dfrac{3}{5-x} + \dfrac{x+2}{x-5}$ $\dfrac{x-1}{x-5}$

Perform the indicated operation. Write each answer in lowest terms. See Examples 5 and 6.

37. $\dfrac{y+1}{y^2-6y+8} - \dfrac{3}{y^2-16}$ $\dfrac{y^2+2y+10}{(y+4)(y-4)(y-2)}$

38. $\dfrac{x+2}{x^2-36} - \dfrac{x}{x^2+9x+18}$ $\dfrac{6+11x}{(x-6)(x+6)(x+3)}$

39. $\dfrac{x+4}{3x^2+11x+6} + \dfrac{x}{2x^2+x-15}$

40. $\dfrac{x+3}{5x^2+12x+4} + \dfrac{6}{x^2-x-6}$ $\dfrac{x^2+30x+3}{(5x+2)(x-3)(x+2)}$

41. $\dfrac{7}{x^2-x-2} + \dfrac{x}{x^2+4x+3}$ $\dfrac{x^2+5x+21}{(x-2)(x+1)(x+3)}$

42. $\dfrac{a}{a^2+10a+25} + \dfrac{4}{a^2+6a+5}$ $\dfrac{a^2+5a+20}{(a+5)^2(a+1)}$

43. $\dfrac{2}{x+1} - \dfrac{3x}{3x+3} + \dfrac{1}{2x+2}$ $\dfrac{-2x+5}{2(x+1)}$

44. $\dfrac{5}{3x-6} - \dfrac{x}{x-2} + \dfrac{3+2x}{5x-10}$ $\dfrac{34-9x}{15(x-2)}$

45. $\dfrac{3}{x+3} + \dfrac{5}{x^2+6x+9} - \dfrac{x}{x^2-9}$ $\dfrac{2(x^2+x-21)}{(x+3)^2(x-3)}$

46. $\dfrac{x+2}{x^2-2x-3} + \dfrac{x}{x-3} - \dfrac{4}{x+1}$ $\dfrac{x^2-2x+14}{(x-3)(x+1)}$

Add or subtract as indicated. Write each answer in lowest terms.

47. $\dfrac{4}{3x^2y^3} + \dfrac{5}{3x^2y^3} - \dfrac{3}{x^2y^3}$

48. $\dfrac{7}{2xy^4} + \dfrac{1}{2xy^4} - \dfrac{4}{xy^4}$

49. $\dfrac{x-5}{2x} - \dfrac{x+5}{2x} - \dfrac{5}{x}$

50. $\dfrac{x+4}{4x} - \dfrac{x-4}{4x} - \dfrac{2}{x}$

51. $\dfrac{3}{2x+10} + \dfrac{8}{3x+15}$

52. $\dfrac{10}{3x-3} + \dfrac{1}{7x-7}$

53. $\dfrac{-2}{x^2-3x} - \dfrac{1}{x^3-3x^2}$

54. $\dfrac{-3}{2a+8} - \dfrac{8}{a^2+4a}$

55. $\dfrac{ab}{a^2-b^2} + \dfrac{b}{a+b}$

56. $\dfrac{x}{25-x^2} + \dfrac{2}{3x-15}$

57. $\dfrac{5}{x^2-4} - \dfrac{3}{x^2+4x+4}$

58. $\dfrac{3z}{z^2-9} - \dfrac{2}{3-z}$

59. $\dfrac{2}{a^2+2a+1} + \dfrac{3}{a^2-1}$ $\dfrac{5a+1}{(a+1)^2(a-1)}$

60. $\dfrac{9x+2}{3x^2-2x-8} + \dfrac{7}{3x^2+x-4}$ $\dfrac{3x-4}{(x-2)(x-1)}$

61. In your own words, explain how to add rational expressions with different denominators.

62. In your own words, explain how to multiply rational expressions.

63. In your own words, explain how to divide rational expressions.

64. In your own words, explain how to subtract rational expressions with different denominators.

---

27. $\dfrac{-13x+4}{(x+4)(x-4)}$

28. $\dfrac{7x+5}{(x-5)(x+5)}$

39. $\dfrac{5(x^2+x-4)}{(3x+2)(x+3)(2x-5)}$

51. $\dfrac{25}{6(x+5)}$

52. $\dfrac{73}{21(x-1)}$

53. $\dfrac{-2x-1}{x^2(x-3)}$

54. $\dfrac{-3a-16}{2a(a+4)}$

55. $\dfrac{2ab-b^2}{(a+b)(a-b)}$

56. $\dfrac{-x+10}{3(x-5)(x+5)}$

57. $\dfrac{2x+16}{(x+2)^2(x-2)}$

58. $\dfrac{5z+6}{(z-3)(z+3)}$

Perform the indicated operation. Write each answer in lowest terms.

65. $\left(\dfrac{2}{3} - \dfrac{1}{x}\right) \cdot \left(\dfrac{3}{x} + \dfrac{1}{2}\right)$

66. $\left(\dfrac{2}{3} - \dfrac{1}{x}\right) \div \left(\dfrac{3}{x} + \dfrac{1}{2}\right)$

67. $\left(\dfrac{1}{x} + \dfrac{2}{3}\right) - \left(\dfrac{1}{x} - \dfrac{2}{3}\right) \dfrac{4}{3}$

68. $\left(\dfrac{1}{2} + \dfrac{2}{x}\right) - \left(\dfrac{1}{2} - \dfrac{1}{x}\right) \dfrac{3}{x}$

69. $\left(\dfrac{2a}{3}\right)^2 \div \left(\dfrac{a^2}{a+1} - \dfrac{1}{a+1}\right) \quad \dfrac{4a^2}{9(a-1)}$

70. $\left(\dfrac{x+2}{2x} - \dfrac{x-2}{2x}\right) \cdot \left(\dfrac{5x}{4}\right)^2 \quad \dfrac{25x}{8}$

71. $\left(\dfrac{2x}{3}\right)^2 \div \left(\dfrac{x}{3}\right)^2 \quad 4$

72. $\left(\dfrac{2x}{3}\right)^2 \cdot \left(\dfrac{3}{x}\right)^2 \quad 4$

73. $\dfrac{x}{x^2 - 9} + \dfrac{3}{x^2 - 6x + 9} - \dfrac{1}{x+3} \quad \dfrac{6x}{(x+3)(x-3)^2}$

74. $\dfrac{3}{x^2 - 9} - \dfrac{x}{x^2 - 6x + 9} + \dfrac{1}{x+3} \quad \dfrac{-6x}{(x-3)^2(x+3)}$

75. $\left(\dfrac{x}{x+1} - \dfrac{x}{x-1}\right) \div \dfrac{x}{2x+2} \quad \dfrac{-4}{x-1}$

76. $\dfrac{x}{2x+2} \div \left(\dfrac{x}{x+1} + \dfrac{x}{x-1}\right) \quad \dfrac{x-1}{4x}$

77. $\dfrac{4}{x} \cdot \left(\dfrac{2}{x+2} - \dfrac{2}{x-2}\right)$

78. $\dfrac{1}{x+1} \cdot \left(\dfrac{5}{x} + \dfrac{2}{x-3}\right)$

Use a grapher to support the results of each exercise.

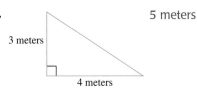 **79.** Exercise 3

 80. Exercise 4

 81. Exercise 27

 82. Exercise 28

If $f(x) = \dfrac{3x}{x^2 - 4}$ and $g(x) = \dfrac{6}{x^2 + 2x}$, find the following.

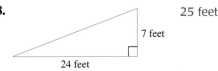 **83.** $(f + g)(x)$

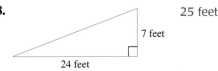 **84.** $(f \cdot g)(x) \quad \dfrac{18}{x^3 + 2x^2 - 4x - 8}$

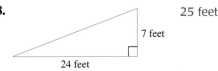 **85.** $\left(\dfrac{f}{g}\right)(x) \quad \dfrac{x^2}{2(x-2)}$

 86. $(f - g)(x) \quad \dfrac{3x^2 - 6x + 12}{x(x+2)(x-2)}$

Review Exercises

Use the distributive property to multiply the following. See Section 1.2.

87. $12\left(\dfrac{2}{3} + \dfrac{1}{6}\right) \quad 10$

88. $14\left(\dfrac{1}{7} + \dfrac{3}{14}\right) \quad 5$

89. $x^2\left(\dfrac{4}{x^2} + 1\right) \quad 4 + x^2$

90. $5y^2\left(\dfrac{1}{y^2} - \dfrac{1}{5}\right) \quad 5 - y^2$

65. $\dfrac{2x^2 + 9x - 18}{6x^2}$

66. $\dfrac{2(2x-3)}{3(6+x)}$

77. $-\dfrac{32}{x(x+2)(x-2)}$

78. $\dfrac{7x-15}{x(x-3)(x+1)}$

83. $\dfrac{3x^2 + 6x - 12}{x(x+2)(x-2)}$

103. $\dfrac{1-3x}{x^3}$

104. $\dfrac{9x+7}{x^4}$

Find each root. See Section 1.3.

91. $\sqrt{100} \quad 10$

92. $\sqrt{25} \quad 5$

93. $\sqrt[3]{8} \quad 2$

94. $\sqrt[3]{27} \quad 3$

95. $\sqrt[4]{81} \quad 3$

96. $\sqrt[4]{16} \quad 2$

Use the Pythagorean theorem to find each unknown length of a right triangle. See Section 5.8.

97.

3 meters · 4 meters · 5 meters

98.

7 feet · 24 feet · 25 feet

A Look Ahead

EXAMPLE

Add $x^{-1} + 3x^{-2}$.

Solution:

$$x^{-1} + 3x^{-2} = \dfrac{1}{x} + \dfrac{3}{x^2}$$

$$= \dfrac{1 \cdot x}{x \cdot x} + \dfrac{3}{x^2}$$

$$= \dfrac{x}{x^2} + \dfrac{3}{x^2}$$

$$= \dfrac{x+3}{x^2}$$

Perform the indicated operation. See the preceding example.

99. $x^{-1} + (2x)^{-1} \quad \dfrac{3}{2x}$

100. $3y^{-1} + (4y)^{-1} \quad \dfrac{13}{4y}$

101. $4x^{-2} - 3x^{-1} \quad \dfrac{4 - 3x}{x^2}$

102. $(4x)^{-2} - (3x)^{-1} \quad \dfrac{3 - 16x}{48x^2}$

103. $x^{-3}(2x + 1) - 5x^{-2}$

104. $4x^{-3} + x^{-4}(5x + 7)$

6.4 | SIMPLIFYING COMPLEX FRACTIONS

TAPE IA 6.4

O B J E C T I V E S

 Identify complex fractions.

 Simplify complex fractions by simplifying the numerator and denominator and then dividing.

3 Simplify complex fractions by multiplying by a common denominator.

4 Simplify expressions with negative exponents.

 A rational expression whose numerator, denominator, or both contain one or more rational expressions is called a **complex rational expression** or a **complex fraction**.

EXAMPLES OF COMPLEX FRACTIONS

$$\frac{\dfrac{1}{a}}{\dfrac{b}{2}} \qquad \frac{\dfrac{x}{2y^2}}{\dfrac{6x-2}{9y}} \qquad \frac{x+\dfrac{1}{y}}{y+1}$$

The parts of a complex fraction are

$$\left.\frac{\dfrac{x}{y+2}}{7+\dfrac{1}{y}}\right.$$

← Numerator of complex fraction.

← Main fraction bar.

← Denominator of complex fraction.

Our goal in this section is to simplify complex fractions. A complex fraction is simplified when it is in the form $\dfrac{P}{Q}$, where P and Q are polynomials that have no common factors. Two methods of simplifying complex fractions are introduced. The first method evolves from the definition of a fraction as a quotient.

TO SIMPLIFY A COMPLEX FRACTION: METHOD I

Step 1. Simplify the numerator and the denominator of the complex fraction so that each is a single fraction.

Step 2. Perform the indicated division by multiplying the numerator of the complex fraction by the reciprocal of the denominator of the complex fraction.

Step 3. Simplify if possible.

EXAMPLE 1 Simplify each complex fraction.

a. $\dfrac{\dfrac{2x}{27y^2}}{\dfrac{6x^2}{9}}$
b. $\dfrac{\dfrac{5x}{x+2}}{\dfrac{10}{x-2}}$
c. $\dfrac{x+\dfrac{1}{y}}{y+\dfrac{1}{x}}$

Solution: **a.** The numerator of the complex fraction is already a single fraction, and so is the denominator. Perform the indicated division by multiplying the numerator, $\dfrac{2x}{27y^2}$, by the reciprocal of the denominator, $\dfrac{6x^2}{9}$. Then simplify.

$$\dfrac{\dfrac{2x}{27y^2}}{\dfrac{6x^2}{9}} = \dfrac{2x}{27y^2} \div \dfrac{6x^2}{9}$$

$$= \dfrac{2x}{27y^2} \cdot \dfrac{9}{6x^2} \qquad \text{Multiply by the reciprocal of } \dfrac{6x^2}{9}.$$

$$= \dfrac{2x \cdot 9}{27y^2 \cdot 6x^2}$$

$$= \dfrac{1}{9xy^2}$$

b. $\dfrac{\dfrac{5x}{x+2}}{\dfrac{10}{x-2}} = \dfrac{5x}{x+2} \cdot \dfrac{x-2}{10} \qquad \text{Multiply by the reciprocal of } \dfrac{10}{x-2}.$

$$= \dfrac{5x(x-2)}{2 \cdot 5(x+2)}$$

$$= \dfrac{x(x-2)}{2(x+2)} \qquad \text{Simplify.}$$

c. First simplify the numerator and the denominator of the complex fraction separately so that each is a single fraction.

$$\dfrac{x+\dfrac{1}{y}}{y+\dfrac{1}{x}} = \dfrac{\dfrac{x \cdot y}{1 \cdot y} + \dfrac{1}{y}}{\dfrac{y \cdot x}{1 \cdot x} + \dfrac{1}{x}} \qquad \begin{array}{l}\text{The LCD is } y.\\[2mm]\text{The LCD is } x.\end{array}$$

$$= \dfrac{\dfrac{xy+1}{y}}{\dfrac{yx+1}{x}} \qquad \begin{array}{l}\text{Add.}\\[2mm]\text{Add.}\end{array}$$

$$= \frac{xy + 1}{y} \cdot \frac{x}{xy + 1} \qquad \text{Multiply by the reciprocal of } \frac{yx + 1}{x}.$$

$$= \frac{x(xy + 1)}{y(xy + 1)}$$

$$= \frac{x}{y}$$

3 Next we look at another method of simplifying complex fractions. With this method we multiply the numerator and the denominator of the complex fraction by the LCD of all fractions in the complex fraction.

To Simplify a Complex Fraction: Method II

Step 1. Multiply the numerator and the denominator of the complex fraction by the LCD of the fractions in both the numerator and the denominator.

Step 2. Simplify.

EXAMPLE 2 Simplify each complex fraction.

a. $\dfrac{\dfrac{5x}{x + 2}}{\dfrac{10}{x - 2}}$

b. $\dfrac{x + \dfrac{1}{y}}{y + \dfrac{1}{x}}$

Solution: **a.** The least common denominator of $\dfrac{5x}{x + 2}$ and $\dfrac{10}{x - 2}$ is $(x + 2)(x - 2)$. Multiply both the numerator, $\dfrac{5x}{x + 2}$, and the denominator, $\dfrac{10}{x - 2}$, by the LCD.

$$\frac{\dfrac{5x}{x + 2}}{\dfrac{10}{x - 2}} = \frac{\left(\dfrac{5x}{x + 2}\right) \cdot (x + 2)(x - 2)}{\left(\dfrac{10}{x - 2}\right) \cdot (x + 2)(x - 2)} \qquad \begin{array}{l}\text{Multiply numerator and} \\ \text{denominator by the LCD.}\end{array}$$

$$= \frac{5x \cdot (x - 2)}{2 \cdot 5 \cdot (x + 2)} \qquad \text{Simplify.}$$

$$= \frac{x(x - 2)}{2(x + 2)} \qquad \text{Simplify.}$$

b. The least common denominator of $\dfrac{1}{y}$ and $\dfrac{1}{x}$ is xy.

$$\frac{x + \dfrac{1}{y}}{y + \dfrac{1}{x}} = \frac{\left(x + \dfrac{1}{y}\right) \cdot xy}{\left(y + \dfrac{1}{x}\right) \cdot xy}$$

Multiply numerator and
denominator by the LCD.

$$= \frac{x \cdot xy + \dfrac{1}{y} \cdot x \; y}{y \cdot xy + \dfrac{1}{x} \cdot x \; y}$$

Apply the distributive property.

$$= \frac{x^2y + x}{xy^2 + y}$$

Simplify.

$$= \frac{x\,(xy + 1)}{y\,(xy + 1)}$$

Factor.

$$= \frac{x}{y}$$

Simplify.

4

If an expression contains negative exponents, write the expression as an equivalent expression with positive exponents.

EXAMPLE 3

Simplify.

$$\frac{x^{-1} + 2xy^{-1}}{x^{-2} - x^{-2}y^{-1}}$$

Solution:

This fraction does not appear to be a complex fraction. If we write it by using only positive exponents, however, we see that it is a complex fraction.

$$\frac{x^{-1} + 2xy^{-1}}{x^{-2} - x^{-2}y^{-1}} = \frac{\dfrac{1}{x} + \dfrac{2x}{y}}{\dfrac{1}{x^2} - \dfrac{1}{x^2y}}$$

The LCD of $\dfrac{1}{x}, \dfrac{2x}{y}, \dfrac{1}{x^2}$, and $\dfrac{1}{x^2y}$ is x^2y. Multiply both the numerator and denominator by x^2y.

$$= \frac{\left(\dfrac{1}{x} + \dfrac{2x}{y}\right) \cdot x^2y}{\left(\dfrac{1}{x^2} - \dfrac{1}{x^2y}\right) \cdot x^2y}$$

$$= \frac{\dfrac{1}{x} \cdot x^2y + \dfrac{2x}{y} \cdot x^2y}{\dfrac{1}{x^2} \cdot x^2y - \left(\dfrac{1}{x^2y}\right) \cdot x^2y}$$ Apply the distributive property.

$$= \frac{xy + 2x^3}{y - 1}$$ Simplify.

EXERCISE SET 6.4

11. $\dfrac{9x - 18}{9x^2 - 4}$ **12.** $\dfrac{15x - 9}{3x^2 + 2}$ **15.** $\dfrac{xy^2}{x^2 + y^2}$ **16.** $\dfrac{b + a^3}{ab}$ **17.** $\dfrac{2b^2 + 3a}{b^2 - ab}$

Simplify each complex fraction. See Examples 1 and 2.

1. $\dfrac{\frac{1}{3}}{\frac{2}{5}}$ $\dfrac{5}{6}$ **2.** $\dfrac{\frac{3}{5}}{\frac{4}{5}}$ $\dfrac{3}{4}$ **3.** $\dfrac{\frac{4}{x}}{\frac{5}{2x}}$ $\dfrac{8}{5}$ **4.** $\dfrac{\frac{5}{2x}}{\frac{4}{x}}$ $\dfrac{5}{8}$

24. $\dfrac{\frac{5}{6} - \frac{1}{2}}{\frac{1}{3} + \frac{1}{8}}$ $\dfrac{8}{11}$ **25.** $\dfrac{\frac{x+1}{3}}{\frac{2x-1}{6}}$ **26.** $\dfrac{\frac{x+3}{12}}{\frac{4x-5}{15}}$

5. $\dfrac{\frac{10}{3x}}{\frac{5}{6x}}$ 4 **6.** $\dfrac{\frac{15}{2x}}{\frac{5}{6x}}$ 9 **7.** $\dfrac{1 + \frac{2}{5}}{2 + \frac{3}{5}}$ $\dfrac{7}{13}$ **8.** $\dfrac{2 + \frac{1}{7}}{3 - \frac{4}{7}}$ $\dfrac{15}{17}$

27. $\dfrac{\frac{x}{3}}{\frac{2}{x+1}}$ $\dfrac{x(x+1)}{6}$ **28.** $\dfrac{\frac{x-1}{5}}{\frac{3}{x}}$ **29.** $\dfrac{\frac{2}{x} + 3}{\frac{4}{x^2} - 9}$ $\dfrac{x}{2 - 3x}$

9. $\dfrac{\frac{4}{x-1}}{\frac{x}{x-1}}$ $\dfrac{4}{x}$ **10.** $\dfrac{\frac{x}{x+2}}{\frac{2}{x+2}}$ $\dfrac{x}{2}$ **11.** $\dfrac{1 - \frac{2}{x}}{x - \frac{4}{9x}}$ **12.** $\dfrac{5 - \frac{3}{x}}{x + \frac{2}{3x}}$

30. $\dfrac{2 + \frac{1}{x}}{4x - \frac{1}{x}}$ **31.** $\dfrac{1 - \frac{x}{y}}{\frac{x^2}{y^2} - 1}$ **32.** $\dfrac{1 - \frac{2}{x}}{x - \frac{4}{x}}$ $\dfrac{1}{x+2}$

13. $\dfrac{\frac{1}{x+1} - 1}{\frac{1}{x-1} + 1}$ $\dfrac{1-x}{1+x}$ **14.** $\dfrac{1 + \frac{1}{x-1}}{1 - \frac{1}{x+1}}$ $\dfrac{x+1}{x-1}$

33. $\dfrac{\frac{-2x}{x-y}}{\frac{y}{x^2}}$ **34.** $\dfrac{\frac{7y}{x^2+xy}}{\frac{y^2}{x^2}}$ **35.** $\dfrac{\frac{2}{x} + \frac{1}{x^2}}{\frac{y}{x^2}}$ $\dfrac{2x+1}{y}$

Simplify. See Example 3.

15. $\dfrac{x^{-1}}{x^{-2} + y^{-2}}$ **16.** $\dfrac{a^{-3} + b^{-1}}{a^{-2}}$ **17.** $\dfrac{2a^{-1} + 3b^{-2}}{a^{-1} - b^{-1}}$

18. $\dfrac{x^{-1} + y^{-1}}{3x^{-2} + 5y^{-2}}$ **19.** $\dfrac{1}{x - x^{-1}}$ **20.** $\dfrac{x^{-2}}{x + 3x^{-1}}$

36. $\dfrac{\frac{5}{x^2} - \frac{2}{x}}{\frac{1}{x} + 2}$ **37.** $\dfrac{\frac{x}{9} - \frac{1}{x}}{1 + \frac{3}{x}}$ $\dfrac{x-3}{9}$ **38.** $\dfrac{\frac{x}{4} - \frac{4}{x}}{1 - \frac{4}{x}}$ $\dfrac{x+4}{4}$

39. $\dfrac{\frac{x-1}{x^2-4}}{1 + \frac{1}{x-2}}$ $\dfrac{1}{x+2}$ **40.** $\dfrac{\frac{2}{x+5} + \frac{4}{x+3}}{\frac{3x+13}{x^2+8x+15}}$ 2

Simplify.

21. $\dfrac{\frac{x+1}{7}}{\frac{x+2}{7}}$ $\dfrac{x+1}{x+2}$ **22.** $\dfrac{\frac{y}{10}}{\frac{x+1}{10}}$ $\dfrac{y}{x+1}$ **23.** $\dfrac{\frac{1}{2} - \frac{1}{3}}{\frac{3}{4} + \frac{2}{5}}$ $\dfrac{10}{69}$

41. $\dfrac{\frac{4}{5-x} + \frac{5}{x-5}}{\frac{2}{x} + \frac{3}{x-5}}$ **42.** $\dfrac{\frac{3}{x-4} - \frac{2}{4-x}}{\frac{2}{x-4} - \frac{2}{x}}$ $\dfrac{5x}{8}$

18. $\dfrac{xy^2 + yx^2}{3y^2 + 5x^2}$ **19.** $\dfrac{x}{x^2 - 1}$ **20.** $\dfrac{1}{x^3 + 3x}$ **25.** $\dfrac{2(x+1)}{2x-1}$ **26.** $\dfrac{5(x+3)}{4(4x-5)}$ **28.** $\dfrac{x(x-1)}{15}$ **30.** $\dfrac{1}{2x-1}$

31. $-\dfrac{y}{x+y}$ **33.** $-\dfrac{2x^3}{y(x-y)}$ **34.** $\dfrac{7x}{y(x+y)}$ **36.** $\dfrac{5-2x}{x(1+2x)}$ **41.** $\dfrac{x}{5x-10}$

43. $\dfrac{\dfrac{x+2}{x} - \dfrac{2}{x-1}}{\dfrac{x+1}{x} + \dfrac{x+1}{x-1}} \dfrac{x-2}{2x-1}$

44. $\dfrac{\dfrac{5}{a+2} - \dfrac{1}{a-2}}{\dfrac{3}{2+a} + \dfrac{6}{2-a}} \dfrac{-4(a-3)}{3(a+6)}$

45. $\dfrac{\dfrac{x-2}{x+2} + \dfrac{x+2}{x-2}}{\dfrac{x-2}{x+2} - \dfrac{x+2}{x-2}} \dfrac{x^2+4}{4x}$

46. $\dfrac{\dfrac{x-1}{x+1} - \dfrac{x+1}{x-1}}{\dfrac{x-1}{x+1} + \dfrac{x+1}{x-1}} \dfrac{2x}{x^2+1}$

47. $\dfrac{\dfrac{2}{y^2} - \dfrac{5}{xy} - \dfrac{3}{x^2}}{\dfrac{2}{y^2} + \dfrac{7}{xy} + \dfrac{3}{x^2}} \dfrac{x-3y}{x+3y}$

48. $\dfrac{\dfrac{2}{x^2} - \dfrac{1}{xy} - \dfrac{1}{y^2}}{\dfrac{1}{x^2} - \dfrac{3}{xy} + \dfrac{2}{y^2}} \dfrac{2y+x}{y-2x}$

49. $\dfrac{a^{-1}+1}{a^{-1}-1} \dfrac{1+a}{1-a}$

50. $\dfrac{a^{-1}-4}{4+a^{-1}} \dfrac{1-4a}{4a+1}$

51. $\dfrac{3x^{-1}+(2y)^{-1}}{x^{-2}} \dfrac{x^2+6xy}{2y}$

52. $\dfrac{5x^{-2}-3y^{-1}}{x^{-1}+y^{-1}} \dfrac{5y-3x^2}{xy+x^2}$

53. $\dfrac{2a^{-1}+(2a)^{-1}}{a^{-1}+2a^{-2}} \dfrac{5a}{2a+4}$

54. $\dfrac{a^{-1}+2a^{-2}}{2a^{-1}+(2a)^{-1}} \dfrac{2a+4}{5a}$

55. $\dfrac{5x^{-1}+2y^{-1}}{x^{-2}y^{-2}} \; 5xy^2+2x^2y$

56. $\dfrac{x^{-2}y^{-2}}{5x^{-1}+2y^{-1}} \dfrac{1}{5xy^2+2yx^2}$

57. $\dfrac{5x^{-1}-2y^{-1}}{25x^{-2}-4y^{-2}} \dfrac{xy}{2x+5y}$

58. $\dfrac{3x^{-1}+3y^{-1}}{4x^{-2}-9y^{-2}} \dfrac{3xy^2+3yx^2}{4y^2-9x^2}$

59. $(x^{-1}+y^{-1})^{-1} \dfrac{xy}{x+y}$

60. $\dfrac{xy}{x^{-1}+y^{-1}} \dfrac{x^2y^2}{y+x}$

61. $\dfrac{x}{1-\dfrac{1}{1+\dfrac{1}{x}}} \; x^2+x$

62. $\dfrac{x}{1-\dfrac{1}{1-\dfrac{1}{x}}} \; -x(x-1)$

In the study of calculus, the difference quotient
$\dfrac{f(a+h)-f(a)}{h}$ *is often found and simplified. Find and simplify this quotient for each function f(x) by following steps a through d.*

a. Find $f(a+h)$.
b. Find $f(a)$.
c. Use steps a and b to find $\dfrac{f(a+h)-f(a)}{h}$.
d. Simplify the result of step c.

63. $f(x) = \dfrac{1}{x}$ **a.** $\dfrac{1}{a+h}$ **63b.** $\dfrac{1}{a}$ **63c.** $\dfrac{\dfrac{1}{a+h}-\dfrac{1}{a}}{h}$ **63d.** $\dfrac{-1}{a(a+h)}$

64. $f(x) = \dfrac{5}{x}$ **64a.** $\dfrac{5}{a+h}$ **64b.** $\dfrac{5}{a}$ **64c.** $\dfrac{\dfrac{5}{a+h}-\dfrac{5}{a}}{h}$ **64d.** $\dfrac{-5}{a(a+h)}$

65. $\dfrac{3}{x+1}$ **a.** $\dfrac{3}{a+h+1}$ **65b.** $\dfrac{3}{a+1}$ **65c.** $\dfrac{\dfrac{3}{a+h+1}-\dfrac{3}{a+1}}{h}$ **65d.** $\dfrac{-3}{(a+h+1)(a+1)}$

66. $\dfrac{2}{x^2}$ **66a.** $\dfrac{2}{(a+h)^2}$ **66b.** $\dfrac{2}{a^2}$ **66c.** $\dfrac{\dfrac{2}{(a+h)^2}-\dfrac{2}{a^2}}{h}$

66d. $\dfrac{-2(2a+h)}{a^2(a+h)^2}$ **79.** $3a^2+4a+4$ **80.** $\dfrac{4x-9y}{4x}$

Review Exercises

Simplify. See Sections 5.1 and 5.2.

67. $\dfrac{3x^3y^2}{12x} \cdot \dfrac{x^2y^2}{4}$

68. $\dfrac{-36xb^3}{9xb^2} \; -4b$

69. $\dfrac{144x^5y^5}{-16x^2y} \; -9x^3y^4$

70. $\dfrac{48x^3y^2}{-4xy} \; -12x^2y$

Solve the following. See Sections 2.6 and 2.7.

71. $|x-5|=9 \;\; \{-4,14\}$ **72.** $|2y+1|=1 \;\; \{-1,0\}$

73. $|x-5|<9 \;\; (-4,14)$ **74.** $|2x+1|\geq 1$

A Look Ahead

$(-\infty,-1] \cup [0,\infty)$

EXAMPLE

Simplify.

$$\dfrac{2(a+b)^{-1}-5(a-b)^{-1}}{4(a^2-b^2)^{-1}}$$

Solution:

$$\dfrac{2(a+b)^{-1}-5(a-b)^{-1}}{4(a^2-b^2)^{-1}} = \dfrac{\dfrac{2}{a+b}-\dfrac{5}{a-b}}{\dfrac{4}{a^2-b^2}}$$

$$= \dfrac{\left(\dfrac{2}{a+b}-\dfrac{5}{a-b}\right)\cdot(a+b)(a-b)}{\left[\dfrac{4}{(a+b)(a-b)}\right]\cdot(a+b)(a-b)}$$

$$= \dfrac{\dfrac{2}{a+b}\cdot(a+b)(a-b)-\dfrac{5}{a-b}\cdot(a+b)(a-b)}{\dfrac{4(a+b)(a-b)}{(a+b)(a-b)}}$$

$$= \dfrac{2(a-b)-5(a+b)}{4}$$

$$= \dfrac{-3a-7b}{4}, \text{ or } -\dfrac{3a+7b}{4}$$

Simplify. See the preceding example.

75. $\dfrac{1}{1-(1-x)^{-1}} \dfrac{x-1}{x}$

76. $\dfrac{1}{1+(1+x)^{-1}} \dfrac{1+x}{2+x}$

77. $\dfrac{(x+2)^{-1}+(x-2)^{-1}}{(x^2-4)^{-1}} \; 2x$

78. $\dfrac{(y-1)^{-1}-(y+4)^{-1}}{(y^2+3y-4)^{-1}} \; 5$

79. $\dfrac{3(a+1)^{-1}+4a^{-2}}{(a^3+a^2)^{-1}}$

80. $\dfrac{9x^{-1}-5(x-y)^{-1}}{4(x-y)^{-1}}$

6.5 | DIVIDING POLYNOMIALS

TAPE IA 6.5

OBJECTIVES

1. Divide a polynomial by a monomial.
2. Divide by a polynomial.

Recall that a rational expression is a quotient of polynomials. An equivalent form of a rational expression can be obtained by performing the indicated division. For example, the rational expression $\dfrac{3x^5y^2 - 15x^3y - 6x}{6x^2y^3}$ can be thought of as the polynomial $3x^5y^2 - 15x^3y - 6x$ divided by the monomial $6x^2y^3$. To perform this division of a polynomial by a monomial (which we do in Example 3), recall the following addition fact for fractions with a common denominator:

$$\frac{a}{c} + \frac{b}{c} = \frac{a + b}{c}$$

If a, b, and c are monomials, we might read this equation from right to left and gain insight into dividing a polynomial by a monomial.

TO DIVIDE A POLYNOMIAL BY A MONOMIAL

Divide each term in the polynomial by the monomial:

$$\frac{a + b}{c} = \frac{a}{c} + \frac{b}{c}, \qquad c \neq 0$$

EXAMPLE 1 Divide $10x^2 - 5x + 20$ by 5.

Solution: Divide each term of $10x^2 - 5x + 20$ by 5 and simplify.

$$\frac{10x^2 - 5x + 20}{5} = \frac{10x^2}{5} - \frac{5x}{5} + \frac{20}{5} = 2x^2 - x + 4$$

To check, see that (quotient) (divisor) = dividend, or

$$(2x^2 - x + 4)\ (5) = 10x^2 - 5x + 20$$

EXAMPLE 2 Find the quotient: $\dfrac{7a^2b - 2ab^2}{2ab^2}$.

Solution: Divide each term of the polynomial in the numerator by $2ab^2$:

$$\frac{7a^2b - 2ab^2}{2ab^2} = \frac{7a^2b}{2ab^2} - \frac{2ab^2}{2ab^2} = \frac{7a}{2b} - 1$$

EXAMPLE 3   Find the quotient: $\dfrac{3x^5y^2 - 15x^3y - 6x}{6x^2y^3}$

Solution: Divide each term in the numerator by $6x^2y^3$:

$$\frac{3x^5y^2 - 15x^3y - 6x}{6x^2y^3} = \frac{3x^5y^2}{6x^2y^3} - \frac{15x^3y}{6x^2y^3} - \frac{6x}{6x^2y^3} = \frac{x^3}{2y} - \frac{5x}{2y^2} - \frac{1}{xy^3}$$

2

To divide a polynomial by a polynomial other than a monomial, we use **long division**. Polynomial long division is similar to long division of real numbers. We review long division of real numbers by dividing 7 into 296.

Divisor: $\quad 7\overline{)296}$

$$
\begin{array}{r}
42 \\
7\overline{)296} \\
-28 \\
\hline
16 \\
-14 \\
\hline
2
\end{array}
$$

$4(7) = 28.$

Subtract and bring down the next digit in the dividend.

$2(7) = 14.$

Subtract. The remainder is 2.

The quotient is $42\dfrac{2\,(\text{remainder})}{7\,(\text{divisor})}$. To check, notice that

$$42(7) + 2 = 296, \qquad \text{the dividend.}$$

This same division process can be applied to polynomials, as shown next.

EXAMPLE 4   Divide $2x^2 - x - 10$ by $x + 2$.

Solution: $2x^2 - x - 10$ is the dividend, and $x + 2$ is the divisor.

Step 1.   Divide $2x^2$ by x.

$$
\begin{array}{r}
2x \\
x + 2\overline{)2x^2 - x - 10}
\end{array}
$$
$\dfrac{2x^2}{x} = 2x$, so $2x$ is the first term of the quotient.

Step 2.   Multiply $2x(x + 2)$.

$$
\begin{array}{r}
2x \\
x + 2\overline{)2x^2 - x - 10} \\
2x^2 + 4x
\end{array}
$$
$2x(x + 2)$

Like terms are lined up vertically.

Step 3.   Subtract $(2x^2 + 4x)$ from $(2x^2 - x - 10)$ by changing the signs of $(2x^2 + 4x)$ and adding.

$$
\begin{array}{r}
2x \\
x + 2\overline{)\ 2x^2 - x - 10} \\
-2x^2 - 4x \\
\hline
-5x
\end{array}
$$

Step 4.   Bring down the next term, -10, and start the process over.

$$
\begin{array}{r}
2x \phantom{{}- x - 10} \\
x + 2 \overline{)\ 2x^2 -\ \ x - 10} \\
\underline{-2x^2 - 4x} \\
-5x - 10
\end{array}
$$

Step 5.   Divide $-5x$ by x.

$$
\begin{array}{r}
2x - 5 \\
x + 2 \overline{)\ 2x^2 -\ \ x - 10} \\
\underline{-2x^2 - 4x} \\
-5x - 10
\end{array}
$$

$\dfrac{-5x}{x} = -5$

so -5 is the second term of the quotient.

Step 6.   Multiply $-5(x + 2)$.

$$
\begin{array}{r}
2x -\ \ 5 \\
x + 2 \overline{)\ 2x^2 -\ \ x - 10} \\
\underline{-2x^2 - 4x} \\
-5x - 10 \\
\underline{-5x - 10}
\end{array}
$$

$-5(x + 2)$

Like terms are lined up vertically.

Step 7.   Subtract $(-5x - 10)$ from $(-5x - 10)$.

$$
\begin{array}{r}
2x -\ \ 5 \\
x + 2 \overline{)\ 2x^2 -\ \ x - 10} \\
\underline{-2x^2 - 4x} \\
-5x - 10 \\
\underline{+5x + 10} \\
0
\end{array}
$$

Then $\dfrac{2x^2 - x - 10}{x + 2} = 2x - 5$. There is no remainder.

Check this result by multiplying $2x - 5$ by $x + 2$. Their product is

$$(2x - 5)(x + 2) = 2x^2 - x - 10, \quad \text{the dividend.}$$

EXAMPLE 5   Find the quotient: $\dfrac{6x^2 - 19x + 12}{3x - 5}$.

Solution:

$$
\begin{array}{r}
2x \\
3x - 5 \overline{)\ 6x^2 - 19x + 12} \\
\underline{6x^2 - 10x} \\
-9x + 12
\end{array}
$$

Divide: $\dfrac{6x^2}{3x} = 2x$.

Multiply: $2x(3x - 5)$.

Subtract by changing the signs of $6x^2 - 10x$ and adding. Bring down the next term, $+ 12$.

$$
\begin{array}{r}
2x -\ \ 3 \\
3x - 5 \overline{)\ 6x^2 - 19x + 12} \\
\underline{6x^2 - 10x} \\
-9x + 12 \\
\underline{-9x + 15} \\
-3
\end{array}
$$

Divide: $\dfrac{-9x}{3x} = -3$.

Multiply: $-3(3x - 5)$.

Subtract.

When checking, we call the **divisor** polynomial $3x - 5$. The **quotient** polynomial is $2x - 3$. The **remainder** polynomial is -3. See that

dividend = divisor · quotient + remainder

or

$$6x^2 - 19x + 12 = (3x - 5)(2x - 3) + (-3)$$
$$= 6x^2 - 19x + 15 - 3$$
$$= 6x^2 - 19x + 12$$

The division checks, so

$$\frac{6x^2 - 19x + 12}{3x - 5} = 2x - 3 - \frac{3}{3x - 5}$$

EXAMPLE 6   Divide $2x^3 + 3x^4 - 8x + 6$ by $x^2 - 1$.

Solution: Before dividing, we write both the divisor and the dividend in descending order of exponents. Any "missing powers" can be represented by the product of 0 and the variable raised to the missing power. There is no x^2 term in the dividend, so include $0x^2$ to represent the missing term. Also, there is no x term in the divisor, so include $0x$ in the divisor.

$$
\begin{array}{r}
3x^2 + 2x + 3 \\
x^2 + 0x - 1 \overline{\smash{)}\, 3x^4 + 2x^3 + 0x^2 - 8x + 6} \\
\underline{3x^4 + 0x^3 - 3x^2} \\
2x^3 + 3x^2 - 8x \\
\underline{2x^3 + 0x^2 - 2x} \\
3x^2 - 6x + 6 \\
\underline{3x^2 + 0x - 3} \\
-6x + 9
\end{array}
$$

$\dfrac{3x^4}{x^2} = 3x^2$.

$3x^2(x^2 + 0x - 1)$.
Subtract. Bring down $-8x$.
$2x^3/x^2 = 2x$, a term of the quotient.
$2x(x^2 + 0x - 1)$.
Subtract. Bring down 6.
$3x^2/x^2 = 3$, a term of the quotient.
$3(x^2 + 0x - 1)$.
Subtract.

The division process is finished when the degree of the remainder polynomial is less than the degree of the divisor. Thus,

$$\frac{3x^4 + 2x^3 - 8x + 6}{x^2 - 1} = 3x^2 + 2x + 3 + \frac{-6x + 9}{x^2 - 1}$$

To check, see that

$$3x^4 + 2x^3 - 8x + 6 = (x^2 - 1)(3x^2 + 2x + 3) + (-6x + 9)$$

EXAMPLE 7   Divide $27x^3 + 8$ by $2 + 3x$.

Solution: Write both the divisor and the dividend in descending order of exponents. Replace the missing terms in the dividend with $0x^2$ and $0x$.

$$
\begin{array}{r}
9x^2 - 6x + 4 \\
3x + 2\overline{)27x^3 + 0x^2 + 0x + 8} \\
\underline{27x^3 + 18x^2} \\
-18x^2 + 0x \\
\underline{-18x^2 - 12x} \\
12x + 8 \\
\underline{12x + 8}
\end{array}
$$

$9x^2(3x + 2)$.

Subtract. Bring down $0x$.

$-6x(3x + 2)$.

Subtract. Bring down 8.

$4(3x + 2)$.

Thus, $\dfrac{27x^3 + 8}{3x + 2} = 9x^2 - 6x + 4$.

EXERCISE SET 6.5   **5.** $2x^2 + 3x - 2$ **6.** $3x^2 + 37x + 12$ **7.** $x^2 + 2x + 1$ **8.** $3x - 1 + \dfrac{2}{x}$

Find each quotient. See Examples 1 through 3.

1. Divide $4a^2 + 8a$ by $2a$. $2a + 4$

2. Divide $6x^4 - 3x^3$ by $3x^2$. $2x^2 - x$

3. $\dfrac{12a^5b^2 + 16a^4b}{4a^4b}$ $3ab + 4$

4. $\dfrac{4x^3y + 12x^2y^2 - 4xy^3}{4xy}$ $x^2 + 3xy - y^2$

5. $\dfrac{4x^2y^2 + 6xy^2 - 4y^2}{2y^2}$

6. $\dfrac{6x^5 + 74x^4 + 24x^3}{2x^3}$

7. $\dfrac{4x^2 + 8x + 4}{4}$

8. $\dfrac{15x^3 - 5x^2 + 10x}{5x^2}$

9. A board of length $3x^4 + 6x^2 - 18$ meters is to be cut into three pieces of the same length. Find the length of each piece. $(x^4 + 2x^2 - 6)$ meters

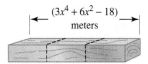

(3x⁴ + 6x² – 18) meters

10. The perimeter of a regular hexagon is given to be $12x^5 - 48x^3 + 3$ miles. Find the length of each side.

$\left(2x^5 - 8x^3 + \dfrac{1}{2}\right)$ miles

Find each quotient. See Examples 4 through 7.

11. $\dfrac{x^2 + 3x + 3}{x + 2}$ **12.** $\dfrac{y^2 + 7y + 10}{y + 5}$ $y + 2$

13. $\dfrac{2x^2 - 6x - 8}{x + 1}$ $2x - 8$ **14.** $\dfrac{3x^2 + 19x + 20}{x + 5}$ $3x + 4$

15. Divide $2x^2 + 3x - 2$ by $2x + 4$ $x - \dfrac{1}{2}$

16. Divide $6x^2 - 17x - 3$ by $3x - 9$ $2x + \dfrac{1}{3}$

17. $\dfrac{4x^3 + 7x^2 + 8x + 20}{2x + 4}$ **18.** $\dfrac{18x^3 + x^2 - 90x - 5}{9x^2 - 45}$

19. If the area of the rectangle is $15x^2 - 29x - 14$ square inches, and its length is $5x + 2$ inches, find its width. $(3x - 7)$ in.

(5x + 2) inches

20. If the area of a parallelogram is $2x^2 - 17x + 35$ square centimeters and its base is $2x - 7$ centimeters, find its height. $(x - 5)$ cm

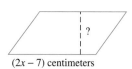

(2x – 7) centimeters

11. $x + 1 + \dfrac{1}{x + 2}$ **17.** $2x^2 - \dfrac{1}{2}x + 5$ **18.** $2x + \dfrac{1}{9}$

Find each quotient.

21. Divide $25a^2b^{12}$ by $10a^5b^7$. $\dfrac{5b^5}{2a^3}$

22. Divide $12a^2b^3$ by $8a^7b$. $\dfrac{3b^2}{2a^5}$

23. $\dfrac{x^6y^6 - x^3y^3}{x^3y^3}$ $x^3y^3 - 1$

24. $\dfrac{25xy^2 + 75xyz + 125x^2yz}{-5x^2y}$ $-\dfrac{5y}{x} - \dfrac{15z}{x} - 25z$

25. $\dfrac{a^2 + 4a + 3}{a + 1}$ $a + 3$ **26.** $\dfrac{3x^2 - 14x + 16}{x - 2}$ $3x - 8$

27. $\dfrac{2x^2 + x - 10}{x - 2}$ $2x + 5$ **28.** $\dfrac{x^2 - 7x + 12}{x - 5}$

29. Divide $-16y^3 + 24y^4$ by $-4y^2$. $4y - 6y^2$

30. Divide $-20a^2b + 12ab^2$ by $-4ab$. $5a - 3b$

31. $\dfrac{2x^2 + 13x + 15}{x - 5}$ **32.** $\dfrac{2x^2 + 13x + 5}{2x + 3}$

33. $\dfrac{20x^2y^3 + 6xy^4 - 12x^3y^5}{2xy^3}$ $10x + 3y - 6x^2y^2$

34. $\dfrac{3x^2y + 6x^2y^2 + 3xy}{3xy}$ $x + 2xy + 1$

35. $\dfrac{6x^2 + 16x + 8}{3x + 2}$ $2x + 4$ **36.** $\dfrac{x^2 - 25}{x + 5}$ $x - 5$

37. $\dfrac{2y^2 + 7y - 15}{2y - 3}$ $y + 5$ **38.** $\dfrac{3x^2 - 4x + 6}{x - 2}$ $3x + 2 + \dfrac{10}{x - 2}$

39. Divide $4x^2 - 9$ by $2x - 3$. $2x + 3$

40. Divide $8x^2 + 6x - 27$ by $4x + 9$. $2x - 3$

41. Divide $2x^3 + 6x - 4$ by $x + 4$. $2x^2 - 8x + 38 - \dfrac{156}{x + 4}$

42. Divide $4x^3 - 5x$ by $2x - 1$.

43. Divide $3x^2 - 4$ by $x - 1$. $3x + 3 - \dfrac{1}{x + 1}$

44. Divide $x^2 - 9$ by $x + 4$. $x - 4 + \dfrac{7}{x + 4}$

45. $\dfrac{-13x^3 + 2x^4 + 16x^2 - 9x + 20}{5 - x}$ $-2x^3 + 3x^2 - x + 4$

46. $\dfrac{5x^2 - 5x + 2x^3 + 20}{4 + x}$ $2x^2 - 3x + 7 - \dfrac{8}{x + 4}$

47. Divide $3x^5 - x^3 + 4x^2 - 12x - 8$ by $x^2 - 2$.

48. Divide $-8x^3 + 2x^4 + 19x^2 - 33x + 15$ by $x^2 - x + 5$. $2x^2 - 6x + 3$

49. $\dfrac{3x^3 - 5}{3x^2}$ $x - \dfrac{5}{3x^2}$ **50.** $\dfrac{14x^3 - 2}{7x - 1}$

51. Find $P(1)$ for the polynomial function $P(x) = 3x^3 + 2x^2 - 4x + 3$. Next divide $3x^3 + 2x^2 - 4x + 3$ by $x - 1$. Compare the remainder with $P(1)$.   4

52. Find $P(-2)$ for the polynomial function $P(x) = x^3 - 4x^2 - 3x + 5$. Next divide $x^3 - 4x^2 - 3x + 5$ by $x + 2$. Compare the remainder with $P(-2)$. -13

53. Find $P(-3)$ for the polynomial $P(x) = 5x^4 - 2x^2 + 3x - 6$. Next, divide $5x^4 - 2x^2 + 3x - 6$ by $x + 3$. Compare the remainder with $P(-3)$.   372

54. Find $P(2)$ for the polynomial function $P(x) = -4x^4 + 2x^3 - 6x + 3$. Next, divide $-4x^4 + 2x^3 - 6x + 3$ by $x - 2$. Compare the remainder with $P(2)$. -57

55. Write down any patterns you noticed from Exercises 51–54.

56. Explain how to check polynomial long division.

57. Try performing the following division without changing the order of the terms. Describe why this makes the process more complicated. Then perform the division again after putting the terms in the dividend in descending order of exponents.

$$\dfrac{4x^2 - 12x - 12 + 3x^3}{x - 2}$$

$$3x^2 + 10x + 8 + \dfrac{4}{x - 2}$$

Review Exercises

Insert $<$, $>$, or $=$ to make each statement true. See Section 1.3.

58. $3^2 \underline{\ =\ } (-3)^2$ **59.** $(-5)^2 \underline{\ =\ } 5^2$

60. $-2^3 \underline{\ =\ } (-2)^3$ **61.** $3^4 \underline{\ =\ } (-3)^4$

Solve each inequality. See Section 2.7.

62. $|x + 5| < 4$ $(-9, -1)$

63. $|x - 1| \le 8$ $[-7, 9]$

64. $|2x + 7| \ge 9$ $(-\infty, -8] \cup [1, \infty)$

65. $|4x + 2| > 10$ $(-\infty, -3) \cup (2, \infty)$

A Look Ahead

EXAMPLE Divide $x^2 - \dfrac{7}{2}x + 4$ by $x + 2$.

28. $x - 2 + \dfrac{2}{x - 5}$ **31.** $2x + 23 + \dfrac{130}{x - 5}$ **32.** $x + 5 - \dfrac{10}{2x + 3}$ **42.** $2x^2 + x - 2 - \dfrac{2}{2x - 1}$

47. $3x^3 + 5x + 4 - \dfrac{2x}{x^2 - 2}$ **50.** $2x^2 + \dfrac{2}{7}x + \dfrac{2}{49} - \dfrac{96}{49(7x - 1)}$

Solution:

$$x + 2 \overline{\smash{\big)}\, x^2 - \frac{7}{2}x + 4} \quad \frac{x - \frac{11}{2}}{}$$

$$\begin{array}{r} x - \frac{11}{2} \\ x + 2 \overline{\smash{\big)}\ x^2 - \frac{7}{2}x + 4} \\ \underline{x^2 + 2x} \\ -\frac{11}{2}x + 4 \\ \underline{-\frac{11}{2}x - 11} \\ 15 \end{array}$$

The quotient is $x - \frac{11}{2} + \frac{15}{x + 2}$.

Find each quotient. See the preceding example.

66. $\dfrac{x^4 + \frac{2}{3}x^3 + x}{x - 1}$

67. $\dfrac{2x^3 + \frac{9}{2}x^2 - 4x - 10}{x + 2}$

68. $\dfrac{3x^4 - x - x^3 + \frac{1}{2}}{2x - 1}$

69. $\dfrac{2x^4 + \frac{1}{2}x^3 + x^2 + x}{x - 2}$

70. $\dfrac{5x^4 - 2x^2 + 10x^3 - 4x}{5x + 10}$ $x^3 - \frac{2}{5}x$

71. $\dfrac{9x^5 + 6x^4 - 6x^2 - 4x}{3x + 2}$ $3x^4 - 2x$

66. $x^3 + \frac{5}{3}x^2 + \frac{5}{3}x + \frac{8}{3} + \frac{8}{3(x - 1)}$ **67.** $2x^2 + \frac{1}{2}x - 5$ **68.** $\frac{3}{2}x^3 + \frac{1}{4}x^2 + \frac{1}{8}x - \frac{7}{16} + \frac{1}{16(2x - 1)}$

69. $2x^3 + \frac{9}{2}x^2 + 10x + 21 + \frac{42}{x - 2}$

6.6 SYNTHETIC DIVISION AND THE REMAINDER THEOREM

TAPE IA 6.6

OBJECTIVES

 Use synthetic division to divide a polynomial by a binomial.

 Use the remainder theorem to evaluate polynomials.

1 When a polynomial is to be divided by a binomial of the form $x - c$, a shortcut process called **synthetic division** may be used. On the left is an example of long division, and on the right the same example showing the coefficients of the variables only.

$$\begin{array}{r} 2x^2 + 5x + 2 \\ x - 3 \overline{\smash{\big)}\ 2x^3 - x^2 - 13x + 1} \\ \underline{2x^3 - 6x^2} \\ 5x^2 - 13x \\ \underline{5x^2 - 15x} \\ 2x + 1 \\ \underline{2x - 6} \\ 7 \end{array} \qquad \begin{array}{r} 2 \quad 5 \quad 2 \\ 1 - 3 \overline{\smash{\big)}\ 2 - 1 - 13 + 1} \\ \underline{2 - 6} \\ 5 - 13 \\ \underline{5 - 15} \\ 2 + 1 \\ \underline{2 - 6} \\ 7 \end{array}$$

Notice that as long as we keep coefficients of powers of x in the same column, we can perform division of polynomials by performing algebraic operations on the coefficients only. This shortcut process of dividing with coefficients only in a special format is called synthetic division. To find $(2x^3 - x^2 - 13x + 1) \div (x - 3)$ by synthetic division, follow the next example.

EXAMPLE 1 Use synthetic division to divide $2x^3 - x^2 - 13x + 1$ by $x - 3$.

Solution: To use synthetic division, the divisor must be in the form $x - c$. Since we are dividing by $x - 3$, c is 3. Write down 3 and the coefficients of the dividend.

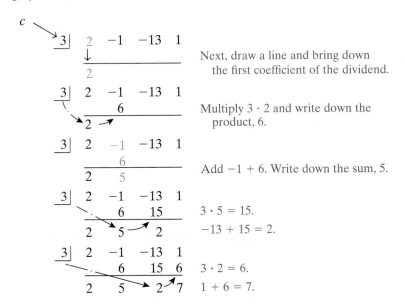

Next, draw a line and bring down the first coefficient of the dividend.

Multiply $3 \cdot 2$ and write down the product, 6.

Add $-1 + 6$. Write down the sum, 5.

$3 \cdot 5 = 15$.
$-13 + 15 = 2$.

$3 \cdot 2 = 6$.
$1 + 6 = 7$.

The quotient is found in the bottom row. The numbers 2, 5, and 2 are the coefficients of the quotient polynomial, and the number 7 is the remainder. The degree of the quotient polynomial is one less than the degree of the dividend. In our example, the degree of the dividend is 3, so the degree of the quotient polynomial is 2. As we found when we performed the long division, the quotient is

$$2x^2 + 5x + 2, \qquad \text{remainder } 7$$

or

$$2x^2 + 5x + 2 + \frac{7}{x - 3}$$

EXAMPLE 2 Use synthetic division to divide $x^4 - 2x^3 - 11x^2 + 5x + 34$ by $x + 2$.

Solution: The divisor is $x + 2$, which we write in the form $x - c$ as $x - (-2)$. Thus, c is -2. The dividend coefficients are $1, -2, -11, 5$, and 34.

$$
\begin{array}{r|rrrrr}
-2 & 1 & -2 & -11 & 5 & 34 \\
   &   & -2 & 8 & 6 & -22 \\
\hline
   & 1 & -4 & -3 & 11 & 12 \\
\end{array}
$$

The dividend is a fourth-degree polynomial, so the quotient polynomial is a third-degree polynomial. The quotient is $x^3 - 4x^2 - 3x + 11$ with a remainder of 12. Thus,

$$\frac{x^4 - 2x^3 - 11x^2 + 5x + 34}{x + 2} = x^3 - 4x^2 - 3x + 11 + \frac{12}{x + 2}$$

REMINDER   Before dividing by synthetic division, write the dividend in descending order of variable exponents. Any "missing powers" of the variable should be represented by 0 times the variable raised to the missing power.

EXAMPLE 3   If $P(x) = 2x^3 - 4x^2 + 5$:

a. Find $P(2)$ by substitution.

b. Use synthetic division to find the remainder when $P(x)$ is divided by $x - 2$.

Solution: **a.** $P(x) = 2x^3 - 4x^2 + 5$

$P(2) = 2(2)^3 - 4(2)^2 + 5$

$= 2(8) - 4(4) + 5 = 16 - 16 + 5 = 5$

Thus, $P(2) = 5$.

b. The coefficients of $P(x)$ are 2, -4, 0, and 5. The number 0 is a coefficient of the missing power of x^1. The divisor is $x - 2$, so c is 2.

$$
\begin{array}{c|rrrr}
c & & & & \\
\searrow & & & & \\
2 & 2 & -4 & 0 & 5 \\
& & 4 & 0 & 0 \\
\hline
& 2 & 0 & 0 & 5 \quad \text{remainder}
\end{array}
$$

The remainder when $P(x)$ is divided by $x - 2$ is 5.

2   Notice in the preceding example that $P(2) = 5$ and that the remainder when $P(x)$ is divided by $x - 2$ is 5. This is no accident. This illustrates the **remainder theorem.**

REMAINDER THEOREM

If a polynomial $P(x)$ is divided by $x - c$, then the remainder is $P(c)$.

EXAMPLE 4 Use the remainder theorem and synthetic division to find $P(4)$ if

$$P(x) = 4x^6 - 25x^5 + 35x^4 + 17x^2.$$

Solution: To find $P(4)$ by the remainder theorem, we divide $P(x)$ by $x - 4$. The coefficients of $P(x)$ are 4, -25, 35, 0, 17, 0, and 0. Also, c is 4.

$$
\begin{array}{c|ccccccc}
c & & & & & & & \\
4 & 4 & -25 & 35 & 0 & 17 & 0 & 0 \\
& & 16 & -36 & -4 & -16 & 4 & 16 \\
\hline
& 4 & -9 & -1 & -4 & 1 & 4 & 16 \quad \text{remainder}
\end{array}
$$

Thus, $P(4) = 16$, the remainder.

5. $x^2 - 5x - 23 - \dfrac{41}{x-2}$ **6.** $x^2 + x - 1 - \dfrac{2}{x+5}$ **15.** $x^2 + \dfrac{2}{x-3}$ **17.** $6x + 7 + \dfrac{1}{x+1}$

18. $x^2 - 2x + 1 - \dfrac{1}{x-3}$ **21.** $3x - 9 + \dfrac{12}{x+3}$ **22.** $3x - 5 + \dfrac{14}{x+4}$ **23.** $3x^2 - \dfrac{9}{2}x + \dfrac{7}{4} + \dfrac{47}{8x-4}$

EXERCISE SET 6.6

24. $8x^2 - 12x + 4$ **25.** $3x^2 + 3x - 3$

Use synthetic division to find each quotient. See Examples 1 through 3.

1. $\dfrac{x^2 + 3x - 40}{x - 5}$ $x + 8$ **2.** $\dfrac{x^2 - 14x + 24}{x - 2}$ $x - 12$

3. $\dfrac{x^2 + 5x - 6}{x + 6}$ $x - 1$ **4.** $\dfrac{x^2 + 12x + 32}{x + 4}$ $x + 8$

5. $\dfrac{x^3 - 7x^2 - 13x + 5}{x - 2}$ **6.** $\dfrac{x^3 + 6x^2 - 4x - 7}{x + 5}$

7. $\dfrac{4x^2 - 9}{x - 2}$ $4x + 8 + \dfrac{7}{x - 2}$ **8.** $\dfrac{3x^2 - 4}{x - 1}$ $3x + 3 - \dfrac{1}{x - 1}$

For the given polynomial $P(x)$ and the given c, find $P(c)$ by (a) direct substitution and (b) the remainder theorem. See Examples 3 and 4.

9. $P(x) = 3x^2 - 4x - 1;\ P(2)$ 3

10. $P(x) = x^2 - x + 3;\ P(5)$ 23

11. $P(x) = 4x^4 + 7x^2 + 9x - 1;\ P(-2)$ 73

12. $P(x) = 8x^5 + 7x + 4;\ P(-3)$ -1961

13. $P(x) = x^5 + 3x^4 + 3x - 7;\ P(-1)$ -8

14. $P(x) = 5x^4 - 4x^3 + 2x - 1;\ P(-1)$ 6

Use synthetic division to find each quotient.

15. $\dfrac{x^3 - 3x^2 + 2}{x - 3}$ **16.** $\dfrac{x^2 + 12}{x + 2}$ $x - 2 + \dfrac{16}{x + 2}$

17. $\dfrac{6x^2 + 13x + 8}{x + 1}$ **18.** $\dfrac{x^3 - 5x^2 + 7x - 4}{x - 3}$

26. $9y^2 + 3y - 3 + \dfrac{4}{y + \dfrac{2}{3}}$ **28.** $x^3 + 6x^2 + 11x + 6 + \dfrac{8}{x - 2}$ **32.** $4x^2 + 1 - \dfrac{15}{x + 3}$

19. $\dfrac{2x^4 - 13x^3 + 16x^2 - 9x + 20}{x - 5}$ $2x^3 - 3x^2 + x - 4$

20. $\dfrac{3x^4 + 5x^3 - x^2 + x - 2}{x + 2}$ $3x^3 - x^2 + x - 1$

21. $\dfrac{3x^2 - 15}{x + 3}$ **22.** $\dfrac{3x^2 + 7x - 6}{x + 4}$

23. $\dfrac{3x^3 - 6x^2 + 4x + 5}{x - \dfrac{1}{2}}$ **24.** $\dfrac{8x^3 - 6x^2 - 5x + 3}{x + \dfrac{3}{4}}$

25. $\dfrac{3x^3 + 2x^2 - 4x + 1}{x - \dfrac{1}{3}}$ **26.** $\dfrac{9y^3 + 9y^2 - y + 2}{y + \dfrac{2}{3}}$

27. $\dfrac{7x^2 - 4x + 12 + 3x^3}{x + 1}$ $3x^2 + 4x - 8 + \dfrac{20}{x + 1}$

28. $\dfrac{x^4 + 4x^3 - x^2 - 16x - 4}{x - 2}$

29. $\dfrac{x^3 - 1}{x - 1}$ $x^2 + x + 1$ **30.** $\dfrac{y^3 - 8}{y - 2}$ $y^2 + 2y + 4$

31. $\dfrac{x^2 - 36}{x + 6}$ $x - 6$ **32.** $\dfrac{4x^3 + 12x^2 + x - 12}{x + 3}$

For the given polynomial $P(x)$ and the given c, use the remainder theorem to find $P(c)$.

33. $P(x) = x^3 + 3x^2 - 7x + 4;\ 1$ 1

34. $P(x) = x^3 + 5x^2 - 4x - 6;\ 2$ 14

35. $P(x) = 3x^3 - 7x^2 - 2x + 5; -3$ -133

36. $P(x) = 4x^3 + 5x^2 - 6x - 4; -2$ -4

37. $P(x) = 4x^4 + x^2 - 2; -1$ 3

38. $P(x) = x^4 - 3x^2 - 2x + 5; -2$ 13

39. $P(x) = 2x^4 - 3x^2 - 2; \dfrac{1}{3}$ $-\dfrac{187}{81}$

40. $P(x) = 4x^4 - 2x^3 + x^2 - x - 4; \dfrac{1}{2}$ $-\dfrac{17}{4}$

41. $P(x) = x^5 + x^4 - x^3 + 3; \dfrac{1}{2}$ $\dfrac{95}{32}$

42. $P(x) = x^5 - 2x^3 + 4x^2 - 5x + 6; \dfrac{2}{3}$ $\dfrac{968}{243}$

43. Explain an advantage of using the remainder theorem instead of direct substitution.

44. Explain an advantage of using synthetic division instead of long division.

We say that 2 is a factor of 8 because 2 divides 8 evenly, or with a remainder of 0. In the same manner, the polynomial $x - 2$ is a factor of the polynomial $x^3 - 18x^2 + 24x$ because the remainder is 0 when $x^3 - 18x^2 + 24x$ is divided by $x - 2$. Use this information for Exercises 45 through 47.

45. Use synthetic division to show that $x + 3$ is a factor of $x^3 + 3x^2 + 4x + 12$.

46. Use synthetic division to show that $x - 2$ is a factor of $x^3 - 2x^2 - 3x + 6$.

47. From the remainder theorem, the polynomial $x - c$ is a factor of a polynomial function $P(x)$ if $P(c)$ is what value? 0

48. If a polynomial is divided by $x - 5$, the quotient is $2x^2 + 5x - 6$ and the remainder is 3. Find the original polynomial. $2x^3 - 5x^2 - 31x + 33$

49. If a polynomial is divided by $x + 3$, the quotient is $x^2 - x + 10$ and the remainder is -2. Find the original polynomial. $x^3 + 2x^2 + 7x + 28$

50. If the area of a parallelogram is $x^4 - 23x^2 + 9x - 5$ square centimeters and its base is $x + 5$ centimeters, find its height. $(x^3 - 5x^2 + 2x - 1)$ cm

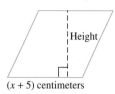

Height

$(x + 5)$ *centimeters*

51. If the volume of a box is $x^4 + 6x^3 - 7x^2$ cubic meters, its height is x^2 meters, and its length is $x + 7$ meters, find its width. $(x - 1)$ meters

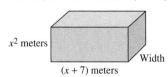

x^2 *meters*

Width

$(x + 7)$ *meters*

Review Exercises

Solve each equation for x. See Sections 2.1 and 5.8.

52. $7x + 2 = x - 3$ $\left\{-\dfrac{5}{6}\right\}$

53. $4 - 2x = 17 - 5x$

54. $x^2 = 4x - 4$ $\{2\}$

55. $5x^2 + 10x = 15$ $\{-3, 1\}$

56. $\dfrac{x}{3} - 5 = 13$ $\{54\}$

57. $\dfrac{2x}{9} + 1 = \dfrac{7}{9}$ $\{-1\}$

Factor the following. See Sections 5.5 and 5.7.

58. $x^3 - 1$

59. $8y^3 + 1$

60. $125z^3 + 8$

61. $a^3 - 27$

62. $xy + 2x + 3y + 6$

63. $x^2 - x + xy - y$

64. $x^3 - 9x$ $x(x + 3)(x - 3)$ **65.** $2x^3 - 32x$

45. $(x + 3)(x^2 + 4) = x^3 + 3x^2 + 4x + 12$ **46.** $(x - 2)(x^2 - 3) = x^3 - 2x^2 - 3x + 6$ **53.** $\left\{\dfrac{13}{3}\right\}$
58. $(x - 1)(x^2 + x + 1)$ **59.** $(2y + 1)(4y^2 - 2y + 1)$ **60.** $(5z + 2)(25z^2 - 10z + 4)$
61. $(a - 3)(a^2 + 3a + 9)$ **62.** $(y + 2)(x + 3)$ **63.** $(x - 1)(x + y)$ **65.** $2x(x + 4)(x - 4)$

6.7   SOLVING EQUATIONS CONTAINING RATIONAL EXPRESSIONS

TAPE IA 6.7

O B J E C T I V E

1 Solve equations containing rational expressions.

1 To solve equations containing rational expressions, we first clear the equation of fractions by multiplying both sides of the equation by the LCD of all rational expressions.

EXAMPLE 1   Solve $\dfrac{8x}{5} + \dfrac{3}{2} = \dfrac{3x}{5}$.

Solution: The LCD of $\dfrac{8x}{5}, \dfrac{3}{2}$, and $\dfrac{3x}{5}$ is 10. Multiply both sides of the equation by 10.

$$\frac{8x}{5} + \frac{3}{2} = \frac{3x}{5}$$

$$10\left(\frac{8x}{5} + \frac{3}{2}\right) = 10\left(\frac{3x}{5}\right) \qquad \text{Multiply by the LCD.}$$

$$10 \cdot \frac{8x}{5} + 10 \cdot \frac{3}{2} = 10 \cdot \frac{3x}{5} \qquad \text{Apply the distributive property.}$$

$$16x + 15 = 6x \qquad \text{Simplify.}$$

$$15 = -10x \qquad \text{Subtract } 16x \text{ from both sides.}$$

$$-\frac{15}{10} = x, \text{ or } x = -\frac{3}{2} \qquad \text{Solve.}$$

Verify this solution by substituting $-\dfrac{3}{2}$ for x in the original equation. The solution set is $\left\{-\dfrac{3}{2}\right\}$.

The important difference in the equations in this section is that the denominator of a rational expression may contain a variable. Recall that a rational expression is undefined for values of the variable that make the denominator 0. Thus, special precautions must be taken when an equation contains rational expressions with variables in the denominator. If a proposed solution makes the denominator 0, then it must be rejected as a solution. Such proposed solutions are called **extraneous solutions**.

EXAMPLE 2   Solve $\dfrac{3}{x} - \dfrac{x + 21}{3x} = \dfrac{5}{3}$.

Solution: The LCD of denominators x, $3x$, and 3 is $3x$. Multiply both sides by $3x$.

$$\frac{3}{x} - \frac{x + 21}{3x} = \frac{5}{3}$$

$$3x\left(\frac{3}{x} - \frac{x + 21}{3x}\right) = 3x\left(\frac{5}{3}\right)$$

$$3x\left(\frac{3}{x}\right) - 3x\left(\frac{x + 21}{3x}\right) = 3x\left(\frac{5}{3}\right) \qquad \text{Apply the distributive property.}$$

$$9 - (x + 21) = 5x \qquad \text{Simplify.}$$

$$9 - x - 21 = 5x$$

$$-12 = 6x$$

$$-2 = x \qquad \text{Solve.}$$

The proposed solution is -2. Check the solution in the original equation.

$$\frac{3}{x} - \frac{x + 21}{3x} = \frac{5}{3}$$

$$\frac{3}{-2} - \frac{-2 + 21}{3(-2)} = \frac{5}{3}$$

$$-\frac{9}{6} + \frac{19}{6} = \frac{5}{3}$$

$$\frac{10}{6} = \frac{5}{3} \qquad \text{True.}$$

The solution set is $\{-2\}$.

The following steps may be used to solve equations containing rational expressions.

TO SOLVE AN EQUATION CONTAINING RATIONAL EXPRESSIONS

Step 1.   Multiply both sides of the equation by the LCD of all rational expressions in the equation.

Step 2.   Simplify both sides.

Step 3.   Determine whether the equation is linear, quadratic, or higher degree and solve accordingly.

Step 4.   Check the solution in the original equation.

EXAMPLE 3   Solve $\dfrac{x + 6}{x - 2} = \dfrac{2(x + 2)}{x - 2}$.

Solution: First multiply both sides of the equation by the LCD, $x - 2$.

$$\frac{x + 6}{x - 2} = \frac{2(x + 2)}{x - 2}$$

$$(x - 2)\left(\frac{x + 6}{x - 2}\right) = (x - 2)\left[\frac{2(x + 2)}{x - 2}\right] \qquad \text{Multiply both sides by } x - 2.$$

This is a linear equation, so we solve by isolating the variable x.

$$x + 6 = 2(x + 2) \qquad \text{Simplify.}$$

$$x + 6 = 2x + 4 \qquad \text{Use the distributive property.}$$

$$x = 2 \qquad \text{Solve.}$$

Now check the proposed solution 2 in the *original equation*.

$$\frac{x + 6}{x - 2} = \frac{2(x + 2)}{x - 2}$$

$$\frac{2 + 6}{2 - 2} = \frac{2(2 + 2)}{2 - 2}$$

$$\frac{8}{0} = \frac{2(4)}{0}$$

The denominators are 0, since 2 is not in the domain of either rational expression in the equation. Therefore, 2 is an extraneous solution. There is no solution to the original equation. The solution set is $\emptyset$ or { }.

EXAMPLE 4   Solve $\dfrac{z}{2z^2 + 3z - 2} - \dfrac{1}{2z} = \dfrac{3}{z^2 + 2z}$.

Solution: Factor the denominators to find that the LCD is $2z(z + 2)(2z - 1)$. Multiply both sides by the LCD.

$$\frac{z}{2z^2 + 3z - 2} - \frac{1}{2z} = \frac{3}{z^2 + 2z}$$

$$\frac{z}{(2z - 1)(z + 2)} - \frac{1}{2z} = \frac{3}{z(z + 2)}$$

$$2z(z + 2)(2z - 1)\left[\frac{z}{(2z - 1)(z + 2)} - \frac{1}{2z}\right]$$

$$= 2z(z + 2)(2z - 1)\left[\frac{3}{z(z + 2)}\right]$$

$$2z(z + 2)(2z - 1)\left[\frac{z}{(2z - 1)(z + 2)}\right] - 2z(z + 2)(2z - 1)\left(\frac{1}{2z}\right)$$

$$= 2z(z + 2)(2z - 1)\left[\frac{3}{z(z + 2)}\right]$$

Apply the distributive property.

$$2z(z) - (z + 2)(2z - 1) = 3 \cdot 2(2z - 1)$$ Simplify.

$$2z^2 - (2z^2 + 3z - 2) = 12z - 6$$

$$2z^2 - 2z^2 - 3z + 2 = 12z - 6$$

$$-3z + 2 = 12z - 6$$

$$-15z = -8$$

$$z = \frac{8}{15}$$ Solve.

The proposed solution $\dfrac{8}{15}$ does not make any denominator 0; the solution set is $\left\{\dfrac{8}{15}\right\}$.

EXAMPLE 5   Solve $\dfrac{2x}{x-3} + \dfrac{6-2x}{x^2-9} = \dfrac{x}{x+3}$.

Solution: Factor the second denominator to find that the LCD is $(x+3)(x-3)$. Multiply both sides of the equation by $(x+3)(x-3)$. By the distributive property, this is the same as multiplying each term by $(x+3)(x-3)$.

$$\frac{2x}{x-3} + \frac{6-2x}{x^2-9} = \frac{x}{x+3}$$

$$(x+3)(x-3)\left(\frac{2x}{x-3}\right) + (x+3)(x-3)\left[\frac{6-2x}{(x+3)(x-3)}\right]$$
$$= (x+3)(x-3)\left(\frac{x}{x+3}\right)$$

$$2x(x+3) + (6-2x) = x(x-3) \qquad \text{Simplify.}$$
$$2x^2 + 6x + 6 - 2x = x^2 - 3x \qquad \text{Apply the distributive property.}$$

Next we solve this quadratic equation by the factoring method. To do so, first write the equation so that one side is 0.

$$x^2 + 7x + 6 = 0$$

Next, see if the trinomial is factorable.

$$(x+6)(x+1) = 0 \qquad \text{Factor.}$$
$$x = -6 \quad \text{or} \quad x = -1 \qquad \text{Set each factor equal to 0.}$$

Neither -6 nor -1 makes any denominator 0. The solution set is $\{-6, -1\}$. ▬▬▬

A graph can be helpful in visualizing solutions of equations. For example, to visualize the solution of the equation $\dfrac{3}{x} - \dfrac{x+21}{3x} = \dfrac{5}{3}$ in Example 2, the graph of the related rational function $f(x) = \dfrac{3}{x} - \dfrac{x+21}{3x}$ is shown. A solution of the equation is an x-value that corresponds to a y-value of $\dfrac{5}{3}$.

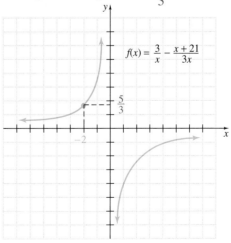

$$f(x) = \frac{3}{x} - \frac{x+21}{3x}$$

Notice that an x-value of -2 corresponds to a y-value of $\frac{5}{3}$. The solution of the equation is indeed -2 as shown in Example 2.

EXERCISE SET 6.7

Solve each equation. See Examples 1 and 2.

1. $\dfrac{x}{2} - \dfrac{x}{3} = 12$ {72}

2. $x = \dfrac{x}{2} - 4$ {-8}

3. $\dfrac{x}{3} = \dfrac{1}{6} + \dfrac{x}{4}$ {2}

4. $\dfrac{x}{2} = \dfrac{21}{10} - \dfrac{x}{5}$ {3}

5. $\dfrac{2}{x} + \dfrac{1}{2} = \dfrac{5}{x}$ {6}

6. $\dfrac{5}{3x} + 1 = \dfrac{7}{6}$ {10}

7. $\dfrac{x + 3}{x} = \dfrac{5}{x}$ {2}

8. $\dfrac{4 - 3x}{2x} = -\dfrac{8}{2x}$ {4}

Solve each equation. See Examples 3 through 5.

9. $\dfrac{x + 5}{x + 3} = \dfrac{8}{x + 3}$ {3}

10. $\dfrac{5}{x - 2} - \dfrac{2}{x + 4} = -\dfrac{4}{x^2 + 2x - 8}$ $\left\{-\dfrac{28}{3}\right\}$

11. $\dfrac{1}{x - 1} + \dfrac{1}{x + 1} = \dfrac{2}{x^2 - 1}$ { }

12. $\dfrac{1}{x - 1} = \dfrac{2}{x + 1}$ {3}

13. $\dfrac{6}{x + 3} = \dfrac{4}{x - 3}$ {15}

14. $\dfrac{1}{x - 4} - \dfrac{3x}{x^2 - 16} = \dfrac{2}{x + 4}$ {3}

15. $\dfrac{3}{2x + 3} - \dfrac{1}{2x - 3} = \dfrac{4}{4x^2 - 9}$ {4}

16. $\dfrac{1}{x - 4} = \dfrac{8}{x^2 - 16}$ { }

17. $\dfrac{2}{x^2 - 4} = \dfrac{1}{2x - 4}$ { }

18. $\dfrac{1}{x - 2} - \dfrac{2}{x^2 - 2x} = 1$ {1}

19. $\dfrac{12}{3x^2 + 12x} = 1 - \dfrac{1}{x + 4}$ {1}

Solve each equation.

20. $\dfrac{5}{x} = \dfrac{20}{12}$ {3}

21. $\dfrac{2}{x} = \dfrac{10}{5}$ {1}

22. $1 - \dfrac{4}{a} = 5$ {-1}

23. $7 + \dfrac{6}{a} = 5$ {-3}

24. $\dfrac{1}{2x} - \dfrac{1}{x + 1} = \dfrac{1}{3x^2 + 3x}$ $\left\{\dfrac{1}{3}\right\}$

25. $\dfrac{2}{x - 5} + \dfrac{1}{2x} = \dfrac{5}{3x^2 - 15x}$ $\left\{\dfrac{5}{3}\right\}$

26. $\dfrac{1}{x} - \dfrac{x}{25} = 0$ {$-5, 5$}

27. $\dfrac{x}{4} + \dfrac{5}{x} = 3$ {10, 2}

28. $5 - \dfrac{2}{2y - 5} = \dfrac{3}{2y - 5}$ {3}

29. $1 - \dfrac{5}{y + 7} = \dfrac{4}{y + 7}$ {2}

30. $\dfrac{x - 1}{x + 2} = \dfrac{2}{3}$ {7}

31. $\dfrac{6x + 7}{2x + 9} = \dfrac{5}{3}$ {3}

32. $\dfrac{x + 3}{x + 2} = \dfrac{1}{x + 2}$ { }

33. $\dfrac{2x + 1}{4 - x} = \dfrac{9}{4 - x}$ { }

34. $\dfrac{1}{a - 3} + \dfrac{2}{a + 3} = \dfrac{1}{a^2 - 9}$ $\left\{\dfrac{4}{3}\right\}$

35. $\dfrac{12}{9 - a^2} + \dfrac{3}{3 + a} = \dfrac{2}{3 - a}$ { }

36. $\dfrac{64}{x^2 - 16} + 1 = \dfrac{2x}{x - 4}$ {-12}

37. $2 + \dfrac{3}{x} = \dfrac{2x}{x + 3}$ {-1}

38. $\dfrac{-15}{4y + 1} + 4 = y$ $\left\{1, \dfrac{11}{4}\right\}$

39. $\dfrac{36}{x^2 - 9} + 1 = \dfrac{2x}{x + 3}$ {9}

40. $\dfrac{28}{x^2 - 9} + \dfrac{2x}{x - 3} + \dfrac{6}{x + 3} = 0$ {$-5, -1$}

41. $\dfrac{x^2 - 20}{x^2 - 7x + 12} = \dfrac{3}{x - 3} + \dfrac{5}{x - 4}$ {1, 7}

42. $\dfrac{x + 2}{x^2 + 7x + 10} = \dfrac{1}{3x + 6} - \dfrac{1}{x + 5}$ $\left\{-\dfrac{7}{5}\right\}$

43. $\dfrac{3}{2x - 5} + \dfrac{2}{2x + 3} = 0$ $\left\{\dfrac{1}{10}\right\}$

44. 3000 game disks   **48.** $\dfrac{2x + 5}{x(x - 3)}$ **52.** $\dfrac{(a + 3)(a + 1)}{a + 2}$ **56.** $\dfrac{4a + 1}{(3a + 1)(3a - 1)}$ **57.** $\dfrac{-a - 8}{4a(a - 2)}$ **58.** $\left(-1, \dfrac{3}{2}\right)$

44. The average cost of producing x game disks for a computer is given by the function $f(x) = 3.3 + \dfrac{5400}{x}$. Find the number of game disks that must be produced for the average cost to be \$5.10.

45. The average cost of producing x electric pencil sharpeners is given by the function $f(x) = 20 + \dfrac{4000}{x}$. Find the number of electric pencil sharpeners that must be produced for the average cost to be \$25.   800 pencil sharpeners

Perform the indicated operation and simplify, or solve the equation for the variable.

46. $\dfrac{2}{x^2 - 4} = \dfrac{1}{x + 2} - \dfrac{3}{x - 2}$ $\{-5\}$

47. $\dfrac{3}{x^2 - 25} = \dfrac{1}{x + 5} + \dfrac{2}{x - 5}$ $\left\{-\dfrac{2}{3}\right\}$

48. $\dfrac{5}{x^2 - 3x} + \dfrac{4}{2x - 6}$ **49.** $\dfrac{5}{x^2 - 3x} \div \dfrac{4}{2x - 6}$ $\dfrac{5}{2x}$

50. $\dfrac{x - 1}{x + 1} + \dfrac{x + 7}{x - 1} = \dfrac{4}{x^2 - 1}$ $\{-2\}$

51. $\left(1 - \dfrac{y}{x}\right) \div \left(1 - \dfrac{x}{y}\right) - \dfrac{y}{x}$ **52.** $\dfrac{a^2 - 9}{a - 6} \cdot \dfrac{a^2 - 5a - 6}{a^2 - a - 6}$

53. $\dfrac{2}{a - 6} + \dfrac{3a}{a^2 - 5a - 6} - \dfrac{a}{5a + 5}$ $\dfrac{-a^2 + 31a + 10}{5(a - 6)(a + 1)}$

54. $\dfrac{2x + 3}{3x - 2} = \dfrac{4x + 1}{6x + 1}$ $\left\{-\dfrac{1}{5}\right\}$ **55.** $\dfrac{5x - 3}{2x} = \dfrac{10x + 3}{4x + 1}$ $\left\{-\dfrac{3}{13}\right\}$

56. $\dfrac{a}{9a^2 - 1} + \dfrac{2}{6a - 2}$ **57.** $\dfrac{3}{4a - 8} - \dfrac{a + 2}{a^2 - 2a}$

58. $\dfrac{-3}{x^2} - \dfrac{1}{x} + 2 = 0$ **59.** $\dfrac{x}{2x + 6} + \dfrac{5}{x^2 - 9}$

60. $\dfrac{x - 8}{x^2 - x - 2} + \dfrac{2}{x - 2}$ $\dfrac{3}{x + 1}$

61. $\dfrac{x - 8}{x^2 - x - 2} + \dfrac{2}{x - 2} = \dfrac{3}{x + 1}$

62. $\dfrac{3}{a} - 5 = \dfrac{7}{a} - 1$ $\{-1\}$ **63.** $\dfrac{7}{3z - 9} + \dfrac{5}{z}$ $\dfrac{22z - 45}{z(3z - 9)}$

Solve each equation. Begin by writing each equation with positive exponents only.

64. $x^{-2} - 19x^{-1} + 48 = 0$ **64.** $\left\{\dfrac{1}{16}, \dfrac{1}{3}\right\}$ **65.** $\left\{\dfrac{1}{9}, -\dfrac{1}{4}\right\}$

65. $x^{-2} - 5x^{-1} - 36 = 0$

66. $p^{-2} + 4p^{-1} - 5 = 0$ **66.** $\left\{-\dfrac{1}{5}, 1\right\}$

67. $6p^{-2} - 5p^{-1} + 1 = 0$ $\{3, 2\}$

59. $\dfrac{x^2 - 3x + 10}{2(x + 3)(x - 3)}$ **61.** $\{x \mid x \text{ is a real number and } x \neq 2, x \neq -1\}$

Solve each equation. Round solutions to two decimal places.   **68.** $\{-0.17\}$ **69.** $\{1.39\}$

68. $\dfrac{1.4}{x - 2.6} = \dfrac{-3.5}{x + 7.1}$ **69.** $\dfrac{-8.5}{x + 1.9} = \dfrac{5.7}{x - 3.6}$

70. $\dfrac{10.6}{y} - 14.7 = \dfrac{9.92}{3.2} + 7.6$ $\{0.42\}$

71. $\dfrac{12.2}{x} + 17.3 = \dfrac{9.6}{x} - 14.7$ $\{-0.08\}$

Use a grapher to verify the solution of each given exercise.

72. Exercise 20   **73.** Exercise 21

74. Exercise 30   **75.** Exercise 31

For Exercises 72–75, see Appendix D.

Review Exercises

Write each sentence as an equation and solve. See Section 2.2.   **77.** 73 and 74   **79.** $\dfrac{1}{2}$ and 2

76. Four more than 3 times a number is 19.   5

77. The sum of two consecutive integers is 147.

78. The length of a rectangle is 5 inches more than the width. Its perimeter is 50 inches. Find the length and width.   length, 15 in.; width, 10 in.

79. The sum of a number and its reciprocal is $\dfrac{5}{2}$.

Simplify the following. See Section 1.4.

80. $-|-6| - (-5)$ -1 **81.** $\sqrt{49} - (10 - 6)^2$ -9

82. $|4 - 8| + (4 - 8)$ 0   **83.** $(-4)^2 - 5^2$ -9

The following is from a survey of state prisons. Use this histogram to answer Exercises 84 through 88.

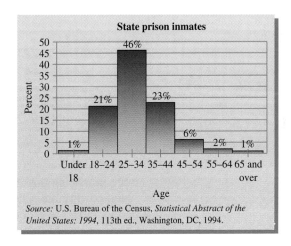

State prison inmates

Source: U.S. Bureau of the Census, *Statistical Abstract of the United States: 1994*, 113th ed., Washington, DC, 1994.

84. What percent of state prison inmates are aged 45 to 54?   6%

85. What percent of state prison inmates are 65 years old or older?   1%

86. What age category shows the highest percent of prison inmates?   25–34 years old

87. What percent of state prison inmates are 18 to 34 years old?   67%

88. A state prison in Louisiana houses 2000 inmates. Approximately how many 25- to 34-year-old inmates might you expect this prison to hold? 920 inmates

A Look Ahead

EXAMPLE

Solve $\left(\dfrac{x}{x+1}\right)^2 - 7\left(\dfrac{x}{x+1}\right) + 10 = 0$.

Solution:

Let $u = \dfrac{x}{x+1}$ and solve for u. Then substitute back and solve for x.

$$\left(\frac{x}{x+1}\right)^2 - 7\left(\frac{x}{x+1}\right) + 10 = 0$$

$$u^2 - 7u + 10 = 0 \quad \text{Let } u = \frac{x}{x+1}.$$

$$(u-5)(u-2) = 0 \quad \text{Factor.}$$

$$u = 5 \quad \text{or} \quad u = 2 \quad \text{Solve.}$$

Since $u = \dfrac{x}{x+1}$, we have that $5 = \dfrac{x}{x+1}$ and $2 = \dfrac{x}{x+1}$.

Thus, there are two rational equations to solve.

1.
$$5 = \frac{x}{x+1}$$
$$5 \cdot (x+1) = x$$
$$5x + 5 = x$$
$$5 = -4x$$
$$x = -\frac{5}{4}$$

2.
$$2 = \frac{x}{x+1}$$
$$2 \cdot (x+1) = x$$
$$2x + 2 = x$$
$$2 = -x$$
$$x = -2$$

Since neither $-\frac{5}{4}$ nor -2 makes the denominator 0, the solution set is $\left\{-\frac{5}{4}, -2\right\}$.

Solve each equation by substitution. See the preceding example.

89. $(x-1)^2 + 3(x-1) + 2 = 0$ $\{-1, 0\}$

90. $(4-x)^2 - 5(4-x) + 6 = 0$ $\{1, 2\}$

91. $\left(\dfrac{3}{x-1}\right)^2 + 2\left(\dfrac{3}{x-1}\right) + 1 = 0$ $\{-2\}$

92. $\left(\dfrac{5}{2+x}\right)^2 + \left(\dfrac{5}{x+x}\right) - 20 = 0$ $\left\{-3, -\dfrac{3}{4}\right\}$

6.8   RATIONAL EQUATIONS AND PROBLEM SOLVING

TAPE IA 6.8

O B J E C T I V E S

1 Solve an equation containing rational expressions for a specified variable.

2 Solve problems by writing equations containing rational expressions.

1   In Section 2.3 we solved equations for a specified variable. In this section, we continue practicing this skill by solving equations containing rational expressions for a specified variable. The steps given in Section 2.3 for solving equations for a specified variable are repeated here.

TO SOLVE EQUATIONS FOR A SPECIFIED VARIABLE

Step 1.   Clear the equation of fractions or rational expressions by multiplying each side of the equation by the least common denominator (LCD) of all denominators in the equation.

Step 2.   Use the distributive property to remove grouping symbols such as parentheses.

Step 3.   Combine like terms on the same side of the equation.

Step 4.   Use the addition property of equality to rewrite the equation as an equivalent equation with terms containing the specified variable on one side and all other terms on the other side.

Step 5.   Use the distributive property and the multiplication property of equality to isolate the specified variable.

EXAMPLE 1   Solve $\dfrac{1}{x} + \dfrac{1}{y} = \dfrac{1}{z}$ for x.

Solution: To clear this equation of fractions, multiply both sides of the equation by xyz, the LCD of $\dfrac{1}{x}, \dfrac{1}{y},$ and $\dfrac{1}{z}$.

$$\frac{1}{x} + \frac{1}{y} = \frac{1}{z}$$

$$xyz\left(\frac{1}{x} + \frac{1}{y}\right) = xyz\left(\frac{1}{z}\right) \qquad \text{Multiply both sides by } xyz.$$

$$xyz\left(\frac{1}{x}\right) + xyz\left(\frac{1}{y}\right) = xyz\left(\frac{1}{z}\right) \qquad \text{Apply the distributive property.}$$

$$yz + xz = xy \qquad \text{Simplify.}$$

Notice the two terms that contain the specified variable x.

Next subtract xz from both sides so that all terms containing the specified variable x are on one side of the equation and all other terms are on the other side.

$$yz = xy - xz$$

Use the distributive property to factor x from $xy - xz$ and then the multiplication property of equality to solve for x.

$$yz = x(y - z)$$

$$\frac{yz}{y - z} = x, \quad \text{or} \quad x = \frac{yz}{y - z} \qquad \text{Divide both sides by } y - z.$$

2

Problem solving sometimes involves modeling a described situation with an equation containing rational expressions. In Examples 2 through 5, we practice solving such problems and use the problem-solving steps first introduced in Section 2.2.

EXAMPLE 2 Find the number that, when subtracted from the numerator and added to the denominator of $\frac{9}{19}$, results in a fraction equivalent to $\frac{1}{3}$.

Solution: **1.** UNDERSTAND the problem. To do so, read and reread the problem and try guessing the solution. For example, if the unknown number is 3, we have the following:

$$\frac{9 - 3}{19 + 3} = \frac{1}{3}$$

To see if this is a true statement, simplify the fraction on the left.

$$\frac{6}{22} = \frac{1}{3}$$

or

$$\frac{3}{11} = \frac{1}{3} \qquad \text{False.}$$

when the fraction on the left is simplified further. Since this is not a true statement, 3 is not the correct number. Remember that the purpose of this step is not to guess the correct solution but to gain an understanding of the problem posed.

2. ASSIGN a variable. Let n be the number to be subtracted from the numerator and added to the denominator.

3. ILLUSTRATE. No illustration is needed.

4. TRANSLATE the problem.

In words:	when the number is subtracted from the numerator and added to the denominator of the fraction	this is equivalent to	$\frac{1}{3}$
Translate:	$\frac{9 - n}{19 + n}$	$=$	$\frac{1}{3}$

5. COMPLETE the work. Here we solve the equation for n.

$$\frac{9 - n}{19 + n} = \frac{1}{3}$$

To solve for n, begin by multiplying both sides by the LCD, $3(19 + n)$.

$$3(19 + n) \cdot \left(\frac{9 - n}{19 + n} \right) = 3(19 + n) \left(\frac{1}{3} \right) \qquad \text{Multiply by the LCD.}$$

$$3(9 - n) = 19 + n \qquad \text{Simplify.}$$

$$27 - 3n = 19 + n$$

$$8 = 4n$$

$$2 = n \qquad \text{Solve.}$$

6. INTERPRET the results. First *check* the stated problem. If we subtract 2 from the numerator and add 2 to the denominator of $\frac{9}{19}$, we have $\frac{9 - 2}{19 + 2} = \frac{7}{21} = \frac{1}{3}$, and the problem checks. Next, *state* the conclusions. The unknown number is 2.

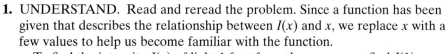

EXAMPLE 3   The intensity of $I(x)$ of light, as measured in foot-candles, that is x feet from its source is given by the rational function

$$I(x) = \frac{320}{x^2}$$

How far away is the source if the intensity of light is 5 foot-candles?

Solution: **1.** UNDERSTAND. Read and reread the problem. Since a function has been given that describes the relationship between $I(x)$ and x, we replace x with a few values to help us become familiar with the function.

To find the intensity $I(x)$ of light 1 foot from the source, we find $I(1)$.

$$I(1) = \frac{320}{1^2} = \frac{320}{1} = 320 \text{ foot-candles}$$

To find the intensity $I(x)$ of light 4 feet from the source, we find $I(4)$.

$$I(4) = \frac{320}{4^2} = \frac{320}{16} = 20 \text{ foot-candles}$$

Notice that as x increases, $I(x)$ decreases. That is, as the number of feet from the light source increases, the intensity decreases, as expected. Steps 2 and 3 are not needed because a formula has been given.

4. TRANSLATE. We are given that the intensity $I(x)$ is 5 foot-candles, and we are asked to find how far away is the light source, x. To do so, let $I(x) = 5$.

$$I(x) = \frac{320}{x^2}$$

$$5 = \frac{320}{x^2} \qquad \text{Let } I(x) = 5.$$

5. COMPLETE. Here we solve the equation for x.

$$5 = \frac{320}{x^2}$$

$$x^2(5) = x^2\left(\frac{320}{x^2}\right) \qquad \text{Multiply both sides by } x^2.$$

$$5x^2 = 320 \qquad \text{Simplify.}$$

$$x^2 = 64 \qquad \text{Divide both sides by 5.}$$

Then, since $8^2 = 64$ and also $(-8)^2 = 64$, we have that

$$x = 8 \qquad \text{or} \qquad x = -8$$

6. INTERPRET. Since x represents distance and distance cannot be negative, the proposed solution -8 must be rejected. *Check* the solution 8 feet in the given formula. Then *state* the solution: The source of light is 8 feet away when the intensity is 5 foot-candles. ▬▬▬

The following work example leads to an equation containing rational expressions.

▬▬▬▬▬▬▬▬▬▬▬▬▬▬▬▬▬▬▬▬▬▬▬▬▬▬▬▬▬▬▬▬▬▬

EXAMPLE 4   Melissa Scarlatti can clean the house in 4 hours, whereas her husband, Zack, can do the same job in 5 hours. They have agreed to clean together so that they can finish in time to watch a movie on TV that starts in 2 hours. How long will it take them to clean the house together? Can they finish before the movie starts?

Solution: **1. UNDERSTAND.** Read and reread the problem. The key idea here is the relationship between the *time* (in hours) it takes to complete the job and the *part of the job* completed in 1 unit of time (1 hour). For example, if the *time* it takes Melissa to complete the job is 4 hours, the *part of the job* she can complete in 1 hour is $\frac{1}{4}$. Similarly, Zack can complete $\frac{1}{5}$ of the job in 1 hour.

2. ASSIGN. Let t represent the *time* in hours it takes Melissa and Zack to clean the house together. Then $\frac{1}{t}$ represents the *part of the job* they complete in 1 hour.

3. ILLUSTRATE. Here we summarize the given information on a chart.

	HOURS TO COMPLETE THE JOB	PART OF JOB COMPLETED IN 1 HOUR
MELISSA ALONE	4	$\frac{1}{4}$
ZACK ALONE	5	$\frac{1}{5}$
TOGETHER	t	$\frac{1}{t}$

4. TRANSLATE.

In words:	part of job Melissa can complete in 1 hour	added to	part of job Zack can complete in 1 hour	is equal to	part of job they can complete together in 1 hour
Translate:	$\dfrac{1}{4}$	$+$	$\dfrac{1}{5}$	$=$	$\dfrac{1}{t}$

5. COMPLETE.

$$\frac{1}{4} + \frac{1}{5} = \frac{1}{t}$$

$$20t\left(\frac{1}{4} + \frac{1}{5}\right) = 20t\left(\frac{1}{t}\right) \qquad \text{Multiply both sides by the LCD, } 20t.$$

$$5t + 4t = 20$$

$$9t = 20$$

$$t = \frac{20}{9} \quad \text{or} \quad 2\frac{2}{9} \qquad \text{Solve.}$$

6. INTERPRET. *Check:* The proposed solution is $2\frac{2}{9}$. That is, Melissa and Zack would take $2\frac{2}{9}$ hours to clean the house together. This proposed solution is reasonable since $2\frac{2}{9}$ hours is more than half of Melissa's time and less than half of Zack's time. Check this solution in the originally stated problem.

State: Can they finish before the movie starts? If they can clean the house together in $2\frac{2}{9}$ hours, they could not complete the job before the movie starts.

EXAMPLE 5   In his boat, Steve Deitmer takes $1\frac{1}{2}$ times as long to go 72 miles upstream as he does to return. If the boat cruises at 30 mph in still water, what is the speed of the current?

Solution: **1.** UNDERSTAND. Read and reread the problem. Next guess a solution. Suppose that the current is 4 mph. The speed of the boat upstream is slowed down by the current: $30 - 4$, or 26 mph, and the speed of the boat downstream is speeded up by the current: $30 + 4$, or 34 mph. Next let's find out how long it takes to travel 72 miles upstream and 72 miles downstream. To do so, we use the formula $d = r \cdot t$, or $\dfrac{d}{r} = t$.

UPSTREAM	DOWNSTREAM
$\dfrac{d}{r} = t$	$\dfrac{d}{r} = t$
$\dfrac{72}{26} = t$	$\dfrac{72}{34} = t$
$2\dfrac{10}{13} = t$	$2\dfrac{2}{17} = t$

Since the time upstream $\left(2\frac{10}{13} \text{ hours}\right)$ is not $1\frac{1}{2}$ times the time downstream $\left(2\frac{2}{17} \text{ hours}\right)$, our guess is not correct. We do, however, have a better understanding of the problem.

2. **ASSIGN.** Since we are asked to find the speed of the current, *let x represent the current's speed.* The speed of the boat traveling downstream is made faster by the current and is represented by $30 + x$. The speed of the boat traveling upstream is made slower by the current and is represented by $30 - x$.

3. **ILLUSTRATE.** This information is summarized in the following chart, where we use the formula $\dfrac{d}{r} = t$:

	DISTANCE	RATE	TIME $\left(\dfrac{d}{r}\right)$
UPSTREAM	72	$30 - x$	$\dfrac{72}{30 - x}$
DOWNSTREAM	72	$30 + x$	$\dfrac{72}{30 + x}$

4. **TRANSLATE.** Since the time spent traveling upstream is $1\frac{1}{2}$ times the time spent traveling downstream, we have

In words:	time upstream	is	$1\dfrac{1}{2}$	times	time downstream
Translate:	$\dfrac{72}{30 - x}$	$=$	$\dfrac{3}{2}$	$\cdot$	$\dfrac{72}{30 + x}$

5. **COMPLETE.**

$$\frac{72}{30 - x} = \frac{3}{2} \cdot \frac{72}{30 + x}$$

Multiply both sides by the LCD, $2(30 + x)(30 - x)$.

$$2(30 + x)(30 - x)\left(\frac{72}{30 - x}\right) = 2(30 + x)(30 - x)\left(\frac{3}{2} \cdot \frac{72}{30 + x}\right)$$

$$72 \cdot 2(30 + x) = 3 \cdot 72 \cdot (30 - x) \qquad \text{Simplify.}$$

$$2(30 + x) = 3(30 - x) \qquad \text{Divide by 72.}$$

$$60 + 2x = 90 - 3x \qquad \text{Apply the distributive property.}$$

$$5x = 30$$

$$x = 6 \qquad \text{Solve.}$$

6. INTERPRET. *Check* this proposed solution of 6 mph in the originally stated problem. *State:* The current's speed is 6 mph.

1. $C = \dfrac{5}{9}(F - 32)$ **3.** $R = \dfrac{R_1 R_2}{R_1 + R_2}$ **4.** $R_1 = \dfrac{RR_2}{(R_2 - R)}$ **5.** $n = \dfrac{2S}{a + L}$ **7.** $b = \dfrac{2A - ah}{h}$

9. $T_2 = \dfrac{P_2 V_2 T_1}{P_1 V_1}$ **10.** $T_2 = T_1 - \dfrac{LH}{kA}$ **11.** $f_2 = \dfrac{f_1 f}{f_1 - f}$ **14.** $a_1 = S(1 - r) + a_n r$

EXERCISE SET 6.8

Solve each equation for the specified variable. See Example 1.

1. $F = \dfrac{9}{5}C + 32;\ C$

2. $V = \dfrac{1}{3}\pi r^2 h;\ h \quad h = \dfrac{3V}{\pi r^2}$

3. $\dfrac{1}{R} = \dfrac{1}{R_1} + \dfrac{1}{R_2};\ R$

4. $\dfrac{1}{R} = \dfrac{1}{R_1} + \dfrac{1}{R_2};\ R_1$

5. $S = \dfrac{n(a + L)}{2};\ n$

6. $S = \dfrac{n(a + L)}{2};\ a \quad a = \dfrac{2S - nL}{n}$

7. $A = \dfrac{h(a + b)}{2};\ b$

8. $A = \dfrac{h(a + b)}{2};\ h \quad h = \dfrac{2A}{a + b}$

9. $\dfrac{P_1 V_1}{T_1} = \dfrac{P_2 V_2}{T_2};\ T_2$

10. $H = \dfrac{kA(T_1 - T_2)}{L};\ T_2$

11. $f = \dfrac{f_1 f_2}{f_1 + f_2};\ f_2$

12. $I = \dfrac{E}{R + r};\ r \quad r = \dfrac{E}{I} - R$

13. $\lambda = \dfrac{2L}{n};\ L \quad L = \dfrac{n\lambda}{2}$

14. $S = \dfrac{a_1 - a_n r}{1 - r};\ a_1$

15. $\dfrac{\theta}{\omega} = \dfrac{2L}{c};\ c \quad c = \dfrac{2L\omega}{\theta}$

16. $F = \dfrac{-GMm}{r^2};\ M \quad M = -\dfrac{Fr^2}{Gm}$

Solve. See Example 2.

17. The sum of a number and 5 times its reciprocal is 6. Find the number(s).   1 and 5

18. The quotient of a number and 9 times its reciprocal is 1. Find the number(s). -3 and 3

19. If a number is added to the numerator of $\frac{12}{41}$ and twice the number is added to the denominator of $\frac{12}{41}$, the resulting fraction is equivalent to $\frac{1}{3}$. Find the number.   5

20. If a number is subtracted from the numerator of $\frac{13}{8}$ and added to the denominator of $\frac{13}{8}$, the resulting fraction is equivalent to $\frac{2}{5}$. Find the number.   7

In electronics, the relationship among the resistances r_1 and r_2 of two resistors wired in a parallel circuit and their combined resistance r is described by the formula $\dfrac{1}{r} = \dfrac{1}{r_1} + \dfrac{1}{r_2}$. *Use this formula to solve Exercises 21 through 23. See Example 3.*

21. If the combined resistance is 2 ohms and one of the two resistances is 3 ohms, find the other resistance.   6 ohms

22. Find the combined resistance of two resistors of 12 ohms each when they are wired in a parallel circuit.   6 ohms

23. The relationship among resistance of two resistors wired in a parallel circuit and their combined resistance may be extended to three resistors of resistances r_1, r_2 and r_3. Write an equation you believe may describe the relationship, and use it to find the combined resistance if r_1 is 5, r_2 is 6 and r_3 is 2. $\dfrac{1}{r} = \dfrac{1}{r_1} + \dfrac{1}{r_2} + \dfrac{1}{r_3};\ r = \dfrac{15}{13}$ ohms

Solve. See Example 4.

24. Alan Cantrell can word process a research paper in 6 hours. With Steve Isaac's help, the paper can be processed in 4 hours. Find how long it takes Steve to word process the paper alone.   12 hrs.

25. An experienced roofer can roof a house in 26 hours. A beginning roofer needs 39 hours to complete the same job. Find how long it takes for the two to do the job together.   15.6 hrs.

26. A new printing press can print newspapers twice as fast as the old one can. The old one can print the afternoon edition in 4 hours. Find how long it takes to print the afternoon edition if both printers are operating. $1\frac{1}{3}$ hrs.

27. Three postal workers can sort a stack of mail in 20 minutes, 30 minutes, and 60 minutes, respectively. Find how long it takes to sort the mail if all three work together.   10 min.

Solve. See Example 5.

28. An F-100 plane and a Toyota Truck leave the same town at sunrise and head for a town 450 miles away. The speed of the plane is three times the speed of the truck, and the plane arrives in 6 hours ahead of the truck. Find the speed of the truck.   50 mph

29. Mattie Evans drove 150 miles in the same amount of time that it took a turbopropeller plane to travel 600 miles. The speed of the plane was 150 mph faster than the speed of the car. Find the speed of the plane.   200 mph

30. The speed of a boat in still water is 24 mph. If the boat travels 54 miles upstream in the same time that it takes to travel 90 miles downstream, find the speed of the current.   6 mph

31. The speed of Lazy River's current is 5 mph. If a boat travels 20 miles downstream in the same time that it takes to travel 10 miles upstream, find the speed of the boat in still water.   15 mph

Solve.

32. The sum of the reciprocals of two consecutive odd integers is $\frac{20}{99}$. Find the two integers.   9 and 11

33. The sum of the reciprocals of two consecutive integers is $-\frac{15}{56}$. Find the two integers.   −8 and −7

34. If Sarah Clark can do a job in 5 hours and Dick Belli and Sarah working together can do the same job in 2 hours, find how long it takes Dick to do the job alone. $3\frac{1}{3}$ hrs.

35. One hose can fill a goldfish pond in 45 minutes, and two hoses can fill the same pond in 20 minutes. Find how long it takes the second hose alone to fill the pond.   36 min.

36. The speed of a bicyclist is 10 mph faster than the speed of a walker. If the bicyclist travels 26 miles in the same amount of time that the walker travels 6 miles, find the speed of the bicyclist.   13 mph

37. Two trains going in opposite directions leave at the same time. One train travels 15 mph faster than the other. In 6 hours the trains are 630 miles apart. Find the speed of each.   45 mph and 60 mph

38. The numerator of a fraction is 4 less than the denominator. If both the numerator and the denominator are increased by 2, the resulting fraction is equivalent to $\frac{2}{3}$. Find the fraction. $\frac{6}{10}$

39. Fabio Casartelli of Italy won the individual road race in cycling during the 1992 Summer Olympics. An amateur cyclist training for a road race rode the first 20-mile portion of his workout at a constant rate. For the 16-mile cooldown portion of his workout, he reduced his speed by 2 miles per hour. Each portion of the workout took equal time. Find the cyclist's rate during the first portion and his rate during the cooldown portion.   10 mph, 8 mph

40. The denominator of a fraction is 1 more than the numerator. If both the numerator and the denominator are decreased by 3, the resulting fraction is equivalent to $\frac{4}{5}$. Find the fraction. $\frac{7}{8}$

41. Moo Dairy has three machines to fill half-gallon milk cartons. The machines can fill the daily quota in 5 hours, 6 hours, and 7.5 hours, respectively. Find how long it takes to fill the daily quota if all three machines are running.   2 hrs.

42. The inlet pipe of an oil tank can fill the tank in 1 hour 30 minutes. The outlet pipe can empty the tank in 1 hour. Find how long it takes to empty a full tank if both pipes are open.   3 hrs.

43. A plane flies 465 miles with the wind and 345 miles against the wind in the same length of time. If the speed of the wind is 20 mph, find the speed of the plane in still air.   135 mph

44. Two rockets are launched. The first travels at 9000 mph. Fifteen minutes later the second is launched at 10,000 mph. Find the distance at which both rockets are an equal distance from Earth.   22,500 miles

45. Two joggers, one averaging 8 mph and one averaging 6 mph, start from a designated initial point.

The slower jogger arrives at the end of the run a half-hour after the other jogger. Find the distance of the run.   12 miles

46. A semi truck travels 300 miles through the flatland in the same amount of time that it travels 180 miles through the Great Smoky Mountains. The rate of the truck is 20 miles per hour slower in the mountains than in the flatland. Find both the flatland rate and mountain rate.

47. Smith Engineering is in the process of reviewing the salaries of their surveyors. During this review, the company found that an experienced surveyor surveys a roadbed in 4 hours. An apprentice surveyor needs 5 hours to survey the same stretch of road. If the two work together, find how long it takes them to complete the job. $2\frac{2}{9}$ hrs.

48. An experienced bricklayer constructs a small wall in 3 hours. An apprentice completes the job in 6 hours. Find how long it takes if they work together.   2 hrs.

49. A marketing manager travels 1080 miles in a corporate jet and then an additional 240 miles by car. If the car ride takes 1 hour longer, and if the rate of the jet is 6 times the rate of the car, find the time the manager travels by jet and find the time she travels by car.   by jet, 3 hrs.; by car, 4 hrs.

50. Gary Marcus and Tony Alva work for Lombardo's Pipe and Concrete. Mr. Lombardo is preparing an estimate for a customer. He knows that Gary lays a slab of concrete in 6 hours. Tony lays the same size slab in 4 hours. If both work on the job and the cost of labor is $45.00 per hour, decide what the labor estimate should be. $108

51. In 2 minutes, a conveyor belt moves 300 pounds of recyclable aluminum from the delivery truck to a storage area. A smaller belt moves the same quantity of cans the same distance in 6 minutes. If both belts are used, find how long it takes to move the cans to the storage area. $1\frac{1}{2}$ min.

52. Mr. Dodson can paint his house by himself in four days. His son needs an additional day to complete the job if he works by himself. If they work together, find how long it takes to paint the house.

53. While road testing a new make of car, the editor of a consumer magazine finds that she can go 10 miles into a 3-mile-per-hour wind in the same amount of time that she can go 11 miles with a 3-mile-per-hour wind behind her. Find the speed of the car in still air.   63 mph

54. The world record for the largest white bass caught is held by Ronald Sprouse of Virginia. The bass weighed 6 pounds 13 ounces. If Ronald rows to his favorite fishing spot 9 miles downstream in the same amount of time that he rows 3 miles upstream and if the current is 6 miles per hour, find how long it takes him to cover the 12 miles.   1 hr.

Review Exercises

Solve the equation for x. See Section 2.1.

55. $\dfrac{x}{5} = \dfrac{x+2}{3}$ $\{-5\}$ **56.** $\dfrac{x}{4} = \dfrac{x+3}{6}$ $\{6\}$

57. $\dfrac{x-3}{2} = \dfrac{x-5}{6}$ $\{2\}$ **58.** $\dfrac{x-6}{4} = \dfrac{x-2}{5}$ $\{22\}$

Use the circle graph to answer Exercises 59 through 64.

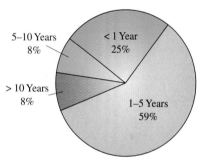

How long have you worn your hair in its current style?

5–10 Years 8%
< 1 Year 25%
> 10 Years 8%
1–5 Years 59%

Source: *Self Magazine*, June 1993.

59. What percent of those polled have worn their hair in its current style for less than 1 year?   25%

60. What percent of those polled have worn their hair in its current style for more than 10 years?   8%

61. What percent of those polled have worn their hair in its current style for 5 years or less?   84%

62. What percent of those polled have worn their hair in its current style for 5 years or more?   16%

63. Poll your algebra class with this same question and compute the percents in each category.

64. Use the results of Exercise 63 and construct a circle graph. (*Hint:* Recall that the number of degrees around a circle is 360°. Then, for example, the number of degrees in a sector representing 8% of a whole should be 0.08(360°) = 28.8°.)

46. flatland rate, 50 mph; mountain rate, 30 mph   **52.** $2\frac{2}{9}$ days   **63.** Answers will vary.   **64.** Answers will vary.

6.9 | VARIATION AND PROBLEM SOLVING

TAPE IA 6.9

O B J E C T I V E S

1. Write an equation expressing direct variation.
2. Write an equation expressing inverse variation.
3. Write an equation expressing joint variation.

In this section, we solve problems that can be modeled by using the concepts of direct variation, inverse variation, or joint variation.

1 A very familiar example of direct variation is the relationship of the circumference C of a circle to its radius r. The formula $C = 2\pi r$ expresses that the circumference is always 2π times the radius. In other words, C is always a constant multiple (2π) of r. Because it is, we say that **C varies directly as r,** that **C varies directly with r,** or that **C is directly proportional to r.**

DIRECT VARIATION

y varies directly as x, or **y is directly proportional to x,** if there is a nonzero constant k such that

$$y = kx$$

The number k is called the **constant of variation** or the **constant of proportionality.**

Recall that the relationship described between x and y is a linear one. In other words, the graph of $y = kx$ is a line. The slope of the line is k, and the line passes through the origin.

The graph of the direct variation example $C = 2\pi r$ is shown next. The horizontal axis represents the radius r, and the vertical axis is the circumference C.

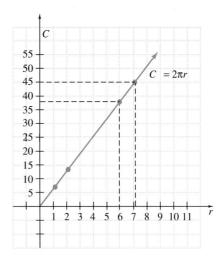

From the graph we can read that when the radius is 6 units, the circumference is approximately 38 units. Also, when the circumference is 45 units, the radius is between 7 and 8 units. Notice that as the radius increases, the circumference increases.

EXAMPLE 1   Suppose that y varies directly as x. If y is 5 when x is 30, find the constant of variation. Also, find y when $x = 90$.

Solution: Since y varies directly as x, we write $y = kx$. If $y = 5$ when $x = 30$, we have that

$$y = kx$$
$$5 = k(30) \qquad \text{Replace } y \text{ with 5 and } x \text{ with 30.}$$
$$\frac{1}{6} = k \qquad \text{Solve for } k.$$

The constant of variation is $\frac{1}{6}$.

After finding the constant of variation k, the direct variation equation can be written as $y = \frac{1}{6}x$. Next find y when x is 90.

$$y = \frac{1}{6}x$$
$$= \frac{1}{6}(90) \qquad \text{Let } x = 90.$$
$$= 15$$

When $x = 90$, y must be 15.

Notice that not only is the direct variation equation $y = kx$ a linear equation, but also y is a function of x. Thus, the notation $y = kx$ or $f(x) = kx$ may be used.

EXAMPLE 2   Hooke's law states that the distance a spring stretches is directly proportional to the weight attached to the spring. If a 40-pound weight attached to the spring stretches the spring 5 inches, find the distance that a 65-pound weight attached to the spring stretches the spring.

Solution: **1.** UNDERSTAND. Read and reread the problem. Notice that we are given that the distance a spring stretches is **directly proportional** to the weight attached.

2. ASSIGN. Let d represent the distance stretched and let w represent the weight attached. The constant of variation is represented by k.

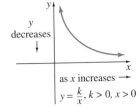

3. ILLUSTRATE. See the illustration shown.

4. TRANSLATE. Because d is directly proportional to w, we write

$$d = kw$$

When a weight of 40 pounds is attached, the spring stretches 5 inches. That is, when $w = 40$, $d = 5$.

$$5 = k(40) \qquad \text{Replace } d \text{ with 5 and } w \text{ with 40.}$$

$$\frac{1}{8} = k \qquad \text{Solve for } k.$$

Replace k with $\frac{1}{8}$ in the equation $d = kw$, and we have

$$d = \frac{1}{8}w.$$

To find the stretch when a weight of 65 pounds is attached, replace w with 65; find d.

$$d = \frac{1}{8}(65)$$

5. COMPLETE.

$$d = \frac{1}{8}(65)$$

$$= \frac{65}{8} = 8\frac{1}{8}, \quad \text{or} \quad 8.125$$

6. INTERPRET. *Check* the proposed solution of 8.125 inches. *State:* The spring stretches 8.125 inches when a 65-pound weight is attached.

When y is proportional to the **reciprocal** of another variable x, we say that **y varies inversely as x,** or that **y is inversely proportional to x.** An example of the inverse variation relationship is the relationship between the pressure that a gas exerts and the volume of its container. As the volume of a container decreases, the pressure of the gas it contains increases.

INVERSE VARIATION

y varies inversely as x, or **y is inversely proportional to x,** if there is a nonzero constant k such that

$$y = \frac{k}{x}$$

The number k is called the **constant of variation** or the **constant of proportionality.**

Notice that $y = \dfrac{k}{x}$, or $f(x) = \dfrac{k}{x}$, is a rational function. Its graph for $k > 0$ and $x > 0$ is shown. From the graph, we can see that as x increases, y decreases.

EXAMPLE 3   Suppose that u varies inversely as w. If u is 3 when w is 5, find u when w is 30.

Solution: Since u varies inversely as w, we have $u = \dfrac{k}{w}$. Let $u = 3$, $w = 5$, and solve for k.

$$u = \frac{k}{w}$$

$$3 = \frac{k}{5} \qquad \text{Let } u = 3 \text{ and } w = 5.$$

$$15 = k \qquad \text{Multiply both sides by 5 or cross multiply.}$$

The constant of variation k is 15. This gives the following inverse variation equation:

$$u = \frac{15}{w}$$

Now find u when $w = 30$.

$$u = \frac{15}{30} \qquad \text{Let } w = 30.$$

$$= \frac{1}{2}$$

Thus, when $w = 30$, $u = \dfrac{1}{2}$.

EXAMPLE 4   Boyle's law says that if the temperature stays the same, the pressure P of a gas is inversely proportional to the volume V. If a cylinder in a steam engine has a pressure of 960 kilopascals when the volume is 1.4 cubic meters, find the pressure when the volume increases to 2.5 cubic meters.

Solution: **1.** UNDERSTAND. Read and reread the problem. Notice that we are given that the pressure of a gas is *inversely proportional* to the volume.

2. ASSIGN. Let P represent the pressure and let V represent the volume. The constant of variation is represented by k.

4. TRANSLATE. Because P is inversely proportional to V, we write

$$P = \frac{k}{V}$$

When $P = 960$ kilopascals, the volume $V = 1.4$ cubic meters. Use this information to find k.

$$960 = \frac{k}{1.4} \qquad \text{Let } P = 960 \text{ and } V = 1.4.$$

$$1344 = k \qquad \text{Cross multiply.}$$

Thus, the value of k is 1344. Replace k with 1344 in the variation equation:

$$P = \frac{1344}{V}$$

Next find P when V is 2.5 cubic meters.

$$P = \frac{1344}{2.5} \qquad \text{Let } V = 2.5.$$

5. COMPLETE.

$$P = \frac{1344}{2.5}$$

$$P = 537.6$$

6. INTERPRET. *Check* the proposed solution. *State:* When the volume is 2.5 cubic meters, the pressure is 537.6 kilopascals. ▬▬▬

3 Sometimes the ratio of a variable to the product of many other variables is constant. For example, the ratio of distance traveled to the product of speed and time traveled is constantly 1:

$$\frac{d}{rt} = 1, \quad \text{or} \quad d = rt$$

Such a relationship is called **joint variation**.

JOINT VARIATION

If the ratio of a variable y to the product of two or more variables is constant, then y **varies jointly as,** or **is jointly proportional to,** the other variables. If

$$y = kxz$$

then the number k is the **constant of variation** or the **constant of proportionality.**

EXAMPLE 5　The surface area of a cylinder varies jointly as its radius and height. Express surface area S in terms of radius r and height h.

Solution:　Because the surface area varies jointly as the radius r and the height h, we equate S to the constant multiple of r and h:

$$S = krh$$
▬▬▬

EXERCISE SET 6.9

Write each statement as an equation. See Examples 1 through 5.

1. *A* is directly proportional to *B*. $A = kB$

2. *C* varies inversely as *D*. $C = \dfrac{k}{D}$

3. *X* is inversely proportional to *Z*. $X = \dfrac{k}{Z}$

4. *G* varies directly with *M*. $G = kM$

5. *N* varies directly with the square of *P*. $N = kP^2$

6. *A* varies jointly with *D* and *E*. $A = kDE$

7. *T* is inversely proportional to *R*. $T = \dfrac{k}{R}$

8. *G* is inversely proportional to *H*. $G = \dfrac{k}{H}$

9. *P* varies directly with *R*. $P = kR$

10. *T* is directly proportional to *S*. $T = kS$

Solve. See Examples 1 and 2.

11. *A* varies directly as *B*. If *A* is 60 when *B* is 12, find *A* when *B* is 9. $A = 45$

12. *C* varies directly as *D*. If *C* is 42 when *D* is 14, find *C* when *D* is 6. $C = 18$ 13. $V = 24$ cu. meters

13. Charles's law states that if the pressure *P* stays the same, the volume *V* of a gas is directly proportional to its temperature *T*. If a balloon is filled with 20 cubic meters of a gas at a temperature of 300 K, find the new volume if the temperature rises to 360 K while the pressure stays the same.

14. The amount *P* of pollution varies directly with the population *N* of people. Kansas City has a population of 450,000 and produces 260,000 tons of pollutants. Find how many tons of pollution we should expect St. Louis to produce, if we know that its population is 980,000. $P = 566,222$ tons

Solve. See Examples 3 and 4.

15. *H* is inversely proportional to *J*. If *H* is 4 when *J* is 5, find *H* when *J* is 2. $H = 10$

16. *D* varies inversely as *A*. If *D* is 16 when *A* is 2, find *D* when *A* is 8. $D = 4$

17. If the voltage *V* in an electric circuit is held constant, the current *I* is inversely proportional to the resistance *R*. If the current is 40 amperes when the resistance is 270 ohms, find the current when the resistance is 150 ohms. 72 amps

18. Pairs of markings a set distance apart are made on highways so that police can detect drivers exceeding the speed limit. Over a fixed distance, the speed *R* varies inversely with the time *T*. In one particular pair of markings, *R* is 45 mph when *T* is 6 seconds. Find the speed of a car that travels the given distance in 5 seconds. $R = 54$ mph

Write each statement as an equation. See Example 5.

19. *x* varies jointly as *y* and *z*. $x = kyz$

20. *P* varies jointly as *R* and the square of *S*. $P = kRS^2$

21. *r* varies jointly as *s* and the cube of *t*. $r = kst^3$

22. *a* varies jointly as *b* and *c*. $a = kbc$

Solve.

23. *Q* is directly proportional to *R*. If *Q* is 4 when *R* is 20, find *Q* when *R* is 35. $Q = 7$

24. *S* is directly proportional to *T*. If *S* is 4 when *T* is 16, find *S* when *T* is 40. $S = 10$

25. *M* varies directly with *P*. If *M* is 8 when *P* is 20, find *M* when *P* is 24.

26. *F* is directly proportional to *G*. If *F* is 18 when *G* is 10, find *F* when *G* is 16.

27. *B* is inversely proportional to *C*. If *B* is 12 when *C* is 3, find *B* when *C* is 18. $B = 2$

28. *U* varies inversely with *V*. If *U* is 14 when *V* is 4, find *U* when *V* is 7. $U = 8$

29. *W* varies inversely with *X*. If *W* is 18 when *X* is 6, find *W* when *X* is 40. $W = 2.7$

30. *Z* is inversely proportional to *Y*. If *Z* is 9 when *Y* is 6, find *Z* when *Y* is 24.

31. The weight of a synthetic ball varies directly with the cube of its radius. A ball with a radius of 2 inches weighs 1.20 pounds. Find the weight of a ball of the same material with a 3-inch radius.

4.05 lbs.

32. At sea, the distance to the horizon is directly proportional to the square root of the elevation of the observer. If a person who is 36 feet above the water can see 7.4 miles, find how far a person 64 feet above the water can see. 9.9 miles

25. $M = \dfrac{48}{5}$ 26. $F = \dfrac{144}{5}$ 30. $Z = \dfrac{9}{4}$

33. Because it is more efficient to produce larger numbers of items, the cost of producing Dysan computer disks is inversely proportional to the number produced. If 4000 can be produced at a cost of $1.20 each, find the cost per disk when 6000 are produced. $0.80

34. The weight of an object on or above the surface of the Earth varies inversely as the square of the distance between the object and the Earth's center. If a person weighs 160 pounds on the Earth's surface, find the individual's weight if he moves 200 miles above Earth. (Assume that the Earth's radius is 4000 miles.) 145 lbs.

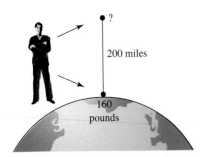

35. The number of cars manufactured on an assembly line at a General Motors plant varies jointly as the number of workers and the time they work. If 200 workers can produce 60 cars in 2 hours, find how many cars 240 workers should be able to make in 3 hours. 108 cars

36. The volume of a cone varies jointly as the square of its radius and its height. If the volume of a cone is 32π when the radius is 4 inches and the height is 6 inches, find the volume of a cone when the radius is 3 inches and the height is 5 inches.

15π cu. in.

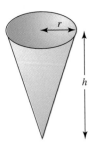

37. When a wind blows perpendicularly against a flat surface, its force is jointly proportional to the surface area and the speed of the wind. A sail whose surface area is 12 square feet experiences a 20-pound force when the wind speed is 10 miles per hour. Find the force on an 8-square-foot sail if the wind speed is 12 miles per hour. 16 lbs.

38. The horsepower that can be safely transmitted to a shaft varies jointly as the shaft's angular speed of rotation (in revolutions per minute) and the cube of its diameter. A 2-inch shaft making 120 revolutions per minute safely transmits 40 horsepower. Find how much horsepower can be safely transmitted by a 3-inch shaft making 80 revolutions per minute. 90 hp

39. A circular column has a *safe load* that is directly proportional to the fourth power of its diameter and inversely proportional to the square of its length. An 8-inch pillar 10 feet long can safely support a 16-ton load. Find the load that a 6-inch pillar made of the same material can support if it is 8 feet long. 7.91 tons

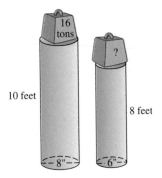

40. The maximum safe load for a rectangular beam varies jointly as its width and the square of its weight and inversely as its length. If a beam 6 inches wide, 4 inches high, and 10 feet long supports 12 tons, find how much a similar beam can support if the beam is 8 inches wide, 5 inches high, and 16 feet long. 15.6 tons

41. The area of a circle is directly proportional to the square of its radius. If the radius is tripled, determine how the area changes. multiplied by 9

42. The horsepower to drive a boat varies directly as the cube of the speed of the boat. If the speed of the boat is to double, determine the corresponding increase in horsepower required. multiplied by 8

43. The intensity I of light varies inversely as the square of the distance d from the light source. If the distance from the light source is doubled (see figure), determine what happens to the intensity of light at the new location.

divided by 4

lamp

24 inches

12 inches

book

□ **44.** The volume of a cylinder varies jointly as the height and the square of the radius. If the height is halved and the radius is doubled, determine what happens to the volume.   multiplied by 2

□ **45.** Suppose that y varies directly as x. If x is doubled, what is the effect on y?   multiplied by 2

□ **46.** Suppose that y varies directly as x^2. If x is doubled, what is the effect on y?   multiplied by 4

Complete the following table for the inverse variation $y = \dfrac{k}{x}$ over each given value of k. Plot the points on a rectangular coordinate system.

x	$\dfrac{1}{4}$	$\dfrac{1}{2}$	1	2	4
$y = \dfrac{k}{x}$					

For Exercises 47–50, see Appendix D.

□ **47.** $k = 1$ □ **48.** $k = 3$ □ **49.** $k = 5$ □ **50.** $k = \dfrac{1}{2}$

47. $4, 2, 1, \dfrac{1}{2}, \dfrac{1}{4}$ **48.** $12, 6, 3, \dfrac{3}{2}, \dfrac{3}{4}$ **49.** $20, 10, 5, \dfrac{5}{2}, \dfrac{5}{4}$ **50.** $2, 1, \dfrac{1}{2}, \dfrac{1}{4}, \dfrac{1}{8}$

Review Exercises

Find the circumference and area of each circle.

51.

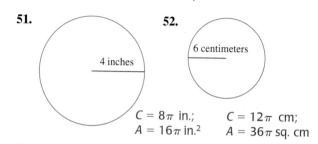

4 inches

$C = 8\pi$ in.;
$A = 16\pi$ in.^{2}

52.

6 centimeters

$C = 12\pi$ cm;
$A = 36\pi$ sq. cm

53.

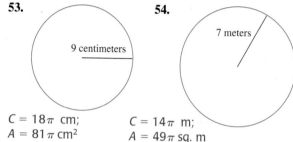

9 centimeters

$C = 18\pi$ cm;
$A = 81\pi$ cm^2

54.

7 meters

$C = 14\pi$ m;
$A = 49\pi$ sq. m

Find the slope of the line containing each pair of points. See Section 3.4.

55. $(-5, -2), (0, 7)$ $\dfrac{9}{5}$ **56.** $(3, 6), (-2, 6)$ 0

57. $(2, 1), (2, -3)$ undefined   **58.** $(4, -1), (5, -2)$ -1

Graph each function. See Sections 3.1 and 3.2.

59. $f(x) = 2x - 3$ **60.** $f(x) = x + 4$

61. $g(x) = |x|$ **62.** $h(x) = |x| + 2$

63. $h(x) = x^2$ **64.** $f(x) = x^2 - 1$

For Exercises 59–64, see Appendix D.

GROUP ACTIVITY

MODELING ELECTRICITY PRODUCTION

MATERIALS:
• Calculator

A progressive community in California is experimenting with electricity generated by windmills. City engineers are analyzing data they have gathered about their field of windmills. The engineers are familiar with other research demonstrating that the amount of electricity that a windmill generates hourly (in watts per hour) is directly proportional to the cube of the wind speed (in miles per hour).

(continued)

1. The city engineers have documented that when the wind speed is exactly 10 miles per hour, a windmill generates electricity at a rate of 15 watts per hour. Find a formula that models the relationship between the wind speed and the amount of electricity generated hourly by a windmill.

2. Complete the following table for the given wind speeds. Estimate from the table the wind speed required to obtain 400 watts per hour.

WIND SPEED (MILES PER HOUR)	ELECTRICITY (WATTS PER HOUR)
15	
17	
19	
21	
23	
25	
27	
29	
31	
33	
35	

3. Construct a bar graph for the data in the table. Describe the trend in the data.

4. The engineers' data shows that for several days the wind speed was more or less steady at 20 miles per hour, and the windmills generated the expected 120 watts per hour. According to the weather forecast for the coming few days, wind speed will fluctuate wildly but will still average 20 miles per hour. Should the engineers still expect the windmills to generate 120 watts per hour? Demonstrate your reasoning with a numerical example.

5. During one three-day period, each windmill generated 150 watts per hour. The forecast predicts the wind speed for the coming few days will drop by half. How many watts per hour should the engineers now expect each windmill to generate? In general, if the wind speed yields c watts per hour, how many watts per hour does half the wind speed yield?

See Appendix D for Group Activity answers and suggestions.

CHAPTER 6   HIGHLIGHTS

DEFINITIONS AND CONCEPTS	EXAMPLES
SECTION 6.1 RATIONAL FUNCTIONS AND SIMPLIFYING RATIONAL EXPRESSIONS	
A **rational expression** is the quotient $\dfrac{P}{Q}$ of two polynomials P and Q as long as Q is not 0.	Rational expressions: $$\frac{2x - 6}{7}, \frac{t^2 - 3t + 5}{t - 1}$$
A **rational function** is a function described by a rational expression.	Rational functions: $$f(x) = \frac{2x - 6}{7}, \quad h(t) = \frac{t^2 - 3t + 5}{t - 1}$$

(continued)

DEFINITIONS AND CONCEPTS	EXAMPLES

SECTION 6.1   RATIONAL FUNCTIONS AND SIMPLIFYING RATIONAL EXPRESSIONS

To simplify a rational expression:

Step 1. Completely factor the numerator and the denominator.

Step 2. Apply the fundamental principle.

Simplify:

$$\frac{2x^2 + 9x - 5}{x^2 - 25} = \frac{(2x - 1)\,(x + 5)}{(x - 5)\,(x + 5)}$$

$$= \frac{2x - 1}{x - 5}$$

SECTION 6.2   MULTIPLYING AND DIVIDING RATIONAL EXPRESSIONS

To multiply rational expressions:

Step 1. Completely factor numerators and denominators.

Step 2. Multiply the numerators and multiply the denominators.

Step 3. Apply the fundamental principle.

Multiply: $\dfrac{x^3 + 8}{12x - 18} \cdot \dfrac{14x^2 - 21x}{x^2 + 2x}$.

$$= \frac{(x + 2)\,(x^2 - 2x + 4)}{6(2x - 3)} \cdot \frac{7x(2x - 3)}{x(x + 2)}$$

$$= \frac{7(x^2 - 2x + 4)}{6}$$

To divide rational expressions:

Multiply the first rational expression by the reciprocal of the second rational expression.

Divide: $\dfrac{x^2 + 6x + 9}{5xy - 5y} \div \dfrac{x + 3}{10y}$

$$= \frac{(x + 3)(x + 3)}{5y(x - 1)} \cdot \frac{2 \cdot 5y}{x + 3}$$

$$= \frac{2(x + 3)}{x - 1}$$

SECTION 6.3   ADDING AND SUBTRACTING RATIONAL EXPRESSIONS

To add or subtract rational expressions:

Step 1. Find the LCD.

Step 2. Write each rational expression as an equivalent rational expression whose denominator is the LCD.

Step 3. Add or subtract numerators and write the sum or difference over the common denominator.

Step 4. Write the result in lowest terms.

Subtract: $\dfrac{3}{x + 2} - \dfrac{x + 1}{x - 3}$

$$= \frac{3 \cdot (x - 3)}{(x + 2) \cdot (x - 3)} - \frac{(x + 1) \cdot (x + 2)}{(x - 3) \cdot (x + 2)}$$

$$= \frac{3(x - 3) - (x + 1)(x + 2)}{(x + 2)(x - 3)}$$

$$= \frac{3x - 9 - (x^2 + 3x + 2)}{(x + 2)(x - 3)}$$

$$= \frac{3x - 9 - x^2 - 3x - 2}{(x + 2)(x - 3)}$$

$$= \frac{-x^2 - 11}{(x + 2)(x - 3)}$$

DEFINITIONS AND CONCEPTS	EXAMPLES

SECTION 6.4 SIMPLIFYING COMPLEX FRACTIONS

Method I: Simplify the numerator and the denominator so that each is a single fraction. Then perform the indicated division and simplify if possible.

Simplify: $\dfrac{\dfrac{x+2}{x}}{x-\dfrac{4}{x}}$.

Method I: $\dfrac{\dfrac{x+2}{x}}{\dfrac{x\cdot x}{1\cdot x}-\dfrac{4}{x}} = \dfrac{\dfrac{x+2}{x}}{\dfrac{x^2-4}{x}}$

$= \dfrac{x+2}{x}\cdot\dfrac{x}{(x+2)(x-2)} = \dfrac{1}{x-2}$

Method II: Multiply the numerator and the denominator of the complex fraction by the LCD of the fractions in both the numerator and the denominator. Then simplify.

Method II: $\dfrac{\left(\dfrac{x+2}{x}\right)\cdot x}{\left(x-\dfrac{4}{x}\right)\cdot x} = \dfrac{x+2}{x\cdot x - \dfrac{4}{x}\cdot x}$

$= \dfrac{x+2}{x^2-4} = \dfrac{x+2}{(x+2)(x-2)} = \dfrac{1}{x-2}$

SECTION 6.5 DIVIDING POLYNOMIALS

To divide a polynomial by a monomial:
Divide each term in the polynomial by the monomial.

$\dfrac{12a^5b^3 - 6a^2b^2 + ab}{6a^2b^2}$

$= \dfrac{12a^5b^3}{6a^2b^2} - \dfrac{6a^2b^2}{6a^2b^2} + \dfrac{ab}{6a^2b^2}$

$= 2a^3b - 1 + \dfrac{1}{6ab}$

To divide a polynomial by a polynomial, other than a monomial:
Use **long division.**

Divide $2x^3 - x^2 - 8x - 1$ by $x - 2$.

$$
\begin{array}{r}
2x^2 + 3x - 2 \\
x-2\overline{)2x^3 - x^2 - 8x - 1} \\
\underline{2x^3 - 4x^2} \\
3x^2 - 8x - 1 \\
\underline{3x^2 - 6x} \\
-2x - 1 \\
\underline{-2x + 4} \\
-5
\end{array}
$$

The quotient is $2x^2 + 3x - 2 - \dfrac{5}{x-2}$.

DEFINITIONS AND CONCEPTS	EXAMPLES

SECTION 6.6 SYNTHETIC DIVISION AND THE REMAINDER THEOREM

A shortcut method called **synthetic division** may be used to divide a polynomial by a binomial of the form $x - c$.

Use synthetic division to divide $2x^3 - x^2 - 8x - 1$ by $x - 2$.

$$
\begin{array}{r|rrrr}
2 & 2 & -1 & -8 & -1 \\
  &   & 4 & 6 & -4 \\
\hline
  & 2 & 3 & -2 & -5
\end{array}
$$

The quotient is $2x^2 + 3x - 2 - \dfrac{5}{x - 2}$.

SECTION 6.7 SOLVING EQUATIONS CONTAINING RATIONAL EXPRESSIONS

To solve an equation containing rational expressions:
Multiply both sides of the equation by the LCD of all rational expressions. Then apply the distributive property and simplify. Solve the resulting equation and then check each proposed solution to see whether it makes the denominator 0. If so, it is an **extraneous solution.**

Solve $x - \dfrac{3}{x} = \dfrac{1}{2}$.

$$2x\left(x - \frac{3}{x}\right) = 2x\left(\frac{1}{2}\right) \qquad \text{The LCD is } 2x.$$

$$2x \cdot x - 2x\left(\frac{3}{x}\right) = 2x\left(\frac{1}{2}\right) \qquad \text{Distribute.}$$

$$2x^2 - 6 = x$$

$$2x^2 - x - 6 = 0 \qquad \text{Subtract } x.$$

$$(2x + 3)(x - 2) = 0 \qquad \text{Factor.}$$

$$x = -\frac{3}{2} \text{ or } x = 2$$

Both $-\dfrac{3}{2}$ and 2 check. The solution set is $\left\{2, -\dfrac{3}{2}\right\}$.

SECTION 6.8 RATIONAL EQUATIONS AND PROBLEM SOLVING

To solve an equation for a specified variable:
Treat the specified variable as the only variable of the equation and solve as usual.

Solve for x.

$$A = \frac{2x + 3y}{5}$$

$$5A = 2x + 3y \qquad \text{Multiply by 5.}$$

$$5A - 3y = 2x \qquad \text{Subtract } 3y.$$

$$\frac{5A - 3y}{2} = x \qquad \text{Divide by 2.}$$

Problem-solving steps:

Jeanee and David Dillon volunteer every year to clean a strip of Lake Ponchatrain beach. Jeanee can clean all the trash in this area of beach in 6 hours; David takes 5 hours. How long will it take them to clean the area of beach together?

1. UNDERSTAND.

1. Read and reread the problem.

2. ASSIGN a variable.

2. Let x = time in hours that it takes Jeanee and David to clean the beach together. (*continued*)

DEFINITIONS AND CONCEPTS	EXAMPLES

SECTION 6.8 RATIONAL EQUATIONS AND PROBLEM SOLVING

3. ILLUSTRATE the problem.

3.

	HOURS TO COMPLETE	PART COMPLETED IN 1 HOUR
JEANEE ALONE	6	$\frac{1}{6}$
DAVID ALONE	5	$\frac{1}{5}$
TOGETHER	x	$\frac{1}{x}$

4. TRANSLATE.

4. In words:

$$\begin{array}{ccccc} \text{part Jeanee can complete in 1 hour} & + & \text{part David can complete in 1 hour} & = & \text{part they can complete together in 1 hour} \end{array}$$

Translate:

$$\frac{1}{6} + \frac{1}{5} = \frac{1}{x}$$

5. COMPLETE.

5.
$$\frac{1}{6} + \frac{1}{5} = \frac{1}{x} \qquad \text{Multiply by } 30x.$$
$$5x + 6x = 30$$
$$11x = 30$$
$$x = \frac{30}{11}, \text{ or } 2\frac{8}{11}$$

6. INTERPRET.

6. *Check* and then *state*. Together, they can clean the beach in $2\frac{8}{11}$ hours.

SECTION 6.9 VARIATION AND PROBLEM SOLVING

y **varies directly** as *x*, or *y* is **directly proportional** to *x*, if there is a nonzero constant *k* such that

$$y = kx$$

y **varies inversely** as *x*, or *y* is **inversely proportional** to *x*, if there is a nonzero constant *k* such that

$$y = \frac{k}{x}$$

The circumference of a circle *C* varies directly as its radius *r*.

$$C = \underset{k}{2\pi} r$$

Pressure *P* varies inversely with volume *V*.

$$P = \frac{k}{V}$$

1. $\{x \mid x$ is a real number$\}$ **2.** $\{x \mid x$ is a real number$\}$

CHAPTER 6 REVIEW

(6.1)   *Find the domain of each rational function.*

1. $f(x) = \dfrac{3 - 5x}{7}$ **2.** $g(x) = \dfrac{2x + 4}{11}$

3. $F(x) = \dfrac{-3x^2}{x - 5}$ **4.** $h(x) = \dfrac{4x}{3x - 12}$

5. $f(x) = \dfrac{x^3 + 2}{x^2 + 8x}$ **6.** $G(x) = \dfrac{20}{3x^2 - 48}$

Write each rational expression in lowest terms.

7. $\dfrac{15x^4}{45x^2}$ $\dfrac{x^2}{3}$ **8.** $\dfrac{x + 2}{2 + x}$ 1   **9.** $\dfrac{18m^6 p^2}{10m^4 p}$ $\dfrac{9m^2 p}{5}$

10. $\dfrac{x - 12}{12 - x}$ -1 **11.** $\dfrac{5x - 15}{25x - 75}$ $\dfrac{1}{5}$ **12.** $\dfrac{22x + 8}{11x + 4}$ 2

13. $\dfrac{2x}{2x^2 - 2x}$ $\dfrac{1}{x - 1}$ **14.** $\dfrac{x + 7}{x^2 - 49}$ $\dfrac{1}{x - 7}$

15. $\dfrac{2x^2 + 4x - 30}{x^2 + x - 20}$ **16.** $\dfrac{xy - 3x + 2y - 6}{x^2 + 4x + 4}$ $\dfrac{y - 3}{x + 2}$

17. The average cost of manufacturing x bookcases is given by the rational function

$$C(x) = \frac{35x + 4200}{x}$$

a. Find the average cost per bookcase of manufacturing 50 bookcases. $\$119$

b. Find the average cost per bookcase of manufacturing 100 bookcases. $\$77$

◻ **c.** As the number of bookcases increases, does the average cost per bookcase increase or decrease? (See parts (a) and (b).)   decrease

(6.2)   *Perform the indicated operation. Write answers in lowest terms.*

18. $\dfrac{5}{x^3} \cdot \dfrac{x^2}{15}$ $\dfrac{1}{3x}$ **19.** $\dfrac{3x^4 yz^3}{15x^2 y^2} \cdot \dfrac{10xy}{z^6}$ $\dfrac{2x^3}{z^3}$

20. $\dfrac{4 - x}{5} \cdot \dfrac{15}{2x - 8}$ $-\dfrac{3}{2}$

21. $\dfrac{x^2 - 6x + 9}{2x^2 - 18} \cdot \dfrac{4x + 12}{5x - 15}$ $\dfrac{2}{5}$

22. $\dfrac{a - 4b}{a^2 + ab} \cdot \dfrac{b^2 - a^2}{8b - 2a}$ $\dfrac{a - b}{2a}$

23. $\dfrac{x^2 - x - 12}{2x^2 - 32} \cdot \dfrac{x^2 + 8x + 16}{3x^2 + 21x + 36}$ $\dfrac{1}{6}$

24. $\dfrac{2x^3 + 54}{5x^2 + 5x - 30} \cdot \dfrac{6x + 12}{3x^2 - 9x + 27}$ $\dfrac{4(x + 2)}{5(x - 2)}$

25. $\dfrac{3}{4x} \div \dfrac{8}{2x^2}$ $\dfrac{3x}{16}$ **26.** $\dfrac{4x + 8y}{3} \div \dfrac{5x + 10y}{9}$

27. $\dfrac{5ab}{14c^3} \div \dfrac{10a^4 b^2}{6ac^5}$ $\dfrac{3c^2}{14a^2 b}$ **28.** $\dfrac{2}{5x} \div \dfrac{4 - 18x}{6 - 27x}$ $\dfrac{3}{5x}$

29. $\dfrac{x^2 - 25}{3} \div \dfrac{x^2 - 10x + 25}{x^2 - x - 20}$ $\dfrac{(x + 4)(x + 5)}{3}$

30. $\dfrac{a - 4b}{a^2 + ab} \div \dfrac{20b - 5a}{b^2 - a^2}$ $\dfrac{a - b}{5a}$

31. $\dfrac{7x + 28}{2x + 4} \div \dfrac{x^2 + 2x - 8}{x^2 - 2x - 8}$ $\dfrac{7(x - 4)}{2(x - 2)}$

32. $\dfrac{3x + 3}{x - 1} \div \dfrac{x^2 - 6x - 7}{x^2 - 1}$ $\dfrac{3(x + 1)}{x - 7}$

33. $\dfrac{2x - x^2}{x^3 - 8} \div \dfrac{x^2}{x^2 + 2x + 4}$ $-\dfrac{1}{x}$

34. $\dfrac{5a^2 - 20}{a^3 + 2a^2 + a + 2} \div \dfrac{7a}{a^3 + a}$ $\dfrac{5(a - 2)}{7}$

35. $\dfrac{2a}{21} \div \dfrac{3a^2}{7} \cdot \dfrac{4}{a}$ $\dfrac{8}{9a^2}$

36. $\dfrac{5x - 15}{3 - x} \cdot \dfrac{x + 2}{10x + 20} \cdot \dfrac{x^2 - 9}{x^2 - x - 6}$ $-\dfrac{x + 3}{2(x + 2)}$

37. $\dfrac{4a + 8}{5a^2 - 20} \cdot \dfrac{3a^2 - 6a}{a + 3} \div \dfrac{2a^2}{5a + 15}$ $\dfrac{6}{a}$

If $f(x) = x - 5$ and $g(x) = 2x + 1$, find:

38. $(f + g)(x)$ $3x - 4$ **39.** $(f - g)(x)$ $-x - 6$

40. $(f \cdot g)(x)$ $2x^2 - 9x - 5$ **41.** $\left(\dfrac{g}{f}\right)(x)$ $\dfrac{2x + 1}{x - 5}$

(6.3)   *Find the LCD of the rational expressions in the list.*   **43.** $60x^2 y^5$ **46.** $10x^3(x - 4)(x + 7)(x - 3)$

42. $\dfrac{4}{9}, \dfrac{5}{2}$ 18   **43.** $\dfrac{5}{4x^2 y^5}, \dfrac{3}{10x^2 y^4}, \dfrac{x}{6y^4}$

44. $\dfrac{5}{2x}, \dfrac{7}{x - 2}$ $2x(x - 2)$ **45.** $\dfrac{3}{5x}, \dfrac{2}{x - 5}$ $5x(x - 5)$

46. $\dfrac{1}{5x^3}, \dfrac{4}{x^2 + 3x - 28}, \dfrac{11}{10x^2 - 30x}$

3. $\{x \mid x$ is a real number and $x \neq 5\}$ **4.** $\{x \mid x$ is a real number and $x \neq 4\}$

5. $\{x \mid x$ is a real number and $x \neq 0, x \neq -8\}$ **6.** $\{x \mid x$ is a real number and $x \neq -4, x \neq 4\}$ **15.** $\dfrac{2(x - 3)}{x - 4}$ **26.** $\dfrac{12}{5}$

Perform the indicated operation. Write answers in lowest terms.

47. $\dfrac{2}{15} + \dfrac{4}{15}$ $\dfrac{2}{5}$

48. $\dfrac{4}{x-4} + \dfrac{x}{x-4}$ $\dfrac{4+x}{x-4}$

49. $\dfrac{4}{3x^2} + \dfrac{2}{3x^2}$ $\dfrac{2}{x^2}$

50. $\dfrac{1}{x-2} - \dfrac{1}{4-2x}$ $\dfrac{3}{2(x-2)}$

51. $\dfrac{2x+1}{x^2+x-6} + \dfrac{2-x}{x^2+x-6}$ $\dfrac{1}{x-2}$

52. $\dfrac{7}{2x} + \dfrac{5}{6x}$ $\dfrac{13}{3x}$

53. $\dfrac{1}{3x^2y^3} - \dfrac{1}{5x^4y}$ $\dfrac{5x^2-3y^2}{15x^4y^3}$

54. $\dfrac{1}{10-x} + \dfrac{x-1}{x-10}$

55. $\dfrac{x-2}{x+1} - \dfrac{x-3}{x-1}$

56. $\dfrac{x}{9-x^2} - \dfrac{2}{5x-15}$

57. $2x+1 - \dfrac{1}{x-3}$

58. $\dfrac{2}{a^2-2a+1} + \dfrac{3}{a^2-1}$ $\dfrac{5a-1}{(a-1)^2(a+1)}$

59. $\dfrac{x}{9x^2+12x+16} - \dfrac{3x+4}{27x^3-64}$ $\dfrac{3x^2-7x-4}{(3x-4)(9x^2+12x+16)}$

Perform the indicated operation. Write answers in lowest terms.

60. $\dfrac{2}{x-1} - \dfrac{3x}{3x-3} + \dfrac{1}{2x-2}$ $\dfrac{5-2x}{2(x-1)}$

61. $\dfrac{3}{2x} \cdot \left(\dfrac{2}{x+1} - \dfrac{2}{x-3}\right)$

62. $\left(\dfrac{2}{x} - \dfrac{1}{5}\right) \cdot \left(\dfrac{2}{x} + \dfrac{1}{3}\right)$

63. $\dfrac{2}{x^2-16} - \dfrac{3x}{x^2+8x+16} + \dfrac{3}{x+4}$ $\dfrac{14x-40}{(x+4)^2(x-4)}$

64. Find the perimeter of the heptagon (polygon with 7 sides). $\dfrac{11}{x}$

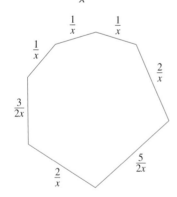

(labels on heptagon:) $\dfrac{1}{x}$, $\dfrac{1}{x}$, $\dfrac{1}{x}$, $\dfrac{2}{x}$, $\dfrac{3}{2x}$, $\dfrac{5}{2x}$, $\dfrac{2}{x}$

(6.4) *Simplify each complex fraction.*

65. $\dfrac{\frac{2}{5}}{\frac{3}{5}}$ $\dfrac{2}{3}$

66. $\dfrac{1 - \frac{3}{4}}{2 + \frac{1}{4}}$ $\dfrac{1}{9}$

67. $\dfrac{\frac{1}{x} - \frac{2}{3x}}{\frac{5}{2x} - \frac{1}{3}}$

68. $\dfrac{\frac{x^2}{15}}{\frac{x+1}{5x}}$

69. $\dfrac{\frac{3}{y^2}}{\frac{6}{y^3}}$ $\dfrac{y}{2}$

70. $\dfrac{\frac{x+2}{3}}{\frac{5}{x-2}}$

71. $\dfrac{2 - \frac{3}{2x}}{x - \frac{2}{5x}}$

72. $\dfrac{1 + \frac{x}{y}}{\frac{x^2}{y^2} - 1}$ $\dfrac{y}{x-y}$

73. $\dfrac{\frac{5}{x} + \frac{1}{xy}}{\frac{3}{x^2}}$ $\dfrac{5xy+x}{3y}$

74. $\dfrac{\frac{x}{3} - \frac{3}{x}}{1 + \frac{3}{x}}$ $\dfrac{x-3}{3}$

75. $\dfrac{\frac{1}{x-1} + 1}{\frac{1}{x+1} - 1}$

76. $\dfrac{\frac{2}{1 - \frac{2}{x}}}{}$ $\dfrac{2x}{x-2}$

77. $\dfrac{1}{1 + \dfrac{2}{1 - \frac{1}{x}}}$ $\dfrac{x-1}{3x-1}$

78. $\dfrac{\frac{x^2+5x-6}{(x+6)^2}}{\frac{4x+3}{8x+6}}$ $\dfrac{2(x-1)}{x+6}$

79. $\dfrac{\frac{x-3}{x+3} + \frac{x+3}{x-3}}{\frac{x-3}{x+3} - \frac{x+3}{x-3}}$

80. $\dfrac{\frac{3}{x-1} - \frac{2}{1-x}}{\frac{2}{x-1} - \frac{2}{x}}$ $\dfrac{5x}{2}$

81. If $f(x) = \dfrac{3}{x}$, find each of the following: **81c.** $\dfrac{\frac{3}{a+h} - \frac{3}{a}}{h}$

a. $f(a+h)$ $\dfrac{3}{a+h}$ **b.** $f(a)$ $\dfrac{3}{a}$

c. Use parts (a) and (b) to find $\dfrac{f(a+h) - f(a)}{h}$.

d. Simplify the results of part (c). $\dfrac{-3}{a(a+h)}$

(6.5) *Find each quotient. Use only positive exponents.*

82. $\dfrac{3x^5yb^9}{9xy^7} \div \dfrac{x^4b^9}{3y^6}$

83. Divide $-9xb^4z^3$ by $-4axb^2$. $\dfrac{9b^2z^3}{4a}$

84. $\dfrac{4xy + 2x^2 - 9}{4xy}$ $1 + \dfrac{x}{2y} - \dfrac{9}{4xy}$

85. Divide $12xb^2 + 16xb^4$ by $4xb^3$. $\dfrac{3}{b} + 4b$

54. $\dfrac{x-2}{x-10}$ **55.** $\dfrac{-x+5}{(x+1)(x-1)}$ **56.** $\dfrac{-7x-6}{5(x-3)(x+3)}$ **57.** $\dfrac{2x^2-5x-4}{x-3}$ **61.** $-\dfrac{12}{x(x+1)(x-3)}$

62. $\dfrac{60+4x-x^2}{15x^2}$ **67.** $\dfrac{2}{15-2x}$ **68.** $\dfrac{x^3}{3(x+1)}$ **70.** $\dfrac{(x+2)(x-2)}{15}$ **71.** $\dfrac{20x-15}{10x^2-4}$ **75.** $\dfrac{1+x}{1-x}$ **79.** $-\dfrac{x^2+9}{6x}$

86. $3x^3 + 9x^2 + 2x + 6 - \dfrac{2}{x-3}$ **87.** $2x^3 + 6x^2 + 17x + 56 + \dfrac{156}{x-3}$ **93.** $3x^2 - \dfrac{5}{2}x - \dfrac{1}{4} - \dfrac{5}{8\left(x + \dfrac{3}{2}\right)}$

Find each quotient.

108. $\dfrac{7}{x} - \dfrac{x}{7} = 0 \quad \{-7, 7\}$

86. $\dfrac{3x^4 - 25x^2 - 20}{x-3}$ **87.** $\dfrac{-x^2 + 2x^4 + 5x - 12}{x-3}$

109. $\dfrac{x-2}{x^2 - 7x + 10} = \dfrac{1}{5x - 10} - \dfrac{1}{x-5} \quad \left\{\dfrac{5}{3}\right\}$

88. $\dfrac{2x^4 - x^3 + 2x^2 - 3x + 1}{x - \dfrac{1}{2}} \quad 2x^3 + 2x - 2$

Solve the equations for x or perform the indicated operation. Simplify. **110.** $\dfrac{2x+5}{x(x-7)}$

89. $\dfrac{x^3 + 3x^2 - 2x + 2}{x - \dfrac{1}{2}} \quad x^2 + \dfrac{7}{2}x - \dfrac{1}{4} + \dfrac{15}{8\left(x - \dfrac{1}{2}\right)}$

110. $\dfrac{5}{x^2 - 7x} + \dfrac{4}{2x - 14}$ **111.** $3 - \dfrac{5}{x} - \dfrac{2}{x^2} = 0 \quad \left\{-\dfrac{1}{3}, 2\right\}$

90. $\dfrac{3x^4 + 5x^3 + 7x^2 + 3x - 2}{x^2 + x + 2} \quad 3x^2 + 2x - 1$

112. $\dfrac{4}{3-x} - \dfrac{7}{2x-6} + \dfrac{5}{x} \quad \dfrac{-5x - 30}{2x(x-3)}$

91. $\dfrac{9x^4 - 6x^3 + 3x^2 - 12x - 30}{3x^2 - 2x - 5} \quad 3x^2 + 6$

(6.8) *Solve the equation for the specified variable.*

(6.6) *Use synthetic division to find each quotient.*

113. $A = \dfrac{h(a+b)}{2}, a$ **114.** $\dfrac{1}{R} = \dfrac{1}{R_1} + \dfrac{1}{R_2}, R_2$

92. $\dfrac{3x^3 + 12x - 4}{x-2}$ **93.** $\dfrac{3x^3 + 2x^2 - 4x - 1}{x + \dfrac{3}{2}}$

115. $I = \dfrac{E}{R+r}, R$ **116.** $A = P + Prt, r$

$3x^2 + 6x + 24 + \dfrac{44}{x-2}$

117. $H = \dfrac{kA(T_1 - T_2)}{L}, A \quad A = \dfrac{HL}{k(T_1 - T_2)}$

94. $\dfrac{x^5 - 1}{x+1}$ **95.** $\dfrac{x^3 - 81}{x-3}$

96. $\dfrac{x^3 - x^2 + 3x^4 - 2}{x-4}$ **97.** $\dfrac{3x^4 - 2x^2 + 10}{x+2}$

Solve.

118. The sum of a number and twice its reciprocal is 3. Find the number(s). $\{1, 2\}$

If $P(x) = 3x^5 - 9x + 7$, use the remainder theorem to find the following.

119. If a number is added to the numerator of $\dfrac{3}{7}$, and twice that number is added to the denominator of $\dfrac{3}{7}$, the result is equivalent to $\dfrac{10}{21}$. Find the number. 7

98. $P(4)$ 3043 **99.** $P(-5)$ -9323

100. $P\left(\dfrac{2}{3}\right)$ $\dfrac{113}{81}$ **101.** $P\left(-\dfrac{1}{2}\right)$ $\dfrac{365}{32}$

120. The denominator of a fraction is 2 more than the numerator. If the numerator is decreased by 3 and the denominator is increased by 5, the resulting fraction is equivalent to $\dfrac{2}{3}$. Find the fraction. $\dfrac{23}{25}$

102. If the area of the rectangle is $x^4 - x^3 - 6x^2 - 6x$ $+ 18$ square miles and its width is $x - 3$ miles, find the length. $(x^3 + 2x^2 - 6)$ miles

$\boxed{x^4 - x^3 - 6x^2 - 6x + 18 \text{ square miles}} \;\; \begin{array}{c} x - 3 \\ \text{miles} \end{array}$

121. The sum of the reciprocals of two consecutive even integers is $-\dfrac{9}{40}$. Find the two integers.

122. Three boys can paint a fence in 4 hours, 5 hours, and 6 hours, respectively. Find how long it will take all three boys to paint the fence. $1\dfrac{23}{37}$ hr.

123. If Sue Katz can type a certain number of mailing labels in 6 hours and Tom Neilson and Sue working together can type the same number of mailing labels in 4 hours, find how long it takes Tom alone to type mailing labels. 12 hrs.

(6.7) *Solve each equation for x.*

103. $\dfrac{2}{5} = \dfrac{x}{15} \quad \{6\}$ **104.** $\dfrac{3}{x} + \dfrac{1}{3} = \dfrac{5}{x} \quad \{6\}$

105. $4 + \dfrac{8}{x} = 8 \quad \{2\}$ **106.** $\dfrac{2x+3}{5x-9} = \dfrac{3}{2} \quad \{3\}$

124. The inlet pipe of a water tank can fill the tank in 2 hours and 30 minutes. The outlet pipe can empty the tank in 2 hours. Find how long it takes to empty a full tank if both pipes are open. 10 hrs.

107. $\dfrac{1}{x-2} - \dfrac{3x}{x^2 - 4} = \dfrac{2}{x+2} \quad \left\{\dfrac{3}{2}\right\}$

94. $x^4 - x^3 + x^2 - x + 1 - \dfrac{2}{x+1}$ **95.** $x^2 + 3x + 9 - \dfrac{54}{x-3}$ **96.** $3x^3 + 13x^2 + 51x + 204 + \dfrac{814}{x-4}$

97. $3x^3 - 6x^2 + 10x - 20 + \dfrac{50}{x+2}$ **113.** $a = \dfrac{2A}{h} - b$ **114.** $R_2 = \dfrac{RR_1}{R_1 - R}$ **115.** $R = \dfrac{E}{I} - r$ **116.** $r = \dfrac{A-P}{Pt}$

121. -10 and -8

125. Timmy Garnica drove 210 miles in the same amount of time that it took a DC 10 jet to travel 1715 miles. The speed of the jet was 430 mph faster than the speed of the car. Find the speed of the jet.   490 mph

126. The combined resistance R of two resistors in parallel with resistances r_1 and r_2 is given by the formula $\dfrac{1}{R} = \dfrac{1}{r_1} + \dfrac{1}{r_2}$. If the combined resistance is $\frac{30}{11}$ ohms and the resistance of one of the two resistors is 5 ohms, find the resistance of the other resistor.   6 ohms

127. The speed of a Ranger boat in still water is 32 mph. If the boat travels 72 miles upstream in the same time that it takes to travel 120 miles downstream, find the current of the stream.   8 mph

128. A B737 jet flies 445 miles with the wind and 355 miles against the wind in the same length of time. If the speed of the jet in still air is 400 mph, find the speed of the wind.   45 mph

129. The speed of a jogger is 3 mph faster than the speed of a walker. If the jogger travels 14 miles in the same amount of time that the walker travels 8 miles, find the speed of the walker.   4 mph

130. Two Amtrak trains traveling on parallel tracks leave Tucson at the same time. In 6 hours the faster train is 382 miles from Tucson and the trains are 112 miles apart. Find how fast each train is traveling. $63\frac{2}{3}$ mph and 45 mph

(6.9)   *Solve each proportion for x.*

131. $\dfrac{3x - 5}{14} = \dfrac{x}{4}$ $\{-10\}$ **132.** $\dfrac{2 - 5x}{12} = \dfrac{x}{8}$ $\left\{\dfrac{4}{13}\right\}$

Solve each of the following variation problems.

133. A is directly proportional to B. If $A = 6$ when $B = 14$, find A when $B = 21$.   9

134. C is inversely proportional to D. If $C = 12$ when $D = 8$, find C when $D = 24$.   4

135. According to Boyle's law, the pressure exerted by a gas is inversely proportional to the volume, as long as the temperature stays the same. If a gas exerts a pressure of 1250 pounds per square inch when the volume is 2 cubic feet, find the volume when the pressure is 800 pounds per square inch.

136. The surface area of a sphere varies directly as the square of its radius. If the surface area is 36 square inches when the radius is 3 inches, find the surface area when the radius is 4 inches. 64 sq. in.

135.   3.125 cubic feet

CHAPTER 6 TEST

Find the domain of each rational function.

1. $f(x) = \dfrac{5x^2}{1 - x}$ **2.** $g(x) = \dfrac{9x^2 - 9}{x^2 + 4x + 3}$

Write each rational expression in lowest terms.

3. $\dfrac{5x^7}{3x^4}$ $\dfrac{5x^3}{3}$ **4.** $\dfrac{7x - 21}{24 - 8x}$ $-\dfrac{7}{8}$

5. $\dfrac{x^2 - 4x}{x^2 + 5x - 36}$ $\dfrac{x}{x + 9}$

Perform the indicated operation. Write answers in lowest terms.

6. $\dfrac{x}{x - 2} \cdot \dfrac{x^2 - 4}{5x}$ $\dfrac{x + 2}{5}$

7. $\dfrac{2x^3 + 16}{6x^2 + 12x} \cdot \dfrac{5}{x^2 - 2x + 4}$ $\dfrac{5}{3x}$

8. $\dfrac{26ab}{7c} \div \dfrac{13a^2c^5}{14a^4b^3}$ $\dfrac{4a^3b^4}{c^6}$

9. $\dfrac{3x^2 - 12}{x^2 + 2x - 8} \div \dfrac{6x + 18}{x + 4}$ $\dfrac{x + 2}{2(x + 3)}$

10. $\dfrac{4x - 12}{2x - 9} \div \dfrac{3 - x}{4x^2 - 81} \cdot \dfrac{x + 3}{5x + 15}$ $\dfrac{-4(2x + 9)}{5}$

11. $\dfrac{5}{4x^3} + \dfrac{7}{4x^3}$ $\dfrac{3}{x^3}$ **12.** $\dfrac{3 + 2x}{10 - x} + \dfrac{13 + x}{x - 10}$ -1

13. $\dfrac{3}{x^2 - x - 6} + \dfrac{2}{x^2 - 5x + 6}$ $\dfrac{5x - 2}{(x - 3)(x + 2)(x - 2)}$

14. $\dfrac{5}{x - 7} - \dfrac{2x}{3x - 21} + \dfrac{x}{2x - 14}$ $\dfrac{-x + 30}{6(x - 7)}$

15. $\dfrac{3x}{5} \cdot \left(\dfrac{5}{x} - \dfrac{5}{2x}\right)$ $\dfrac{3}{2}$

1. $\{x \mid x \text{ is a real number and } x \ne 1\}$ **2.** $\{x \mid x \text{ is a real number and } x \ne -3, x \ne -1\}$

Simplify each complex fraction.

16. $\dfrac{\dfrac{4x}{13}}{\dfrac{20x}{13}} \quad \dfrac{1}{5}$

17. $\dfrac{\dfrac{5}{x} - \dfrac{7}{3x}}{\dfrac{9}{8x} - \dfrac{1}{x}} \quad \dfrac{64}{3}$

18. $\dfrac{\dfrac{x^2 - 5x + 6}{x + 3}}{\dfrac{x^2 - 4x + 4}{x^2 - 9}} \quad \dfrac{(x-3)^2}{x-2}$

Divide.

19. $\dfrac{4x^2y + 9x + z}{3xz} \quad \dfrac{4xy}{3z} + \dfrac{3}{z} + \dfrac{1}{3x}$

20. $\dfrac{x^6 + 3x^5 - 2x^4 + x^2 - 3x + 2}{x - 2}$

21. Use synthetic division to divide $4x^4 - 3x^3 + 2x^2 - x - 1$ by $x + 3$.

22. If $P(x) = 4x^4 + 7x^2 - 2x - 5$, use the remainder theorem to find $P(-2)$.   91

If $f(x) = x$, $g(x) = x - 7$, and $h(x) = x^2 - 6x + 5$, find the following.

23. $(h - g)(x)$
$(h - g)(x) = x^2 - 7x + 12$

24. $(h \cdot f)(x) \quad (h \cdot f)(x) = x^3 - 6x^2 + 5x$

Solve each equation for x.

25. $\dfrac{5x + 3}{3x - 7} = \dfrac{19}{7} \quad \{7\}$

26. $\dfrac{5}{x - 5} + \dfrac{x}{x + 5} = -\dfrac{29}{21} \quad \{2, -2\}$

27. $\dfrac{x}{x - 4} = 3 - \dfrac{4}{x - 4} \quad \{8\}$

20. $x^5 + 5x^4 + 8x^3 + 16x^2 + 33x + 63 + \dfrac{128}{x - 2}$ **21.** $4x^3 - 15x^2 + 47x - 142 + \dfrac{425}{x + 3}$

28. Solve for x: $\dfrac{x + b}{a} = \dfrac{4x - 7a}{b} \quad x = \dfrac{7a^2 + b^2}{4a - b}$

29. The product of one more than a number and twice the reciprocal of the number is $\frac{12}{5}$. Find the number.   5

30. If Jan can weed the garden in 2 hours and her husband can weed it in 1 hour and 30 minutes, find how long it takes them to weed the garden together. $\frac{6}{7}$ hr.

31. Suppose that W is inversely proportional to V. If $W = 20$ when $V = 12$, find W when $V = 15$.   16

32. Suppose that Q is jointly proportional to R and the square of S. If $Q = 24$ when $R = 3$ and $S = 4$, find Q when $R = 2$ and $S = 3$.   9

33. When an anvil is dropped into a gorge, the speed with which it strikes the ground is directly proportional to the square root of the distance it falls. An anvil that falls 400 feet hits the ground at a speed of 160 feet per second. Find the height of a cliff over the gorge if a dropped anvil hits the ground at a speed of 128 feet per second.   256 ft.

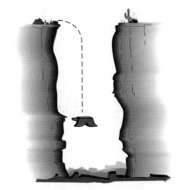

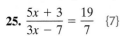

CHAPTER 6   CUMULATIVE REVIEW

1. Use mathematical symbols to write each sentence.

 a. The sum of x and 5 is 20. $x + 5 = 20$

 b. Two times the sum of 3 and y amounts to 4.

 c. Subtract 8 from x, and the difference is the same as the product of 2 and x. $x - 8 = 2x$

 d. The quotient of z and 9 is 3 times the difference of z and 5. $\dfrac{z}{9} = 3(z - 5)$; *Sec. 1.2, Ex. 1*

1b. $2(3 + y) = 4$

2. Find the following roots.   *Sec. 1.3, Ex. 9*

 a. $\sqrt[3]{27}$ 3   **b.** $\sqrt[5]{1}$ 1   **c.** $\sqrt[4]{16}$ 2

3. Pennants in the shape of an isosceles triangle are to be constructed for the Slidell High School Athletic Club and sold as a fund raiser. The company manufacturing the pennants charges according to perimeter. The club has determined that a perimeter of 149 centimeters each should make a nice profit. If each equal side of the triangle is twice the length of the third side, increased by 12 centimeters, find the lengths of the sides of each triangular pennant.   25 cm, 62 cm, 62 cm; *Sec. 2.2, Ex. 4*

4. Graph the equation $5x - 2y = 10$.   *Sec. 3.1, Ex. 4*

5. Which of the following graphs are graphs of functions?

a.
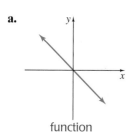
function

b.
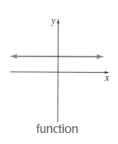
function

c.
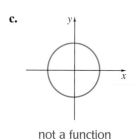
not a function

d.
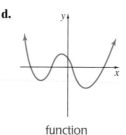
function

e.
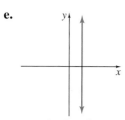
not a function

f.
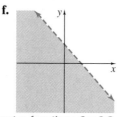
not a function; *Sec 3.2, Ex. 5*

6. Graph $y = -3$.   See Appendix D; *Sec. 3.3, Ex. 7*

7. Find an equation of the line that contains the point $(4, 4)$ and is parallel to the line $2x + 3y = -6$. Write the equation in standard form.

8. Solve the system: { }; *Sec. 4.2, Ex. 2*

$$\begin{cases} 2x - 4y + 8z = 2 & (1) \\ -x - 3y + z = 11 & (2) \\ x - 2y + 4z = 0 & (3) \end{cases}$$

9. Use the product rule to multiply.

a. $(3x^6)(5x)$ $15x^7$

b. $(-2x^3p^2)(4xp^{10})$

10. Simplify each expression. Use positive exponents to write answers.

a. $\left(\dfrac{3x^2y}{y^{-9}z}\right)^{-2}$ $\dfrac{z^2}{9x^4y^{20}}$

b. $\left(\dfrac{3a^2}{2x^{-1}}\right)^3\left(\dfrac{x^{-3}}{4a^{-2}}\right)^{-1}$ $\dfrac{27a^4x^6}{2}$; *Sec. 5.2, Ex. 4*

11. Multiply $(2x - 7)(3x - 4)$.

12. Factor $2n^2 - 38n + 80$.

13. Solve $x^3 = 4x$. $\{-2, 0, 2\}$; *Sec. 5.8, Ex. 6*

14. Given the graph of the function,
14a. domain $(-\infty, \infty)$, range $(-\infty, 4]$
14b. $(-3, 0), (-1, 0), (1, 0), (5, 0); (0, -2)$

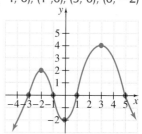

14f. between $x = -3$ and $x = -1$ and between $x = 1$ and $x = 5$; *Sec 5.9, Ex. 1*

a. Find the domain and the range of the function.

b. List the x- and y-intercept points.

c. Find the coordinates of the point with the greatest y-value. $(3, 4)$

d. Find the coordinates of the point with the least y-value. There is no point with the least y-value.

e. List the x-values where the y-values are equal to 0. $-3, -1, 1, 5$

f. List the x-values where the y-values are greater than 0.

15. For the ICL Production Company, the rational function $C(x) = \dfrac{2.6x + 10,000}{x}$ describes the company's cost per disc of pressing x compact discs. Find the cost per disc for pressing:

a. 100 compact discs $102.60

b. 1000 compact discs $ 12.60; *Sec. 6.1, Ex. 1*

16. Multiply.   **b.** $-(a + b)^2$; *Sec. 6.2, Ex. 2*

a. $\dfrac{2x^2 + 3x - 2}{-4x - 8} \cdot \dfrac{16x^2}{4x^2 - 1}$ $-\dfrac{4x^2}{2x + 1}$

b. $(ac - ad + bc - bd) \cdot \dfrac{a + b}{d - c}$

17. Simplify.

$$\dfrac{x^{-1} + 2xy^{-1}}{x^{-2} - x^{-2}y^{-1}} \qquad \dfrac{xy + 2x^3}{y - 1}; \text{ Sec. 6.4, Ex. 3}$$

4. See Appendix D.   **7.** $2x + 3y = 20$; *Sec. 3.5, Ex. 7*   **9b.** $-8x^4p^{12}$; *Sec. 5.1, Ex. 2*
11. $6x^2 - 29x + 28$; *Sec. 5.4, Ex. 6*   **12.** $2(n^2 - 19n + 40)$; *Sec. 5.6, Ex. 4*

18. Find the quotient: $\dfrac{7a^2b - 2ab^2}{2ab^2} \cdot \dfrac{7a}{2b} - 1$; *Sec. 6.5, Ex. 2*

19. If $P(x) = 2x^3 - 4x^2 + 5$,

    a. Find $P(2)$ by substitution. $P(2) = 5$

    b. Use synthetic division to find the remainder when $P(x)$ is divided by $x - 2$.

20. Solve $\dfrac{x + 6}{x - 2} = \dfrac{2(x + 2)}{x - 2}$. $\{\,\}$; *Sec. 6.7, Ex. 3*

21. The intensity $I(x)$ of light, in foot-candles, that is x feet from its source is given by the rational function

$$I(x) = \frac{320}{x^2}$$

19b. The remainder is 5; *Sec. 6.6, Ex. 3*

How far away is the source if the intensity of light is 5 foot-candles? The source of light is 8 feet away. *Sec. 6.8, Ex. 3*

22. Suppose that y varies directly as x. If y is 5 when x is 30, find the constant of variation. Also, find y when $x = 90$. $k = \frac{1}{6}$; when $x = 90$, $y = 15$; *Sec. 6.9, Ex. 1*

RATIONAL EXPONENTS, RADICALS, AND COMPLEX NUMBERS

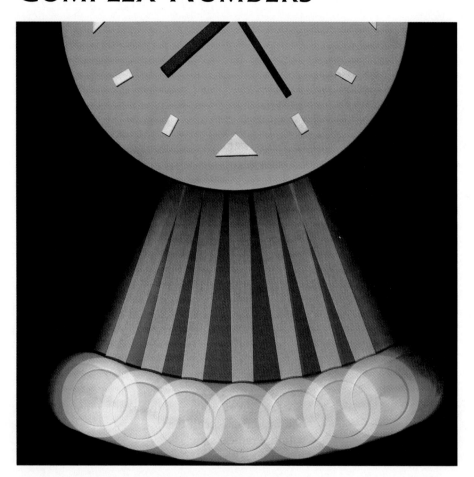

CALCULATING THE LENGTH AND PERIOD OF A PENDULUM

A simple pendulum, like the kind found in a clock, of a given length has a unique property. The time required to complete one full back-and-forth swing is the same regardless of the mass of the pendulum or the distance it travels. The time to complete one full swing *does*, however, depend on the pendulum's length.

IN THE CHAPTER GROUP ACTIVITY ON PAGE 459 YOU WILL HAVE THE OPPORTUNITY TO INVESTIGATE THE RELATIONSHIP BETWEEN THE LENGTH OF A PENDULUM AND ITS PERIOD.

I n this chapter, radical notation is reviewed, and then rational exponents are introduced. As the name implies, rational exponents are exponents that are rational numbers. We present an interpretation of rational exponents that is consistent with the meaning and rules already established for integer exponents, and we present two forms of notation for roots: radical and exponent. We conclude this chapter with complex numbers, a natural extension of the real number system.

7.1 RADICALS AND RADICAL FUNCTIONS

O B J E C T I V E S

TAPE IA 7.1

1 Find square roots.
2 Find cube roots.
3 Find nth roots.
4 Graph square root functions.
5 Graph cube root functions.

Recall from Section 1.3 that if some number a is the square of some other number b, such that

$$a = b^2$$

then we also say that b is a square root of a.

Thus, because

$$25 = 5^2$$

we say that 5 is a square root of 25.

Also, because

$$25 = (-5)^2$$

we say that -5 is a square root of 25.

Like every other positive real number, 25 has two square roots—one positive, one negative. Recall that we symbolize the **nonnegative,** or **principal, square root** with the **radical sign:**

$$\sqrt{25} = 5$$

We denote the **negative square root** with the **negative radical sign:**

$$-\sqrt{25} = -5$$

An expression containing a radical sign is called a **radical expression.** An expression within, or "under," a radical sign is called a **radicand.**

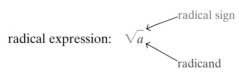

radical expression: $\sqrt{a}$

radical sign

radicand

> ### SQUARE ROOT
>
> The **principal square root** of a nonnegative number a is its nonnegative square root. The principal square root is written as $\sqrt{a}$. The **negative square root** of a is written as $-\sqrt{a}$. Thus,
>
> $$\sqrt{a} = b \text{ only if } b^2 = a \text{ and } b \geq 0$$

EXAMPLE 1   Simplify. Assume that all variables represent positive numbers.

  a. $\sqrt{36}$ **b.** $\sqrt{0}$ **c.** $\sqrt{\dfrac{4}{49}}$ **d.** $\sqrt{0.25}$ **e.** $\sqrt{x^6}$ **f.** $\sqrt{9x^{10}}$ **g.** $-\sqrt{81}$

Solution: **a.** $\sqrt{36} = 6$ because $6^2 = 36$ and 6 is not negative.

  b. $\sqrt{0} = 0$ because $0^2 = 0$ and 0 is not negative.

  c. $\sqrt{\dfrac{4}{49}} = \dfrac{2}{7}$ because $\left(\dfrac{2}{7}\right)^2 = \dfrac{4}{49}$ and $\dfrac{2}{7}$ is not negative.

  d. $\sqrt{0.25} = 0.5$ because $(0.5)^2 = 0.25$.

  e. $\sqrt{x^6} = x^3$ because $(x^3)^2 = x^6$.

  f. $\sqrt{9x^{10}} = 3x^5$ because $(3x^5)^2 = 9x^{10}$.

  g. $-\sqrt{81} = -9$. The negative in front of the radical indicates the negative square root of 81.

  Can we find the square root of a negative number, say $\sqrt{-4}$? That is, can we find a real number whose square is -4? No, there is no real number whose square is -4, and we say that $\sqrt{-4}$ is not a real number. In general,

  The square root of a negative number is not a real number.

  Recall that numbers such as 1, 4, 9, and 25 are called **perfect squares,** since $1 = 1^2$, $4 = 2^2$, $9 = 3^2$, and $25 = 5^2$. Square roots of perfect square radicands simplify to rational numbers. What happens when we try to simplify a root such as $\sqrt{3}$? Since 3 is not a perfect square, $\sqrt{3}$ is not a rational number. It is called an **irrational number,** and we can find a decimal **approximation** of it. To find decimal approximations, use the table in the Appendix or a calculator. For example, an approximation for $\sqrt{3}$ is

$$\sqrt{3} \approx 1.732$$
$$\uparrow$$

approximation symbol

2   Finding roots can be extended to other roots such as cube roots. For example, since $2^3 = 8$, we call 2 the **cube root** of 8. In symbols, we write

$$\sqrt[3]{8} = 2$$

> ### CUBE ROOT
> The **cube root** of a real number a is written as $\sqrt[3]{a}$, and
> $$\sqrt[3]{a} = b \text{ only if } b^3 = a$$

From this definition, we have

$$\sqrt[3]{64} = 4 \text{ since } 4^3 = 64$$
$$\sqrt[3]{-27} = -3 \text{ since } (-3)^3 = -27$$

Notice that, unlike with square roots, **it is possible to have a negative radicand when finding a cube root.** This is so because the *cube* of a negative number is a negative number. Therefore, the *cube root* of a negative number is a negative number.

EXAMPLE 2   Find the cube roots.

a. $\sqrt[3]{1}$ **b.** $\sqrt[3]{-64}$ **c.** $\sqrt[3]{\dfrac{8}{125}}$ **d.** $\sqrt[3]{x^6}$ **e.** $\sqrt[3]{-8x^9}$

Solution: **a.** $\sqrt[3]{1} = 1$ because $1^3 = 1$.

b. $\sqrt[3]{-64} = -4$ because $(-4)^3 = -64$.

c. $\sqrt[3]{\dfrac{8}{125}} = \dfrac{2}{5}$ because $\left(\dfrac{2}{5}\right)^3 = \dfrac{8}{125}$.

d. $\sqrt[3]{x^6} = x^2$ because $\left(x^2\right)^3 = x^6$.

e. $\sqrt[3]{-8x^9} = -2x^3$ because $\left(-2x^3\right)^3 = -8x^9$.

3   Just as we can raise a real number to powers other than 2 or 3, we can find roots other than square roots and cube roots. In fact, we can find the **nth root** of a number, where n is any natural number. In symbols, the nth root of a is written as $\sqrt[n]{a}$, where n is called the **index.** The index 2 is usually omitted for square roots.

> **R E M I N D E R**   If the index is even, such as $\sqrt{}, \sqrt[4]{}, \sqrt[6]{}$, and so on, the radicand must be nonnegative for the root to be a real number. For example,
>
> $$\sqrt[4]{16} = 2, \text{ but } \sqrt[4]{-16} \text{ is not a real number.}$$
> $$\sqrt[6]{64} = 2, \text{ but } \sqrt[6]{-64} \text{ is not a real number.}$$
>
> If the index is odd, such as $\sqrt[3]{}, \sqrt[5]{}$, and so on, the radicand may be any real number. For example,
>
> $$\sqrt[3]{64} = 4 \text{ and } \sqrt[3]{-64} = -4$$
> $$\sqrt[5]{32} = 2 \text{ and } \sqrt[5]{-32} = -2$$

EXAMPLE 3   Simplify the following expressions.

        a. $\sqrt[4]{81}$ **b.** $\sqrt[5]{-243}$ **c.** $-\sqrt{25}$ **d.** $\sqrt[4]{-81}$ **e.** $\sqrt[3]{64x^3}$

Solution: **a.** $\sqrt[4]{81} = 3$ because $3^4 = 81$ and 3 is positive.

        b. $\sqrt[5]{-243} = -3$ because $(-3)^5 = -243$.

        c. $-\sqrt{25} = -5$ because -5 is the opposite of $\sqrt{25}$.

        d. $\sqrt[4]{-81}$ is not a real number. There is no real number that, when raised to the fourth power, is -81.

        e. $\sqrt[3]{64x^3} = 4x$ because $(4x)^3 = 64x^3$.

        Recall that the notation $\sqrt{a^2}$ indicates the positive square root of a^2 only. For example,

$$\sqrt{(-5)^2} = \sqrt{25} = 5$$

When variables are present in the radicand and it is unclear whether the variable represents a positive number or a negative number, absolute value bars are sometimes needed to ensure that the result is a positive number. For example,

$$\sqrt{x^2} = |x|$$

This ensures that the result is positive. This same situation may occur when the index is any *even* positive integer. When the index is any *odd* positive integer, absolute value bars are not necessary.

If n is an even positive integer, then $\sqrt[n]{a^n} = |a|$.

If n is an odd positive integer, then $\sqrt[n]{a^n} = a$.

EXAMPLE 4   Simplify.

        a. $\sqrt{(-3)^2}$ **b.** $\sqrt{x^2}$ **c.** $\sqrt[4]{(x-2)^4}$ **d.** $\sqrt[3]{(-5)^3}$ **e.** $\sqrt[5]{(2x-7)^5}$

Solution: **a.** $\sqrt{(-3)^2} = |-3| = 3$ When the index is even, the absolute value bars ensure us

        b. $\sqrt{x^2} = |x|$ that our result is not negative.

        c. $\sqrt[4]{(x-2)^4} = |x-2|$

        d. $\sqrt[3]{(-5)^3} = -5$

        e. $\sqrt[5]{(2x-7)^5} = 2x - 7$ Absolute value bars are not needed when the index is odd.

Recall that an equation in x and y describes a function if each x-value is paired with exactly one y-value. With this in mind, does the equation

$$y = \sqrt{x}$$

describe a function? First, notice that replacement values for x must be nonnegative real numbers, since $\sqrt{x}$ is not a real number if $x < 0$. The notation $\sqrt{x}$ denotes the principal square root of x, so for every nonnegative number x, there is exactly one number, $\sqrt{x}$. Therefore, $y = \sqrt{x}$ describes a function, and we may write it as

$$f(x) = \sqrt{x}$$

Recall that the domain of a function in x is the set of all possible replacement values for x. This means that the domain of this function is the set of all nonnegative numbers, or $\{x \mid x \geq 0\}$.

We find function values for $f(x)$ as usual. For example,

$$f(0) = \sqrt{0} = 0$$
$$f(1) = \sqrt{1} = 1$$
$$f(4) = \sqrt{4} = 2$$
$$f(9) = \sqrt{9} = 3$$

Choosing perfect squares for x ensures us that $f(x)$ is a rational number, but it is important to stress that $f(x) = \sqrt{x}$ is defined for all nonnegative real numbers. For example,

$$f(3) = \sqrt{3} \approx 1.7$$

EXAMPLE 5 Graph the square root function $f(x) = \sqrt{x}$.

Solution: To graph, we identify the domain, evaluate the function for several values of x, plot the resulting points, and connect the points with a smooth curve. The domain of this function is the set of all nonnegative numbers or $\{x \mid x \geq 0\}$. The table comes from the function values obtained earlier.

x	$f(x) = \sqrt{x}$
0	0
1	1
3	$\sqrt{3} \approx 1.7$
4	2
9	3

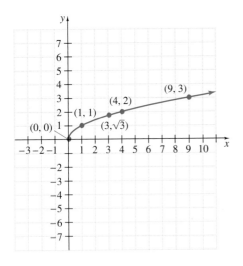

Notice that the graph of this function passes the vertical line test, as expected.

⑤ The equation $f(x) = \sqrt[3]{x}$ also describes a function. Here x may be any real number, so the domain of this function is the set of all real numbers. A few function values are given next.

$$f(0) = \sqrt[3]{0} = 0$$
$$f(1) = \sqrt[3]{1} = 1$$
$$f(-1) = \sqrt[3]{-1} = -1$$
$$f(6) = \sqrt[3]{6}$$
$$f(-6) = \sqrt[3]{-6}$$
$$f(8) = \sqrt[3]{8} = 2$$
$$f(-8) = \sqrt[3]{-8} = -2$$

} Here, the radicands are not perfect cubes. The radicals do not simplify to rational numbers.

EXAMPLE 6 Graph the function $f(x) = \sqrt[3]{x}$.

Solution: To graph, we identify the domain, plot points, and connect the points with a smooth curve. The domain of this function is the set of all real numbers. The table comes from the function values obtained earlier. We have approximated $\sqrt[3]{6}$ and $\sqrt[3]{-6}$ for graphing purposes.

x	$f(x) = \sqrt[3]{x}$
0	0
1	1
-1	-1
6	$\sqrt[3]{6} \approx 1.8$
-6	$\sqrt[3]{-6} \approx -1.8$
8	2
-8	-2

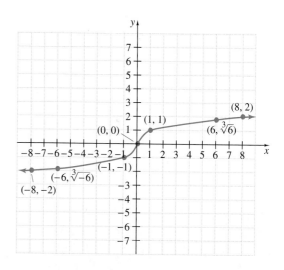

The graph of this function passes the vertical line test, as expected.

EXERCISE SET 7.1

Simplify. Assume that variables represent positive real numbers. See Example 1.

1. $\sqrt{100}$ 10

2. $\sqrt{400}$ 20

3. $\sqrt{\dfrac{1}{4}}$ $\dfrac{1}{2}$

4. $\sqrt{\dfrac{9}{25}}$ $\dfrac{3}{5}$

5. $\sqrt{0.0001}$ 0.01

6. $\sqrt{0.04}$ 0.2

7. $-\sqrt{36}$ −6

8. $-\sqrt{9}$ −3

9. $\sqrt{x^{10}}$ x^5

10. $\sqrt{x^{16}}$ x^8

11. $\sqrt{16y^6}$ $4y^3$

12. $\sqrt{64y^{20}}$ $8y^{10}$

Simplify. If the radical is not a real number, state so. Assume that variables represent positive real numbers. See Examples 1 through 3.

13. $\sqrt[3]{\dfrac{1}{8}}$ $\dfrac{1}{2}$

14. $\sqrt[3]{\dfrac{27}{64}}$ $\dfrac{3}{4}$

15. $-\sqrt[4]{16}$ -2

16. $\sqrt[5]{-243}$ -3

17. $\sqrt[4]{-16}$

18. $\sqrt{-16}$

19. $\sqrt[3]{-125x^9}$ $-5x^3$

20. $\sqrt[3]{a^{12}b^3}$ a^4b

21. $\sqrt[5]{32x^5}$ $2x$

22. $\sqrt[5]{-32y^{10}}$ $-2y^2$

Simplify each root. Assume that variables represent any real numbers. See Example 4.

23. $\sqrt{z^2}$ $|z|$

24. $\sqrt{y^{10}}$ $|y^5|$

25. $\sqrt[3]{x^3}$ x

26. $\sqrt[3]{z^6}$ z^2

27. $\sqrt{(x-5)^2}$ $|x-5|$

28. $\sqrt{(y-6)^2}$ $|y-6|$

29. $\sqrt[4]{(2z)^4}$ $2|z|$

30. $\sqrt{(9x)^2}$ $9|x|$

31. $\sqrt{100(2x-y)^6}$

32. $\sqrt[4]{81(y-10z)^{12}}$

33. When does a square root simplify to a rational number? Answers will vary.

34. When does a cube root simplify to a rational number? Answers will vary.

Simplify each radical. Assume that all variables represent positive real numbers.

35. $-\sqrt{121}$ -11

36. $-\sqrt[3]{125}$ -5

37. $\sqrt[3]{8x^3}$ $2x$

38. $\sqrt{16x^8}$ $4x^4$

39. $\sqrt{y^{12}}$ y^6

40. $\sqrt[3]{y^{12}}$ y^4

41. $\sqrt{25a^2b^{20}}$ $5ab^{10}$

42. $\sqrt{9x^4y^6}$ $3x^2y^3$

43. $\sqrt[3]{-27x^9}$ $-3x^3$

44. $\sqrt[3]{-8a^{21}b^6}$ $-2a^7b^2$

45. $\sqrt[4]{a^{16}b^4}$ a^4b

46. $\sqrt[4]{x^8y^{12}}$ x^2y^3

47. $\sqrt[5]{-32x^{10}y^5}$ $-2x^2y$

48. $\sqrt[5]{-243z^{15}}$ $-3z^3$

49. $\sqrt{\dfrac{25}{49}}$ $\dfrac{5}{7}$

50. $\sqrt{\dfrac{4}{81}}$ $\dfrac{2}{9}$

51. $\sqrt{\dfrac{x^2}{4y^2}}$ $\dfrac{x}{2y}$

52. $\sqrt{\dfrac{y^{10}}{9x^6}}$ $\dfrac{y^5}{3x^3}$

53. $-\sqrt[3]{\dfrac{z^{21}}{27x^3}}$ $-\dfrac{z^7}{3x}$

54. $-\sqrt[3]{\dfrac{64a^3}{b^9}}$ $-\dfrac{4a}{b^3}$

55. $\sqrt[4]{\dfrac{x^4}{16}}$ $\dfrac{x}{2}$

56. $\sqrt[4]{\dfrac{y^4}{81x^4}}$ $\dfrac{y}{3x}$

*Determine whether each square root is rational or irrational. If it is rational, find its **exact value**. If it is irra-tional, use a calculator to write a three-decimal-place approximation.*

57. $\sqrt{9}$ rational, 3

58. $\sqrt{8}$ irrational, 2.828

59. $\sqrt{37}$ irrational, 6.083

60. $\sqrt{36}$ rational, 6

61. $\sqrt{169}$ rational, 13

62. $\sqrt{160}$ irrational, 12.649

63. $\sqrt{4}$ rational, 2

64. $\sqrt{27}$ irrational, 5.196

Simplify each root. Assume that variables represent any real numbers.

65. $\sqrt[3]{x^{15}}$ x^5

66. $\sqrt[3]{y^{12}}$ y^4

67. $\sqrt{x^{12}}$ x^6

68. $\sqrt{z^{16}}$ z^8

69. $\sqrt{81x^2}$ $9|x|$

70. $\sqrt{100z^4}$ $10z^2$

71. $-\sqrt{144y^{14}}$ $-12|y^7|$

72. $-\sqrt{121z^{22}}$ $-11|z^{11}|$

73. $\sqrt{x^2+4x+4}$ $|x+2|$

(*Hint*: Factor the polynomial first.)

74. $\sqrt{x^2-6x+9}$ $|x-3|$

(*Hint*: Factor the polynomial first.)

If $f(x) = \sqrt{2x+3}$ and $g(x) = \sqrt[3]{x-8}$, find the fol-lowing function values.

75. $f(0)$ $\sqrt{3}$

76. $g(0)$ -2

77. $g(7)$ -1

78. $f(-1)$ 1

79. $g(-19)$ -3

80. $f(3)$ 3

81. $f(2)$ $\sqrt{7}$

82. $g(1)$ $\sqrt[3]{-7}$

Identify the domain and then graph each function. See Example 5. For Exercises 83–86, see Appendix D.

83. $f(x) = \sqrt{x} + 2$ $[0, \infty)$

84. $f(x) = \sqrt{x} - 2$ $[0, \infty)$

85. $f(x) = \sqrt{x-3}$; use the following table. $[3, \infty)$

x	$f(x)$
3	0
4	1
7	2
12	3

86. $f(x) = \sqrt{x+1}$; use the following table. $[-1, \infty)$

x	$f(x)$
-1	0
0	1
3	2
8	3

17. not a real number **18.** not a real number **31.** $10\left|(2x-y)^3\right|$ **32.** $3\left|(y-10z)^3\right|$

Identify the domain and then graph each function. See Example 6. For Exercises 87–98, see Appendix D.

87. $f(x) = \sqrt[3]{x} + 1$ $(-\infty, \infty)$ **88.** $f(x) = \sqrt[3]{x} - 2$ $(-\infty, \infty)$

89. $g(x) = \sqrt[3]{x} - 1$; use the following table. $(-\infty, \infty)$

x	$g(x)$
1	0
2	1
0	-1
9	2
-7	-2

90. $g(x) = \sqrt[3]{x} + 1$; use the following table. $(-\infty, \infty)$

x	$g(x)$
-1	0
0	1
-2	-1
7	2
-9	-2

Use a grapher to verify the domain of each function and its graph.

91. Exercise 83   **92.** Exercise 84

93. Exercise 85   **94.** Exercise 86

95. Exercise 87   **96.** Exercise 88

97. Exercise 89   **98.** Exercise 90

Review Exercises

Simplify each exponential expression. See Sections 5.1 and 5.2.

99. $(-2x^3y^2)^5$ $-32x^{15}y^{10}$ **100.** $(4y^6z^7)^3$ $64y^{18}z^{21}$

101. $(-3x^2y^3z^5)(20x^5y^7)$ **102.** $(-14a^5bc^2)(2abc^4)$

103. $\dfrac{7x^{-1}y}{14(x^5y^2)^{-2}}$ $\dfrac{x^9y^5}{2}$ **104.** $\dfrac{(2a^{-1}b^2)^3}{(8a^2b)^{-2}}$ $512ab^8$

101. $-60x^7y^{10}z^5$ **102.** $-28a^6b^2c^6$

7.2   RATIONAL EXPONENTS

O B J E C T I V E S

TAPE IA 7.2

1 Understand the meaning of a raised to the $\dfrac{1}{n}$th power.

2 Understand the meaning of a raised to the $\dfrac{m}{n}$th power.

3 Understand the meaning of a raised to the $-\dfrac{m}{n}$th power.

4 Use rules for exponents to simplify expressions that contain rational exponents.

 We introduce now an alternate, equivalent notation for the nth root of a number a. Recall that if b is the nth root of a, we write with radicals

$$b = \sqrt[n]{a}$$

Alternately, we write with **rational exponents**

$$b = a^{1/n}$$

Thus, we define $a^{1/n}$ to be $\sqrt[n]{a}$. The two symbols have identical meaning and can be used interchangeably to denote the positive number whose nth power is a.

> **DEFINITION OF $a^{1/n}$**
>
> If n is a positive integer greater than 1 and $\sqrt[n]{a}$ is a real number, then $\sqrt[n]{a}$ can also be written as $a^{1/n}$:
>
> $$a^{1/n} = \sqrt[n]{a}$$

Notice that the denominator of the rational exponent corresponds to the index of the radical.

EXAMPLE 1   Use radical notation to write the following. Simplify if possible.

    a. $4^{1/2}$ **b.** $64^{1/3}$ **c.** $x^{1/4}$ **d.** $0^{1/6}$ **e.** $-9^{1/2}$ **f.** $(81x^8)^{1/4}$ **g.** $(5y)^{1/3}$

Solution: **a.** $4^{1/2} = \sqrt{4} = 2$

    b. $64^{1/3} = \sqrt[3]{64} = 4$

    c. $x^{1/4} = \sqrt[4]{x}$

    d. $0^{1/6} = \sqrt[6]{0} = 0$

    e. $-9^{1/2} = -\sqrt{9} = -3$

    f. $(81x^8)^{1/4} = \sqrt[4]{81x^8} = 3x^2$

    g. $(5y)^{1/3} = \sqrt[3]{5y}$

As we expand our use of exponents to include $\dfrac{m}{n}$, we define their meaning so that rules for exponents still hold true. For example, by properties of exponents,

$$8^{1/3} \cdot 8^{1/3} = 8^{1/3 + 1/3} = 8^{2/3}$$

For this to be true, we give meaning to the expression $8^{2/3}$ in such a way that the product and power rules do indeed hold true.

> **DEFINITION OF $a^{m/n}$**
>
> If m and n are positive integers greater than 1 with $\dfrac{m}{n}$ in lowest terms, then
>
> $$a^{m/n} = \sqrt[n]{a^m} = \left(\sqrt[n]{a}\right)^m$$
>
> as long as $\sqrt[n]{a}$ is a real number.

Notice that the denominator n of the rational exponent corresponds to the index of the radical. The numerator m of the rational exponent indicates that the base is to be raised to the mth power.

This means that

$$8^{2/3} = \sqrt[3]{8^2} = \sqrt[3]{64} = 4 \quad \text{or}$$

$$8^{2/3} = \left(\sqrt[3]{8}\right)^2 = 2^2 = 4$$

Most of the time, $\left(\sqrt[n]{a}\right)^m$ will be easier to calculate.

EXAMPLE 2   Use radical notation to write the following. Then simplify if possible.

a. $4^{3/2}$ **b.** $-16^{3/4}$ **c.** $(-27)^{2/3}$ **d.** $\left(\dfrac{1}{9}\right)^{3/2}$ **e.** $(4x - 1)^{3/5}$

Solution:
   a. $4^{3/2} = \left(\sqrt{4}\right)^3 = 2^3 = 8$
   b. $-16^{3/4} = -\left(\sqrt[4]{16}\right)^3 = -(2)^3 = -8$
   c. $(-27)^{2/3} = \left(\sqrt[3]{-27}\right)^2 = (-3)^2 = 9$
   d. $\left(\dfrac{1}{9}\right)^{3/2} = \left(\sqrt{\dfrac{1}{9}}\right)^3 = \left(\dfrac{1}{3}\right)^3 = \dfrac{1}{27}$
   e. $(4x - 1)^{3/5} = \sqrt[5]{(4x - 1)^3}$

> **R E M I N D E R**   The *denominator* of a rational exponent is the index of the corresponding radical. For example, $x^{1/5} = \sqrt[5]{x}$ and $z^{2/3} = \sqrt[3]{z^2}$, or $z^{2/3} = \left(\sqrt[3]{z}\right)^2$.

3 The rational exponents we have given meaning to exclude negative rational numbers. To complete the set of definitions, we define $a^{-m/n}$.

DEFINITION OF $a^{-m/n}$

$$a^{-m/n} = \frac{1}{a^{m/n}}$$

as long as $a^{m/n}$ is a nonzero real number.

EXAMPLE 3   Write each expression with a positive exponent, and then simplify.

   a. $16^{-3/4}$ **b.** $(-27)^{-2/3}$

Solution: **a.** $16^{-3/4} = \dfrac{1}{16^{3/4}} = \dfrac{1}{(\sqrt[4]{16})^3} = \dfrac{1}{2^3} = \dfrac{1}{8}$

b. $(-27)^{-2/3} = \dfrac{1}{(-27)^{2/3}} = \dfrac{1}{(\sqrt[3]{-27})^2} = \dfrac{1}{(-3)^2} = \dfrac{1}{9}$

REMINDER   Notice that the sign of the base is not affected by the sign of its exponent. For example,

$$9^{-3/2} = \dfrac{1}{9^{3/2}} = \dfrac{1}{(\sqrt{9})^3} = \dfrac{1}{27}$$

Also,

$$(-27)^{-1/3} = \dfrac{1}{(-27)^{1/3}} = -\dfrac{1}{3}$$

---

4   It can be shown that the properties of integer exponents hold for rational exponents. By using these properties and definitions, we can now simplify expressions that contain rational exponents.

These rules are repeated here for review.

SUMMARY OF EXPONENT RULES

If m and n are rational numbers, and a, b, and c are numbers for which the expressions below exist, then

Product rule for exponents: $a^m \cdot a^n = a^{m+n}$

Power rule for exponents: $(a^m)^n = a^{m \cdot n}$

Power rules for products and quotients: $(ab)^n = a^n b^n$ and

$$\left(\dfrac{a}{c}\right)^n = \dfrac{a^n}{c^n}, \; c \neq 0$$

Quotient rule for exponents: $\dfrac{a^m}{a^n} = a^{m-n}, \; a \neq 0$

Zero exponent: $a^0 = 1, \; a \neq 0$

Negative exponent: $a^{-n} = \dfrac{1}{a^n}, \; a \neq 0$

EXAMPLE 4   Write with rational exponents. Then simplify if possible.

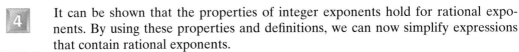

a. $\sqrt[5]{x}$ **b.** $\sqrt[3]{17x^2y^5}$ **c.** $\sqrt{x-5a}$ **d.** $3\sqrt{2p} - 5\sqrt[3]{p^2}$

Solution: **a.** $\sqrt[5]{x} = x^{1/5}$

b. $\sqrt[3]{17x^2y^5} = (17x^2y^5)^{1/3}$. We can further simplify this expression by using a power rule for exponents.

$$(17x^2y^5)^{1/3} = 17^{1/3}x^{2/3}y^{5/3}$$

c. $\sqrt{x - 5a} = (x - 5a)^{1/2}$

d. $3\sqrt{2p} - 5\sqrt[3]{p^2} = 3(2p)^{1/2} - 5(p^2)^{1/3} = 3(2p)^{1/2} - 5p^{2/3}$

We can use rational exponents to help us simplify radicals.

EXAMPLE 5   Use rational exponents to simplify. Assume that variables represent positive numbers.

  a. $\sqrt[8]{x^4}$ **b.** $\sqrt[4]{r^2s^6}$

Solution: **a.** $\sqrt[8]{x^4} = x^{4/8} = x^{1/2} = \sqrt{x}$

  b. $\sqrt[4]{r^2s^6} = (r^2s^6)^{1/4} = r^{2/4}s^{6/4} = r^{1/2}s^{3/2} = (rs^3)^{1/2} = \sqrt{rs^3}$

EXAMPLE 6   Use properties of exponents to simplify the following expressions. Write results with only positive exponents.

  a. $x^{1/2}x^{1/3}$ **b.** $\dfrac{7^{1/3}}{7^{4/3}}$ **c.** $\dfrac{(2x^{2/5}y^{-1/3})^5}{x^2y}$

Solution: **a.** $x^{1/2}x^{1/3} = x^{(1/2 + 1/3)} = x^{3/6 + 2/6} = x^{5/6}$ **b.** $\dfrac{7^{1/3}}{7^{4/3}} = 7^{1/3 - 4/3} = 7^{-3/3} = 7^{-1} = \dfrac{1}{7}$

  c. We begin by using the power rule $(ab)^m = a^m b^m$ to simplify the numerator.

$$\frac{(2x^{2/5}y^{-1/3})^5}{x^2y} = \frac{2^5(x^{2/5})^5(y^{-1/3})^5}{x^2y} = \frac{32x^2y^{-5/3}}{x^2y}$$

$$= 32x^{2-2}y^{-5/3 - 3/3}\quad \text{Apply the quotient rule.}$$

$$= 32x^0y^{-8/3}$$

$$= \frac{32}{y^{8/3}}$$

EXAMPLE 7   Multiply.

  a. $z^{2/3}(z^{1/3} - z^5)$ **b.** $(x^{1/3} - 5)(x^{1/3} + 2)$

Solution: **a.** $z^{2/3}(z^{1/3} - z^5) = z^{2/3}z^{1/3} - z^{2/3}z^5$ Apply the distributive property.

$$= z^{(2/3 + 1/3)} - z^{(2/3 + 5)}\quad \text{Use the product rule.}$$

$$= z^{3/3} - z^{(2/3 + 15/3)}$$

$$= z - z^{17/3}$$

  b. $(x^{1/3} - 5)(x^{1/3} + 2) = x^{2/3} + 2x^{1/3} - 5x^{1/3} - 10$

$$= x^{2/3} - 3x^{1/3} - 10$$

EXAMPLE 8 Factor $x^{-1/2}$ from the expression $3x^{-1/2} - 7x^{5/2}$. Assume that all variables represent positive numbers.

Solution: $3x^{-1/2} - 7x^{5/2} = (x^{-1/2})(3) - (x^{-1/2})(7x^{6/2})$

$$= x^{-1/2}(3 - 7x^3)$$

To check, multiply $x^{-1/2}(3 - 7x^3)$ to see that the product is $3x^{-1/2} - 7x^{5/2}$. ▬▬▬

---

21. $(\sqrt[4]{-16})^3$, not a real number **22.** $(\sqrt{-9})^3$, not a real number **23.** $\sqrt[5]{(2x)^3}$, or $(\sqrt[5]{2x})^3$

25. $\sqrt[3]{(7x + 2)^2}$, or $(\sqrt[3]{7x + 2})^2$ **26.** $\sqrt[4]{(x - 4)^3}$, or $(\sqrt[4]{x - 4})^3$ **31.** $\dfrac{1}{(-64)^{2/3}} = \dfrac{1}{16}$ **32.** $\dfrac{1}{(-8)^{4/3}} = \dfrac{1}{16}$

EXERCISE SET 7.2

33. $\dfrac{1}{(-4)^{3/2}}$, not a real number **34.** $\dfrac{1}{(-16)^{5/4}}$, not a real number

Use radical notation to write the following. Simplify if possible.

39. $\dfrac{5}{7x^{-3/4}}$ $\dfrac{5x^{3/4}}{7}$ **40.** $\dfrac{2}{3y^{-5/7}}$ $\dfrac{2y^{5/7}}{3}$

1. $49^{1/2}$ $\sqrt{49} = 7$ **2.** $64^{1/3}$ $\sqrt[3]{64} = 4$

3. $27^{1/3}$ $\sqrt[3]{27} = 3$ **4.** $8^{1/3}$ $\sqrt[3]{8} = 2$

📝 41. Explain how writing x^{-7} with positive exponents is similar to writing $x^{-1/4}$ with positive exponents.

5. $\left(\dfrac{1}{16}\right)^{1/4}$ $\sqrt[4]{\dfrac{1}{16}} = \dfrac{1}{2}$ **6.** $\left(\dfrac{1}{64}\right)^{1/2}$ $\sqrt{\dfrac{1}{64}} = \dfrac{1}{8}$

📝 42. Explain how writing $2x^{-5}$ with positive exponents is similar to writing $2x^{-3/4}$ with positive exponents.

7. $169^{1/2}$ $\sqrt{169} = 13$ **8.** $81^{1/4}$ $\sqrt[4]{81} = 3$

Write with rational exponents. Then simplify if possible. See Example 4.

9. $2m^{1/3}$ $2\sqrt[3]{m}$ **10.** $(2m)^{1/3}$ $\sqrt[3]{2m}$

43. $\sqrt{3}$ $3^{1/2}$ **44.** $\sqrt[3]{y}$ $y^{1/3}$

11. $(9x^4)^{1/2}$ $\sqrt{9x^4} = 3x^2$ **12.** $(16x^8)^{1/2}$ $\sqrt{16x^8} = 4x^4$

45. $\sqrt[3]{y^5}$ $y^{5/3}$ **46.** $\sqrt[4]{x^3}$ $x^{3/4}$

13. $(-27)^{1/3}$ $\sqrt[3]{-27} = -3$ **14.** $-64^{1/2}$ $-\sqrt{64} = -8$

47. $\sqrt[5]{4y^7}$ $4^{1/5}y^{7/5}$ **48.** $\sqrt[4]{11x^5}$ $11^{1/4}x^{5/4}$

15. $-16^{1/4}$ $-\sqrt[4]{16} = -2$ **16.** $(-32)^{1/5}$ $\sqrt[5]{-32} = -2$

49. $\sqrt{(y + 1)^3}$ $(y + 1)^{3/2}$ **50.** $\sqrt{(3 + y^2)^5}$ $(3 + y^2)^{5/2}$

17. $16^{3/4}$ $(\sqrt[4]{16})^3 = 8$ **18.** $4^{5/2}$ $(\sqrt{4})^5 = 32$

51. $2\sqrt{x} - 3\sqrt{y}$ **52.** $4\sqrt{2x} + \sqrt{xy}$

19. $(-64)^{2/3}$ $(\sqrt[3]{-64})^2 = 16$ **20.** $(-8)^{4/3}$ $(\sqrt[3]{-8})^4 = 16$

21. $(-16)^{3/4}$ **22.** $(-9)^{3/2}$

Use rational exponents to simplify each radical. (Variables represent positive numbers.) See Example 5.

23. $(2x)^{3/5}$ **24.** $2x^{3/5}$ $2\sqrt[5]{x^3}$

53. $\sqrt[6]{x^3}$ $\sqrt{x}$ **54.** $\sqrt[9]{a^3}$ $\sqrt[3]{a}$

25. $(7x + 2)^{2/3}$ **26.** $(x - 4)^{3/4}$

55. $\sqrt[4]{16x^2}$ $2\sqrt{x}$ **56.** $\sqrt[8]{4y^2}$ $\sqrt[4]{2y}$

57. $\sqrt[4]{(x + 3)^2}$ $\sqrt{x + 3}$ **58.** $\sqrt[6]{a^3b^6}$ $b\sqrt{a}$

27. $\left(\dfrac{16}{9}\right)^{3/2}$ $\left(\sqrt{\dfrac{16}{9}}\right)^3 = \dfrac{64}{27}$ **28.** $\left(\dfrac{49}{25}\right)^{3/2}$ $\left(\sqrt{\dfrac{49}{25}}\right)^3 = \dfrac{343}{125}$

59. $\sqrt[8]{x^4y^4}$ $\sqrt{xy}$ **60.** $\sqrt[9]{y^6z^3}$ $\sqrt[3]{y^2z}$

Write with positive exponents. Simplify if possible. See Example 3.

Use properties of exponents to simplify each expression. Write with positive exponents. See Example 6.

29. $8^{-4/3}$ $\dfrac{1}{8^{4/3}} = \dfrac{1}{16}$ **30.** $64^{-2/3}$ $\dfrac{1}{64^{2/3}} = \dfrac{1}{16}$

61. $a^{2/3}a^{5/3}$ $a^{7/3}$ **62.** $b^{9/5}b^{8/5}$ $b^{17/5}$

31. $(-64)^{-2/3}$ **32.** $(-8)^{-4/3}$

63. $(4u^2v^{-6})^{3/2}$ $\dfrac{8u^3}{v^9}$ **64.** $(32^{1/5}x^{2/3}y^{1/3})^3$ $8x^2y$

33. $(-4)^{-3/2}$ **34.** $(-16)^{-5/4}$

65. $\dfrac{b^{1/2}b^{3/4}}{-b^{1/4}}$ $-b$ **66.** $\dfrac{a^{1/4}a^{-1/2}}{a^{2/3}}$ $\dfrac{1}{a^{11/12}}$

35. $x^{-1/4}$ $\dfrac{1}{x^{1/4}}$ **36.** $y^{-1/6}$ $\dfrac{1}{y^{1/6}}$

Multiply. See Example 7.

37. $\dfrac{1}{a^{-2/3}}$ $a^{2/3}$ **38.** $\dfrac{1}{n^{-8/9}}$ $n^{8/9}$

67. $y^{1/2}(y^{1/2} - y^{2/3})$ $y - y^{7/6}$ **68.** $x^{1/2}(x^{1/2} + x^{3/2})$ $x + x^2$

41. Answers will vary. **42.** Answers will vary. **51.** $2x^{1/2} - 3y^{1/2}$ **52.** $4(2x)^{1/2} + (xy)^{1/2} = 2^{5/2}x^{1/2} + x^{1/2}y^{1/2}$

69. $2x^{5/3} - 2x^{2/3}$ **70.** $3x^{3/2} + 3x^{1/2}y$

69. $x^{2/3}(2x - 2)$ **70.** $3x^{1/2}(x + y)$

71. $(2x^{1/3} + 3)(2x^{1/3} - 3)$ **72.** $(y^{1/2} + 5)(y^{1/2} + 5)$

$\quad 4x^{2/3} - 9$ $\quad y + 10y^{1/2} + 25$

Factor the common factor from the given expression. See Example 8. **73.** $x^{8/3}(1 + x^{2/3})$

73. $x^{8/3}; x^{8/3} + x^{10/3}$ **74.** $x^{3/2}; x^{5/2} - x^{3/2}$ $x^{3/2}(x - 1)$

75. $x^{1/5}; x^{2/5} - 3x^{1/5}$ **76.** $x^{2/7}; x^{3/7} - 2x^{2/7}$

77. $x^{-1/3}; 5x^{-1/3} + x^{2/3}$ **78.** $x^{-3/4}; x^{-3/4} + 3x^{1/4}$

Fill in the box with the correct expression.

79. $\boxed{} \cdot a^{2/3} = a^{3/3}$, or a **80.** $\boxed{} \cdot x^{1/8} = x^{4/8}$, or $x^{1/2}$ $x^{3/8}$

81. $\dfrac{\boxed{}}{x^{-2/5}} = x^{3/5}$ $x^{1/5}$ **82.** $\dfrac{\boxed{}}{y^{-3/4}} = y^{4/4}$, or y $y^{1/4}$

75. $x^{1/5}(x^{1/5} - 3)$ **76.** $x^{2/7}(x^{1/7} - 2)$ **77.** $x^{-1/3}(5 + x)$ **78.** $x^{-3/4}(1 + 3x)$ **79.** $a^{1/3}$

Use a calculator to write a four-decimal-place approximation of each.

83. $8^{1/4}$ 1.6818   **84.** $20^{1/5}$ 1.8206

85. $18^{3/5}$ 5.6645   **86.** $76^{5/7}$ 22.0515

Review Exercises

Write each integer as a product of two integers such that one of the factors is a perfect square. For example, write 18 as $9 \cdot 2$, because 9 is a perfect square.

87. 75 $25 \cdot 3$ **88.** 20 $4 \cdot 5$ **89.** 48 $16 \cdot 3$ **90.** 45 $9 \cdot 5$

Write each integer as a product of two integers such that one of the factors is a perfect cube. For example, write 24 as $8 \cdot 3$, because 8 is a perfect cube.

91. 16 $8 \cdot 2$ **92.** 56 $8 \cdot 7$ **93.** 54 $27 \cdot 2$ **94.** 80 $8 \cdot 10$

7.3 | SIMPLIFYING RADICAL EXPRESSIONS

TAPE IA 7.3

O B J E C T I V E S

1 Understand the product rule for radicals.

2 Understand the quotient rule for radicals.

3 Simplify radicals.

4 Use rational exponents to simplify radical expressions.

1 It is possible to simplify some radicals that do not evaluate to rational numbers. To do so, we use a product rule and a quotient rule for radicals. To discover the product rule, notice the following pattern:

$$\sqrt{9} \cdot \sqrt{4} = 3 \cdot 2 = 6$$
$$\sqrt{9 \cdot 4} = \sqrt{36} = 6$$

Since both expressions simplify to 6, it is true that

$$\sqrt{9} \cdot \sqrt{4} = \sqrt{9 \cdot 4}$$

This pattern suggests the following product rule for exponents.

PRODUCT RULE FOR RADICALS

If $\sqrt[n]{a}$ and $\sqrt[n]{b}$ are real numbers, then

$$\sqrt[n]{a} \cdot \sqrt[n]{b} = \sqrt[n]{ab}$$

To see that this is true, notice that the product rule is the relationship $a^{1/n} \cdot b^{1/n} = (ab)^{1/n}$ stated in radical notation.

EXAMPLE 1 Multiply.

a. $\sqrt{3} \cdot \sqrt{5}$ **b.** $\sqrt{21} \cdot \sqrt{x}$ **c.** $\sqrt[3]{4} \cdot \sqrt[3]{2}$ **d.** $\sqrt[4]{5y^2} \cdot \sqrt[4]{2x^3}$ **e.** $\sqrt{\dfrac{2}{a}} \cdot \sqrt{\dfrac{b}{3}}$

Solution:

a. $\sqrt{3} \cdot \sqrt{5} = \sqrt{3 \cdot 5} = \sqrt{15}$

b. $\sqrt{21} \cdot \sqrt{x} = \sqrt{21x}$

c. $\sqrt[3]{4} \cdot \sqrt[3]{2} = \sqrt[3]{4 \cdot 2} = \sqrt[3]{8} = 2$

d. $\sqrt[4]{5y^2} \cdot \sqrt[4]{2x^3} = \sqrt[4]{5y^2 \cdot 2x^3} = \sqrt[4]{10y^2x^3}$

e. $\sqrt{\dfrac{2}{a}} \cdot \sqrt{\dfrac{b}{3}} = \sqrt{\dfrac{2}{a} \cdot \dfrac{b}{3}} = \sqrt{\dfrac{2b}{3a}}$

To discover a quotient rule for radicals, notice the following pattern:

$$\sqrt{\dfrac{4}{9}} = \dfrac{2}{3}$$

$$\dfrac{\sqrt{4}}{\sqrt{9}} = \dfrac{2}{3}$$

Since both expressions simplify to $\dfrac{2}{3}$, it is true that

$$\sqrt{\dfrac{4}{9}} = \dfrac{\sqrt{4}}{\sqrt{9}}$$

This pattern suggests the following quotient rule for radicals.

> **QUOTIENT RULE FOR RADICALS**
> If $\sqrt[n]{a}$ and $\sqrt[n]{b}$ are real numbers and $\sqrt[n]{b}$ is not zero, then
> $$\sqrt[n]{\dfrac{a}{b}} = \dfrac{\sqrt[n]{a}}{\sqrt[n]{b}}$$

To see that this is true, notice that the quotient rule is the relationship $\left(\dfrac{a}{b}\right)^{1/n} = \dfrac{a^{1/n}}{b^{1/n}}$ stated in radical notation. We can use the quotient rule to simplify radical expressions by reading the rule from left to right or from right to left as shown in Example 2.

For the remainder of this chapter, we will assume that variables represent positive real numbers. If this is so, we need not insert absolute value bars when we simplify even roots.

EXAMPLE 2   Use the quotient rule to simplify.

a. $\sqrt{\dfrac{25}{49}}$ b. $\sqrt[3]{\dfrac{8}{27}}$ c. $\sqrt{\dfrac{x}{9}}$ d. $\sqrt[4]{\dfrac{3}{16y^4}}$

Solution: a. $\sqrt{\dfrac{25}{49}} = \dfrac{\sqrt{25}}{\sqrt{49}} = \dfrac{5}{7}$ b. $\sqrt[3]{\dfrac{8}{27}} = \dfrac{\sqrt[3]{8}}{\sqrt[3]{27}} = \dfrac{2}{3}$

c. $\sqrt{\dfrac{x}{9}} = \dfrac{\sqrt{x}}{\sqrt{9}} = \dfrac{\sqrt{x}}{3}$ d. $\sqrt[4]{\dfrac{3}{16y^4}} = \dfrac{\sqrt[4]{3}}{\sqrt[4]{16y^4}} = \dfrac{\sqrt[4]{3}}{2y}$

3   Both the product and quotient rules can be used to simplify a radical. If the product rule is read from right to left, we have that $\sqrt[n]{ab} = \sqrt[n]{a} \cdot \sqrt[n]{b}$. This is used to simplify the following radicals.

EXAMPLE 3   Simplify the following.

a. $\sqrt{50}$ b. $\sqrt[3]{24}$ c. $\sqrt{26}$ d. $\sqrt[4]{32}$

Solution: **a.** Factor 50 such that one factor is the largest perfect square that divides 50. The largest perfect square factor of 50 is 25, so we write 50 as $25 \cdot 2$ and use the product rule for radicals to simplify.

$$\sqrt{50} = \sqrt{25 \cdot 2} = \sqrt{25} \cdot \sqrt{2} = 5\sqrt{2}$$

b. Since the index is 3, we find the largest perfect cube factor of 24. We write 24 as $8 \cdot 3$ since 8 is the largest perfect cube factor of 24.

$$\sqrt[3]{24} = \sqrt[3]{8 \cdot 3} = \sqrt[3]{8} \cdot \sqrt[3]{3} = 2\sqrt[3]{3}$$

c. The largest perfect square factor of 26 is 1, so $\sqrt{26}$ cannot be simplified further.

d. $\sqrt[4]{32} = \sqrt[4]{16 \cdot 2} = \sqrt[4]{16} \cdot \sqrt[4]{2} = 2\sqrt[4]{2}$

After simplifying a radical such as a square root, always check the radicand to see that it contains no other perfect square factors. It may, if the largest perfect square factor of the radicand was not originally recognized. For example,

$$\sqrt{200} = \sqrt{4 \cdot 50} = \sqrt{4} \cdot \sqrt{50} = 2\sqrt{50}$$

Notice that the radicand 50 still contains the perfect square factor of 25. This is because 4 is not the largest perfect square factor of 200. We continue as follows:

$$2\sqrt{50} = 2\sqrt{25 \cdot 2} = 2 \cdot \sqrt{25} \cdot \sqrt{2} = 2 \cdot 5 \cdot \sqrt{2} = 10\sqrt{2}$$

The radical is now simplified since 2 contains no perfect square factors (other than 1).

In general, we say that a radicand of the form $\sqrt[n]{a}$ is simplified when a contains no factors that are perfect nth powers (other than 1 or −1).

EXAMPLE 4   Use the product rule to simplify.

a. $\sqrt{25x^3}$

b. $\sqrt[3]{54x^6y^8}$

c. $\sqrt[4]{81z^{11}}$

Solution: **a.** $\sqrt{25x^3} = \sqrt{25x^2 \cdot x}$ Find the largest perfect square factor.

$= \sqrt{25x^2} \cdot \sqrt{x}$ Apply the product rule.

$= 5x\sqrt{x}$ Simplify.

b. $\sqrt[3]{54x^6y^8} = \sqrt[3]{27 \cdot 2 \cdot x^6 \cdot y^6 \cdot y^2}$ Factor the radicand and identify perfect cube

$= \sqrt[3]{27x^6y^6 \cdot 2y^2}$ factors.

$= \sqrt[3]{27x^6y^6} \cdot \sqrt[3]{2y^2}$ Apply the product rule.

$= 3x^2y^2\sqrt[3]{2y^2}$ Simplify.

c. $\sqrt[4]{81z^{11}} = \sqrt[4]{81 \cdot z^8 \cdot z^3}$ Factor the radicand and identify perfect fourth

$= \sqrt[4]{81z^8 \cdot z^3}$ power factors.

$= \sqrt[4]{81z^8} \cdot \sqrt[4]{z^3}$ Apply the product rule.

$= 3z^2\sqrt[4]{z^3}$ Simplify.

EXAMPLE 5   Use the quotient rule to divide, and simplify if possible.

a. $\dfrac{\sqrt{20}}{\sqrt{5}}$ **b.** $\dfrac{\sqrt{50x}}{2\sqrt{2}}$ **c.** $\dfrac{7\sqrt[3]{48x^4y^8}}{\sqrt[3]{6y^2}}$

Solution: **a.** $\dfrac{\sqrt{20}}{\sqrt{5}} = \sqrt{\dfrac{20}{5}}$ Apply the quotient rule.

$= \sqrt{4}$ Simplify.

$= 2$ Simplify.

b. $\dfrac{\sqrt{50x}}{2\sqrt{2}} = \dfrac{1}{2} \cdot \sqrt{\dfrac{50x}{2}}$ Apply the quotient rule.

$= \dfrac{1}{2} \cdot \sqrt{25x}$ Simplify.

$= \dfrac{1}{2} \cdot \sqrt{25} \cdot \sqrt{x}$ Factor 25x.

$= \dfrac{1}{2} \cdot 5 \cdot \sqrt{x}$ Simplify.

$= \dfrac{5}{2}\sqrt{x}$

c. $\dfrac{7\sqrt[3]{48x^4y^8}}{\sqrt[3]{6y^2}} = 7 \cdot \sqrt[3]{\dfrac{48x^4y^8}{6y^2}}$ Apply the quotient rule.

$= 7 \cdot \sqrt[3]{8x^4y^6}$ Simplify.

$= 7\sqrt[3]{8x^3y^6 \cdot x}$ Factor.

$= 7 \cdot \sqrt[3]{8x^3y^6} \cdot \sqrt[3]{x}$ Apply the product rule.

$= 7 \cdot 2xy^2 \cdot \sqrt[3]{x}$ Simplify.

$= 14xy^2\sqrt[3]{x}$

Rational exponents may be useful when multiplying radicands with different indices.

EXAMPLE 6   Use rational exponents to write $\sqrt{5} \cdot \sqrt[3]{2}$ as a single radical.

Solution: Notice that the indices 2 and 3 have a least common multiple of 6. Use rational exponents to write each radical as a sixth root:

$$\sqrt{5} = 5^{1/2} = 5^{3/6} = \sqrt[6]{5^3} = \sqrt[6]{125}$$

$$\sqrt[3]{2} = 2^{1/3} = 2^{2/6} = \sqrt[6]{2^2} = \sqrt[6]{4}$$

Thus,

$$\sqrt{5} \cdot \sqrt[3]{2} = \sqrt[6]{125} \cdot \sqrt[6]{4} = \sqrt[6]{500}$$

35. $2y^2\sqrt[3]{2y}$ **37.** $a^2b\sqrt[4]{b^3}$ **38.** $2z^2\sqrt[5]{z^2}$ **41.** $5ab\sqrt{b}$ **42.** $3x^2y^3\sqrt{xy}$ **43.** $-2x^2\sqrt[5]{y}$ **44.** $-3z\sqrt[5]{z^4}$

EXERCISE SET 7.3

45. $x^4\sqrt[3]{50x^2}$ **46.** $2y^3\sqrt[3]{5y}$ **47.** $-4a^4b^3\sqrt{2b}$ **48.** $-2b^3\sqrt{5a}$

Use the product rule to multiply. See Example 1.

1. $\sqrt{7} \cdot \sqrt{2}$ $\sqrt{14}$ **2.** $\sqrt{11} \cdot \sqrt{10}$ $\sqrt{110}$

3. $\sqrt[3]{4} \cdot \sqrt[3]{9}$ $\sqrt[3]{36}$ **4.** $\sqrt[3]{10} \cdot \sqrt[3]{5}$ $\sqrt[3]{50}$

5. $\sqrt{2} \cdot \sqrt{3x}$ $\sqrt{6x}$ **6.** $\sqrt{3y} \cdot \sqrt{5x}$ $\sqrt{15yx}$

7. $\sqrt{\dfrac{7}{x}} \cdot \sqrt{\dfrac{2}{y}}$ $\sqrt{\dfrac{14}{xy}}$ **8.** $\sqrt{\dfrac{6}{m}} \cdot \sqrt{\dfrac{n}{5}}$ $\sqrt{\dfrac{6n}{5m}}$

9. $\sqrt[4]{4x^3} \cdot \sqrt[4]{5}$ $\sqrt[4]{20x^3}$ **10.** $\sqrt[3]{ab^2} \cdot \sqrt[3]{27ab}$ $\sqrt[3]{27a^2b^3}$

Use the quotient rule to divide, and simplify if possible. See Example 5.

11. $\dfrac{\sqrt{14}}{\sqrt{7}}$ $\sqrt{2}$ **12.** $\dfrac{\sqrt{45}}{\sqrt{9}}$ $\sqrt{5}$ **13.** $\dfrac{\sqrt[3]{24}}{\sqrt[3]{3}}$ $\sqrt[3]{8} = 2$

14. $\dfrac{\sqrt[3]{10}}{\sqrt[3]{2}}$ $\sqrt[3]{5}$ **15.** $\dfrac{\sqrt{x^5y^3}}{\sqrt{xy}}$ x^2y **16.** $\dfrac{\sqrt{a^7b^6}}{\sqrt{a^3b^2}}$ a^2b^2

17. $\dfrac{8\sqrt[3]{54m^7}}{\sqrt[3]{2m}}$ $24m^2$ **18.** $\dfrac{\sqrt[3]{128x^3}}{-3\sqrt[3]{2x}}$ $-\dfrac{4\sqrt[3]{x^2}}{3}$

Simplify. See Examples 2–4.

19. $\sqrt{32}$ $4\sqrt{2}$ **20.** $\sqrt{27}$ $3\sqrt{3}$ **21.** $\sqrt[3]{192}$ $4\sqrt[3]{3}$

22. $\sqrt[3]{108}$ $3\sqrt[3]{4}$ **23.** $5\sqrt{75}$ $25\sqrt{3}$ **24.** $3\sqrt{8}$ $6\sqrt{2}$

25. $\sqrt{\dfrac{6}{49}}$ $\dfrac{\sqrt{6}}{7}$ **26.** $\sqrt{\dfrac{8}{81}}$ $\dfrac{2\sqrt{2}}{9}$ **27.** $\sqrt{20}$ $2\sqrt{5}$

28. $\sqrt{24}$ $2\sqrt{6}$ **29.** $\sqrt[3]{\dfrac{4}{27}}$ $\dfrac{\sqrt[3]{4}}{3}$ **30.** $\sqrt[3]{\dfrac{3}{64}}$ $\dfrac{\sqrt[3]{3}}{4}$

31. $\sqrt{\dfrac{2}{49}}$ $\dfrac{\sqrt{2}}{7}$ **32.** $\sqrt{\dfrac{5}{121}}$ $\dfrac{\sqrt{5}}{11}$ **33.** $\sqrt{100x^5}$ $10x^2\sqrt{x}$

34. $\sqrt{64y^9}$ $8y^4\sqrt{y}$ **35.** $\sqrt[3]{16y^7}$ **36.** $\sqrt[3]{64y^9}$ $4y^3$

37. $\sqrt[4]{a^8b^7}$ **38.** $\sqrt[5]{32z^{12}}$ **39.** $\sqrt{y^5}$ $y^2\sqrt{y}$

40. $\sqrt[3]{y^5}$ $y\sqrt[3]{y^2}$ **41.** $\sqrt{25a^2b^3}$ **42.** $\sqrt{9x^5y^7}$

43. $\sqrt[5]{-32x^{10}y}$ **44.** $\sqrt[5]{-243z^9}$ **45.** $\sqrt[3]{50x^{14}}$

46. $\sqrt[3]{40y^{10}}$ **47.** $-\sqrt{32a^8b^7}$ **48.** $-\sqrt{20ab^6}$

49. $\sqrt{\dfrac{5x^2}{4y^2}}$ $\dfrac{\sqrt{5x}}{2y}$ **50.** $\sqrt{\dfrac{y^{10}}{9x^6}}$ $\dfrac{y^5}{3x^3}$ **51.** $-\sqrt[3]{\dfrac{z^7}{27x^3}}$ $-\dfrac{z^2\sqrt[3]{z}}{3x}$

52. $-\sqrt[3]{\dfrac{64a}{b^9}}$

53. $\sqrt[4]{\dfrac{x^7}{16}} \quad \dfrac{x\sqrt[4]{x^3}}{2}$

54. $\sqrt[4]{\dfrac{y}{81x^4}} \quad \dfrac{\sqrt[4]{y}}{3x}$

55. $\sqrt{9x^7y^9}$

56. $\sqrt{12r^9s^{12}}$

57. $\sqrt[3]{125r^9s^{12}} \quad 5r^3s^4$

58. $\sqrt[3]{8a^6b^9}$

59. $\sqrt{\dfrac{x^2y}{100}} \quad \dfrac{x\sqrt{y}}{10}$

60. $\sqrt{\dfrac{y^2z}{36}} \quad \dfrac{y\sqrt{z}}{6}$

61. $\sqrt[4]{\dfrac{8}{x^8}} \quad \dfrac{\sqrt[4]{8}}{x^2}$

62. $\sqrt[4]{\dfrac{a^3}{81}} \quad \dfrac{\sqrt[4]{a^3}}{3}$

Use rational exponents to write each radical with the same index. Then multiply. See Example 6.

63. $\sqrt{2} \cdot \sqrt[3]{3} \quad \sqrt[6]{72}$

64. $\sqrt[3]{5} \cdot \sqrt{2} \quad \sqrt[6]{200}$

65. $\sqrt[5]{7} \cdot \sqrt[3]{y} \quad \sqrt[15]{343y^5}$

66. $\sqrt[3]{x} \cdot \sqrt[4]{5} \quad \sqrt[12]{125x^4}$

67. $\sqrt[3]{x} \cdot \sqrt{x} \quad \sqrt[6]{x^5}$

68. $\sqrt[4]{y} \cdot \sqrt[7]{y} \quad \sqrt[28]{y^{11}}$

69. $\sqrt{5r} \cdot \sqrt[3]{s} \quad \sqrt[6]{125r^3s^2}$

70. $\sqrt[5]{4a} \cdot \sqrt[3]{b} \quad \sqrt[15]{64a^3b^5}$

71. The formula for the surface area A of a cone with height h and radius r is given by
$$A = \pi r\sqrt{r^2 + h^2} \quad \textbf{71a.} \ 20\pi \text{ sq. cm}$$

a. Find the surface area of a cone whose height is 3 centimeters and whose radius is 4 centimeters.

b. Approximate to two decimal places the surface area of a cone whose height is 7.2 feet and whose radius is 6.8 feet. 211.57 sq. ft

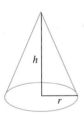

72. The formula for the radius r of a sphere, given the surface area A, is
$$r = \sqrt{\dfrac{A}{4\pi}}$$

a. The Safety First Company has decided to manufacture a new rubber ball for children. Both the marketing and production departments have determined that the surface area of the ball should be 300 square inches. Approximate the radius of the ball to two decimal places. 4.89 in.

b. Why do you think that the marketing department of a company might be involved in the planning of a new product such as the one above? Answers will vary.

73. Before Mount Vesuvius, a volcano in Italy, erupted violently in A.D. 79, its height was 4190 feet. Vesuvius was roughly cone-shaped, and its base had a radius of approximately 25,200 feet. Use the formula for the surface area of a cone, given in Exercise 71, to approximate the surface area this volcano had before it erupted. (*Source:* Global Volcanism Network) 2,022,426,050 square feet

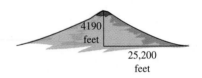

74. The surface area of Earth is approximately 1.97×10^8 square miles. Use the formula for the radius of a sphere, given in Exercise 72, to find the radius of Earth. Round to the nearest whole mile.

75. The management team of GoodDay Tire Company has determined that the demand equation for their Premium GoodDay tires is given by
$$F(x) = \sqrt{130 - x}$$

where x is the unit price of the tire and $F(x)$ is the quantity demanded per week in thousands.

a. Approximate to two decimal places the quantity of Premium GoodDay tires demanded per week if the price per tire is $86. 6.63 thousand tires

b. Approximate to two decimal places the quantity of Premium GoodDay tires demanded per week if the tires go on sale and the price per tire is $79.99. 7.07 thousand tires

c. From the results of parts (a) and (b), explain why companies have sales. Answers will vary.

52. $-\dfrac{4\sqrt[3]{a}}{b^3}$ **55.** $3x^3y^4\sqrt{xy}$ **56.** $2r^4s^6\sqrt{3r}$ **58.** $2a^2b^3$ **74.** 3959 miles

76. The owner of Knightime Video has determined that the demand equation for renting older released tapes is given by the equation

$$F(x) = 0.6\sqrt{49 - x^2}$$

where x is the price in dollars per two-day rental and $F(x)$ is the quantity of the video demanded per week.

a. Approximate to one decimal place the demand per week of an older released video if the rental price is $3 per two-day rental. 3.8

b. Approximate to one decimal place the demand per week of an older released video if the rental price is $5 per two-day rental. 2.9

c. Explain how the owner of the video store can use this equation to predict the number of copies of each tape that should be in stock. Answers will vary.

Review Exercises

Perform the following operations. See Sections 1.4 and 5.4.

77. $6x + 8x$ $14x$

78. $(6x)(8x)$ $48x^2$

79. $(2x + 3)(x - 5)$

80. $(2x + 3) + (x - 5)$

81. $9y^2 - 8y^2$ y^2

82. $(9y^2)(-8y^2)$ $-72y^4$

83. $-3(x + 5)$ $-3x - 15$ **84.** $-3 + x + 5$ $x + 2$

79. $2x^2 - 7x - 15$ **80.** $3x - 2$

7.4 | ADDING, SUBTRACTING, AND MULTIPLYING RADICAL EXPRESSIONS

TAPE IA 7.4

O B J E C T I V E S

 Add or subtract radical expressions.

2 Multiply radical expressions.

1 We have learned that the sum or difference of like terms can be simplified. To simplify these sums or differences, we use the distributive property. For example,

$$2x + 3x = (2 + 3)x = 5x \quad \text{and} \quad 7x^2y - 4x^2y = (7 - 4)x^2y = 3x^2y$$

The distributive property can also be used to add **like radicals.**

> **LIKE RADICALS**
> Radicals with the same index and the same radicand are like radicals.

For example, $2\sqrt{7} + 3\sqrt{7} = (2 + 3)\sqrt{7} = 5\sqrt{7}$. Also,

$$5\sqrt{3x} - 7\sqrt{3x} = (5 - 7)\sqrt{3x} = -2\sqrt{3x}$$

The expression $2\sqrt{7} + 2\sqrt[3]{7}$ cannot be simplified since $2\sqrt{7}$ and $2\sqrt[3]{7}$ are not like radicals.

EXAMPLE 1 Add or subtract. Assume that variables represent positive real numbers.

a. $\sqrt{20} + 2\sqrt{45}$ **b.** $\sqrt[3]{54} - 5\sqrt[3]{16} + \sqrt[3]{2}$ **c.** $\sqrt{27x} - 2\sqrt{9x} + \sqrt{72x}$

d. $\sqrt[3]{98} + \sqrt{98}$ **e.** $\sqrt[3]{48y^4} + \sqrt[3]{6y^4}$

Solution: First, simplify each radical. Then add or subtract any like radicals.

a. $\sqrt{20} + 2\sqrt{45} = \sqrt{4 \cdot 5} + 2\sqrt{9 \cdot 5}$

$\qquad\qquad\quad = \sqrt{4} \cdot \sqrt{5} + 2 \cdot \sqrt{9} \cdot \sqrt{5}$

$\qquad\qquad\quad = 2 \cdot \sqrt{5} + 2 \cdot 3 \cdot \sqrt{5}$

$\qquad\qquad\quad = 2\sqrt{5} + 6\sqrt{5} = 8\sqrt{5}$

b. $\sqrt[3]{54} - 5\sqrt[3]{16} + \sqrt[3]{2} = \sqrt[3]{27} \cdot \sqrt[3]{2} - 5 \cdot \sqrt[3]{8} \cdot \sqrt[3]{2} + \sqrt[3]{2}$

$\qquad\qquad\qquad\qquad\quad = 3 \cdot \sqrt[3]{2} - 5 \cdot 2 \cdot \sqrt[3]{2} + \sqrt[3]{2}$

$\qquad\qquad\qquad\qquad\quad = 3\sqrt[3]{2} - 10\sqrt[3]{2} + \sqrt[3]{2}$

$\qquad\qquad\qquad\qquad\quad = -6\sqrt[3]{2}$

c. $\sqrt{27x} - 2\sqrt{9x} + \sqrt{72x} = \sqrt{9} \cdot \sqrt{3x} - 2 \cdot \sqrt{9} \cdot \sqrt{x} + \sqrt{36} \cdot \sqrt{2x}$

$\qquad\qquad\qquad\qquad\qquad = 3 \cdot \sqrt{3x} - 2 \cdot 3 \cdot \sqrt{x} + 6 \cdot \sqrt{2x}$

$\qquad\qquad\qquad\qquad\qquad = 3\sqrt{3x} - 6\sqrt{x} + 6\sqrt{2x}$

d. We can simplify $\sqrt{98}$, but since the indexes are different, these radicals cannot be added:

$$\sqrt[3]{98} + \sqrt{98} = \sqrt[3]{98} + \sqrt{49} \cdot \sqrt{2} = \sqrt[3]{98} + 7\sqrt{2}$$

e. $\sqrt[3]{48y^4} + \sqrt[3]{6y^4} = \sqrt[3]{8y^3} \cdot \sqrt[3]{6y} + \sqrt[3]{y^3} \cdot \sqrt[3]{6y}$

$\qquad\qquad\qquad\quad = 2y\sqrt[3]{6y} + y\sqrt[3]{6y}$

$\qquad\qquad\qquad\quad = 3y\sqrt[3]{6y}$

The following may be used to simplify radical expressions.

TO SIMPLIFY RADICAL EXPRESSIONS

1. Write each radical term in simplest form.

2. Add or subtract any like radicals.

EXAMPLE 2 Simplify.

a. $\dfrac{\sqrt{45}}{4} - \dfrac{\sqrt{5}}{3}$ **b.** $\sqrt[3]{\dfrac{7x}{8}} + 2\sqrt[3]{7x}$

Solution: **a.** $\dfrac{\sqrt{45}}{4} - \dfrac{\sqrt{5}}{3} = \dfrac{3\sqrt{5}}{4} - \dfrac{\sqrt{5}}{3}$ To subtract, notice that the LCD is 12.

$\qquad\qquad\qquad = \dfrac{3\sqrt{5} \cdot 3}{4 \cdot 3} - \dfrac{\sqrt{5} \cdot 4}{3 \cdot 4}$ Write each expression as an equivalent expression with a denominator of 12.

$\qquad\qquad\qquad = \dfrac{9\sqrt{5}}{12} - \dfrac{4\sqrt{5}}{12}$ Multiply factors in the numerator and the denominator.

$\qquad\qquad\qquad = \dfrac{5\sqrt{5}}{12}$ Subtract.

b. $\sqrt[3]{\dfrac{7x}{8}} + 2\sqrt[3]{7x} = \dfrac{\sqrt[3]{7x}}{\sqrt[3]{8}} + 2\sqrt[3]{7x}$ Apply the quotient property for radicals.

$\qquad = \dfrac{\sqrt[3]{7x}}{2} + 2\sqrt[3]{7x}$ Simplify.

$\qquad = \dfrac{\sqrt[3]{7x}}{2} + \dfrac{2\sqrt[3]{7x} \cdot 2}{2}$ Write each expression as an equivalent expression with a denominator of 2.

$\qquad = \dfrac{\sqrt[3]{7x}}{2} + \dfrac{4\sqrt[3]{7x}}{2}$

$\qquad = \dfrac{5\sqrt[3]{7x}}{2}$ Add.

2 Radical expressions are multiplied by using many of the same properties used to multiply polynomial expressions. For instance, to multiply $\sqrt{2}(\sqrt{6} - 3\sqrt{2})$, we use the distributive property and multiply $\sqrt{2}$ by each term inside the parentheses:

$$\sqrt{2}(\sqrt{6} - 3\sqrt{2}) = \sqrt{2}(\sqrt{6}) - \sqrt{2}(3\sqrt{2})$$
$$= \sqrt{12} - 3\sqrt{2 \cdot 2}$$
$$= \sqrt{4 \cdot 3} - 3 \cdot 2 \qquad \text{Apply the product rule for radicals.}$$
$$= 2\sqrt{3} - 6$$

EXAMPLE 3   Multiply.

a. $\sqrt{3}(5 + \sqrt{30})$ **b.** $(\sqrt{5} - \sqrt{6})(\sqrt{7} + 1)$ **c.** $(7\sqrt{x} + 5)(3\sqrt{x} - \sqrt{5})$

d. $(4\sqrt{3} - 1)^2$ **e.** $(\sqrt{2x} - 5)(\sqrt{2x} + 5)$

Solution: **a.** $\sqrt{3}(5 + \sqrt{30}) = \sqrt{3}(5) + \sqrt{3}(\sqrt{30})$

$\qquad = 5\sqrt{3} + \sqrt{3 \cdot 30}$

$\qquad = 5\sqrt{3} + \sqrt{3 \cdot 3 \cdot 10}$

$\qquad = 5\sqrt{3} + 3\sqrt{10}$

b. To multiply, we can use the FOIL method.

$$\begin{array}{cccc} & \textbf{First} & \textbf{Outside} & \textbf{Inside} & \textbf{Last} \end{array}$$
$$(\sqrt{5} - \sqrt{6})(\sqrt{7} + 1) = \sqrt{5} \cdot \sqrt{7} + \sqrt{5} \cdot 1 - \sqrt{6} \cdot \sqrt{7} - \sqrt{6} \cdot 1$$
$$= \sqrt{35} + \sqrt{5} - \sqrt{42} - \sqrt{6}$$

c. $(7\sqrt{x} + 5)(3\sqrt{x} - \sqrt{5}) = 7\sqrt{x}(3\sqrt{x}) - 7\sqrt{x}(\sqrt{5}) + 5(3\sqrt{x}) - 5(\sqrt{5})$

$\qquad = 21x - 7\sqrt{5x} + 15\sqrt{x} - 5\sqrt{5}$

d. $(4\sqrt{3} - 1)^2 = (4\sqrt{3} - 1)(4\sqrt{3} - 1)$

$\qquad = 4\sqrt{3}(4\sqrt{3}) - 4\sqrt{3}(1) - 1(4\sqrt{3}) - 1(-1)$

$\qquad = 16 \cdot 3 - 4\sqrt{3} - 4\sqrt{3} + 1$

$\qquad = 48 - 8\sqrt{3} + 1$

$\qquad = 49 - 8\sqrt{3}$

e. $(\sqrt{2x} - 5)(\sqrt{2x} + 5) = \sqrt{2x} \cdot \sqrt{2x} + 5\sqrt{2x} - 5\sqrt{2x} - 5 \cdot 5$

$\qquad = 2x - 25$

MENTAL MATH

Simplify. Assume that all variables represent positive real numbers.

1. $2\sqrt{3} + 4\sqrt{3}$ $6\sqrt{3}$

2. $5\sqrt{7} + 3\sqrt{7}$ $8\sqrt{7}$

3. $8\sqrt{x} - 5\sqrt{x}$ $3\sqrt{x}$

4. $3\sqrt{y} + 10\sqrt{y}$ $13\sqrt{y}$

5. $7\sqrt[3]{x} + 5\sqrt[3]{x}$ $12\sqrt[3]{x}$

6. $8\sqrt[3]{z} - 2\sqrt[3]{z}$ $6\sqrt[3]{z}$

EXERCISE SET 7.4

Add or subtract. Assume that all variables represent positive numbers. See Examples 1 and 2.

1. $\sqrt{8} - \sqrt{32}$ $-2\sqrt{2}$

2. $\sqrt{27} - \sqrt{75}$ $-2\sqrt{3}$

3. $2\sqrt{2x^3} + 4x\sqrt{8x}$

4. $3\sqrt{45x^3} + x\sqrt{5x}$ $10x\sqrt{5x}$

5. $2\sqrt{50} - 3\sqrt{125} + \sqrt{98}$ $17\sqrt{2} - 15\sqrt{5}$

6. $4\sqrt{32} - \sqrt{18} + 2\sqrt{128}$ $29\sqrt{2}$

7. $\sqrt[3]{16x} - \sqrt[3]{54x}$ $-\sqrt[3]{2x}$

8. $2\sqrt[3]{3a^4} - 3a\sqrt[3]{81a}$ $-7a\sqrt[3]{3a}$

9. $\sqrt{9b^3} - \sqrt{25b^3} + \sqrt{49b^3}$ $5b\sqrt{b}$

10. $\sqrt{4x^7} + 9x^2\sqrt{x^3} - 5x\sqrt{x^5}$ $6x^3\sqrt{x}$

11. $\dfrac{5\sqrt{2}}{3} + \dfrac{2\sqrt{2}}{5}$ $\dfrac{31\sqrt{2}}{15}$

12. $\dfrac{\sqrt{3}}{2} + \dfrac{4\sqrt{3}}{3}$ $\dfrac{11\sqrt{3}}{6}$

13. $\sqrt[3]{\dfrac{11}{8}} - \dfrac{\sqrt[3]{11}}{6}$ $\dfrac{\sqrt[3]{11}}{3}$

14. $\dfrac{2\sqrt[3]{4}}{7} - \dfrac{\sqrt[3]{4}}{14}$ $\dfrac{3\sqrt[3]{4}}{14}$

15. $\dfrac{\sqrt{20x}}{9} + \sqrt{\dfrac{5x}{9}}$ $\dfrac{5\sqrt{5x}}{9}$

16. $\dfrac{3x\sqrt{7}}{5} + \sqrt{\dfrac{7x^2}{100}}$ $\dfrac{7x\sqrt{7}}{10}$

17. $7\sqrt{9} - 7 + \sqrt{3}$

18. $\sqrt{16} - 5\sqrt{10} + 7$

19. $2 + 3\sqrt{y^2} - 6\sqrt{y^2} + 5$ $7 - 3y$

20. $3\sqrt{7} - \sqrt[3]{x} + 4\sqrt{7} - 3\sqrt[3]{x}$ $7\sqrt{7} - 4\sqrt[3]{x}$

21. $3\sqrt{108} - 2\sqrt{18} - 3\sqrt{48}$ $6\sqrt{3} - 6\sqrt{2}$

22. $-\sqrt{75} + \sqrt{12} - 3\sqrt{3}$ $-6\sqrt{3}$

23. $-5\sqrt[3]{625} + \sqrt[3]{40}$ $-23\sqrt[3]{5}$

24. $-2\sqrt[3]{108} - \sqrt[3]{32}$ $-8\sqrt[3]{4}$

25. $\sqrt{9b^3} - \sqrt{25b^3} + \sqrt{16b^3}$ $2b\sqrt{b}$

26. $\sqrt{4x^7y^5} + 9x^2\sqrt{x^3y^5} - 5xy\sqrt{x^5y^3}$ $6x^3y^2\sqrt{xy}$

27. $5y\sqrt{8y} + 2\sqrt{50y^3}$

28. $3\sqrt{8x^2y^3} - 2x\sqrt{32y^3}$

29. $\sqrt[3]{54xy^3} - 5\sqrt[3]{2xy^3} + y\sqrt[3]{128x}$ $2y\sqrt[3]{2x}$

30. $2\sqrt[3]{24x^3y^4} + 4x\sqrt[3]{81y^4}$ $16xy\sqrt[3]{3y}$

31. $6\sqrt[3]{11} + 8\sqrt{11} - 12\sqrt{11}$ $6\sqrt[3]{11} - 4\sqrt{11}$

32. $3\sqrt[3]{5} + 4\sqrt{5}$ $3\sqrt[3]{5} + 4\sqrt{5}$

33. $-2\sqrt[4]{x^7} + 3\sqrt[4]{16x^7}$ $4x\sqrt[4]{x^3}$

34. $6\sqrt[3]{24x^3} - 2\sqrt[3]{81x^3} - x\sqrt[3]{3}$ $5x\sqrt[3]{3}$

35. $\dfrac{4\sqrt{3}}{3} - \dfrac{\sqrt{12}}{3}$ $\dfrac{2\sqrt{3}}{3}$

36. $\dfrac{\sqrt{45}}{10} + \dfrac{7\sqrt{5}}{10}$ $\sqrt{5}$

37. $\dfrac{\sqrt[3]{8x^4}}{7} + \dfrac{3x\sqrt[3]{x}}{7}$ $\dfrac{5x\sqrt[3]{x}}{7}$

38. $\dfrac{\sqrt[4]{48}}{5x} - \dfrac{2\sqrt[3]{3}}{10x}$ $\dfrac{\sqrt[4]{3}}{5x}$

39. $\sqrt{\dfrac{28}{x^2}} + \sqrt{\dfrac{7}{4x^2}}$ $\dfrac{5\sqrt{7}}{2x}$

40. $\dfrac{\sqrt{99}}{5x} - \sqrt{\dfrac{44}{x^2}}$ $-\dfrac{7\sqrt{11}}{5x}$

41. $\sqrt[3]{\dfrac{16}{27}} - \dfrac{\sqrt[3]{54}}{6}$ $\dfrac{\sqrt[3]{2}}{6}$

42. $\dfrac{\sqrt[3]{3}}{10} + \sqrt[3]{\dfrac{24}{125}}$ $\dfrac{\sqrt[3]{3}}{2}$

43. $-\dfrac{\sqrt[3]{2x^4}}{9} + \sqrt[3]{\dfrac{250x^4}{27}}$

44. $\dfrac{\sqrt[3]{y^5}}{8} + \dfrac{5y\sqrt[3]{y^2}}{4}$ $\dfrac{11y\sqrt[3]{y^2}}{8}$

45. Find the perimeter of the trapezoid. $15\sqrt{3}$ in.

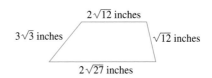

$2\sqrt{12}$ inches

$3\sqrt{3}$ inches $\sqrt{12}$ inches

$2\sqrt{27}$ inches

46. Find the perimeter of the triangle.

$(6\sqrt{2} + 3\sqrt{5})$ meters

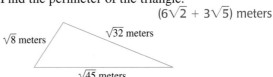

$\sqrt{8}$ meters $\sqrt{32}$ meters

$\sqrt{45}$ meters

Multiply, and then simplify if possible. Assume that all variables represent positive real numbers. See Example 3.

47. $\sqrt{7}(\sqrt{5} + \sqrt{3})$

48. $\sqrt{5}(\sqrt{15} - \sqrt{35})$

49. $(\sqrt{5} - \sqrt{2})^2$

50. $(3x - \sqrt{2})(3x - \sqrt{2})$

3. $10x\sqrt{2x}$ **17.** $14 + \sqrt{3}$ **18.** $11 - 5\sqrt{10}$ **27.** $20y\sqrt{2y}$ **28.** $-2xy\sqrt{2y}$ **43.** $\dfrac{14x\sqrt[3]{2x}}{9}$ **47.** $\sqrt{35} + \sqrt{21}$

48. $5\sqrt{3} - 5\sqrt{7}$ **49.** $7 - 2\sqrt{10}$ **50.** $9x^2 - 6x\sqrt{2} + 2$

51. $3\sqrt{x} - x\sqrt{3}$ **52.** $y\sqrt{5} + 5\sqrt{y}$ **53.** $6x - 13\sqrt{x} - 5$ **54.** $32y - 8\sqrt{y} + 4z\sqrt{y} - z$ **55.** $\sqrt[3]{a^2} + \sqrt[3]{a} - 20$

51. $\sqrt{3x}(\sqrt{3} - \sqrt{x})$ **52.** $\sqrt{5y}(\sqrt{y} + \sqrt{5})$

53. $(2\sqrt{x} - 5)(3\sqrt{x} + 1)$ **54.** $(8\sqrt{y} + z)(4\sqrt{y} - 1)$

55. $(\sqrt[3]{a} - 4)(\sqrt[3]{a} + 5)$ **56.** $(\sqrt[3]{a} + 2)(\sqrt[3]{a} + 7)$

57. $6(\sqrt{2} - 2)$ $6\sqrt{2} - 12$ **58.** $\sqrt{5}(6 - \sqrt{5})$ $6\sqrt{5} - 5$

59. $\sqrt{2}(\sqrt{2} + x\sqrt{6})$ **60.** $\sqrt{3}(\sqrt{3} - 2\sqrt{5x})$ $3 - 2\sqrt{15x}$

61. $(2\sqrt{7} + 3\sqrt{5})(\sqrt{7} - 2\sqrt{5})$ $-16 - \sqrt{35}$

62. $(\sqrt{6} - 4\sqrt{2})(3\sqrt{6} + 1)$ $18 + \sqrt{6} - 24\sqrt{3} - 4\sqrt{2}$

63. $(\sqrt{x} - y)(\sqrt{x} + y)$ $x - y^2$

64. $(3\sqrt{x} + 2)(\sqrt{3x} - 2)$ $3x\sqrt{3} - 6\sqrt{x} + 2\sqrt{3x} - 4$

65. $(\sqrt{3} + x)^2$ **66.** $(\sqrt{y} - 3x)^2$ $y - 6x\sqrt{y} + 9x^2$

67. $(\sqrt{5x} - 3\sqrt{2})(\sqrt{5x} - 3\sqrt{3})$ $5x - 3\sqrt{10x} - 3\sqrt{15x} + 9\sqrt{6}$

68. $(5\sqrt{3x} - \sqrt{y})(4\sqrt{x} + 1)$ $20x\sqrt{3} + 5\sqrt{3x} - 4\sqrt{xy} - \sqrt{y}$

69. $(\sqrt[3]{4} + 2)(\sqrt[3]{2} - 1)$ $2\sqrt[3]{2} - \sqrt[3]{4}$

70. $(\sqrt[3]{3} + \sqrt[3]{2})(\sqrt[3]{9} - \sqrt[3]{4})$ $1 - \sqrt[3]{12} + \sqrt[3]{18}$

71. $(\sqrt[3]{x} + 1)(\sqrt[3]{x} - 4\sqrt{x} + 7)$

72. $(\sqrt[3]{3x} + 3)(\sqrt[3]{2x} - 3x - 1)$

 73. Baseboard needs to be installed around the perimeter of a rectangular room.

a. Find how much baseboard should be ordered by finding the perimeter of the room. $22\sqrt{5}$ ft

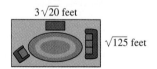

$3\sqrt{20}$ feet

$\sqrt{125}$ feet

b. Find the area of the room.   150 sq. ft

56. $\sqrt[3]{a^2} + 9\sqrt[3]{a} + 14$ **59.** $2 + 2x\sqrt{3}$ **65.** $3 + 2x\sqrt{3} + x^2$ **71.** $-4\sqrt[6]{x^5} + \sqrt[3]{x^2} + 8\sqrt[3]{x} - 4\sqrt{x} + 7$
72. $\sqrt[3]{6x^2} - 3x\sqrt[3]{3x} - \sqrt[3]{3x} + 3\sqrt[3]{2x} - 9x - 3$

 74. A border of wallpaper is to be used around the perimeter of the odd-shaped room shown.

a. Find how much wallpaper border is needed by finding the perimeter of the room.
$(12\sqrt{3} + 13\sqrt{7})$ meters

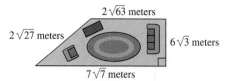

$2\sqrt{63}$ meters

$2\sqrt{27}$ meters

$6\sqrt{3}$ meters

$7\sqrt{7}$ meters

b. Find the area of the room. (*Hint:* The area of a trapezoid is the product of half the height $6\sqrt{3}$ meters and the sum of the bases $2\sqrt{63}$ and $7\sqrt{7}$ meters.) $39\sqrt{21}$ sq. meters

Review Exercises

Factor each numerator and denominator. Then simplify if possible. See Section 6.1.

75. $\dfrac{2x - 14}{2}$ $x - 7$ **76.** $\dfrac{8x - 24y}{4}$ $2(x - 3y)$

77. $\dfrac{7x - 7y}{x^2 - y^2}$ $\dfrac{7}{x + y}$ **78.** $\dfrac{x^3 - 8}{4x - 8}$ $\dfrac{x^2 + 2x + 4}{4}$

79. $\dfrac{6a^2b - 9ab}{3ab}$ $2a - 3$ **80.** $\dfrac{14r - 28r^2s^2}{7rs}$ $\dfrac{2(1 - 2rs^2)}{s}$

81. $\dfrac{-4 + 2\sqrt{3}}{6}$ $\dfrac{-2 + \sqrt{3}}{3}$ **82.** $\dfrac{-5 + 10\sqrt{7}}{5}$ $-1 + 2\sqrt{7}$

7.5 RATIONALIZING NUMERATORS AND DENOMINATORS OF RADICAL EXPRESSIONS

TAPE IA 7.5

O B J E C T I V E S

1. Rationalize denominators.
2. Rationalize numerators.

Often in mathematics, it is helpful to write a radical expression such as $\dfrac{\sqrt{3}}{\sqrt{2}}$ either without a radical in the denominator or without a radical in the numerator. The process of writing this expression as an equivalent expression but without a radical in the denominator is called **rationalizing the denominator.** To rationalize the denominator of $\frac{\sqrt{3}}{\sqrt{2}}$, we use the fundamental principle of fractions and multiply the numerator and the denominator by $\sqrt{2}$. Recall that this is the same as multiplying by $\frac{\sqrt{2}}{\sqrt{2}}$, which simplifies to 1.

$$\frac{\sqrt{3}}{\sqrt{2}} = \frac{\sqrt{3} \cdot \sqrt{2}}{\sqrt{2} \cdot \sqrt{2}} = \frac{\sqrt{6}}{\sqrt{4}} = \frac{\sqrt{6}}{2}$$

EXAMPLE 1   Rationalize the denominator of each expression.

a. $\dfrac{\sqrt{27}}{\sqrt{5}}$ **b.** $\dfrac{2\sqrt{16}}{\sqrt{9x}}$ **c.** $\sqrt[3]{\dfrac{1}{2}}$

Solution: **a.** First, we simplify $\sqrt{27}$; then we rationalize the denominator.

$$\frac{\sqrt{27}}{\sqrt{5}} = \frac{\sqrt{9 \cdot 3}}{\sqrt{5}} = \frac{3\sqrt{3}}{\sqrt{5}}$$

Recall that to rationalize the denominator, multiply the numerator and denominator by $\sqrt{5}$.

$$\frac{3\sqrt{3}}{\sqrt{5}} = \frac{3\sqrt{3} \cdot \sqrt{5}}{\sqrt{5} \cdot \sqrt{5}} = \frac{3\sqrt{15}}{5}$$

b. First, we simplify the radicals and then rationalize the denominator.

$$\frac{2\sqrt{16}}{\sqrt{9x}} = \frac{2(4)}{3\sqrt{x}} = \frac{8}{3\sqrt{x}}$$

To rationalize the denominator, multiply the numerator and denominator by $\sqrt{x}$. Then

$$\frac{8}{3\sqrt{x}} = \frac{8 \cdot \sqrt{x}}{3\sqrt{x} \cdot \sqrt{x}} = \frac{8\sqrt{x}}{3x}$$

c. $\sqrt[3]{\dfrac{1}{2}} = \dfrac{\sqrt[3]{1}}{\sqrt[3]{2}} = \dfrac{1}{\sqrt[3]{2}}$. Now we rationalize the denominator. Since $\sqrt[3]{2}$ is a cube root, we want to multiply by a value that will make the radicand 2 a perfect cube. If we multiply by $\sqrt[3]{2^2}$, we get $\sqrt[3]{2^3} = \sqrt[3]{8} = 2$.

$$\frac{1 \cdot \sqrt[3]{2^2}}{\sqrt[3]{2} \cdot \sqrt[3]{2^2}} = \frac{\sqrt[3]{4}}{\sqrt[3]{8}} = \frac{\sqrt[3]{4}}{2}$$

A different method is needed to rationalize a denominator that is a sum or difference of two terms. For example, to rationalize the denominator of the expression $\frac{5}{\sqrt{3}-2}$, multiply both the numerator and denominator by $\sqrt{3} + 2$, the

conjugate of the denominator $\sqrt{3} - 2$. In general, the conjugate of $a + b$ is $a - b$. Then

$$\frac{5}{\sqrt{3} - 2} = \frac{5(\sqrt{3} + 2)}{(\sqrt{3} - 2)(\sqrt{3} + 2)}$$

$$= \frac{5(\sqrt{3} + 2)}{\sqrt{3} \cdot \sqrt{3} + 2\sqrt{3} - 2\sqrt{3} - 4}$$

$$= \frac{5(\sqrt{3} + 2)}{3 - 4}$$

$$= \frac{5(\sqrt{3} + 2)}{-1}$$

$$= -5(\sqrt{3} + 2), \quad \text{or} \quad -5\sqrt{3} - 10$$

EXAMPLE 2   Rationalize each denominator.

a. $\dfrac{2}{3\sqrt{2} + 4}$ **b.** $\dfrac{\sqrt{6} + 2}{\sqrt{5} - \sqrt{3}}$ **c.** $\dfrac{7\sqrt{y}}{\sqrt{12x}}$ **d.** $\dfrac{2\sqrt{m}}{3\sqrt{x} + \sqrt{m}}$

Solution: **a.** Multiply the numerator and denominator by the conjugate of the denominator, $3\sqrt{2} + 4$:

$$\frac{2}{3\sqrt{2} + 4} = \frac{2(3\sqrt{2} - 4)}{(3\sqrt{2} + 4)(3\sqrt{2} - 4)}$$

$$= \frac{2(3\sqrt{2} - 4)}{(3\sqrt{2})^2 - 4^2}$$

$$= \frac{2(3\sqrt{2} - 4)}{18 - 16}$$

$$= \frac{2(3\sqrt{2} - 4)}{2}, \quad \text{or} \quad 3\sqrt{2} - 4$$

It is often helpful to leave a numerator in factored form to help determine whether the expression can be simplified.

b. Multiply the numerator and denominator by the conjugate of $\sqrt{5} - \sqrt{3}$:

$$\frac{\sqrt{6} + 2}{\sqrt{5} - \sqrt{3}} = \frac{(\sqrt{6} + 2)(\sqrt{5} + \sqrt{3})}{(\sqrt{5} - \sqrt{3})(\sqrt{5} + \sqrt{3})}$$

$$= \frac{\sqrt{6}\sqrt{5} + \sqrt{6}\sqrt{3} + 2\sqrt{5} + 2\sqrt{3}}{(\sqrt{5})^2 - (\sqrt{3})^2}$$

$$= \frac{\sqrt{30} + \sqrt{18} + 2\sqrt{5} + 2\sqrt{3}}{5 - 3}$$

$$= \frac{\sqrt{30} + 3\sqrt{2} + 2\sqrt{5} + 2\sqrt{3}}{2}$$

c. Notice that the denominator of this example is *not the sum or difference of two terms*. For this reason, we simplify the radical expression and then multiply the numerator and denominator by a factor so that the resulting denominator is a rational expression:

$$\frac{7\sqrt{y}}{\sqrt{12x}} = \frac{7\sqrt{y}}{\sqrt{4}\sqrt{3x}} = \frac{7\sqrt{y} \cdot \sqrt{3x}}{2\sqrt{3x} \cdot \sqrt{3x}} = \frac{7\sqrt{3xy}}{2 \cdot 3x} = \frac{7\sqrt{3xy}}{6x}$$

d. Multiply by the conjugate of $3\sqrt{x} + \sqrt{m}$ to eliminate the radicals from the denominator:

$$\frac{2\sqrt{m}}{3\sqrt{x} + \sqrt{m}} = \frac{2\sqrt{m}(3\sqrt{x} - \sqrt{m})}{(3\sqrt{x} + \sqrt{m})(3\sqrt{x} - \sqrt{m})} = \frac{6\sqrt{mx} - 2m}{(3\sqrt{x})^2 - (\sqrt{m})^2}$$

$$= \frac{6\sqrt{mx} - 2m}{9x - m}$$

2

As mentioned earlier, it is also often helpful to write an expression such as $\frac{\sqrt{3}}{\sqrt{2}}$ as an equivalent expression without a radical in the numerator. This process is called **rationalizing the numerator.** To rationalize the numerator of $\frac{\sqrt{3}}{\sqrt{2}}$, we multiply the numerator and denominator by $\sqrt{3}$.

$$\frac{\sqrt{3}}{\sqrt{2}} = \frac{\sqrt{3} \cdot \sqrt{3}}{\sqrt{2} \cdot \sqrt{3}} = \frac{\sqrt{9}}{\sqrt{6}} = \frac{3}{\sqrt{6}}$$

EXAMPLE 3   Rationalize the numerator of each expression.

a. $\dfrac{\sqrt{28}}{\sqrt{45}}$ **b.** $\dfrac{\sqrt[3]{2x^2}}{\sqrt[3]{5y}}$

Solution: **a.** First, we simplify $\sqrt{28}$ and $\sqrt{45}$:

$$\frac{\sqrt{28}}{\sqrt{45}} = \frac{\sqrt{4 \cdot 7}}{\sqrt{9 \cdot 5}} = \frac{2\sqrt{7}}{3\sqrt{5}}$$

Next, we rationalize the numerator by multiplying the numerator and denominator by $\sqrt{7}$:

$$\frac{2\sqrt{7}}{3\sqrt{5}} = \frac{2\sqrt{7} \cdot \sqrt{7}}{3\sqrt{5} \cdot \sqrt{7}} = \frac{2 \cdot 7}{3\sqrt{5} \cdot 7} = \frac{14}{3\sqrt{35}}$$

b. The numerator and the denominator of this expression are already simplified. To rationalize the numerator, $\sqrt[3]{2x^2}$, we multiply the numerator and denominator by a factor that will make the radicand a perfect cube. If we multiply $\sqrt[3]{2x^2}$ by $\sqrt[3]{4x}$, we get $\sqrt[3]{8x^3} = 2x$:

$$\frac{\sqrt[3]{2x^2}}{\sqrt[3]{5y}} = \frac{\sqrt[3]{2x^2} \cdot \sqrt[3]{4x}}{\sqrt[3]{5y} \cdot \sqrt[3]{4x}} = \frac{\sqrt[3]{8x^3}}{\sqrt[3]{20xy}} = \frac{2x}{\sqrt[3]{20xy}}$$

EXAMPLE 4   Rationalize the numerator of $\dfrac{\sqrt{x}+2}{5}$.

Solution: Multiply the numerator and denominator by the conjugate of $\sqrt{x}+2$, the numerator:

$$\frac{\sqrt{x}+2}{5} = \frac{(\sqrt{x}+2)(\sqrt{x}-2)}{5(\sqrt{x}-2)}$$

$$= \frac{x - 2\sqrt{x} + 2\sqrt{x} - 4}{5(\sqrt{x}-2)}$$

$$= \frac{x-4}{5(\sqrt{x}-2)}$$

13. $\dfrac{7(3+\sqrt{x})}{9-x}$ **14.** $\dfrac{32-8\sqrt{y}}{y-16}$ **15.** $-5+2\sqrt{6}$ **16.** $3-\sqrt{6}-2\sqrt{2}+2\sqrt{3}$ **17.** $\dfrac{2a+2\sqrt{a}+\sqrt{ab}+\sqrt{b}}{4a-b}$

EXERCISE SET 7.5   **18.** $\dfrac{4a+2\sqrt{ab}-6\sqrt{a}-3\sqrt{b}}{4a-b}$ **25.** $-\dfrac{8(1-\sqrt{10})}{9}$ **26.** $\dfrac{-3\sqrt{6}-6}{2}$

Simplify by rationalizing the denominator. Assume that all variables represent positive numbers. See Examples 1 and 2.

1. $\dfrac{1}{\sqrt{3}}\quad\dfrac{\sqrt{3}}{3}$ **2.** $\dfrac{\sqrt{2}}{\sqrt{6}}\quad\dfrac{\sqrt{3}}{3}$ **3.** $\sqrt{\dfrac{1}{5}}\quad\dfrac{\sqrt{5}}{5}$

4. $\sqrt{\dfrac{1}{2}}\quad\dfrac{\sqrt{2}}{2}$ **5.** $\dfrac{4}{\sqrt[3]{3}}\quad\dfrac{4\sqrt[3]{9}}{3}$ **6.** $\dfrac{6}{\sqrt[3]{9}}\quad 2\sqrt[3]{3}$

7. $\dfrac{3}{\sqrt{8x}}\quad\dfrac{3\sqrt{2x}}{4x}$ **8.** $\dfrac{5}{\sqrt{27a}}\quad\dfrac{5\sqrt{3a}}{9a}$ **9.** $\dfrac{3}{\sqrt[3]{4x^2}}\quad\dfrac{3\sqrt[3]{2x}}{2x}$

10. $\dfrac{5}{\sqrt[3]{3y}}\quad\dfrac{5\sqrt[3]{9y^2}}{3y}$ **11.** $\dfrac{6}{2-\sqrt{7}}$ **12.** $\dfrac{3}{\sqrt{7}-4}\quad-\dfrac{\sqrt{7}+4}{3}$

13. $\dfrac{-7}{\sqrt{x}-3}$ **14.** $\dfrac{-8}{\sqrt{y}+4}$ **15.** $\dfrac{\sqrt{2}-\sqrt{3}}{\sqrt{2}+\sqrt{3}}$

16. $\dfrac{\sqrt{3}+\sqrt{4}}{\sqrt{2}+\sqrt{3}}$ **17.** $\dfrac{\sqrt{a}+1}{2\sqrt{a}-\sqrt{b}}$ **18.** $\dfrac{2\sqrt{a}-3}{2\sqrt{a}-\sqrt{b}}$

19. $\dfrac{9}{\sqrt{3a}}\quad\dfrac{3\sqrt{3a}}{a}$ **20.** $\dfrac{x}{\sqrt{5}}\quad\dfrac{x\sqrt{5}}{5}$ **21.** $\dfrac{3}{\sqrt[3]{2}}\quad\dfrac{3\sqrt[3]{4}}{2}$

22. $\dfrac{5}{\sqrt[3]{9}}\quad\dfrac{5\sqrt[3]{3}}{3}$ **23.** $\dfrac{2\sqrt{3}}{\sqrt{7}}\quad\dfrac{2\sqrt{21}}{7}$ **24.** $\dfrac{-5\sqrt{2}}{\sqrt{11}}\quad\dfrac{-5\sqrt{22}}{11}$

25. $\dfrac{8}{1+\sqrt{10}}$ **26.** $\dfrac{-3}{\sqrt{6}-2}$ **27.** $\dfrac{\sqrt{x}}{\sqrt{x}+\sqrt{y}}$

28. $\dfrac{2\sqrt{a}}{2\sqrt{x}-\sqrt{y}}\quad\dfrac{4\sqrt{ax}+2\sqrt{ay}}{4x-y}$

Rationalize each numerator. See Examples 3 and 4.

29. $\sqrt{\dfrac{5}{3}}\quad\dfrac{5}{\sqrt{15}}$ **30.** $\sqrt{\dfrac{3}{2}}\quad\dfrac{3}{\sqrt{6}}$ **31.** $\sqrt{\dfrac{18}{5}}\quad\dfrac{6}{\sqrt{10}}$

32. $\sqrt{\dfrac{12}{7}}\quad\dfrac{6}{\sqrt{21}}$ **33.** $\dfrac{\sqrt{4x}}{7}\quad\dfrac{4x}{7\sqrt{4x}}$ **34.** $\dfrac{\sqrt{3x^5}}{6}\quad\dfrac{x^3}{2\sqrt{3x}}$

35. $\dfrac{\sqrt[3]{5y^2}}{\sqrt[3]{4x}}$ **36.** $\dfrac{\sqrt[3]{4x}}{\sqrt[3]{z^4}}$ **37.** $\dfrac{2-\sqrt{7}}{-5}$

38. $\dfrac{\sqrt{5}+2}{\sqrt{2}}$ **39.** $\dfrac{\sqrt{x}+3}{\sqrt{x}}$ **40.** $\dfrac{5+\sqrt{2}}{\sqrt{2x}}$

41. $\sqrt{\dfrac{2}{5}}\quad\dfrac{2}{\sqrt{10}}$ **42.** $\sqrt{\dfrac{3}{7}}\quad\dfrac{3}{\sqrt{21}}$ **43.** $\dfrac{\sqrt{2x}}{11}\quad\dfrac{2x}{11\sqrt{2x}}$

44. $\dfrac{\sqrt{y}}{7}\quad\dfrac{y}{7\sqrt{y}}$ **45.** $\sqrt[3]{\dfrac{7}{8}}\quad\dfrac{7}{2\sqrt[3]{49}}$ **46.** $\sqrt[3]{\dfrac{25}{2}}\quad\dfrac{5}{\sqrt[3]{10}}$

47. $\dfrac{2-\sqrt{11}}{6}$ **48.** $\dfrac{\sqrt{15}+1}{2}$ **49.** $\dfrac{\sqrt[3]{3x^5}}{10}\quad\dfrac{3x^2}{10\sqrt[3]{9x}}$

50. $\sqrt[3]{\dfrac{9y}{7}}\quad\dfrac{3y}{\sqrt[3]{21y^2}}$ **51.** $\sqrt{\dfrac{18x^4y^6}{3z}}$ **52.** $\sqrt{\dfrac{8x^5y}{2z}}\quad\dfrac{2x^3y}{\sqrt{xyz}}$

53. $\dfrac{\sqrt{2}-1}{\sqrt{2}+1}\quad\dfrac{1}{3-2\sqrt{2}}$ **54.** $\dfrac{\sqrt{8}-\sqrt{3}}{\sqrt{2}+\sqrt{3}}\quad\dfrac{5}{7+3\sqrt{6}}$

55. $\dfrac{\sqrt{x}+1}{\sqrt{x}-1}\quad\dfrac{x-1}{x-2\sqrt{x}+1}$ **56.** $\dfrac{\sqrt{x}+\sqrt{y}}{\sqrt{x}-\sqrt{y}}$

57. When rationalizing the denominator of $\dfrac{\sqrt{5}}{\sqrt{7}}$, explain why both the numerator and the denominator must be multiplied by $\sqrt{7}$.

11. $-2(2+\sqrt{7})$ **27.** $\dfrac{x-\sqrt{xy}}{x-y}$ **35.** $\dfrac{5y}{\sqrt[3]{100xy}}$ **36.** $\dfrac{2x}{z\sqrt[3]{2x^2z}}$ **37.** $\dfrac{3}{10+5\sqrt{7}}$ **38.** $\dfrac{1}{\sqrt{10}-2\sqrt{2}}$ **39.** $\dfrac{x-9}{x-3\sqrt{x}}$

40. $\dfrac{23}{5\sqrt{2x}-2\sqrt{x}}$ **47.** $\dfrac{-7}{12+6\sqrt{11}}$ **48.** $\dfrac{7}{\sqrt{15}-1}$ **51.** $\dfrac{6x^2y^3}{\sqrt{6z}}$ **56.** $\dfrac{x-y}{x-2\sqrt{xy}+y}$

58. When rationalizing the numerator of $\frac{\sqrt{5}}{\sqrt{7}}$, explain why both the numerator and the denominator must be multiplied by $\sqrt{5}$.

59. The formula of the radius of a sphere r with surface area A is given by the formula

$$r = \sqrt{\frac{A}{4\pi}}$$

Rationalize the denominator of the radical expression in this formula.

$$r = \frac{\sqrt{A\pi}}{2\pi}$$

60. The formula for the radius of a cone r with height 7 centimeters and volume V is given by the formula

$$r = \sqrt{\frac{3V}{7\pi}}$$

Rationalize the numerator of the radical expression in this formula.

$$r = \frac{3V}{\sqrt{21\pi V}}$$

61. Explain why rationalizing the denominator does not change the value of the original expression.

62. Explain why rationalizing the numerator does not change the value of the original expression.

61. Answers will vary. **62.** Answers will vary.

Review Exercises   **65.** $\left\{-\frac{1}{2}, 6\right\}$ **66.** $\left\{-2, -\frac{4}{5}\right\}$

Solve each equation.

63. $2x - 7 = 3(x - 4)$ $\{5\}$ **64.** $9x - 4 = 7(x - 2)$ $\{-5\}$

65. $(x - 6)(2x + 1) = 0$ **66.** $(y + 2)(5y + 4) = 0$

67. $x^2 - 8x = -12$ $\{2, 6\}$ **68.** $x^3 = x$ $\{0, 1, -1\}$

TAPE IA 7.6

7.6   RADICAL EQUATIONS AND PROBLEM SOLVING

OBJECTIVES

1. Solve equations that contain radical expressions.
2. Use the Pythagorean theorem to model problems.

1 In this section, we present techniques to solve equations containing radical expressions such as

$$\sqrt{2x - 3} = 9$$

We use the power rule to help us solve these radical equations.

POWER RULE

If both sides of an equation are raised to the same power, **all** solutions of the original equation are **among** the solutions of the new equation.

This property *does not* say that raising both sides of an equation to a power yields an equivalent equation. A solution of the new equation *may or may not* be a solution of the original equation. Thus, *each solution of the new equation must be checked* to make sure it is a solution of the original equation.

EXAMPLE 1   Solve $\sqrt{2x - 3} = 9$ for x.

Solution: Use the power rule to square both sides of the equation to eliminate the radical:

$$\sqrt{2x - 3} = 9$$
$$(\sqrt{2x - 3})^2 = 9^2$$
$$2x - 3 = 81$$
$$2x = 84$$
$$x = 42$$

Now check the solution in the original equation.

$$\sqrt{2x - 3} = 9$$
$$\sqrt{2(42) - 3} = 9 \qquad \text{Let } x = 42.$$
$$\sqrt{84 - 3} = 9$$
$$\sqrt{81} = 9$$
$$9 = 9 \qquad \text{True.}$$

The solution checks, so we conclude that the solution set is {42}.

EXAMPLE 2   Solve $\sqrt{-10x - 1} + 3x = 0$ for x.

Solution: First, isolate the radical on one side of the equation. To do this, we subtract $3x$ from both sides.

$$\sqrt{-10x - 1} + 3x = 0$$
$$\sqrt{-10x - 1} + 3x - 3x = 0 - 3x$$
$$\sqrt{-10x - 1} = -3x$$

Next, use the power rule to eliminate the radical:

$$(\sqrt{-10x - 1})^2 = (-3x)^2$$
$$-10x - 1 = 9x^2$$

Since this is a quadratic equation, set the equation equal to 0 and try to solve by factoring.

$$9x^2 + 10x + 1 = 0$$
$$(9x + 1)(x + 1) = 0 \qquad \text{Factor.}$$
$$9x + 1 = 0 \quad \text{or} \quad x + 1 = 0 \qquad \text{Set each factor equal to 0.}$$
$$x = -\frac{1}{9} \quad \text{or} \quad x = -1$$

Possible solutions are $-\dfrac{1}{9}$ and -1. Now we check our work in the original equation.

$$\text{Let } x = -\frac{1}{9}$$

$$\sqrt{-10x - 1} + 3x = 0$$

$$\sqrt{-10\left(-\frac{1}{9}\right) - 1} + 3\left(-\frac{1}{9}\right) = 0$$

$$\sqrt{\frac{10}{9} - \frac{9}{9}} - \frac{3}{9} = 0$$

$$\sqrt{\frac{1}{9}} - \frac{1}{3} = 0$$

$$\frac{1}{3} - \frac{1}{3} = 0 \qquad \text{True.}$$

$$\text{Let } x = -1$$

$$\sqrt{-10x - 1} + 3x = 0$$

$$\sqrt{-10(-1) - 1} + 3(-1) = 0$$

$$\sqrt{10 - 1} - 3 = 0$$

$$\sqrt{9} - 3 = 0$$

$$3 - 3 = 0 \qquad \text{True.}$$

Both solutions check. The solution set is $\left\{-\dfrac{1}{9}, -1\right\}$.

TO SOLVE A RADICAL EQUATION

Step 1. Write the equation so that one radical with variables is by itself on one side of the equation.

Step 2. Raise each side of the equation to a power equal to the index of the radical.

Step 3. Simplify each side of the equation.

Step 4. If the equation still contains a radical term, repeat *steps 1 through 3*.

Step 5. Solve the equation.

Step 6. Check all proposed solutions in the original equation for extraneous solutions.

EXAMPLE 3 Solve for x: $\sqrt[3]{x + 1} + 5 = 3$.

Solution: First, we isolate the radical by subtracting 5 from both sides of the equation.

$$\sqrt[3]{x + 1} + 5 = 3$$

$$\sqrt[3]{x + 1} = -2$$

Next, raise both sides of the equation to the third power to eliminate the radical.

$$(\sqrt[3]{x + 1})^3 = (-2)^3$$
$$x + 1 = -8$$
$$x = -9$$

The solution checks in the original equation, so the solution set is $\{-9\}$. ▬▬▬

EXAMPLE 4   Solve $\sqrt{4 - x} = x - 2$ for x.

Solution:
$$\sqrt{4 - x} = x - 2$$
$$(\sqrt{4 - x})^2 = (x - 2)^2$$
$$4 - x = x^2 - 4x + 4 \qquad \text{Write the quadratic equation}$$
$$x^2 - 3x = 0 \qquad\qquad\quad \text{in standard form.}$$
$$x(x - 3) = 0 \qquad\qquad\quad \text{Factor.}$$
$$x = 0 \quad \text{or} \quad x - 3 = 0$$
$$x = 3$$

Check the possible solutions:

$$\sqrt{4 - x} = x - 2 \qquad\qquad\qquad \sqrt{4 - x} = x - 2$$
$$\sqrt{4 - 0} = 0 - 2 \quad \text{Let } x = 0. \qquad \sqrt{4 - 3} = 3 - 2 \quad \text{Let } x = 3.$$
$$2 = -2 \quad\quad \text{False.} \qquad\qquad\quad 1 = 1 \quad\quad \text{True.}$$

The proposed solution 3 checks, but 0 does not. When a proposed solution does not check, it is an **extraneous root or solution.** Since 0 is an extraneous solution, the solution set is $\{3\}$. ▬▬▬

> **REMINDER**   In Example 4, notice that $(x - 2)^2 = x^2 - 4x + 4$. Make sure binomials are squared correctly.

EXAMPLE 5   Solve $\sqrt{2x + 5} + \sqrt{2x} = 3$.

Solution: Isolate a radical by subtracting $\sqrt{2x}$ from both sides:

$$\sqrt{2x + 5} + \sqrt{2x} = 3$$
$$\sqrt{2x + 5} = 3 - \sqrt{2x}$$

Use the power rule to begin eliminating the radicals. Square both sides:

$$(\sqrt{2x + 5})^2 = (3 - \sqrt{2x})^2$$
$$2x + 5 = 9 - 6\sqrt{2x} + 2x \qquad \text{Multiply: } (3 - \sqrt{2x})(3 - \sqrt{2x}).$$

There is still a radical in the equation, so isolate the radical again. Then square both sides:

$$2x + 5 = 9 - 6\sqrt{2x} + 2x$$

$$6\sqrt{2x} = 4 \qquad \text{Isolate the radical.}$$

$$(6\sqrt{2x})^2 = 4^2 \qquad \text{Square both sides of the equation}$$
$$\text{to eliminate the radical.}$$

$$36(2x) = 16$$

$$72x = 16 \qquad \text{Multiply.}$$

$$x = \frac{16}{72} \qquad \text{Solve.}$$

$$x = \frac{2}{9} \qquad \text{Simplify.}$$

The proposed solution, $\frac{2}{9}$, does check in the original equation, so the solution set is $\left\{\frac{2}{9}\right\}$.

> **R E M I N D E R**   Make sure expressions are squared correctly. In Example 5, we squared $(3 - \sqrt{2x})$:
>
> $$(3 - \sqrt{2x})^2 = (3 - \sqrt{2x})(3 - \sqrt{2x})$$
> $$= 3 \cdot 3 - 3\sqrt{2x} - 3\sqrt{2x} + \sqrt{2x} \cdot \sqrt{2x}$$
> $$= 9 - 6\sqrt{2x} + 2x$$

2   Recall that the Pythagorean theorem states that in a right triangle, the length of the hypotenuse squared equals the sum of the lengths of each of the legs squared.

> **PYTHAGOREAN THEOREM**
>
> If a and b are the lengths of the legs of a right triangle and c is the length of the hypotenuse, then $a^2 + b^2 = c^2$.
>
>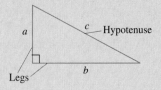

EXAMPLE 6   Find the length of the unknown leg of the following right triangle.

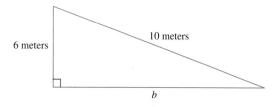

Solution: In the formula $a^2 + b^2 = c^2$, c is the hypotenuse. Let $c = 10$, the length of the hypotenuse, let $a = 6$, and solve for b. Then $a^2 + b^2 = c^2$ becomes

$$6^2 + b^2 = 10^2$$
$$36 + b^2 = 100$$
$$b^2 = 64 \qquad \text{Subtract 36 from both sides.}$$
$$b = 8 \quad \text{and} \quad b = -8 \qquad \text{Because } b^2 = 64.$$

Since we are solving for a length, we will list the positive solution only. The unknown leg of the triangle is 8 meters long.

EXAMPLE 7   A 50-foot supporting wire is to be attached to a 75-foot antenna. Because of surrounding buildings, sidewalks, and roadways, the wire must be anchored exactly 20 feet from the base of the antenna.

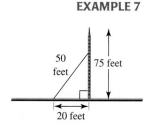

a. How high from the base of the antenna is the wire attached?

b. Local regulations require that a supporting wire be attached at a height no less than $\dfrac{3}{5}$ of the total height of the antenna. From part (a), have local regulations been met?

Solution: **1.** UNDERSTAND. Read and reread the problem. From the diagram, notice that a right triangle is formed with hypotenuse 50 feet and one leg 20 feet.

2. ASSIGN. Let x = the height from the base of the antenna to the attached wire.

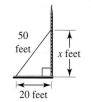

3. ILLUSTRATE.

4. TRANSLATE. Use the Pythagorean theorem.

$$(\text{leg})^2 + (\text{leg})^2 = (\text{hypotenuse})^2$$
$$(20)^2 + \quad x^2 \quad = (50)^2$$

5. COMPLETE.
$$(20)^2 + x^2 = (50)^2$$
$$400 + x^2 = 2500$$
$$x^2 = 2100 \qquad \text{Subtract 400 from both sides.}$$
$$x = \sqrt{2100}$$

6. INTERPRET. *Check* the work and *state* the solution.
   a. The wire is attached exactly $\sqrt{2100}$ feet from the base of the pole, or approximately 45.8 feet.
   b. The supporting wire must be attached at a height no less than $\frac{3}{5}$ of the total height of the antenna. This height is $\frac{3}{5}$(75 feet), or 45 feet. Since we know from part (a) that the wire is to be attached at a height of approximately 45.8 feet, local regulations have been met.

GRAPHING CALCULATOR EXPLORATIONS

We can use a grapher to solve equations such as radical equations. For example, to use a grapher to approximate the solutions of the equation solved in Example 4, we graph the following:

$$Y_1 = \sqrt{4 - x} \quad \text{and} \quad Y_2 = x - 2$$

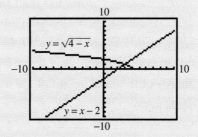

The x-value of the point of intersection is the solution. Use the Intersect feature or the Zoom and Trace features of your grapher to see that the solution is 3.

Use a grapher to solve each radical equation. Round all solutions to the nearest hundredth.

1. $\sqrt{x + 7} = x$ {3.19}

2. $\sqrt{3x + 5} = 2x$ {1.55}

3. $\sqrt{2x + 1} = \sqrt{2x + 2}$ { }

4. $\sqrt{10x - 1} = \sqrt{-10x + 10} - 1$ {0.34}

5. $1.2x = \sqrt{3.1x + 5}$ {3.23}

6. $\sqrt{1.9x^2 - 2.2} = -0.8x + 3$ {−5.44, 1.63}

EXERCISE SET 7.6

4. {24}   **11.** $\left\{-\dfrac{9}{2}\right\}$ **17.** {−4}   **33.** $\left\{\dfrac{15}{4}\right\}$ **39.** {−12}   **49.** $\left\{\dfrac{1}{2}\right\}$

Solve. See Examples 1 and 2.

1. $\sqrt{2x} = 4$ {8}   **2.** $\sqrt{3x} = 3$ {3}   **3.** $\sqrt{x-3} = 2$ {7}

4. $\sqrt{x+1} = 5$ **5.** $\sqrt{2x} = -4$ { }   **6.** $\sqrt{5x} = -5$ { }

7. $\sqrt{4x-3} - 5 = 0$ {7}   **8.** $\sqrt{x-3} - 1 = 0$ {4}

9. $\sqrt{2x-3} - 2 = 1$ {6}   **10.** $\sqrt{3x+3} - 4 = 8$ {47}

Solve. See Example 3.

11. $\sqrt[3]{6x} = -3$ **12.** $\sqrt[3]{4x} = -2$ {−2}

13. $\sqrt[3]{x-2} - 3 = 0$ {29}   **14.** $\sqrt[3]{2x-6} - 4 = 0$ {35}

Solve. See Examples 4 and 5.

15. $\sqrt{13-x} = x - 1$ {4}   **16.** $\sqrt{2x-3} = 3 - x$ {2}

17. $x - \sqrt{4-3x} = -8$ **18.** $2x + \sqrt{x+1} = 8$ {3}

19. $\sqrt{y+5} = 2 - \sqrt{y-4}$ { }

20. $\sqrt{x+3} + \sqrt{x-5} = 3$ $\left\{\dfrac{181}{36}\right\}$

21. $\sqrt{x-3} + \sqrt{x+2} = 5$ {7}

22. $\sqrt{2x-4} - \sqrt{3x+4} = -2$ {4, 20}

Find the length of the unknown side in each triangle. See Example 6.

23.

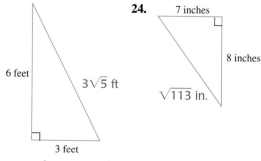

6 feet $3\sqrt{5}$ ft   3 feet

24.

7 inches   8 inches $\sqrt{113}$ in.

25.

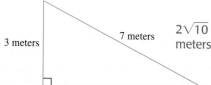

3 meters   7 meters $2\sqrt{10}$ meters

26.

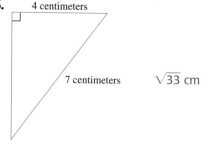

4 centimeters   7 centimeters $\sqrt{33}$ cm

Solve.

27. $\sqrt{3x-2} = 5$ {9}   **28.** $\sqrt{5x-4} = 9$ {17}

29. $-\sqrt{2x} + 4 = -6$ {50}   **30.** $-\sqrt{3x+9} = -12$ {45}

31. $\sqrt{3x+1} + 2 = 0$ { }   **32.** $\sqrt{3x+1} - 2 = 0$ {1}

33. $\sqrt[4]{4x+1} - 2 = 0$ **34.** $\sqrt[4]{2x-9} - 3 = 0$ {45}

35. $\sqrt{4x-3} = 5$ {7}   **36.** $\sqrt{3x+9} = 12$ {45}

37. $\sqrt[3]{6x-3} - 3 = 0$ {5}   **38.** $\sqrt[3]{3x} + 4 = 7$ {9}

39. $\sqrt[3]{2x-3} - 2 = -5$ **40.** $\sqrt[3]{x-4} - 5 = -7$ {−4}

41. $\sqrt{x+4} = \sqrt{2x-5}$ {9}   **42.** $\sqrt{3y+6} = \sqrt{7y-6}$ {3}

43. $x - \sqrt{1-x} = -5$ {−3}   **44.** $x - \sqrt{x-2} = 4$ {6}

45. $\sqrt[3]{-6x-1} = \sqrt[3]{-2x-5}$ {1}

46. $x + \sqrt{x+5} = 7$ {4}

47. $\sqrt{5x-1} - \sqrt{x} + 2 = 3$ {1}

48. $\sqrt{2x-1} - 4 = -\sqrt{x-4}$ {5}

49. $\sqrt{2x-1} = \sqrt{1-2x}$

50. $\sqrt{3x+4} - 1 = \sqrt{2x+1}$ {0, 4}

51. Explain why proposed solutions of radical equations must be checked. Answers will vary.

52. Consider the equations $\sqrt{2x} = 4$ and $\sqrt[3]{2x} = 4$.

    a. Explain the difference in solving these equations.

    b. Explain the similarity in solving these equations. **52a, b.** Answers will vary.

Find the length of the unknown side of each triangle. Give the exact length and a one-decimal-place approximation.

53.

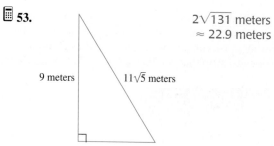

9 meters $11\sqrt{5}$ meters $2\sqrt{131}$ meters ≈ 22.9 meters

54.

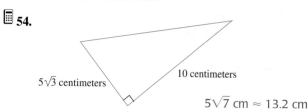

$5\sqrt{3}$ centimeters   10 centimeters $5\sqrt{7}$ cm ≈ 13.2 cm

55.

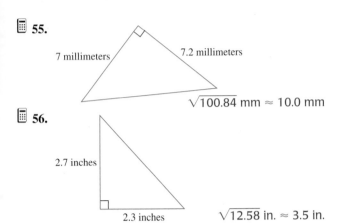

7 millimeters 7.2 millimeters

$\sqrt{100.84}$ mm ≈ 10.0 mm

56.

2.7 inches

2.3 inches $\sqrt{12.58}$ in. ≈ 3.5 in.

Solve. See Example 7. Give exact answers and two-decimal-place approximations where appropriate.

57. A wire is needed to support a vertical pole 15 feet high. The cable will be anchored to a stake 8 feet from the base of the pole. How much cable is needed?   17 ft

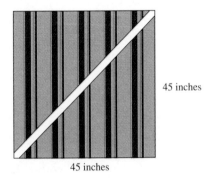

15 feet

8 feet

58. A furniture upholsterer wished to cut a strip from a piece of fabric that is 45 inches by 45 inches. The strip must be cut on the bias of the fabric. What is the longest strip that can be cut? $45\sqrt{2}$ in. ≈ 63.64 in.

45 inches

45 inches

59. A spotlight is mounted on the eaves of a house 12 feet above the ground. A flower bed runs between the house and the sidewalk, so the closest the ladder can be placed to the house is 5 feet. How long a ladder is needed so that an electrician can reach the place where the light is mounted?   13 ft

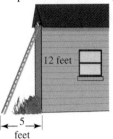

12 feet

5 feet

60. A wire is to be attached to support a telephone pole. Because of surrounding buildings, sidewalks, and roadway, the wire must be anchored exactly 15 feet from the base of the pole. Telephone company workers have only 30 feet of cable, and 2 feet of that must be used to attach the cable to the pole and to the stake on the ground. How high from the base of the pole can the wire be attached?

$\sqrt{559}$ ft ≈ 23.64 ft

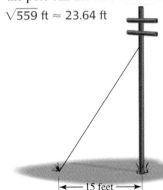

15 feet

61. The tallest structure in the United States is a TV tower in Blanchard, North Dakota. Its height is 2063 feet. A 2382-foot length of wire is to be used as a guy wire attached to the top of the tower. Approximate to the nearest foot how far from the base of the tower the guy wire must be anchored. (*Source:* U.S. Geological Survey)   1191 ft

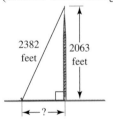

2382 feet   2063 feet

?

62. The radius of the Moon is 1080 miles. Use the formula for the radius r of a sphere given its surface area A,

$$r = \sqrt{\frac{A}{4\pi}}$$

to find the surface area of the Moon. Round to the nearest square mile. 14,657,415 square miles

63. The cost $C(x)$ in dollars per day to operate a small delivery service is given by $C(x) = 80\sqrt[3]{x} + 500$, where x is the number of deliveries per day. In July, the manager decides that it is necessary to keep delivery costs below $1620.00. Find the greatest number of deliveries this company can make per day and still keep overhead below $1620.00. 2743 deliveries

64. The formula $v = \sqrt{2gh}$ relates the velocity v, in feet per second, of an object after it falls h feet accelerated by gravity g, in feet per second squared. If g is approximately 32 feet per second squared, find how far an object has fallen if its velocity is 80 feet per second. 100 ft

65. Two tractors are pulling a tree stump from a field. If two forces A and B pull at right angles ($90°$) to each other, the size of the resulting force R is given by the formula

$$R = \sqrt{A^2 + B^2}$$

If tractor A is exerting 600 pounds of force and the resulting force is 850 pounds, find how much force tractor B is exerting. $50\sqrt{145}$ lb ≈ 602.08 lb

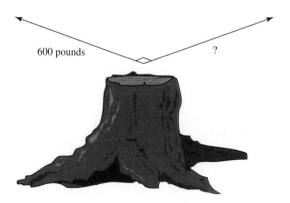

600 pounds   ?

66. Police departments find it very useful to be able to approximate the speed of a car when they are given the distance that the car skidded before it came to a stop. If the road surface is wet concrete, the function $S(x) = \sqrt{10.5x}$ is used, where $S(x)$ is the speed of the car in miles per hour and x is the distance skidded in feet. Find how fast a car was

moving if it skidded 280 feet on wet concrete. (See the accompanying photograph.) $14\sqrt{5}$ mph ≈ 54.22 mph

67. The maximum distance $D(h)$ that a person can see from a height h kilometers above the ground is given by the function $D(h) = 111.7\sqrt{h}$. Find the height that would allow us to see 80 kilometers. 0.51 km

Review Exercises

Use the vertical line test to determine whether each graph represents the graph of a function. See Section 3.2.

68.

function

69.

not a function

70.

function

71.

not a function

72.

not a function

73.

not a function

Simplify. See Section 6.4.

74. $\dfrac{\dfrac{x}{6}}{\dfrac{2x}{3}+\dfrac{1}{2}}$ $\dfrac{x}{4x+3}$

75. $\dfrac{\dfrac{1}{y}+\dfrac{4}{5}}{\dfrac{-3}{20}}$ $\dfrac{20+16y}{3y}$

76. $\dfrac{\dfrac{z}{5}+\dfrac{1}{10}}{\dfrac{z}{20}-\dfrac{z}{5}}$ $\dfrac{4z+2}{3z}$

77. $\dfrac{\dfrac{1}{y}+\dfrac{1}{x}}{\dfrac{1}{y}-\dfrac{1}{x}}$ $\dfrac{x+y}{x-y}$

A Look Ahead

EXAMPLE

Solve $(t^2 - 3t) - 2\sqrt{t^2 - 3t} = 0$.

Solution:

Substitution can be used to make this problem somewhat simpler. Since $t^2 - 3t$ occurs more than once, let $x = t^2 - 3t$.

$$(t^2 - 3t) - 2\sqrt{t^2 - 3t} = 0$$
$$x - 2\sqrt{x} = 0 \qquad \text{Let } x = t^2 - 3t.$$
$$x = 2\sqrt{x}$$
$$x^2 = (2\sqrt{x})^2$$
$$x^2 = 4x$$
$$x^2 - 4x = 0$$
$$x(x - 4) = 0$$
$$x = 0 \quad \text{or} \quad x - 4 = 0$$
$$x = 4$$

Now we "undo" the substitution.

$x = 0$ Replace x with $t^2 - 3t$.
$$t^2 - 3t = 0$$
$$t(t - 3) = 0$$
$$t = 0 \quad \text{or} \quad t - 3 = 0$$
$$t = 3$$

$x = 4$ Replace x with $t^2 - 3t$.
$$t^2 - 3t = 4$$
$$t^2 - 3t - 4 = 0$$
$$(t - 4)(t + 1) = 0$$
$$t - 4 = 0 \quad \text{or} \quad t + 1 = 0$$
$$t = 4 \quad \text{or} \qquad t = -1$$

In this problem, we have four possible solutions: 0, 3, 4, and -1. All four solutions check in the original equation, so the solution set is $\{-1, 0, 3, 4\}$.

Solve. See the preceding example.

78. $3\sqrt{x^2 - 8x} = x^2 - 8x$ $\{-1, 0, 8, 9\}$

79. $\sqrt{(x^2 - x) + 7} = 2(x^2 - x) - 1$ $\{-1, 2\}$

80. $7 - (x^2 - 3x) = \sqrt{(x^2 - 3x) + 5}$ $\{-1, 4\}$

81. $x^2 + 6x = 4\sqrt{x^2 + 6x}$ $\{-8, -6, 0, 2\}$

7.7 | COMPLEX NUMBERS

O B J E C T I V E S

1   Define imaginary and complex numbers.
2   Add or subtract complex numbers.
3   Multiply complex numbers.
4   Divide complex numbers.
5   Raise i to powers.

TAPE IA 7.7

1   Our work with radical expressions has excluded expressions such as $\sqrt{-16}$ because $\sqrt{-16}$ is not a real number; there is no real number whose square is -16. In this section, we discuss a number system that includes roots of negative numbers. This number system is the **complex number system,** and it includes the set of real

numbers as a subset. The complex number system allows us to solve equations such as $x^2 + 1 = 0$ that have no real number solutions. The set of complex numbers includes the **imaginary unit.**

IMAGINARY UNIT

The imaginary unit, written i, is the number whose square is -1. That is,

$$i^2 = -1 \quad \text{and} \quad i = \sqrt{-1}$$

To write the square root of a negative number in terms of i, use the property that if a is a positive number, then

$$\sqrt{-a} = \sqrt{-1} \cdot \sqrt{a}$$
$$= i \cdot \sqrt{a}$$

Using i, we can write $\sqrt{-16}$ as

$$\sqrt{-16} = \sqrt{-1 \cdot 16} = \sqrt{-1} \cdot \sqrt{16} = i \cdot 4, \text{ or } 4i$$

EXAMPLE 1   Write with i notation.

a. $\sqrt{-36}$

b. $\sqrt{-5}$

Solution: **a.** $\sqrt{-36} = \sqrt{-1 \cdot 36} = \sqrt{-1} \cdot \sqrt{36} = i \cdot 6$, or $6i$

b. $\sqrt{-5} = \sqrt{-1(5)} = \sqrt{-1} \cdot \sqrt{5} = i\sqrt{5}$. Since $\sqrt{5}i$ can easily be confused with $\sqrt{5}i$, we write $\sqrt{5}i$ as $i\sqrt{5}$.

Now that we have practiced working with the imaginary unit, complex numbers are defined.

COMPLEX NUMBERS

A complex number is a number that can be written in the form $a + bi$, where a and b are real numbers.

The number a is the **real part** of $a + bi$, and the number b is the **imaginary part** of $a + bi$. Notice that the set of real numbers is a subset of the complex numbers since any real number can be written in the form of a complex number. For example,

$$16 = 16 + 0i$$

In general, a complex number $a + bi$ is a real number if $b = 0$. Also, a complex number is called a **pure imaginary number** if $a = 0$. For example,

$$3i = 0 + 3i \quad \text{and} \quad i\sqrt{7} = 0 + i\sqrt{7}$$

are pure imaginary numbers.

The following diagram shows the relationship between complex numbers and their subsets.

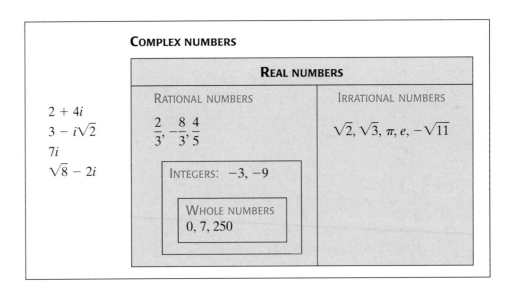

Two complex numbers $a + bi$ and $c + di$ are equal if and only if $a = c$ and $b = d$. Complex numbers can be added or subtracted by adding or subtracting their real parts and then adding or subtracting their imaginary parts.

SUM OR DIFFERENCE OF COMPLEX NUMBERS

If $a + bi$ and $c + di$ are complex numbers, then their sum is

$$(a + bi) + (c + di) = (a + c) + (b + d)i$$

Their difference is

$$(a + bi) - (c + di) = a + bi - c - di = (a - c) + (b - d)i$$

EXAMPLE 2   Add or subtract the complex numbers. Write the sum or difference in the form $a + bi$.

   a. $(2 + 3i) + (-3 + 2i)$ **b.** $(5i) - (1 - i)$ **c.** $(-3 - 7i) - (-6)$

Solution: **a.** $(2 + 3i) + (-3 + 2i) = (2 - 3) + (3 + 2)i = -1 + 5i$

b. $5i - (1 - i) = 5i - 1 + i$
$$= -1 + (5 + 1)i$$
$$= -1 + 6i$$

c. $(-3 - 7i) - (-6) = -3 - 7i + 6$
$$= (-3 + 6) - 7i$$
$$= 3 - 7i$$

3

We will use the relationship $i^2 = -1$ to simplify when we multiply two complex numbers.

EXAMPLE 3   Multiply the complex numbers. Write the product in the form $a + bi$.

a. $(2 - 5i)(4 + i)$ **b.** $(2 - i)^2$ **c.** $(7 + 3i)(7 - 3i)$

Solution: Multiply complex numbers as though they were binomials.

a. $(2 - 5i)(4 + i) = 2(4) + 2(i) - 5i(4) - 5i(i)$
$$= 8 + 2i - 20i - 5i^2$$
$$= 8 - 18i - 5(-1) \qquad \text{Replace } i^2 = -1.$$
$$= 8 - 18i + 5$$
$$= 13 - 18i$$

b. $(2 - i)^2 = (2 - i)(2 - i)$
$$= 2(2) - 2(i) - 2(i) + i^2$$
$$= 4 - 4i + (-1) \qquad \text{Replace } i^2 = -1.$$
$$= 3 - 4i$$

c. $(7 + 3i)(7 - 3i) = 7(7) - 7(3i) + 3i(7) - 3i(3i)$
$$= 49 - 21i + 21i - 9i^2$$
$$= 49 - 9(-1) \qquad \text{Replace } i^2 = -1.$$
$$= 49 + 9$$
$$= 58$$

Notice that if you add, subtract, or multiply two complex numbers, just like real numbers, the result is a complex number.

4

From Example 3c, notice that the product of $7 + 3i$ and $7 - 3i$ is a real number. These two complex numbers are called **complex conjugates** of one another. In general, we have the following definition.

COMPLEX CONJUGATES

The complex numbers $(a + bi)$ and $(a - bi)$ are called **complex conjugates** of each other, and $(a + bi)(a - bi) = a^2 + b^2$.

To see that the product of a complex number $a + bi$ and its conjugate $a - bi$ is the real number $a^2 + b^2$, we multiply:

$$(a + bi)(a - bi) = a^2 - abi + abi - b^2i^2$$
$$= a^2 - b^2(-1)$$
$$= a^2 + b^2$$

We use complex conjugates to divide by a complex number.

EXAMPLE 4   Find each quotient. Write in the form $a + bi$.

a. $\dfrac{2 + i}{1 - i}$

b. $\dfrac{7}{3i}$

Solution: **a.** Multiply the numerator and denominator by the complex conjugate of $1 - i$ to eliminate the imaginary number in the denominator.

$$\frac{2 + i}{1 - i} = \frac{(2 + i)(1 + i)}{(1 - i)(1 + i)}$$

$$= \frac{2(1) + 2(i) + 1(i) + i^2}{1^2 - i^2}$$

$$= \frac{2 + 3i - 1}{1 + 1}$$

$$= \frac{1 + 3i}{2}, \quad \text{or} \quad \frac{1}{2} + \frac{3}{2}i$$

To check that $\dfrac{2 + i}{1 - i} = \dfrac{1}{2} + \dfrac{3}{2}i$, multiply $\left(\dfrac{1}{2} + \dfrac{3}{2}i\right)(1 - i)$ to verify that the product is $2 + i$.

b. Multiply the numerator and denominator by the conjugate of $3i$. Note that $3i = 0 + 3i$, so its conjugate is $0 - 3i$ or $-3i$.

$$\frac{7}{3i} = \frac{7(-3i)}{(3i)(-3i)} = \frac{-21i}{-9i^2} = \frac{-21i}{-9(-1)} = \frac{-21i}{9} = \frac{-7i}{3}, \quad \text{or} \quad 0 - \frac{7}{3}i$$

The product rule for radicals does not necessarily hold true for imaginary numbers. *To multiply imaginary numbers, first write each number in terms of the imaginary unit i.* For example, to multiply $\sqrt{-4}$ and $\sqrt{-9}$, first write each number in the form bi:

$$\sqrt{-4}\,\sqrt{-9} = 2i(3i) = 6i^2 = -6$$

EXAMPLE 5   Multiply or divide the following as indicated.

a. $\sqrt{-3}\,\sqrt{-5}$ b. $\sqrt{-36}\,\sqrt{-1}$ c. $\sqrt{8}\,\sqrt{-2}$ d. $\dfrac{\sqrt{-125}}{\sqrt{5}}$

Solution: First, write each imaginary number in the form bi.

a. $\sqrt{-3}\sqrt{-5} = i\sqrt{3}\,(i\sqrt{5}) = i^2\sqrt{15} = -\sqrt{15}$

b. $\sqrt{-36}\sqrt{-1} = 6i(i) = 6i^2 = 6(-1) = -6$

c. $\sqrt{8}\sqrt{-2} = 2\sqrt{2}\,(i\sqrt{2}) = 2i(\sqrt{2}\sqrt{2}) = 2i(2) = 4i$

d. $\dfrac{\sqrt{-125}}{\sqrt{5}} = \dfrac{i\sqrt{125}}{\sqrt{5}} = i\sqrt{25} = 5i$

5

We can use the fact that $i^2 = -1$ to find higher powers of i. To find i^3, we rewrite it as the product of i^2 and i:

$$i^3 = i^2 \cdot i = (-1)i = -i$$

Continue this process and find higher powers of i:

$$i^4 = i^2 \cdot i^2 = (-1)(-1) = 1$$
$$i^5 = i^4 \cdot i = 1 \cdot i = i$$

If we continue finding powers of i, we generate a pattern:

$i^1 = i$	$i^5 = i$	$i^9 = i$
$i^2 = -1$	$i^6 = -1$	$i^{10} = -1$
$i^3 = -i$	$i^7 = -i$	$i^{11} = -i$
$i^4 = 1$	$i^8 = 1$	$i^{12} = 1$

The values i, -1, $-i$, and 1 repeat as i is raised to higher and higher powers. This pattern allows us to find other powers of i.

EXAMPLE 6   Find the following powers of i.

a. i^7 **b.** i^{20} **c.** i^{46} **d.** i^{-12}

Solution: **a.** $i^7 = i^4 \cdot i^3 = 1(-i) = -i$ **b.** $i^{20} = (i^4)^5 = 1^5 = 1$

c. $i^{46} = i^{44} \cdot i^2 = (i^4)^{11} \cdot i^2 = 1^{11}(-1) = -1$

d. $i^{-12} = \dfrac{1}{i^{12}} = \dfrac{1}{(i^4)^3} = \dfrac{1}{(1)^3} = \dfrac{1}{1} = 1$

MENTAL MATH

Simplify. See Example 1.

1. $\sqrt{-81}$ $9i$ **2.** $\sqrt{-49}$ $7i$ **3.** $\sqrt{-7}$ $i\sqrt{7}$ **4.** $\sqrt{-3}$ $i\sqrt{3}$

5. $-\sqrt{16}$ -4 **6.** $-\sqrt{4}$ -2 **7.** $\sqrt{-64}$ $8i$ **8.** $\sqrt{-100}$ $10i$

EXERCISE SET 7.7

Write in terms of i. See Example 1.

1. $\sqrt{-24}$ $2i\sqrt{6}$ **2.** $\sqrt{-32}$ $4i\sqrt{2}$ **3.** $-\sqrt{-36}$ $-6i$

4. $-\sqrt{-121}$ $-11i$ **5.** $8\sqrt{-63}$ $24i\sqrt{7}$ **6.** $4\sqrt{-20}$ $8i\sqrt{5}$

7. $-\sqrt{54}$ $-3\sqrt{6}$ **8.** $\sqrt{-63}$ $3i\sqrt{7}$

Add or subtract. Write the sum or difference in the form a + bi. See Example 2.

9. $(4 - 7i) + (2 + 3i)$ **10.** $(2 - 4i) - (2 - i)$ $-3i$

11. $(6 + 5i) - (8 - i)$ **12.** $(8 - 3i) + (-8 + 3i)$ 0

13. $6 - (8 + 4i)$ $-2 - 4i$ **14.** $(9 - 4i) - 9$ $-4i$

Multiply. Write the product in the form a + bi. See Example 3.

15. $6i(2 - 3i)$ $18 + 12i$ **16.** $5i(4 - 7i)$ $35 + 20i$

17. $(\sqrt{3} + 2i)(\sqrt{3} - 2i)$ 7 **18.** $(\sqrt{5} - 5i)(\sqrt{5} + 5i)$ 30

19. $(4 - 2i)^2$ $12 - 16i$ **20.** $(6 - 3i)^2$ $27 - 36i$

Write each quotient in the form a + bi. See Example 4.

21. $\dfrac{4}{i}$ $-4i$ **22.** $\dfrac{5}{6i}$ $-\dfrac{5i}{6}$ **23.** $\dfrac{7}{4 + 3i}$ $\dfrac{28}{25} - \dfrac{21}{25}i$

24. $\dfrac{9}{1 - 2i}$ **25.** $\dfrac{3 + 5i}{1 + i}$ $4 + i$ **26.** $\dfrac{6 + 2i}{4 - 3i}$ $\dfrac{18}{25} + \dfrac{26}{25}i$

27. Describe how to find the conjugate of a complex number. Answers will vary.

28. Explain why the product of a complex number and its complex conjugate is a real number.
Answers will vary.

Multiply or divide. See Example 5.

29. $\sqrt{-2} \cdot \sqrt{-7}$ $-\sqrt{14}$ **30.** $\sqrt{-11} \cdot \sqrt{-3}$ $-\sqrt{33}$

31. $\sqrt{-5} \cdot \sqrt{-10}$ $-5\sqrt{2}$ **32.** $\sqrt{-2} \cdot \sqrt{-6}$ $-2\sqrt{3}$

33. $\sqrt{16} \cdot \sqrt{-1}$ $4i$ **34.** $\sqrt{3} \cdot \sqrt{-27}$ $9i$

35. $\dfrac{\sqrt{-9}}{\sqrt{3}}$ $i\sqrt{3}$ **36.** $\dfrac{\sqrt{49}}{\sqrt{-10}}$ $-\dfrac{7i\sqrt{10}}{10}$

37. $\dfrac{\sqrt{-80}}{\sqrt{-10}}$ $2\sqrt{2}$ **38.** $\dfrac{\sqrt{-40}}{\sqrt{-8}}$ $\sqrt{5}$

Find each power of i. See Example 6.

39. i^8 1 **40.** i^{10} -1 **41.** i^{21} i **42.** i^{15} $-i$

43. i^{11} $-i$ **44.** i^{40} 1 **45.** i^{-6} -1 **46.** i^{-9} $-i$

9. $6 - 4i$ **11.** $-2 + 6i$ **24.** $\dfrac{9}{5} + \dfrac{18}{5}i$ **49.** $2 - i$ **50.** $-8 + 4i$ **71.** $x^2 - 5x - 2 - \dfrac{6}{x - 1}$

72. $5x^3 - 10x^2 + 17x - 34 + \dfrac{70}{x + 2}$

Perform the indicated operation. Write the result in the form a + bi.

47. $(7i)(-9i)$ 63 **48.** $(-6i)(-4i)$ -24

49. $(6 - 3i) - (4 - 2i)$ **50.** $(-2 - 4i) - (6 - 8i)$

51. $(6 - 2i)(3 + i)$ 20 **52.** $(2 - 4i)(2 - i)$ $-10i$

53. $(8 - \sqrt{-3}) - (2 + \sqrt{-12})$ $6 - 3i\sqrt{3}$

54. $(8 - \sqrt{-4}) - (2 + \sqrt{-16})$ $6 - 6i$

55. $(1 - i)(1 + i)$ 2 **56.** $(6 + 2i)(6 - 2i)$ 40

57. $\dfrac{16 + 15i}{-3i}$ $-5 + \dfrac{16}{3}i$ **58.** $\dfrac{2 - 3i}{-7i}$ $\dfrac{3}{7} + \dfrac{2}{7}i$

59. $(9 + 8i)^2$ $17 + 144i$ **60.** $(4 - 7i)^2$ $-33 - 56i$

61. $\dfrac{2}{3 + i}$ $\dfrac{3}{5} - \dfrac{1}{5}i$ **62.** $\dfrac{5}{3 - 2i}$ $\dfrac{15}{13} + \dfrac{10}{13}i$

63. $(5 - 6i) - 4i$ $5 - 10i$ **64.** $(6 - 2i) + 7i$ $6 + 5i$

65. $\dfrac{2 - 3i}{2 + i}$ $\dfrac{1}{5} - \dfrac{8}{5}i$ **66.** $\dfrac{6 + 5i}{6 - 5i}$ $\dfrac{11}{61} + \dfrac{60}{61}i$

67. $(2 + 4i) + (6 - 5i)$ **68.** $(5 - 3i) + (7 - 8i)$
$8 - i$ $12 - 11i$

Review Exercises

Recall that the sum of the measures of the angles of a triangle is 180°. Find the unknown angle in each triangle. See Section 4.3.

69. 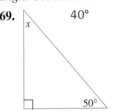 **70.**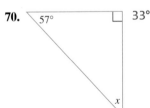

Use synthetic division to divide the following. See Section 6.6.

71. $(x^3 - 6x^2 + 3x - 4) \div (x - 1)$

72. $(5x^4 - 3x^2 + 2) \div (x + 2)$

Thirty people were recently polled about their average monthly balance in their checking account. The results of this poll are shown in the following histogram. Use this graph to answer Exercises 73 through 78. See Section 1.1.

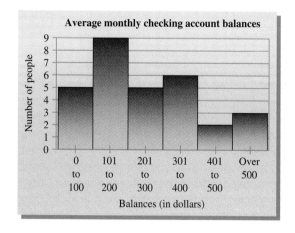

Average monthly checking account balances

Number of people (vertical axis): 0 to 9

Balances (in dollars): 0 to 100, 101 to 200, 201 to 300, 301 to 400, 401 to 500, Over 500

73. How many people polled reported an average checking balance of $201 to $300?   5

74. How many people polled reported an average checking balance of 0 to $100?   5

75. How many people polled reported an average checking balance of $200 or less?   14

76. How many people polled reported an average checking balance of $301 or more?   11

77. What percent of people polled reported an average checking balance of $201 to $300?   16.7%

78. What percent of people polled reported an average checking balance of 0 to $100?   16.7%

GROUP ACTIVITY

CALCULATING THE LENGTH AND PERIOD OF A PENDULUM

MATERIALS:
- string (at least 1 meter long), weight, meter stick, stopwatch, calculator

Make a simple pendulum by securely tying the string to a weight.

The formula relating a pendulum's period T (in seconds) to its length l (in centimeters) is

$$T = 2\pi\sqrt{\frac{l}{980}}$$

The **period** of a pendulum is defined as the time it takes the pendulum to complete one full back-and-forth swing. In this activity, you will be measuring your simple pendulum's period with a stopwatch. Because the periods will be only a few seconds long, it will be more accurate for you to time a total of five complete swings and then find the average time of one complete swing.

(continued)

1. For each of the pendulum (string) lengths given in Table 1, measure the time required for 5 complete swings and record it in the appropriate column. Next, divide this value by 5 to find the measured period of the pendulum for the given length and record it in the Measured Period T_m column in the table. Use the given formula to calculate the theoretical period T for the same pendulum length and record it in the appropriate column. (Round to two decimal places.) Find and record in the last column the difference between the measured period and the theoretical period.

2. For each of the periods T given in Table 2, use the given formula and calculate the theoretical pendulum length l required to yield the given period. Record l in the appropriate column; round to one decimal place. Next, use this length l and measure and record the time for 5 complete swings. Divide this value by 5 to find the measured period T_m, and record it. Then find and record in the last column the difference between the theoretical period and the measured period.

3. Use the general trends you find in the tables to describe the relationship between a pendulum's period and its length.

4. Discuss the differences you found between the values of the theoretical period and the measured period. What factors contributed to these differences?

Table 1

Length l (centimeters)	Time for 5 Swings (seconds)	Measured Period T_m (seconds)	Theoretical Period T (seconds)	Difference $\lvert T - T_m \rvert$
30				
55				
70				

Table 2

Period T (seconds)	Theoretical Length l (centimeters)	Time for 5 Swings (seconds)	Measured Period T_m (seconds)	Difference $\lvert T - T_m \rvert$
1				
1.25				
2				

See Appendix D for Group Activity answers and suggestions.

CHAPTER 7   HIGHLIGHTS

DEFINITIONS AND CONCEPTS	EXAMPLES
SECTION 7.1 RADICALS AND RADICAL FUNCTIONS	
The **positive,** or **principal, square root** of a nonnegative number a is written as $\sqrt{a}$.	$\sqrt{36} = 6$ $\qquad$ $\sqrt{\dfrac{9}{100}} = \dfrac{3}{10}$
The **negative square root** of a is written as $-\sqrt{a}$. $\sqrt{a} = b$ only if $b^2 = a$ and $b \geq 0$	$-\sqrt{36} = -6$ $\qquad$ $\sqrt{0.04} = 0.2$ *(continued)*

DEFINITIONS AND CONCEPTS	**EXAMPLES**

SECTION 7.1   RADICALS AND RADICAL FUNCTIONS

The **cube root** of a real number a is written as $\sqrt[3]{a}$:

$$\sqrt[3]{a} = b \text{ only if } b^3 = a$$

If n is an even positive integer, then $\sqrt[n]{a^n} = |a|$.

If n is an odd positive integer, then $\sqrt[n]{a^n} = a$.

A **radical function** in x is a function defined by an expression containing a root of x.

$$\sqrt[3]{27} = 3 \qquad \sqrt[3]{-\frac{1}{8}} = -\frac{1}{2}$$

$$\sqrt[3]{y^6} = y^2 \qquad \sqrt[3]{64x^9} = 4x^3$$

$$\sqrt{(-3)^2} = |-3| = 3$$

$$\sqrt[3]{(-7)^3} = -7$$

If $f(x) = \sqrt{x} + 2$,

$$f(1) = \sqrt{1} + 2 = 1 + 2 = 3$$

$$f(3) = \sqrt{3} + 2 \approx 3.73$$

SECTION 7.2   RATIONAL EXPONENTS

$a^{1/n} = \sqrt[n]{a}$ if $\sqrt[n]{a}$ is a real number.

If m and n are positive integers, where $n \geq 2$ and $\sqrt[n]{a}$ is a real number, then

$$a^{m/n} = \left(a^{1/n}\right)^m = \left(\sqrt[n]{a}\right)^m$$

$a^{-m/n} = \dfrac{1}{a^{m/n}}$ as long as $a^{m/n}$ is a nonzero number.

Exponent rules are true for rational exponents.

$$81^{1/2} = \sqrt{81} = 9$$

$$(-8x^3)^{1/3} = \sqrt[3]{-8x^3} = -2x$$

$$4^{5/2} = \left(\sqrt{4}\right)^5 = 2^5 = 32$$

$$27^{2/3} = \left(\sqrt[3]{27}\right)^2 = 3^2 = 9$$

$$16^{-3/4} = \frac{1}{16^{3/4}} = \frac{1}{\left(\sqrt[4]{16}\right)^3} = \frac{1}{2^3} = \frac{1}{8}$$

$$x^{2/3} \cdot x^{-5/6} = x^{2/3 - 5/6} = x^{-1/6} = \frac{1}{x^{1/6}}$$

$$(8^{14})^{1/7} = 8^2 = 64$$

$$\frac{a^{4/5}}{a^{-2/5}} = a^{4/5 - (-2/5)} = a^{6/5}$$

SECTION 7.3   SIMPLIFYING RADICAL EXPRESSIONS

Product and quotient rules:

If $\sqrt[n]{a}$ and $\sqrt[n]{b}$ are real numbers,

$$\sqrt[n]{a} \cdot \sqrt[n]{b} = \sqrt[n]{a \cdot b}$$

$$\frac{\sqrt[n]{a}}{\sqrt[n]{b}} = \sqrt[n]{\frac{a}{b}}, \text{ provided } \sqrt[n]{b} \neq 0$$

A radical of the form $\sqrt[n]{a}$ is **simplified** when a contains no factors that are perfect nth powers.

Multiply or divide as indicated:

$$\sqrt{11} \cdot \sqrt{3} = \sqrt{33}$$

$$\frac{\sqrt[3]{40x}}{\sqrt[3]{5x}} = \sqrt[3]{8} = 2$$

$$\sqrt{40} = \sqrt{4 \cdot 10} = 2\sqrt{10}$$

$$\sqrt{36x^5} = \sqrt{36x^4 \cdot x} = 6x^2\sqrt{x}$$

$$\sqrt[3]{24x^7y^3} = \sqrt[3]{8x^6y^3 \cdot 3x} = 2x^2y\sqrt[3]{3x}$$

(continued)

DEFINITIONS AND CONCEPTS	EXAMPLES

SECTION 7.4 ADDING, SUBTRACTING, AND MULTIPLYING RADICAL EXPRESSIONS

Radicals with the same index and the same radicand are **like radicals.**

$$5\sqrt{6} + 2\sqrt{6} = (5 + 2)\sqrt{6} = 7\sqrt{6}$$

The distributive property can be used to add like radicals.

$$-\sqrt[3]{3x} - 10\sqrt[3]{3x} + 3\sqrt[3]{10x} = (-1 - 10)\sqrt[3]{3x}$$
$$+ 3\sqrt[3]{10x} = -11\sqrt[3]{3x} + 3\sqrt[3]{10x}$$

Radical expressions are multiplied by using many of the same properties used to multiply polynomials.

Multiply:
$$(\sqrt{5} - \sqrt{2x})(\sqrt{2} + \sqrt{2x})$$
$$= \sqrt{10} + \sqrt{10x} - \sqrt{4x} - 2x$$
$$= \sqrt{10} + \sqrt{10x} - 2\sqrt{x} - 2x$$
$$(2\sqrt{3} - \sqrt{8x})(2\sqrt{3} + \sqrt{8x})$$
$$= 4\sqrt{9} - 8x = 12 - 8x$$

SECTION 7.5 RATIONALIZING NUMERATORS AND DENOMINATORS OF RADICAL EXPRESSIONS

The **conjugate** of $a + b$ is $a - b$.

The conjugate of $\sqrt{7} + \sqrt{3}$ is $\sqrt{7} - \sqrt{3}$.

The process of writing the denominator of a radical expression without a radical is called **rationalizing the denominator.**

Rationalize each denominator:
$$\frac{\sqrt{5}}{\sqrt{3}} = \frac{\sqrt{5} \cdot \sqrt{3}}{\sqrt{3} \cdot \sqrt{3}} = \frac{\sqrt{15}}{3}$$

$$\frac{6}{\sqrt{7} + \sqrt{3}} = \frac{6(\sqrt{7} - \sqrt{3})}{(\sqrt{7} + \sqrt{3})(\sqrt{7} - \sqrt{3})}$$
$$= \frac{6(\sqrt{7} - \sqrt{3})}{7 - 3}$$
$$= \frac{6(\sqrt{7} - \sqrt{3})}{4} = \frac{3(\sqrt{7} - \sqrt{3})}{2}$$

The process of writing the numerator of a radical expression without a radical is called **rationalizing the numerator.**

Rationalize each numerator:
$$\frac{\sqrt[3]{9}}{\sqrt[3]{5}} = \frac{\sqrt[3]{9} \cdot \sqrt[3]{3}}{\sqrt[3]{5} \cdot \sqrt[3]{3}} = \frac{\sqrt[3]{27}}{\sqrt[3]{15}} = \frac{3}{\sqrt[3]{15}}$$

$$\frac{\sqrt{9} + \sqrt{3x}}{12} = \frac{(\sqrt{9} + \sqrt{3x})(\sqrt{9} - \sqrt{3x})}{12(\sqrt{9} - \sqrt{3x})}$$
$$= \frac{9 - 3x}{12(\sqrt{9} - \sqrt{3x})}$$
$$= \frac{3(3 - x)}{3 \cdot 4(3 - \sqrt{3x})} = \frac{3 - x}{4(3 - \sqrt{3x})}$$

SECTION 7.6 RADICAL EQUATIONS AND PROBLEM SOLVING

To solve a radical equation:

Solve: $x = \sqrt{4x + 9} + 3$.

Step 1. Write the equation so that one radical is by itself on one side of the equation.

1. $x - 3 = \sqrt{4x + 9}$

Step 2. Raise each side of the equation to a power equal to the index of the radical.

2. $(x - 3)^2 = 4x + 9$

(continued)

DEFINITIONS AND CONCEPTS	EXAMPLES

SECTION 7.6 RADICAL EQUATIONS AND PROBLEM SOLVING

Step 3. Simplify each side of the equation.	3. $x^2 - 6x + 9 = 4x + 9$
Step 4. If the equation still contains a radical, repeat *steps 1* through *3*.	5. $x^2 - 10x = 0$
	$x(x - 10) = 0$
Step 5. Solve the equation.	$x = 0$ or $x = 10$
Step 6. Check proposed solutions in the original equation for extraneous solutions.	6. The proposed solution 10 checks, but 0 does not. The solution set is {10}.

SECTION 7.7 COMPLEX NUMBERS

A **complex number** is a number that can be written in the form $a + bi$, where a and b are real numbers.	Simplify $\sqrt{-9}$.
	$\sqrt{-9} = \sqrt{-1 \cdot 9} = \sqrt{-1} \cdot \sqrt{9} = i \cdot 3$, or $3i$
$i^2 = -1$ and $i = \sqrt{-1}$	

COMPLEX NUMBERS	WRITTEN IN FORM $a + bi$
12	$12 + 0i$
$-5i$	$0 + (-5)i$
$-2 - 3i$	$-2 + (-3)i$

Multiply or divide as indicated.
$$\sqrt{-3} \cdot \sqrt{-7} = i\sqrt{3} \cdot i\sqrt{7}$$
$$= i^2\sqrt{21}$$
$$= -\sqrt{21}$$

To add or subtract complex numbers, add or subtract their real parts and then add or subtract their imaginary parts.	Perform the indicated operations:
	$(-3 + 2i) - (7 - 4i) = -3 + 2i - 7 + 4i$
	$= -10 + 6i$
To multiply complex numbers, multiply as though they are binomials.	$(-7 - 2i)(6 + i) = -42 - 7i - 12i - 2i^2$
	$= -42 - 19i - 2(-1)$
	$= -42 - 19i + 2$
	$= -40 - 19i$
The complex numbers $(a + bi)$ and $(a - bi)$ are called **complex conjugates.**	The complex conjugate of
	$(3 + 6i)$ is $(3 - 6i)$
	Their product is a real number:
	$(3 - 6i)(3 + 6i) = 9 - 36i^2$
	$= 9 - 36(-1) = 9 + 36 = 45$
To divide complex numbers, multiply the numerator and the denominator by the conjugate of the denominator.	Divide: $\dfrac{4}{2 - i} = \dfrac{4(2 + i)}{(2 - i)(2 + i)}$
	$= \dfrac{4(2 + i)}{4 - i^2}$
	$= \dfrac{4(2 + i)}{5}$
	$= \dfrac{8 + 4i}{5}$, or $\dfrac{8}{5} + \dfrac{4}{5}i$

4. not a real number **9.** $-a^2b^3$ **10.** $4a^2b^6$ **12.** $-2x^3y^4$ **16.** $|x^2 - 4|$ **20.** $2|(2y + z)^3|$
25. $(-\infty, \infty)$; see Appendix D

CHAPTER 7 REVIEW

(7.1) *Take the root. Assume that all variables represent positive numbers.*

1. $\sqrt{81}$ 9   **2.** $\sqrt[4]{81}$ 3   **3.** $\sqrt[3]{-8}$ -2 **4.** $\sqrt[4]{-16}$

5. $-\sqrt{\dfrac{1}{49}}$ $-\dfrac{1}{7}$ **6.** $\sqrt{x^{64}}$ x^{32} **7.** $-\sqrt{36}$ -6 **8.** $\sqrt[3]{64}$ 4

9. $\sqrt[3]{-a^6b^9}$ **10.** $\sqrt{16a^4b^{12}}$ **11.** $\sqrt[5]{32a^5b^{10}}$ $2ab^2$

12. $\sqrt[5]{-32x^{15}y^{20}}$ **13.** $\sqrt{\dfrac{x^{12}}{36y^2}}$ $\dfrac{x^6}{6y}$ **14.** $\sqrt[3]{\dfrac{27y^3}{z^{12}}}$ $\dfrac{3y}{z^4}$

Simplify. Use absolute value bars when necessary.
15. $\sqrt{(-x)^2}$ $|-x|$ **16.** $\sqrt[4]{(x^2 - 4)^4}$ **17.** $\sqrt[3]{(-27)^3}$ -27
18. $\sqrt[5]{(-5)^5}$ -5 **19.** $-\sqrt[5]{x^5}$ $-x$
20. $\sqrt[4]{16(2y + z)^{12}}$ **21.** $\sqrt{25(x - y)^{10}}$ $5|(x - y)^5|$
22. $\sqrt[5]{-y^5}$ $-y$ **23.** $\sqrt[9]{-x^9}$ $-x$

Identify the domain and then graph each function.
24. $f(x) = \sqrt{x} + 3$ $[0, \infty)$; see Appendix D
25. $g(x) = \sqrt[3]{x} - 3$; use the accompanying table.

x	-5	2	3	4	11
$g(x)$	-2	-1	0	1	2

(7.2) *Evaluate the following.*

26. $\left(\dfrac{1}{81}\right)^{1/4}$ $\dfrac{1}{3}$ **27.** $\left(-\dfrac{1}{27}\right)^{1/3}$ $-\dfrac{1}{3}$ **28.** $(-27)^{-1/3}$ $-\dfrac{1}{3}$

29. $(-64)^{-1/3}$ $-\dfrac{1}{4}$ **30.** $-9^{3/2}$ -27 **31.** $64^{-1/3}$ $\dfrac{1}{4}$

32. $(-25)^{5/2}$ not a real number   **33.** $\left(\dfrac{25}{49}\right)^{-3/2}$ $\dfrac{343}{125}$

34. $\left(\dfrac{8}{27}\right)^{-2/3}$ $\dfrac{9}{4}$ **35.** $\left(-\dfrac{1}{36}\right)^{-1/4}$ not a real number

Write with rational exponents.
36. $\sqrt[3]{x^2}$ $x^{2/3}$ **37.** $\sqrt[5]{5x^2y^3}$ $5^{1/5}x^{2/5}y^{3/5}$

Write with radical notation.
38. $y^{4/5}$ $\sqrt[5]{y^4}$ **39.** $5(xy^2z^5)^{1/3}$ **40.** $(x + 2y)^{-1/2}$

Simplify each expression. Assume that all variables represent positive numbers. Write with only positive exponents.

41. $a^{1/3}a^{4/3}a^{1/2}$ $a^{13/6}$ **42.** $\dfrac{b^{1/3}}{b^{4/3}}$ $\dfrac{1}{b}$

39. $5\sqrt[3]{xy^2z^5}$ **40.** $\dfrac{1}{\sqrt{x + 2y}}$

43. $(a^{1/2}a^{-2})^3$ $\dfrac{1}{a^{9/2}}$ **44.** $(x^{-3}y^6)^{1/3}$ $\dfrac{y^2}{x}$

45. $\left(\dfrac{b^{3/4}}{a^{-1/2}}\right)^8$ a^4b^6 **46.** $\dfrac{x^{1/4}x^{-1/2}}{x^{2/3}}$ $\dfrac{1}{x^{11/12}}$

47. $\left(\dfrac{49c^{5/3}}{a^{-1/4}b^{5/6}}\right)^{-1}$ $\dfrac{b^{5/6}}{49a^{1/4}c^{5/3}}$ **48.** $a^{-1/4}(a^{5/4} - a^{9/4})$ $a - a^2$

Use a calculator and write a three-decimal-place approximation.
49. $\sqrt{20}$ 4.472   **50.** $\sqrt[3]{-39}$ -3.391 **51.** $\sqrt[4]{726}$ 5.191
52. $56^{1/3}$ 3.826   **53.** $-78^{3/4}$ -26.246 **54.** $105^{-2/3}$ 0.045

(7.3) *Perform the indicated operations and then simplify if possible. For the remainder of this review, assume that variables represent positive numbers only.*
55. $\sqrt[3]{3} \cdot \sqrt[3]{8}$ $2\sqrt[3]{6}$ **56.** $\sqrt[3]{7y} \cdot \sqrt[3]{x^2z}$ $\sqrt[3]{7x^2yz}$

57. $\dfrac{\sqrt{44x^3}}{\sqrt{11x}}$ $2x$ **58.** $\dfrac{\sqrt[4]{a^6b^{13}}}{\sqrt[4]{a^2b}}$ ab^3

Simplify.
59. $\sqrt{60}$ $2\sqrt{15}$ **60.** $-\sqrt{75}$ $-5\sqrt{3}$
61. $\sqrt[3]{162}$ $3\sqrt[3]{6}$ **62.** $\sqrt[3]{-32}$ $-2\sqrt[3]{4}$
63. $\sqrt{36x^7}$ $6x^3\sqrt{x}$ **64.** $\sqrt[3]{24a^5b^7}$ $2ab^2\sqrt[3]{3a^2b}$
65. $\sqrt{\dfrac{p^{17}}{121}}$ $\dfrac{p^8\sqrt{p}}{11}$ **66.** $\sqrt[3]{\dfrac{y^5}{27x^6}}$ $\dfrac{y\sqrt[3]{y^2}}{3x^2}$
67. $\sqrt[4]{\dfrac{xy^6}{81}}$ $\dfrac{y\sqrt[4]{xy^2}}{3}$ **68.** $\sqrt{\dfrac{2x^3}{49y^4}}$ $\dfrac{x\sqrt{2x}}{7y^2}$

Use rational exponents to write each radical with the same index. Then multiply.
69. $\sqrt[3]{2} \cdot \sqrt{7}$ $\sqrt[6]{1372}$ **70.** $\sqrt[3]{3} \cdot \sqrt[4]{x}$ $\sqrt[12]{81x^3}$

71. The formula for the radius r of a circle of area A is
$$r = \sqrt{\dfrac{A}{\pi}}$$
71a. $\dfrac{5}{\sqrt{\pi}}$ meters, or $\dfrac{5\sqrt{\pi}}{\pi}$ meters
a. Find the exact radius of a circle whose area is 25 square meters.
b. Approximate to two decimal places the radius of a circle whose area is 104 square inches. 5.75 in.

(7.4) *Perform the indicated operation.*
72. $x\sqrt{75xy} - \sqrt{27x^3y}$ $2x\sqrt{3xy}$ **73.** $2\sqrt{32x^2y^3} - xy\sqrt{98y}$ $xy\sqrt{2y}$

74. $\sqrt[3]{128} + \sqrt[3]{250}$ $9\sqrt[3]{2}$ **75.** $3\sqrt[4]{32a^5} - a\sqrt[4]{162a}$

76. $\dfrac{5}{\sqrt{4}} + \dfrac{\sqrt{3}}{3}$ $\dfrac{15 + 2\sqrt{3}}{6}$ **77.** $\sqrt{\dfrac{8}{x^2}} - \sqrt{\dfrac{50}{16x^2}}$ $\dfrac{3\sqrt{2}}{4x}$

78. $2\sqrt{50} - 3\sqrt{125} + \sqrt{98}$ $17\sqrt{2} - 15\sqrt{5}$

79. $2a\sqrt[4]{32b^5} - 3b\sqrt[4]{162a^4b} + \sqrt[4]{2a^4b^5}$ $-4ab\sqrt[4]{2b}$

Multiply and then simplify if possible.

80. $\sqrt{3}(\sqrt{27} - \sqrt{3})$ 6 **81.** $(\sqrt{x} - 3)^2$ $x - 6\sqrt{x} + 9$

82. $(\sqrt{5} - 5)(2\sqrt{5} + 2)$ $-8\sqrt{5}$

83. $(2\sqrt{x} - 3\sqrt{y})(2\sqrt{x} + 3\sqrt{y})$ $4x - 9y$

84. $(\sqrt{a} + 3)(\sqrt{a} - 3)$ **85.** $(\sqrt[3]{a} + 2)^2$

86. $(\sqrt[3]{5x} + 9)(\sqrt[3]{5x} - 9)$ $\sqrt[3]{25x^2} - 81$

87. $(\sqrt[3]{a} + 4)(\sqrt[3]{a^2} - 4\sqrt[3]{a} + 16)$ $a + 64$

(7.5) *Rationalize each denominator.*

88. $\dfrac{3}{\sqrt{7}}$ $\dfrac{3\sqrt{7}}{7}$ **89.** $\sqrt{\dfrac{x}{12}}$ $\dfrac{\sqrt{3x}}{6}$

90. $\dfrac{5}{\sqrt[3]{4}}$ $\dfrac{5\sqrt[3]{2}}{2}$ **91.** $\sqrt{\dfrac{24x^5}{3y^2}}$ $\dfrac{2x^2\sqrt{2x}}{y}$

92. $\sqrt[3]{\dfrac{15x^6y^7}{z^2}}$ $\dfrac{x^2y^2\sqrt[3]{15yz}}{z}$ **93.** $\dfrac{5}{2 - \sqrt{7}}$ $\dfrac{10 + 5\sqrt{7}}{3}$

94. $\dfrac{3}{\sqrt{y} - 2}$ $\dfrac{3\sqrt{y} + 6}{y - 4}$ **95.** $\dfrac{\sqrt{2} - \sqrt{3}}{\sqrt{2} + \sqrt{3}}$ $-5 + 2\sqrt{6}$

Rationalize each numerator.

96. $\dfrac{\sqrt{11}}{3}$ $\dfrac{11}{3\sqrt{11}}$ **97.** $\sqrt{\dfrac{18}{y}}$ $\dfrac{6}{\sqrt{2y}}$

98. $\dfrac{\sqrt[3]{9}}{7}$ $\dfrac{3}{7\sqrt[3]{3}}$ **99.** $\sqrt{\dfrac{24x^5}{3y^2}}$ $\dfrac{4x^3}{y\sqrt{2x}}$

100. $\sqrt[3]{\dfrac{xy^2}{10z}}$ $\dfrac{xy}{\sqrt[3]{10x^2yz}}$ **101.** $\dfrac{\sqrt{x} + 5}{-3}$ $\dfrac{x - 25}{-3\sqrt{x} + 15}$

(7.6) *Solve each equation for the variable.*

102. $\sqrt{y - 7} = 5$ $\{32\}$ **103.** $\sqrt{2x} + 10 = 4$ $\{\ \}$

104. $\sqrt[3]{2x - 6} = 4$ $\{35\}$ **105.** $\sqrt{x + 6} = \sqrt{x + 2}$ $\{\ \}$

106. $2x - 5\sqrt{x} = 3$ $\{9\}$ **107.** $\sqrt{x + 9} = 2 + \sqrt{x - 7}$

  $\{16\}$

Find each unknown length.

108.

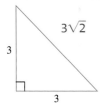

$3\sqrt{2}$

3

3

109.

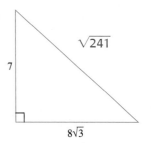

$\sqrt{241}$

7

$8\sqrt{3}$

110. Beverly Hillis wants to determine the distance x across a pond on her property. She is able to measure the distances shown on the following diagram. Find how wide the lake is at the crossing point indicated by the triangle to the nearest tenth of a foot.   51.2 ft

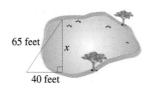

65 feet

x

40 feet

111. A pipefitter needs to connect two underground pipelines that are offset by 3 feet, as pictured in the diagram. Neglecting the joints needed to join the pipes, find the length of the shortest possible connecting pipe rounded to the nearest hundredth of a foot.   4.24 ft

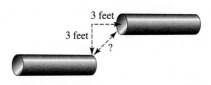

3 feet

3 feet

?

(7.7) *Perform the indicated operation and simplify. Write the result in the form $a + bi$.*

112. $\sqrt{-8}$ $2i\sqrt{2}$ **113.** $-\sqrt{-6}$ $-i\sqrt{6}$

114. $\sqrt{-4} + \sqrt{-16}$ $6i$ **115.** $\sqrt{-2} \cdot \sqrt{-5}$ $-\sqrt{10}$

116. $(12 - 6i) + (3 + 2i)$ **117.** $(-8 - 7i) - (5 - 4i)$

118. $(\sqrt{3} + \sqrt{2}) + (3\sqrt{2} - \sqrt{-8})$ $\sqrt{3} + 4\sqrt{2} - 2i\sqrt{2}$

119. $2i(2 - 5i)$ $10 + 4i$ **120.** $-3i(6 - 4i)$ $-12 - 18i$

121. $(3 + 2i)(1 + i)$ **122.** $(2 - 3i)^2$ $-5 - 12i$

123. $(\sqrt{6} - 9i)(\sqrt{6} + 9i)$ 87

124. $\dfrac{2 + 3i}{2i}$ $\dfrac{3}{2} - i$ **125.** $\dfrac{1 + i}{-3i}$ $-\dfrac{1}{3} + \dfrac{1}{3}i$

75. $3a\sqrt[4]{2a}$ **84.** $a - 9$ **85.** $\sqrt[3]{a^2} + 4\sqrt[3]{a} + 4$ **116.** $15 - 4i$ **117.** $-13 - 3i$ **121.** $1 + 5i$

CHAPTER 7 TEST

Raise to the power or take the root. Assume that all variables represent positive numbers. Write with only positive exponents. **1.** $6\sqrt{6}$ **2.** $-x^{16}$

1. $\sqrt{216}$ **2.** $-\sqrt[4]{x^{64}}$ **3.** $\left(\dfrac{1}{125}\right)^{1/3}$ $\dfrac{1}{5}$ **4.** $\left(\dfrac{1}{125}\right)^{-1/3}$ 5

5. $\left(\dfrac{8x^3}{27}\right)^{2/3}$ $\dfrac{4x^2}{9}$ **6.** $\sqrt[3]{-a^{18}b^9}$ $-a^6b^3$

7. $\left(\dfrac{64c^{4/3}}{a^{-2/3}b^{5/6}}\right)^{1/2}$ $\dfrac{8a^{1/3}c^{2/3}}{b^{5/12}}$ **8.** $a^{-2/3}(a^{5/4} - a^3)$ $a^{7/12} - a^{7/3}$

Take the root. Use absolute value bars when necessary.
9. $\sqrt[4]{(4xy)^4}$ $|4xy|$ **10.** $\sqrt[3]{(-27)^3}$ -27

Rationalize the denominator. Assume that all variables represent positive numbers.

11. $\sqrt{\dfrac{9}{y}}$ $\dfrac{3\sqrt{y}}{y}$ **12.** $\dfrac{4 - \sqrt{x}}{4 + 2\sqrt{x}}$ **13.** $\dfrac{\sqrt[3]{ab}}{\sqrt[3]{ab^2}}$ $\dfrac{\sqrt[3]{b^2}}{b}$

14. Rationalize the numerator of $\dfrac{\sqrt{6} + x}{8}$ and simplify.

Perform the indicated operations. Assume that all variables represent positive numbers.
15. $\sqrt{125x^3} - 3\sqrt{20x^3}$ **16.** $\sqrt{3}(\sqrt{16} - \sqrt{2})$ $4\sqrt{3} - \sqrt{6}$
17. $(\sqrt{x} + 1)^2$ **18.** $(\sqrt{2} - 4)(\sqrt{3} + 1)$
19. $(\sqrt{5} + 5)(\sqrt{5} - 5)$ -20

Use a calculator to approximate each to three decimal places.
20. $\sqrt{561}$ 23.685 **21.** $386^{-2/3}$ 0.019

Solve.
22. $x = \sqrt{x - 2} + 2$ {2, 3} **23.** $\sqrt{x^2 - 7} + 3 = 0$ { }
24. $\sqrt{x + 5} = \sqrt{2x - 1}$ {6}

12. $\dfrac{8 - 6\sqrt{x} + x}{8 - 2x}$ **14.** $\dfrac{6 - x^2}{8(\sqrt{6} - x)}$ **15.** $-x\sqrt{5x}$ **17.** $x + 2\sqrt{x} + 1$ **18.** $\sqrt{6} - 4\sqrt{3} + \sqrt{2} - 4$ **27.** $-3i$

Perform the indicated operation and simplify. Write the result in the form a + bi.
25. $\sqrt{-2} \cdot i\sqrt{2}$ **26.** $-\sqrt{-8}$ $-2i\sqrt{2}$
27. $(12 - 6i) - (12 - 3i)$ **28.** $(6 - 2i)(6 + 2i)$ 40
29. $(4 + 3i)^2$ $7 + 24i$ **30.** $\dfrac{1 + 4i}{1 - i}$ $-\dfrac{3}{2} + \dfrac{5}{2}i$

31. Find x.

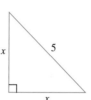

$$x = \dfrac{5\sqrt{2}}{2}$$

32. Identify the domain of $g(x)$. Then complete the accompanying table and graph $g(x)$.
$g(x) = \sqrt{x + 2}$ $[-2, \infty)$; see Appendix D.

x	-2	-1	2	7
$g(x)$	0	1	2	3

Solve.
33. The function $V(r) = \sqrt{2.5r}$ can be used to estimate the maximum safe velocity, V, in miles per hour, at which a car can travel if it is driven along a curved road with a *radius of curvature*, r, in feet. To the nearest whole number, find the maximum safe speed if a cloverleaf exit on an expressway has a radius of curvature of 300 feet. 27 mph

34. Use the formula from Exercise 33 to find the radius of curvature if the safe velocity is 30 mph. 360 ft

CHAPTER 7 CUMULATIVE REVIEW

1. Use the associative property of multiplication to write an expression equivalent to $4 \cdot (9y)$. Then simplify this equivalent expression.

2. Write the following as algebraic expressions. Then simplify.

   a. The sum of two consecutive integers, if x is the first consecutive integer. $2x + 1$

b. The perimeter of the triangle with sides x, $5x$, and $6x - 3$. $12x - 3$; *Sec. 2.1, Ex. 9*

1. $4 \cdot (9y) = (4 \cdot 9)y = 36y$; *Sec. 1.2, Ex. 7*

3. $(-\infty, -3] \cup [9, \infty)$; *Sec. 2.7, Ex. 7*

3. Solve $\left| \dfrac{x}{3} - 1 \right| - 7 \geq -5$.

4. Graph $g(x) = 2x + 1$. Compare this graph with the graph of $f(x) = 2x$.

5. Use the elimination method to solve the system
$\begin{cases} 3x - 2y = 10 \\ 4x - 3y = 15 \end{cases}$ $\{(0, -5)\}$; *Sec. 4.1, Ex. 6*

6. Lynn Pike, a pharmacist, needs 70 liters of 50% alcohol solution. She has available a 30% alcohol solution and an 80% alcohol solution. How many liters of each solution should she mix to obtain 70 liters of a 50% alcohol solution?

7. If $P(x) = 3x^2 - 2x - 5$, find the following:
a. $P(1)$ $P(1) = -4$ **b.** $P(-2)$

8. Factor $ab - 6a + 2b - 12$.

9. Factor $2x^2 + 11x + 15$.

10. Factor each polynomial completely.
a. $5p^2 + 5 + qp^2 + q$ $(p^2 + 1)(5 + q)$
b. $9x^2 + 24x + 16$ $(3x + 4)^2$
c. $y^2 + 25$ prime; *Sec. 5.7, Ex. 11*

11. Simplify each rational expression.
a. $\dfrac{24x^6y^5}{8x^7y}$ $\dfrac{3y^4}{x}$ **b.** $\dfrac{2x^2}{10x^3 - 2x^2}$

12. Divide.
a. $\dfrac{3x}{5y} \div \dfrac{9y}{x^5}$ $\dfrac{x^6}{15y^2}$ **b.** $\dfrac{8m^2}{3m^2 - 12} \div \dfrac{40}{2 - m}$

13. Add or subtract.
a. $\dfrac{5}{7} + \dfrac{x}{7}$ $\dfrac{5 + x}{7}$ **b.** $\dfrac{x}{4} + \dfrac{5x}{4} - \dfrac{3x}{2}$
c. $\dfrac{x^2}{x + 7} - \dfrac{49}{x + 7}$ $x - 7$ **d.** $\dfrac{x}{3y^2} - \dfrac{x + 1}{3y^2}$

14. Simplify each complex fraction.
a. $\dfrac{\dfrac{5x}{x + 2}}{\dfrac{10}{x - 2}}$ $\dfrac{x(x - 2)}{2(x + 2)}$ **b.** $\dfrac{x + \dfrac{1}{y}}{y + \dfrac{1}{x}}$ $\dfrac{x}{y}$; *Sec. 6.4, Ex. 2*

15. Find the quotient: $\dfrac{6x^2 - 19x + 12}{3x - 5}$.

16. Solve $\dfrac{8x}{5} + \dfrac{3}{2} = \dfrac{3x}{5}$. $\left\{ -\dfrac{3}{2} \right\}$; *Sec. 6.7, Ex. 1*

17. Melissa Scarlatti can clean the house in 4 hours, and her husband, Zack, can do the same job in 5 hours. They have agreed to clean together so that they can finish in time to watch a movie on TV that starts in 2 hours. How long will it take them to clean the house together? Can they finish before the movie starts? $2\dfrac{2}{9}$ hours, no; *Sec. 6.8, Ex. 4*

18. Simplify. Assume that all variables represent positive numbers.
a. $\sqrt{36}$ 6 **b.** $\sqrt{0}$ 0 **c.** $\sqrt{\dfrac{4}{49}}$ $\dfrac{2}{7}$
d. $\sqrt{0.25}$ 0.5 **e.** $\sqrt{x^6}$ x^3 **f.** $\sqrt{9x^{10}}$ $3x^5$
g. $-\sqrt{81}$ -9; *Sec. 7.1, Ex. 1*

19. Use radical notation to write the following. Simplify if possible.
a. $4^{1/2}$ 2 **b.** $64^{1/3}$ 4 **c.** $x^{1/4}$ $\sqrt[4]{x}$
d. $0^{1/6}$ 0 **e.** $-9^{1/2}$ -3 **f.** $(81x^8)^{1/4}$ $3x^2$
g. $(5y)^{1/3}$ $\sqrt[3]{5y}$; *Sec. 7.2, Ex. 1*

20. Simplify the following.
a. $\sqrt{50}$ $5\sqrt{2}$ **b.** $\sqrt[3]{24}$ $2\sqrt[3]{3}$ **c.** $\sqrt{26}$ $\sqrt{26}$
d. $\sqrt[4]{32}$ $2\sqrt[4]{2}$; *Sec. 7.3, Ex. 3*

21. Multiply.
a. $\sqrt{3}(5 + \sqrt{30})$ $5\sqrt{3} + 3\sqrt{10}$
b. $(\sqrt{5} - \sqrt{6})(\sqrt{7} + 1)$ $\sqrt{35} + \sqrt{5} - \sqrt{42} - \sqrt{6}$
c. $(7\sqrt{x} + 5)(3\sqrt{x} - \sqrt{5})$
d. $(4\sqrt{3} - 1)^2$ $49 - 8\sqrt{3}$
e. $(\sqrt{2x} - 5)(\sqrt{2x} + 5)$ $2x - 25$; *Sec. 7.4, Ex. 3*

22. Rationalize the denominator of each expression.
a. $\dfrac{\sqrt{27}}{\sqrt{5}}$ $\dfrac{3\sqrt{15}}{5}$ **b.** $\dfrac{2\sqrt{16}}{\sqrt{9x}}$ $\dfrac{8\sqrt{x}}{3x}$ **c.** $\sqrt[3]{\dfrac{1}{2}}$

23. Solve $\sqrt{4 - x} = x - 2$ for x. $\{3\}$; *Sec. 7.6, Ex. 4*

4. The graph of $g(x) = 2x + 1$ is the graph of $f(x) = 2x$ shifted upward 1 unit (see Appendix D). *Sec. 3.3, Ex. 1*

6. 42 liters of 30% solution, 28 liters of 80% solution; *Sec. 4.3, Ex. 3* **7b.** $P(-2) = 11$; *Sec. 5.3, Ex. 3*

8. $(b - 6)(a + 2)$; *Sec. 5.5, Ex. 7* **9.** $(2x + 5)(x + 3)$; *Sec. 5.6, Ex. 5* **11b.** $\dfrac{1}{5x - 1}$; *Sec. 6.1, Ex. 3*

12b. $-\dfrac{m^2}{15(m + 2)}$; *Sec. 6.2, Ex. 3* **13d.** $-\dfrac{1}{3y^2}$; *Sec. 6.3, Ex. 1* **15.** $2x - 3 - \dfrac{3}{3x - 5}$; *Sec. 6.5, Ex. 5*

21c. $21x - 7\sqrt{5x} + 15\sqrt{x} - 5\sqrt{5}$ **22c.** $\dfrac{\sqrt[3]{4}}{2}$; *Sec. 7.5, Ex. 1*

QUADRATIC EQUATIONS AND FUNCTIONS

MODELING THE POSITION OF A FREELY FALLING OBJECT

In physics, the position of an object in free fall is modeled by a quadratic function. Once an object is in free fall, its position depends on its initial position, its initial velocity, and the acceleration due to gravity that it experiences.

IN THE CHAPTER GROUP ACTIVITY ON PAGE 516, YOU WILL HAVE THE OPPORTUNITY TO MATCH A DESCRIPTION OF A FREELY FALLING OBJECT TO ITS MATHEMATICAL MODEL AND ITS GRAPH.

An important part of the study of algebra is learning to model and solve problems. Often, the model of a problem is a quadratic equation or a function containing a second-degree polynomial. In this chapter, we continue the work begun in Chapter 5, when we solved polynomial equations in one variable by factoring. Two additional methods of solving quadratic equations are analyzed, as well as methods of solving nonlinear inequalities in one variable.

8.1 SOLVING QUADRATIC EQUATIONS BY COMPLETING THE SQUARE

TAPE IA 8.1

OBJECTIVES

1. Use the square root property to solve quadratic equations.
2. Write perfect square trinomials.
3. Solve quadratic equations by completing the square.
4. Use quadratic equations to solve problems.

In Chapter 5, we solved quadratic equations by factoring. Recall that a **quadratic,** or **second-degree, equation** is an equation that can be written in the form $ax^2 + bx + c = 0$, where a, b, and c are real numbers and a is not 0. To solve a quadratic equation such as $x^2 = 9$ by factoring, we use the zero-factor theorem. To use the zero-factor theorem, the equation must first be written in standard form, $ax^2 + bx + c = 0$.

$$x^2 = 9$$
$$x^2 - 9 = 0 \qquad \text{Subtract 9 from both sides.}$$
$$(x + 3)(x - 3) = 0 \qquad \text{Factor.}$$
$$x + 3 = 0 \quad \text{or} \quad x - 3 = 0 \qquad \text{Set each factor equal to 0.}$$
$$x = -3 \qquad\qquad x = 3 \qquad \text{Solve.}$$

The solution set is $\{-3, 3\}$, the positive and negative square roots of 9. Not all quadratic equations can be solved by factoring, so we need to explore other methods. Notice that the solutions of the equation $x^2 = 9$ are two numbers whose square is 9.

$$3^2 = 9 \quad \text{and} \quad (-3)^2 = 9$$

Thus, we can solve the equation $x^2 = 9$ by taking the square root of both sides. Be sure to include both $\sqrt{9}$ and $-\sqrt{9}$ as solutions since both $\sqrt{9}$ and $-\sqrt{9}$ are numbers whose square is 9.

$$x^2 = 9$$
$$x = \pm\sqrt{9}$$
$$x = \pm 3 \qquad \text{The notation } \pm 3 \text{ (read as plus or minus 3)}$$
$$\text{indicates the pair of numbers } +3 \text{ and } -3.$$

This illustrates the square root property.

> **SQUARE ROOT PROPERTY**
> If b is a real number and if $a^2 = b$, then $a = \pm\sqrt{b}$.

EXAMPLE 1   Use the square root property to solve $x^2 = 50$.

Solution:

$$x^2 = 50$$
$$x = \pm\sqrt{50} \qquad \text{Apply the square root property.}$$
$$x = \pm 5\sqrt{2} \qquad \text{Simplify the radical.}$$

The solution set is $\{5\sqrt{2}, -5\sqrt{2}\}$.

EXAMPLE 2   Use the square root property to solve $(x + 1)^2 = 12$ for x.

Solution: By the square root property, we have

$$(x + 1)^2 = 12$$
$$x + 1 = \pm\sqrt{12} \qquad \text{Apply the square root property.}$$
$$x + 1 = \pm 2\sqrt{3} \qquad \text{Simplify the radical.}$$
$$x = -1 \pm 2\sqrt{3} \qquad \text{Subtract 1 from both sides.}$$

The solution set is $\{-1 + 2\sqrt{3}, -1 - 2\sqrt{3}\}$.

EXAMPLE 3   Solve $(2x - 5)^2 = -16$.

Solution:

$$(2x - 5)^2 = -16$$
$$2x - 5 = \pm\sqrt{-16} \qquad \text{Apply the square root property.}$$
$$2x - 5 = \pm 4i \qquad \text{Simplify the radical.}$$
$$2x = 5 \pm 4i \qquad \text{Add 5 to both sides.}$$
$$x = \frac{5 \pm 4i}{2} \qquad \text{Divide both sides by 2.}$$

The solution set is $\left\{\dfrac{5 + 4i}{2}, \dfrac{5 - 4i}{2}\right\}$.

Notice from Examples 2 and 3 that, if we write a quadratic equation so that one side is the square of a binomial, we can solve by using the square root property. To write the square of a binomial, we write perfect square trinomials. Recall that a perfect square trinomial is a trinomial that can be factored into two identical binomial factors.

PERFECT SQUARE TRINOMIALS | FACTORED FORM

$$x^2 + 8x + 16 \qquad (x + 4)^2$$
$$x^2 - 6x + 9 \qquad (x - 3)^2$$
$$x^2 + 3x + \frac{9}{4} \qquad \left(x + \frac{3}{2}\right)^2$$

Notice that for each perfect square trinomial, **the constant term of the trinomial is the square of half the coefficient of the x-term.** For example,

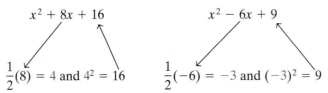

$$\frac{1}{2}(8) = 4 \text{ and } 4^2 = 16 \qquad \frac{1}{2}(-6) = -3 \text{ and } (-3)^2 = 9$$

The process of writing a quadratic equation so that one side is a perfect square trinomial is called **completing the square.**

EXAMPLE 4   Solve $p^2 + 2p = 4$ by completing the square.

Solution: First, add the square of half the coefficient of p to both sides so that the resulting trinomial will be a perfect square trinomial. The coefficient of p is 2.

$$\frac{1}{2}(2) = 1 \quad \text{and} \quad 1^2 = 1$$

Add 1 to both sides of the original equation:

$$p^2 + 2p = 4$$
$$p^2 + 2p + 1 = 4 + 1 \qquad \text{Add 1 to both sides.}$$
$$(p + 1)^2 = 5 \qquad \text{Factor the trinomial; simplify the right side.}$$

We may now apply the square root property and solve for p:

$$p + 1 = \pm\sqrt{5} \qquad \text{Use the square root property.}$$
$$p = -1 \pm \sqrt{5} \qquad \text{Subtract 1 from both sides.}$$

Notice that there are two solutions: $-1 + \sqrt{5}$ and $-1 - \sqrt{5}$. The solution set is $\{-1 + \sqrt{5}, -1 - \sqrt{5}\}$.

EXAMPLE 5   Solve $m^2 - 7m - 1 = 0$ for m by completing the square.

Solution: First, add 1 to both sides of the equation so that the left side has no constant term.

$$m^2 - 7m - 1 = 0$$
$$m^2 - 7m = 1$$

Now find the constant term that makes the left side a perfect square trinomial by squaring half the coefficient of m. Add this constant to both sides of the equation:

$$\frac{1}{2}(-7) = -\frac{7}{2} \quad \text{and} \quad \left(-\frac{7}{2}\right)^2 = \frac{49}{4}$$

$$m^2 - 7m + \frac{49}{4} = 1 + \frac{49}{4}$$ Add $\frac{49}{4}$ to both sides of the equation.

$$\left(m - \frac{7}{2}\right)^2 = \frac{53}{4}$$ Factor the perfect square trinomial and simplify the right side.

$$m - \frac{7}{2} = \pm\sqrt{\frac{53}{4}}$$ Apply the square root property.

$$m = \frac{7}{2} \pm \frac{\sqrt{53}}{2}$$ Add $\frac{7}{2}$ to both sides and simplify $\sqrt{\frac{53}{4}}$.

$$m = \frac{7 \pm \sqrt{53}}{2}$$ Simplify.

The solution set is $\left\{\dfrac{7 + \sqrt{53}}{2}, \dfrac{7 - \sqrt{53}}{2}\right\}$.

EXAMPLE 6   Solve $2x^2 - 8x + 3 = 0$.

Solution: Our procedure for finding the constant term to complete the square works only if the coefficient of the squared variable term is 1. Therefore, to solve this equation, the first step is to divide both sides by 2, the coefficient of x^2.

$$2x^2 - 8x + 3 = 0$$

$$x^2 - 4x + \frac{3}{2} = 0$$ Divide both sides by 2.

$$x^2 - 4x = -\frac{3}{2}$$ Subtract $\frac{3}{2}$ from both sides.

Next find the square of half of -4:

$$\frac{1}{2}(-4) = -2 \quad \text{and} \quad (-2)^2 = 4$$

Add 4 to both sides of the equation to complete the square:

$$x^2 - 4x + 4 = -\frac{3}{2} + 4$$

$$(x - 2)^2 = \frac{5}{2}$$ Factor the perfect square and simplify the right side.

$$x - 2 = \pm\sqrt{\frac{5}{2}}$$ Apply the square root property.

$$x - 2 = \pm\frac{\sqrt{10}}{2}$$ Rationalize the denominator.

$$x = 2 \pm \frac{\sqrt{10}}{2}$$ Add 2 to both sides.

$$= \frac{4}{2} \pm \frac{\sqrt{10}}{2}$$ Find the common denominator.

$$= \frac{4 \pm \sqrt{10}}{2}$$ Simplify.

The solution set is $\left\{\dfrac{4 + \sqrt{10}}{2}, \dfrac{4 - \sqrt{10}}{2}\right\}$.

The following steps may be used to solve a quadratic equation such as $ax^2 + bx + c = 0$ by completing the square. This method may be used whether or not the polynomial $ax^2 + bx + c$ is factorable.

TO SOLVE A QUADRATIC EQUATION IN x BY COMPLETING THE SQUARE

Step 1. If the coefficient of x^2 is 1, go to *step 2*. Otherwise, divide both sides of the equation by the coefficient of x^2.

Step 2. Isolate all variable terms on one side of the equation.

Step 3. Complete the square for the resulting binomial by adding the square of half of the coefficient of x to both sides of the equation.

Step 4. Factor the resulting perfect square trinomial and write it as the square of a binomial.

Step 5. Apply the square root property to solve for x.

EXAMPLE 7 Solve $3x^2 - 9x + 8 = 0$ by completing the square.

Solution: $\qquad\qquad 3x^2 - 9x + 8 = 0$

Step 1. $\quad x^2 - 3x + \dfrac{8}{3} = 0$ $\qquad$ Divide both sides of the equation by 3.

Step 2. $\qquad x^2 - 3x = -\dfrac{8}{3}$ $\qquad$ Subtract $\dfrac{8}{3}$ from both sides.

Since $\dfrac{1}{2}(-3) = -\dfrac{3}{2}$ and $\left(-\dfrac{3}{2}\right)^2 = \dfrac{9}{4}$, we add $\dfrac{9}{4}$ to both sides of the equation.

Step 3. $\quad x^2 - 3x + \dfrac{9}{4} = -\dfrac{8}{3} + \dfrac{9}{4}$

Step 4. $\qquad \left(x - \dfrac{3}{2}\right)^2 = -\dfrac{5}{12}$ $\qquad$ Factor the perfect square trinomial.

Step 5.

$$x - \frac{3}{2} = \pm\sqrt{-\frac{5}{12}} \qquad \text{Apply the square root property.}$$

$$x - \frac{3}{2} = \pm\frac{i\sqrt{5}}{2\sqrt{3}} \qquad \text{Simplify the radical.}$$

$$x - \frac{3}{2} = \pm\frac{i\sqrt{15}}{6} \qquad \text{Rationalize the denominator.}$$

$$x = \frac{3}{2} \pm \frac{i\sqrt{15}}{6} \qquad \text{Add } \frac{3}{2} \text{ to both sides.}$$

$$= \frac{9}{6} \pm \frac{i\sqrt{15}}{6} \qquad \text{Find a common denominator.}$$

$$= \frac{9 \pm i\sqrt{15}}{6} \qquad \text{Simplify.}$$

The solution set is $\left\{ \dfrac{9 + i\sqrt{15}}{6}, \dfrac{9 - i\sqrt{15}}{6} \right\}$.

4

Recall the **simple interest** formula $I = Prt$, where I is the interest earned, P is the principal, r is the rate of interest, and t is time. If \$100 is invested at a simple interest rate of 5% annually, at the end of 3 years the total interest I earned is

$$I = P \cdot r \cdot t$$

or

$$I = 100 \cdot 0.05 \cdot 3 = \$15$$

and the new principal is

$$\$100 + \$15 = \$115$$

Most of the time, the interest computed on money borrowed or money deposited is **compound interest.** Compound interest, unlike simple interest, is computed on original principal *and* on interest already earned. To see the difference between simple interest and compound interest, suppose that \$100 is invested at a rate of 5% compounded annually. To find the total amount of money at the end of 3 years, we calculate as follows:

$$I \quad = \quad P \quad \cdot \quad r \quad \cdot t$$

First year: Interest $= \$100 \cdot 0.05 \cdot 1 = \$5.00$

New principal $= \$100.00 + \$5.00 = \$105.00$

Second year: Interest $= \$105.00 \cdot 0.05 \cdot 1 = \$5.25$

New principal $= \$105.00 + \$5.25 = \$110.25$

Third year: Interest $= \$110.25 \cdot 0.05 \cdot 1 \approx \$5.51$

New principal $= \$110.25 + \$5.51 = \$115.76$

At the end of the third year, the total compound interest earned is \$15.76, whereas the total simple interest earned is \$15.

It is tedious to calculate compound interest as we did above, so we use a compound interest formula. The formula for calculating the total amount of money when interest is compounded annually is

$$A = P(1 + r)^t$$

where P is the original investment, r is the interest rate per compounding period, and t is the number of periods. For example, the amount of money A at the end of 3 years if $100 is invested at 5% compounded annually is

$$A = \$100(1 + 0.05)^3 \approx \$100(1.1576) = \$115.76$$

as expected.

EXAMPLE 8 Find the interest rate r if $2000, compounded annually, grows to $2420 in 2 years.

Solution: **1.** UNDERSTAND the problem.   For this example, make sure that you understand the formula for compounding interest annually. Since a formula is known, we go to step 4.

4. TRANSLATE.   Here we substitute given values into the formula:

$$A = P(1 + r)^t$$
$$2420 = 2000(1 + r)^2 \qquad \text{Let } A = 2420, P = 2000, \text{ and } t = 2.$$

5. COMPLETE.   Solve the equation for r.

$$2420 = 2000(1 + r)^2$$

$$\frac{2420}{2000} = (1 + r)^2 \qquad \text{Divide both sides by 2000.}$$

$$\frac{121}{100} = (1 + r)^2 \qquad \text{Simplify the fraction.}$$

$$\pm\sqrt{\frac{121}{100}} = 1 + r \qquad \text{Apply the square root property.}$$

$$\pm\frac{11}{10} = 1 + r \qquad \text{Simplify.}$$

$$-1 \pm \frac{11}{10} = r$$

$$-\frac{10}{10} \pm \frac{11}{10} = r$$

$$\frac{1}{10} = r \quad \text{or} \quad -\frac{21}{10} = r$$

6. INTERPRET.   The rate cannot be negative, so we reject $-\dfrac{21}{10}$ and *check*

$\dfrac{1}{10} = 0.10 = 10\%$ per year. If we invest $2000 at 10% compounded annually, in

2 years the amount in the account would be $2000(1 + 0.10)^2 = 2420$ dollars, the desired amount.

State: The interest rate is 10% compounded annually.

GRAPHING CALCULATOR EXPLORATIONS

In Section 5.8, we showed how we can use a grapher to approximate real number solutions of a quadratic equation written in standard form. We can also use a grapher to solve a quadratic equation when it is not written in standard form. For example, to solve $(x + 1)^2 = 12$, the quadratic equation in Example 2, we graph the following on the same set of axes. Use Xmin $= -10$, Xmax $= 10$, Ymin $= -13$, and Ymax $= 13$.

$$Y_1 = (x + 1)^2 \quad \text{and} \quad Y_2 = 12$$

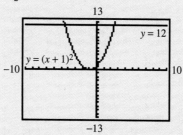

Use the Intersect feature or the Zoom and Trace features to locate the points of intersection of the graphs. The x-values of these points are the solutions of $(x + 1)^2 = 12$. The solutions, rounded to two decimal points, are 2.46 and -4.46.

Check to see that these numbers are approximations of the exact solutions $-1 \pm 2\sqrt{3}$.

Use a grapher to solve each quadratic equation. Round all solutions to the nearest hundredth.

1. $x(x - 5) = 8$ $\{-1.27, 6.27\}$ **2.** $x(x + 2) = 5$ $\{-3.45, 1.45\}$

3. $x^2 + 0.5x = 0.3x + 1$ $\{-1.10, 0.90\}$ **4.** $x^2 - 2.6x = -2.2x + 3$ $\{-1.54, 1.94\}$

5. Use a grapher and solve $(2x - 5)^2 = -16$, Example 3 in this section, using the window

    Xmin $= -20$

    Xmax $= 20$

    Xscl $= 1$

    Ymin $= -20$

    Ymax $= 20$

    Yscl $= 1$

Explain the results. Compare your results with the solution found in Example 3.   { }

6. What are the advantages and disadvantages of using a grapher to solve quadratic equations?   Answers will vary.

7. $\{-\sqrt{10}, \sqrt{10}\}$ **11.** $\{6 - 3\sqrt{2}, 6 + 3\sqrt{2}\}$ **12.** $\{-4 - 3\sqrt{3}, -4 + 3\sqrt{3}\}$ **13.** $\left\{\dfrac{3 - 2\sqrt{2}}{2}, \dfrac{3 + 2\sqrt{2}}{2}\right\}$

14. $\left\{\dfrac{-9 - \sqrt{6}}{4}, \dfrac{-9 + \sqrt{6}}{4}\right\}$ **19.** $\{-2i\sqrt{2}, 2i\sqrt{2}\}$ **20.** $\{-2i\sqrt{3}, 2i\sqrt{3}\}$ **21.** $\{1 - 4i, 1 + 4i\}$

EXERCISE SET 8.1

22. $\{-2 - 5i, -2 + 5i\}$ **23.** $\{-7 - \sqrt{5}, -7 + \sqrt{5}\}$ **24.** $\{-10 - \sqrt{11}, -10 + \sqrt{11}\}$

Use the square root property to solve each equation. These equations have real number solutions. See Examples 1 and 2.

1. $x^2 = 16$ $\{-4, 4\}$ **2.** $x^2 = 49$ $\{-7, 7\}$

3. $x^2 - 7 = 0$ $\{-\sqrt{7}, \sqrt{7}\}$ **4.** $x^2 - 11 = 0$ $\{-\sqrt{11}, \sqrt{11}\}$

5. $x^2 = 18$ $\{-3\sqrt{2}, 3\sqrt{2}\}$ **6.** $y^2 = 20$ $\{-2\sqrt{5}, 2\sqrt{5}\}$

7. $3z^2 - 30 = 0$ **8.** $2x^2 = 4$ $\{-\sqrt{2}, \sqrt{2}\}$

9. $(x + 5)^2 = 9$ $\{-8, -2\}$ **10.** $(y - 3)^2 = 4$ $\{1, 5\}$

11. $(z - 6)^2 = 18$ **12.** $(y + 4)^2 = 27$

13. $(2x - 3)^2 = 8$ **14.** $(4x + 9)^2 = 6$

Use the square root property to solve each equation. See Examples 1 and 3.

15. $x^2 + 9 = 0$ $\{-3i, 3i\}$ **16.** $x^2 + 4 = 0$ $\{-2i, 2i\}$

17. $x^2 - 6 = 0$ $\{-\sqrt{6}, \sqrt{6}\}$ **18.** $y^2 - 10 = 0$ $\{-\sqrt{10}, \sqrt{10}\}$

19. $2z^2 + 16 = 0$ **20.** $3p^2 + 36 = 0$

21. $(x - 1)^2 = -16$ **22.** $(y + 2)^2 = -25$

23. $(z + 7)^2 = 5$ **24.** $(x + 10)^2 = 11$

25. $(x + 3)^2 = -8$ **26.** $(y - 4)^2 = -18$
$\{-3 - 2i\sqrt{2}, -3 + 2i\sqrt{2}\}$ $\{4 - 3i\sqrt{2}, 4 + 3i\sqrt{2}\}$

Add the proper constant to each binomial so that the resulting trinomial is a perfect square trinomial. Then factor the trinomial. **27.** $x^2 + 16x + 64 = (x + 8)^2$

27. $x^2 + 16x$ **28.** $y^2 + 2y$

29. $z^2 - 12z$ **30.** $x^2 - 8x$

31. $p^2 + 9p$ **32.** $n^2 + 5n$

33. $x^2 + x$ **34.** $y^2 - y$

Find two possible missing terms so that each is a perfect square trinomial.

❏ **35.** $x^2 + \blacksquare + 16$ $-8x, 8x$ ❏ **36.** $y^2 + \blacksquare + 9$ $-6y, 6y$

❏ **37.** $z^2 + \blacksquare + \dfrac{25}{4}$ $-5z, 5z$ ❏ **38.** $x^2 + \blacksquare + \dfrac{1}{4}$ $-x, x$

Solve each equation by completing the square. These equations have real number solutions. See Examples 4 through 7. **39.** $\{-5, -3\}$ **41.** $\{-3 - \sqrt{7}, -3 + \sqrt{7}\}$

39. $x^2 + 8x = -15$ **40.** $y^2 + 6y = -8$ $\{-4, -2\}$

41. $x^2 + 6x + 2 = 0$ **42.** $x^2 - 2x - 2 = 0$

43. $x^2 + x - 1 = 0$ **44.** $x^2 + 3x - 2 = 0$

45. $x^2 + 2x - 5 = 0$ **46.** $y^2 + y - 7 = 0$

47. $3p^2 - 12p + 2 = 0$ **48.** $2x^2 + 14x - 1 = 0$

49. $4y^2 - 12y - 2 = 0$ **50.** $6x^2 - 3 = 6x$

51. $2x^2 + 7x = 4$ $\left\{-4, \dfrac{1}{2}\right\}$ **52.** $3x^2 - 4x = 4$ $\left\{-\dfrac{2}{3}, 2\right\}$

53. $x^2 - 4x - 5 = 0$ **54.** $y^2 + 6y - 8 = 0$

55. $x^2 + 8x + 1 = 0$ **56.** $x^2 - 10x + 2 = 0$

57. $3y^2 + 6y - 4 = 0$ **58.** $2y^2 + 12y + 3 = 0$

59. $2x^2 - 3x - 5 = 0$ **60.** $5x^2 + 3x - 2 = 0$ $\left\{-1, \dfrac{2}{5}\right\}$

For additional answers, see p. 479.
Solve each equation by completing the square. See Examples 4 through 7. **61.** $\{-1 - i, -1 + i\}$

61. $y^2 + 2y + 2 = 0$ **62.** $x^2 + 4x + 6 = 0$

63. $x^2 - 6x + 3 = 0$ **64.** $x^2 - 7x - 1 = 0$

65. $2a^2 + 8a = -12$ **66.** $3x^2 + 12x = -14$

67. $5x^2 + 15x - 1 = 0$ **68.** $16y^2 + 16y - 1 = 0$

69. $2x^2 - x + 6 = 0$ **70.** $4x^2 - 2x + 5 = 0$

71. $x^2 + 10x + 28 = 0$ **72.** $y^2 + 8y + 18 = 0$

73. $z^2 + 3z - 4 = 0$ $\{-4, 1\}$**74.** $y^2 + y - 2 = 0$ $\{-2, 1\}$

75. $2x^2 - 4x + 3 = 0$ **76.** $9x^2 - 36x = -40$

77. $3x^2 + 3x = 5$ **78.** $5y^2 - 15y = 1$

For additional answers, see p. 479.
Use the formula $A = P(1 + r)^t$ to solve Exercises 79–82. See Example 8. **83.** Answers will vary.

79. Find the rate r at which \$3000 grows to \$4320 in 2 years.   20%

80. Find the rate r at which \$800 grows to \$882 in 2 years.   5%

81. Find the rate at which \$810 grows to \$1000 in 2 years.   11%

82. Find the rate at which \$2000 grows to \$2880 in 2 years.   20%

❏ **83.** In your own words, what is the difference between simple interest and compound interest?

❏ **84.** If you are depositing money in an account that pays 4%, would you prefer the interest to be simple or compound?   compound

❏ **85.** If you are borrowing money at a rate of 10%, would you prefer the interest to be simple or compound?   simple

28. $y^2 + 2y + 1 = (y + 1)^2$ **29.** $z^2 - 12z + 36 = (z - 6)^2$ **30.** $x^2 - 8x + 16 = (x - 4)^2$ **31.** $p^2 + 9p + \dfrac{81}{4} = \left(p + \dfrac{9}{2}\right)^2$

32. $n^2 + 5n + \dfrac{25}{4} = \left(n + \dfrac{5}{2}\right)^2$ **33.** $x^2 + x + \dfrac{1}{4} = \left(x + \dfrac{1}{2}\right)^2$ **34.** $y^2 - y + \dfrac{1}{4} = \left(y - \dfrac{1}{2}\right)^2$ **42.** $\{1 - \sqrt{3}, 1 + \sqrt{3}\}$

43. $\left\{\dfrac{-1 - \sqrt{5}}{2}, \dfrac{-1 + \sqrt{5}}{2}\right\}$ **44.** $\left\{\dfrac{-3 - \sqrt{17}}{2}, \dfrac{-3 + \sqrt{17}}{2}\right\}$ **45.** $\{-1 - \sqrt{6}, -1 + \sqrt{6}\}$ **46.** $\left\{\dfrac{-1 - \sqrt{29}}{2}, \dfrac{-1 + \sqrt{29}}{2}\right\}$

47. $\left\{\dfrac{6 - \sqrt{30}}{3}, \dfrac{6 + \sqrt{30}}{3}\right\}$ **48.** $\left\{\dfrac{-7 - \sqrt{51}}{2}, \dfrac{-7 + \sqrt{51}}{2}\right\}$ **49.** $\left\{\dfrac{3 - \sqrt{11}}{2}, \dfrac{3 + \sqrt{11}}{2}\right\}$ **50.** $\left\{\dfrac{1 - \sqrt{3}}{2}, \dfrac{1 + \sqrt{3}}{2}\right\}$

Neglecting air resistance, the distance s(t) traveled by a freely falling object is given by the function $s(t) = 16t^2$, *where t is time in seconds. Use this formula to solve Exercises 86–89. Round answers to 2 decimal places.*

86. The height of the Columbia Seafirst Center in Seattle is 954 feet. How long would it take an object to fall to the ground from the top of the building? 7.72 sec.

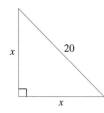

53. $\{-1, 5\}$
54. $\{-3 - \sqrt{17}, -3 + \sqrt{17}\}$
55. $\{-4 - \sqrt{15}, -4 + \sqrt{15}\}$
56. $\{5 - \sqrt{23}, 5 + \sqrt{23}\}$
57. $\left\{\dfrac{-3 - \sqrt{21}}{3}, \dfrac{-3 + \sqrt{21}}{3}\right\}$

87. The height of the First Interstate World Center in Los Angeles is 1017 feet. How long would it take an object to fall from the top of the building to the ground? 7.97 sec.

88. The height of the John Hancock Center in Chicago is 1127 feet. How long would it take an object to fall from the top of the building to the ground? 8.39 sec.

89. The height of the Carew Tower in Cincinnati is 568 feet. How long would it take an object to fall from the top of the building to the ground? 5.96 sec.

Solve.

90. The area of a square room is 225 square feet. Find the dimensions of the room. 15 ft by 15 ft

91. The area of a circle is 36π square inches. Find the radius of the circle. 6 in.

92. An isosceles right triangle has legs of equal length. If the hypotenuse is 20 centimeters long, find the length of each leg. $10\sqrt{2}$ cm

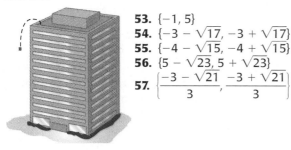

58. $\left\{\dfrac{-6 - \sqrt{30}}{2}, \dfrac{-6 + \sqrt{30}}{2}\right\}$

59. $\left\{-1, \dfrac{5}{2}\right\}$

62. $\{-2 - i\sqrt{2}, -2 + i\sqrt{2}\}$

93. A 27-inch TV is advertised in the *Daily Sentry* newspaper. If 27 inches is the measure of the diag-onal of the picture tube, find the measure of the side of the picture tube. $\dfrac{27\sqrt{2}}{2}$ in.

27 inches

63. $\{3 - \sqrt{6}, 3 + \sqrt{6}\}$

64. $\left\{\dfrac{7 - \sqrt{53}}{2}, \dfrac{7 + \sqrt{53}}{2}\right\}$

65. $\{-2 - i\sqrt{2}, -2 + i\sqrt{2}\}$

66. $\left\{\dfrac{-6 - i\sqrt{6}}{3}, \dfrac{-6 + i\sqrt{6}}{3}\right\}$

A common equation used in business is a demand equation. It expresses the relationship between the unit price of some commodity and the quantity demanded. For Exercises 94 and 95, p represents the unit price and x represents the quantity demanded in thousands.

94. A manufacturing company has found that the demand equation for a certain type of scissors is given by the equation $p = -x^2 + 47$. Find the demand for the scissors if the price is \$11 per pair.

95. Acme, Inc., sells desk lamps and has found that the demand equation for a certain style of desk lamp is given by the equation $p = -x^2 + 15$. Find the demand for the desk lamp if the price is \$7 per lamp. 2.828 thousand units

94. 6 thousand scissors

Review Exercises

Simplify each expression. See Section 7.1.

96. $\dfrac{3}{4} - \sqrt{\dfrac{25}{16}}$ $-\dfrac{1}{2}$ **97.** $\dfrac{3}{5} + \sqrt{\dfrac{16}{25}}$ $\dfrac{7}{5}$

98. $\dfrac{1}{2} - \sqrt{\dfrac{9}{4}}$ -1 **99.** $\dfrac{9}{10} - \sqrt{\dfrac{49}{100}}$ $\dfrac{1}{5}$

Simplify each expression. See Section 7.5.

100. $\dfrac{6 + 4\sqrt{5}}{2}$ $3 + 2\sqrt{5}$ **101.** $\dfrac{10 - 20\sqrt{3}}{2}$ $5 - 10\sqrt{3}$

102. $\dfrac{3 - 9\sqrt{2}}{6}$ $\dfrac{1 - 3\sqrt{2}}{2}$ **103.** $\dfrac{12 - 8\sqrt{7}}{16}$ $\dfrac{3 - 2\sqrt{7}}{4}$

Evaluate $\sqrt{b^2 - 4ac}$ *for each set of values. See Section 7.3.*

104. $a = 2, b = 4, c = -1$ $2\sqrt{6}$

105. $a = 1, b = 6, c = 2$ $2\sqrt{7}$

106. $a = 3, b = -1, c = -2$ 5

107. $a = 1, b = -3, c = -1$ $\sqrt{13}$

67. $\left\{\dfrac{-15 - 7\sqrt{5}}{10}, \dfrac{-15 + 7\sqrt{5}}{10}\right\}$ **68.** $\left\{\dfrac{-2 - \sqrt{5}}{4}, \dfrac{-2 + \sqrt{5}}{4}\right\}$ **69.** $\left\{\dfrac{1 - i\sqrt{47}}{4}, \dfrac{1 + i\sqrt{47}}{4}\right\}$

70. $\left\{\dfrac{1 - i\sqrt{19}}{4}, \dfrac{1 + i\sqrt{19}}{4}\right\}$ **71.** $\{-5 - i\sqrt{3}, -5 + i\sqrt{3}\}$ **72.** $\{-4 - i\sqrt{2}, -4 + i\sqrt{2}\}$ **75.** $\left\{\dfrac{2 - i\sqrt{2}}{2}, \dfrac{2 + i\sqrt{2}}{2}\right\}$

76. $\left\{\dfrac{6 - 2i}{3}, \dfrac{6 + 2i}{3}\right\}$ **77.** $\left\{\dfrac{-3 - \sqrt{69}}{6}, \dfrac{-3 + \sqrt{69}}{6}\right\}$ **78.** $\left\{\dfrac{15 - 7\sqrt{5}}{10}, \dfrac{15 + 7\sqrt{5}}{10}\right\}$

8.2 | SOLVING QUADRATIC EQUATIONS BY THE QUADRATIC FORMULA

TAPE IA 8.2

O B J E C T I V E S

1. Solve quadratic equations by using the quadratic formula.
2. Determine the number and type of solutions of a quadratic equation by using the discriminant.
3. Solve geometric problems that lead to quadratic equations.

1

Any quadratic equation can be solved by completing the square. Since the same sequence of steps is repeated each time we complete the square, let's complete the square for a general quadratic equation, $ax^2 + bx + c = 0$. By doing so, we find a pattern for the solutions of a quadratic equation known as the **quadratic formula.**

Recall that to complete the square for an equation such as $ax^2 + bx + c = 0$, we first divide both sides by the coefficient of x^2.

$$ax^2 + bx + c = 0$$

$$x^2 + \frac{b}{a}x + \frac{c}{a} = 0 \qquad \text{Divide both sides by } a, \text{ the coefficient of } x^2.$$

$$x^2 + \frac{b}{a}x = -\frac{c}{a} \qquad \text{Subtract the constant } \frac{c}{a} \text{ from both sides.}$$

Next, find the square of half $\dfrac{b}{a}$, the coefficient of x.

$$\frac{1}{2}\left(\frac{b}{a}\right) = \frac{b}{2a} \quad \text{and} \quad \left(\frac{b}{2a}\right)^2 = \frac{b^2}{4a^2}$$

Add this result to both sides of the equation:

$$x^2 + \frac{b}{a}x + \frac{b^2}{4a^2} = -\frac{c}{a} + \frac{b^2}{4a^2} \qquad \text{Add } \frac{b^2}{4a^2} \text{ to both sides.}$$

$$x^2 + \frac{b}{a}x + \frac{b^2}{4a^2} = \frac{-c \cdot 4a}{a \cdot 4a} + \frac{b^2}{4a^2} \qquad \begin{array}{l}\text{Find a common denominator on the} \\ \text{right side.}\end{array}$$

$$x^2 + \frac{b}{a}x + \frac{b^2}{4a^2} = \frac{b^2 - 4ac}{4a^2} \qquad \text{Simplify the right side.}$$

$$\left(x + \frac{b}{2a}\right)^2 = \frac{b^2 - 4ac}{4a^2} \qquad \begin{array}{l}\text{Factor the perfect square trinomial} \\ \text{on the left side.}\end{array}$$

$$x + \frac{b}{2a} = \pm\sqrt{\frac{b^2 - 4ac}{4a^2}} \qquad \text{Apply the square root property.}$$

$$x + \frac{b}{2a} = \pm\frac{\sqrt{b^2 - 4ac}}{2a} \qquad \text{Simplify the radical.}$$

$$x = -\frac{b}{2a} \pm \frac{\sqrt{b^2 - 4ac}}{2a} \qquad \text{Subtract } \frac{b}{2a} \text{ from both sides.}$$

$$x = \frac{-b \pm \sqrt{b^2 - 4ac}}{2a} \qquad \text{Simplify.}$$

This equation identifies the solutions of the general quadratic equation in standard form and is called the quadratic formula. It can be used to solve any equation written in standard form $ax^2 + bx + c = 0$ as long as a is not 0.

QUADRATIC FORMULA

A quadratic equation written in the form $ax^2 + bx + c = 0$ has the solutions

$$x = \frac{-b \pm \sqrt{b^2 - 4ac}}{2a}$$

EXAMPLE 1   Solve $3x^2 + 16x + 5 = 0$ for x.

Solution: This equation is in standard form, so $a = 3$, $b = 16$, and $c = 5$. Substitute these values into the quadratic formula:

$$x = \frac{-b \pm \sqrt{b^2 - 4ac}}{2a} \qquad \text{Quadratic formula.}$$

$$= \frac{-16 \pm \sqrt{16^2 - 4(3)(5)}}{2(3)} \qquad \text{Let } a = 3, b = 16, \text{ and } c = 5.$$

$$= \frac{-16 \pm \sqrt{256 - 60}}{6}$$

$$= \frac{-16 \pm \sqrt{196}}{6} = \frac{-16 \pm 14}{6}$$

$$x = \frac{-16 + 14}{6} = -\frac{1}{3} \quad \text{or} \quad x = \frac{-16 - 14}{6} = -\frac{30}{6} = -5$$

The solution set is $\left\{ -\dfrac{1}{3}, -5 \right\}$.

R E M I N D E R   To replace a, b, and c correctly in the quadratic formula, write the quadratic equation in standard form $ax^2 + bx + c = 0$.

EXAMPLE 2   Solve $2x^2 - 4x = 3$.

Solution: First write the equation in standard form by subtracting 3 from both sides:

$$2x^2 - 4x - 3 = 0$$

Now $a = 2, b = -4$, and $c = -3$. Substitute these values into the quadratic formula:

$$x = \frac{-b \pm \sqrt{b^2 - 4ac}}{2a}$$

$$= \frac{-(-4) \pm \sqrt{(-4)^2 - 4(2)(-3)}}{2(2)}$$

$$= \frac{4 \pm \sqrt{16 + 24}}{4}$$

$$= \frac{4 \pm \sqrt{40}}{4} = \frac{4 \pm 2\sqrt{10}}{4}$$

$$= \frac{\boxed{2}\,(2 \pm \sqrt{10})}{\boxed{2} \cdot 2} = \frac{2 \pm \sqrt{10}}{2}$$

The solution set is $\left\{ \dfrac{2 + \sqrt{10}}{2}, \dfrac{2 - \sqrt{10}}{2} \right\}$.

REMINDER   To simplify the expression $\dfrac{4 \pm 2\sqrt{10}}{4}$ in the preceding example, note that 2 is factored out of both terms of the numerator *before* simplifying.

$$\frac{4 \pm 2\sqrt{10}}{4} = \frac{\boxed{2}\,(2 \pm \sqrt{10})}{\boxed{2} \cdot 2} = \frac{2 \pm \sqrt{10}}{2}$$

EXAMPLE 3   Solve $\dfrac{1}{4}m^2 - m + \dfrac{1}{2} = 0$.

Solution: We could use the quadratic formula with $a = \dfrac{1}{4}, b = -1$, and $c = \dfrac{1}{2}$. Instead, we find a simpler, equivalent standard form equation whose coefficients are not fractions.

Multiply both sides of the equation by 4 to clear fractions:

$$4\left(\frac{1}{4}m^2 - m + \frac{1}{2}\right) = 4 \cdot 0$$

$$m^2 - 4m + 2 = 0 \qquad \text{Simplify.}$$

Substitute $a = 1, b = -4$, and $c = 2$ into the quadratic formula and simplify:

$$m = \frac{-(-4) \pm \sqrt{(-4)^2 - 4(1)(2)}}{2(1)} = \frac{4 \pm \sqrt{16 - 8}}{2}$$

$$= \frac{4 \pm \sqrt{8}}{2} = \frac{4 \pm 2\sqrt{2}}{2} = \frac{\boxed{2}\,(2 \pm \sqrt{2})}{\boxed{2}} = 2 \pm \sqrt{2}$$

The solution set is $\{2 + \sqrt{2}, 2 - \sqrt{2}\}$.

EXAMPLE 4   Solve $p = -3p^2 - 3$.

Solution: The equation in standard form is $3p^2 + p + 3 = 0$. Thus, let $a = 3$, $b = 1$, and $c = 3$ in the quadratic formula:

$$p = \frac{-1 \pm \sqrt{1^2 - 4(3)(3)}}{2 \cdot 3} = \frac{-1 \pm \sqrt{1 - 36}}{6} = \frac{-1 \pm \sqrt{-35}}{6} = \frac{-1 \pm i\sqrt{35}}{6}$$

The solution set is $\left\{\dfrac{-1 + i\sqrt{35}}{6}, \dfrac{-1 - i\sqrt{35}}{6}\right\}$.

In the quadratic formula, $x = \dfrac{-b \pm \sqrt{b^2 - 4ac}}{2a}$, the radicand $b^2 - 4ac$ is called the **discriminant** because, by knowing its value, we can **discriminate** among the possible number and type of solutions of a quadratic equation. Possible values of the discriminant and their meanings are summarized next.

> **DISCRIMINANT**
>
> The following table corresponds the discriminant $b^2 - 4ac$ of a quadratic equation of the form $ax^2 + bx + c = 0$ with the number and type of solutions of the equation.
>
> | $b^2 - 4ac$ | NUMBER AND TYPE OF SOLUTIONS |
> | --- | --- |
> | Positive | Two real solutions |
> | Zero | One real solution |
> | Negative | Two complex but not real solutions |

EXAMPLE 5   Use the discriminant to determine the number and type of solutions of each quadratic equation.

    a. $x^2 + 2x + 1 = 0$ **b.** $3x^2 + 2 = 0$ **c.** $2x^2 - 7x - 4 = 0$

Solution: **a.** In $x^2 + 2x + 1 = 0$, $a = 1$, $b = 2$, and $c = 1$. Thus,

$$b^2 - 4ac = 2^2 - 4\,(1)(1) = 0$$

Since $b^2 - 4ac = 0$, this quadratic equation has one real solution.

b. In this equation, $a = 3$, $b = 0$, $c = 2$. Then $b^2 - 4ac = 0 - 4(3)(2) = -24$. Since $b^2 - 4ac$ is negative, the quadratic equation has two complex but not real solutions.

c. In this equation, $a = 2$, $b = -7$, and $c = -4$. Then

$$b^2 - 4ac = 49 - 4(2)(-4) = 81$$

Since $b^2 - 4ac$ is positive, the quadratic equation has two real solutions.

3 The quadratic formula is useful in solving problems that are modeled by quadratic equations.

EXAMPLE 6 The hypotenuse of an isosceles right triangle is 2 centimeters longer than either of its legs. Find the perimeter of the triangle.

Solution: **1.** UNDERSTAND.   Read and reread the problem. Recall that an isosceles right triangle has legs of equal length. Also recall the Pythagorean theorem for right triangles:

$$a^2 + b^2 = c^2$$

2. ASSIGN.

Let

$$x = \text{the length of each leg of the triangle}$$

so that

$$x + 2 = \text{the length of the hypotenuse of the triangle}$$

3. ILLUSTRATE.   Draw a triangle and label it with the assigned variables.

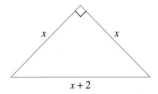

4. TRANSLATE.   By the Pythagorean theorem,

In words: $(\text{leg})^2 + (\text{leg})^2 = (\text{hypotenuse})^2$
Translate: $x^2 + x^2 = (x + 2)^2$

5. COMPLETE.   Solve the quadratic equation:

$$x^2 + x^2 = (x + 2)^2$$
$$x^2 + x^2 = x^2 + 4x + 4 \qquad \text{Square } (x + 2).$$
$$x^2 - 4x - 4 = 0 \qquad\qquad \text{Set the equation equal to 0.}$$

Next substitute $a = 1$, $b = -4$, and $c = -4$ in the quadratic formula.

$$x = \frac{4 \pm \sqrt{16 - 4(1)(-4)}}{2} = \frac{4 \pm \sqrt{32}}{2} = \frac{4 \pm 4\sqrt{2}}{2}$$

$$= \frac{2 \cdot (2 \pm 2\sqrt{2})}{2} = 2 \pm 2\sqrt{2}$$

6. INTERPRET. Since the length of a side cannot be negative, we reject the solution $2 - 2\sqrt{2}$. The length of each leg is $(2 + 2\sqrt{2})$ centimeters. Since the hypotenuse is 2 centimeters longer than the legs, the hypotenuse is $(4 + 2\sqrt{2})$ centimeters. *Check* the lengths of the sides in the Pythagorean theorem. *State:* The perimeter of the triangle is $(2 + 2\sqrt{2})$ cm $+ (2 + 2\sqrt{2})$ cm $+ (4 + 2\sqrt{2})$ cm $= (8 + 6\sqrt{2})$ centimeters. ▬▬▬

7. $\left\{\dfrac{-7 - \sqrt{33}}{2}, \dfrac{-7 + \sqrt{33}}{2}\right\}$ **8.** $\left\{\dfrac{-5 - \sqrt{13}}{2}, \dfrac{-5 + \sqrt{13}}{2}\right\}$ **9.** $\left\{\dfrac{1 - \sqrt{57}}{8}, \dfrac{1 + \sqrt{57}}{8}\right\}$ **10.** $\left\{\dfrac{9 - 5\sqrt{5}}{22}, \dfrac{9 + 5\sqrt{5}}{22}\right\}$

11. $\left\{\dfrac{7 - \sqrt{85}}{6}, \dfrac{7 + \sqrt{85}}{6}\right\}$ **12.** $\left\{\dfrac{5 - \sqrt{77}}{2}, \dfrac{5 + \sqrt{77}}{2}\right\}$ **13.** $\{1 - \sqrt{3}, 1 + \sqrt{3}\}$ **14.** $\{-3 - \sqrt{7}, -3 + \sqrt{7}\}$

17. $\left\{\dfrac{3 - \sqrt{11}}{2}, \dfrac{3 + \sqrt{11}}{2}\right\}$ **18.** $\{1 - \sqrt{2}, 1 + \sqrt{2}\}$ **21.** $\left\{\dfrac{3 - i\sqrt{87}}{8}, \dfrac{3 + i\sqrt{87}}{8}\right\}$

EXERCISE SET 8.2

22. $\left\{\dfrac{-1 - i\sqrt{71}}{18}, \dfrac{-1 + i\sqrt{71}}{18}\right\}$ **23.** $\{-2 - \sqrt{11}, -2 + \sqrt{11}\}$ **24.** $\{-3 - \sqrt{11}, -3 + \sqrt{11}\}$

Use the quadratic formula to solve each equation. These equations have real number solutions. See Examples 1 through 3. **1.** $\{-6, 1\}$

1. $m^2 + 5m - 6 = 0$ **2.** $p^2 + 11p - 12 = 0$ $\{-12, 1\}$

3. $2y = 5y^2 - 3$ $\left\{-\dfrac{3}{5}, 1\right\}$ **4.** $5x^2 - 3 = 14x$ $\left\{-\dfrac{1}{5}, 3\right\}$

5. $x^2 - 6x + 9 = 0$ $\{3\}$ **6.** $y^2 + 10y + 25 = 0$ $\{-5\}$

7. $x^2 + 7x + 4 = 0$ **8.** $y^2 + 5y + 3 = 0$

9. $8m^2 - 2m = 7$ **10.** $11n^2 - 9n = 1$

11. $3m^2 - 7m = 3$ **12.** $x^2 - 13 = 5x$

13. $\dfrac{1}{2}x^2 - x - 1 = 0$ **14.** $\dfrac{1}{6}x^2 + x + \dfrac{1}{3} = 0$

15. $\dfrac{2}{5}y^2 + \dfrac{1}{5}y = \dfrac{3}{5}$ $\left\{-\dfrac{3}{2}, 1\right\}$ **16.** $\dfrac{1}{8}x^2 + x = \dfrac{5}{2}$ $\{-10, 2\}$

17. $\dfrac{1}{3}y^2 - y - \dfrac{1}{6} = 0$ **18.** $\dfrac{1}{2}y^2 = y + \dfrac{1}{2}$

19. Solve Exercise 1 by factoring. Explain the result.

20. Solve Exercise 2 by factoring. Explain the result.
19, 20. Answers will vary.
Use the quadratic formula to solve each equation. See Example 4.

21. $6 = -4x^2 + 3x$ **22.** $9x^2 + x + 2 = 0$

23. $(x + 5)(x - 1) = 2$ **24.** $x(x + 6) = 2$

25. $10y^2 + 10y + 3 = 0$

26. $3y^2 + 6y + 5 = 0$

The solutions of the quadratic equation $ax^2 + bx + c = 0$ are

$$\frac{-b + \sqrt{b^2 - 4ac}}{2a} \quad and \quad \frac{-b - \sqrt{b^2 - 4ac}}{2a}$$

27. Show that the sum of these solutions is $\dfrac{-b}{a}$.

28. Show that the product of these solutions is $\dfrac{c}{a}$.

Use the discriminant to determine the number and types of solutions of each equation. See Example 5.

29. $9x - 2x^2 + 5 = 0$ **30.** $5 - 4x + 12x^2 = 0$

31. $4x^2 + 12x = -9$ **32.** $9x^2 + 1 = 6x$

33. $3x = -2x^2 + 7$ **34.** $3x^2 = 5 - 7x$

35. $6 = 4x - 5x^2$ **36.** $8x = 3 - 9x^2$

Use the quadratic formula to solve each equation. These equations have real number solutions.

37. $x^2 + 5x = -2$ **38.** $y^2 - 8 = 4y$

39. $(m + 2)(2m - 6) = 5(m - 1) - 12$ $\left\{\dfrac{5}{2}, 1\right\}$

40. $7p(p - 2) + 2(p + 4) = 3$ $\left\{\dfrac{5}{7}, 1\right\}$

25. $\left\{\dfrac{-5 - i\sqrt{5}}{10}, \dfrac{-5 + i\sqrt{5}}{10}\right\}$ **26.** $\left\{\dfrac{-3 - i\sqrt{6}}{3}, \dfrac{-3 + i\sqrt{6}}{3}\right\}$ **29.** two real solutions **30.** two complex solutions

31. one real solution **32.** one real solution **33.** two real solutions **34.** two real solutions **35.** two complex solutions

36. two real solutions **37.** $\left\{\dfrac{-5 - \sqrt{17}}{2}, \dfrac{-5 + \sqrt{17}}{2}\right\}$ **38.** $\{2 - 2\sqrt{3}, 2 + 2\sqrt{3}\}$

41. $\left\{\dfrac{3 - \sqrt{29}}{2}, \dfrac{3 + \sqrt{29}}{2}\right\}$ **42.** $\left\{\dfrac{-9 - \sqrt{105}}{2}, \dfrac{-9 + \sqrt{105}}{2}\right\}$ **43.** $\left\{\dfrac{-1 - \sqrt{19}}{6}, \dfrac{-1 + \sqrt{19}}{6}\right\}$

41. $\dfrac{x^2}{3} - x = \dfrac{5}{3}$

42. $\dfrac{x^2}{2} - 3 = -\dfrac{9}{2}x$

43. $x(6x + 2) - 3 = 0$ **44.** $x(7x + 1) = 2$

Use the quadratic formula to solve each equation.

45. $x^2 + 6x + 13 = 0$ $\{-3 - 2i, -3 + 2i\}$

46. $x^2 + 2x + 2 = 0$ $\{-1 - i, -1 + i\}$

47. $\dfrac{2}{5}y^2 + \dfrac{1}{5}y + \dfrac{3}{5} = 0$ $\left\{\dfrac{-1 - i\sqrt{23}}{4}, \dfrac{-1 + i\sqrt{23}}{4}\right\}$

48. $\dfrac{1}{8}x^2 + x + \dfrac{5}{2} = 0$ $\{-4 - 2i, -4 + 2i\}$

49. $\dfrac{1}{2}y^2 = y - \dfrac{1}{2}$ $\{1\}$ **50.** $\dfrac{2}{3}x^2 - \dfrac{20}{3}x = -\dfrac{100}{6}$ $\{5\}$

51. $(n - 2)^2 = 15n$ **52.** $\left(p - \dfrac{1}{2}\right)^2 = \dfrac{p}{2}$

$\left\{\dfrac{3 - \sqrt{5}}{4}, \dfrac{3 + \sqrt{5}}{4}\right\}$

Solve. See Example 6.

53. Uri Chechov's rectangular dog pen for his Irish setter must have an area of 400 square feet. Also, the length must be $2\frac{1}{2}$ times the width. Find the dimensions of the pen. width, $4\sqrt{10}$ ft; length, $10\sqrt{10}$ ft

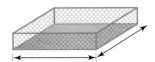

54. The hypotenuse of an isosceles right triangle is 5 inches longer than either of the legs. Find the length of the legs and the length of the hypotenuse.

55. The base of a triangle is twice its height. If the area of the triangle is 42 square centimeters, find its base and height. base, $2\sqrt{42}$ cm; height, $\sqrt{42}$ cm

56. The width of a rectangle is $\frac{1}{3}$ its length. If its area is 12 square inches, find its length and width.

57. An entry in the Peach Festival Poster Contest must be rectangular and have an area of 1200 square inches. Furthermore, its length must be $1\frac{1}{2}$ times its width. Find the dimensions each entry must have. width, $20\sqrt{2}$ in.; height, $30\sqrt{2}$ in.

58. A holding pen for cattle must be square and have a diagonal length of 100 meters.

   a. Find the length of a side of the pen. $50\sqrt{2}$ meters

   b. Find the area of the pen. 5000 sq. meters

59. A rectangle is three times longer than it is wide. It has a diagonal of length 50 centimeters.

44. $\left\{\dfrac{-1 - \sqrt{57}}{14}, \dfrac{-1 + \sqrt{57}}{14}\right\}$ **51.** $\left\{\dfrac{19 - \sqrt{345}}{2}, \dfrac{19 + \sqrt{345}}{2}\right\}$

56. length, 6 in.; width, 2 in.

   a. Find the dimensions of the rectangle.

   b. Find the perimeter of the rectangle. $40\sqrt{10}$ cm

59a. width, $5\sqrt{10}$ cm; length, $15\sqrt{10}$ cm

60. If a point B divides a line segment such that the smaller portion is to the larger portion as the larger is to the whole, the whole is the length of the *golden ratio*.

The golden ratio was thought by the Greeks as the most pleasing to the eye, and many of their buildings contained numerous examples of the golden ratio. The value of the golden ratio is the positive solution of

$$\underset{\text{(larger)}}{\underset{\text{(smaller)}}{\dfrac{x - 1}{1}}} = \underset{\text{(smaller)}}{\underset{\text{(larger)}}{\dfrac{1}{x}}}$$

Find this value. $\dfrac{1 + \sqrt{5}}{2}$

Use the quadratic formula and a calculator to approximate each solution to the nearest tenth.

61. $2x^2 - 6x + 3 = 0$ $\{0.6, 2.4\}$

62. $5x^2 - 8x + 2 = 0$ $\{0.3, 1.3\}$

63. $1.3x^2 - 2.5x - 7.9 = 0$ $\{-1.7, 3.6\}$

64. $3.6x^2 + 1.8x - 4.3 = 0$ $\{-1.4, 0.9\}$

A ball is thrown downward from the top of a 180-foot building with an initial velocity of -20 feet per second. The height of the ball $h(t)$ after t seconds is given by the function

$$h(t) = -16t^2 - 20t + 180$$

65. How long after the ball is thrown will it strike the ground? Round the result to the nearest tenth of a second. 2.8 sec.

66. How long after the ball is thrown will it be 50 feet from the ground? Round the result to the nearest tenth of a second. 2.3 sec.

54. leg, $5 + 5\sqrt{2}$ in.; hypotenuse, $10 + 5\sqrt{2}$ in.

72. $x \approx 4.3$ or $x \approx 1.7$. The temperature was 35° on Monday and Thursday. Yes.

The accompanying graph shows the daily low temperatures for one week in New Orleans, Louisiana.

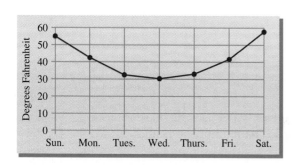

67. Which day of the week shows the greatest decrease in temperature low?   Sunday to Monday

68. Which day of the week shows the greatest increase in temperature low?   Friday to Saturday

69. Which day of the week had the lowest temperature?   Wednesday

70. Use the graph to estimate the low temperature on Thursday.   33°F

Notice that the shape of the temperature graph is similar to a parabola (see Section 5.9). In fact, this graph can be modeled by the quadratic function $f(x) = 3x^2 - 18x + 57$, where $f(x)$ is the temperature in degrees Fahrenheit and x is the number of days from Sunday. Use this function to answer Exercises 71 and 72.

71. Use the quadratic function given to approximate the temperature on Thursday. Does your answer agree with the graph above? $f(4) = 33$; yes

72. Use the function given and the quadratic formula to find when the temperature was 35°F. [*Hint:* Let $f(x) = 35$ and solve for x.] Round your answer to one decimal place and interpret your result. Does your answer agree with the graph above?

73. Use a grapher to solve Exercises 61 and 63.

74. Use a grapher to solve Exercises 62 and 64.
73, 74. See Appendix D.

Review Exercises

Solve each equation. See Sections 6.7 and 7.6.

75. $\sqrt{5x - 2} = 3$ $\left\{\dfrac{11}{5}\right\}$ **76.** $\sqrt{y + 2} + 7 = 12$ $\{23\}$

77. $\dfrac{1}{x} + \dfrac{2}{5} = \dfrac{7}{x}$ $\{15\}$ **78.** $\dfrac{10}{z} = \dfrac{5}{z} - \dfrac{1}{3}$ $\{-15\}$

Factor. See Section 5.7.

79. $x^4 + x^2 - 20$ **80.** $2y^4 + 11y^2 - 6$

81. $z^4 - 13z^2 + 36$ **82.** $x^4 - 1$

79. $(x^2 + 5)(x + 2)(x - 2)$
80. $(2y^2 - 1)(y^2 + 6)$
81. $(z + 3)(z - 3)(z + 2)(z - 2)$
82. $(x^2 + 1)(x - 1)(x + 1)$

A Look Ahead

EXAMPLE

Solve $x^2 - 3\sqrt{2}x + 2 = 0$.

Solution:

In this equation, $a = 1$, $b = -3\sqrt{2}$, and $c = 2$. By the quadratic formula, we have

$$x = \frac{-b \pm \sqrt{b^2 - 4ac}}{2a}$$

$$= \frac{3\sqrt{2} \pm \sqrt{(-3\sqrt{2})^2 - 4(1)(2)}}{2(1)}$$

$$= \frac{3\sqrt{2} \pm \sqrt{18 - 8}}{2} = \frac{3\sqrt{2} \pm \sqrt{10}}{2}$$

The solution set is $\left\{\dfrac{3\sqrt{2} + \sqrt{10}}{2}, \dfrac{3\sqrt{2} - \sqrt{10}}{2}\right\}$.

Use the quadratic formula to solve each quadratic equation. See the preceding example.   **83.** $\left\{\dfrac{\sqrt{3}}{3}\right\}$ **84.** $\left\{\dfrac{-\sqrt{5}}{5}\right\}$

83. $3x^2 - \sqrt{12}x + 1 = 0$

84. $5x^2 + \sqrt{20}x + 1 = 0$ **85.** $\left\{\dfrac{-\sqrt{2} - i\sqrt{2}}{2}, \dfrac{-\sqrt{2} + i\sqrt{2}}{2}\right\}$

85. $x^2 + \sqrt{2}x + 1 = 0$

86. $x^2 - \sqrt{2}x + 1 = 0$ **86.** $\left\{\dfrac{\sqrt{2} - i\sqrt{2}}{2}, \dfrac{\sqrt{2} + i\sqrt{2}}{2}\right\}$

87. $2x^2 - \sqrt{3}x - 1 = 0$

88. $7x^2 + \sqrt{7}x - 2 = 0$ **87.** $\left\{\dfrac{\sqrt{3} - \sqrt{11}}{4}, \dfrac{\sqrt{3} + \sqrt{11}}{4}\right\}$

88. $\left\{\dfrac{-2\sqrt{7}}{7}, \dfrac{\sqrt{7}}{7}\right\}$

8.3   SOLVING EQUATIONS BY USING QUADRATIC METHODS

TAPE IA 8.3

O B J E C T I V E S

1 Solve various equations that are quadratic in form.

2 Solve problems that lead to quadratic equations.

In this section, we discuss various types of equations that can be solved in part by using the methods for solving quadratic equations.

The first example is a rational equation that simplifies to a quadratic equation.

EXAMPLE 1 Solve $\dfrac{3x}{x-2} - \dfrac{x+1}{x} = \dfrac{6}{x(x-2)}$.

Solution: In this equation, x cannot be either 2 or 0, because these values cause denominators to equal zero. To solve for x, first multiply both sides of the equation by $x(x-2)$ to clear fractions. By the distributive property, this means that we multiply each term by $x(x-2)$:

$$x(x-2)\left(\frac{3x}{x-2}\right) - x(x-2)\left(\frac{x+1}{x}\right) = x(x-2)\left[\frac{6}{x(x-2)}\right]$$

$3x^2 - (x-2)(x+1) = 6$	Simplify.
$3x^2 - (x^2 - x - 2) = 6$	Multiply.
$3x^2 - x^2 + x + 2 = 6$	
$2x^2 + x - 4 = 0$	Simplify.
$x = \dfrac{-1 \pm \sqrt{1^2 - 4(2)(-4)}}{2 \cdot 2}$	Let $a = 2$, $b = 1$, and $c = -4$ in the quadratic formula.
$= \dfrac{-1 \pm \sqrt{1 + 32}}{4}$	Simplify.
$= \dfrac{-1 \pm \sqrt{33}}{4}$	

The solution set is $\left\{\dfrac{-1 + \sqrt{33}}{4}, \dfrac{-1 - \sqrt{33}}{4}\right\}$.

EXAMPLE 2 Solve $x^3 = 8$ for x.

Solution: Begin by subtracting 8 from both sides; then factor the resulting difference of cubes.

$x^3 = 8$	
$x^3 - 8 = 0$	
$(x-2)(x^2 + 2x + 4) = 0$	Factor.
$x - 2 = 0$ or $x^2 + 2x + 4 = 0$	Set each factor equal to 0.

The solution of $x - 2 = 0$ is 2. Solve the second equation by the quadratic formula.

$$x = \frac{-2 \pm \sqrt{2^2 - 4(1)(4)}}{2 \cdot 1} = \frac{-2 \pm \sqrt{-12}}{2} \qquad \text{Let } a = 1, b = 2, c = 4.$$

$$= \frac{-2 \pm 2i\sqrt{3}}{2} = \frac{2(-1 \pm i\sqrt{3})}{2}$$

$$= -1 \pm i\sqrt{3}$$

The solution set is $\{2, -1 + i\sqrt{3}, -1 - i\sqrt{3}\}$.

Notice that solutions of $x^3 = 8$ in Example 2 are numbers whose cube is 8. As expected, the real number 2 is a solution since $2^3 = 8$. Unexpectedly, we also found two nonreal numbers whose cube is 8, $-1 + i\sqrt{3}$ and $-1 - i\sqrt{3}$.

EXAMPLE 3   Solve $p^4 - 3p^2 - 4 = 0$.

Solution: First, factor the trinomial:

$$p^4 - 3p^2 - 4 = 0$$
$$(p^2 - 4)(p^2 + 1) = 0 \qquad \text{Factor.}$$
$$(p - 2)(p + 2)(p^2 + 1) = 0 \qquad \text{Factor further.}$$
$$p - 2 = 0 \quad \text{or} \quad p + 2 = 0 \quad \text{or} \quad p^2 + 1 = 0 \qquad \text{Set each factor equal to 0.}$$
$$p = 2 \quad \text{or} \qquad p = -2 \quad \text{or} \qquad p^2 = -1$$
$$p = \pm\sqrt{-1} = \pm i$$

The solution set is $\{2, -2, i, -i\}$.

EXAMPLE 4   Solve $(x - 3)^2 - 3(x - 3) - 4 = 0$.

Solution: Notice that the quantity $(x - 3)$ is repeated in this equation. Sometimes it is helpful to substitute a variable (in this case other than x) for the repeated quantity.
Let $y = x - 3$. Then

$$(x - 3)^2 - 3(x - 3) - 4 = 0$$

becomes

$$y^2 - 3y - 4 = 0 \qquad \text{Let } x - 3 = y.$$
$$(y - 4)(y + 1) = 0 \qquad \text{Factor.}$$

To solve, use the zero factor property:

$$y - 4 = 0 \quad \text{or} \quad y + 1 = 0 \qquad \text{Set each factor equal to 0.}$$
$$y = 4 \quad \text{or} \qquad y = -1 \qquad \text{Solve.}$$

To find values of x, substitute back. That is, let $y = x - 3$:

$$x - 3 = 4 \quad \text{or} \quad x - 3 = -1$$
$$x = 7 \quad \text{or} \quad x = 2$$

The solution set is $\{2, 7\}$.

EXAMPLE 5 Solve $x^{2/3} - 5x^{1/3} + 6 = 0$.

Solution: The key to solving this equation is recognizing that $x^{2/3} = (x^{1/3})^2$. Replace $x^{1/3}$ with m so that

$$(x^{1/3})^2 - 5x^{1/3} + 6 = 0$$

becomes

$$m^2 - 5m + 6 = 0$$

Now solve by factoring:

$$m^2 - 5m + 6 = 0$$
$$(m - 3)(m - 2) = 0 \qquad \text{Factor.}$$
$$m - 3 = 0 \quad \text{or} \quad m - 2 = 0 \qquad \text{Set each factor equal to 0.}$$
$$m = 3 \quad \text{or} \quad m = 2$$

Since $m = x^{1/3}$, we have

$$x^{1/3} = 3 \qquad \text{or} \quad x^{1/3} = 2$$
$$x = 3^3 = 27 \quad \text{or} \quad x = 2^3 = 8.$$

The solution set is $\{8, 27\}$.

The next example is a work problem. This problem is modeled by a rational equation that simplifies to a quadratic equation.

EXAMPLE 6 Together, an experienced typist and an apprentice typist can process a document in 6 hours. Alone, the experienced typist can process the document 2 hours faster than the apprentice typist can. Find the time that each person can process the document alone.

Solution: 1. **UNDERSTAND.** Read and reread the problem. The key idea here is the relationship between the *time* (hours) it takes to complete the job and the *part of the job* completed in one unit of time (hour). For example, because they can complete the job together in 6 hours, the *part of the job* they can complete in 1 hour is $\frac{1}{6}$.

2. **ASSIGN.** Let x represent the *time* in hours it takes the apprentice typist to complete the job alone. Then $x - 2$ represents the time in hours it takes the experienced typist to complete the job alone.

3. ILLUSTRATE.   Here, we summarize in a chart the information discussed.

	TOTAL HOURS TO COMPLETE JOB	**PART OF JOB COMPLETED IN 1 HOUR**
APPRENTICE TYPIST	x	$\dfrac{1}{x}$
EXPERIENCED TYPIST	$x - 2$	$\dfrac{1}{x - 2}$
TOGETHER	6	$\dfrac{1}{6}$

4. TRANSLATE.

In words:	part of job completed by apprentice typist in 1 hour	added to	part of job completed by experienced typist in 1 hour	is equal to	part of job completed together in 1 hour
Translate:	$\dfrac{1}{x}$	$+$	$\dfrac{1}{x - 2}$	$=$	$\dfrac{1}{6}$

5. COMPLETE.

$$\frac{1}{x} + \frac{1}{x - 2} = \frac{1}{6}$$

$$6x(x - 2)\left(\frac{1}{x} + \frac{1}{x - 2}\right) = 6x(x - 2)\left(\frac{1}{6}\right) \quad \text{Multiply both sides by the LCD, } 6x(x - 2).$$

$$6x(x - 2)\left(\frac{1}{x}\right) + 6x(x - 2)\left(\frac{1}{x - 2}\right) = 6x(x - 2)\left(\frac{1}{6}\right) \quad \text{Apply the distributive property.}$$

$$6(x - 2) + 6x = x(x - 2)$$

$$6x - 12 + 6x = x^2 - 2x$$

$$0 = x^2 - 14x + 12$$

Substitute $a = 1$, $b = -14$, and $c = 12$ into the quadratic formula and simplify:

$$x = \frac{-(-14) \pm \sqrt{(-14)^2 - 4(1)(12)}}{2(1)} = \frac{14 \pm \sqrt{148}}{2}$$

Using a calculator or a square root table, we see that $\sqrt{148} \approx 12.2$ rounded to one decimal place. Thus,

$$x \approx \frac{14 \pm 12.2}{2}$$

$$x \approx \frac{14 + 12.2}{2} = 13.1 \quad \text{or} \quad x \approx \frac{14 - 12.2}{2} = 0.9$$

1. $\{3 - \sqrt{7}, 3 + \sqrt{7}\}$ **2.** $\{1 - \sqrt{3}, 1 + \sqrt{3}\}$

3. $\left\{\dfrac{3 - \sqrt{57}}{4}, \dfrac{3 + \sqrt{57}}{4}\right\}$

6. INTERPRET.   *Check:* If the apprentice typist completes the job alone in 0.9 hours, the experienced typist completes the job alone in $x - 2 = 0.9 - 2 = -1.1$ hours. Since this is not possible we reject the solution of 0.9. The approximate solution is thus 13.1 hours.

4. $\left\{\dfrac{9 - \sqrt{105}}{2}, \dfrac{9 + \sqrt{105}}{2}\right\}$

State: The apprentice typist can complete the job alone in approximately 13.1 hours, and the experienced typist completes the job alone in approximately $x - 2 = 13.1 - 2 = 11.1$ hours.

7. $\left\{1, \dfrac{-1 - i\sqrt{3}}{2}, \dfrac{-1 + i\sqrt{3}}{2}\right\}$

9. $\left\{0, -3, \dfrac{3 - 3i\sqrt{3}}{2}, \dfrac{3 + 3i\sqrt{3}}{2}\right\}$ **10.** $\left\{0, -1, \dfrac{1 - i\sqrt{3}}{2}, \dfrac{1 + i\sqrt{3}}{2}\right\}$ **12.** $\left\{-5, \dfrac{5 - 5i\sqrt{3}}{2}, \dfrac{5 + 5i\sqrt{3}}{2}\right\}$

14. $\{-1, 1, -i\sqrt{3}, i\sqrt{3}\}$ **15.** $\left\{-\dfrac{1}{2}, \dfrac{1}{2}, -i\sqrt{3}, i\sqrt{3}\right\}$ **17.** $\{-3, 3, -2, 2\}$ **18.** $\left\{-\dfrac{2}{3}, \dfrac{2}{3}, -i, i\right\}$ **23.** $\left\{-\dfrac{1}{8}, 27\right\}$

EXERCISE SET 8.3

32. $\{-1, 1, -\sqrt{11}, \sqrt{11}\}$

Solve. See Example 1.

1. $\dfrac{2}{x} + \dfrac{3}{x - 1} = 1$ **2.** $\dfrac{6}{x^2} = \dfrac{3}{x + 1}$

3. $\dfrac{3}{x} + \dfrac{4}{x + 2} = 2$ **4.** $\dfrac{5}{x - 2} + \dfrac{4}{x + 2} = 1$

5. $\dfrac{7}{x^2 - 5x + 6} = \dfrac{2x}{x - 3} - \dfrac{x}{x - 2}$ $\left\{\dfrac{1 - \sqrt{29}}{2}, \dfrac{1 + \sqrt{29}}{2}\right\}$

6. $\dfrac{11}{2x^2 + x - 15} = \dfrac{5}{2x - 5} - \dfrac{x}{x + 3}$ $\left\{\dfrac{5 - \sqrt{33}}{2}, \dfrac{5 + \sqrt{33}}{2}\right\}$

Solve. See Example 2.   **8.** $\{-2, 1 - i\sqrt{3}, 1 + i\sqrt{3}\}$

7. $y^3 - 1 = 0$ **8.** $x^3 + 8 = 0$

9. $x^4 + 27x = 0$ **10.** $y^5 + y^2 = 0$

11. $z^3 = 64$ **12.** $z^3 = -125$
$\{4, -2 - 2i\sqrt{3}, -2 + 2i\sqrt{3}\}$

Solve. See Example 3.   **13.** $\{-2, 2, -2i, 2i\}$

13. $p^4 - 16 = 0$ **14.** $x^4 + 2x^2 - 3 = 0$

15. $4x^4 + 11x^2 = 3$ **16.** $z^4 = 81$ $\{-3, 3, -3i, 3i\}$

17. $z^4 - 13z^2 + 36 = 0$ **18.** $9x^4 + 5x^2 - 4 = 0$

Solve. See Examples 4 and 5.   **19.** $\{125, -8\}$

19. $x^{2/3} - 3x^{1/3} - 10 = 0$ **20.** $x^{2/3} + 2x^{1/3} + 1 = 0$ $\{-1\}$

21. $(5n + 1)^2 + 2(5n + 1) - 3 = 0$ $\left\{-\dfrac{4}{5}, 0\right\}$

22. $(m - 6)^2 + 5(m - 6) + 4 = 0$ $\{2, 5\}$

23. $2x^{2/3} - 5x^{1/3} = 3$ **24.** $3x^{2/3} + 11x^{1/3} = 4$ $\left\{-64, \dfrac{1}{27}\right\}$

25. $1 + \dfrac{2}{3t - 2} = \dfrac{8}{(3t - 2)^2}$ $\left\{-\dfrac{2}{3}, \dfrac{4}{3}\right\}$

26. $2 - \dfrac{7}{x + 6} = \dfrac{15}{(x + 6)^2}$ $\left\{-\dfrac{15}{2}, -1\right\}$

27. $20x^{2/3} - 6x^{1/3} - 2 = 0$ **27.** $\left\{-\dfrac{1}{125}, \dfrac{1}{8}\right\}$

28. $4x^{2/3} + 16x^{1/3} = -15$ $\left\{-\dfrac{125}{8}, -\dfrac{27}{8}\right\}$

29. Write a polynomial equation that has three solutions: 2, 5, and -7.   Answers will vary.

30. Write a polynomial equation that has three solutions: 0, $2i$, and $-2i$.   Answers will vary.

Solve each equation.   **31.** $\{-\sqrt{2}, \sqrt{2}, -\sqrt{3}, \sqrt{3}\}$

31. $a^4 - 5a^2 + 6 = 0$ **32.** $x^4 - 12x^2 + 11 = 0$

33. $\dfrac{2x}{x - 2} + \dfrac{x}{x + 3} = \dfrac{-5}{x + 3}$ $\left\{\dfrac{-9 - \sqrt{201}}{6}, \dfrac{-9 + \sqrt{201}}{6}\right\}$

34. $\dfrac{5}{x - 3} + \dfrac{x}{x + 3} = \dfrac{19}{x^2 - 9}$ $\{-1 - \sqrt{5}, -1 + \sqrt{5}\}$

35. $(p + 2)^2 = 9(p + 2) - 20$ $\{2, 3\}$

36. $2(4m - 3)^2 - 9(4m - 3) = 5$ $\left\{\dfrac{5}{8}, 2\right\}$

37. $x^3 + 64 = 0$ **38.** $y^3 - 27 = 0$

39. $x^{2/3} - 8x^{1/3} + 15 = 0$ $\{27, 125\}$

40. $x^{2/3} - 2x^{1/3} - 8 = 0$ $\{-8, 64\}$

41. $y^3 + 9y - y^2 - 9 = 0$ $\{1, -3i, 3i\}$

42. $x^3 + x - 3x^2 - 3 = 0$ $\{3, -i, i\}$

43. $2x^{2/3} + 3x^{1/3} - 2 = 0$ $\left\{\dfrac{1}{8}, -8\right\}$

44. $6x^{2/3} - 25x^{1/3} - 25 = 0$ $\left\{\dfrac{-125}{216}, 125\right\}$

45. $x^{-2} - x^{-1} - 6 = 0$ $\left\{-\dfrac{1}{2}, \dfrac{1}{3}\right\}$

46. $y^{-2} - 8y^{-1} + 7 = 0$ $\left\{\dfrac{1}{7}, 1\right\}$

47. $2x^3 - 250 = 0$

48. $8y^3 + 8 = 0$ $\left\{-1, \dfrac{1 - i\sqrt{3}}{2}, \dfrac{1 + i\sqrt{3}}{2}\right\}$

49. $\dfrac{x}{x - 1} + \dfrac{1}{x + 1} = \dfrac{2}{x^2 - 1}$ $\{-3\}$

50. $\dfrac{x}{x - 5} + \dfrac{5}{x + 5} = \dfrac{-1}{x^2 - 25}$ $\{-12, 2\}$

51. $p^4 - p^2 - 20 = 0$ $\{-\sqrt{5}, \sqrt{5}, -2i, 2i\}$

52. $x^4 - 10x^2 + 9 = 0$ $\{-1, 1, -3, 3\}$

37. $\{-4, 2 - 2i\sqrt{3}, 2 + 2i\sqrt{3}\}$ **38.** $\left\{3, \dfrac{-3 - 3i\sqrt{3}}{2}, \dfrac{-3 + 3i\sqrt{3}}{2}\right\}$ **47.** $\left\{5, \dfrac{-5 - 5i\sqrt{3}}{2}, \dfrac{-5 + 5i\sqrt{3}}{2}\right\}$

53. $\left\{-3, \dfrac{3 - 3i\sqrt{3}}{2}, \dfrac{3 + 3i\sqrt{3}}{2}\right\}$ **54.** $\{6, -3 - 3i\sqrt{3}, -3 + 3i\sqrt{3}\}$

53. $2x^3 = -54$ **54.** $y^3 - 216 = 0$

55. $1 = \dfrac{4}{x - 7} + \dfrac{5}{(x - 7)^2}$ $\{6, 12\}$

56. $3 + \dfrac{1}{(2p + 4)} = \dfrac{10}{(2p + 4)^2}$ $\left\{-\dfrac{7}{6}, -3\right\}$

57. $27y^4 + 15y^2 = 2$ **58.** $8z^4 + 14z^2 = -5$

Solve. If appropriate, use a calculator to approximate each solution to the nearest tenth. See Example 6.

59. Bill Shaughnessy and his son Billy can clean the house together in 4 hours. When the son works alone, it takes him an hour longer to clean than it takes his dad alone. Find how long to the nearest hundredth hour it takes the son to clean alone.   8.53 hr.

60. Together, Scratchy and Freckles eat a 50-pound bag of dog food in 30 days. Scratchy by himself eats a 50-pound bag in 2 weeks less time than Freckles does by himself. How many days to the nearest whole day would a 50-pound bag of dog food last Freckles?   68 days

61. The product of a number and 4 less than the number is 96. Find the number.   12 or -8

62. A whole number increased by its square is two more than twice itself. Find the number.   2

63. An IBM computer and a Toshiba computer can complete a job together in 8 hours. The IBM computer alone can complete the job in 1 hour less time than the Toshiba computer alone. Find the time each computer can complete the job alone.

64. Two fax machines working together can fax a lengthy document in 75 minutes. Fax machine A alone can fax the document in 30 minutes less time than fax machine B alone. Find the time each machine can fax the document alone.

65. Use a grapher to solve Exercise 31. Compare the solution with the solution from Exercise 31. Explain any differences.   See Appendix D.

66. Use a grapher to solve Exercise 48. Compare the solution with the solution from Exercise 48. Explain any differences.   See Appendix D.

Review Exercises

Solve each inequality. See Section 2.4.

67. $\dfrac{5x}{3} + 2 \le 7$ $(-\infty, 3]$ **68.** $\dfrac{2x}{3} + \dfrac{1}{6} \ge 2$ $\left[\dfrac{11}{4}, \infty\right)$

69. $\dfrac{y - 1}{15} > \dfrac{-2}{5}$ $(-5, \infty)$ **70.** $\dfrac{z - 2}{12} < \dfrac{1}{4}$ $(-\infty, 5)$

Find the domain and range of each relation graphed. Decide which relations are also functions. See Section 3.2.

71.

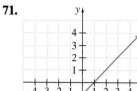

domain: $\{x \,|\, x \text{ is a real number}\}$
range: $\{y \,|\, y \text{ is a real number}\}$
function

72.

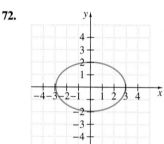

domain: $\{x \,|\, -3 \le x \le 3\}$
range: $\{y \,|\, -2 \le y \le 2\}$
not a function

73.

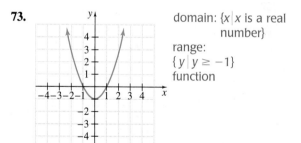

domain: $\{x \,|\, x \text{ is a real number}\}$
range: $\{y \,|\, y \ge -1\}$
function

74.

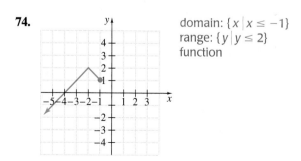

domain: $\{x \,|\, x \le -1\}$
range: $\{y \,|\, y \le 2\}$
function

57. $\left\{-\dfrac{1}{3}, \dfrac{1}{3}, \dfrac{-i\sqrt{6}}{3}, \dfrac{i\sqrt{6}}{3}\right\}$ **58.** $\left\{\dfrac{-i\sqrt{2}}{2}, \dfrac{i\sqrt{2}}{2}, \dfrac{-i\sqrt{5}}{2}, \dfrac{i\sqrt{5}}{2}\right\}$ **63.** Toshiba, 16.5 hr.; IBM, 15.5 hr.

64. machine A, 136.5 min.; machine B, 166.5 min.

TAPE IA 8.4

8.4   NONLINEAR INEQUALITIES IN ONE VARIABLE

O B J E C T I V E S

1. Solve polynomial inequalities of degree 2 or greater.
2. Solve inequalities that contain rational expressions with variables in the denominator.

Just as we can solve linear inequalities in one variable, so we can also solve quadratic inequalities in one variable. A **quadratic inequality** is an inequality that can be written so that one side is a quadratic expression and the other side is 0. Here are examples of quadratic inequalities in one variable. Each is written in **standard form.**

$$x^2 - 10x + 7 \le 0 \qquad 3x^2 + 2x - 6 > 0$$
$$2x^2 + 9x - 2 < 0 \qquad x^2 - 3x + 11 \ge 0$$

A solution of a quadratic inequality in one variable is a value of the variable that makes the inequality a true statement.

The value of an expression such as $x^2 - 3x - 10$ will sometimes be positive, sometimes negative, and sometimes 0, depending on the value substituted for x. To solve the inequality $x^2 - 3x - 10 < 0$, we are looking for all values of x that make the expression $x^2 - 3x - 10$ **less than 0,** or **negative.** To understand how we find these values, we'll study the graph of the quadratic function $y = x^2 - 3x - 10$.

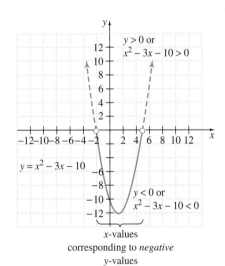

x-values corresponding to *negative* y-values

Notice that the x-values for which y or $x^2 - 3x - 10$ is positive are separated from the x values for which y or $x^2 - 3x - 10$ is negative by the values for which y or $x^2 - 3x - 10$ is 0, the x-intercepts. Thus, the solution set of $x^2 - 3x - 10 < 0$ consists of all real numbers from -2 to 5, or $(-2, 5)$.

It is not necessary to graph $y = x^2 - 3x - 10$ to solve the related inequality $x^2 - 3x - 10 < 0$. Instead, we can draw a number line representing the x-axis and keep the following in mind: **A region on the number line for which the value of $x^2 - 3x - 10$ is positive is separated from a region on the number line for which the value of $x^2 - 3x - 10$ is negative by a value for which the expression is 0.** Find these values for which the expression is 0 by solving the related equation:

$$x^2 - 3x - 10 = 0$$
$$(x - 5)(x + 2) = 0 \qquad \text{Factor.}$$
$$x - 5 = 0 \quad \text{or} \quad x + 2 = 0 \qquad \text{Set each factor equal to 0.}$$
$$x = 5 \quad \text{or} \qquad x = -2 \qquad \text{Solve.}$$

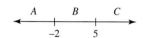

These two numbers divide the number line into three regions. We will call the regions A, B, and C. These regions are important because, if the value of $x^2 - 3x - 10$ is negative when a number from a region is substituted for x, then $x^2 - 3x - 10$ is negative when any number in the region is substituted for x. The same is true if the value of $x^2 - 3x - 10$ is positive for a particular value of x in a region.

To see whether the inequality $x^2 - 3x - 10 < 0$ is true or false in each region, choose a test point from each region and substitute its value for x in the inequality $x^2 - 3x - 10 < 0$. If the resulting inequality is true, the region containing the test point is a solution region.

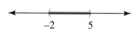

	TEST POINT VALUE	$(x - 5)(x + 2) < 0$	
REGION A	-3	$(-8)(-1) < 0$	False
REGION B	0	$(-5)(2) < 0$	True
REGION C	6	$(1)(8) < 0$	False

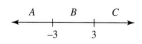

The values in region B satisfy the inequality. The numbers -2 and 5 are not included in the solution set since the inequality symbol is $<$. The solution set is $(-2, 5)$, and its graph is shown.

EXAMPLE 1   Solve $(x + 3)(x - 3) \geq 0$.

Solution: First, solve the related equation $(x + 3)(x - 3) = 0$.

$$(x + 3)(x - 3) = 0$$
$$x + 3 = 0 \quad \text{or} \quad x - 3 = 0$$
$$x = -3 \quad \text{or} \qquad x = 3$$

The two numbers -3 and 3 separate the number line into three regions.

Substitute the value of a test point from each region. If the test value satisfies the inequality, every value in the region containing the test value is a solution.

	TEST POINT VALUE	$(x + 3)(x - 3) \geq 0$	
REGION A	−4	$(-1)(-7) \geq 0$	True
REGION B	0	$(3)(-3) \geq 0$	False
REGION C	4	$(7)(1) \geq 0$	True

The points in regions A and C satisfy the inequality. The numbers -3 and 3 are included in the solution since the inequality symbol is $\geq$. The solution set is $(-\infty, -3] \cup [3, \infty)$, and its graph is shown.

The following steps may be used to solve a polynomial inequality.

TO SOLVE A POLYNOMIAL INEQUALITY

Step 1.   Write the inequality in standard form.

Step 2.   Solve the related equation.

Step 3.   Separate the number line into regions with the solutions from *step 2*.

Step 4.   For each region, choose a test point and determine whether its value satisfies the **original inequality.**

Step 5.   Write the solution set as the union of regions whose test point value is a solution.

EXAMPLE 2   Solve $x^2 - 4x \geq 0$.

Solution: First, solve the related equation $x^2 - 4x = 0$.

$$x^2 - 4x = 0$$
$$x(x - 4) = 0$$
$$x = 0 \quad \text{or} \quad x = 4$$

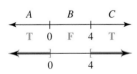

The numbers 0 and 4 separate the number line into three regions.

Check a test value in each region in the original inequality. Values in regions A and C satisfy the inequality. The numbers 0 and 4 are included in the solution since the inequality symbol is $\geq$. The solution set is $(-\infty, 0] \cup [4, \infty)$, and its graph is shown.

EXAMPLE 3   Solve $(x + 2)(x - 1)(x - 5) \leq 0$.

Solution: First, solve $(x + 2)(x - 1)(x - 5) = 0$. By inspection, the solutions are $-2, 1$, and 5. They separate the number line into four regions. Next check test points from each region.

	TEST POINT VALUE	$(x + 2)(x - 1)(x - 5) \leq 0$	
REGION *A*	-3	$(-1)(-4)(-8) \leq 0$	True
REGION *B*	0	$(2)(-1)(-5) \leq 0$	False
REGION *C*	2	$(4)(1)(-3) \leq 0$	True
REGION *D*	6	$(8)(5)(1) \leq 0$	False

The solution set is $(-\infty, -2] \cup [1, 5]$, and its graph is shown. We include the numbers -2, 1, and 5 because the inequality symbol is $\leq$.

2   Inequalities containing rational expressions with variables in the denominator are solved by using a similar procedure.

EXAMPLE 4   Solve $\dfrac{x + 2}{x - 3} \leq 0$.

Solution: First, find all values that make the denominator equal to 0. To do this, solve $x - 3 = 0$, or $x = 3$.

Next, solve the related equation $\dfrac{x + 2}{x - 3} = 0$.

$$\frac{x + 2}{x - 3} = 0 \qquad \text{Multiply both sides by the LCD, } x - 3.$$

$$x + 2 = 0$$

$$x = -2$$

Place these numbers on a number line and proceed as before, checking test point values in the original inequality.

Choose -3 from region *A*.   Choose 0 from region *B*.

$$\frac{x + 2}{x - 3} \leq 0 \qquad\qquad\qquad \frac{x + 2}{x - 3} \leq 0$$

$$\frac{-3 + 2}{-3 - 3} \leq 0 \qquad\qquad\qquad \frac{0 + 2}{0 - 3} \leq 0$$

$$\frac{-1}{-6} \leq 0 \qquad\qquad\qquad\qquad -\frac{2}{3} \leq 0 \qquad \text{True.}$$

$$\frac{1}{6} \leq 0 \qquad \text{False.}$$

Choose 4 from region C.

$$\frac{x + 2}{x - 3} \leq 0$$

$$\frac{4 + 2}{4 - 3} \leq 0$$

$$6 \leq 0 \qquad \text{False.}$$

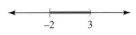

The solution set is $[-2, 3)$. This interval includes -2 because -2 satisfies the original inequality. This interval does not include 3, because 3 would make the denominator 0.

The following steps may be used to solve a rational inequality with variables in the denominator.

TO SOLVE A RATIONAL INEQUALITY

Step 1. Solve for values that make all denominators 0.

Step 2. Solve the related equation.

Step 3. Separate the number line into regions with the solutions from *steps 1* and 2.

Step 4. For each region, choose a test point and determine whether its value satisfies the **original inequality.**

Step 5. Write the solution set as the union of regions whose test point value is a solution.

EXAMPLE 5 Solve $\dfrac{5}{x+1} < -2$.

Solution: First, find values for x that make the denominator equal to 0.

$$x + 1 = 0$$
$$x = -1$$

Next, solve $\dfrac{5}{x+1} = -2$.

$$(x+1) \cdot \frac{5}{x+1} = (x+1) \cdot -2 \qquad \text{Multiply both sides by the LCD, } x+1.$$
$$5 = -2x - 2 \qquad\qquad \text{Simplify.}$$
$$7 = -2x$$
$$-\frac{7}{2} = x$$

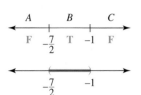

Use these two solutions to divide a number line into three regions and choose test points.

Only a test point value from region B satisfies the **original inequality.** The solution set is $\left(-\dfrac{7}{2}, -1\right)$, and its graph is shown.

EXERCISE SET 8.4

For answer graphs, please see page 526.

Solve each quadratic inequality. Write the solution set in interval notation and graph the solution set. See Examples 1 through 3.

1. $(x + 1)(x + 5) > 0$

2. $(x + 1)(x + 5) \leq 0$

3. $(x - 3)(x + 4) \leq 0$

4. $(x + 4)(x - 1) > 0$

5. $x^2 - 7x + 10 \leq 0$

6. $x^2 + 8x + 15 \geq 0$

7. $3x^2 + 16x < -5$

8. $2x^2 - 5x < 7$

9. $(x - 6)(x - 4)(x - 2) > 0$

10. $(x - 6)(x - 4)(x - 2) \leq 0$

11. $x(x - 1)(x + 4) \leq 0$

12. $x(x - 6)(x + 2) > 0$

13. $(x^2 - 9)(x^2 - 4) > 0$

14. $(x^2 - 16)(x^2 - 1) \leq 0$

Solve each inequality. Write the solution set in interval notation and graph the solution. See Example 4.

15. $\dfrac{x + 7}{x - 2} < 0$

16. $\dfrac{x - 5}{x - 6} > 0$

17. $\dfrac{5}{x + 1} > 0$

18. $\dfrac{3}{y - 5} < 0$

19. $\dfrac{x + 1}{x - 4} \geq 0$

20. $\dfrac{x + 1}{x - 4} \leq 0$

21. Explain why $\dfrac{x + 2}{x - 3} > 0$ and $(x + 2)(x - 3) > 0$ have the same solutions. Answers will vary.

22. Explain why $\dfrac{x + 2}{x - 3} \geq 0$ and $(x + 2)(x - 3) \geq 0$ do not have the same solutions. Answers will vary.

Solve each inequality. Write the solution set in interval notation and graph the solution. See Example 5.

23. $\dfrac{3}{x - 2} < 4$

24. $\dfrac{-2}{y + 3} > 2$

25. $\dfrac{x^2 + 6}{5x} \geq 1$

26. $\dfrac{y^2 + 15}{8y} \leq 1$

Solve each inequality. Write the solution set in interval notation and graph the solution.

27. $(x - 8)(x + 7) > 0$

28. $(x - 5)(x + 1) < 0$

29. $(2x - 3)(4x + 5) \leq 0$

30. $(6x + 7)(7x - 12) > 0$

31. $x^2 > x$

32. $x^2 < 25$

33. $(2x - 8)(x + 4)(x - 6) \leq 0$

34. $(3x - 12)(x + 5)(2x - 3) \geq 0$

35. $6x^2 - 5x \geq 6$

36. $12x^2 + 11x \leq 15$

37. $4x^3 + 16x^2 - 9x - 36 > 0$

38. $x^3 + 2x^2 - 4x - 8 < 0$

39. $x^4 - 26x^2 + 25 \geq 0$

40. $16x^4 - 40x^2 + 9 \leq 0$

41. $(2x - 7)(3x + 5) > 0$

42. $(4x - 9)(2x + 5) < 0$

43. $\dfrac{x}{x - 10} < 0$

44. $\dfrac{x + 10}{x - 10} > 0$

45. $\dfrac{x - 5}{x + 4} \geq 0$

46. $\dfrac{x - 3}{x + 2} \leq 0$

47. $\dfrac{x(x + 6)}{(x - 7)(x + 1)} \geq 0$

48. $\dfrac{(x - 2)(x + 2)}{(x + 1)(x - 4)} \leq 0$

49. $\dfrac{-1}{x - 1} > -1$

50. $\dfrac{4}{y + 2} < -2$

51. $\dfrac{x}{x + 4} \leq 2$

52. $\dfrac{4x}{x - 3} \geq 5$

53. $\dfrac{z}{z - 5} \geq 2z$

54. $\dfrac{p}{p + 4} \leq 3p$

55. $\dfrac{(x + 1)^2}{5x} > 0$

56. $\dfrac{(2x - 3)^2}{x} < 0$

Find all numbers that satisfy each of the following.

57. A number minus its reciprocal is less than zero. Find the numbers.

58. Twice a number added to its reciprocal is nonnegative. Find the numbers.

59. The total profit function $P(x)$ for a company producing x thousand units is given by
$$P(x) = -2x^2 + 26x - 44$$
Find the values of x for which the company makes a profit. [*Hint:* The company makes a profit when $P(x) > 0$.] x is between 2 and 11.

60. A projectile is fired straight up from the ground with an initial velocity of 80 feet per second. Its height $s(t)$ in feet at any time t is given by the function   between 2 and 3 sec.
$$s(t) = -16t^2 + 80t$$
Find the interval of time for which the height of the projectile is greater than 96 feet.

Use a graphing calculator to check the exercise.

61. Exercise 27.

62. Exercise 28.

63. Exercise 39.

64. Exercise 40.

57. any number less than -1 or between 0 and 1   **58.** any number greater than or equal to 0
61–64. See Appendix D.

Review Exercises

Recall that the graph of $f(x) + K$ is the same as the graph of $f(x)$ shifted $|K|$ units upward if $K > 0$ and $|K|$ units downward if $K < 0$. Use the graph of $f(x) = |x|$ below to sketch the graph of each function. (See Sections 3.1 and 3.3.) For Exercises 65–72, see Appendix D.

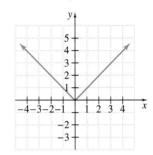

65. $g(x) = |x| + 2$

66. $H(x) = |x| - 2$

67. $F(x) = |x| - 1$

68. $h(x) = |x| + 5$

Use the graph of $f(x) = x^2$ below to sketch the graph of each function.

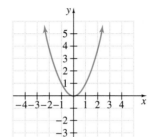

69. $F(x) = x^2 - 3$

70. $h(x) = x^2 - 4$

71. $H(x) = x^2 + 1$

72. $g(x) = x^2 + 3$

8.5 QUADRATIC FUNCTIONS AND THEIR GRAPHS

TAPE IA 8.5

OBJECTIVES

1. Graph quadratic functions of the form $f(x) = x^2 + k$.
2. Graph quadratic functions of the form $f(x) = (x - h)^2$.
3. Graph quadratic functions of the form $f(x) = ax^2$.
4. Graph quadratic functions of the form $f(x) = a(x - h)^2 + k$.

We first graphed the quadratic equation $y = x^2$ in Section 3.1. In Section 3.2, we learned that this graph defines a function, and we wrote $y = x^2$ as $f(x) = x^2$. Quadratic functions and their graphs were studied further in Section 5.9. Throughout these sections, we discovered that the graph of a quadratic function is a parabola opening upward or downward. In this section, we continue our study of quadratic functions and their graphs.

First, let's recall the definition of a quadratic function.

> **QUADRATIC FUNCTION**
>
> A quadratic function is a function that can be written in the form $f(x) = ax^2 + bx + c$, where a, b, and c are real numbers and $a \neq 0$.

Notice that equations of the form $y = ax^2 + bx + c$, where $a \neq 0$, define quadratic functions, since y is a function of x or $y = f(x)$.

Recall that if $a > 0$, the parabola opens upward and if $a < 0$, the parabola opens downward. Also, the vertex of a parabola is the lowest point if the parabola opens upward and the highest point if the parabola opens downward. The axis of symmetry is the vertical line that passes through the vertex.

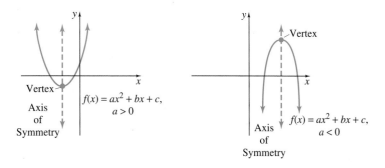

EXAMPLE 1   Sketch the graphs of $f(x) = x^2$ and $g(x) = x^2 + 3$ on the same set of axes.

Solution: Construct a table of values for $f(x)$ and plot the points. Notice that for each x-value, the corresponding value of $g(x)$ must be 3 more than the corresponding value of $f(x)$ since $f(x) = x^2$ and $g(x) = x^2 + 3$. In other words, the graph of $g(x) = x^2 + 3$ is the same as the graph of $f(x) = x^2$ shifted upward 3 units. The axis of symmetry for each graph is the y-axis.

x	$f(x) = x^2$	$g(x) = x^2 + 3$
-2	4	7
-1	1	4
0	0	3
1	1	4
2	4	7

each y-value
is increased
by 3

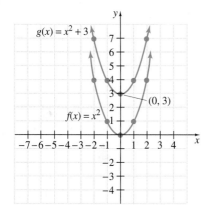

In general, we have the following properties.

GRAPHING THE PARABOLA DEFINED BY $f(x) = x^2 + k$

If k is positive, the graph of $f(x) = x^2 + k$ is the graph of $y = x^2$ shifted upward k units.

If k is negative, the graph of $f(x) = x^2 + k$ is the graph of $y = x^2$ shifted downward $|k|$ units.

The vertex is $(0, k)$, and the axis of symmetry is the y-axis.

EXAMPLE 2   Sketch the graph of each function.

a. $F(x) = x^2 + 2$ **b.** $g(x) = x^2 - 3$

Solution: **a.** The graph of $F(x) = x^2 + 2$ is obtained by shifting the graph of $y = x^2$ upward 2 units.

     b. The graph of $g(x) = x^2 - 3$ is obtained by shifting the graph of $y = x^2$ downward 3 units.

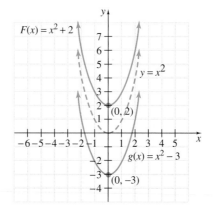

EXAMPLE 3   Sketch the graph of $f(x) = x^2$ and $g(x) = (x - 2)^2$ on the same set of axes.

Solution: By plotting points, we see that for each x-value, the corresponding value of $g(x)$ is the same as the value of $f(x)$ when the x-value is increased by 2. Thus, the graph of $g(x) = (x - 2)^2$ is the graph of $f(x) = x^2$ shifted to the right 2 units. The axis of symmetry for the graph of $g(x) = (x - 2)^2$ is also shifted 2 units to the right and is the line $x = 2$.

x	$f(x) = x^2$
-2	4
-1	1
0	0
1	1
2	4

x	$g(x) = (x - 2)^2$
0	4
1	1
2	0
3	1
4	4

each x-value
increased by 2
corresponds to
same y-value

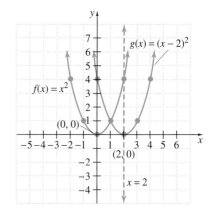

In general, we have the following properties.

GRAPHING THE PARABOLA DEFINED BY $f(x) = (x - h)^2$

If h is positive, the graph of $f(x) = (x - h)^2$ is the graph of $y = x^2$ shifted to the right h units.

If h is negative, the graph of $f(x) = (x - h)^2$ is the graph of $y = x^2$ shifted to the left $|h|$ units.

The vertex is $(h, 0)$, and the line of symmetry has the equation $x = h$.

EXAMPLE 4   Sketch the graph of each function.

    a. $G(x) = (x - 3)^2$ **b.** $F(x) = (x + 1)^2$

Solution: **a.** The graph of $G(x) = (x - 3)^2$ is obtained by shifting the graph of $y = x^2$ to the right 3 units.

    b. The equation $F(x) = (x + 1)^2$ can be written as $F(x) = [x - (-1)]^2$. The graph of $F(x) = [x - (-1)]^2$ is obtained by shifting the graph of $y = x^2$ to the left 1 unit.

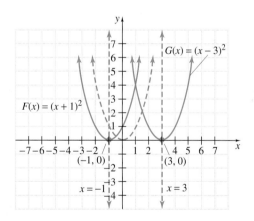

It is possible to combine vertical and horizontal shifts.

EXAMPLE 5   Sketch the graph of $F(x) = (x - 3)^2 + 1$.

Solution: The graph of $F(x) = (x - 3)^2 + 1$ is the graph of $y = x^2$ shifted 3 units to the right and 1 unit up. The vertex is then $(3, 1)$, and the axis of symmetry is $x = 3$. A few ordered pair solutions are plotted to aid in graphing.

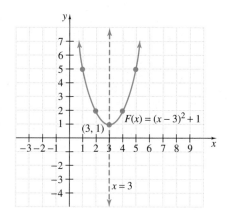

Next, we discover the change in the shape of the graph when the coefficient of x^2 is not 1.

EXAMPLE 6   Sketch the graph of $f(x) = x^2$, $g(x) = 3x^2$, and $h(x) = \dfrac{1}{2}x^2$ on the same set of axes.

Solution: Compare the table of values. We see that for each x-value, the corresponding value of $g(x)$ is triple the corresponding value of $f(x)$. Similarly, the value of $h(x)$ is half the value of $f(x)$. The result is that the graph of $g(x) = 3x^2$ is narrower than the graph of $f(x) = x^2$ and that the graph of $h(x) = \dfrac{1}{2}x^2$ is wider. The vertex for each graph is $(0, 0)$, and the axis of symmetry is the y-axis.

x	$f(x) = x^2$
-2	4
-1	1
0	0
1	1
2	4

x	$g(x) = 3x^2$
-2	12
-1	3
0	0
1	3
2	12

x	$h(x) = \dfrac{1}{2}x^2$
-2	2
-1	$\dfrac{1}{2}$
0	0
1	$\dfrac{1}{2}$
2	2

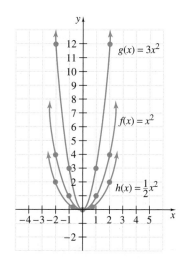

> **GRAPHING THE PARABOLA DEFINED BY $f(x) = ax^2$**
>
> If a is positive, the parabola opens upward, and if a is negative, the parabola opens downward.
>
> If $|a| > 1$, the graph of the parabola is narrower than the graph of $y = x^2$.
>
> If $|a| < 1$, the graph of the parabola is wider than the graph of $y = x^2$.

EXAMPLE 7   Sketch the graph of $f(x) = -2x^2$.

Solution: Because $a = -2$, a negative value, this parabola opens downward. Since $|-2| = 2$ and $2 > 1$, the parabola is narrower than the graph of $y = x^2$. The vertex is $(0, 0)$, and the axis of symmetry is the y-axis. We verify this by plotting a few points.

x	$f(x) = -2x^2$
-2	-8
-1	-2
0	0
1	-2
2	-8

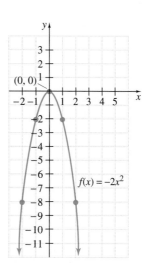

4

EXAMPLE 8   Sketch the graph of $g(x) = \dfrac{1}{2}(x + 2)^2 + 5$. Find the vertex and the axis of symmetry.

Solution: The function $g(x) = \dfrac{1}{2}(x + 2)^2 + 5$ may be written as $g(x) = \dfrac{1}{2}[x - (-2)]^2 + 5$. Thus, this graph is the same as the graph of $y = x^2$ shifted 2 units to the left and 5 units up, and it is wider because a is $\dfrac{1}{2}$. The vertex is $(-2, 5)$, and the axis of symmetry is $x = -2$. We plot a few points to verify.

x	$g(x) = \frac{1}{2}(x + 2)^2 + 5$
-4	7
-3	$5\frac{1}{2}$
-2	5
-1	$5\frac{1}{2}$
0	7

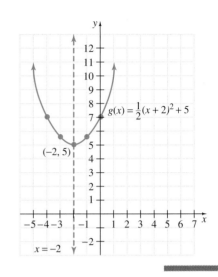

In general, the following holds.

GRAPH OF A QUADRATIC FUNCTION

The graph of a quadratic function written in the form
$f(x) = a(x - h)^2 + k$ is a parabola with vertex (h, k). If $a > 0$, the
parabola opens upward, and if $a < 0$, the parabola opens downward. The
axis of symmetry is the line whose equation is $x = h$.

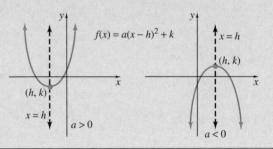

GRAPHING CALCULATOR EXPLORATIONS

*Use a grapher to graph the first function of each pair that follows. Then use its graph to
predict the graph of the second function. Check your prediction by graphing both on the same
set of axes.*   For Exercises 1–6, see Appendix D.

1. $F(x) = \sqrt{x}; G(x) = \sqrt{x} + 1$

2. $g(x) = x^3; H(x) = x^3 - 2$

3. $H(x) = |x|; f(x) = |x - 5|$

4. $h(x) = x^3 + 2; g(x) = (x - 3)^3 + 2$

5. $f(x) = |x + 4|; F(x) = |x + 4| + 3$

6. $G(x) = \sqrt{x} - 2; g(x) = \sqrt{x - 4} - 2$

MENTAL MATH

State the vertex of the graph of each quadratic function.   **7.** $(-1, 5)$ **8.** $(10, -7)$

1. $f(x) = x^2$ $(0, 0)$ **2.** $f(x) = -5x^2$ $(0, 0)$ **3.** $g(x) = (x - 2)^2$ $(2, 0)$ **4.** $g(x) = (x + 5)^2$ $(-5, 0)$

5. $f(x) = 2x^2 + 3$ $(0, 3)$ **6.** $h(x) = x^2 - 1$ $(0, -1)$ **7.** $g(x) = (x + 1)^2 + 5$ **8.** $h(x) = (x - 10)^2 - 7$

EXERCISE SET 8.5   For Exercises 1–54, see Appendix D.

Sketch the graph of each quadratic function. Label the vertex and sketch and label the axis of symmetry. See Examples 1 through 5.

1. $f(x) = x^2 - 1$

2. $g(x) = x^2 + 3$

3. $h(x) = x^2 + 5$

4. $h(x) = x^2 - 4$

5. $g(x) = x^2 + 7$

6. $f(x) = x^2 - 2$

7. $f(x) = (x - 5)^2$

8. $g(x) = (x + 5)^2$

9. $h(x) = (x + 2)^2$

10. $H(x) = (x - 1)^2$

11. $G(x) = (x + 3)^2$

12. $f(x) = (x - 6)^2$

13. $f(x) = (x - 2)^2 + 5$

14. $g(x) = (x - 6)^2 + 1$

15. $h(x) = (x + 1)^2 + 4$

16. $G(x) = (x + 3)^2 + 3$

17. $g(x) = (x + 2)^2 - 5$

18. $h(x) = (x + 4)^2 - 6$

Sketch the graph of each quadratic function. Label the vertex, and sketch and label the axis of symmetry. See Examples 6 and 7.

19. $g(x) = -x^2$

20. $f(x) = 5x^2$

21. $h(x) = \frac{1}{3}x^2$

22. $g(x) = -3x^2$

23. $H(x) = 2x^2$

24. $f(x) = -\frac{1}{4}x^2$

Sketch the graph of each quadratic function. Label the vertex, and sketch and label the axis of symmetry. See Example 8.

25. $f(x) = 2(x - 1)^2 + 3$

26. $g(x) = 4(x - 4)^2 + 2$

27. $h(x) = -3(x + 3)^2 + 1$

28. $f(x) = -(x - 2)^2 - 6$

29. $H(x) = \frac{1}{2}(x - 6)^2 - 3$

30. $G(x) = \frac{1}{5}(x + 4)^2 + 3$

Sketch the graph of each quadratic function. Label the vertex and sketch and label the axis of symmetry.

31. $f(x) = -(x - 2)^2$

32. $g(x) = -(x + 6)^2$

33. $F(x) = -x^2 + 4$

34. $H(x) = -x^2 + 10$

35. $F(x) = 2x^2 - 5$

36. $g(x) = \frac{1}{2}x^2 - 2$

37. $h(x) = (x - 6)^2 + 4$

38. $f(x) = (x - 5)^2 + 2$

39. $F(x) = \left(x + \frac{1}{2}\right)^2 - 2$

40. $H(x) = \left(x + \frac{1}{4}\right)^2 - 3$

41. $F(x) = \frac{3}{2}(x + 7)^2 + 1$

42. $g(x) = -\frac{3}{2}(x - 1)^2 - 5$

43. $f(x) = \frac{1}{4}x^2 - 9$

44. $H(x) = \frac{3}{4}x^2 - 2$

45. $G(x) = 5\left(x + \frac{1}{2}\right)^2$

46. $F(x) = 3\left(x - \frac{3}{2}\right)^2$

47. $h(x) = -(x - 1)^2 - 1$

48. $f(x) = -3(x + 2)^2 + 2$

49. $g(x) = \sqrt{3}(x + 5)^2 + \frac{3}{4}$

50. $G(x) = \sqrt{5}(x - 7)^2 - \frac{1}{2}$

51. $h(x) = 10(x + 4)^2 - 6$

52. $h(x) = 8(x + 1)^2 + 9$

53. $f(x) = -2(x - 4)^2 + 5$

54. $G(x) = -4(x + 9)^2 - 1$

Write the equation of the parabola that has the same shape as $f(x) = 5x^2$ but with the following vertex.

 55. $(2, 3)$

56. $(1, 6)$

 57. $(-3, 6)$

58. $(4, -1)$

The shifting properties covered in this section apply to the graphs of all functions. Given the accompanying graph of $y = f(x)$, sketch the graph of each of the following. For Exercises 59–64, see Appendix D.

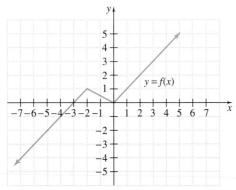

59. $y = f(x) + 1$

60. $y = f(x) - 2$

61. $y = f(x - 3)$

62. $y = f(x + 3)$

63. $y = f(x + 2) + 2$

64. $y = f(x - 1) + 1$

Review Exercises

Add the proper constant to each binomial so that the resulting trinomial is a perfect square trinomial. See Section 8.1.

65. $x^2 + 8x$ $x^2 + 8x + 16$ **66.** $y^2 + 4y$ $y^2 + 4y + 4$

67. $z^2 - 16z$ **68.** $x^2 - 10x$

69. $y^2 + y$ $y^2 + y + \dfrac{1}{4}$ **70.** $z^2 - 3z$ $z^2 - 3z + \dfrac{9}{4}$

Solve by completing the square. See Section 8.1.

71. $x^2 + 4x = 12$ $\{-6, 2\}$ **72.** $y^2 + 6y = -5$ $\{-5, -1\}$

73. $z^2 + 10z - 1 = 0$ **74.** $x^2 + 14x + 20 = 0$

75. $z^2 - 8z = 2$ **76.** $y^2 - 10y = 3$

55. $f(x) = 5(x - 2)^2 + 3$ **56.** $f(x) = 5(x - 1)^2 + 6$ **57.** $f(x) = 5(x + 3)^2 + 6$ **58.** $f(x) = 5(x - 4)^2 - 1$
67. $z^2 - 16z + 64$ **68.** $x^2 - 10x + 25$ **73.** $\{-5 - \sqrt{26}, -5 + \sqrt{26}\}$ **74.** $\{-7 - \sqrt{29}, -7 + \sqrt{29}\}$
75. $\{4 - 3\sqrt{2}, 4 + 3\sqrt{2}\}$ **76.** $\{5 - 2\sqrt{7}, 5 + 2\sqrt{7}\}$

8.6   FURTHER GRAPHING OF QUADRATIC FUNCTIONS

TAPE IA 8.6

O B J E C T I V E S

1   Write quadratic functions in the form $y = a(x - h)^2 + k$.

2   Derive a formula for finding the vertex of a parabola.

3   Find the minimum or maximum value of a quadratic function.

1   We know that the graph of a quadratic function is a parabola. If a quadratic function is written in the form

$$f(x) = a(x - h)^2 + k$$

we can easily find the vertex (h, k) and graph the parabola. To write a quadratic function in this form, complete the square. See Section 8.1 for a review of completing the square.

EXAMPLE 1   Graph $f(x) = x^2 - 4x - 12$.

Solution: The graph of this quadratic function is a parabola. To find the vertex of the parabola, we complete the square on the binomial $x^2 - 4x$. To simplify our work, let $f(x) = y$:

$$y = x^2 - 4x - 12 \qquad \text{Let } f(x) = y.$$
$$y + 12 = x^2 - 4x \qquad \text{Add 12 to both sides to isolate the } x\text{-variable terms.}$$

Now add the square of half of -4 to both sides:

$$\frac{1}{2}(-4) = -2 \quad \text{and} \quad (-2)^2 = 4$$

$$y + 12 + 4 = x^2 - 4x + 4 \qquad \text{Add 4 to both sides.}$$
$$y + 16 = (x - 2)^2 \qquad \text{Factor the trinomial.}$$
$$y = (x - 2)^2 - 16 \qquad \text{Subtract 16 from both sides.}$$
$$f(x) = (x - 2)^2 - 16 \qquad \text{Replace } y \text{ with } f(x).$$

From this equation, we can see that the vertex of the parabola is $(2, -16)$, and the axis of symmetry is the line $x = 2$.

Since $a > 0$, the parabola opens upward. This parabola opening upward with vertex $(2, -16)$ will have two x-intercepts. To find them, let $f(x)$ or $y = 0$:

$$0 = x^2 - 4x - 12$$
$$0 = (x - 6)(x + 2)$$
$$0 = x - 6 \quad \text{or} \quad 0 = x + 2$$
$$6 = x \quad \text{or} \quad -2 = x$$

The two x-intercepts are 6 and -2. The sketch of $f(x) = x^2 - 4x - 12$ is shown.

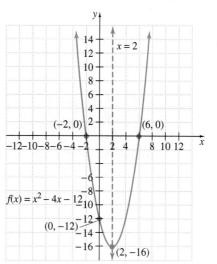

EXAMPLE 2   Graph $f(x) = 3x^2 + 3x + 1$. Find the vertex and any intercepts.

Solution: Replace $f(x)$ with y and complete the square on x to write the equation in the form $y = a(x - h)^2 + k$:

$$y = 3x^2 + 3x + 1 \qquad \text{Replace } f(x) \text{ with } y.$$
$$y - 1 = 3x^2 + 3x \qquad \text{Isolate } x\text{-variable terms.}$$

Factor 3 from the terms $3x^2 + 3x$ so that the coefficient of x^2 is 1:

$$y - 1 = 3(x^2 + x) \qquad \text{Factor out 3.}$$

The coefficient of x is 1. Then $\dfrac{1}{2}(1) = \dfrac{1}{2}$ and $\left(\dfrac{1}{2}\right)^2 = \dfrac{1}{4}$. Since we are adding $\dfrac{1}{4}$ inside the parentheses, we are really adding $3\left(\dfrac{1}{4}\right)$, so we *must* add $3\left(\dfrac{1}{4}\right)$ to the left side:

$$y - 1 + 3\left(\frac{1}{4}\right) = 3\left(x^2 + x + \frac{1}{4}\right)$$
$$y - \frac{1}{4} = 3\left(x + \frac{1}{2}\right)^2 \qquad \begin{array}{c}\text{Simplify the left side and factor the}\\ \text{right side.}\end{array}$$
$$y = 3\left(x + \frac{1}{2}\right)^2 + \frac{1}{4} \qquad \text{Add } \frac{1}{4} \text{ to both sides.}$$
$$f(x) = 3\left(x + \frac{1}{2}\right)^2 + \frac{1}{4} \qquad \text{Replace } y \text{ with } f(x).$$

Then $a = 3$, $h = -\dfrac{1}{2}$, and $k = \dfrac{1}{4}$. This means that the parabola opens upward with vertex $\left(-\dfrac{1}{2}, \dfrac{1}{4}\right)$ and that the axis of symmetry is the line $x = -\dfrac{1}{2}$.

To find the y-intercept, let $x = 0$. Then

$$f(0) = 3(0)^2 + 3(0) + 1 = 1$$

This parabola has no x-intercepts since the vertex is in the second quadrant and opens upward. Use the vertex, axis of symmetry, and y-intercept to sketch the parabola.

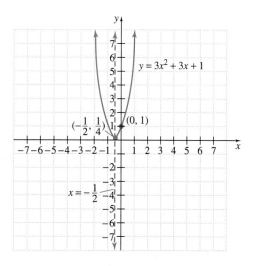

EXAMPLE 3   Graph $f(x) = -x^2 - 2x + 3$. Find the vertex and any intercepts.

Solution: Write $f(x)$ in the form $a(x - h)^2 + k$ by completing the square: First, replace $f(x)$ with y.

$$f(x) = -x^2 - 2x + 3$$
$$y = -x^2 - 2x + 3$$
$$y - 3 = -x^2 - 2x \qquad \text{Subtract 3 from both sides to isolate } x\text{-variable terms.}$$
$$y - 3 = -1(x^2 + 2x) \qquad \text{Factor } -1 \text{ from the terms } -x^2 - 2x.$$

The coefficient of x is 2. Then $\dfrac{1}{2}(2) = 1$ and $1^2 = 1$. Add 1 to the right side inside the parentheses and add $-1(1)$ to the left side:

$$y - 3 - 1(1) = -1(x^2 + 2x + 1)$$
$$y - 4 = -1(x + 1)^2 \qquad \text{Simplify the left side and factor the right side.}$$
$$y = -1(x + 1)^2 + 4 \qquad \text{Add 4 to both sides.}$$
$$f(x) = -1(x + 1)^2 + 4 \qquad \text{Replace } y \text{ with } f(x).$$

Since $a = -1$, the parabola opens downward with vertex $(-1, 4)$ and axis of symmetry $x = -1$.

To find the y-intercept, let $x = 0$ and solve for y. Then

$$f(0) = -0^2 - 2(0) + 3 = 3$$

Thus, 3 is the y-intercept.

To find the x-intercepts, let y or $f(x) = 0$ and solve for x:

$$f(x) = -x^2 - 2x + 3$$
$$0 = -x^2 - 2x + 3 \qquad \text{Let } f(x) = 0.$$

Divide both sides by -1 so that the coefficient of x^2 is 1:

$$\frac{0}{-1} = \frac{-x^2}{-1} - \frac{2x}{-1} + \frac{3}{-1} \qquad \text{Divide both sides by } -1.$$

$$0 = x^2 + 2x - 3 \qquad \text{Simplify.}$$

$$0 = (x + 3)(x - 1) \qquad \text{Factor.}$$

$$x + 3 = 0 \quad \text{or} \quad x - 1 = 0 \qquad \text{Set each factor equal to 0.}$$

$$x = -3 \qquad\qquad x = 1 \qquad \text{Solve.}$$

The x-intercepts are -3 and 1. Use these points to sketch the parabola.

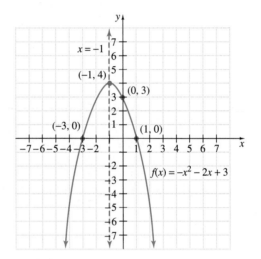

2 Recall from Section 5.9 that we introduced a formula for finding the vertex of a parabola. Now that we have practiced completing the square, we will show that the x-coordinate of the vertex of the graph of $f(x)$ or $y = ax^2 + bx + c$ can be found by the formula $x = \dfrac{-b}{2a}$. To do so, we complete the square on x and write the equation in the form $y = a(x - h)^2 + k$.

First, isolate the x-variable terms by subtracting c from both sides:

$$y = ax^2 + bx + c$$

$$y - c = ax^2 + bx$$

Next, factor a from the terms $ax^2 + bx$:

$$y - c = a\left(x^2 + \frac{b}{a}x\right)$$

Next, add the square of half of $\dfrac{b}{a}$, or $\left(\dfrac{b}{2a}\right)^2 = \dfrac{b^2}{4a^2}$, to the right side inside the parentheses. Because of the factor a, what we really added is $a\left(\dfrac{b^2}{4a^2}\right)$ and this must be added to the left side:

$$y - c + a\left(\frac{b^2}{4a^2}\right) = a\left(x^2 + \frac{b}{a}x + \frac{b^2}{4a^2}\right)$$

$$y - c + \frac{b^2}{4a} = a\left(x + \frac{b}{2a}\right)^2 \qquad \begin{array}{l}\text{Simplify the left side and factor} \\ \text{the right side.}\end{array}$$

$$y = a\left(x + \frac{b}{2a}\right)^2 + c - \frac{b^2}{4a} \qquad \begin{array}{l}\text{Add } c \text{ to both sides and subtract} \\ \dfrac{b^2}{4a} \text{ from both sides.}\end{array}$$

Compare this form with $f(x)$ or $y = a(x - h)^2 + k$ and see that h is $\dfrac{-b}{2a}$, which means that the x-coordinate of the vertex of the graph of $f(x) = ax^2 + bx + c$ is $\dfrac{-b}{2a}$.

VERTEX FORMULA

The graph of $f(x) = ax^2 + bx + c$, when $a \neq 0$, is a parabola with vertex

$$\left(\frac{-b}{2a},\ f\left(\frac{-b}{2a}\right)\right)$$

Let's use this formula to find the vertex of the parabola we graphed in Example 1. In the quadratic function $f(x) = x^2 - 4x - 12$, notice that $a = 1, b = -4$, and $c = -12$. Then

$$\frac{-b}{2a} = \frac{-(-4)}{2(1)} = 2$$

The x-value of the vertex is 2. To find the corresponding $f(x)$ or y-value, find $f(2)$. Then

$$f(2) = 2^2 - 4(2) - 12 = 4 - 8 - 12 = -16$$

The vertex is $(2, -16)$, and the axis of symmetry is the line $x = 2$. These results agree with our findings in Example 1.

3 The quadratic function whose graph is a parabola that opens upward has a minimum value, and the quadratic function whose graph is a parabola that opens downward has a maximum value. The $f(x)$ or y-value of the vertex is the minimum or maximum value of the function.

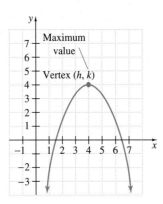 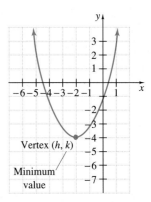

EXAMPLE 4   A rock is thrown upward from the ground. Its height in feet above ground after t seconds is given by the equation $f(t) = -16t^2 + 20t$. Find the maximum height of the rock and the number of seconds it took for the rock to reach its maximum height.

Solution: **1. UNDERSTAND.** The maximum height of the rock is the largest value of $f(t)$. Since the equation $f(t) = -16t^2 + 20t$ is a quadratic function, its graph is a parabola. It opens downward since $-16 < 0$. Thus, the maximum value of $f(t)$ is the $f(t)$ or y-value of the vertex of its graph.

4. TRANSLATE. To find the vertex (h, k), notice that for $f(t) = -16t^2 + 20t$, $a = -16$, $b = 20$, and $c = 0$. Use these values and the vertex formula

$$\left(\frac{-b}{2a}, f\left(\frac{-b}{2a} \right) \right)$$

5. COMPLETE. $h = \dfrac{-b}{2a} = \dfrac{-20}{-32} = \dfrac{5}{8}$

$$f\left(\frac{5}{8} \right) = -16\left(\frac{5}{8} \right)^2 + 20\left(\frac{5}{8} \right) = -16\left(\frac{25}{64} \right) + \frac{25}{2} = -\frac{25}{4} + \frac{50}{4} = \frac{25}{4}$$

6. INTERPRET. The graph of $f(t)$ is a parabola opening downward with vertex $\left(\frac{5}{8}, \frac{25}{4} \right)$. This means that the rock's maximum height is $\frac{25}{4}$ feet, or $6\frac{1}{4}$ feet, which was reached in $\frac{5}{8}$ second.

EXERCISE SET 8.6

3. (5, 30)   **4.** (−4, 18)   **5.** (1, −2)   **6.** (1, 7)   **7.** $\left(\frac{1}{2}, \frac{5}{4} \right)$ **8.** $\left(\frac{9}{2}, -\frac{49}{4} \right)$

Find the vertex of the graph of each quadratic function.
See Example 1.

1. $f(x) = x^2 + 8x + 7$ **2.** $f(x) = x^2 + 6x + 5$
    (−4, −9)   (−3, −4)

3. $f(x) = -x^2 + 10x + 5$ **4.** $f(x) = -x^2 - 8x + 2$
5. $f(x) = 5x^2 - 10x + 3$ **6.** $f(x) = -3x^2 + 6x + 4$
7. $f(x) = -x^2 + x + 1$ **8.** $f(x) = x^2 - 9x + 8$

Find the vertex of the graph of each quadratic function. Determine whether the graph opens upward or downward, find any intercepts, and sketch the graph. See Examples 1 through 3. For Exercises 9–40, see Appendix D.

9. $f(x) = x^2 + 4x - 5$ **10.** $f(x) = x^2 + 2x - 3$

11. $f(x) = -x^2 + 2x - 1$ **12.** $f(x) = -x^2 + 4x - 4$

13. $f(x) = x^2 - 4$ **14.** $f(x) = x^2 - 1$

15. $f(x) = 4x^2 + 4x - 3$ **16.** $f(x) = 2x^2 - x - 3$

17. $f(x) = x^2 + 8x + 15$ **18.** $f(x) = x^2 + 10x + 9$

19. $f(x) = x^2 - 6x + 5$ **20.** $f(x) = x^2 - 4x + 3$

21. $f(x) = x^2 - 4x + 5$ **22.** $f(x) = x^2 - 6x + 11$

23. $f(x) = 2x^2 + 4x + 5$ **24.** $f(x) = 3x^2 + 12x + 16$

25. $f(x) = -2x^2 + 12x$ **26.** $f(x) = -4x^2 + 8x$

27. $f(x) = x^2 + 1$ **28.** $f(x) = x^2 + 4$

29. $f(x) = x^2 - 2x - 15$ **30.** $f(x) = x^2 - 4x + 3$

31. $f(x) = -5x^2 + 5x$ **32.** $f(x) = 3x^2 - 12x$

33. $f(x) = -x^2 + 2x - 12$

34. $f(x) = -x^2 + 8x - 17$

35. $f(x) = 3x^2 - 12x + 15$

36. $f(x) = 2x^2 - 8x + 11$ **37.** $f(x) = x^2 + x - 6$

38. $f(x) = x^2 + 3x - 18$

39. $f(x) = -2x^2 - 3x + 35$

40. $f(x) = 3x^2 - 13x - 10$

Solve. See Example 4.

41. If a projectile is fired straight upward from the ground with an initial speed of 96 feet per second, then its height h in feet after t seconds is given by the equation

$$h(t) = -16t^2 + 96t$$

Find the maximum height of the projectile.   144 ft

42. The cost C in dollars of manufacturing x bicycles at Holladay's Production Plant is given by the function $C(x) = 2x^2 - 800x + 92,000$.

a. Find the number of bicycles that must be manufactured to minimize the cost.   200 bicycles

b. Find the minimum cost. $12,000

43. If Rheam Gaspar throws a ball upward with an initial speed of 32 feet per second, then its height h in feet after t seconds is given by the equation

$$h(t) = -16t^2 + 32t$$

Find the maximum height of the ball.   16 ft

44. The Utah Ski Club sells calendars to raise money. The profit P, in cents, from selling x calendars is given by the equation $P(x) = 360x - x^2$.

a. Find how many calendars must be sold to maximize profit.   180 calendars

b. Find the maximum profit. $32,400

45. Find two positive numbers whose sum is 60 and whose product is as large as possible. [*Hint:* Let x and $60 - x$ be the two positive numbers. Their product can be described by the function $f(x) = x(60 - x)$.]   30 and 30

46. The length and width of a rectangle must have a sum of 40. Find the dimensions of the rectangle that will have maximum area. (Use the hint for Exercise 45.)   length, 20; width, 20

Find the vertex of the graph of each quadratic function. Determine whether the graph opens upward or downward, find the y-intercept, approximate the x-intercepts to one decimal place, and sketch the graph.

47. $f(x) = x^2 + 10x + 15$ For Exercises 47–54, see Appendix D.

48. $f(x) = x^2 - 6x + 4$

49. $f(x) = 3x^2 - 6x + 7$

50. $f(x) = 2x^2 + 4x - 1$

Use a grapher to check each exercise.

51. Exercise 23.   **52.** Exercise 24.

53. Exercise 33.   **54.** Exercise 34.

Find the maximum or minimum value of each function. Approximate to two decimal places.

55. $f(x) = 2.3x^2 - 6.1x + 3.2$ −0.84

56. $f(x) = 7.6x^2 + 9.8x - 2.1$ −5.26

57. $f(x) = -1.9x^2 + 5.6x - 2.7$ 1.43

58. $f(x) = -5.2x^2 - 3.8x + 5.1$ 5.79

Review Exercises

For Exercises 59–68, see Appendix D.
Sketch the graph of each function. See Section 8.5.

59. $f(x) = x^2 + 2$ **60.** $f(x) = (x - 3)^2$

61. $g(x) = x + 2$ **62.** $h(x) = x - 3$

63. $f(x) = (x + 5)^2 + 2$ **64.** $f(x) = 2(x - 3)^2 + 2$

65. $f(x) = 3(x - 4)^2 + 1$ **66.** $f(x) = (x + 1)^2 + 4$

67. $f(x) = -(x - 4)^2 + \dfrac{3}{2}$ **68.** $f(x) = -2(x + 7)^2 + \dfrac{1}{2}$

GROUP ACTIVITY

MODELING THE POSITION OF A FREELY FALLING OBJECT

The quadratic model that describes the position of a freely falling object as a function of time (neglecting air resistance) is

$$s(t) = \frac{1}{2}gt^2 + v_0 t + s_0$$

where t is time in seconds, s_0 is the object's initial height or position in either meters or feet, v_0 is the object's initial velocity in either meters per second or feet per second, and g is the acceleration due to gravity. This function $s(t)$ is also known as a *position function*. Velocity is directional. If the velocity is directed upward, it has a positive value; if it is directed downward, it has a negative value. The acceleration due to gravity on Earth is -9.8 meters per second per second, or -32 feet per second per second. The acceleration due to gravity on the Moon is -1.617 meters per second per second, or -5.28 feet per second per second.

For each of the following descriptions of a freely falling object, (a) match the description to the position function that models it, (b) match the description to the graph that represents it, and (c) find how long the object is in the air (approximate to two decimal places).

1. A rock is thrown *down* a 60-foot-deep well at a velocity of 15 feet per second.

2. A golf ball is thrown *upward* into the air at a velocity of 100 feet per second from a height of 5 feet on the Moon.

3. A bottle rocket is *launched* from sea level at a velocity of 50 meters per second.

4. A forest ranger *drops* a bundle of mail from a 100-meter-tall fire tower.

5. A bottle rocket is *launched* at a velocity of 50 meters per second from the top of a 100-meter-tall fire tower.

6. A sandbag is thrown *toward the ground* with a velocity of 50 feet per second from a hot air balloon at a height of 400 feet.

(a) $s(t) = -4.9t^2 + 100$

(b) $s(t) = -16t^2 + 15t + 60$

(c) $s(t) = -4.9t^2 + 50t + 100$

(d) $s(t) = -4.9t^2 + 50t$

(e) $s(t) = -16t^2 - 50t + 400$

(f) $s(t) = -2.64t^2 + 100t + 5$

(g) $s(t) = -16t^2 - 15t + 60$

(h) $s(t) = -16t^2 + 50t + 400$

(i) $s(t) = -16t^2 + 100$

(i)

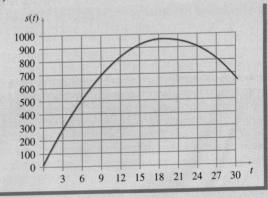

(continued)

(ii)

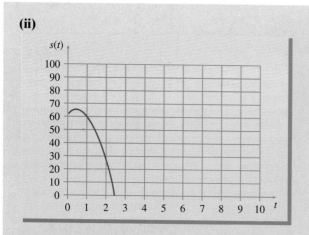

(iii)

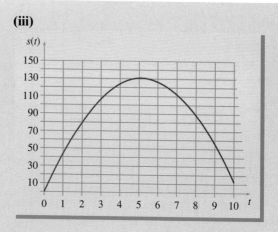

(iv)

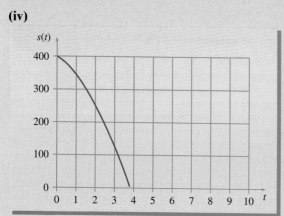

(v)

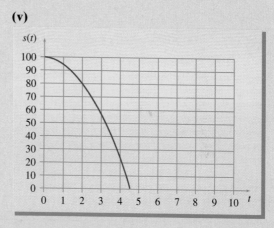

(vi)

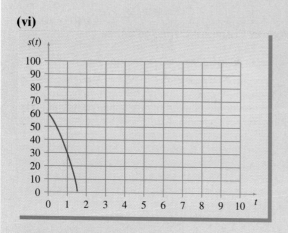

(vii)

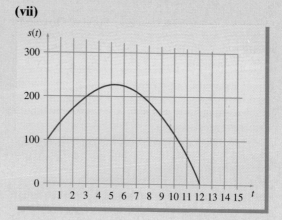

(continued)

(viii)

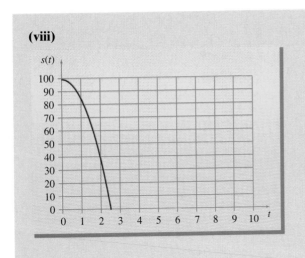

(ix)

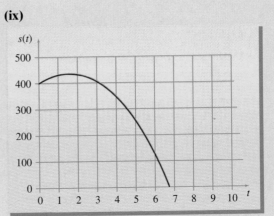

For Group Activity answers and suggestions, see Appendix D.

CHAPTER 8   HIGHLIGHTS

DEFINITIONS AND CONCEPTS	EXAMPLES
SECTION 8.1 SOLVING QUADRATIC EQUATIONS BY COMPLETING THE SQUARE	

Square root property:

If b is a real number and if $a^2 = b$, then $a = \pm\sqrt{b}$.

Solve: $(x + 3)^2 = 14$

$$x + 3 = \pm\sqrt{14}$$

$$x = -3 \pm \sqrt{14}$$

To solve a quadratic equation in x by completing the square:

Step 1. If the coefficient of x^2 is not 1, divide both sides of the equation by the coefficient of x^2.

Step 2. Isolate the variable terms.

Step 3. Complete the square by adding the square of half of the coefficient of x to both sides.

Step 4. Write the resulting trinomial as the square of a binomial.

Step 5. Apply the square root property.

Solve $3x^2 - 12x - 18 = 0$.

1. $x^2 - 4x - 6 = 0$

2. $x^2 - 4x = 6$

3. $\frac{1}{2}(-4) = -2$ and $(-2)^2 = 4$

   $x^2 - 4x + 4 = 6 + 4$

4. $(x - 2)^2 = 10$

5. $x - 2 = \pm\sqrt{10}$

      $x = 2 \pm \sqrt{10}$

DEFINITIONS AND CONCEPTS	EXAMPLES

SECTION 8.2 SOLVING QUADRATIC EQUATIONS BY THE QUADRATIC FORMULA

A quadratic equation written in the form $ax^2 + bx + c = 0$ has solutions

$$x = \frac{-b \pm \sqrt{b^2 - 4ac}}{2a}$$

Solve $x^2 - x - 3 = 0$.

$$a = 1, b = -1, c = -3$$

$$x = \frac{-(-1) \pm \sqrt{(-1)^2 - 4(1)(-3)}}{2 \cdot 1}$$

$$x = \frac{1 \pm \sqrt{13}}{2}$$

SECTION 8.3 SOLVING EQUATIONS USING QUADRATIC METHODS

Substitution is often helpful in solving an equation that contains a repeated variable expression.

Solve $(2x + 1)^2 - 5(2x + 1) + 6 = 0$.

Let $m = 2x + 1$. Then

$$m^2 - 5m + 6 = 0 \quad \text{Let } m = 2x + 1.$$
$$(m - 3)(m - 2) = 0$$

$$m = 3 \quad \text{or} \quad m = 2$$
$$2x + 1 = 3 \quad \text{or} \quad 2x + 1 = 2 \quad \text{Substitute}$$
$$x = 1 \quad \text{or} \quad x = \frac{1}{2} \quad \text{back.}$$

SECTION 8.4 NONLINEAR INEQUALITIES IN ONE VARIABLE

To solve a polynomial inequality:

Step 1. Write the inequality in standard form.

Step 2. Solve the related equation.

Step 3. Use solutions from *step 2* to separate the number line into regions.

Step 4. Use a test point to determine whether values in each region satisfy the original inequality.

Step 5. Write the solution set as the union of regions whose test point value is a solution.

Solve $x^2 \geq 6x$.

1. $x^2 - 6x \geq 0$

2. $x^2 - 6x = 0$

$$x(x - 6) = 0$$
$$x = 0 \quad \text{or} \quad x = 6$$

3.

4.

REGION	TEST POINT VALUE	$x^2 \geq 6x$	
A	-2	$(-2)^2 \geq 6(-2)$	True
B	1	$1^2 \geq 6(1)$	False
C	7	$7^2 \geq 6(7)$	True

5.

The solution set is $(-\infty, 0] \cup [6, \infty)$.

(continued)

DEFINITIONS AND CONCEPTS	EXAMPLES

SECTION 8.4 NONLINEAR INEQUALITIES IN ONE VARIABLE

To solve a rational inequality:

Step 1. Solve for values that make all denominators 0.

Step 2. Solve the related equation.

Step 3. Use solutions from *steps 1* and *2* to separate the number line into regions.

Step 4. Use a test point to determine whether values in each region satisfy the original inequality.

Step 5. Write the solution set as the union of regions whose test point value is a solution.

Solve $\dfrac{6}{x-1} < -2$.

1. $x - 1 = 0$ Set denominator equal to 0.

$\qquad x = 1$

2. $\dfrac{6}{x-1} = -2$

$\qquad 6 = -2(x-1)$ Multiply by $(x-1)$.

$\qquad 6 = -2x + 2$

$\qquad 4 = -2x$

$\qquad -2 = x$

3.

$$\overset{A \qquad\qquad B \qquad\qquad C}{\underset{-2 \qquad\qquad 1}{\longleftarrow\!\!|\!\!\longleftrightarrow\!\!|\!\!\longrightarrow}}$$

4. Only a test value from region B satisfies the original inequality.

5.

$$\underset{-2 \qquad\qquad 1}{\longleftarrow\!\!(\!\!\rule{2cm}{0.4pt}\!\!)\!\!\longrightarrow}$$

The solution set is $(-2, 1)$.

SECTION 8.5 QUADRATIC FUNCTIONS AND THEIR GRAPHS

Graph of a quadratic function

The graph of a quadratic function written in the form $f(x) = a(x-h)^2 + k$ is a parabola with vertex (h, k). If $a > 0$, the parabola opens upward; if $a < 0$, the parabola opens downward. The axis of symmetry is the line whose equation is $x = h$.

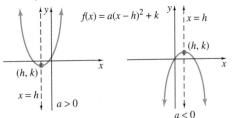

Graph $g(x) = 3(x-1)^2 + 4$.

The graph is a parabola with vertex $(1, 4)$ and axis of symmetry $x = 1$. Since $a = 3$ is positive, the graph opens upward.

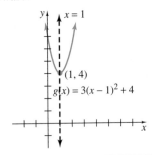

SECTION 8.6 FURTHER GRAPHING OF QUADRATIC FUNCTIONS

The graph of $f(x) = ax^2 + bx + c, a \neq 0$, is a parabola with vertex

$$\left(\frac{-b}{2a}, f\left(\frac{-b}{2a}\right) \right)$$

Graph $f(x) = x^2 - 2x - 8$. Find the vertex and x- and y-intercepts.

$$\frac{-b}{2a} = \frac{-(-2)}{2 \cdot 1} = 1$$

(continued)

DEFINITIONS AND CONCEPTS	EXAMPLES

SECTION 8.6 FURTHER GRAPHING OF QUADRATIC FUNCTIONS

$$f(1) = 1^2 - 2(1) - 8 = -9$$

The vertex is $(1, -9)$.

$$0 = x^2 - 2x - 8$$
$$0 = (x - 4)(x + 2)$$
$$x = 4 \quad \text{or} \quad x = -2$$

The x-intercept points are $(4, 0)$ and $(-2, 0)$.

$$f(0) = 0^2 - 2 \cdot 0 - 8 = -8$$

The y-intercept point is $(0, -8)$.

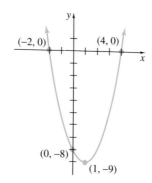

CHAPTER 8 REVIEW

(8.1) *Solve by factoring.*

1. $x^2 - 15x + 14 = 0$ $\{14, 1\}$

2. $x^2 - x - 30 = 0$ $\{-5, 6\}$

3. $10x^2 = 3x + 4$ $\left\{\dfrac{4}{5}, -\dfrac{1}{2}\right\}$ **4.** $7a^2 = 29a + 30$ $\left\{-\dfrac{6}{7}, 5\right\}$

Solve by using the square root property.

5. $4m^2 = 196$ $\{-7, 7\}$ **6.** $9y^2 = 36$ $\{-2, 2\}$

7. $(9n + 1)^2 = 9$ $\left\{-\dfrac{4}{9}, \dfrac{2}{9}\right\}$ **8.** $5x - 2)^2 = 2$ $\left\{\dfrac{2 - \sqrt{2}}{5}, \dfrac{2 + \sqrt{2}}{5}\right\}$

Solve by completing the square.

9. $z^2 + 3z + 1 = 0$ **10.** $x^2 + x + 7 = 0$

11. $(2x + 1)^2 = x$ **12.** $(3x - 4)^2 = 10x$ $\left\{\dfrac{17 - \sqrt{145}}{9}, \dfrac{17 + \sqrt{145}}{9}\right\}$

9. $\left\{\dfrac{-3 - \sqrt{5}}{2}, \dfrac{-3 + \sqrt{5}}{2}\right\}$ **10.** $\left\{\dfrac{-1 - 3i\sqrt{3}}{2}, \dfrac{-1 + 3i\sqrt{3}}{2}\right\}$ **11.** $\left\{\dfrac{-3 - i\sqrt{7}}{8}, \dfrac{-3 + i\sqrt{7}}{8}\right\}$

13. Two ships leave a port at the same time and travel at the same speed. One ship is traveling due north and the other due east. In a few hours, the ships are 150 miles apart. How many miles has each ship traveled? Give an exact answer and a one-decimal-place approximation. $75\sqrt{2}$ miles; 106.1 miles

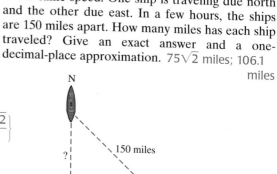

20. $\{-i\sqrt{11}, i\sqrt{11}\}$ **22.** $\left\{\dfrac{5 - i\sqrt{143}}{12}, \dfrac{5 + i\sqrt{143}}{12}\right\}$ **23.** $\left\{\dfrac{1 - i\sqrt{35}}{9}, \dfrac{1 + i\sqrt{35}}{9}\right\}$ **26b.** $\dfrac{15 + \sqrt{321}}{16}$ sec.; 2.1 sec.

(8.2) *If the discriminant of a quadratic equation has the given value, determine the number and type of solutions of the equation.*

14. -8 two complex solutions   **15.** 48   two real solutions

16. 100   two real solutions   **17.** 0   one real solution

Solve by using the quadratic formula.

18. $x^2 - 16x + 64 = 0$ $\{8\}$ **19.** $x^2 + 5x = 0$ $\{-5, 0\}$

20. $x^2 + 11 = 0$ **21.** $2x^2 + 3x = 5$ $\left\{-\dfrac{5}{2}, 1\right\}$

22. $6x^2 + 7 = 5x$ **23.** $9a^2 + 4 = 2a$

24. $(5a - 2)^2 - a = 0$ $\left\{\dfrac{21 - \sqrt{41}}{50}, \dfrac{21 + \sqrt{41}}{50}\right\}$

25. $(2x - 3)^2 = x$ $\left\{1, \dfrac{9}{4}\right\}$

26. Cadets graduating from military school usually toss their hats high into the air at the end of the ceremony. One cadet threw his hat so that its distance $d(t)$ in feet above the ground t seconds after it was thrown was $d(t) = -16t^2 + 30t + 6$.

  a. Find the distance above the ground of the hat 1 second after it was thrown.   20 ft

  b. Find the time it takes the hat to hit the ground. Give an exact time and a one-decimal-place approximation.

27. The hypotenuse of an isosceles right triangle is 6 centimeters longer than either of the legs. Find the length of the legs. $(6 + 6\sqrt{2})$ cm

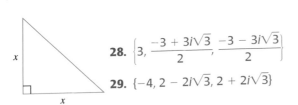

28. $\left\{3, \dfrac{-3 + 3i\sqrt{3}}{2}, \dfrac{-3 - 3i\sqrt{3}}{2}\right\}$

29. $\{-4, 2 - 2i\sqrt{3}, 2 + 2i\sqrt{3}\}$

(8.3) *Solve each equation for the variable.*

28. $x^3 = 27$ **29.** $y^3 = -64$

30. $\dfrac{5}{x} + \dfrac{6}{x - 2} = 3$ $\left\{\dfrac{2}{3}, 5\right\}$ **31.** $\dfrac{7}{8} = \dfrac{8}{x^2}$ $\left\{\dfrac{-8\sqrt{7}}{7}, \dfrac{8\sqrt{7}}{7}\right\}$

32. $x^4 - 21x^2 - 100 = 0$ $\{-5, 5, -2i, 2i\}$

33. $5(x + 3)^2 - 19(x + 3) = 4$ $\left\{-\dfrac{16}{5}, 1\right\}$

34. $x^{2/3} - 6x^{1/3} + 5 = 0$ **35.** $x^{2/3} - 6x^{1/3} = -8$ $\{8, 64\}$

36. $a^6 - a^2 = a^4 - 1$ **37.** $y^{-2} + y^{-1} = 20$ $\left\{-\dfrac{1}{5}, \dfrac{1}{4}\right\}$

38. Two postal workers, Al MacDonell and Tim Bozik, can sort a stack of mail in 5 hours. Working alone, Tim can sort the mail in 1 hour less time

than Al can. Find the time that each postal worker can sort the mail alone. Round the result to one decimal place.   Al, 10.5 hr; Tim, 9.5 hr

39. A negative number decreased by its reciprocal is $-\dfrac{24}{5}$. Find the number.   The number is -5.

40. The formula $A = P(1 + r)^2$ gives the amount A in an account paying interest rate r compounded annually after 2 years of P dollars were originally invested. Find the interest rate r such that $2500 increases to $2717 in 2 years. Round the result to the nearest hundredth of a percent.   4.25%

(8.4) *Solve each inequality for x. Write each solution set in interval notation and then graph the solution set.*

41. $2x^2 - 50 \le 0$ **42.** $\dfrac{1}{4}x^2 < \dfrac{1}{16}$ $\left(-\dfrac{1}{2}, \dfrac{1}{2}\right)$

43. $(2x - 3)(4x + 5) \ge 0$ $\left(-\infty, -\dfrac{5}{4}\right] \cup \left[\dfrac{3}{2}, \infty\right)$

44. $(x^2 - 16)(x^2 - 1) > 0$ $(-\infty, -4) \cup (-1, 1) \cup (4, \infty)$

45. $\dfrac{x - 5}{x - 6} < 0$ **46.** $\dfrac{x(x + 5)}{4x - 3} \ge 0$

47. $\dfrac{(4x + 3)(x - 5)}{x(x + 6)} > 0$ $(-\infty, -6) \cup \left(-\dfrac{3}{4}, 0\right) \cup (5, \infty)$

48. $(x + 5)(x - 6)(x + 2) \le 0$ $(-\infty, -5] \cup [-2, 6]$

49. $x^3 + 3x^2 - 25x - 75 > 0$ $(-5, -3) \cup (5, \infty)$

50. $\dfrac{x^2 + 4}{3x} \le 1$ **51.** $\dfrac{(5x + 6)(x - 3)}{x(6x - 5)} < 0$

52. $\dfrac{3}{x - 2} > 2$ $\left(2, \dfrac{7}{2}\right)$

For Exercises 53–65, see Appendix D.
(8.5) *Sketch the graph of each function. Label the vertex and the axis of symmetry.*

53. $f(x) = x^2 - 4$ **54.** $g(x) = x^2 + 7$

55. $H(x) = 2x^2$ **56.** $h(x) = -\dfrac{1}{3}x^2$

57. $F(x) = (x - 1)^2$ **58.** $G(x) = (x + 5)^2$

59. $f(x) = (x - 4)^2 - 2$

60. $f(x) = -3(x - 1)^2 + 1$

(8.6) *Sketch the graph of each function. Find the vertex and the intercepts.*

61. $f(x) = x^2 + 10x + 25$ **62.** $f(x) = -x^2 + 6x - 9$

34. $\{1, 125\}$ **36.** $\{-1, 1, -i, i\}$ **41.** $[-5, 5]$ **45.**

46. $[-5, 0] \cup \left(\dfrac{3}{4}, \infty\right)$ **50.** $(-\infty, 0)$ **51.** $\left(-\dfrac{6}{5}, 0\right) \cup \left(\dfrac{5}{6}, 3\right)$

45. (5, 6)

63. $f(x) = 4x^2 - 1$

64. $f(x) = -5x^2 + 5$

65. Find the vertex of the graph of $f(x) = -3x^2 - 5x + 4$. Determine whether the graph opens upward or downward, find the y-intercept, approximate the x-intercepts to one decimal place, and sketch the graph.

66. The function $h(t) = -16t^2 + 120t + 300$ gives the height in feet of a projectile fired from the top of a building in t seconds.

66a. 0.4 sec. and 7.1 sec.

a. When will the object reach a height of 350 feet? Round your answer to one decimal place.

b. Explain why part (a) has two answers.

67. Find two numbers whose product is as large as possible, given that their sum is 420.

67. The numbers are both 210.

68. Write an equation of a quadratic function whose graph is a parabola that has vertex $(-3, 7)$ and that passes through the origin.

68. $y = -\frac{7}{9}(x + 3)^2 + 7$

CHAPTER 8 TEST

1. $\left\{\frac{7}{5}, -1\right\}$

2. $\{-1 - \sqrt{10}, -1 + \sqrt{10}\}$

3. $\left\{\frac{1 + i\sqrt{31}}{2}, \frac{1 - i\sqrt{31}}{2}\right\}$

4. $\{3 - \sqrt{7}, 3 + \sqrt{7}\}$

Solve each equation for the variable.

1. $5x^2 - 2x = 7$

2. $(x + 1)^2 = 10$

3. $m^2 - m + 8 = 0$

4. $u^2 - 6u + 2 = 0$

5. $7x^2 + 8x + 1 = 0$

6. $a^2 - 3a = 5$ $\left\{\frac{3 + \sqrt{29}}{2}, \frac{3 - \sqrt{29}}{2}\right\}$

7. $\frac{4}{x + 2} + \frac{2x}{x - 2} = \frac{6}{x^2 - 4}$ $\{-2 - \sqrt{11}, -2 + \sqrt{11}\}$

8. $x^4 - 8x^2 - 9 = 0$ $\{-3, 3, -i, i\}$

9. $x^6 + 1 = x^4 + x^2$ $\{-1, 1, -i, i\}$

10. $(x + 1)^2 - 15(x + 1) + 56 = 0$ $\{6, 7\}$

Solve the equation for the variable by completing the square.

11. $x^2 - 6x = -2$ $\{3 - \sqrt{7}, 3 + \sqrt{7}\}$

12. $2a^2 + 5 = 4a$ $\left\{\frac{2 - i\sqrt{6}}{2}, \frac{2 + i\sqrt{6}}{2}\right\}$

Solve each inequality for x. Graph the solution set and then write the solution set in interval notation.

13. $2x^2 - 7x > 15$

14. $(x^2 - 16)(x^2 - 25) > 0$

15. $\frac{5}{x + 3} < 1$

16. $\frac{7x - 14}{x^2 - 9} \le 0$

Graph each function. Label the vertex. For Exercises 17–20, see Appendix D.

17. $f(x) = 3x^2$

18. $G(x) = -2(x - 1)^2 + 5$

Graph each function. Find and label the vertex, y-intercept, and x-intercepts (if any).

19. $h(x) = x^2 - 4x + 4$

20. $F(x) = 2x^2 - 8x + 9$

21. A 10-foot ladder is leaning against a house. The distance from the bottom of the ladder to the house is 4 feet less than the distance from the top of the ladder to the ground. Find how far the top of the ladder is from the ground. Give an exact answer and a one-decimal-place approximation.

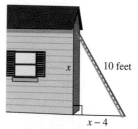

$(2 + \sqrt{46})$ ft $\approx$ 8.8 ft

22. Dave and Sandy Hartranft can paint a room together in 4 hours. Working alone, Dave can paint the room in 2 hours less time than Sandy can. Find how long it takes Sandy to paint the room alone. $(5 + \sqrt{17})$ hr $\approx$ 9.12 hr

23. A stone is thrown upward from a bridge. The stone's height in feet, $s(t)$, above the water t seconds after the stone is thrown is a function given by the equation $s(t) = -16t^2 + 32t + 256$.

a. Find the maximum height of the stone. 272 ft

b. Find the time it takes the stone to hit the water. 5.12 sec.

5. $\left\{-\frac{1}{7}, -1\right\}$

13. $\left(-\infty, -\frac{3}{2}\right] \cup (5, \infty)$

14. $(-\infty, -5) \cup (-4, 4) \cup (5, \infty)$

15. $(-\infty, -3) \cup (2, \infty)$

16. $(-\infty, -3) \cup [2, 3)$

CHAPTER 8 CUMULATIVE REVIEW

1. Find each product.

  a. $(-8)(-1)$ 8 **b.** $(-2)\dfrac{1}{6}$ $-\dfrac{1}{3}$ **c.** $3(-3)$ -9

  d. $(0)(11)$ 0 **e.** $\left(\dfrac{1}{5}\right)\left(-\dfrac{10}{11}\right)$ $-\dfrac{2}{11}$

  f. $(7)(1)(-2)(-3)$ 42 **g.** $8(-2)(0)$ 0; Sec. 1.3, Ex. 5

2. Solve $|2x| + 5 = 7$. $\{-1, 1\}$; Sec. 2.6, Ex. 4

3. If $f(x) = 7x^2 - 3x + 1$ and $g(x) = 3x - 2$, find the following.

  a. $f(1) f(1) = 5$ **b.** $g(1) g(1) = 1$ **c.** $f(-2) f(-2) = 35$

  d. $g(0) g(0) = -2$; Sec. 3.2, Ex. 7

4. Simplify each expression. Write answers with positive exponents.

  a. $\dfrac{x^{-9}}{x^2}$ $\dfrac{1}{x^{11}}$ **b.** $\dfrac{p^4}{p^{-3}}$ p^7 **c.** $\dfrac{2^{-3}}{2^{-1}}$ $\dfrac{1}{4}$

  d. $\dfrac{2x^{-7}y^2}{10xy^{-5}}$ $\dfrac{y^7}{5x^8}$ **e.** $\dfrac{(3x^{-3})(x^2)}{x^6}$ $\dfrac{3}{x^7}$; Sec. 5.1, Ex. 6

5. Simplify by combining like terms.

  a. $-12x^2 + 7x^2 - 6x$ $-5x^2 - 6x$

  b. $3xy - 2x + 5xy - x$ $8xy - 3x$; Sec. 5.3, Ex. 5

6. Factor $16x^2 + 24xy + 9y^2$. $(4x + 3y)^2$; Sec. 5.6, Ex. 8

7. Graph $f(x) = 3x^2 - 12x + 13$. Find the vertex and any intercepts. See Appendix D; Sec. 5.9, Ex. 4

8. Write each rational expression as an equivalent rational expression with the given denominator.

  a. $\dfrac{3x}{2y}$, denominator $10xy^3$ $\dfrac{15x^2y^2}{10xy^3}$

  b. $\dfrac{3x + 1}{x - 5}$, denominator $2x^2 - 11x + 5$ $\dfrac{6x^2 - x - 1}{(x - 5)(2x - 1)}$; Sec. 6.1, Ex. 6

9. If $f(x) = x - 1$ and $g(x) = 2x - 3$, find the following. **9c.** $2x^2 - 5x + 3$

  a. $(f + g)(x)$ $3x - 4$ **b.** $(f - g)(x)$ $-x + 2$

  c. $(f \cdot g)(x)$ **d.** $\left(\dfrac{f}{g}\right)(x)$ $\dfrac{x - 1}{2x - 3}$, $x \neq \dfrac{3}{2}$; Sec. 6.2, Ex. 5

10. Perform the indicated operation. **10a.** $\dfrac{-4x + 15}{x - 3}$

  10b. $\dfrac{4}{x - y}$; Sec. 6.3, Ex. 4

  a. $\dfrac{x}{x - 3} - 5$ **b.** $\dfrac{7}{x - y} + \dfrac{3}{y - x}$

11. Divide $2x^2 - x - 10$ by $x + 2$. $2x - 5$; Sec. 6.5, Ex. 4

12. Use synthetic division to divide $2x^3 - x^2 - 13x + 1$ by $x - 3$.

13. Simplify the following expressions.

  a. $\sqrt[4]{81}$ 3 **b.** $\sqrt[5]{-32}$ -2

  c. $-\sqrt{25}$ -5 **d.** $\sqrt[4]{-81}$ not a real number

  e. $\sqrt[3]{64x^3}$ $4x$; Sec. 7.1, Ex. 3

14. Write with rational exponents.

  a. $\sqrt[5]{x}$ $x^{1/5}$ **b.** $\sqrt[3]{17x^2y^5}$ $17^{1/3}x^{2/3}y^{5/3}$

  c. $\sqrt{x - 5a}$ $(x - 5a)^{1/2}$

  d. $3\sqrt{2p} - 5\sqrt[3]{p^2}$ $3(2p)^{1/2} - 5p^{2/3}$; Sec. 7.2, Ex. 4

15. Simplify.

  a. $\dfrac{\sqrt{45}}{4} - \dfrac{\sqrt{5}}{3}$ $\dfrac{5\sqrt{5}}{12}$

  b. $\sqrt[3]{\dfrac{7x}{8}} + 2\sqrt[3]{7x}$ $\dfrac{5\sqrt[3]{7x}}{2}$; Sec. 7.4, Ex. 2

16. Rationalize the numerator of $\dfrac{\sqrt{x} + 2}{5}$.

17. Solve $\sqrt{2x + 5} + \sqrt{2x} = 3$. $\left\{\dfrac{2}{9}\right\}$; Sec. 7.6, Ex. 5

18. A 50-foot supporting wire is to be attached to a 75-foot antenna. Because of surrounding buildings, sidewalks, and roadways, the wire must be anchored exactly 20 feet from the base of the antenna.

  a. How high from the base of the antenna is the wire attached? $\sqrt{2100}$ feet ≈ 45.8 feet

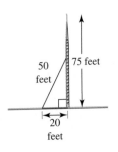

12. $2x^2 + 5x + 2 + \dfrac{7}{x - 3}$; Sec. 6.6, Ex. 1   **16.** $\dfrac{x - 4}{5(\sqrt{x} - 2)}$; Sec. 7.5, Ex. 4

b. Local regulations require that a supporting wire be attached at a height no less than $\frac{3}{5}$ of the total height of the antenna. From part (a), have local regulations been met? yes; *Sec. 7.6, Ex. 7*

19. Use the square root property to solve $(x + 1)^2 = 12$ for x. $\{-1 + 2\sqrt{3}, -1 - 2\sqrt{3}\}$; *Sec. 8.1, Ex. 2*

20. Solve $2x^2 - 4x = 3$.

21. Solve $p^4 - 3p^2 - 4 = 0$. $\{2, -2, i, -i\}$; *Sec. 8.3, Ex. 3*

20. $\left\{\dfrac{2 + \sqrt{10}}{2}, \dfrac{2 - \sqrt{10}}{2}\right\}$; *Sec. 8.2, Ex. 2*

22. Solve $\dfrac{x + 2}{x - 3} \le 0$. $[-2, 3)$; *Sec. 8.4, Ex. 4*

23. Sketch the graph of $F(x) = (x - 3)^2 + 1$.

24. A rock is thrown upward from the ground. Its height in feet above ground after t seconds is given by the equation $f(t) = -16t^2 + 20t$. Find the maximum height of the rock and the number of seconds it took for the rock to reach that maximum height.

23. See Appendix D; *Sec. 8.5, Ex. 5*

24. maximum height is $6\dfrac{1}{4}$ feet in $\dfrac{5}{8}$ second; *Sec. 8.6, Ex. 4*

The following are answer graphs for Exercise Set 8.4.

1. −5 −1 $(-\infty, -5) \cup (-1, \infty)$ **2.** −5 −1 $[-5, -1]$ **3.** −4 3 $[-4, 3]$

4. −4 1 $(-\infty, -4) \cup (1, \infty)$ **5.** 2 5 $[2, 5]$ **6.** −5 −3 $(-\infty, -5] \cup [-3, \infty)$

7. −5 $-\frac{1}{3}$ $\left(-5, -\frac{1}{3}\right)$ **8.** −1 $\frac{7}{2}$ $\left(-1, \frac{7}{2}\right)$ **9.** 2 4 6 $(2, 4) \cup (6, \infty)$

10. 2 4 6 $(-\infty, 2] \cup [4, 6]$ **11.** −4 0 1 $(-\infty, -4] \cup [0, 1]$ **12.** −2 0 6 $(-2, 0) \cup (6, \infty)$

13. −3 −2 2 3 $(-\infty, -3) \cup (-2, 2) \cup (3, \infty)$ **14.** −4 −1 1 4 $[-4, -1] \cup [1, 4]$

15. −7 2 $(-7, 2)$ **16.** 5 6 $(-\infty, 5) \cup (6, \infty)$ **17.** −1 $(-1, \infty)$

18. 5 $(-\infty, 5)$ **19.** −1 4 $(-\infty, -1] \cup (4, \infty)$ **20.** −1 4 $[-1, 4)$

23. 2 $\frac{11}{4}$ $(-\infty, 2) \cup \left(\frac{11}{4}, \infty\right)$ **24.** −4 −3 $(-4, -3)$ **25.** 0 2 3 $(0, 2] \cup [3, \infty)$

26. 0 3 5 $(-\infty, 0) \cup [3, 5]$ **27.** −7 8 $(-\infty, -7) \cup (8, \infty)$ **28.** −1 5 $(-1, 5)$

29. $-\frac{5}{4}$ $\frac{3}{2}$ $\left[-\frac{5}{4}, \frac{3}{2}\right]$ **30.** $-\frac{7}{6}$ $\frac{12}{7}$ $\left(-\infty, -\frac{7}{6}\right) \cup \left(\frac{12}{7}, \infty\right)$ **31.** 0 1 $(-\infty, 0) \cup (1, \infty)$

32. −5 5 $(-5, 5)$ **33.** −4 4 6 $(-\infty, -4] \cup [4, 6]$ **34.** −5 $\frac{3}{2}$ 4 $\left[-5, \frac{3}{2}\right] \cup [4, \infty)$

35. $-\frac{2}{3}$ $\frac{3}{2}$ $\left(-\infty, -\frac{2}{3}\right] \cup \left[\frac{3}{2}, \infty\right)$ **36.** $-\frac{5}{3}$ $\frac{3}{4}$ $\left[-\frac{5}{3}, \frac{3}{4}\right]$ **37.** −4 $-\frac{3}{2}$ $\frac{3}{2}$ $\left(-4, -\frac{3}{2}\right) \cup \left(\frac{3}{2}, \infty\right)$

38. 2 $(-\infty, 2)$ **39.** −5 −1 1 5 $(-\infty, -5] \cup [-1, 1] \cup [5, \infty)$

40. $-\frac{3}{2}$ $-\frac{1}{2}$ $\frac{1}{2}$ $\frac{3}{2}$ $\left[-\frac{3}{2}, -\frac{1}{2}\right] \cup \left[\frac{1}{2}, \frac{3}{2}\right]$ **41.** $-\frac{5}{3}$ $\frac{7}{2}$ $\left(-\infty, -\frac{5}{3}\right) \cup \left(\frac{7}{2}, \infty\right)$ **42.** $-\frac{5}{2}$ $\frac{9}{4}$ $\left(-\frac{5}{2}, \frac{9}{4}\right)$

43. 0 10 $(0, 10)$ **44.** −10 10 $(-\infty, -10) \cup (10, \infty)$ **45.** −4 5 $(-\infty, -4) \cup [5, \infty)$

46. −2 3 $(-2, 3]$ **47.** −6 −1 0 7 $(-\infty, -6] \cup (-1, 0] \cup (7, \infty)$

48. −2 −1 2 4 $[-2, -1) \cup [2, 4)$ **49.** 1 2 $(-\infty, 1) \cup (2, \infty)$ **50.** −4 −2 $(-4, -2)$

51. −8 −4 $(-\infty, -8] \cup (-4, \infty)$ **52.** 3 15 $(3, 15]$ **53.** 0 5 $\frac{11}{2}$ $(-\infty, 0] \cup \left(5, \frac{11}{2}\right]$

54. −4 $-\frac{11}{3}$ 0 $\left(-4, -\frac{11}{3}\right] \cup [0, \infty)$ **55.** 0 $(0, \infty)$ **56.** 0 $(-\infty, 0)$

CONIC SECTIONS

MODELING CONIC SECTIONS

The shapes of conic sections are used in a variety of applications. They are used in architecture in the design of bridges, arches, and vaults. They are used in astronomy to model the orbits of planets, comets, and satellites. The are used also in engineering in the design of certain gears and reflectors. In blueprints or diagrams of these situations, the exact shape of the conic section involved must be depicted.

IN THE CHAPTER GROUP ACTIVITY ON PAGE 557, YOU WILL HAVE THE OPPORTUNITY TO DRAW THE SHAPES OF SEVERAL CONIC SECTIONS.

In Chapter 8, we analyzed some of the important connections between a parabola and its equation. Parabolas are interesting in their own right but are more interesting still because they are part of a collection of curves known as conic sections. This chapter is devoted to quadratic equations in two variables and their conic section graphs: the parabola, circle, ellipse, and hyperbola. ▬▬▬

9.1 THE PARABOLA AND THE CIRCLE

O B J E C T I V E S

1. Graph parabolas of the forms $x = a(y - k)^2 + h$ and $y = a(x - h)^2 + k$.
2. Use the distance formula and the midpoint formula.
3. Graph circles of the form $(x - h)^2 + (y - k)^2 = r^2$.
4. Write the equation of a circle given its center and radius.
5. Find the center and the radius of a circle, given its equation.

TAPE IA 9.1

Conic sections derive their name because each conic section is the intersection of a right circular cone and a plane. The circle, parabola, ellipse, and hyperbola are the conic sections.

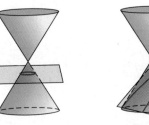

Circle

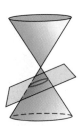

Parabola

Ellipse

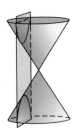

Hyperbola

1

Thus far, we have seen that $f(x)$ or $y = a(x - h)^2 + k$ is the equation of a parabola that opens upward if $a > 0$ or downward if $a < 0$. Parabolas can also open left or right, or even on a slant. Equations of these parabolas are not functions of x, of course, since a parabola opening any way other than upward or downward fails the vertical line test. In this section, we introduce parabolas that open to the left and to the right. Parabolas opening on a slant will not be developed in this book.

Just as $y = a(x - h)^2 + k$ is the equation of a parabola that opens upward or downward, $x = a(y - k)^2 + h$ is the equation of a parabola that opens to the right or to the left. The parabola opens to the right if $a > 0$ and to the left if $a < 0$. The parabola has vertex (h, k), and its axis of symmetry is the line $y = k$.

PARABOLAS

$$y = a(x - h)^2 + k$$

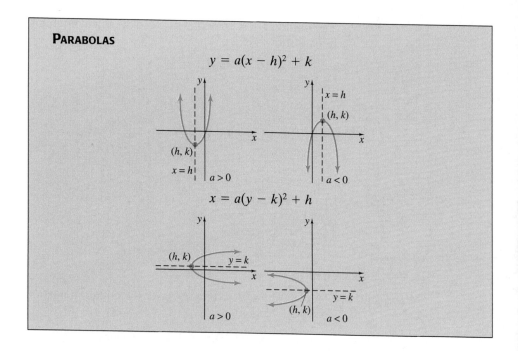

$$x = a(y - k)^2 + h$$

The forms $y = a(x - h)^2 + k$ and $x = a(y - k)^2 + h$ are called **standard forms**.

EXAMPLE 1   Graph the parabola $x = 2y^2$.

Solution: Written in standard form, the equation $x = 2y^2$ is $x = 2(y - 0)^2 + 0$ with $a = 2, h = 0$, and $k = 0$. Its graph is a parabola with vertex $(0, 0)$, and its axis of symmetry is the line $y = 0$. Since $a > 0$, this parabola opens to the right. The table shows a few more ordered pair solutions of $x = 2y^2$. Its graph is also shown.

x	y
8	-2
2	-1
0	0
2	1
8	2

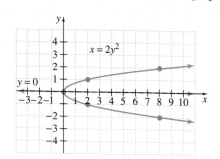

EXAMPLE 2   Graph the parabola $x = -3(y - 1)^2 + 2$.

Solution: The equation $x = -3(y - 1)^2 + 2$ is in the form $x = a(y - k)^2 + h$ with $a = -3$, $k = 1$, and $h = 2$. Since $a < 0$, the parabola opens to the left. The vertex (h, k) is $(2, 1)$, and the axis of symmetry is the line $y = 1$. When $y = 0, x = -1$, the x-intercept.

The parabola is sketched next. Also, a table containing a few ordered pair solutions is shown.

x	y
2	1
−1	0
−1	2

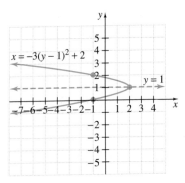

EXAMPLE 3   Graph $y = -x^2 - 2x + 15$.

Solution: Complete the square on x to write the equation in standard form:

$$y - 15 = -x^2 - 2x \qquad \text{Subtract 15 from both sides.}$$
$$y - 15 = -1(x^2 + 2x) \qquad \text{Factor } -1 \text{ from the terms } -x^2 - 2x.$$

The coefficient of x is 2. Find the square of half of 2.

$$\frac{1}{2}(2) = 1 \text{ and } 1^2 = 1$$

$$y - 15 - 1(1) = -1(x^2 + 2x + 1) \qquad \text{Add } -1(1) \text{ to both sides.}$$
$$y - 16 = -1(x + 1)^2 \qquad \begin{array}{l}\text{Simplify the left side and factor the} \\ \text{right side.}\end{array}$$

$$y = -(x + 1)^2 + 16 \qquad \text{Add 16 to both sides.}$$

The equation is now in the form $y = a(x - h)^2 + k$ with $a = -1$, $h = -1$, and $k = 16$. The parabola opens downward since $a < 0$ and has vertex $(-1, 16)$. Its axis of symmetry is the line $x = -1$. The y-intercept is 15. The graph is next, along with a few more ordered pair solutions.

x	y
−1	16
0	15
−2	15
1	12
−3	12
3	0
−5	0

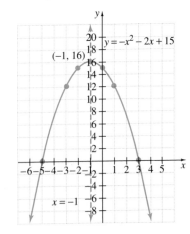

EXAMPLE 4   Graph $x = 2y^2 + 4y + 5$.

Solution: Complete the square on y to write the equation in standard form:

$$x - 5 = 2y^2 + 4y \qquad \text{Subtract 5 from both sides.}$$
$$x - 5 = 2(y^2 + 2y)$$
$$x - 5 + 2(1) = 2(y^2 + 2y + 1) \qquad \text{Add } 2(1) \text{ to both sides.}$$
$$x - 3 = 2(y + 1)^2 \qquad \text{Simplify the left side and factor the right side.}$$
$$x = 2(y + 1)^2 + 3 \qquad \text{Add 3 to both sides.}$$

The equation is now in the form $x = a(y - k)^2 + h$ with $a = 2$, $k = -1$, and $h = 3$. The parabola opens to the right since $a > 0$ and has vertex $(3, -1)$. Its axis of symmetry is the line $y = -1$. The x-intercept is 5.

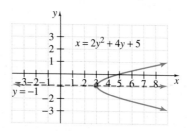

The Cartesian coordinate system helps us visualize a distance between points. To find the distance between two points, we use the distance formula, which is derived from the Pythagorean theorem.

To find the distance d between two points (x_1, y_1) and (x_2, y_2) as shown next, notice that the length of leg a is $x_2 - x_1$ and that the length of leg b is $y_2 - y_1$.

Thus, the Pythagorean theorem tells us that

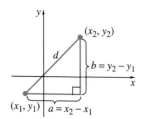

$$d^2 = a^2 + b^2$$

or

$$d^2 = (x_2 - x_1)^2 + (y_2 - y_1)^2$$

or

$$d = \sqrt{(x_2 - x_1)^2 + (y_2 - y_1)^2}$$

This formula gives us the distance between any two points on the real plane.

DISTANCE FORMULA

The distance d between two points (x_1, y_1) and (x_2, y_2) is given by

$$d = \sqrt{(x_2 - x_1)^2 + (y_2 - y_1)^2}$$

EXAMPLE 5   Find the distance between the pair of points $(2, -5)$ and $(1, -4)$.

Solution: Use the distance formula. It makes no difference which point we call (x_1, y_1) and which point we call (x_2, y_2). Let $(x_1, y_1) = (2, -5)$ and $(x_2, y_2) = (1, -4)$:

$$d = \sqrt{(x_2 - x_1)^2 + (y_2 - y_1)^2}$$
$$= \sqrt{(1 - 2)^2 + [-4 - (-5)]^2}$$
$$= \sqrt{(-1)^2 + (1)^2}$$
$$= \sqrt{1 + 1}$$
$$= \sqrt{2}$$

The distance between the two points is $\sqrt{2}$ units.

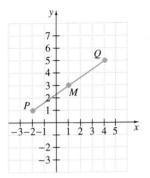

The **midpoint** of a line segment is the **point** located exactly halfway between the two end points of the line segment. On the graph to the left, the point M is the midpoint of line segment PQ. Thus, the distance between M and P equals the distance between M and Q.

The x-coordinate of M is at half the distance between the x-coordinates of P and Q, and the y-coordinate of M is at half the distance between the y-coordinates of P and Q. That is, the x-coordinate of M is the average of the x-coordinates of P and Q; the y-coordinate of M is the average of the y-coordinates of P and Q.

> **MIDPOINT FORMULA**
>
> The midpoint of the line segment whose end points are (x_1, y_1) and (x_2, y_2) is the point with coordinates
>
> $$\left(\frac{x_1 + x_2}{2}, \frac{y_1 + y_2}{2} \right)$$

EXAMPLE 6   Find the midpoint of the line segment that joins points $P(-3, 3)$ and $Q(1, 0)$.

Solution: Use the midpoint formula. It makes no difference which point we call (x_1, y_1) or which point we call (x_2, y_2). Let $(x_1, y_1) = (-3, 3)$ and $(x_2, y_2) = (1, 0)$.

$$\text{midpoint} = \left(\frac{x_1 + x_2}{2}, \frac{y_1 + y_2}{2} \right)$$
$$= \left(\frac{-3 + 1}{2}, \frac{3 + 0}{2} \right)$$
$$= \left(\frac{-2}{2}, \frac{3}{2} \right)$$
$$= \left(-1, \frac{3}{2} \right)$$

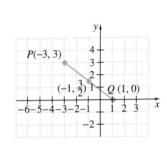

The midpoint of the segment is $\left(-1, \frac{3}{2} \right)$.

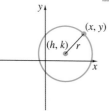

Another conic section is the **circle.** A circle is the set of all points in a plane that are the same distance from a fixed point called the **center.** The distance is called the **radius** of the circle. To find a standard equation for a circle, let (h, k) represent the center of the circle, and let (x, y) represent any point on the circle. The distance between (h, k) and (x, y) is defined to be the radius, r units. We can find this distance r by using the distance formula.

$$r = \sqrt{(x - h)^2 + (y - k)^2}$$
$$r^2 = (x - h)^2 + (y - k)^2 \qquad \text{Square both sides.}$$

CIRCLE

The graph of $(x - h)^2 + (y - k)^2 = r^2$ is a circle with center (h, k) and radius r.

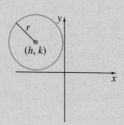

The form $(x - h)^2 + (y - k)^2 = r^2$ is called **standard form.**

If an equation can be written in the standard form

$$(x - h)^2 + (y - k)^2 = r^2$$

then its graph is a circle, which we can draw by graphing the center (h, k) and using the radius r.

EXAMPLE 7   Graph $x^2 + y^2 = 4$.

Solution: The equation can be written in standard form as

$$(x - 0)^2 + (y - 0)^2 = 2^2$$

The center of the circle is $(0, 0)$, and the radius is 2. Its graph is shown at left.

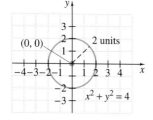

REMINDER   Notice the difference between the equation of a circle and the equation of a parabola. The equation of a circle contains both x^2 and y^2 terms on the same side of the equation with equal coefficients. The equation of a parabola has either an x^2 term or a y^2 term but not both.

EXAMPLE 8   Graph $(x + 1)^2 + y^2 = 8$.

Solution: The equation can be written as $(x + 1)^2 + (y - 0)^2 = 8$ with $h = -1, k = 0$, and $r = \sqrt{8}$. The center is $(-1, 0)$, and the radius is $\sqrt{8} = 2\sqrt{2} \approx 2.8$.

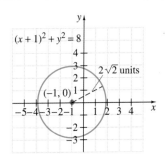

4   Since a circle is determined entirely by its center and radius, this information is all we need to write the equation of a circle.

EXAMPLE 9   Find an equation of the circle with the center $(-7, 3)$ and radius 10.

Solution: Using the given values $h = -7, k = 3$, and $r = 10$, we write the equation

$$(x - h)^2 + (y - k)^2 = r^2$$

or

$$[x - (-7)]^2 + (y - 3)^2 = 10^2$$

or

$$(x + 7)^2 + (y - 3)^2 = 100$$

5   To find the center and the radius of a circle from its equation, write the equation in standard form. To write the equation of a circle in standard form, we complete the square on both x and y.

EXAMPLE 10   Graph $x^2 + y^2 + 4x - 8y = 16$.

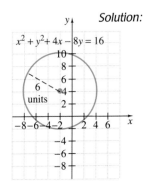

Solution: Since this equation contains x^2 and y^2 terms on the same side of the equation with equal coefficients, its graph is a circle. To write the equation in standard form, group the terms involving x and the terms involving y, and then complete the square on each variable.

$$(x^2 + 4x) + (y^2 - 8y) = 16$$

Thus, $\dfrac{1}{2}(4) = 2$ and $2^2 = 4$. Also, $\dfrac{1}{2}(-8) = -4$ and $(-4)^2 = 16$. Add 4 and then 16 to both sides.

$$(x^2 + 4x + 4) + (y^2 - 8y + 16) = 16 + 4 + 16$$
$$(x + 2)^2 + (y - 4)^2 = 36 \qquad \text{Factor.}$$

This circle has the center $(-2, 4)$ and radius 6, as shown.

GRAPHING CALCULATOR EXPLORATIONS

To graph an equation such as $x^2 + y^2 = 25$ with a grapher, we first solve the equation for y.

$$x^2 + y^2 = 25$$
$$y^2 = 25 - x^2$$
$$y = \pm\sqrt{25 - x^2}$$

The graph of $y = \sqrt{25 - x^2}$ will be the top half of the circle, and the graph of $y = -\sqrt{25 - x^2}$ will be the bottom half of the circle.

To graph, press $\boxed{Y=}$ and enter $Y_1 = \sqrt{25 - x^2}$ and $Y_2 = -\sqrt{25 - x^2}$. Insert parentheses about $25 - x^2$ so that $\sqrt{25 - x^2}$ and not $\sqrt{25} - x^2$ is graphed.

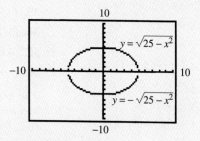

The graph does not appear to be a circle because we are currently using a standard window and the screen is rectangular. This causes the tick marks on the x-axis to be farther apart than the tick marks on the y-axis and thus creates the distorted circle. If we want the graph to appear circular, define a square window by using a feature of your grapher or redefine your window to show the x-axis from -15 to 15 and the y-axis from -10 to 10. Using a square window, the graph appears as follows:

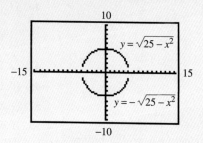

Use a grapher to graph each circle.   For Exercises 1–4, see Appendix D.

1. $x^2 + y^2 = 55$

2. $x^2 + y^2 = 20$

3. $7x^2 + 7y^2 - 89 = 0$

4. $3x^2 + 3y^2 - 35 = 0$

MENTAL MATH

The graph of each equation is a parabola. Determine whether the parabola opens upward, downward, to the left, or to the right.

1. $y = x^2 - 7x + 5$ upward

2. $y = -x^2 + 16$ downward

3. $x = -y^2 - y + 2$ to the left

4. $x = 3y^2 + 2y - 5$ to the right

5. $y = -x^2 + 2x + 1$ downward

6. $x = -y^2 + 2y - 6$ to the left

17. $\sqrt{10}$ units   **18.** $\sqrt{17}$ units   **27.** $\left(-5, \frac{5}{2}\right)$ **30.** $\left(-\frac{3}{2}, \frac{11}{2}\right)$ **33.** $\left(\sqrt{2}, \frac{\sqrt{5}}{2}\right)$ **34.** $\left(\frac{5\sqrt{2}}{2}, \frac{5\sqrt{3}}{2}\right)$

EXERCISE SET 9.1

The graph of each equation is a parabola. Find the vertex of the parabola and sketch its graph. See Examples 1 through 4.   For Exercises 1–12, see Appendix D.

1. $x = 3y^2$

2. $x = -2y^2$

3. $x = (y - 2)^2 + 3$

4. $x = (y - 4)^2 - 1$

5. $y = 3(x - 1)^2 + 5$

6. $x = -4(y - 2)^2 + 2$

7. $x = y^2 + 6y + 8$

8. $x = y^2 - 6y + 6$

9. $y = x^2 + 10x + 20$

10. $y = x^2 + 4x - 5$

11. $x = -2y^2 + 4y + 6$

12. $x = 3y^2 + 6y + 7$

Find the distance between each pair of points. Approximate the distance in Exercises 21 and 22 to two decimal places. See Example 5.

13. $(5, 1)$ and $(8, 5)$ 5 units

14. $(2, 3)$ and $(14, 8)$ 13 units

15. $(-3, 2)$ and $(1, -3)$ $\sqrt{41}$ units

16. $(3, -2)$ and $(-4, 1)$ $\sqrt{58}$ units

17. $(-9, 4)$ and $(-8, 1)$

18. $(-5, -2)$ and $(-6, -6)$

19. $(0, -\sqrt{2})$ and $(\sqrt{3}, 0)$ $\sqrt{5}$ units

20. $(-\sqrt{5}, 0)$ and $(0, \sqrt{7})$ $2\sqrt{3}$ units

21. $(1.7, -3.6)$ and $(-8.6, 5.7)$ 13.88 units

22. $(9.6, 2.5)$ and $(-1.9, -3.7)$ 13.06 units

23. $(2\sqrt{3}, \sqrt{6})$ and $(-\sqrt{3}, 4\sqrt{6})$ 9 units

24. $(5\sqrt{2}, -4)$ and $(-3\sqrt{2}, -8)$ 12 units

Find the midpoint of the line segment whose end points are given. See Example 6.

25. $(6, -8), (2, 4)$ $(4, -2)$

26. $(3, 9), (7, 11)$ $(5, 10)$

27. $(-2, -1), (-8, 6)$

28. $(-3, -4), (6, -8)$ $\left(\frac{3}{2}, -6\right)$

29. $(7, 3), (-1, -3)$ $(3, 0)$

30. $(-2, 5), (-1, 6)$

31. $\left(\frac{1}{2}, \frac{3}{8}\right), \left(-\frac{3}{2}, \frac{5}{8}\right)$ $\left(-\frac{1}{2}, \frac{1}{2}\right)$

32. $\left(-\frac{2}{5}, \frac{7}{15}\right), \left(-\frac{2}{5}, -\frac{4}{15}\right)$

33. $(\sqrt{2}, 3\sqrt{5}), (\sqrt{2}, -2\sqrt{5})$

32. $\left(-\frac{2}{5}, \frac{1}{10}\right)$

34. $(\sqrt{8}, -\sqrt{12}), (3\sqrt{2}, 7\sqrt{3})$

35. $(4.6, -3.5), (7.8, -9.8)$ $(6.2, -6.65)$

36. $(-4.6, 2.1), (-6.7, 1.9)$ $(-5.65, 2)$

The graph of each equation is a circle. Find the center and the radius, and then sketch. See Examples 7, 8, and 10.   For Exercises 37–46, see Appendix D.

37. $x^2 + y^2 = 9$

38. $x^2 + y^2 = 25$

39. $x^2 + (y - 2)^2 = 1$

40. $(x - 3)^2 + y^2 = 9$

41. $(x - 5)^2 + (y + 2)^2 = 1$

42. $(x + 3)^2 + (y + 3)^2 = 4$

43. $x^2 + y^2 + 6y = 0$

44. $x^2 + 10x + y^2 = 0$

45. $x^2 + y^2 + 2x - 4y = 4$

46. $x^2 + 6x - 4y + y^2 = 3$

Write an equation of the circle with the given center and radius. See Example 9.   **47.** $(x - 2)^2 + (y - 3)^2 = 36$

47. $(2, 3); 6$

48. $(-7, 6); 2$

49. $(0, 0); \sqrt{3}$ $x^2 + y^2 = 3$ **50.** $(0, -6); \sqrt{2}$

51. $(-5, 4); 3\sqrt{5}$

52. the origin; $4\sqrt{7}$

53. If you are given a list of equations of circles and parabolas and none are in standard form, explain how you would determine which is an equation of a circle and which is an equation of a parabola. Explain also how you would distinguish the upward or downward parabolas from the left-opening or right-opening parabolas.   Answers will vary.

48. $(x + 7)^2 + (y - 6)^2 = 4$ **50.** $x^2 + (y + 6)^2 = 2$ **51.** $(x + 5)^2 + (y - 4)^2 = 45$ **52.** $x^2 + y^2 = 112$

Sketch the graph of each equation. If the graph is a parabola, find its vertex. If the graph is a circle, find its center and radius. For Exercises 54–85, see Appendix D.

54. $x = y^2 + 2$

55. $x = y^2 - 3$

56. $y = (y + 3)^2 + 3$

57. $y = (x - 2)^2 - 2$

58. $x^2 + y^2 = 49$

59. $x^2 + y^2 = 1$

60. $x = (y - 1)^2 + 4$

61. $x = (y + 3)^2 - 1$

62. $(x + 3)^2 + (y - 1)^2 = 9$

63. $(x - 2)^2 + (y - 2)^2 = 16$

64. $x = -2(y + 5)^2$

65. $x = -(y - 1)^2$

66. $x^2 + (y + 5)^2 = 5$

67. $(x - 4)^2 + y^2 = 7$

68. $y = 3(x - 4)^2 + 2$

69. $y = 5(x + 5)^2 + 3$

70. $2x^2 + 2y^2 = \dfrac{1}{2}$

71. $\dfrac{x^2}{8} + \dfrac{y^2}{8} = 2$

72. $y = x^2 - 2x - 15$

73. $y = x^2 + 7x + 6$

74. $x^2 + y^2 + 6x + 10y - 2 = 0$

75. $x^2 + y^2 + 2x + 12y - 12 = 0$

76. $x = y^2 + 6y + 2$

77. $x = y^2 + 8y - 4$

78. $x^2 + y^2 - 8y + 5 = 0$

79. $x^2 - 10y + y^2 + 4 = 0$

80. $x = -2y^2 - 4y$

81. $x = -3y^2 + 30y$

82. $\dfrac{x^2}{3} + \dfrac{y^2}{3} = 2$

83. $5x^2 + 5y^2 = 25$

84. $y = 4x^2 - 40x + 105$

85. $y = 5x^2 - 20x + 16$

Solve.

86. Two surveyors need to find the distance across a lake. They place a reference pole at point A in the diagram. Point B is 3 meters east and 1 meter north of the reference point A. Point C is 19 meters east and 13 meters north of point A. Find the distance across the lake, from B to C.   20 meters

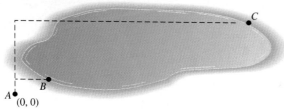

87. Determine whether the triangle with vertices $(2, 6)$, $(0, -2)$, and $(5, 1)$ is an isosceles triangle.   Yes, it is.

88.c. $x^2 + y^2 = 25$; See Appendix D.

88. Cindy Brown, an architect, is drawing plans on grid paper for a circular pool with a fountain in the middle. The paper is marked off in centimeters, and each centimeter represents 1 foot. On the paper, the diameter of the "pool" is 20 centimeters, and "fountain" is the point $(0, 0)$.

  a. Sketch the architect's drawing. Be sure to label the axes.   See Appendix D.

  b. Write an equation that describes the circular pool. $x^2 + y^2 = 100$

  c. Cindy plans to place a circle of lights around the fountain such that each light is 5 feet from the fountain. Write an equation for the circle of lights and sketch the circle on your drawing.

89. A bridge constructed over a bayou has a supporting arch in the shape of a parabola. Find an equation of the parabolic arch if the length of the road over the arch is 100 meters and the maximum height of the arch is 40 meters. $y = -\dfrac{2}{125}x^2 + 40$

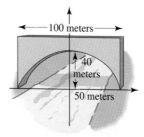

For Exercises 90–97, see Appendix D.
Use a grapher to verify each exercise. Use a square viewing window.

90. Exercise 82.

91. Exercise 83.

92. Exercise 84.

93. Exercise 85.

Review Exercises

Graph each equation. See Section 3.3.

94. $y = 2x + 5$

95. $y = -3x + 3$

96. $y = 3$

97. $x = -2$

Rationalize each denominator and simplify if possible. See Section 7.4

98. $\dfrac{1}{\sqrt{3}}$ $\dfrac{\sqrt{3}}{3}$

99. $\dfrac{\sqrt{5}}{\sqrt{8}}$ $\dfrac{\sqrt{10}}{4}$

100. $\dfrac{4\sqrt{7}}{\sqrt{6}}$ $\dfrac{2\sqrt{42}}{3}$

101. $\dfrac{10}{\sqrt{5}}$ $2\sqrt{5}$

9.2 | THE ELLIPSE AND THE HYPERBOLA

O B J E C T I V E S

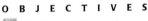

1 Define and graph an ellipse.
2 Define and graph a hyperbola.

TAPE IA 9.2

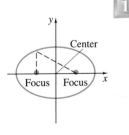

An **ellipse** can be thought of as the set of points in a plane such that the sum of the distances of those points from two fixed points is constant. Each of the two fixed points is called a **focus.** The plural of focus of **foci.** The point midway between the foci is called the **center.**

An ellipse may be drawn by hand by using two tacks, a piece of string, and a pencil. Secure the two tacks into a piece of cardboard, for example, and tie each end of the string to a tack. Use your pencil to pull the string tight and draw the ellipse. The two tacks are the foci of the drawn ellipse.

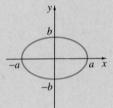

It can be shown that the **standard form** of an ellipse with center $(0, 0)$ is

$$\frac{x^2}{a^2} + \frac{y^2}{b^2} = 1.$$

ELLIPSE WITH CENTER (0, 0)

The graph of an equation of the form $\dfrac{x^2}{a^2} + \dfrac{y^2}{b^2} = 1$ is an ellipse with center $(0, 0)$. The x-intercepts are a and $-a$, and the y-intercepts are b and $-b$.

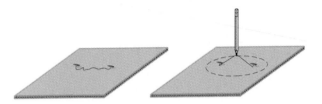

EXAMPLE 1   Graph $\dfrac{x^2}{9} + \dfrac{y^2}{16} = 1$.

Solution: The equation is of the form $\dfrac{x^2}{a^2} + \dfrac{y^2}{b^2} = 1$, with $a = 3$ and $b = 4$, so its graph is an ellipse with center $(0, 0)$, x-intercepts 3 and -3, and y-intercepts 4 and -4 as graphed next.

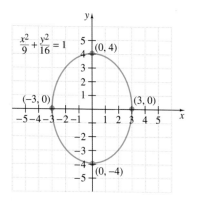

EXAMPLE 2   Graph the equation $4x^2 + 16y^2 = 64$.

Solution: Although this equation contains a sum of squared terms in x and y on the same side of an equation, this is not the equation of a circle since the coefficients of x^2 and y^2 are not the same. When this happens, the graph is an ellipse. Since the standard form of the equation of an ellipse has 1 on one side, divide both sides of this equation by 64.

$$4x^2 + 16y^2 = 64$$
$$\frac{4x^2}{64} + \frac{16y^2}{64} = \frac{64}{64} \qquad \text{Divide both sides by 64.}$$
$$\frac{x^2}{16} + \frac{y^2}{4} = 1 \qquad \text{Simplify.}$$

We now recognize the equation of an ellipse with center $(0, 0)$, x-intercepts 4 and -4, and y-intercepts 2 and -2.

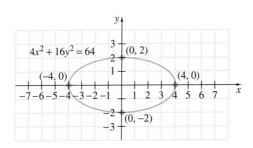

The center of an ellipse is not always $(0, 0)$, as shown in the next example.

EXAMPLE 3 Graph $\dfrac{(x+3)^2}{25} + \dfrac{(y-2)^2}{36} = 1.$

Solution: This ellipse has center $(-3, 2)$. Notice that $a = 5$ and $b = 6$. To find four points on the graph of the ellipse, first graph the center, $(-3, 2)$. Since $a = 5$, count 5 units right and then 5 units left of the point with coordinates $(-3, 2)$. Next, since $b = 6$, start at $(-3, 2)$ and count 6 units up and then 6 units down to find two more points on the ellipse.

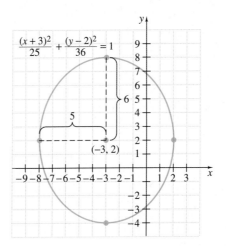

The final conic section is the **hyperbola.** A hyperbola is the set of points in a plane such that the absolute value of the difference of the distance from two fixed points is constant. Each of the two fixed points is called a **focus.** The point midway between the foci is called the **center.**

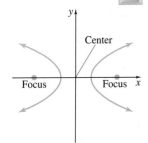

Using the distance formula, we can show that the graph of $\dfrac{x^2}{a^2} - \dfrac{y^2}{b^2} = 1$ is a hyperbola with center $(0, 0)$ and x-intercepts a and $-a$. Also, the graph of $\dfrac{y^2}{b^2} - \dfrac{x^2}{a^2} = 1$ is a hyperbola with center $(0, 0)$ and y-intercepts b and $-b$. These equations are called **standard form** equations for a hyperbola.

HYPERBOLA WITH CENTER (0, 0)

The graph of an equation of the form $\dfrac{x^2}{a^2} - \dfrac{y^2}{b^2} = 1$ is a hyperbola with center $(0, 0)$ and x-intercepts a and $-a$.

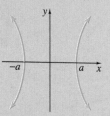

The graph of an equation of the form $\dfrac{y^2}{b^2} - \dfrac{x^2}{a^2} = 1$ is a hyperbola with center $(0, 0)$ and y-intercepts b and $-b$.

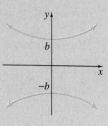

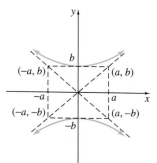

Graphing a hyperbola such as $\dfrac{y^2}{b^2} - \dfrac{x^2}{a^2} = 1$ is made easier by recognizing one of its important characteristics. Examining the figure to the left, notice how the sides of the branches of the hyperbola extend indefinitely and seem to approach the dashed lines in the figure. These dashed lines are called the **asymptotes** of the hyperbola.

To sketch these lines, or asymptotes, draw a rectangle with vertices (a, b), $(-a, b)$, $(a, -b)$, and $(-a, -b)$. The asymptotes of the hyperbola are the extended diagonals of this rectangle.

EXAMPLE 4   Sketch the graph of $\dfrac{x^2}{16} - \dfrac{y^2}{25} = 1$.

Solution: This equation has the form $\dfrac{x^2}{a^2} - \dfrac{y^2}{b^2} = 1$, with $a = 4$ and $b = 5$. Thus, its graph is a hyperbola with center $(0, 0)$ and x-intercepts 4 and -4. To aid in graphing the hyperbola, we first sketch its asymptotes. The extended diagonals of the rectangle with coordinates $(4, 5)$, $(4, -5)$, $(-4, 5)$, and $(-4, -5)$ are the asymptotes of the hyperbola. Then use the asymptotes to aid in sketching the hyperbola.

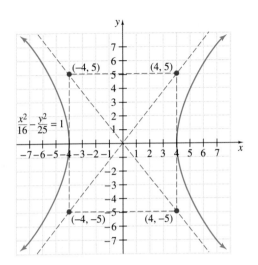

EXAMPLE 5   Sketch the graph of the equation $4y^2 - 9x^2 = 36$.

Solution: Since this is a difference of squared terms in x and y on the same side of the equation, its graph is a hyperbola, as opposed to an ellipse or a circle. The standard form of the equation of a hyperbola has a 1 on one side, so divide both sides of the equation by 36.

$$4y^2 - 9x^2 = 36$$

$$\frac{4y^2}{36} - \frac{9x^2}{36} = \frac{36}{36} \qquad \text{Divide both sides by 36.}$$

$$\frac{y^2}{9} - \frac{x^2}{4} = 1 \qquad \text{Simplify.}$$

The equation is of the form $\dfrac{y^2}{b^2} - \dfrac{x^2}{a^2} = 1$, with $a = 2$ and $b = 3$, so the hyperbola is centered at $(0, 0)$ with y-intercepts 3 and -3. The sketch of the hyperbola is shown.

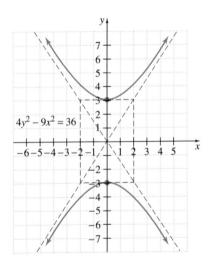

The following box provides a summary of conic sections.

CONIC SECTIONS

	STANDARD FORM	GRAPH
Parabola	$y = a(x - h)^2 + k$	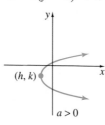
Parabola	$x = a(y - k)^2 + h$	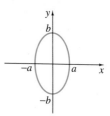
Circle	$(x - h)^2 + (y - k)^2 = r^2$	
Ellipse	$\dfrac{x^2}{a^2} + \dfrac{y^2}{b^2} = 1$	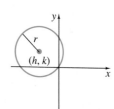
Hyperbola	$\dfrac{x^2}{a^2} - \dfrac{y^2}{b^2} = 1$	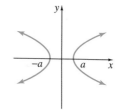

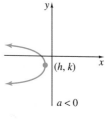

(continued)

CONIC SECTIONS (CONT.)		
	STANDARD FORM	**GRAPH**
Hyperbola	$\dfrac{y^2}{b^2} - \dfrac{x^2}{a^2} = 1$	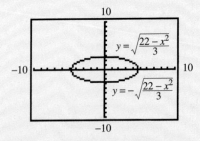

GRAPHING CALCULATOR EXPLORATIONS

To find the graph of an ellipse by using a grapher, use the same procedure as for graphing a circle. For example, to graph $x^2 + 3y^2 = 22$, first solve for y.

$$3y^2 = 22 - x^2$$

$$y^2 = \frac{22 - x^2}{3}$$

$$y = \pm\sqrt{\frac{22 - x^2}{3}}$$

Next press the $\boxed{Y=}$ key and enter $Y_1 = \sqrt{\dfrac{22 - x^2}{3}}$ and $Y_2 = -\sqrt{\dfrac{22 - x^2}{3}}$. (Insert two sets of parentheses in the radicand as $\sqrt{((22 - x^2)/3)}$ so that the desired graph is obtained.) The graph appears as follows:

Use a grapher to graph each ellipse. For Exercises 1–4, see Appendix D.

1. $10x^2 + y^2 = 32$

2. $20x^2 + 5y^2 = 100$

3. $7.3x^2 + 15.5y^2 = 95.2$

4. $18.8x^2 + 36.1y^2 = 205.8$

EXERCISE SET 9.2

Sketch the graph of each equation. See Examples 1 and 2.
For Exercises 1–20, see Appendix D.

1. $\dfrac{x^2}{4} + \dfrac{y^2}{25} = 1$ **2.** $\dfrac{x^2}{9} + y^2 = 1$

3. $\dfrac{x^2}{16} + \dfrac{y^2}{9} = 1$ **4.** $x^2 + \dfrac{y^2}{4} = 1$

5. $9x^2 + 4y^2 = 36$ **6.** $x^2 + 4y^2 = 16$

7. $4x^2 + 25y^2 = 100$ **8.** $36x^2 + y^2 = 36$

Sketch the graph of each equation. See Example 3.

9. $\dfrac{(x+1)^2}{36} + \dfrac{(y-2)^2}{49} = 1$

10. $\dfrac{(x-3)^2}{9} + \dfrac{(y+3)^2}{16} = 1$

11. $\dfrac{(x-1)^2}{4} + \dfrac{(y-1)^2}{25} = 1$

12. $\dfrac{(x+3)^2}{16} + \dfrac{(y+2)^2}{4} = 1$

Sketch the graph of each equation. See Examples 4 and 5.

13. $\dfrac{x^2}{4} - \dfrac{y^2}{9} = 1$ **14.** $\dfrac{x^2}{36} - \dfrac{y^2}{36} = 1$

15. $\dfrac{y^2}{25} - \dfrac{x^2}{16} = 1$ **16.** $\dfrac{y^2}{25} - \dfrac{x^2}{49} = 1$

Sketch the graph of each equation. See Example 5.

17. $x^2 - 4y^2 = 16$ **18.** $4x^2 - y^2 = 36$

19. $16y^2 - x^2 = 16$ **20.** $4y^2 - 25x^2 = 100$

21. If you are given a list of equations of circles, parabolas, ellipses, and hyperbolas, explain how you could distinguish the different conic sections from their equations.

Identify whether each equation, when graphed, will be a parabola, circle, ellipse, or hyperbola. Sketch the graph of each equation.
For Exercises 22–37, see Appendix D.

22. $(x-7)^2 + (y-2)^2 = 4$ circle

23. $y = x^2 + 4$ parabola

24. $y = x^2 + 12x + 36$ parabola

25. $\dfrac{x^2}{4} + \dfrac{y^2}{9} = 1$ ellipse   **26.** $\dfrac{y^2}{9} - \dfrac{x^2}{9} = 1$ hyperbola

27. $\dfrac{x^2}{16} - \dfrac{y^2}{4} = 1$ hyperbola   **28.** $\dfrac{x^2}{16} + \dfrac{y^2}{4} = 1$ ellipse

29. $x^2 + y^2 = 16$ circle   **30.** $x = y^2 + 4y - 1$ parabola

31. $x = -y^2 + 6y$ parabola   **32.** $9x^2 - 4y^2 = 36$ hyperbola

33. $9x^2 + 4y^2 = 36$ ellipse

34. $\dfrac{(x-1)^2}{49} + \dfrac{(y+2)^2}{25} = 1$ ellipse

35. $y^2 = x^2 + 16$ hyperbola

36. $\left(x + \dfrac{1}{2}\right)^2 + \left(y - \dfrac{1}{2}\right)^2 = 1$ circle

37. $y = -2x^2 + 4x - 3$ parabola

Solve.

38. A planet's orbit about the Sun can be described as an ellipse. Consider the Sun as the origin of a rectangular coordinate system. Suppose that the x-intercepts of the elliptical path of the planet are $\pm130{,}000{,}000$ and that the y-intercepts are $\pm125{,}000{,}000$. Write the equation of the elliptical path of the planet.

39. Comets orbit the Sun in elongated ellipses. Consider the Sun as the origin of a rectangular coordinate system. Suppose that the equation of the path of the comet is

$$\frac{(x - 1{,}782{,}000{,}000)^2}{(3.42)(10^{23})} + \frac{(y - 356{,}400{,}000)^2}{(1.368)(10^{22})} = 1$$

Find the center of the path of the comet.

40. Use a grapher to verify Exercise 5.

41. Use a grapher to verify Exercise 6.
For Exercises 40–41, see Appendix D.

Review Exercises

Solve each inequality. See Section 2.5.

42. $x < 5$ and $x < 1$ $(-\infty, 1)$

43. $x < 5$ or $x < 1$ $(-\infty, 5)$

44. $2x - 1 \geq 7$ or $-3x \leq -6$ $[2, \infty)$

45. $2x - 1 \geq 7$ and $-3x \leq -6$ $[4, \infty)$

Perform the indicated operations. See Sections 5.3 and 5.4.

46. $(2x^3)(-4x^2)$ $-8x^5$ **47.** $2x^3 - 4x^3$ $-2x^3$

48. $-5x^2 + x^2$ $-4x^2$ **49.** $(-5x^2)(x^2)$ $-5x^4$

38. $\dfrac{x^2}{1.69 \cdot 10^{16}} + \dfrac{y^2}{1.5625 \cdot 10^{16}} = 1$ **39.** $(1{,}782{,}000{,}000,\ 356{,}400{,}000)$

A Look Ahead

EXAMPLE Sketch the graph of $\dfrac{(x-2)^2}{25} - \dfrac{(y-1)^2}{9} = 1$.

Solution:

This hyperbola has center $(2, 1)$. Notice that $a = 5$ and $b = 3$.

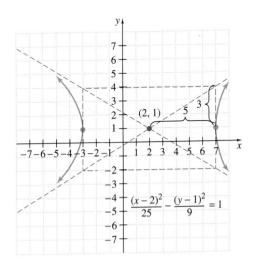

Sketch the graph of each equation. See the preceding example. For Exercises 50–55, see Appendix D.

50. $\dfrac{(x-1)^2}{4} - \dfrac{(y+1)^2}{25} = 1$

51. $\dfrac{(x+2)^2}{9} - \dfrac{(y-1)^2}{4} = 1$

52. $\dfrac{y^2}{16} - \dfrac{(x+3)^2}{9} = 1$

53. $\dfrac{(y+4)^2}{4} - \dfrac{x^2}{25} = 1$

54. $\dfrac{(x+5)^2}{16} - \dfrac{(y+2)^2}{25} = 1$

55. $\dfrac{(x-3)^2}{9} - \dfrac{(y-2)^2}{4} = 1$

$\dfrac{9.3}{}$ | SOLVING NONLINEAR SYSTEMS OF EQUATIONS

TAPE IA 9.3

O B J E C T I V E S

1 Solve a nonlinear system by substitution.

2 Solve a nonlinear system by elimination.

In Section 4.1, we used graphing, substitution, and elimination methods to find solutions of systems of linear equations in two variables. We now apply these same methods to nonlinear systems of equations in two variables. A **nonlinear system of equations** is a system of equations at least one of which is not linear. Since we will be graphing the equations in each system, we are interested in real number solutions only.

First, nonlinear systems are solved by the substitution method.

EXAMPLE 1   Solve the system

$$\begin{cases} x^2 - 3y = 1 \\ x - y = 1 \end{cases}$$

Solution: We can solve this system by substitution if we solve one equation for one of the variables. Solving the first equation for x is not the best choice since doing so introduces a radical. Also, solving for y in the first equation introduces a fraction. We solve the second equation for y.

$$x - y = 1 \qquad \text{Second equation.}$$
$$x - 1 = y \qquad \text{Solve for } y.$$

Replace y with $x - 1$ in the first equation, and then solve for x:

$$x^2 - 3y = 1 \qquad \text{First equation.}$$

$$x^2 - 3(\overbrace{x - 1}) = 1 \qquad \text{Replace } y \text{ with } x - 1.$$
$$x^2 - 3x + 3 = 1$$
$$x^2 - 3x + 2 = 0$$
$$(x - 2)(x - 1) = 0$$
$$x = 2 \quad \text{or} \quad x = 1$$

Let $x = 2$ and then $x = 1$ in the equation $y = x - 1$ to find corresponding y-values:

Let $x = 2$.   Let $x = 1$.

$$y = x - 1 \qquad\qquad y = x - 1$$
$$y = 2 - 1 = 1 \qquad\quad y = 1 - 1 = 0$$

The solution set is $\{(2, 1), (1, 0)\}$. Check both solutions in both equations. Both solutions satisfy both equations, so both are solutions of the system. The graph of each equation in the system is next.

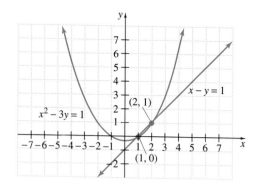

EXAMPLE 2   Solve the system

$$\begin{cases} y = \sqrt{x} \\ x^2 + y^2 = 6 \end{cases}$$

Solution: This system is ideal for substitution since y is expressed in terms of x in the first equation. Notice that if $y = \sqrt{x}$, then both x and y must be nonnegative if they are real numbers. Substitute $\sqrt{x}$ for y in the second equation, and solve for x:

$$x^2 + y^2 = 6$$
$$x^2 + (\sqrt{x})^2 = 6 \qquad \text{Let } y = \sqrt{x}.$$
$$x^2 + x = 6$$
$$x^2 + x - 6 = 0$$
$$(x + 3)(x - 2) = 0$$
$$x = -3 \text{ or } x = 2$$

The solution -3 is discarded because we have noted that x must be nonnegative. To see this, let $x = -3$ in the first equation. Then let $x = 2$ in the first equation to find a corresponding y-value.

Let $x = -3$.
$$y = \sqrt{x}$$
$$y = \sqrt{-3} \quad \text{Not a real number.}$$

Let $x = 2$.
$$y = \sqrt{x}$$
$$y = \sqrt{2}$$

Since we are interested only in real number solutions, the only solution is $(2, \sqrt{2})$. The solution set is $\{(2, \sqrt{2})\}$. Check to see that this solution satisfies both equations. The graph of each equation in this system is shown next.

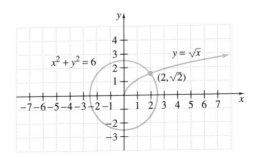

Some nonlinear systems may be solved by the elimination method.

2

EXAMPLE 3 Solve the system

$$\begin{cases} x^2 + 2y^2 = 10 \\ x^2 - y^2 = 1 \end{cases}$$

Solution: Use elimination, or addition, to solve this system. To eliminate x^2 when we add the two equations, multiply both sides of the second equation by -1. Then

$$\begin{cases} x^2 + 2y^2 = 10 \\ (-1(x^2 - y^2) = -1 \cdot 1 \end{cases} \quad \begin{matrix} \text{is} \\ \text{equivalent} \\ \text{to} \end{matrix} \quad \begin{cases} x^2 + 2y^2 = 10 \\ -x^2 + y^2 = -1 \end{cases}$$

$$\begin{array}{ll} 3y^2 = 9 & \text{Add.} \\ y^2 = 3 & \text{Divide by 3.} \\ y = \pm\sqrt{3} \end{array}$$

To find the corresponding x-values, let $y = \sqrt{3}$ and $y = -\sqrt{3}$ in either original equation. We choose the second equation.

Let $y = \sqrt{3}$.

$$x^2 - y^2 = 1$$
$$x^2 - (\sqrt{3})^2 = 1$$
$$x^2 - 3 = 1$$
$$x^2 = 4$$
$$x = \pm\sqrt{4} = \pm 2$$

Let $y = -\sqrt{3}$.

$$x^2 - y^2 = 1$$
$$x^2 - (-\sqrt{3})^2 = 1$$
$$x^2 - 3 = 1$$
$$x^2 = 4$$
$$x = \pm\sqrt{4} = \pm 2$$

The solution set is $\{(2, \sqrt{3}), (-2, \sqrt{3}), (2, -\sqrt{3}), (-2, -\sqrt{3})\}$. Check all four ordered pairs in both equations of the system. The graph of each equation in this system is shown next.

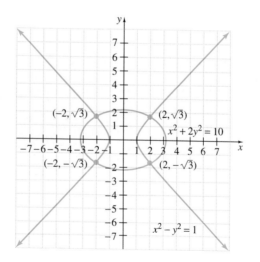

EXAMPLE 4   Solve the system

$$\begin{cases} x^2 + y^2 = 4 \\ x + y = 3 \end{cases}$$

Solution: The elimination method is not a good choice here, since x and y are squared in the first equation but not in the second equation. Use the substitution method and solve the second equation for x.

$$x + y = 3 \qquad \text{Second equation.}$$
$$x = 3 - y$$

Let $x = 3 - y$ in the first equation:

$$x^2 + y^2 = 4 \qquad \text{First equation.}$$

$$(3 - y)^2 + y^2 = 4$$
$$9 - 6y + y^2 + y^2 = 4$$
$$2y^2 - 6y + 5 = 0$$

By the quadratic formula, where $a = 2$, $b = -6$, and $c = 5$, we have

$$y = \frac{6 \pm \sqrt{(-6)^2 - 4 \cdot 2 \cdot 5}}{2 \cdot 2} = \frac{6 \pm \sqrt{-4}}{4}$$

Since $\sqrt{-4}$ is not a real number, there is no solution. Graphically, the circle and the line do not intersect, as shown.

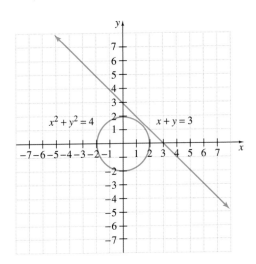

EXERCISE SET 9.3

Solve each nonlinear system of equations. See Examples 1 through 4.

1. $\begin{cases} x^2 + y^2 = 25 \\ 4x + 3y = 0 \end{cases}$
   $\{(3, -4), (-3, 4)\}$

2. $\begin{cases} x^2 + y^2 = 25 \\ 3x + 4y = 0 \end{cases}$
   $\{(-4, 3), (4, -3)\}$

3. $\begin{cases} x^2 + 4y^2 = 10 \\ y = x \end{cases}$

4. $\begin{cases} 4x^2 + y^2 = 10 \\ y = x \end{cases}$

5. $\begin{cases} y^2 = 4 - x \\ x - 2y = 4 \end{cases}$
   $\{(4, 0), (0, -2)\}$

6. $\begin{cases} x^2 + y^2 = 4 \\ x + y = -2 \end{cases}$
   $\{(-2, 0), (0, -2)\}$

3. $\{(\sqrt{2}, \sqrt{2}), (-\sqrt{2}, -\sqrt{2})\}$ 4. $\{(-\sqrt{2}, -\sqrt{2}), (\sqrt{2}, \sqrt{2})\}$

7. $\{(-\sqrt{5}, -2), (-\sqrt{5}, 2), (\sqrt{5}, -2), (\sqrt{5}, 2)\}$ **8.** $\{(2\sqrt{2}, 1), (-2\sqrt{2}, 1), (2\sqrt{2}, -1), (-2\sqrt{2}, -1)\}$ **11.** $\{(1, -2), (3, 6)\}$

7. $\begin{cases} x^2 + y^2 = 9 \\ 16x^2 - 4y^2 = 64 \end{cases}$

8. $\begin{cases} 4x^2 + 3y^2 = 35 \\ 5x^2 + 2y^2 = 42 \end{cases}$

9. $\begin{cases} x^2 + 2y^2 = 2 \\ x - y = 2 \end{cases}$ $\{\}$

10. $\begin{cases} x^2 + 2y^2 = 2 \\ x^2 - 2y^2 = 6 \end{cases}$ $\{\}$

11. $\begin{cases} y = x^2 - 3 \\ 4x - y = 6 \end{cases}$

12. $\begin{cases} y = x + 1 \\ x^2 - y^2 = 1 \end{cases}$ $\{(-1, 0)\}$

13. $\begin{cases} y = x^2 \\ 3x + y = 10 \end{cases}$

14. $\begin{cases} 6x - y = 5 \\ xy = 1 \end{cases}$

15. $\begin{cases} y = 2x^2 + 1 \\ x + y = -1 \end{cases}$ $\{\}$

16. $\begin{cases} x^2 + y^2 = 9 \\ x + y = 5 \end{cases}$ $\{\}$

17. $\begin{cases} y = x^2 - 4 \\ y = x^2 - 4x \end{cases}$ $\{(1, -3)\}$ **18.** $\begin{cases} x = y^2 - 3 \\ x = y^2 - 3y \end{cases}$ $\{(-2, 1)\}$

19. $\begin{cases} 2x^2 + 3y^2 = 14 \\ -x^2 + y^2 = 3 \end{cases}$ **20.** $\begin{cases} 4x^2 - 2y^2 = 2 \\ -x^2 + y^2 = 2 \end{cases}$

21. $\begin{cases} x^2 + y^2 = 1 \\ x^2 + (y + 3)^2 = 4 \end{cases}$ $\{(0, -1)\}$

22. $\begin{cases} x^2 + 2y^2 = 4 \\ x^2 - y^2 = 4 \end{cases}$ $\{(-2, 0), (2, 0)\}$

23. $\begin{cases} y = x^2 + 2 \\ y = -x^2 + 4 \end{cases}$ **24.** $\begin{cases} x = -y^2 - 3 \\ x = y^2 - 5 \end{cases}$ $\{(-4, -1)(-4, 1)\}$

25. $\begin{cases} 3x^2 + y^2 = 9 \\ 3x^2 - y^2 = 9 \end{cases}$ **26.** $\begin{cases} x^2 + y^2 = 25 \\ x = y^2 - 5 \end{cases}$

27. $\begin{cases} x^2 + 3y^2 = 6 \\ x^2 - 3y^2 = 10 \end{cases}$ $\{\}$ **28.** $\begin{cases} x^2 + y^2 = 1 \\ y = x^2 - 9 \end{cases}$ $\{\}$

29. $\begin{cases} x^2 + y^2 = 36 \\ y = \dfrac{1}{6}x^2 - 6 \end{cases}$ **30.** $\begin{cases} x^2 + y^2 = 16 \\ y = -\dfrac{1}{4}x^2 + 4 \end{cases}$

31. How many real solutions are possible for a system of equations whose graphs are a circle and a parabola?   0, 1, 2, 3, or 4

32. How many real solutions are possible for a system of equations whose graphs are an ellipse and a line?   0, 1, or 2

33. The sum of the squares of two numbers is 130. The difference of the squares of the two numbers is 32. Find the two numbers.

34. The sum of the squares of two numbers is 20. Their product is 8. Find the two numbers.

35. During the development stage of a new rectangular keypad for a security system, it was decided that the area of the rectangle should be 285 square centime-

ters and the perimeter should be 68 centimeters. Find the dimensions of the keypad. 15 cm by 19 cm

36. A rectangular holding pen for cattle is to be designed so that its perimeter is 92 feet and its area is 525 feet. Find the dimensions of the holding pen.
21 ft by 25 ft

*Recall that in business, a demand function expresses the quantity of a commodity demanded as a function of the commodity's unit price. A supply function expresses the quantity of a commodity supplied as a function of the commodity's unit price. When the quantity produced and supplied is equal to the quantity demanded, then we have what is called **market equilibrium**.*

37. The demand function for a certain compact disc is given by the function

$$p = -0.01x^2 - 0.2x + 9$$

and the corresponding supply function is given by

$$p = 0.01x^2 - 0.1x + 3$$ 15 thousand compact discs; price, $3.75

where p is in dollars and x is in thousands of units. Find the equilibrium quantity and the corresponding price by solving the system consisting of the two given equations.

38. The demand function for a certain style of picture frame is given by the function

$$p = -2x^2 + 90$$

and the corresponding supply function is given by

$$p = 9x + 34$$ 3.5 thousand frames; price, $65.50

where p is in dollars and x is in thousands of units. Find the equilibrium quantity and the corresponding price by solving the system consisting of the two given equations.

For Exercises 39–46, see Appendix D.

Use a grapher to verify the results of each exercise.

39. Exercise 3.   **40.** Exercise 4.

41. Exercise 23.   **42.** Exercise 24.

Review Exercises

Graph each inequality in two variables. See Section 3.6.

43. $x > -3$ **44.** $y \le 1$

45. $y < 2x - 1$ **46.** $3x - y \le 4$

13. $\{(2, 4), (-5, 25)\}$ **14.** $\left\{\left(-\dfrac{1}{6}, -6\right), (1, 1)\right\}$ **19.** $\{(-1, -2), (-1, 2), (1, -2), (1, 2)\}$
20. $\{(\sqrt{3}, \sqrt{5}), (\sqrt{3}, -\sqrt{5}), (-\sqrt{3}, \sqrt{5}), (-\sqrt{3}, -\sqrt{5})\}$ **23.** $\{(-1, 3), (1, 3)\}$ **25.** $\{(\sqrt{3}, 0), (-\sqrt{3}, 0)\}$
26. $\{(-5, 0), (4, -3), (4, 3)\}$ **29.** $\{(-6, 0), (6, 0), (0, -6)\}$ **30.** $\{(0, 4), (-4, 0), (4, 0)\}$
33. 9 and 7; 9 and -7; -9 and 7; -9 and -7 **34.** 2 and 4; -2 and -4

Find the perimeter of each geometric figure. See Section 5.3.

47.

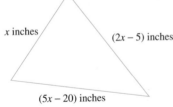

(8x − 25) in.

x inches

(2x − 5) inches

(5x − 20) inches

48.

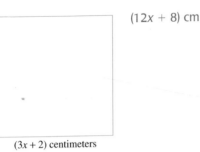

(12x + 8) cm

(3x + 2) centimeters

49.

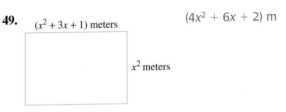

(x² + 3x + 1) meters

(4x² + 6x + 2) m

x² meters

50.

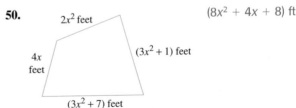

2x² feet

(8x² + 4x + 8) ft

4x feet

(3x² + 1) feet

(3x² + 7) feet

9.4 NONLINEAR INEQUALITIES AND SYSTEMS OF INEQUALITIES

O B J E C T I V E S

1 Sketch the graph of a nonlinear inequality.

2 Sketch the solution set of a system of nonlinear inequalities.

TAPE IA 9.4

We can graph a nonlinear equality in two variables such as $\dfrac{x^2}{9} + \dfrac{y^2}{16} \le 1$ in a way similar to the way we graphed a linear inequality in two variables in Section 3.6. First, we graph the related equation $\dfrac{x^2}{9} + \dfrac{y^2}{16} = 1$. The graph of the equation is our boundary. Then, using test points, we determine and shade the region whose points satisfy the inequality.

EXAMPLE 1 Graph $\dfrac{x^2}{9} + \dfrac{y^2}{16} \le 1$.

Solution: First, graph the equation $\dfrac{x^2}{9} + \dfrac{y^2}{16} = 1$. Sketch a solid curve since the graph of $\dfrac{x^2}{9} + \dfrac{y^2}{16} \le 1$ includes the graph of $\dfrac{x^2}{9} + \dfrac{y^2}{16} = 1$. The graph is an ellipse, and it divides the plane into two regions, the "inside" and the "outside" of the ellipse. To determine which region contains the solutions, select a test point in either region and determine whether the coordinates of the point satisfy the inequality. We choose $(0, 0)$ as the test point.

$$\frac{x^2}{9} + \frac{y^2}{16} \le 1$$

$$\frac{0^2}{9} + \frac{0^2}{16} \le 1 \qquad \text{Let } x = 0 \text{ and } y = 0.$$

$$0 \le 1 \qquad \text{True.}$$

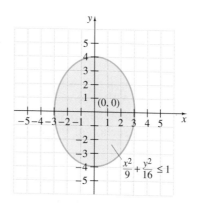

Since this statement is true, the solution set is the region containing $(0, 0)$. The graph of the solution set includes the points on and inside the ellipse, as shaded in the figure.

EXAMPLE 2   Graph $4y^2 > x^2 + 16$.

Solution: The related equation is $4y^2 = x^2 + 16$, or $\frac{y^2}{4} - \frac{x^2}{16} = 1$, which is a hyperbola.

Graph the hyperbola as a dashed curve since the graph of $4y^2 > x^2 + 16$ does *not* include the graph of $4y^2 = x^2 + 16$. The hyperbola divides the plane into three regions. Select a test point in each region—not on a boundary line—to determine whether that region contains solutions of the inequality.

Test region A with $(0, 4)$	Test region B with $(0, 0)$	Test region C with $(0, -4)$
$4y^2 > x^2 + 16$	$4y^2 > x^2 + 16$	$4y^2 > x^2 + 16$
$4(4)^2 > 0^2 + 16$	$4(0)^2 > 0^2 + 16$	$4(-4)^2 > 0^2 + 16$
$64 > 16$ True.	$0 > 16$ False.	$64 > 16$ True.

The graph of the solution set includes the shaded regions only, not the boundary.

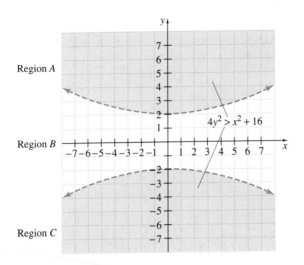

In the Section 3.6, we looked at the intersection of graphs of inequalities in two variables. Although we did not identify them as such, we now can recognize these sets of inequalities as systems of inequalities. The graph of a system of inequalities is the intersection of the graphs of the inequalities.

EXAMPLE 3   Graph the system

$$\begin{cases} x \le 1 - 2y \\ y \le x^2 \end{cases}$$

Solution: Graph each inequality on the same set of axes. The intersection is the darkest shaded region along with its boundary lines. The coordinates of the points of intersection can be found by solving the related system

$$\begin{cases} x = 1 - 2y \\ y = x^2 \end{cases}$$

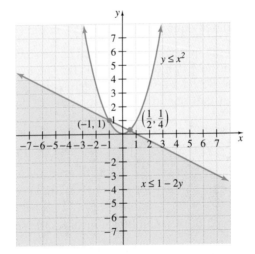

EXAMPLE 4   Graph the system

$$\begin{cases} x^2 + y^2 < 25 \\ \dfrac{x^2}{9} - \dfrac{y^2}{25} < 1 \\ \qquad y < x + 3 \end{cases}$$

Solution: Graph each inequality. The graph of $x^2 + y^2 < 25$ contains points "inside" the circle that has center $(0,0)$ and radius 5. The graph of $\dfrac{x^2}{9} - \dfrac{y^2}{25} < 1$ is the region between the two branches of the hyperbola with x-intercepts -3 and 3 and center $(0,0)$. The graph of $y < x + 3$ is the region "below" the line with the slope 1 and y-intercept 3. The graph of the solution set of the system is the intersection of all the graphs, the darkest shaded region shown. The boundary of this region is not part of the solution.

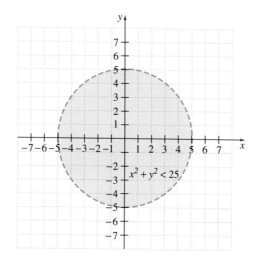

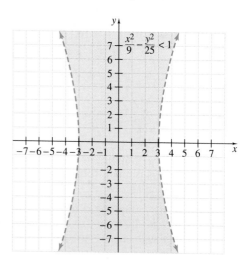

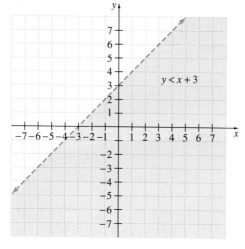

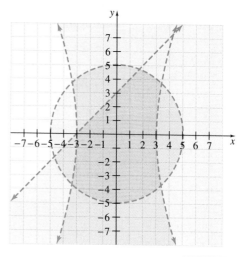

EXERCISE SET 9.4 For Exercises 1–20, see Appendix D.

Graph each inequality. See Examples 1 and 2.

1. $y < x^2$

2. $y < -x^2$

3. $x^2 + y^2 \geq 16$

4. $x^2 + y^2 < 36$

5. $\dfrac{x^2}{4} - y^2 < 1$

6. $x^2 - \dfrac{y^2}{9} \geq 1$

7. $y > (x - 1)^2 - 3$

8. $y > (x + 3)^2 + 2$

9. $x^2 + y^2 \leq 9$

10. $x^2 + y^2 > 4$

11. $y > -x^2 + 5$

12. $y < -x^2 + 5$

13. $\dfrac{x^2}{4} + \dfrac{y^2}{9} \leq 1$

14. $\dfrac{x^2}{25} + \dfrac{y^2}{4} \geq 1$

15. $\dfrac{y^2}{4} - x^2 \leq 1$

16. $\dfrac{y^2}{16} - \dfrac{x^2}{9} > 1$

17. $y < (x - 2)^2 + 1$

18. $y > (x - 2)^2 + 1$

19. $y \leq x^2 + x - 2$

20. $y > x^2 + x - 2$

21. Discuss how graphing a linear inequality such as $x + y < 9$ is similar to graphing a nonlinear inequality such as $x^2 + y^2 < 9$.

22. Discuss how graphing a linear inequality such as $x + y < 9$ is different from graphing a nonlinear inequality such as $x^2 + y^2 < 9$.

Graph the solution of each system. See Examples 3 and 4.
For Exercises 23–42, see Appendix D.

23. $\begin{cases} 2x - y < 2 \\ y \leq -x \end{cases}$

24. $\begin{cases} x - 2y > 4 \\ y > -x^2 \end{cases}$

25. $\begin{cases} 4x + 3y \geq 12 \\ x^2 + y^2 < 16 \end{cases}$

26. $\begin{cases} 3x - 4y \leq 12 \\ x^2 + y^2 < 16 \end{cases}$

27. $\begin{cases} x^2 + y^2 \leq 9 \\ x^2 + y^2 \geq 1 \end{cases}$

28. $\begin{cases} x^2 + y^2 \geq 9 \\ x^2 + y^2 \geq 16 \end{cases}$

29. $\begin{cases} y > x^2 \\ y \geq 2x + 1 \end{cases}$

30. $\begin{cases} y \leq -x^2 + 3 \\ y \leq 2x - 1 \end{cases}$

31. $\begin{cases} x > y^2 \\ y > 0 \end{cases}$

32. $\begin{cases} x < (y + 1)^2 + 2 \\ x + y \geq 3 \end{cases}$

33. $\begin{cases} x^2 + y^2 > 9 \\ y > x^2 \end{cases}$

34. $\begin{cases} x^2 + y^2 \leq 9 \\ y < x^2 \end{cases}$

35. $\begin{cases} \dfrac{x^2}{4} + \dfrac{y^2}{9} \geq 1 \\ x^2 + y^2 \geq 4 \end{cases}$

36. $\begin{cases} x^2 + (y - 2)^2 \geq 9 \\ \dfrac{x^2}{4} + \dfrac{y^2}{25} < 1 \end{cases}$

37. $\begin{cases} x^2 - y^2 \geq 1 \\ y \geq 0 \end{cases}$

38. $\begin{cases} x^2 - y^2 \geq 1 \\ x \geq 0 \end{cases}$

39. $\begin{cases} x + y \geq 1 \\ 2x + 3y < 1 \\ x > -3 \end{cases}$

40. $\begin{cases} x - y < -1 \\ 4x - 3y > 0 \\ y > 0 \end{cases}$

41. $\begin{cases} x^2 - y^2 < 1 \\ \dfrac{x^2}{16} + y^2 \leq 1 \\ x \geq -2 \end{cases}$

42. $\begin{cases} x^2 - y^2 \geq 1 \\ \dfrac{x^2}{16} + \dfrac{y^2}{4} \leq 1 \\ y \geq 1 \end{cases}$

Review Exercises

Determine which graph is the graph of a function. See Section 3.2.

43.
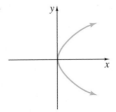
not a function

44.
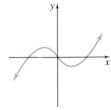
function

45.
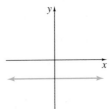
function

46.
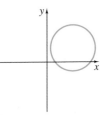
not a function

Find each function value if $f(x) = 3x^2 - 2$. See Section 3.2.

47. $f(-1)$ 1

48. $f(-3)$ 25

49. $f(a)$ $3a^2 - 2$

50. $f(b)$ $3b^2 - 2$

GROUP ACTIVITY

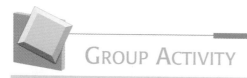

MODELING CONIC SECTIONS

MATERIALS:
- Two thumbtacks (or nails), graph paper, cardboard, tape, string, pencil, ruler.

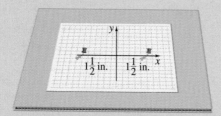

FIGURE 1

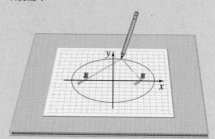

FIGURE 2

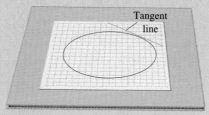

FIGURE 3

Try constructing the following model and investigating the results.

1. Draw an *x*-axis and a *y*-axis on the graph paper as shown in Figure 1.

2. Place the graph paper on the cardboard and use tape to attach.

3. Locate two points on the *x*-axis each about $1\frac{1}{2}$ inches from the origin and on opposite sides of the origin (see Figure 1). Insert thumbtacks (or nails) at each of these locations.

4. Fasten a 9-inch piece of string to the thumbtacks as shown in Figure 2. Use your pencil to draw and keep the string taut while you carefully move the pencil in a path all around the thumbtacks.

5. Using the grid of the graph paper as a guide, find an approximate equation of the ellipse you drew.

6. Experiment by moving the tacks closer together or farther apart and drawing new ellipses. What do you observe?

7. Write a paragraph explaining why the figure drawn by the pencil is an ellipse. How might you use the same materials to draw a circle?

8. (Optional) Choose one of the ellipses you drew with the string and pencil. Use a ruler to draw any six tangent lines to the ellipse. (A line is tangent to the ellipse if it intersects, or just touches, the ellipse at only one point. See Figure 3.) Extend the tangent lines to yield six points of intersection among the tangents. Use a straight edge to draw a line connecting each pair of opposite points of intersection. What do you observe? Repeat with a different ellipse. Can you make a conjecture about the relationship among the lines that connect opposite points of intersection?

See Appendix D for Group Activity answers and suggestions.

Chapter 9 Highlights

Definitions and Concepts	Examples

Section 9.1 The Parabola and the Circle

Parabolas $y = a(x - h)^2 + k$

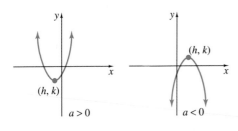

$$a > 0 \qquad a < 0$$

$$x = a(y - k)^2 + h$$

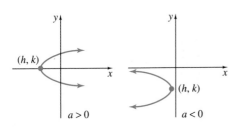

$$a > 0 \qquad a < 0$$

Graph $x = 3y^2 - 12y + 13$.

$$x + 3(4) = 3(y^2 - 4y + 4) + 13$$
$$x = 3(y - 2)^2 + 1$$

Since $a = 3$, this parabola opens to the right with vertex $(1, 2)$. Its axis of symmetry is $y = 2$. The x-intercept is 13.

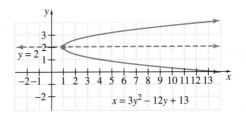

$$x = 3y^2 - 12y + 13$$

Distance formula

The distance d between two points (x_1, y_1) and (x_2, y_2) is given by

$$d = \sqrt{(x_2 - x_1)^2 + (y_2 - y_1)^2}$$

Find the distance between points $(-1, 6)$ and $(-2, -4)$.
Let $(x_1, y_1) = (-1, 6)$ and $(x_2, y_2) = (-2, -4)$.

$$\begin{aligned} d &= \sqrt{(x_2 - x_1)^2 + (y_2 - y_1)^2} \\ &= \sqrt{(-2 - (-1))^2 + (-4 - 6)^2} \\ &= \sqrt{1 + 100} = \sqrt{101} \end{aligned}$$

Midpoint formula

The midpoint of the line segment whose end points are (x_1, y_1) and (x_2, y_2) is the point with coordinates

$$\left(\frac{x_1 + x_2}{2}, \frac{y_1 + y_2}{2} \right)$$

Find the midpoint of the line segment whose end-points are $(-1, 6)$ and $(-2, -4)$.

$$\left(\frac{-1 + (-2)}{2}, \frac{6 + (-4)}{2} \right)$$

The midpoint is

$$\left(-\frac{3}{2}, 1 \right)$$

(continued)

DEFINITIONS AND CONCEPTS	EXAMPLES

SECTION 9.1   THE PARABOLA AND THE CIRCLE

Circle

The graph of $(x - h)^2 + (y - k)^2 = r^2$ is a circle with center (h, k) and radius r.

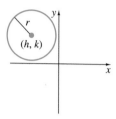

Graph $x^2 + (y + 3)^2 = 5$.

This equation can be written as

$$(x - 0)^2 + (y + 3)^2 = 5 \text{ with } h = 0, k = -3,$$
$$\text{and } r = \sqrt{5}$$

The center of this circle is $(0, -3)$, and the radius is $\sqrt{5}$.

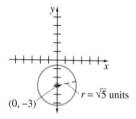

SECTION 9.2   THE ELLIPSE AND THE HYPERBOLA

Ellipse with center (0, 0)

The graph of an equation of the form $\dfrac{x^2}{a^2} + \dfrac{y^2}{b^2} = 1$

is an ellipse with center $(0, 0)$. The x-intercepts are a and $-a$, and the y-intercepts are b and $-b$.

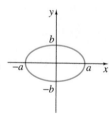

Graph $4x^2 + 9y^2 = 36$.

$$\frac{x^2}{9} + \frac{y^2}{4} = 1 \qquad \text{Divide by 36.}$$

$$\frac{x^2}{3^2} + \frac{y^2}{2^2} = 1$$

The ellipse has center $(0, 0)$, x-intercepts 3 and -3, and y-intercepts 2 and -2.

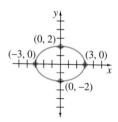

(continued)

DEFINITIONS AND CONCEPTS	**EXAMPLES**

SECTION 9.2 THE ELLIPSE AND THE HYPERBOLA

Hyperbola with center $(0, 0)$

The graph of an equation of the form $\dfrac{x^2}{a^2} - \dfrac{y^2}{b^2} = 1$ is a hyperbola with center $(0, 0)$ and x-intercepts a and $-a$.

Graph $\dfrac{x^2}{9} - \dfrac{y^2}{4} = 1$. Here $a = 3$ and $b = 2$.

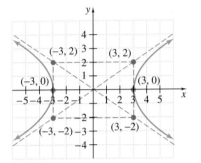

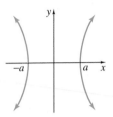

The graph of an equation of the form $\dfrac{y^2}{b^2} - \dfrac{x^2}{a^2} = 1$ is a hyperbola with center $(0, 0)$ and y-intercepts b and $-b$.

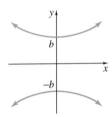

SECTION 9.3 SOLVING NONLINEAR SYSTEMS OF EQUATIONS

A **nonlinear system of equations** is a system of equations at least one of which is not linear. Both the substitution method and the elimination method may be used to solve a nonlinear system of equations.

Solve the nonlinear system $\begin{cases} y = x + 2 \\ 2x^2 + y^2 = 3 \end{cases}$

Substitute $x + 2$ for y in the second equation:

$$2x^2 + y^2 = 3$$
$$2x^2 + (x + 2)^2 = 3$$
$$2x^2 + x^2 + 4x + 4 = 3$$
$$3x^2 + 4x + 1 = 0$$
$$(3x + 1)(x + 1) = 0$$
$$x = -\frac{1}{3}, x = -1$$

(continued)

DEFINITIONS AND CONCEPTS	EXAMPLES
SECTION 9.3 SOLVING NONLINEAR SYSTEMS OF EQUATIONS	

$$\text{If } x = -\frac{1}{3}, y = x + 2 = -\frac{1}{3} + 2 = \frac{5}{3}.$$

$$\text{If } x = -1, y = x + 2 = -1 + 2 = 1.$$

$$\text{The solution set is } \left\{ \left(-\frac{1}{3}, \frac{5}{3} \right), (-1, 1) \right\}.$$

SECTION 9.4 NONLINEAR INEQUALITIES AND SYSTEMS OF INEQUALITIES

The graph of a system of inequalities is the intersection of the graphs of the inequalities.

Graph the system $\begin{cases} x \geq y^2 \\ x + y \leq 4 \end{cases}$

The graph of the system is the darkest shaded region along with its boundary lines.

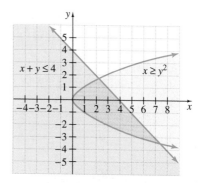

(9.1) *Find the distance between each pair of points.*

1. $(-6, 3)$ and $(8, 4)$ **2.** $(3, 5)$ and $(8, 9)$

3. $(-4, -6)$ and $(-1, 5)$ **4.** $(-1, 5)$ and $(2, -3)$

5. $(-\sqrt{2}, 0)$ and $(0, -4\sqrt{6})$ $7\sqrt{2}$ units

6. $(-\sqrt{5}, -\sqrt{11})$ and $(-\sqrt{5}, -3\sqrt{11})$ $2\sqrt{11}$ units

7. $(7.4, -8.6)$ and $(-1.2, 5.6)$ 16.60 units

8. $(2.3, 1.8)$ and $(10.7, -9.2)$ 13.84 units

Find the midpoint of the line segment whose end points are given.

9. $(2, 6)$ and $(-12, 4)$ **10.** $(-3, 8)$ and $(11, 24)$ (4, 16)

1. $\sqrt{197}$ units   **2.** $\sqrt{41}$ units   **3.** $\sqrt{130}$ units   **4.** $\sqrt{73}$ units

11. $(-6, -5)$ and $(-9, 7)$ **12.** $(4, -6)$ and $(-15, 2)$

13. $\left(0, -\frac{3}{8} \right)$ and $\left(\frac{1}{10}, 0 \right)$ **14.** $\left(\frac{3}{4}, -\frac{1}{7} \right)$ and $\left(-\frac{1}{4}, -\frac{3}{7} \right)$

15. $(\sqrt{3}, -2\sqrt{6})$ and $(\sqrt{3}, -4\sqrt{6})$ $(\sqrt{3}, -3\sqrt{6})$

16. $(-5\sqrt{3}, 2\sqrt{7})$ and $(-3\sqrt{3}, 10\sqrt{7})$ $(-4\sqrt{3}, 6\sqrt{7})$

Write an equation of the circle with the given center and radius.

17. center $(-4, 4)$, *radius* 3 $(x + 4)^2 + (y - 4)^2 = 9$

18. center $(5, 0)$, *diameter* 10 $(x - 5)^2 + y^2 = 25$

19. center $(-7, -9)$, *radius* $\sqrt{11}$ $(x + 7)^2 + (y + 9)^2 = 11$

20. center $(0, 0)$, *diameter* 7 $x^2 + y^2 = \dfrac{49}{4}$

9. $(-5, 5)$ **11.** $\left(-\dfrac{15}{2}, 1 \right)$ **12.** $\left(-\dfrac{11}{2}, -2 \right)$ **13.** $\left(\dfrac{1}{20}, -\dfrac{3}{16} \right)$ **14.** $\left(\dfrac{1}{4}, -\dfrac{2}{7} \right)$

For Exercises 21–34, see Appendix D.

Sketch the graph of the equation. If the graph is a circle, find its center. If the graph is a parabola, find its vertex.

21. $x^2 + y^2 = 7$

22. $x = 2(y - 5)^2 + 4$

23. $x = -(y + 2)^2 + 3$

24. $(x - 1)^2 + (y - 2)^2 = 4$

25. $y = -x^2 + 4x + 10$

26. $x = -y^2 - 4y + 6$

27. $x = \frac{1}{2}y^2 + 2y + 1$

28. $y = -3x^2 + \frac{1}{2}x + 4$

29. $x^2 + y^2 + 2x + y = \frac{3}{4}$

30. $x^2 + y^2 + 3y = \frac{7}{4}$

31. $4x^2 + 4y^2 + 16x + 8y = 1$

32. $3x^2 + 6x + 3y^2 = 9$

33. $y = x^2 + 6x + 9$

34. $x = y^2 + 6y + 9$

35. Write an equation of the circle centered at $(5.6, -2.4)$ with diameter 6.2.
$(x - 5.6)^2 + (y + 2.4)^2 = 9.61$

(9.2) *Sketch the graph of each equation.*
For Exercises 36–66, see Appendix D.

36. $x^2 + \frac{y^2}{4} = 1$

37. $x^2 - \frac{y^2}{4} = 1$

38. $\frac{y^2}{4} - \frac{x^2}{16} = 1$

39. $\frac{y^2}{4} + \frac{x^2}{16} = 1$

40. $\frac{x^2}{5} + \frac{y^2}{5} = 1$

41. $\frac{x^2}{5} - \frac{y^2}{5} = 1$

42. $-5x^2 + 25y^2 = 125$

43. $4y^2 + 9x^2 = 36$

44. $\frac{(x - 2)^2}{4} + (y - 1)^2 = 1$

45. $\frac{(x + 3)^2}{9} + \frac{(y - 4)^2}{25} = 1$

46. $x^2 - y^2 = 1$

47. $36y^2 - 49x^2 = 1764$

48. $y^2 = x^2 + 9$

49. $x^2 = 4y^2 - 16$

50. $100 - 25x^2 = 4y^2$

Sketch the graph of each equation.

51. $y = x^2 + 4x + 6$

52. $y^2 = x^2 + 6$

53. $y^2 + x^2 = 4x + 6$

54. $y^2 + 2x^2 = 4x + 6$

55. $x^2 + y^2 - 8y = 0$

56. $x - 4y = y^2$

57. $x^2 - 4 = y^2$

58. $x^2 = 4 - y^2$

59. $6(x - 2)^2 + 9(y + 5)^2 = 36$

60. $36y^2 = 576 + 16x^2$

61. $\frac{x^2}{16} - \frac{y^2}{25} = 1$

62. $3(x - 7)^2 + 3(y + 4)^2 = 1$

Use a grapher to verify the results of each exercise.

63. Exercise 39.

64. Exercise 40.

65. Exercise 51.

66. Exercise 58.

(9.3) *Solve each system of equations.*

67. $\begin{cases} y = 2x - 4 \\ y^2 = 4x \end{cases}$

68. $\begin{cases} x^2 + y^2 = 4 \\ x - y = 4 \end{cases}$ $\{\}$

69. $\begin{cases} y = x + 2 \\ y = x^2 \end{cases}$

70. $\begin{cases} y = x^2 - 5x + 1 \\ y = -x + 6 \end{cases}$

71. $\begin{cases} 4x - y^2 = 0 \\ 2x^2 + y^2 = 16 \end{cases}$

72. $\begin{cases} x^2 + 4y^2 = 16 \\ x^2 + y^2 = 4 \end{cases}$

73. $\begin{cases} x^2 + y^2 = 10 \\ 9x^2 + y^2 = 18 \end{cases}$

74. $\begin{cases} x^2 + 2y = 9 \\ 5x - 2y = 5 \end{cases}$

75. $\begin{cases} y = 3x^2 + 5x - 4 \\ y = 3x^2 - x + 2 \end{cases}$ $\{(1, 4)\}$

76. $\begin{cases} x^2 - 3y^2 = 1 \\ 4x^2 + 5y^2 = 21 \end{cases}$

77. Find the length and the width of a room whose area is 150 square feet and whose perimeter is 50 feet. 15 ft by 10 ft

78. What is the greatest number of real solutions possible for a system of two equations whose graphs are an ellipse and a hyperbola? 4

(9.4) *Graph the inequality or system of inequalities.*
For Exercises 79–88, see Appendix D.

79. $y \le -x^2 + 3$

80. $x^2 + y^2 < 9$

81. $x^2 - y^2 < 1$

82. $\frac{x^2}{4} + \frac{y^2}{9} \ge 1$

83. $\begin{cases} 2x \le 4 \\ x + y \ge 1 \end{cases}$

84. $\begin{cases} 3x + 4y \le 12 \\ x - 2y > 6 \end{cases}$

85. $\begin{cases} y > x^2 \\ x + y \ge 3 \end{cases}$

86. $\begin{cases} x^2 + y^2 \le 16 \\ x^2 + y^2 \ge 4 \end{cases}$

87. $\begin{cases} x^2 + y^2 < 4 \\ x^2 - y^2 \le 1 \end{cases}$

88. $\begin{cases} x^2 + y^2 < 4 \\ y \ge x^2 - 1 \\ x \ge 0 \end{cases}$

67. $\{(1, -2), (4, 4)\}$ **69.** $\{(-1, 1), (2, 4)\}$ **70.** $\{(5, 1), (-1, 7)\}$ **71.** $\{(2, 2\sqrt{2}), (2, -2\sqrt{2})\}$ **72.** $\{(0, 2), (0, -2)\}$

73. $\{(-1, 3), (-1, -3), (1, 3), (1, -3)\}$ **74.** $\left\{\left(2, \frac{5}{2}\right), (-7, -20)\right\}$ **76.** $\{(-2, -1), (-2, 1), (2, -1), (2, 1)\}$

CHAPTER 9 TEST

For Exercises 5–12, see Appendix D.
For Exercises 17–20, see Appendix D.

1. Find the distance between the points $(-6, 3)$ and $(-8, -7)$. $2\sqrt{26}$ units

2. Find the distance between the points $(-2\sqrt{5}, \sqrt{10})$ and $(-\sqrt{5}, 4\sqrt{10})$. $\sqrt{95}$ units

3. Find the midpoint of the line segment whose end points are $(-2, -5)$ and $(-6, 12)$. $\left(-4, \frac{7}{2}\right)$

4. Find the midpoint of the line segment whose end points are $\left(-\frac{2}{3}, -\frac{1}{5}\right)$ and $\left(-\frac{1}{3}, \frac{4}{5}\right)$. $\left(-\frac{1}{2}, \frac{3}{10}\right)$

Sketch the graph of each equation.

5. $x^2 + y^2 = 36$

6. $x^2 - y^2 = 36$

7. $16x^2 + 9y^2 = 144$

8. $y = x^2 - 8x + 16$

9. $x^2 + y^2 + 6x = 16$

10. $x = y^2 + 8y - 3$

11. $\dfrac{(x - 4)^2}{16} + \dfrac{(y - 3)^2}{9} = 1$

12. $y^2 - x^2 = 0$

Solve each system.

13. $\begin{cases} x^2 + y^2 = 169 \\ 5x + 12y = 0 \end{cases}$

14. $\begin{cases} x^2 + y^2 = 26 \\ x^2 - y^2 = 24 \end{cases}$

15. $\begin{cases} y = x^2 - 5x + 6 \\ y = 2x \end{cases}$ $\{(6, 12), (1, 2)\}$

16. $\begin{cases} x^2 + 4y^2 = 5 \\ y = x \end{cases}$ $\{(1, 1), (-1, -1)\}$

Graph the solution of each system.

17. $\begin{cases} 2x + 5y \geq 10 \\ y \geq x^2 + 1 \end{cases}$

18. $\begin{cases} \dfrac{x^2}{4} + y^2 \leq 1 \\ x + y > 1 \end{cases}$

19. $\begin{cases} x^2 + y^2 > 1 \\ \dfrac{x^2}{4} - y^2 \geq 1 \end{cases}$

20. $\begin{cases} x^2 + y^2 \geq 4 \\ x^2 + y^2 < 16 \\ y \geq 0 \end{cases}$

13. $\{(-12, 5), (12, -5)\}$ **14.** $\{(-5, -1), (-5, 1), (5, -1), (5, 1)\}$

21. Which graph best resembles the graph of $x = a(y - k)^2 + h$ if $a > 0, h < 0$, and $k > 0$? B

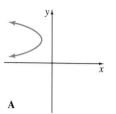
A

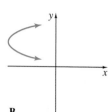
B

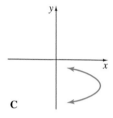
C

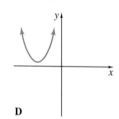
D

22. A bridge has an arch in the shape of half an ellipse. If the equation of the ellipse, measured in feet, is $100x^2 + 225y^2 = 22,500$, find the height of the arch from the road and the width of the arch. height, 10 ft; width, 30 ft

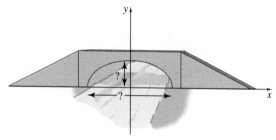

CHAPTER 9 CUMULATIVE REVIEW

1. Simplify each expression.

a. 3^2 9

b. $\left(\dfrac{1}{2}\right)^4$ $\dfrac{1}{16}$

c. -5^2 -25

d. $(-5)^2$ 25

e. -5^3 -125

f. $(-5)^3$ -125; *Sec. 1.3, Ex. 7*

2. Solve for x: $\dfrac{x + 5}{2} + \dfrac{1}{2} = 2x - \dfrac{x - 3}{8}$. $\left\{\dfrac{21}{11}\right\}$; *Sec. 2.1, Ex. 6*

3. Marial Callier just received an inheritance of $10,000 and plans to place all the money in a savings account that pays 5% compounded quarterly to help her son go to college in 3 years. How much money will be in the account in 3 years? $11,607.55; *Sec. 2.3, Ex. 4*

4. Solve $2|x| + 25 = 23$. $\{ \ \}$; *Sec. 2.6, Ex. 6*

6. See Appendix D; *Sec. 3.1, Ex. 7*

5. Solve for x: $\left| 2x - \dfrac{1}{10} \right| < -13$. { }; *Sec. 2.7, Ex. 4*

6. Graph the equation $y = |x|$.

7. The following graph shows the research and development expenditures by the Pharmaceutical Manufacturers Association as a function of time.

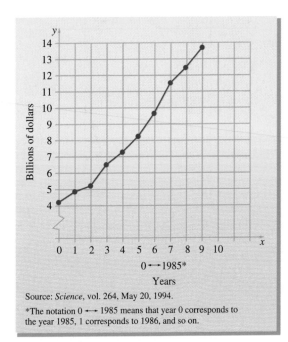

Source: *Science*, vol. 264, May 20, 1994.

*The notation $0 \longleftrightarrow 1985$ means that year 0 corresponds to the year 1985, 1 corresponds to 1986, and so on.

(Sec. 3.2, Ex. 8)

a. Approximate the money spent on research and development in 1992. $11.5 billion

b. In 1958, research and development expenditures were $200 million. Find the increase in expenditures from 1958 to 1994. $13,600 million

8. Write an equation of the line with y-intercept -3 and slope of $\frac{1}{4}$. $y = \frac{1}{4}x - 3$; *Sec. 3.5, Ex. 1*

9. Use substitution to solve the system

$$\begin{cases} -\dfrac{x}{6} + \dfrac{y}{2} = \dfrac{1}{2} \\ \dfrac{x}{3} - \dfrac{y}{6} = -\dfrac{3}{4} \end{cases} \left\{ \left(-\dfrac{21}{10}, \dfrac{3}{10} \right) \right\}; \textit{Sec. 4.1, Ex. 3}$$

10. The measure of the largest angle of a triangle is 80° more than the measure of the smallest angle, and the measure of the remaining angle is 10° more than the measure of the smallest angle. Find the measure of each angle. 30°, 40°, 110°; *Sec. 4.3, Ex. 5*

11. Evaluate the following. *Sec. 5.1, Ex. 3*

a. $7°$ 1 **b.** $-7°$ -1 **c.** $(2x + 5)°$ 1 **d.** $2x°$ 2

12. Simplify each expression. Write answers with positive exponents.

a. $(2x^0 y^{-3})^{-2}$ $\dfrac{y^6}{4}$ **b.** $\left(\dfrac{x^{-5}}{x^{-2}} \right)^{-3}$ x^9

c. $\left(\dfrac{2}{7} \right)^{-2}$ $\dfrac{49}{4}$ **d.** $\dfrac{5^{-2}x^{-3}y^{11}}{x^2 y^{-5}}$ $\dfrac{y^{16}}{25x^5}$; *Sec. 5.2, Ex. 3*

13. Add $11x^3 - 12x^2 + x - 3$ and $x^3 - 10x + 5$.

14. Factor $x^2 - 12x + 35$. $(x - 5)(x - 7)$; *Sec. 5.6, Ex. 2*

15. Divide $\dfrac{8x^3 + 125}{x^4 + 5x^2 + 4} \div \dfrac{2x + 5}{2x^2 + 8}$.

16. Divide $10x^2 - 5x + 20$ by 5.

17. Find the cube roots.

a. $\sqrt[3]{1}$ 1 **b.** $\sqrt[3]{-64}$ -4 **c.** $\sqrt[3]{\dfrac{8}{125}}$ $\dfrac{2}{5}$

d. $\sqrt[3]{x^6}$ x^2 **e.** $\sqrt[3]{-8x^9}$ $-2x^3$; *Sec. 7.1, Ex. 2*

18. Rationalize the numerator of each expression.

a. $\dfrac{\sqrt{28}}{\sqrt{45}}$ $\dfrac{14}{3\sqrt{35}}$ **b.** $\dfrac{\sqrt[3]{2x^2}}{\sqrt[3]{5y}}$

19. Use the square root property to solve $x^2 = 50$.

20. Solve $(x + 3)(x - 3) \geq 0$.

21. Sketch the graph of each function.

a. $F(x) = x^2 + 2$

b. $g(x) = x^2 - 3$ See Appendix D; *Sec. 8.5, Ex. 2*

22. Find the distance between the pair of points $(2, -5)$ and $(1, -4)$. $\sqrt{2}$; *Sec. 9.1, Ex. 5*

23. Solve the system

$$\begin{cases} y = \sqrt{x} \\ x^2 + y^2 = 6 \end{cases} \{(2, \sqrt{2})\}; \textit{Sec. 9.3, Ex. 2}$$

Find real number solutions only.

---

13. $12x^3 - 12x^2 - 9x + 2$; *Sec. 5.3, Ex. 7* **15.** $\dfrac{2(4x^2 - 10x + 25)}{x^2 + 1}$; *Sec. 6.2, Ex. 4* **16.** $2x^2 - x + 4$; *Sec. 6.5, Ex. 1*

18b. $\dfrac{2x}{\sqrt[3]{20xy}}$; *Sec. 7.5, Ex. 3* **19.** $x = \pm 5\sqrt{2}$; *Sec. 8.1, Ex. 1* **20.** $(-\infty, -3] \cup [3, \infty)$; *Sec. 8.4, Ex. 1*

EXPONENTIAL AND LOGARITHMIC FUNCTIONS

MODELING TEMPERATURE

When a cold object is placed in a warm room, the object's temperature gradually rises until it becomes, or nearly becomes, room temperature. Similarly, if a hot object is placed in a cooler room, the object's temperature gradually falls to room temperature. The way in which a cold or hot object warms up or cools off is modeled by an exact mathematical relationship.

IN THE CHAPTER GROUP ACTIVITY ON PAGE 610, YOU WILL HAVE THE OPPORTUNITY TO INVESTIGATE THIS MODEL OF COOLING AND WARMING.

I n this chapter, we discuss two closely related functions: exponential and logarithmic functions. These functions are vital in applications in economics, finance, engineering, the sciences, education, and other fields. Models of tumor growth and learning curves are two examples of the uses of exponential and logarithmic functions.

10.1   COMPOSITE AND INVERSE FUNCTIONS

TAPE IA 10.1

OBJECTIVES

1. Compose functions.
2. Determine whether a function is a one-to-one function.
3. Use the horizontal line test to test whether a function is a one-to-one function.
4. Define the inverse of a function.
5. Find the equation of the inverse of a function.

Thus far in this text, we have seen several ways to combine functions and create new functions. Functions have been added, subtracted, multiplied, and divided, and the result each time has been a function.

Another way to combine functions is called **function composition.** To understand this new way of combining functions, study the tables below. They show degrees Fahrenheit converted to equivalent degrees Celsius, and then degrees Celsius converted to equivalent degrees Kelvin. (The Kelvin scale is a temperature scale devised by Lord Kelvin in 1848.)

x = DEGREES FAHRENHEIT (INPUT)	-31	-13	32	68	149	212
$C(x)$ = DEGREES CELSIUS (OUTPUT)	-35	-25	0	20	65	100

C = DEGREES CELSIUS (INPUT)	-35	-25	0	20	65	100
$K(C)$ = KELVINS (OUTPUT)	238.15	248.15	273.15	293.15	338.15	373.15

Suppose that we want a table that shows a direct conversion from degrees Fahrenheit to kelvins. In other words, suppose that a table is needed that shows kelvins as a function of degrees Fahrenheit. This can easily be done because in the tables, the output of the first table is the same as the input of the second table. The new table is as follows.

x = DEGREES FAHRENHEIT (INPUT)	-31	-13	32	68	149	212
$K(C(x))$ = KELVINS (OUTPUT)	238.15	248.15	273.15	293.15	338.15	373.15

Since the output of the first table is used as the input of the second table, we write the new function as $K(C(x))$. This new function is formed from the composition of the other two functions and the mathematical symbol for this composition is $(K \circ C)(x)$. Thus, $(K \circ C)(x) = K(C(x))$.

It is possible to find an equation for the composition of the two functions $C(x)$ and $K(x)$. In other words, we can find a function that converts degrees Fahrenheit directly to kelvins. The function $C(x) = \frac{5}{9}(x - 32)$ converts degrees Fahrenheit to degrees Celsius, and the function $K(C) = C + 273.15$ converts degrees Celsius to kelvins. Thus,

$$(K \circ C)(x) = K(C(x)) = K\left(\frac{5}{9}(x - 32)\right) = \frac{5}{9}(x - 32) + 273.15$$

In general, the notation $f(g(x))$ means "f composed with g" and can be written as $(f \circ g)(x)$. Also, $g(f(x))$, or $(g \circ f)(x)$, means "g composed with f."

EXAMPLE 1   If $f(x) = x^2$ and $g(x) = x + 3$, find the following.

    a. $(f \circ g)(2)$ and $(g \circ f)(2)$ **b.** $(f \circ g)(x)$ and $(g \circ f)(x)$

Solution: **a.** $(f \circ g)(2) = f(g(2))$

$\qquad\qquad\quad = f(5)$ Since $g(x) = x + 3$, then $g(2) = 2 + 3 = 5$.

$\qquad\qquad\quad = 5^2 = 25$

$\qquad\quad (g \circ f)(2) = g(f(2))$

$\qquad\qquad\quad = g(4)$ Since $f(x) = x^2$, then $f(2) = 2^2 = 4$.

$\qquad\qquad\quad = 4 + 3 = 7$

    b. $(f \circ g)(x) = f(g(x))$

$\qquad\qquad\quad = f(x + 3)$ Replace $g(x)$ with $x + 3$.

$\qquad\qquad\quad = (x + 3)^2$ $f(x + 3) = (x + 3)^2$.

$\qquad\qquad\quad = x^2 + 6x + 9$ Square $(x + 3)$.

$\qquad\quad (g \circ f)(x) = g(f(x))$

$\qquad\qquad\quad = g(x^2)$ Replace $f(x)$ with x^2.

$\qquad\qquad\quad = x^2 + 3$ $g(x^2) = x^2 + 3$.

EXAMPLE 2   If $f(x) = 5x$, $g(x) = x - 2$, and $h(x) = \sqrt{x}$, write each function as a composition with f, g, or h.

    a. $F(x) = \sqrt{x - 2}$ **b.** $G(x) = 5x - 2$

Solution: **a.** Notice the order in which the function F operates on an input value x. First, 2 is subtracted from x, and then the square root of that result is taken. This means that $F = h \circ g$. To check, find $h \circ g$:

$$(h \circ g)(x) = h(g(x))$$
$$= h(x - 2)$$
$$= \sqrt{x - 2}$$

b. Notice the order in which the function G operates on an input value x. First, x is multiplied by 5, and then 2 is subtracted from the result. This means that $G = g \circ f$. To check, find $g \circ f$:

$$(g \circ f)(x) = g(f(x))$$
$$= g(5x)$$
$$= 5x - 2$$

---

2 Study the following table. Determine whether the table describes a function.

STATE (INPUT)	ALABAMA	DELAWARE	MARYLAND	LOUISIANA	HAWAII	MONTANA
RECORD-LOW TEMPERATURE IN DEGREES FAHRENHEIT (OUTPUT)	-27	-17	-40	-16	12	-70

Since each state (input) corresponds to exactly one record-low temperature (output), this table of inputs and outputs does describe a function. Also notice that each output corresponds to a different input. This type of function is given a special name—a one-to-one function.

Does the set $f = \{(0, 1), (2, 2), (-3, 5), (7, 6)\}$ describe a one-to-one function? It is a function since each x-value corresponds to a unique y-value. For this particular function f, each y-value also corresponds to a unique x-value. Thus, this function is also a **one-to-one function.**

ONE-TO-ONE FUNCTION

For a one-to-one function, each x-value (input) corresponds to only one y-value (output) and each y-value (output) corresponds to only one x-value (input).

EXAMPLE 3 Determine whether each function described is one-to-one.

a. $f = \{(6, 2), (5, 4), (-1, 0), (7, 3)\}$
b. $g = \{(3, 9), (-4, 2), (-3, 9), (0, 0)\}$
c. $h = \{(1, 1), (2, 2), (10, 10), (-5, -5)\}$

d.

MINERAL (INPUT)	TALC	GYPSUM	DIAMOND	TOPAZ	STIBNITE
HARDNESS ON THE MOHS SCALE (OUTPUT)	1	2	10	8	2

e.

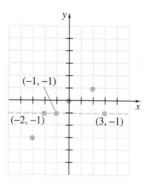

Solution: **a.** *f* is one-to-one since each *y*-value corresponds to only one *x*-value.

b. *g* is not one-to-one because the *y*-value 9 in (3, 9) and (−3, 9) corresponds to two different *x*-values.

c. *h* is a one-to-one function since each *y*-value corresponds to only one *x*-value.

d. This table does not describe a one-to-one function since the output 2 corresponds to two different inputs, gypsum and stibnite.

e. This graph does not describe a one-to-one function since the *y*-value −1 corresponds to three different *x*-values, −2, −1, and 3.

Recall that we recognize the graph of a function when it passes the vertical line test. Since every *x*-value of the function corresponds to exactly one *y*-value, each vertical line intersects the function's graph at most once. The graph shown next, for instance, is the graph of a function.

Is this function a *one-to-one* function? The answer is no. To see why not, notice that the *y*-value of the ordered pair (−3, 3), for example, is the same as the *y*-value of the ordered pair (3, 3). This function is therefore not one-to-one.

To test whether a graph is the graph of a one-to-one function, apply the vertical line test to see if it is a function, and then apply a similar **horizontal line test** to see if it is a one-to-one function.

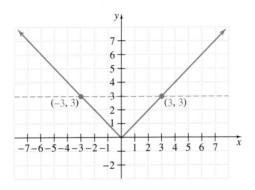

HORIZONTAL LINE TEST

If every horizontal line intersects the graph of a function at most once, then the function is a one-to-one function.

EXAMPLE 4 Determine whether each graph is the graph of a one-to-one function.

a.

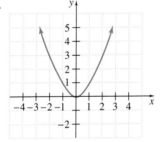

b.

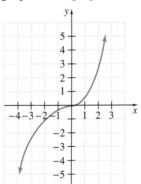

c.

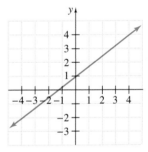

d.

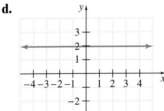

e.

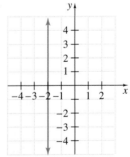

Solution: Graphs **a, b, c,** and **d** all pass the vertical line test, so only these graphs are graphs of functions. But, of these, only **b** and **c** pass the horizontal line test, so only **b** and **c** are graphs of one-to-one-functions.

> R E M I N D E R   All linear equations are one-to-one functions except those whose graphs are horizontal or vertical lines. A vertical line does not pass the vertical line test and hence is not the graph of a function. A horizontal line is the graph of a function but does not pass the horizontal line test and hence is not the graph of a one-to-one function.

4   One-to-one functions are special in that their graphs pass the vertical and horizontal line tests. They are special, too, in another sense: For each one-to-one function we can find its **inverse function** by switching the coordinates of the ordered pairs of the function, or the inputs and the outputs. For example, the inverse of the one-to-one function

STATE (INPUT)	**ALABAMA**	**DELAWARE**	**MARYLAND**	**LOUISIANA**	**HAWAII**	**MONTANA**
RECORD-LOW TEMPERATURE (OUTPUT)	-27	-17	-40	-16	12	-70

is the function

RECORD-LOW TEMPERATURE (INPUT)	**-27**	**-17**	**-40**	**-16**	**12**	**-70**
STATE (OUTPUT)	Alabama	Delaware	Maryland	Louisiana	Hawaii	Montana

Notice that the ordered pair (Alabama, -27) of the function f, for example, becomes the ordered pair (-27, Alabama) of its inverse.

Also, the inverse of the one-to-one function $f = \{(2, -3), (5, 10), (9, 1)\}$ is $\{(-3, 2), (10, 5), (1, 9)\}$. For a function f, we use the notation f^{-1}, read "f inverse," to denote its inverse function. Notice that since the coordinates of each ordered pair have been switched, the domain (set of inputs) of f is the range (set of outputs) of f^{-1}, and the range of f is the domain of f^{-1}.

> **INVERSE FUNCTION**
>
> The inverse of a one-to-one function f is the one-to-one function f^{-1} that consists of the set of all ordered pairs (y, x) where (x, y) belongs to f.

REMINDER   The symbol f^{-1} is the single symbol used to denote the inverse of the function f. It is read as "f inverse." This symbol *does not mean* $\dfrac{1}{f}$.

5   If a one-to-one function f is defined as a set of ordered pairs, we can find f^{-1} by interchanging the x and y coordinates of the ordered pairs. If a one-to-one function f is given in the form of an equation, we can find f^{-1} by using a similar procedure.

TO FIND THE INVERSE OF A ONE-TO-ONE FUNCTION $f(x)$

Step 1.   Replace $f(x)$ with y.
Step 2.   Interchange x and y.
Step 3.   Solve the equation for y.
Step 4.   Replace y with the notation $f^{-1}(x)$.

EXAMPLE 5   Find the equation of the inverse of $f(x) = 3x - 5$. Graph f and f^{-1} on the same set of axes.

Solution: First, let $y = f(x)$.

$$f(x) = 3x - 5$$
$$y = 3x - 5$$

Next, interchange x and y and solve for y:

$$x = 3y - 5 \qquad \text{Interchange } x \text{ and } y.$$
$$3y = x + 5$$
$$y = \frac{x + 5}{3} \qquad \text{Solve for } y.$$

Let $y = f^{-1}(x)$.

$$f^{-1}(x) = \frac{x + 5}{3}$$

Now graph $f(x)$ and $f^{-1}(x)$ on the same set of axes. Both $f(x) = 3x - 5$ and $f^{-1}(x) = \dfrac{x + 5}{3}$ are linear functions, so each graph is a line.

$$f(x) = 3x - 5 \qquad\qquad f^{-1}(x) = \frac{x + 5}{3}$$

x	$y = f(x)$
1	−2
0	−5
$\frac{5}{3}$	0

x	$y = f^{-1}(x)$
−2	1
−5	0
0	$\frac{5}{3}$

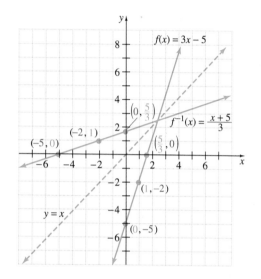

Notice that the graphs of f and f^{-1} are mirror images of each other, and the "mirror" is the dashed line $y = x$. This is true for every function and its inverse. For this reason, we say that **the graphs of f and f^{-1} are symmetric about the line $y = x$.**

Notice also in the preceding table of values that $f(0) = -5$ and $f^{-1}(-5) = 0$, as expected. Also, for example, $f(1) = -2$ and $f^{-1}(-2) = 1$. In words, we say that for some input x, the function f^{-1} takes the output of x called $f(x)$ back to x:

$$x \to f(x) \text{ and } f^{-1}(f(x)) \to x$$
$$\downarrow \quad\; \downarrow \qquad\qquad \downarrow \qquad\; \downarrow$$
$$f(0) = -5 \text{ and } f^{-1}(-5) = 0$$
$$f(1) = -2 \text{ and } f^{-1}(-2) = 1$$

In general,

If f is a one-to-one function, then the inverse of f is the function f^{-1} such that

$$(f^{-1} \circ f)(x) = x \text{ and } (f \circ f^{-1})(x) = x$$

EXAMPLE 6   Show that if $f(x) = 3x + 2$, then $f^{-1}(x) = \dfrac{x-2}{3}$.

Solution: See that $f^{-1}(f(x)) = x$ and $f(f^{-1}(x)) = x$.

$$(f^{-1} \circ f)(x) = f^{-1}(f(x))$$
$$= f^{-1}(3x + 2) \qquad \text{Replace } f(x) \text{ with } 3x + 2.$$
$$= \frac{3x + 2 - 2}{3}$$
$$= \frac{3x}{3}$$
$$= x$$

$$(f \circ f^{-1})(x) = f(f^{-1}(x))$$
$$= f\!\left(\frac{x-2}{3}\right) \qquad \text{Replace } f^{-1}(x) \text{ with } \frac{x-2}{3}.$$
$$= 3\!\left(\frac{x-2}{3}\right) + 2$$
$$= x - 2 + 2$$
$$= x$$

GRAPHING CALCULATOR EXPLORATIONS

A grapher can be used to visualize the results of Example 6. Recall that the graph of a function f and its inverse f^{-1} are mirror images of each other across the line $y = x$. To see this for the function from Example 6, use a square window and graph

the given function: $Y_1 = 3x + 2$

its inverse: $Y_2 = \dfrac{x-2}{3}$

and the line: $Y_3 = x$

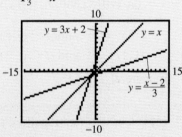

Exercises will follow in Exercise Set 10.1.

7. $(f \circ g)(x) = 25x^2 + 1$; $(g \circ f)(x) = 5x^2 + 5$; $(f \circ f)(x) = x^4 + 2x^2 + 2$

EXERCISE SET 10.1

8. $(f \circ g)(x) = x^2 - 3$; $(g \circ f)(x) = x^2 - 6x + 9$; $(f \circ f)(x) = x - 6$

9. $(f \circ g)(x) = 2x + 11$; $(g \circ f)(x) = 2x + 4$; $(f \circ f)(x) = 4x - 9$

If $f(x) = x^2 - 6x + 2$, $g(x) = -2x$, and $h(x) = \sqrt{x}$, find the following. See Example 1.

1. $(f \circ g)(2)$ 42

2. $(h \circ f)(-2)$ $3\sqrt{2}$

3. $(g \circ f)(-1)$ -18

4. $(f \circ h)(1)$ -3

5. $(g \circ h)(0)$ 0

6. $(h \circ g)(0)$ 0

Find $(f \circ g)(x)$, $(g \circ f)(x)$, and $(f \circ f)(x)$. See Example 1.

7. $f(x) = x^2 + 1$ $g(x) = 5x$

8. $f(x) = x - 3$ $g(x) = x^2$

9. $f(x) = 2x - 3$ $g(x) = x + 7$

10. $f(x) = x + 10$ $g(x) = 3x + 1$

$(f \circ g)(x) = 3x + 11$; $(g \circ f)(x) = 3x + 31$; $(f \circ f)(x) = x + 20$

Find $(f \circ g)(x)$ and $(g \circ f)(x)$. See Example 1.

11. $f(x) = x^3 + x - 2$ $g(x) = -2x$

12. $f(x) = -4x$ $g(x) = x^3 + x^2 - 6$

13. $f(x) = \sqrt{x}$ $g(x) = -5x + 2$

14. $f(x) = 7x - 1$ $g(x) = \sqrt[3]{x}$

$(f \circ g)(x) = 7\sqrt[3]{x} - 1$; $(g \circ f)(x) = \sqrt[3]{7x - 1}$

If $f(x) = 3x$, $g(x) = \sqrt{x}$, and $h(x) = x^2 + 2$, write each of the following functions as a composition of f, g, and h. See Example 2.

15. $H(x) = \sqrt{x^2 + 2}$ $H(x) = (g \circ h)(x)$

16. $G(x) = \sqrt{3x}$ $G(x) = (g \circ f)(x)$

17. $F(x) = 9x^2 + 2$ $F(x) = (h \circ f)(x)$

18. $H(x) = 3x^2 + 6$ $H(x) = (f \circ h)(x)$

19. $G(x) = 3\sqrt{x}$ $G(x) = (f \circ g)(x)$

20. $F(x) = x + 2$ $F(x) = (h \circ g)(x)$

Determine whether each function is a one-to-one function. If it is one-to-one, list the inverse function by switching coordinates or inputs and outputs. See Example 3.

21. $f = \{(-1, -1), (1, 1), (0, 2), (2, 0)\}$

22. $g = \{(8, 6), (9, 6), (3, 4), (-4, 4)\}$ not one-to-one

23. $h = \{(10, 10)\}$ one-to-one; $h^{-1} = \{(10, 10)\}$

24. $r = \{(1, 2), (3, 4), (5, 6), (6, 7)\}$

25. $f = \{(11, 12), (4, 3), (3, 4), (6, 6)\}$

26. $g = \{(0, 3), (3, 7), (6, 7), (-2, -2)\}$ not one-to-one

27.

Year (Input)	1970	1975	1980	1985	1990
Average Composite Score on the ACT (Output)	19.9	18.6	18.5	18.6	20.6

not one-to-one

28.

State (Input)	Washington	Ohio	Georgia	Colorado	California	Arizona
Electoral Votes (Output)	11	21	13	8	54	8

not one-to-one

29.

Rank in Population (Input)	1	49	12	2	45
State (Output)	California	Vermont	Virginia	Texas	South Dakota

29.

State (Input)	California	Vermont	Virginia	Texas	South Dakota
Rank in Population (Output)	1	49	12	2	45

one-to-one

11. $(f \circ g)(x) = -8x^3 - 2x - 2$; $(g \circ f)(x) = -2x^3 - 2x + 4$

12. $(f \circ g)(x) = -4x^3 - 4x^2 + 24$; $(g \circ f)(x) = -64x^3 + 16x^2 - 6$ **13.** $(f \circ g)(x) = \sqrt{-5x + 2}$; $(g \circ f)(x) = -5\sqrt{x} + 2$

21. one-to-one; $f^{-1} = \{(-1, -1), (1, 1), (2, 0), (0, 2)\}$ **24.** one-to-one; $r^{-1} = \{(2, 1), (4, 3), (6, 5), (7, 6)\}$

25. one-to-one; $f^{-1} = \{(12, 11), (3, 4), (4, 3), (6, 6)\}$

30.

SHAPE (INPUT)	TRIANGLE	PENTAGON	QUADRILATERAL	HEXAGON	DECAGON
NUMBER OF SIDES (OUTPUT)	3	5	4	6	10

one-to-one

Given the one-to-one function $f(x) = x^3 + 2$, *find the following.*

31. a. $f(1)$ 3   **b.** $f^{-1}(3)$ 1

32. a. $f(0)$ 2   **b.** $f^{-1}(2)$ 0

33. a. $f(-1)$ 1   **b.** $f^{-1}(1)$ −1

34. a. $f(-2)$ −6   **b.** $f^{-1}(-6)$ −2

Determine whether the graph of each function is the graph of a one-to-one function. See Example 4.

35.

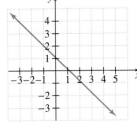

one-to-one

36.

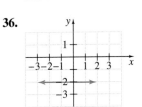

not one-to-one

37.

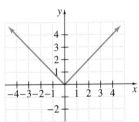

not one-to-one

38.

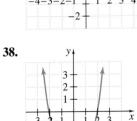

not one-to-one

39.

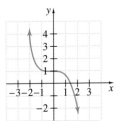

one-to-one

40.

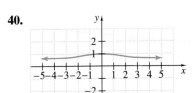

not one-to-one

41.

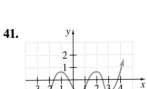

not one-to-one

42.

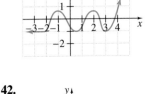

one-to-one

Each of the following functions is one-to-one. Find the inverse of each function and graph the function and its inverse on the same set of axes. See Example 5.
For Exercises 43–54, see Appendix D.

43. $f(x) = x + 4$ **44.** $f(x) = x - 5$

45. $f(x) = 2x - 3$ **46.** $f(x) = 4x + 9$

47. $f(x) = \dfrac{12x - 4}{3}$ **48.** $f(x) = \dfrac{7x + 5}{11}$

30.

NUMBER OF SIDES (INPUT)	3	5	4	6	10
SHAPE (OUTPUT)	TRIANGLE	PENTAGON	QUADRILATERAL	HEXAGON	DECAGON

43. $f^{-1}(x) = x - 4$ **44.** $f^{-1}(x) = x + 5$ **45.** $f^{-1}(x) = \dfrac{x + 3}{2}$ **46.** $f^{-1}(x) = \dfrac{x - 9}{4}$ **47.** $f^{-1}(x) = \dfrac{3x + 4}{12}$ **48.** $f^{-1}(x) = \dfrac{11x - 5}{7}$

50. $f^{-1}(x) = \sqrt[3]{x+1}$ **51.** $f^{-1}(x) = 5x + 2$ **52.** $f^{-1}(x) = \dfrac{2x+3}{4}$ **55.** $(f \circ f^{-1})(x) = x;\ (f^{-1} \circ f)(x) = x$

49. $f(x) = x^3\ f^{-1}(x) = \sqrt[3]{x}$ **50.** $f(x) = x^3 - 1$ **60.**

51. $f(x) = \dfrac{x-2}{5}$ **52.** $f(x) = \dfrac{4x-3}{2}$

53. $g(x) = x^2 + 5,\ x \geq 0$ **54.** $g(x) = x^2 - 2,\ x \geq 0$
$g^{-1}(x) = \sqrt{x-5},\ x \geq 5$ $g^{-1}(x) = \sqrt{x+2},\ x \geq -2$

Solve. See Example 6.

55. If $f(x) = 2x + 1$, show that $f^{-1}(x) = \dfrac{x-1}{2}$.

56. If $f(x) = 3x - 10$, show that $f^{-1}(x) = \dfrac{x+10}{3}$.

57. If $f(x) = x^3 + 6$, show that $f^{-1}(x) = \sqrt[3]{x-6}$.

58. If $f(x) = x^3 - 5$, show that $f^{-1}(x) = \sqrt[3]{x+5}$.

For Exercises 59 and 60,

a. Write the ordered pairs for $f(x)$ whose points are highlighted. (Include the points whose coordinates are given.)

b. Write the corresponding ordered pairs for the inverse of f, f^{-1}.

c. Graph the ordered pairs for f^{-1} found in part (b).

d. Graph $f^{-1}(x)$ by drawing a smooth curve through the plotted points.

59.

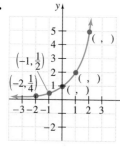

a. $(0, 1), (1, 2), (2, 5)$
b. $(\tfrac{1}{4}, -2), (\tfrac{1}{2}, -1), (1, 0),$
$(2, 1), (5, 2)$
c. See Appendix D.
d. See Appendix D.

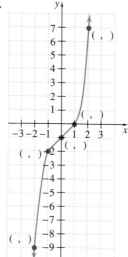

a. $(-2, -9), (-1, -2),$
$(0, -1), (1, 0), (2, 7)$
b. $(-9, -2), (-2, -1),$
$(-1, 0), (0, 1), (7, 2)$
c. See Appendix D.
d. See Appendix D.

Find the inverse of each given one-to-one function. Then graph the function and its inverse on a square window. For Exercises 61–64, see Appendix D.

 61. $f(x) = 3x + 1$ **62.** $f(x) = -2x - 6$

63. $f(x) = \sqrt[3]{x+3}$ **64.** $f(x) = x^3 - 3$
$f^{-1}(x) = x^3 - 1$ $f^{-1}(x) = \sqrt[3]{x+3}$

Review Exercises

Evaluate each of the following. See Section 7.2.

65. $25^{1/2}$ 5   **66.** $49^{1/2}$ 7   **67.** $16^{3/4}$ 8

68. $27^{2/3}$ 9   **69.** $9^{-3/2}$ $\dfrac{1}{27}$ **70.** $81^{-3/4}$ $\dfrac{1}{27}$

If $f(x) = 3^x$, find the following. In Exercises 73 and 74, give an exact answer and a two-decimal place approximation. See Sections 3.2 and 5.1.

71. $f(2)$ 9   **72.** $f(0)$ 1
73. $f(\tfrac{1}{2})$ $3^{1/2} \approx 1.73$ **74.** $f(\tfrac{2}{3})$ $3^{2/3} \approx 2.08$

56. $(f \circ f^{-1})(x) = x;\ (f^{-1} \circ f)(x) = x$ **57.** $(f \circ f^{-1})(x) = x;\ (f^{-1} \circ f)(x) = x$ **58.** $(f \circ f^{-1})(x) = x;\ (f^{-1} \circ f)(x) = x$

61. $f^{-1}(x) = \dfrac{x-1}{3}$

62. $f^{-1}(x) = \dfrac{-x-6}{2}$

10.2 | EXPONENTIAL FUNCTIONS

TAPE IA 10.2

O B J E C T I V E S

1. Identify exponential functions.
2. Graph exponential functions
3. Solve equations of the form $b^x = b^y$.

In earlier chapters, we gave meaning to exponential expressions such as 2^x, where x is a rational number. For example,

$$2^3 = 2 \cdot 2 \cdot 2 \qquad \text{Three factors, each factor is 2.}$$
$$2^{3/2} = (2^{1/2})^3 = \sqrt{2} \cdot \sqrt{2} \cdot \sqrt{2} \qquad \text{Three factors, each factor is } \sqrt{2}.$$

When x is an irrational number (for example, $\sqrt{3}$), what meaning can we give to $2^{\sqrt{3}}$?

It is beyond the scope of this book to give precise meaning to 2^x if x is irrational. We can confirm your intuition and say that $2^{\sqrt{3}}$ is a real number, and since $1 < \sqrt{3} < 2$, then $2^1 < 2^{\sqrt{3}} < 2^2$. We can also use a calculator and approximate $2^{\sqrt{3}}$: $2^{\sqrt{3}} \approx 3.321997$. In fact, as long as the base b is positive, b^x is a real number for all real numbers x. Finally, the rules of exponents apply whether x is rational or irrational, as long as b is positive. In this section, we are interested in functions of the form $f(x) = b^x$, where $b > 0$. A function of this form is called an **exponential function.**

EXPONENTIAL FUNCTION

A function of the form

$$f(x) = b^x$$

is called an exponential function if $b > 0$, b is not 1, and x is a real number.

---

2   Next, we practice graphing exponential functions.

EXAMPLE 1   Graph the exponential functions defined by $f(x) = 2^x$ and $g(x) = 3^x$ on the same set of axes.

Solution: Graph each function by plotting points. Set up a table of values for each of the two functions:

$f(x) = 2^x$

x	0	1	2	3	-1	-2
$f(x)$	1	2	4	8	$\dfrac{1}{2}$	$\dfrac{1}{4}$

$g(x) = 3^x$

x	0	1	2	3	-1	-2
$g(x)$	1	3	9	27	$\dfrac{1}{3}$	$\dfrac{1}{9}$

If each set of points is plotted and connected with a smooth curve, the following graphs result:

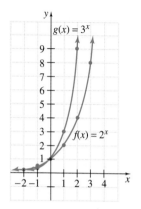

A number of things should be noted about these two graphs of exponential functions. First, the graphs show that $f(x) = 2^x$ and $g(x) = 3^x$ are one-to-one functions since each graph passes the vertical and horizontal line tests. The y-intercept of each graph is 1, but neither graph has an x-intercept. From the graph, we can also see that the domain of each function is all real numbers and that the range is $(0, \infty)$. We can also see that as x values are increasing, y values are increasing also.

EXAMPLE 2   Graph the exponential functions $y = \left(\dfrac{1}{2}\right)^x$ and $y = \left(\dfrac{1}{3}\right)^x$ on the same set of axes.

Solution: As before, plot points and connect them with a smooth curve.

$y = \left(\dfrac{1}{2}\right)^x$

x	0	1	2	3	-1	-2
y	1	$\dfrac{1}{2}$	$\dfrac{1}{4}$	$\dfrac{1}{8}$	2	4

$y = \left(\dfrac{1}{3}\right)^x$

x	0	1	2	3	-1	-2
y	1	$\dfrac{1}{3}$	$\dfrac{1}{9}$	$\dfrac{1}{27}$	3	9

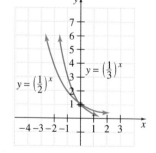

Each function again is a one-to-one function. The y-intercept of both is 1. The domain is the set of all real numbers, and the range is $(0, \infty)$.

Notice the difference between the graphs of Example 1 and the graphs of Example 2. An exponential function is always increasing if the base is greater than

1. When the base is between 0 and 1, the graph is decreasing. The following figures summarize these characteristics of exponential functions.

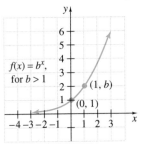

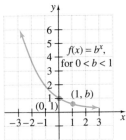

3

We have seen that an exponential function $y = b^x$ is a one-to-one function. Another way of stating this fact is a property that we can use to solve exponential equations.

UNIQUENESS OF b^x

Let $b > 0$ and $b \neq 1$. Then $b^x = b^y$ is equivalent to $x = y$.

EXAMPLE 3 Solve each equation for x.

a. $2^x = 16$ **b.** $9^x = 27$ **c.** $4^{x+3} = 8^x$

Solution: **a.** We write 16 as a power of 2 and then use the uniqueness of b^x to solve.

$$2^x = 16$$
$$2^x = 2^4$$

Since the bases are the same and are nonnegative, by the uniqueness of b^x, we then have that the exponents are equal. Thus,

$$x = 4$$

The solution set is $\{4\}$.

b. Notice that both 9 and 27 are powers of 3.

$$9^x = 27$$
$$(3^2)^x = 3^3 \qquad \text{Write 9 and 27 as powers of 3.}$$
$$3^{2x} = 3^3$$
$$2x = 3 \qquad \text{Apply the uniqueness of } b^x.$$
$$x = \frac{3}{2} \qquad \text{Divide by 2.}$$

To check, replace x with $\dfrac{3}{2}$ in the original expression, $9^x = 27$. The solution set is $\left\{\dfrac{3}{2}\right\}$.

c. Write both 4 and 8 as powers of 2.

$$4^{x+3} = 8^x$$
$$(2^2)^{x+3} = (2^3)^x$$
$$2^{2x+6} = 2^{3x}$$
$$2x + 6 = 3x \qquad \text{Apply the uniqueness of } b^x.$$
$$6 = x \qquad \text{Subtract } 2x \text{ from both sides.}$$

The solution set is {6}.

There is one major problem with the preceding technique. Often the two sides of an equation cannot easily be written as powers of a common base. We explore how to solve an equation such as $4 = 3^x$ with the help of **logarithms** later.

The world abounds with patterns that can be modeled by exponential functions. To make these applications realistic, we use numbers that warrant a calculator. The first application deals with exponential decay. When an application involves direct substitution into a given formula, problem-solving steps are not shown.

EXAMPLE 4   As a result of the Chernobyl nuclear accident, radioactive debris was carried through the atmosphere. One immediate concern was the impact that debris had on the milk supply. The percent y of radioactive material in raw milk after t days is estimated by $y = 100\,(2.7)^{-0.1t}$. Estimate the expected percent of radioactive material in the milk after 30 days.

Solution: Replace t with 30 in the given equation.

$$y = 100(2.7)^{-0.1t}$$
$$= 100(2.7)^{-0.1(30)} \qquad \text{Let } t = 30.$$
$$= 100(2.7)^{-3}$$

To approximate the percent y, the following keystrokes may be used on a scientific calculator.

| 2.7 | | y^x | | 3 | | +/− | | = | | × | | 100 | | = |

The display should read

| 5.0805263 |

Thus, approximately 5% of the radioactive material still contaminated the milk supply after 30 days.

The exponential function defined by $A = P\left(1 + \dfrac{r}{n}\right)^{nt}$ models the pattern relating the dollars A accrued (or owed) after P dollars are invested (or loaned) at an annual rate of interest r compounded n times each year for t years.

EXAMPLE 5   Find the amount owed at the end of 5 years if $1600 is loaned at a rate of 9% compounded monthly.

Solution: Use the formula $A = P\left(1 + \dfrac{r}{n}\right)^{nt}$, with the following values:

$P = \$1600$ (the amount of the loan)

$r = 9\% = 0.09$ (the annual rate of interest)

$n = 12$ (the number of times interest is compounded each year)

$t = 5$ (the duration of the loan, in years)

$$A = P\left(1 + \frac{r}{n}\right)^{nt} \qquad \text{Compound interest formula.}$$

$$= 1600\left(1 + \frac{0.09}{12}\right)^{12(5)} \qquad \text{Substitute known values.}$$

$$= 1600(1.0075)^{60}$$

To approximate A, press the following keys on your calculator:

$$\boxed{1.0075}\ \boxed{y^x}\ \boxed{60}\ \boxed{=}\ \boxed{\times}\ \boxed{1600}\ \boxed{=}$$

The display should read

$$\boxed{2505.0896}$$

Thus, the amount A owed is approximately $2505.09.

GRAPHING CALCULATOR EXPLORATIONS

We can use a graphing calculator and its TRACE feature to solve Example 4 graphically.

To estimate the expected percent of radioactive material in the milk after 30 days, enter $Y_1 = 100(2.7)^{-0.1x}$. The graph does not appear on a standard viewing window, so we need to determine an appropriate viewing window. Because it doesn't make sense to look at radioactivity *before* the Chernobyl nuclear accident, use Xmin = 0. We are interested in finding the percent of radioactive material in the milk when $x = 30$, so choose Xmax = 35 to leave enough space to see the graph at $x = 30$. Because the values of y are percents, it seems appropriate that $0 \le y \le 100$. (Also, use Xscl = 1 and Yscl = 10.) Now, graph the function.

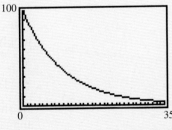

(continued)

Use the TRACE feature to obtain an approximation of the expected percent of radioactive material in the milk when $x = 30$. (A TABLE feature may also be used to approximate the percent.) To obtain a better approximation, use the ZOOM feature several times to zoom in near $x = 30$.

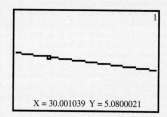

X = 30.001039 Y = 5.0800021

The percent of radioactive material in the milk 30 days after the Chernobyl accident was 5.08%, accurate to two decimal places.

Use a grapher to find each of the following graphically. Approximate your solutions so that they are accurate to two decimal places.

1. Estimate the expected percent of radioactive material in the milk 2 days after the Chernobyl nuclear accident.   81.98%

2. Estimate the expected percent of radioactive material in the milk 10 days after the Chernobyl nuclear accident.   37.04%

3. Estimate the expected percent of radioactive material in the milk 15 days after the Chernobyl nuclear accident.   22.54%

4. Estimate the expected percent of radioactive material in the milk 25 days after the Chernobyl nuclear accident.   8.35%

EXERCISE SET 10.2

Graph each exponential function. See Examples 1 and 2.
For Exercises 1-16, see Appendix D.

1. $y = 4^x$

2. $y = 5^x$

3. $y = 1 + 2^x$

4. $y = 3^x - 1$

5. $y = \left(\frac{1}{4}\right)^x$

6. $y = \left(\frac{1}{5}\right)^x$

7. $y = \left(\frac{1}{2}\right)^x - 2$

8. $y = \left(\frac{1}{3}\right)^x + 2$

9. $y = -2^x$

10. $y = -3^x$

11. $y = 3^x - 2$

12. $y = 2^x - 3$

13. $y = -\left(\frac{1}{4}\right)^x$

14. $y = -\left(\frac{1}{5}\right)^x$

15. $y = \left(\frac{1}{3}\right)^x + 1$

16. $y = \left(\frac{1}{2}\right)^x - 2$

17. Explain why the graph of an exponential function $y = b^x$ contains the point $(1, b)$.

18. Explain why an exponential function $y = b^x$ has a y-intercept of 1.

Solve each equation for x. See Example 3.

19. $3^x = 27$ 3

20. $6^x = 36$ 2

21. $16^x = 8$ $\frac{3}{4}$

22. $64^x = 16$ $\frac{2}{3}$

23. $32^{2x-3} = 2$ $\frac{8}{5}$

24. $9^{2x+1} = 81$ $\frac{1}{2}$

25. $\frac{1}{4} = 2^{3x}$ $-\frac{2}{3}$

26. $\frac{1}{27} = 3^{2x}$ $-\frac{3}{2}$

27. $5^x = 625$ 4   **28.** $2^x = 64$ 6

29. $4^x = 8$ $\dfrac{3}{2}$ **30.** $32^x = 4$ $\dfrac{2}{5}$

31. $27^{x+1} = 9$ $-\dfrac{1}{3}$ **32.** $125^{x-2} = 25$ $\dfrac{8}{3}$

33. $81^{x-1} = 27^{2x}$ -2 **34.** $4^{3x-7} = 32^{2x}$ $-\dfrac{7}{2}$

Write an exponential equation defining the function whose graph is given.

35.

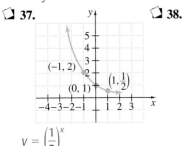

$y = 3^x$

36.

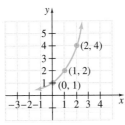

$y = 2^x$

37.

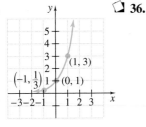

$y = \left(\dfrac{1}{2}\right)^x$

38.

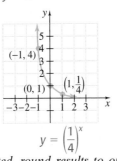

$y = \left(\dfrac{1}{4}\right)^x$

Solve. Unless otherwise indicated, round results to one decimal place. See Example 4.

39. One type of uranium has a daily radioactive decay rate of 0.4%. If 30 pounds of this uranium is available today, find how much will still remain after 50 days. Use $y = 30(2.7)^{-0.004t}$, and let t be 50.   24.6 lb

40. The nuclear waste from an atomic energy plant decays at a rate of 3% each century. If 150 pounds of nuclear waste is disposed of, find how much of it will still remain after 10 centuries. Use $y = 150(2.7)^{-0.03t}$, and let t be 10.   111.3 lb

41. The size of the rat population of a wharf area grows at a rate of 8% monthly. If there are 200 rats in January, find how many rats rounded to the nearest whole should be expected by next January. Use $y = 200(2.7)^{0.08t}$.   519 rats

42. National Park Service personnel are trying to increase the size of the bison population of Theodore Roosevelt National Park. If 260 bison currently live in the park, and if the population's rate of growth is 2.5% annually, find how many bison (rounded to the nearest whole) there should be in 10 years. Use $y = 260(2.7)^{0.025t}$.   333 bison

52. 20.16 lb   **53.** 18.62 lb

43. A rare isotope of a nuclear material is very unstable, decaying at a rate of 15% each second. Find how much isotope remains 10 seconds after 5 grams of the isotope is created. Use $y = 5(2.7)^{-0.15t}$.   1.1 g

44. An accidental spill of 75 grams of radioactive material in a local stream has led to the presence of radioactive debris decaying at a rate of 4% each day. Find how much debris still remains after 14 days. Use $y = 75(2.7)^{-0.04t}$.   43.0 g

45. Mexico City is growing at a rate of 0.7% annually. If there were 15,525,000 residents of Mexico City in 1994, predict how many (to the nearest ten-thousand) will be living in the city in 2000. Use $y = 15,525,000(2.7)^{0.007t}$.   16,190,000 residents

46. An unusually wet spring has caused the size of the Cape Cod mosquito population to increase by 8% each day. If an estimated 200,000 mosquitoes are on Cape Cod on May 12, find how many thousands of mosquitoes will inhabit the Cape on May 25. Use $y = 200,000(2.7)^{0.08t}$.   562,000 mosquitoes

Solve. Use $A = P\left(1 + \dfrac{r}{n}\right)^{nt}$. Round answers to two decimal places. See Example 5.

47. Find the amount Erica owes at the end of 3 years if $6000 is loaned to her at a rate of 8% compounded monthly. $7621.42

48. Find the amount owed at the end of 5 years if $3000 is loaned at a rate of 10% compounded quarterly. $4915.85

49. Find the total amount Janina has in a college savings account if $2000 was invested and earned 6% compounded semiannually for 12 years. $4065.59

50. Find the amount accrued if $500 is invested and earns 7% compounded monthly for 4 years. $661.03

Use a grapher to solve. Estimate the result to two decimal places.

51. Verify the results of Exercise 39.   See Appendix D.

52. From Exercise 39, estimate the number of pounds of uranium that will be available after 100 days.

53. From Exercise 39, estimate the number of pounds of uranium that will be available after 120 days.

54. Verify the results of Exercise 44.   See Appendix D.

55. From Exercise 44, estimate the amount of debris that remains after 10 days.   50.41 g

56. From Exercise 44, estimate the amount of debris that remains after 20 days.   33.88 g

Review Exercises

Solve each equation. See Sections 2.1 and 5.8.

57. $5x - 2 = 18$ {4}   **58.** $3x - 7 = 11$ {6}

59. $3x - 4 = 3(x + 1)$ { }

60. $2 - 6x = 6(1 - x)$ { }

61. $x^2 + 6 = 5x$ {2, 3}   **62.** $18 = 11x - x^2$ {2, 9}

By inspection, find the value for x that makes each statement true.

63. $2^x = 8$ {3}   **64.** $3^x = 9$ {2}

65. $5^x = \dfrac{1}{5}$ {−1}   **66.** $4^x = 1$ {0}

10.3 LOGARITHMIC FUNCTIONS

TAPE IA 10.3

O B J E C T I V E S

1 Write exponential equations with logarithmic notation and write logarithmic equations with exponential notation.

2 Solve logarithmic equations by using exponential notation.

3 Identify and graph logarithmic functions.

1 Since the exponential function $f(x) = 2^x$ is a one-to-one function, it has an inverse. We can create a table of values for f^{-1} by switching the coordinates in the accompanying table of values for $f(x) = 2^x$.

x	$y = f(x)$	x	$y = f^{-1}(x)$
-3	$\dfrac{1}{8}$	$\dfrac{1}{8}$	-3
-2	$\dfrac{1}{4}$	$\dfrac{1}{4}$	-2
-1	$\dfrac{1}{2}$	$\dfrac{1}{2}$	-1
0	1	1	0
1	2	2	1
2	4	4	2
3	8	8	3

The graphs of $f(x)$ and its inverse are shown next. Notice that the graphs of f and f^{-1} are symmetric about the line $y = x$, as expected.

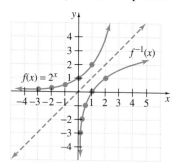

Finally, we would like to be able to write an equation for f^{-1}. To do so, we follow the steps for finding an inverse:

$$f(x) = 2^x$$

Step 1.	Replace $f(x)$ by y.	$y = 2^x$
Step 2.	Interchange x and y.	$x = 2^y$
Step 3.	Solve for y.	

At this point, we are stuck. To solve this equation for y, a new notation, the **logarithmic notation,** is needed. The symbol $\log_b x$ means "the power to which b is raised in order to produce a result of x."

$$\log_b x = y \text{ means } b^y = x$$

We say that $\log_b x$ is "the logarithm of x to the base b" or "the log of x to the base b."

LOGARITHMIC DEFINITION

If $b > 0$ and $b \neq 1$, then

$$y = \log_b x \text{ means } x = b^y$$

for every $x > 0$ and every real number y.

Before returning to the function $x = 2^y$ and solving it for y in terms of x, let's practice using the new notation $\log_b x$.

It is important to be able to write exponential equations with logarithmic notation, and vice versa. The following table shows examples of both forms.

LOGARITHMIC STATEMENT	CORRESPONDING EXPONENTIAL STATEMENT
$\log_3 9 = 2$	$3^2 = 9$
$\log_6 1 = 0$	$6^0 = 1$
$\log_2 8 = 3$	$2^3 = 8$
$\log_4 \dfrac{1}{16} = -2$	$4^{-2} = \dfrac{1}{16}$
$\log_8 2 = \dfrac{1}{3}$	$8^{1/3} = 2$

R E M I N D E R   Notice that the base of the logarithmic notation is the base of the exponential notation.

EXAMPLE 1   Find the value of each logarithmic expression.

a. $\log_4 16$ **b.** $\log_{10} \dfrac{1}{10}$ **c.** $\log_9 3$

Solution: **a.** $\log_4 16 = 2$ because $4^2 = 16$.   **b.** $\log_{10} \dfrac{1}{10} = -1$ because $10^{-1} = \dfrac{1}{10}$.

c. $\log_9 3 = \dfrac{1}{2}$ because $9^{1/2} = \sqrt{9} = 3$.

2   The ability to interchange the logarithmic and exponential forms of a statement is often the key to solving logarithmic equations.

EXAMPLE 2   Solve each equation for x.

a. $\log_4 \dfrac{1}{4} = x$ **b.** $\log_5 x = 3$ **c.** $\log_x 25 = 2$

d. $\log_3 1 = x$ **e.** $\log_b 1 = x$

Solution: **a.** $\log_4 \dfrac{1}{4} = x$ means $4^x = \dfrac{1}{4}$. Solve $4^x = \dfrac{1}{4}$ for x.

$$4^x = \dfrac{1}{4}$$
$$4^x = 4^{-1}$$

Since the bases are the same, by the uniqueness of b^x, we have that

$$x = -1$$

The solution set is $\{-1\}$. To check, see that $\log_4 \dfrac{1}{4} = -1$, since $4^{-1} = \dfrac{1}{4}$.

b. $\log_5 x = 3$ means $5^3 = x$ or

$$x = 125.$$

The solution set is $\{125\}$.

c. $\log_x 25 = 2$ means $x^2 = 25$ and $x > 0$ and $x \neq 1$.

$$x = 5$$

Even though $(-5)^2 = 25$, the base b of a logarithm must be positive. The solution set is $\{5\}$.

d. $\log_3 1 = x$ means $3^x = 1$. Either solve this equation by inspection or solve by writing 1 as 3^0 as shown:

$3^x = 3^0$ Write 1 as 3^0.

$x = 0$ Apply the uniqueness of b^x.

The solution set is $\{0\}$.

e. $\log_b 1 = x$ means $b^x = 1$ and $b > 0$ and $b \neq 1$.

$$b^x = b^0 \qquad \text{Write 1 as } b^0.$$

$$x = 0 \qquad \text{Apply the uniqueness of } b^x.$$

The solution set is $\{0\}$.

In Example 2(e) we proved an important property of logarithms. That is, $\log_b 1$ is always 0. This property as well as two important others are given next.

PROPERTIES OF LOGARITHMS

If b is a real number, $b > 0$ and $b \neq 1$, then

1. $\log_b 1 = 0$
2. $\log_b b^x = x$
3. $b^{\log_b x} = x$

To see that $\log_b b^x = x$, change the logarithmic form to exponential form. Then, $\log_b b^x = x$ means $b^x = b^x$. In exponential form, the statement is true, so in logarithmic form, the statement is also true.

EXAMPLE 3   Simplify.

    a. $\log_3 3^2$ **b.** $\log_7 7^{-1}$ **c.** $5^{\log_5 3}$ **d.** $2^{\log_2 6}$

Solution: **a.** From property 2, $\log_3 3^2 = 2$.
           b. From property 2, $\log_7 7^{-1} = -1$.
           c. From property 3, $5^{\log_5 3} = 3$.
           d. From property 3, $2^{\log_2 6} = 6$.

Having gained proficiency with the notation $\log_b x$, we return to the function $f(x) = 2^x$ and write an equation for its inverse, $f^{-1}(x)$. Recall our earlier work.

$$f(x) = 2^x$$

Step 1.	Replace $f(x)$ by y.	$y = 2^x$
Step 2.	Interchange x and y.	$x = 2^y$
Step 3.	Solve for y.	$y = \log_2 x$
Step 4.	Replace y with $f^{-1}(x)$.	$f^{-1}(x) = \log_2 x$

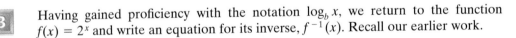

Thus, $f^{-1}(x) = \log_2 x$ defines a function that is the inverse function of the function $f(x) = 2^x$. The function $f^{-1}(x)$ or $y = \log_2 x$ is called a **logarithmic function**.

LOGARITHMIC FUNCTION

If x is a positive real number, b is a constant positive real number, and b is not 1, then a **logarithmic function** is a function that can be defined by

$$f(x) = \log_b x$$

The domain of f is the set of positive real numbers, and the range of f is the set of real numbers.

We can explore logarithmic functions by graphing them.

EXAMPLE 4   Graph the logarithmic function $y = \log_2 x$.

Solution: Write the equation with exponential notation as $2^y = x$. Find some ordered pair solutions that satisfy this equation. Plot the points and connect them with a smooth curve. The domain of this function is $(0, \infty)$, and the range of the function is all real numbers, **R.**

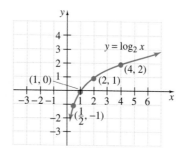

EXAMPLE 5   Graph the logarithmic function $f(x) = \log_{1/3} x$.

Solution: Replace $f(x)$ with y, and write the result with exponential notation:

$$f(x) = \log_{1/3} x$$
$$y = \log_{1/3} x \qquad \text{Replace } f(x) \text{ with } y.$$
$$\left(\frac{1}{3}\right)^y = x \qquad \text{Write in exponential form.}$$

Find ordered pair solutions that satisfy $\left(\frac{1}{3}\right)^y = x$, plot these points, and connect them with a smooth curve as shown in the figure.

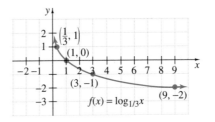

The domain of this function is $(0, \infty)$, and the range is the set of all real numbers.

The following figures summarize characteristics of logarithmic functions.

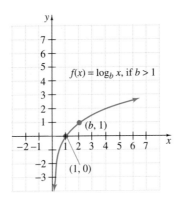

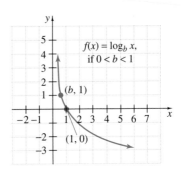

EXERCISE SET 10.3

Find the value of each logarithmic expression. See Example 1.

1. $\log_2 8$ 3

2. $\log_4 64$ 3

3. $\log_3 \frac{1}{9}$ −2

4. $\log_2 \frac{1}{32}$ −5

5. $\log_{25} 5$ $\frac{1}{2}$

6. $\log_8 \frac{1}{2}$ $-\frac{1}{3}$

7. $\log_{1/2} 2$ −1

8. $\log_{2/3} \frac{4}{9}$ 2

9. $\log_7 1$ 0

10. $\log_9 9$ 1

11. $\log_2 2^4$ 4

12. $\log_6 6^{-2}$ −2

13. $\log_{10} 100$ 2

14. $\log_{10} \frac{1}{10}$ −1

15. $3^{\log_3 5}$ 5

16. $5^{\log_5 7}$ 7

17. $\log_3 81$ 4

18. $\log_2 16$ 4

19. $\log_4 \frac{1}{64}$ −3

20. $\log_3 \frac{1}{9}$ −2

21. Explain why negative numbers are not included as logarithmic bases.

22. Explain why 1 is not included as a logarithmic base.

Solve each equation for x. See Example 2.

23. $\log_3 9 = x$ {2}

24. $\log_2 8 = x$ {3}

25. $\log_3 x = 4$ {81}

26. $\log_2 x = 3$ {8}

27. $\log_x 49 = 2$ {7}

28. $\log_x 8 = 3$ {2}

29. $\log_2 \frac{1}{8} = x$ {−3}

30. $\log_3 \frac{1}{81} = x$ {−4}

Solve each equation for x.

31. $\log_3 \frac{1}{27} = x$ {−3}

32. $\log_5 \frac{1}{125} = x$ {−3}

33. $\log_8 x = \frac{1}{3}$ {2}

34. $\log_9 x = \frac{1}{2}$ {3}

35. $\log_4 16 = x$ {2}

36. $\log_2 16 = x$ {4}

37. $\log_{3/4} x = 3$ $\left\{\frac{27}{64}\right\}$

38. $\log_{2/3} x = 2$ $\left\{\frac{4}{9}\right\}$

39. $\log_x 100 = 2$ {10}

40. $\log_x 27 = 3$ {3}

Simplify. See Example 3.

41. $\log_5 5^3$ 3

42. $\log_6 6^2$ 2

43. $2^{\log_2 3}$ 3

44. $7^{\log_7 4}$ 4

45. $\log_9 9$ 1

46. $\log_8 (8)^{-1}$ −1

Graph each logarithmic function. Label any intercepts. See Example 4.

47. $y = \log_3 x$

48. $y = \log_2 x$

49. $f(x) = \log_{1/4} x$

50. $f(x) = \log_{1/2} x$

51. $f(x) = \log_5 x$

52. $f(x) = \log_6 x$

53. $f(x) = \log_{1/6} x$

54. $f(x) = \log_{1/5} x$

For Exercises 47–58, see Appendix D.

Graph each function and its inverse function on the same set of axes. Label any intercepts. See Example 5.

55. $y = 4^x$; $y = \log_4 x$

56. $y = 3^x$; $y = \log_3 x$

57. $y = \left(\frac{1}{3}\right)^x$; $y = \log_{1/3} x$

58. $y = \left(\frac{1}{2}\right)^x$; $y = \log_{1/2} x$

59. The formula $\log_{10}(1 - k) = \frac{-0.3}{H}$ models the relationship between the half-life H of a radioactive material and its rate of decay k. Find the rate of decay of the iodine isotope I-131 if its half-life is 8 days. Round to 4 decimal places.   0.0827

60. Explain why the graph of the function $y = \log_b x$ contains the point $(1, 0)$ no matter what b is.

61. $\text{Log}_3 10$ is between which two integers? Explain your answer.   2 and 3

Review Exercises

Simplify each rational expression. See Section 6.1.

62. $\frac{x + 3}{3 + x}$ 1

63. $\frac{x - 5}{5 - x}$ −1

64. $\frac{x^2 - 8x + 16}{2x - 8}$ $\frac{x - 4}{2}$

65. $\frac{x^2 - 3x - 10}{2 + x}$ $x - 5$

Add or subtract as indicated. See Section 6.3.

66. $\frac{2}{x} + \frac{3}{x^2}$ $\frac{2x + 3}{x^2}$

67. $\frac{3x}{x + 3} + \frac{9}{x + 3}$ 3

68. $\frac{m^2}{m + 1} - \frac{1}{m + 1}$ $m - 1$

69. $\frac{5}{y + 1} - \frac{4}{y - 1}$ $\frac{y - 9}{y^2 - 1}$

10.4 | PROPERTIES OF LOGARITHMS

TAPE IA 10.4

O B J E C T I V E S

1 Apply the product property of logarithms.

2 Apply the quotient property of logarithms.

3 Apply the power property of logarithms.

In the previous section we explored some basic properties of logarithms. We now introduce and explore additional properties. Because a logarithm is an exponent, logarithmic properties are just restatements of exponential properties.

The first of these properties is called the **product property of logarithms,** because it deals with the logarithm of a product.

> **PRODUCT PROPERTY OF LOGARITHMS**
>
> If x, y, and b are positive real numbers and $b \neq 1$, then
> $$\log_b xy = \log_b x + \log_b y$$

To prove this, let $\log_b x = M$ and $\log_b y = N$. Now write each logarithm with exponential notation:

$$\log_b x = M \quad \text{is equivalent to} \quad b^M = x$$
$$\log_b y = N \quad \text{is equivalent to} \quad b^N = y$$

Multiply the left sides and the right sides of the exponential equations, and we have that

$$xy = (b^M)(b^N) = b^{M+N}$$

Now, write the equation $xy = b^{M+N}$ in equivalent logarithmic form:

$$xy = b^{M+N} \quad \text{is equivalent to} \quad \log_b xy = M + N$$

But since $M = \log_b x$ and $N = \log_b y$, we can write

$$\log_b xy = M + N$$

as

$$\log_b xy = \log_b x + \log_b y \qquad \text{Let } M = \log_b x \text{ and } N = \log_b y.$$

The logarithm of a product is the sum of the logarithms of the factors. This property is sometimes used to simplify logarithmic expressions.

EXAMPLE 1   Use the product property to write each sum as the logarithm of a single expression. Assume that variables represent positive numbers.

a. $\log_{11} 10 + \log_{11} 3$ **b.** $\log_3 \dfrac{1}{2} + \log_3 12$ **c.** $\log_2 (x + 2) + \log_2 x$

Solution: In each case, both terms have a common logarithmic base.

a. $\log_{11} 10 + \log_{11} 3 = \log_{11} (10 \cdot 3)$ Apply the product property.
$$= \log_{11} 30$$

b. $\log_3 \dfrac{1}{2} + \log_3 12 = \log_3 \left(\dfrac{1}{2} \cdot 12\right) = \log_3 6$

c. $\log_2 (x + 2) + \log_2 x = \log_2 [(x + 2) \cdot x] = \log_2 (x^2 + 2x)$

2 The second property is the **quotient property of logarithms.**

> **QUOTIENT PROPERTY OF LOGARITHMS**
> If x, y, and b are positive real numbers and $b \neq 1$, then
> $$\log_b \frac{x}{y} = \log_b x - \log_b y$$

The proof of the quotient property of logarithms is similar to the proof of the product rule. Notice that the quotient property says that the logarithm of a quotient is the difference of the logarithms of the dividend and divisor.

EXAMPLE 2 Use the quotient property to write each difference as the logarithm of a single expression. Assume that x represents positive numbers.

a. $\log_{10} 27 - \log_{10} 3$ **b.** $\log_5 8 - \log_5 x$ **c.** $\log_4 25 + \log_4 3 - \log_4 5$
d. $\log_3 (x^2 + 5) - \log_3 (x^2 + 1)$

Solution: All terms have a common logarithmic base.

a. $\log_{10} 27 - \log_{10} 3 = \log_{10} \dfrac{27}{3} = \log_{10} 9$

b. $\log_5 8 - \log_5 x = \log_5 \dfrac{8}{x}$

c. Use both the product and quotient properties:

$$\log_4 25 + \log_4 3 - \log_4 5 = \log_4 (25 \cdot 3) - \log_4 5 \qquad \text{Apply the product property.}$$
$$= \log_4 75 - \log_4 5 \qquad \text{Simplify.}$$
$$= \log_4 \frac{75}{5} \qquad \text{Apply the quotient property.}$$
$$= \log_4 15 \qquad \text{Simplify.}$$

d. $\log_3 (x^2 + 5) - \log_3 (x^2 + 1) = \log_3 \dfrac{x^2 + 5}{x^2 + 1}$ Apply the quotient property.

3 The third and final property we introduce is the **power property of logarithms.**

> **POWER PROPERTY OF LOGARITHMS**
> If x and b are positive real numbers, $b \neq 1$, and r is a real number, then
> $$\log_b x^r = r \log_b x$$

For example,

$$\log_3 2^4 = 4 \log_3 2, \quad \log_5 x^3 = 3 \log_5 x, \quad \log_4 \sqrt{x} = \log_4 (x)^{1/2} = \frac{1}{2} \log_4 x$$

EXAMPLE 3   Write each sum as the logarithm of a single expression. Assume that x is a positive number.

 a. $2 \log_5 3 + 3 \log_5 2$ **b.** $2 \log_9 x + \log_9 (x + 1)$

Solution: In each case, both terms have a common logarithmic base.

 a. $2 \log_5 3 + 3 \log_5 2 = \log_5 3^2 + \log_5 2^3$ Apply the power property.
$$= \log_5 9 + \log_5 8$$
$$= \log_5 (9 \cdot 8) \qquad \text{Apply the product property.}$$
$$= \log_5 72$$

 b. $2 \log_9 x + \log_9 (x + 1) = \log_9 x^2 + \log_9 (x + 1)$ Apply the power property.
$$= \log_9 [x^2(x + 1)] \qquad \text{Apply the product property.}$$

EXAMPLE 4   Use properties of logarithms to write each expression as a sum or difference of multiples of logarithms.

 a. $\log_3 \dfrac{5 \cdot 7}{4}$ **b.** $\log_2 \dfrac{x^5}{y^2}$

Solution: **a.** $\log_3 \dfrac{5 \cdot 7}{4} = \log_3 (5 \cdot 7) - \log_3 4$ Apply the quotient property.
$$= \log_3 5 + \log_3 7 - \log_3 4 \qquad \text{Apply the product property.}$$

 b. $\log_2 \dfrac{x^5}{y^2} = \log_2 (x^5) - \log_2 (y^2)$ Apply the quotient property.
$$= 5 \log_2 x - 2 \log_2 y \qquad \text{Apply the power property.}$$

> **REMINDER**   Notice that we are not able to simplify further a logarithmic expression such as $\log_5 (2x - 1)$. None of the basic properties gives a way to write the logarithm of a difference in some equivalent form.

EXAMPLE 5   If $\log_b 2 = 0.43$ and $\log_b 3 = 0.68$, use the properties of logarithms to evaluate.

 a. $\log_b 6$ **b.** $\log_b 9$ **c.** $\log_b \sqrt{2}$

Solution: **a.** $\log_b 6 = \log_b (2 \cdot 3)$ Write 6 as $2 \cdot 3$.

$= \log_b 2 + \log_b 3$ Apply the product property.

$= 0.43 + 0.68$ Substitute given values.

$= 1.11$ Simplify.

b. $\log_b 9 = \log_b 3^2$ Write 9 as 3^2.

$= 2 \log_b 3$

$= 2(0.68)$ Substitute 0.68 for $\log_b 3$.

$= 1.36$ Simplify.

c. First, recall that $\sqrt{2} = 2^{1/2}$. Then

$\log_b \sqrt{2} = \log_b 2^{1/2}$ Write $\sqrt{2}$ as $2^{1/2}$.

$= \dfrac{1}{2} \log_b 2$ Apply the power property.

$= \dfrac{1}{2}(0.43)$ Substitute the given value.

$= 0.215$ Simplify.

A summary of the basic properties of logarithms that we have developed so far is given next.

> ## PROPERTIES OF LOGARITHMS
>
> If x, y, and b are positive real numbers, $b \neq 1$, and r is a real number, then:
>
> **1.** $\log_b 1 = 0$ **4.** $\log_b xy = \log_b x + \log_b y$ Product property.
>
> **2.** $\log_b b^x = x$ **5.** $\log_b \dfrac{x}{y} = \log_b x - \log_b y$ Quotient property.
>
> **3.** $b^{\log_b x} = x$ **6.** $\log_b x^r = r \log_b x$ Power property.

EXERCISE SET 10.4

Write each sum as the logarithm of a single expression. Assume that variables represent positive numbers. See Example 1.

1. $\log_5 2 + \log_5 7$ $\log_5 14$ **2.** $\log_3 8 + \log_3 4$ $\log_3 32$

3. $\log_4 9 + \log_4 x$ $\log_4 9x$ **4.** $\log_2 x + \log_2 y$ $\log_2 xy$

5. $\log_{10} 5 + \log_{10} 2 + \log_{10} (x^2 + 2)$ $\log_{10} (10x^2 + 20)$

6. $\log_6 3 + \log_6 (x + 4) + \log_6 5$ $\log_6 (15x + 60)$

9. $\log_2 \dfrac{x}{y}$ **10.** $\log_3 \dfrac{12}{z}$

Write each difference as the logarithm of a single expression. Assume that variables represent positive numbers. See Example 2.

7. $\log_5 12 - \log_5 4$ $\log_5 3$ **8.** $\log_7 20 - \log_7 4$ $\log_7 5$

9. $\log_2 x - \log_2 y$ **10.** $\log_3 12 - \log_3 z$

11. $\log_4 2 + \log_4 10 - \log_4 5$ $\log_4 4$, or 1

12. $\log_6 18 + \log_6 2 - \log_6 9$ $\log_6 4$

20. $\log_4 2 - \log_4 9 - \log_4 z$ **21.** $3\log_2 x - \log_2 y$

Write each of the following as the logarithm of a single expression. Assume that variables represent positive numbers. See Example 3.

13. $2\log_2 5$ $\log_2 25$ **14.** $3\log_5 2$ $\log_5 8$

15. $3\log_5 x + 6\log_5 z$ $\log_5 x^3 z^6$

16. $2\log_7 y + 6\log_7 z$ $\log_7 y^2 z^6$

17. $\log_{10} x - \log_{10}(x+1) + \log_{10}(x^2-2)$ $\log_{10}\dfrac{x^3-2x}{x+1}$

18. $\log_9(4x) - \log_9(x-3) + \log_9(x^3+1)$ $\log_9\dfrac{4x^4+4x}{x-3}$

Write each expression as a sum or difference of multiples of logarithms. Assume that variables represent positive numbers. See Example 4.   **19.** $\log_3 4 + \log_3 y - \log_3 5$

19. $\log_3\dfrac{4y}{5}$

20. $\log_4\dfrac{2}{9z}$

21. $\log_2\dfrac{x^3}{y}$

22. $\log_5\dfrac{x}{y^4}$ $\log_5 x - 4\log_5 y$

23. $\log_b\sqrt{7x}$

24. $\log_b\sqrt{\dfrac{3}{y}}$

$\dfrac{1}{2}\log_b 7 + \dfrac{1}{2}\log_b x$. $\dfrac{1}{2}\log_b 3 - \dfrac{1}{2}\log_b y$

If $\log_b 3 \approx 0.5$ and $\log_b 5 \approx 0.7$, approximate the following. See Example 5.

25. $\log_b\dfrac{5}{3}$ 0.2

26. $\log_b 25$ 1.4

27. $\log_b 15$ 1.2

28. $\log_b\dfrac{3}{5}$ -0.2

29. $\log_b\sqrt[3]{5}$ 0.23

30. $\log_b\sqrt[4]{3}$ 0.125

Write each of the following as the logarithm of a single expression. Assume that variables represent positive numbers.

31. $\log_4 5 + \log_4 7$ $\log_4 35$ **32.** $\log_3 2 + \log_3 5$ $\log_3 10$

33. $\log_3 8 - \log_3 2$ $\log_3 4$ **34.** $\log_5 12 - \log_5 3$ $\log_5 4$

35. $\log_7 6 + \log_7 3 - \log_7 4$ $\log_7\dfrac{9}{2}$

36. $\log_8 5 + \log_8 15 - \log_8 20$ $\log_8\dfrac{15}{4}$

37. $3\log_4 2 + \log_4 6$ $\log_4 48$

38. $2\log_3 5 + \log_3 2$ $\log_3 50$

39. $3\log_2 x + \dfrac{1}{2}\log_2 x - 2\log_2(x+1)$ $\log_2\dfrac{x^{7/2}}{(x+1)^2}$

40. $2\log_5 x + \dfrac{1}{3}\log_5 x - 3\log_5(x+5)$ $\log_5\dfrac{x^{7/3}}{(x+5)^3}$

45. $3\log_5 x + \log_5(x+1)$ **46.** $3\log_2 y + \log_2 z$ **47.** $2\log_6 x - \log_6(x+3)$ **48.** $2\log_3(x+5) - \log_3 x$

41. $2\log_8 x - \dfrac{2}{3}\log_8 x + 4\log_8 x$ $\log_8 x^{16/3}$

42. $5\log_6 x - \dfrac{3}{4}\log_6 x + 3\log_6 x$ $\log_6 x^{29/4}$

Write each expression as a sum or difference of multiples of logarithms. Assume that variables represent positive numbers.   **43.** $\log_7 5 + \log_7 x - \log_7 4$

43. $\log_7\dfrac{5x}{4}$

44. $\log_9\dfrac{7}{y}$ $\log_9 7 - \log_9 y$

45. $\log_5 x^3(x+1)$

46. $\log_2 y^3 z$

47. $\log_6\dfrac{x^2}{x+3}$

48. $\log_3\dfrac{(x+5)^2}{x}$

If $\log_b 2 = 0.43$ and $\log_b 3 = 0.68$, evaluate the following.

49. $\log_b 8$ 1.29

50. $\log_b 81$ 2.72

51. $\log_b\dfrac{3}{9}$ -0.68

52. $\log_b\dfrac{4}{32}$ -1.29

53. $\log_b\sqrt{\dfrac{2}{3}}$ -0.125

54. $\log_b\sqrt{\dfrac{3}{2}}$ 0.125

Answer the following true or false.

55. $\log_2 x^3 = 3\log_2 x$ true

56. $\log_3(x+y) = \log_3 x + \log_3 y$ false

57. $\dfrac{\log_7 10}{\log_7 5} = \log_7 2$ false

58. $\log_7\dfrac{14}{8} = \log_7 14 - \log_7 8$ true

59. $\dfrac{\log_7 x}{\log_7 y} = (\log_7 x) - (\log_7 y)$ false

60. $(\log_3 6)\cdot(\log_3 4) = \log_3 24$ false

Review Exercises

61. Graph the functions $y = 10^x$ and $y = \log_{10} x$ on the same set of axes. See Section 10.3.   See Appendix D.

Evaluate each expression. See Section 10.3.

62. $\log_{10} 100$ 2

63. $\log_{10}\dfrac{1}{10}$ -1

64. $\log_7 7^2$ 2

65. $\log_7\sqrt{7}$ $\dfrac{1}{2}$

10.5 | COMMON LOGARITHMS, NATURAL LOGARITHMS, AND CHANGE OF BASE

TAPE IA 10.5

O B J E C T I V E S

1. Identify common logarithms and approximate them by calculator.
2. Evaluate common logarithms of powers of 10.
3. Identify natural logarithms and approximate them by calculator.
4. Evaluate natural logarithms of powers of e.
5. Apply the change of base formula.

In this section we look closely at two particular logarithmic bases. These two logarithmic bases are used so frequently that logarithms to their bases are given special names. **Common logarithms** are logarithms to base 10. **Natural logarithms** are logarithms to base e, which we introduce in this section. Because of the wide availability and low cost of calculators today, the work in this section is based on the use of calculators, which typically have both the common "log" and the natural "log" keys.

Logarithms to base 10, common logarithms, are used frequently because our number system is a base 10 decimal system. The notation $\log x$ means the same as $\log_{10} x$.

COMMON LOGARITHMS

$$\log x \text{ means } \log_{10} x$$

EXAMPLE 1   Use a calculator to approximate $\log 7$ to four decimal places.

Solution: Press the following sequence of keys:

$$\boxed{7} \quad \boxed{\log}$$

(Some calculators require $\boxed{7}$ $\boxed{\log}$ $\boxed{=}$.) The number 0.845098 should appear in the display. To four decimal places,

$$\log 7 \approx 0.8451$$

Some scientific calculators do not have a $\boxed{\log}$ key but do have a $\boxed{10^x}$ key. If this is the case, then $\log 7$ is approximated by pressing

$$\boxed{7} \quad \boxed{\text{INV}} \quad \boxed{10^x}$$

This sequence is based on the fact that the functions $y = \log_{10} x$ or $y = \log x$ and $y = 10^x$ are inverses of each other.

2 To evaluate the common log of a power of 10, a calculator is not needed. According to the property of logarithms,

$$\log_b b^x = x$$

It follows that

$$\log 10^x = x$$

because the base of this logarithm is understood to be 10.

EXAMPLE 2   Find the exact value of each logarithm.

a. $\log 10$ **b.** $\log 1000$ **c.** $\log \dfrac{1}{10}$ **d.** $\log \sqrt{10}$

Solution: **a.** $\log 10 = \log 10^1 = 1$ **b.** $\log 1000 = \log 10^3 = 3$

   c. $\log \dfrac{1}{10} = \log 10^{-1} = -1$ **d.** $\log \sqrt{10} = \log 10^{1/2} = \dfrac{1}{2}$

As we will soon see, equations containing common logs are useful models of many natural phenomena.

EXAMPLE 3   Solve $\log x = 1.2$ for x. Give an exact solution, and then approximate the solution to four decimal places.

Solution: Write the logarithmic equation with exponential notation. Keep in mind that the base of a common log is understood to be 10.

$$\log x = 1.2$$
$$10^{1.2} = x \qquad \text{Write with exponential notation.}$$

The exact solution is $10^{1.2}$. To approximate, use a calculator.

$$15.848932 \approx x$$

To four decimal places, $x \approx 15.8489$.

The Richter scale measures the intensity, or magnitude, of an earthquake. The formula for the magnitude R of an earthquake is $R = \log \left(\dfrac{a}{T}\right) + B$, where a is the amplitude in micrometers of the vertical motion of the ground at the recording station, T is the number of seconds between successive seismic waves, and B is an adjustment factor that takes into account the weakening of the seismic wave as the distance increases from the epicenter of the earthquake.

EXAMPLE 4   Find an earthquake's magnitude on the Richter scale if a recording station measures an amplitude of 300 micrometers and 2.5 seconds between waves. Assume that B is 4.2. Approximate the solution to the nearest tenth.

Solution: Substitute the known values into the formula for earthquake intensity.

$$R = \log\left(\frac{a}{T}\right) + B \qquad \text{Richter scale formula.}$$

$$= \log\left(\frac{300}{2.5}\right) + 4.2 \qquad \text{Let } a = 300, T = 2.5, \text{ and } B = 4.2.$$

$$= \log(120) + 4.2$$

$$\approx 2.1 + 4.2 \qquad \text{Approximate log 120 by 2.1.}$$

$$= 6.3$$

This earthquake had a magnitude of 6.3 on the Richter scale.

3   **Natural logarithms** are also frequently used, especially to describe natural events; hence the label "natural logarithm." Natural logarithms are logarithms to the base e, which is a constant approximately equal to 2.7183. The number e is an irrational number, as is π. The notation $\log_e x$ is usually appreviated to $\ln x$. (The abbreviation ln is read "el en.")

NATURAL LOGARITHMS

$\ln x$ means $\log_e x$

The graph of $y = \ln x$ is shown next.

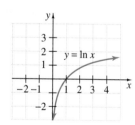

EXAMPLE 5   Use a calculator to approximate ln 8 to four decimal places.

Solution: Press the following sequence of keys:

$\boxed{8}$ $\boxed{\ln}$

The display should show 2.0794415. To four decimal places,

$$\ln 8 \approx 2.0794$$

Some scientific calculators do not have a $\boxed{\ln}$ key, but they have an $\boxed{e^x}$ key instead. If this is the case, ln 8 is approximated by pressing

$$\boxed{8}\ \ \boxed{\text{INV}}\ \ \boxed{e^x}$$

This sequence is based on the fact that the functions $y = \log_e x$ or $y = \ln x$ and $y = e^x$ are inverses of each other. ■

As a result of the property $\log_b b^x = x$, we know that $\log_e e^x = x$, or $\ln e^x = x$.

EXAMPLE 6   Find the exact value of each natural logarithm.

a. $\ln e^3$ **b.** $\ln \sqrt[5]{e}$

Solution: **a.** $\ln e^3 = 3$ **b.** $\ln \sqrt[5]{e} = \ln e^{1/5} = \dfrac{1}{5}$ ■

EXAMPLE 7   Solve the equation $\ln 3x = 5$ for x. Give an exact solution and then approximate the solution to four decimal places.

Solution: Write the equation with exponential notation. Keep in mind that the base of a natural logarithm is understood to be e:

$$\ln 3x = 5$$
$$e^5 = 3x \qquad \text{Write with exponential notation.}$$
$$\frac{e^5}{3} = x$$

The exact solution is $\dfrac{e^5}{3}$. To four decimal places, $x \approx 49.4711$.   ■

Recall from Section 10.2 the formula $A = P\left(1 + \dfrac{r}{n}\right)^{nt}$ for compound interest, where n represents the number of compoundings per year. When interest is compounded continuously, the formula $A = Pe^{rt}$ is used, where r is the annual interest rate and interest is compounded continuously for t years.

EXAMPLE 8   Find the amount owed at the end of 5 years if $1600 is loaned at a rate of 9% compounded continuously.

Solution: Use the formula $A = Pe^{rt}$, where

$$P = \$1600 \text{ (the size of the loan)}$$
$$r = 9\% = 0.09 \text{ (the rate of interest)}$$
$$t = 5 \text{ (the 5-year duration of the loan)}$$
$$A = Pe^{rt}$$
$$= 1600e^{0.09(5)} \quad \text{Substitute in known values.}$$
$$= 1600e^{0.45}$$

Now use a calculator to approximate the solution.

$$A \approx 2509.30$$

The total amount of money owed is approximately $2509.30.

5 Calculators are handy tools for approximating natural and common logarithms. Unfortunately, some calculators cannot be used to approximate logarithms to bases other than e or 10—at least not directly. In such cases, we use the change of base formula.

CHANGE OF BASE

If a, b, and c are positive real numbers and neither b nor c is 1, then

$$\log_b a = \frac{\log_c a}{\log_c b}$$

EXAMPLE 9   Approximate $\log_5 3$ to four decimal places.

Solution: Use the change of base property to write $\log_5 3$ as a quotient of logarithms to base 10:

$$\log_5 3 = \frac{\log 3}{\log 5} \qquad \text{Use the change of base property.}$$

$$\approx \frac{0.4771213}{0.69897} \qquad \text{Approximate logarithms by calculator.}$$

$$\approx 0.6826063 \qquad \text{Simplify by calculator.}$$

To four decimal places, $\log_5 3 \approx 0.6826$.

EXERCISE SET 10.5

Use a calculator to approximate each logarithm to four decimal places. See Examples 1 and 5.

1. log 8 0.9031   **2.** log 6 0.7782   **3.** log 2.31 0.3636

4. log 4.86   **5.** ln 2 0.6931   **6.** ln 3 1.0986

7. ln 0.0716   **8.** ln 0.0032   **9.** log 12.6 1.1004

10. log 25.9   **11.** ln 5 1.6094 **12.** ln 7 1.9459

4. 0.6866   **7.** −2.6367   **8.** −5.7446   **10.** 1.4133

35. $10^{1.3} \approx 19.9526$ **36.** $10^{2.1} \approx 125.8925$ **37.** $\dfrac{10^{1.1}}{2} \approx 6.2946$ **38.** $\dfrac{10^{1.3}}{3} \approx 6.6509$ **41.** $\dfrac{4 + e^{2.3}}{3} \approx 4.6581$

13. log 41.5   1.6180

14. ln 41.5   3.7257

15. Use a calculator and try to approximate log 0. Describe what happens and explain why.

16. Use a calculator and try to approximate ln 0. Describe what happens and explain why.

Find the exact value. See Examples 2 and 6.

17. log 100   2

18. log 10,000   4

19. $\log\left(\dfrac{1}{1000}\right)$ -3

20. $\log\left(\dfrac{1}{100}\right)$ -2

21. $\ln e^2$ 2

22. $\ln e^4$ 4

23. $\ln \sqrt[4]{e}$ $\dfrac{1}{4}$

24. $\ln \sqrt[5]{e}$ $\dfrac{1}{5}$

25. $\log 10^3$ 3

26. $\ln e^5$ 5

27. $\ln e^2$ 2

28. $\log 10^7$ 7

29. log 0.0001 -4

30. log 0.001 -3

31. $\ln \sqrt{e}$ $\dfrac{1}{2}$

32. $\log \sqrt{10}$ $\dfrac{1}{2}$

33. Without using a calculator, explain which of log 50 or ln 50 must be larger.

34. Without using a calculator, explain which of log 50^{-1} or ln 50^{-1} must be larger.

Solve each equation for x. Give an exact solution and a four-decimal-place approximation. See Examples 3 and 7.

35. log $x = 1.3$

36. log $x = 2.1$

37. log $2x = 1.1$

38. log $3x = 1.3$

39. $\ln x = 1.4$ $e^{1.4} \approx 4.0552$ **40.** $\ln x = 2.1$ $e^{2.1} \approx 8.1662$

41. $\ln(3x - 4) = 2.3$

42. $\ln(2x + 5) = 3.4$

43. log $x = 2.3$

44. log $x = 3.1$

45. $\ln x = -2.3$

46. $\ln x = -3.7$ $e^{-3.7} \approx 0.0247$

47. log $(2x + 1) = -0.5$

48. log $(3x - 2) = -0.8$

49. $\ln 4x = 0.18$

50. $\ln 3x = 0.76$

Approximate each logarithm to four decimal places. See Example 9.

51. $\log_2 3$ 1.5850

52. $\log_3 2$ 0.6309

53. $\log_{1/2} 5$ -2.3219

54. $\log_{1/3} 2$ -0.6309

55. $\log_4 9$ 1.5850

56. $\log_9 4$ 0.6309

57. $\log_3 \dfrac{1}{6}$ -1.6309

58. $\log_6 \dfrac{2}{3}$ -0.2263

59. $\log_8 6$ 0.8617

60. $\log_6 8$ 1.1606

Use the formula $R = \log\left(\dfrac{a}{T}\right) + B$ to find the intensity R on the Richter scale of the earthquakes that fit the de-

scriptions given. Round answers to one decimal place. See Example 4.

61. Amplitude a is 200 micrometers, time T between waves is 1.6 seconds, and B is 2.1.   4.2

62. Amplitude a is 150 micrometers, time T between waves is 3.6 seconds, and B is 1.9.   3.5

63. Amplitude a is 400 micrometers, time T between waves is 2.6 seconds, and B is 3.1.   5.3

64. Amplitude a is 450 micrometers, time T between waves is 4.2 seconds, and B is 2.7.   4.7

Use the formula $A = Pe^{rt}$ to solve. See Example 8.

65. Find how much money Dana Jones has after 12 years if \$1400 is invested at 8% interest compounded continuously.   \$3656.38

66. Determine the size of an account in which \$3500 earns 6% interest compounded continuously for 1 year.   \$3716.43

67. Find the amount of money Barbara Mack owes at the end of 4 years if 6% interest is compounded continuously on her \$2000 debt.   \$2542.50

68. Find the amount of money for which a \$2500 certificate of deposit is redeemable if it has been paying 10% interest compounded continuously for 3 years.   \$3374.65

Graph each function by finding ordered pair solutions, plotting the solutions, and then drawing a smooth curve through the plotted points. For Exercises 69–88, see Appendix D.

69. $f(x) = e^x$

70. $f(x) = e^{2x}$

71. $f(x) = e^{-3x}$

72. $f(x) = e^{-x}$

73. $f(x) = e^x + 2$

74. $f(x) = e^x - 3$

75. $f(x) = e^{x-1}$

76. $f(x) = e^{x+4}$

77. $f(x) = 3e^x$

78. $f(x) = -2e^x$

79. $f(x) = \ln x$

80. $f(x) = \log x$

81. $f(x) = -2 \log x$

82. $f(x) = 3 \ln x$

83. $f(x) = \log(x + 2)$

84. $f(x) = \log(x - 2)$

85. $f(x) = \ln x - 3$

86. $f(x) = \ln x + 3$

87. Graph $f(x) = e^x$ (Exercise 69), $f(x) = e^x + 2$ (Exercise 73), and $f(x) = e^x - 3$ (Exercise 74) on the same screen. Discuss any trends shown on the graphs.

88. Graph $f(x) = \ln x$ (Exercise 79), $f(x) = \ln x - 3$ (Exercise 85), and $f(x) = \ln x + 3$ (Exercise 86). Discuss any trends shown on the graphs.

42. $\dfrac{e^{3.4} - 5}{2} \approx 12.4821$ **43.** $10^{2.3} \approx 199.5262$ **44.** $10^{3.1} \approx 1258.9254$ **45.** $e^{-2.3} \approx 0.1003$

47. $\dfrac{10^{-0.5} - 1}{2} \approx -0.3419$ **48.** $\dfrac{10^{-0.8} + 2}{3} \approx 0.7195$ **49.** $\dfrac{e^{0.18}}{4} \approx 0.2993$ **50.** $\dfrac{e^{0.76}}{3} \approx 0.7128$

Review Exercises

Solve each equation for x. See Sections 2.1 and 5.8.

89. $6x - 3(2 - 5x) = 6$ $\quad\left\{\dfrac{4}{7}\right\}$

90. $2x + 3 = 5 - 2(3x - 1)$ $\quad\left\{\dfrac{1}{2}\right\}$

91. $2x + 3y = 6x$ $\quad x = \dfrac{3y}{4}$

92. $4x - 8y = 10x$ $\quad x = -\dfrac{4y}{3}$

93. $x^2 + 7x = -6$ $\{-6, -1\}$ **94.** $x^2 + 4x = 12$ $\quad \{-6, 2\}$

Solve each system of equations. See Section 4.1.

95. $\begin{cases} x + 2y = -4 \\ 3x - y = 9 \end{cases}$ $\{(2, -3)\}$

96. $\begin{cases} 5x + y = 5 \\ -3x - 2y = -10 \end{cases}$ $\{(0, 5)\}$

10.6 EXPONENTIAL AND LOGARITHMIC EQUATIONS AND PROBLEM SOLVING

TAPE IA 10.6

O B J E C T I V E S

 1 Solve exponential equations.

2 Solve logarithmic equations.

3 Solve problems that can be modeled by exponential and logarithmic equations.

1 In Section 10.2 we solved exponential equations such as $2^x = 16$ by writing 16 as a power of 2 and applying the uniqueness of b^x:

$$2^x = 16$$
$$2^x = 2^4 \qquad \text{Write 16 as } 2^4.$$
$$x = 4 \qquad \text{Apply the uniqueness of } b^x.$$

To solve an equation such as $3^x = 7$, we use the fact that $f(x) = \log_b x$ is a one-to-one function. Another way of stating this fact is as a property of equality.

LOGARITHM PROPERTY OF EQUALITY

Let a, b, and c be real numbers such that $\log_b a$ and $\log_b c$ are real numbers and b is not 1. Then

$$\log_b a = \log_b c \text{ is equivalent to } a = c$$

EXAMPLE 1 Solve $3^x = 7$.

Solution: To solve, use the logarithm property of equality and take the logarithm of both sides. For this example, we use the common logarithm:

$$3^x = 7$$
$$\log 3^x = \log 7 \qquad \text{Take the common log of both sides.}$$
$$x \log 3 = \log 7 \qquad \text{Apply the power property of logarithms.}$$
$$x = \frac{\log 7}{\log 3} \qquad \text{Divide both sides by log 3.}$$

The exact solution is $\dfrac{\log 7}{\log 3}$. If a decimal approximation is preferred,

$$\frac{\log 7}{\log 3} \approx \frac{0.845098}{0.4771213} \approx 1.7712 \text{ to four decimal places. The solution set is } \left\{\frac{\log 7}{\log 3}\right\}, \text{ or}$$

approximately $\{1.7712\}$.

2   By applying the appropriate properties of logarithms, we can solve a broad variety of logarithmic equations.

EXAMPLE 2   Solve $\log_4 (x - 2) = 2$ for x.

Solution: Notice that $x - 2$ must be positive, so x must be greater than 2. First, write the equation with exponential notation:

$$\log_4 (x - 2) = 2$$
$$4^2 = x - 2$$
$$16 = x - 2$$
$$18 = x \qquad \text{Add 2 to both sides.}$$

To check, replace x with 18 in the *original equation*:

$$\log_4 (x - 2) = 2$$
$$\log_4 (18 - 2) = 2 \qquad \text{Let } x = 18.$$
$$\log_4 16 = 2$$
$$4^2 = 16 \qquad \text{True.}$$

The solution set is $\{18\}$.

EXAMPLE 3   Solve $\log_2 x + \log_2 (x - 1) = 1$ for x.

Solution: Notice that $x - 1$ must be positive, so x must be greater than 1. Apply the product rule to the left side of the equation:

$$\log_2 x + \log_2 (x - 1) = 1$$
$$\log_2 x(x - 1) = 1 \qquad \text{Use the product rule.}$$
$$\log_2 (x^2 - x) = 1$$

Next, write the equation with exponential notation and solve for x:

$$2^1 = x^2 - x$$
$$0 = x^2 - x - 2 \qquad \text{Subtract 2 from both sides.}$$
$$0 = (x - 2)(x + 1) \qquad \text{Factor.}$$
$$0 = x - 2 \quad \text{or} \quad 0 = x + 1 \qquad \text{Set each factor equal to 0.}$$
$$2 = x \qquad\qquad -1 = x$$

Verify that 2 satisfies the original equation.

Recall that -1 cannot be a solution because x must be greater than 1. If this is not noticed, -1 will be rejected after checking. To see this, replace x with -1 in the original equation.

$$\log_2 x + \log_2 (x - 1) = 1$$
$$\log_2 (-1) + \log_2 (-1 - 1) = 1 \qquad \text{Let } x = -1.$$

Because the logarithm of a negative number is undefined, -1 is an extraneous solution. The solution set is $\{2\}$.

EXAMPLE 4   Solve $\log (x + 2) - \log x = 2$ for x.

Solution: Apply the quotient property of logarithms to the left side of the equation:

$$\log (x + 2) - \log x = 2$$

$$\log \frac{x + 2}{x} = 2 \qquad \text{Apply the quotient property.}$$

$$10^2 = \frac{x + 2}{x} \qquad \text{Write with exponential notation.}$$

$$100 = \frac{x + 2}{x}$$

$$100x = x + 2 \qquad \text{Multiply both sides by } x.$$

$$99x = 2 \qquad \text{Subtract } x \text{ from both sides.}$$

$$x = \frac{2}{99} \qquad \text{Divide both sides by 99.}$$

Verify that the solution set is $\left\{\frac{2}{99}\right\}$.

3   Throughout this chapter we have emphasized that logarithmic and exponential functions are used in a variety of scientific, technical, and business settings. A few examples follow.

EXAMPLE 5   The population size y of a community of lemmings varies according to the relationship $y = y_0 e^{0.15t}$. In this formula, t is time in months, and y_0 is some initial population at time 0. Estimate the population after 6 months if there were originally 5000 lemmings.

Solution: Substitute 5000 for y_0 and 6 for t:

$$y = y_0 e^{0.15t}$$
$$= 5000 e^{0.15(6)} \qquad \text{Let } t = 6 \text{ and } y_0 = 5000.$$
$$= 5000 e^{0.9} \qquad \text{Multiply.}$$

Using a calculator, we find that $y \approx 12{,}298.016$. In 6 months the population will be approximately 12,300 lemmings.

EXAMPLE 6 How long does it take an investment of $2000 to double if it is invested at 5% interest compounded quarterly? The necessary formula is $A = P\left(1 + \dfrac{r}{n}\right)^{nt}$, where A is the accrued (or owed) amount, P is the principal invested, r is the annual rate of interest, n is the number of compounding periods per year, and t is the number of years.

Solution: We are given that $P = \$2000$ and $r = 5\% = 0.05$. Compounding quarterly means 4 times a year, so $n = 4$. The investment is to double, so A must be $4000. Substitute these values and solve for t.

$$A = P\left(1 + \frac{r}{n}\right)^{nt}$$

$$4000 = 2000\left(1 + \frac{0.05}{4}\right)^{4t} \qquad \text{Substitute in known values.}$$

$$4000 = 2000(1.0125)^{4t} \qquad \text{Simplify } 1 + \frac{0.05}{4}.$$

$$2 = (1.0125)^{4t} \qquad \text{Divide both sides by 2000.}$$

$$\log 2 = \log 1.0125^{4t} \qquad \text{Take the logarithm of both sides.}$$

$$\log 2 = 4t(\log 1.0125) \qquad \text{Apply the power property.}$$

$$\frac{\log 2}{(4 \log 1.0125)} = t \qquad \text{Divide both sides by 4 log 1.0125.}$$

$$13.949408 \approx t \qquad \text{Approximate by calculator.}$$

Thus, it takes nearly 14 years for the money to double in value.

GRAPHING CALCULATOR EXPLORATIONS

Use a grapher to find how long it takes an investment of $1500 to triple if it is invested at 8% interest compounded monthly.

First, let $P = \$1500$, $r = 0.08$, and $n = 12$ (for 12 months) in the formula

$$A = P\left(1 + \frac{r}{n}\right)^{nt}$$

(continued)

Notice that when the investment has tripled, the accrued amount A is $4500. Thus,

$$4500 = 1500\left(1 + \frac{0.08}{12}\right)^{12t}$$

Determine an appropriate viewing window and enter and graph the equations

$$Y_1 = 1500\left(1 + \frac{0.08}{12}\right)^{12t}$$

and

$$Y_2 = 4500$$

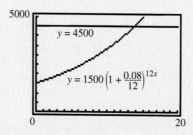

The point of intersection of the two curves is the solution. The x-coordinate tells how long it takes for the investment to triple.

Use a TRACE feature or an INTERSECT feature to approximate the coordinates of the point of intersection of the two curves. It takes approximately 13.78 years, or 13 years and 9 months, for the investment to triple in value to $4500.

Use this graphical solution method to solve each problem. Round each answer to the nearest hundredth.

1. Find how long it takes an investment of $5,000 to grow to $6000 if it is invested at 5% interest compounded quarterly.   3.67 years, or 3 years and 8 months

2. Find how long it takes an investment of $1000 to double if it is invested at 4.5% interest compounded daily. (Use 365 days in a year.)   15.40 years, or 15 years and 5 months

3. Find how long it takes an investment of $10,000 to quadruple if it is invested at 6% interest compounded monthly.   23.16 years, or 23 years and 2 months

4. Find how long it takes $500 to grow to $800 if it is invested at 4% interest compounded semiannually.   11.87 years, or 11 years and 10 months

EXERCISE SET 10.6

Solve each equation. Give an exact solution, and also approximate the solution to four decimal places. See Example 1.

1. $3^x = 6$

2. $4^x = 7$

3. $3^{2x} = 3.8$

4. $5^{3x} = 5.6$

5. $2^{x-3} = 5$

6. $8^{x-2} = 12$

7. $9^x = 5$

8. $3^x = 11$

9. $4^{x+7} = 3$

10. $6^{x+3} = 2$

11. $7^{3x-4} = 11$

12. $5^{2x-6} = 12$

13. $e^{6x} = 5$

14. $e^{2x} = 8$

1. $\left\{\dfrac{\log 6}{\log 3}\right\}$; {1.6309} **2.** $\left\{\dfrac{\log 7}{\log 4}\right\}$; {1.4037}

3. $\left\{\dfrac{\log 3.8}{2 \log 3}\right\}$; {0.6076}

4. $\left\{\dfrac{\log 5.6}{3 \log 5}\right\}$; {0.3568}

5. $\left\{3 + \dfrac{\log 5}{\log 2}\right\}$; {5.3219}

6. $\left\{\dfrac{\log 12}{\log 8} + 2\right\}$; {3.1950}

7. $\left\{\dfrac{\log 5}{\log 9}\right\}$; {0.7325} **8.** $\left\{\dfrac{\log 11}{\log 3}\right\}$; {2.1827}

9. $\left\{\dfrac{\log 3}{\log 4} - 7\right\}$; {−6.2075}

10. $\left\{\dfrac{\log 2}{\log 6} - 3\right\}$; {−2.6131}

11. $\left\{\dfrac{1}{3}\left(4 + \dfrac{\log 11}{\log 7}\right)\right\}$; {1.7441}

Solve each equation. See Examples 2 through 4.

15. $\log_2 (x + 5) = 4$ {11}

16. $\log_6 (x^2 - x) = 1$ {−2, 3}

17. $\log_3 x^2 = 4$ {9, −9}

18. $\log_2 x^2 = 6$ {−8, 8}

19. $\log_4 2 + \log_4 x = 0$ $\left\{\dfrac{1}{2}\right\}$

20. $\log_3 5 + \log_3 x = 1$ $\left\{\dfrac{3}{5}\right\}$

21. $\log_2 6 - \log_2 x = 3$ $\left\{\dfrac{3}{4}\right\}$

22. $\log_4 10 - \log_4 x = 2$ $\left\{\dfrac{5}{8}\right\}$

23. $\log_4 x + \log_4 (x + 6) = 2$ {2}

24. $\log_3 x + \log_3 (x + 6) = 3$ {3}

25. $\log_5 (x + 3) - \log_5 x = 2$ $\left\{\dfrac{1}{8}\right\}$

26. $\log_6 (x + 2) - \log_6 x = 2$ $\left\{\dfrac{2}{35}\right\}$

27. $\log_3 (x - 2) = 2$ {11}

28. $\log_2 (x - 5) = 3$ {13}

29. $\log_4 (x^2 - 3x) = 1$ {4, −1}

30. $\log_8 (x^2 - 2x) = 1$ {−2, 4}

31. $\ln 5 + \ln x = 0$ $\left\{\dfrac{1}{5}\right\}$

32. $\ln 3 + \ln (x - 1) = 0$ $\left\{\dfrac{4}{3}\right\}$

33. $3 \log x - \log x^2 = 2$ {100}

34. $2 \log x - \log x = 3$ {1000}

35. $\log_2 x + \log_2 (x + 5) = 1$ $\left\{\dfrac{-5 + \sqrt{33}}{2}\right\}$

36. $\log_4 x + \log_4 (x + 7) = 1$

37. $\log_4 x - \log_4 (2x - 3) = 3$ $\left\{\dfrac{192}{127}\right\}$

38. $\log_2 x - \log_2 (3x + 5) = 4$ { }

39. $\log_2 x + \log_2 (3x + 1) = 1$ $\left\{\dfrac{2}{3}\right\}$

40. $\log_3 x + \log_3 (x - 8) = 2$ {9}

Solve. See Example 5.   **41.** 103 wolves

41. The size of the wolf population at Isle Royale National Park increases at a rate of 4.3% per year. If the size of the current population is 83 wolves, find how many there should be in 5 years. Use $y = y_0\, e^{0.043t}$ and round to the nearest whole.

42. The number of victims of a flu epidemic is increasing at a rate of 7.5% per week. If 20,000 persons are currently infected, find in how many days we can expect 45,000 to have the flu. Use $y = y_0\, e^{0.075t}$ and round to the nearest whole.   76 days

43. The size of the population of Senegal is increasing at a rate of 2.6% per year. If 9,000,000 people lived in Senegal in 1995, find how many inhabitants there will be by 2000. Round to the nearest ten thousand. Use $y = y_0 e^{0.026t}$.   10,250,000 inhabitants

44. In 1995, 937 million people were citizens of India. Find how long it will take India's population to reach a size of 1500 million (that is, 1.5 billion) if the population size is growing at a rate of 2.1% per year. Round to the nearest tenth. Use $y = y_0 e^{0.021t}$.   22.4 years

Use the formula $A = P\left(1 + \dfrac{r}{n}\right)^{nt}$ to solve these compound interest problems. Round to the nearest tenth. See Example 6.

45. Find how long it takes $600 to double if it is invested at 7% interest compounded monthly.   10 years

46. Find how long it takes $600 to double if it is invested at 12% interest compounded monthly.   5.8 years

47. Find how long it takes a $1200 investment to earn $200 interest if it is invested at 9% interest compounded quarterly.   1.7 years

48. Find how long it takes a $1500 investment to earn $200 interest if it is invested at 10% compounded semiannually.   1.3 years

12. $\left\{\dfrac{1}{2}\left(6 + \dfrac{\log 12}{\log 5}\right)\right\}$; {3.7720} **13.** $\left\{\dfrac{\ln 5}{6}\right\}$; {0.2682} **14.** $\left\{\dfrac{\ln 8}{2}\right\}$; {1.0397} **36.** $\left\{\dfrac{-7 + \sqrt{65}}{2}\right\}$

49. 8.8 years
49. Find how long it takes $1000 to double if it is invested at 8% interest compounded semiannually.

50. Find how long it takes $1000 to double if it is invested at 8% interest compounded monthly.
8.7 years

The formula $w = 0.00185h^{2.67}$ is used to estimate the normal weight w of a boy h inches tall. Use this formula to solve the height–weight problems. Round to the nearest tenth.

51. Find the expected height of a boy who weighs 85 pounds.   55.7 inches

52. Find the expected height of a boy who weighs 140 pounds.   67.2 inches

The formula $P = 14.7e^{-0.21x}$ gives the average atmospheric pressure P, in pounds per square inch, at an altitude x, in miles above sea level. Use this formula to solve these pressure problems. Round answers to the nearest tenth.

53. Find the average atmospheric pressure of Denver, which is 1 mile above sea level.   11.9 lb/sq. in.

54. Find the average atmospheric pressure of Pikes Peak, which is 2.7 miles above sea level. 8.3 lb/sq. in.

55. Find the elevation of a Delta jet if the atmospheric pressure outside the jet is 7.5 lb/in.^{2}.   3.2 miles

56. Find the elevation of a remote Himalayan peak if the atmospheric pressure atop the peak is 6.5 lb/in.^{2}.   3.9 miles

Psychologists call the graph of the formula
$$t = \frac{1}{c} \ln\left(\frac{A}{A - N}\right)$$ *the learning curve, since the formula relates time t passed, in weeks, to a measure N of learning achieved, to a measure A of maximum learning possible, and to a measure c of an individual's learning style. Round to the nearest week.*

57. Norman is learning to type. If he wants to type at a rate of 50 words per minute (N is 50) and his expected maximum rate is 75 words per minute (A is 75), find how many weeks it should take him to achieve his goal. Assume that c is 0.09.   13 weeks

58. An experiment with teaching chimpanzees sign language shows that a typical chimp can master a maximum of 65 signs. Find how many weeks it should take a chimpanzee to master 30 signs if c is 0.03.   21 weeks

59. Janine is working on her dictation skills. She wants to take dictation at a rate of 150 words per minute and believes that the maximum rate she can hope for is 210 words per minute. Find how many weeks it should take her to achieve the 150-word level if c is 0.07.   18 weeks

60. A psychologist is measuring human capability to memorize nonsense syllables. Find how many weeks it should take a subject to learn 15 nonsense syllables if the maximum possible to learn is 24 syllables and c is 0.17.   6 weeks

Use a graphing calculator to solve each equation. For example, to solve Exercise 61, let $Y_1 = e^{0.3x}$, $Y_2 = 8$, and graph the equations. The x-value of the point of intersection is the solution. Round all solutions to two decimal places.   For Exercises 61–68, see Appendix D.

61. $e^{0.3x} = 8$ $\{6.93\}$

62. $10^{0.5x} = 7$ $\{1.69\}$

63. $2 \log(-5.6x + 1.3) = -x - 1$ $\{-3.68\}$

64. $\ln(1.3x - 2.1) = -3.5x + 5$ $\{1.81\}$

65. Check Exercise 11. $\{1.74\}$

66. Check Exercise 12. $\{3.77\}$

67. Check Exercise 31. $\{0.2\}$

68. Check Exercise 32. $\{1.33\}$

Review Exercises

If $x = -2, y = 0,$ and $z = 3$, find the value of each expression. See Section 1.4.

69. $\dfrac{x^2 - y + 2z}{3x}$ $-\dfrac{5}{3}$

70. $\dfrac{x^3 - 2y + z}{2z}$ $-\dfrac{5}{6}$

71. $\dfrac{3z - 4x + y}{x + 2z}$ $\dfrac{17}{4}$

72. $\dfrac{4y - 3x + z}{2x + y}$ $-\dfrac{9}{4}$

Find the inverse function of each one-to-one function. See Section 10.1.

73. $f(x) = 5x + 2$ $f^{-1}(x) = \dfrac{x - 2}{5}$

74. $f(x) = \dfrac{x - 3}{4}$ $f^{-1}(x) = 4x + 3$

GROUP ACTIVITY

MODELING TEMPERATURE

METHOD 1 MATERIALS:
- A container of either cold or hot liquid, thermometer, stopwatch, grapher with curve-fitting capabilities (optional).

METHOD 2 MATERIALS:
- A container of either cold or hot liquid, TI-82, TI-83, or TI-85 graphing calculator with unit-to-unit link cable, TI-CBL (Calculator-Based Laboratory) unit with temperature probe.

Newton's law of cooling relates the temperature of an object to the time elapsed since its warming or cooling began. In this activity you will investigate experimental data to find a mathematical model for this relationship. You may collect the temperature data by using either Method 1 (stopwatch and thermometer) or Method 2 (CBL).

Method 1:

a. Insert the thermometer into the liquid and allow a thermometer reading to register. Take a temperature reading T as you start the stopwatch (at $t = 0$) and record it in the accompanying data table.

b. Continue taking temperature readings at uniform intervals anywhere between 5 and 10 minutes long. At each reading use the stopwatch to measure the length of time that has elapsed *since the temperature readings started*. Record your time t and liquid temperature T in the data table. Gather data for six to twelve readings.

c. Plot the data from the data table. Plot t on the horizontal axis and T on the vertical axis.

Method 2:

a. Prepare the CBL unit and TI-82, TI-83, or TI-85 graphing calculator. Insert the temperature probe into the liquid.

b. Start the HEAT program on the TI graphing calculator and follow its instructions to begin collecting data. The program will collect 36 temperature readings in degrees Celsius and plot them in real time with t on the horizontal axis and T on the vertical axis.

1. Which of the following mathematical models best fits the data you collected? Explain your reasoning. (Assume $a > 0$.)

   a. $T = ab^t + c$

   b. $T = ab^{-t} + c$

   c. $T = -ab^{-t} + c$

   d. $T = \ln(-ax + b) + c$

   e. $T = -\ln(-ax + b) + c$

2. What does the constant c represent in the model you chose? What is the value of c in this activity?

3. (Optional) Subtract the value of c from each of your observations of T. Enter the new ordered pairs $(t, T - c)$ into a grapher. Use the exponential or logarithmic curve-fitting feature to find a model for your experimental data. Graph the ordered pairs $(t, T - c)$ with the model you found. How well does the model fit the data? How does the model compare with your selection from question 1?

See Appendix D for Group Activity answers and suggestions.

CHAPTER 10   HIGHLIGHTS

DEFINITIONS AND CONCEPTS	EXAMPLES

SECTION 10.1   COMPOSITE AND INVERSE FUNCTIONS

The notation $(f \circ g)(x)$ means "f composed with g":

$$(f \circ g)(x) = f(g(x))$$
$$(g \circ f)(x) = g(f(x))$$

Let $f(x) = x^2 + 1$ and let $g(x) = x - 5$.

$$(f \circ g)(x) = f(g(x))$$
$$= f(x - 5)$$
$$= (x - 5)^2 + 1$$
$$= x^2 - 10x + 26$$

If f is a function, then f is a **one-to-one function** only if each y-value (output) corresponds to only one x-value (input).

Determine whether each graph is a one-to-one function.

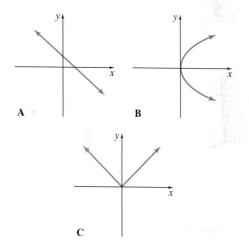

Horizontal line test:

If every horizontal line intersects the graph of a function at most once, then the function is a one-to-one function.

The **inverse** of a one-to-one function f is the one-to-one function f^{-1} that is the set of all ordered pairs (b, a) such that (a, b) belongs to f.

To find the inverse of a one-to-one function $f(x)$,

Step 1.   Replace $f(x)$ with y.

Step 2.   Interchange x and y.

Step 3.   Solve for y.

Step 4.   Replace y with $f^{-1}(x)$.

Graphs A and C pass the vertical line test, so only these are graphs of functions. Of graphs A and C, only graph A passes the horizontal line test, so only graph A is the graph of a one-to-one function.

Find the inverse of $f(x) = 2x + 7$.

$$y = 2x + 7 \qquad \text{Let } f(x) \text{ be y.}$$
$$x = 2y + 7 \qquad \text{Switch } x \text{ and } y.$$
$$2y = x - 7$$
$$y = \frac{x - 7}{2} \qquad \text{Solve for } y.$$
$$f^{-1}(x) = \frac{x - 7}{2} \qquad \text{Let } y \text{ be } f^{-1}(x).$$

The inverse of $f(x) = 2x + 7$ is $f^{-1}(x) = \dfrac{x - 7}{2}$.

DEFINITIONS AND CONCEPTS	EXAMPLES

SECTION 10.2 EXPONENTIAL FUNCTIONS

A function of the form $f(x) = b^x$ is an **exponential function,** where $b > 0, b \neq 1$, and x is a real number.

Graph the exponential function $y = 4^x$.

x	y
-2	$\dfrac{1}{16}$
-1	$\dfrac{1}{4}$
0	1
1	4
2	16

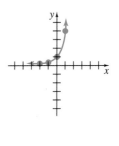

Solve $2^{x+5} = 8$.

Uniqueness of b^x

If $b > 0$ and $b \neq 1$, then $b^x = b^y$ is equivalent to $x = y$.

$2^{x+5} = 2^3$	Write 8 as 2^3.
$x + 5 = 3$	Apply uniqueness of b^x.
$x = -2$	Subtract 5.

SECTION 10.3 LOGARITHMIC FUNCTIONS

Logarithmic definition:

If $b > 0$ and $b \neq 1$, then

$$y = \log_b x \text{ means } x = b^y$$

for positive x and real number y.

LOGARITHMIC FORM	CORRESPONDING EXPONENTIAL STATEMENT
$\log_5 25 = 2$	$5^2 = 25$
$\log_9 3 = \dfrac{1}{2}$	$9^{1/2} = 3$

Properties of logarithms:

If b is a real number, $b > 0$, and $b \neq 1$, then

$$\log_b 1 = 0, \log_b b^x = x, b^{\log_b x} = x$$

$$\log_5 1 = 0, \log_7 7^2 = 2, 3^{\log_3 6} = 6$$

Logarithmic function:

If $b > 0$ and $b \neq 1$, then a **logarithmic function** is a function that can be defined as

$$f(x) = \log_b x$$

The domain of f is the set of positive real numbers, and the range of f is the set of real numbers.

Graph $y = \log_3 x$.

Write $y = \log_3 x$ as $3^y = x$. Some ordered pair solutions are listed in the table.

x	y
3	1
1	0
$\dfrac{1}{3}$	-1
$\dfrac{1}{9}$	-2

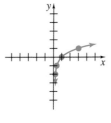

DEFINITIONS AND CONCEPTS	EXAMPLES

SECTION 10.4 PROPERTIES OF LOGARITHMS

Let x, y, and b be positive numbers and $b \neq 1$.

Product property:

$$\log_b xy = \log_b x + \log_b y$$

Quotient property:

$$\log_b \frac{x}{y} = \log_b x - \log_b y$$

Power property:

$$\log_b x^r = r \log_b x$$

Write as a single logarithm:

$2 \log_5 6 + \log_5 x - \log_5 (y + 2)$

$= \log_5 6^2 + \log_5 x - \log_5 (y + 2)$ Power property.

$= \log_5 36 \cdot x - \log_5 (y + 2)$ Product property.

$= \log_5 \dfrac{36x}{y + 2}$ Quotient property.

SECTION 10.5 COMMON LOGARITHMS, NATURAL LOGARITHMS, AND CHANGE OF BASE

Common logarithms:

$\log x$ means $\log_{10} x$

Natural logarithms:

$\ln x$ means $\log_e x$

Continuously compounded interest formula:

$$A = Pe^{rt}$$

where r is the annual interest rate for P dollars invested for t years.

$\log 5 = \log_{10} 5 \approx 0.69897$

$\ln 7 = \log_e 7 \approx 1.94591$

Find the amount in an account at the end of 3 years if \$1000 is invested at an interest rate of 4% compounded continuously.

Here, $t = 3$ years, $P = \$1000$, and $r = 0.04$:

$A = Pe^{rt}$

$\quad = 1000e^{0.04(3)}$

$\quad \approx \$1127.50$

SECTION 10.6 EXPONENTIAL AND LOGARITHMIC EQUATIONS AND PROBLEM SOLVING

Logarithm property of equality:

Let $\log_b a$ and $\log_b c$ be real numbers and $b \neq 1$. Then

$$\log_b a = \log_b c \text{ is equivalent to } a = c$$

Solve $2^x = 5$.

$\log 2^x = \log 5$ Log property of equality.

$x \log 2 = \log 5$ Power property.

$x = \dfrac{\log 5}{\log 2}$ Divide by log 2.

$x \approx 2.3219$

CHAPTER 10 REVIEW

(10.1) If $f(x) = x^2 - 2$, $g(x) = x + 1$, and $h(x) = x^3 - x^2$, find the following.

1. $(f \circ g)(x)$ **2.** $(g \circ f)(x)$

3. $(h \circ g)(2)$ 18   **4.** $(f \circ f)(x)$

5. $(f \circ g)(-1)$ -2 **6.** $(h \circ h)(2)$ 48

Determine whether each of the following functions is a one-to-one function. If it is one-to-one, list the elements of its inverse.

7. $h = \{(-9, 14), (6, 8), (-11, 12), (15, 15)\}$

8. $f = \{(-5, 5), (0, 4), (13, 5), (11, -6)\}$ not one-to-one

1. $(f \circ g)(x) = x^2 + 2x - 1$ **2.** $(g \circ f)(x) = x^2 - 1$ **4.** $(f \circ f)(x) = x^4 - 4x^2 + 2$
7. one-to-one; $h^{-1} = \{(14, -9), (8, 6), (12, -11), (15, 15)\}$

17. $f^{-1}(x) = \dfrac{x - 11}{6}$ **19.** $q^{-1}(x) = \dfrac{x - b}{m}$ **20.** $g^{-1}(x) = \dfrac{6x + 7}{12}$

9.

U.S. REGION (INPUT)	WEST	MIDWEST	SOUTH	NORTHEAST
RANK IN AUTOMOBILE THEFTS (OUTPUT)	2	4	1	3

one-to-one

10.

SHAPE (INPUT)	SQUARE	TRIANGLE	PARALLELOGRAM	RECTANGLE
NUMBER OF SIDES (OUTPUT)	4	3	4	4

not one-to-one

Given that $f(x) = \sqrt{x + 2}$ is a one-to-one function, find the following.

11. a. $f(7)$ 3 **b.** $f^{-1}(3)$ 7

12. a. $f(-1)$ 1 **b.** $f^{-1}(1)$ -1

Determine whether each function is a one-to-one function.

13. 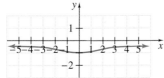 not one-to-one

14. 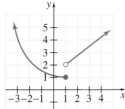 not one-to-one

15. 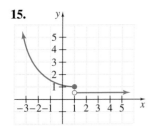 **16.**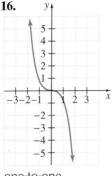

not one-to-one

one-to-one

Find an equation defining the inverse function of the given one-to-one function.

17. $f(x) = 6x + 11$ **18.** $f(x) = 12x$ $f^{-1}(x) = \dfrac{x}{12}$

9.

RANK IN AUTOMOBILE THEFTS (INPUT)	2	4	1	3
U.S. REGION (OUTPUT)	WEST	MIDWEST	SOUTH	NORTHEAST

19. $q(x) = mx + b$ **20.** $g(x) = \dfrac{12x - 7}{6}$

21. $r(x) = \dfrac{13}{2}x - 4$ $r^{-1}(x) = \dfrac{2(x + 4)}{13}$

For Exercises 22–24, see Appendix D.
On the same set of axes, graph the given one-to-one function and its inverse.

22. $g(x) = \sqrt{x}$ **23.** $h(x) = 5x - 5$

24. Find the inverse of the one-to-one function $f(x) = 2x - 3$. Then graph both $f(x)$ and $f^{-1}(x)$ with a square window. $f^{-1}(x) = \dfrac{x + 3}{2}$

(10.2) *Solve each equation for x.*

25. $4^x = 64$ 3 **26.** $3^x = \dfrac{1}{9}$ $\{-2\}$

27. $2^{3x} = \dfrac{1}{16}$ $-\dfrac{4}{3}$ **28.** $5^{2x} = 125$ $\left\{\dfrac{3}{2}\right\}$

29. $9^{x+1} = 243$ $\dfrac{3}{2}$ **30.** $8^{3x-2} = 4$ $\left\{\dfrac{8}{9}\right\}$

Graph each exponential function. For Exercises 31–34, see Appendix D.

31. $y = 3^x$ **32.** $y = \left(\dfrac{1}{3}\right)^x$

33. $y = 4 \cdot 2^x$ **34.** $y = 2^x + 4$

Use the formula $A = P\left(1 + \dfrac{r}{n}\right)^{nt}$ to solve the interest problems. In this formula,

A = amount accrued (or owed)
P = principal invested (or loaned)
r = rate of interest
n = number of compounding periods per year
t = time in years **35.** $2963.11

35. Find the amount accrued if $1600 is invested at 9% interest compounded semiannually for 7 years.

36. A total of $800 is invested in a 7% certificate of deposit for which interest is compounded quarterly. Find the value that this certificate will have at the end of 5 years. $1131.82

37. Use a grapher to verify the results of Exercise 33. See Appendix D.

(10.3)   *Write each equation with logarithmic notation.*

38. $49 = 7^2$ $\log_7 49 = 2$ **39.** $2^{-4} = \dfrac{1}{16}$ $\log_2 \left(\dfrac{1}{16}\right) = -4$

Write each logarithmic equation with exponential notation.
$\left(\dfrac{1}{2}\right)^{-4} = 16$

40. $\log_{1/2} 16 = -4$

41. $\log_{0.4} 0.064 = 3$
$0.4^3 = 0.064$

Solve for x.

42. $\log_4 x = -3$ $\left\{\dfrac{1}{64}\right\}$ **43.** $\log_3 x = 2$ $\{9\}$

44. $\log_3 1 = x$ $\{0\}$ **45.** $\log_4 64 = x$ $\{3\}$

46. $\log_x 64 = 2$ $\{8\}$ **47.** $\log_x 81 = 4$ $\{3\}$

48. $\log_4 4^5 = x$ $\{5\}$ **49.** $\log_7 7^{-2} = x$ $\{-2\}$

50. $5^{\log_5 4} = x$ $\{4\}$ **51.** $2^{\log_2 9} = x$ $\{9\}$

52. $\log_2 (3x - 1) = 4$ $\left\{\dfrac{17}{3}\right\}$ **53.** $\log_3 (2x + 5) = 2$ $\{2\}$

54. $\log_4 (x^2 - 3x) = 1$ **55.** $\log_8 (x^2 + 7x) = 1$
$\{-1, 4\}$ $\{-8, 1\}$

Graph each pair of equations on the same coordinate system.   For Exercises 56–57, see Appendix D.

56. $y = 2^x$ and $y = \log_2 x$ **57.** $y = \left(\dfrac{1}{2}\right)^x$ and $y = \log_{1/2} x$

(10.4)   *Write each of the following as single logarithms.*
58. $\log_3 8 + \log_3 4$ $\log_3 32$
59. $\log_2 6 + \log_2 3$ $\log_2 18$
60. $\log_7 15 - \log_7 20$ $\log_7 \dfrac{3}{4}$
61. $\log 18 - \log 12$ $\log \left(\dfrac{3}{2}\right)$
62. $\log_{11} 8 + \log_{11} 3 - \log_{11} 6$ $\log_{11} 4$
63. $\log_5 14 + \log_5 3 - \log_5 21$ $\log_5 2$
64. $2 \log_5 x - 2 \log_5 (x + 1) + \log_5 x$ $\log_5 \dfrac{x^3}{(x + 1)^2}$
65. $4 \log_3 x - \log_3 x + \log_3 (x + 2)$ $\log_3 (x^4 + 2x^3)$

Use properties of logarithms to write each expression as a sum or difference of multiples of logarithms.

66. $\log_3 \dfrac{x^3}{x + 2}$ **67.** $\log_4 \dfrac{x + 5}{x^2}$

68. $\log_2 \dfrac{3x^2 y}{z}$ **69.** $\log_7 \dfrac{yz^3}{x}$

66. $3 \log_3 x - \log_3 (x + 2)$ **67.** $\log_4 (x + 5) - 2 \log_4 x$ **68.** $\log_2 3 + 2 \log_2 x + \log_2 y - \log_2 z$

69. $\log_7 y + 3 \log_7 z - \log_7 x$ **82.** $\left\{\dfrac{e^{-1} + 3}{2}\right\}$ **83.** $\left\{\dfrac{e^2 - 1}{3}\right\}$

If $\log_b 2 = 0.36$ *and* $\log_b 5 = 0.83$, *find the following.*

70. $\log_b 50$ 2.02   **71.** $\log_b \dfrac{4}{5}$ −0.11

(10.5)   *Use a calculator to approximate the logarithm to four decimal places.*

72. $\log 3.6$ 0.5563   **73.** $\log 0.15$ −0.8239

74. $\ln 1.25$ 0.2231   **75.** $\ln 4.63$ 1.5326

Find the exact value.

76. $\log 1000$ 3   **77.** $\log \dfrac{1}{10}$ −1

78. $\ln \left(\dfrac{1}{e}\right)$ −1   **79.** $\ln (e^4)$ 4

Solve each equation for x.

80. $\ln (2x) = 2$ $\left\{\dfrac{e^2}{2}\right\}$ **81.** $\ln (3x) = 1.6$ $\left\{\dfrac{e^{1.6}}{3}\right\}$

82. $\ln (2x - 3) = -1$ **83.** $\ln (3x + 1) = 2$

Use the formula $\ln \dfrac{I}{I_0} = -kx$ *to solve radiation problems. In this formula,*

   $x =$ depth in millimeters
   $I =$ intensity of radiation
   $I_0 =$ initial intensity
   $k =$ a constant measure dependent on the
        material

Round answers to two decimal places.   **84.** 1.67 mm

84. Find the depth at which the intensity of the radiation passing through a lead shield is reduced to 3% of the original intensity if the value of k is 2.1.

85. If k is 3.2, find the depth at which 2% of the original radiation will penetrate.   1.22 mm

Approximate the logarithm to four decimal places.

86. $\log_5 1.6$ 0.2920   **87.** $\log_3 4$ 1.2619

Use the formula $A = Pe^{rt}$ *to solve the interest problems in which interest is compounded continuously. In this formula,*

   $A =$ amount accrued (or owed)
   $P =$ principal invested (or loaned)
   $r =$ rate of interest
   $t =$ time in years

88. Bank of New York offers a 5-year 6% continuously compounded investment option. Find the amount accrued if $1450 is invested. $1957.30

89. Find the amount to which a $940 investment grows if it is invested at 11% compounded continuously for 3 years. $1307.51

(10.6) *Solve each exponential equation for x. Give an exact solution and also approximate the solution to four decimal places.*

90. $3^{2x} = 7$

91. $6^{3x} = 5$

92. $3^{2x+1} = 6$

93. $4^{3x+2} = 9$

94. $5^{3x-5} = 4$

95. $8^{4x-2} = 3$

96. $2 \cdot 5^{x-1} = 1$

97. $3 \cdot 4^{x+5} = 2$

Solve the equation for x.

98. $\log_5 2 + \log_5 x = 2$ $\left(\dfrac{25}{2}\right)$

99. $\log_3 x + \log_3 10 = 2$ $\left(\dfrac{9}{10}\right)$

100. $\log (5x) - \log (x + 1) = 4$ $\{\}$

101. $\ln (3x) - \ln (x - 3) = 2$

102. $\log_2 x + \log_2 2x - 3 = 1$ $\{2\sqrt{2}\}$

103. $-\log_6 (4x + 7) + \log_6 x = 1$ $\{\}$

Use the formula $y = y_0 e^{kt}$ to solve the population growth problems. In this formula,

y = size of population

y_0 = initial count of population

k = rate of growth

t = time

Round each answer to the nearest whole.

104. The population of mallard ducks in Nova Scotia is expected to grow at a rate of 6% per week during the spring migration. If 155,000 ducks are already in Nova Scotia, find how many are expected by the end of 4 weeks.   197,044 ducks

105. The population of Indonesia is growing at a rate of 1.7% per year. If the population in 1995 was 203,583,886, find the expected population by the year 2005.   241,308,968

106. Anaheim, California, is experiencing an annual growth rate of 3.16%. If 230,000 people now live in Anaheim, find how long it will take for the size of the population to be 500,000.   25 years

107. Memphis, Tennessee, is growing at a rate of 0.36% per year. Find how long it will take the population of Memphis to increase from 650,000 to 700,000.   21 years

108. Egypt's population is increasing at a rate of 2.1% per year. Find how long it will take for its 50,500,000-person population to double in size.

109. The greater Mexico City area had a population of 15.5 million in 1994. How long will it take the city to triple in population if its growth rate is 0.7% annually?   157 years

Use the compound interest equation $A = P\left(1 + \dfrac{r}{n}\right)^{nt}$ to solve the following. (See the directions for Exercises 35 and 36 for an explanation of this formula. Round answers to the nearest tenth.)

110. Find how long it will take a $5000 investment to grow to $10,000 if it is invested at 8% interest compounded quarterly.   8.8 years

111. An investment of $6000 has grown to $10,000 while the money was invested at 6% interest compounded monthly. Find how long it was invested.   8.5 years

For Exercises 112–113, see Appendix D.
Use a graphing calculator to solve each equation. Round all solutions to two decimal places.

112. $e^x = 2$ $\{0.69\}$

113. $10^{0.3x} = 7$ $\{2.82\}$

90. $\left\{\dfrac{\log 7}{2 \log 3}\right\}$; $\{0.8856\}$ **91.** $\left\{\dfrac{\log 5}{3 \log 6}\right\}$; $\{0.2994\}$ **92.** $\left\{\dfrac{1}{2}\left(\dfrac{\log 6}{\log 3} - 1\right)\right\}$; $\{0.3155\}$ **93.** $\left\{\dfrac{1}{3}\left(\dfrac{\log 9}{\log 4} - 2\right)\right\}$; $\{-0.1383\}$

94. $\left\{\dfrac{1}{3}\left(\dfrac{\log 4}{\log 5} + 5\right)\right\}$; $\{1.9538\}$ **95.** $\left\{\dfrac{1}{4}\left(\dfrac{\log 3}{\log 8} + 2\right)\right\}$; $\{0.6321\}$ **96.** $\left\{-\dfrac{\log 2}{\log 5} + 1\right\}$; $\{0.5693\}$

97. $\left\{\dfrac{\log \frac{2}{3}}{\log 4} - 5\right\}$; $\{-5.2925\}$ **101.** $\left\{\dfrac{3e^2}{e^2 - 3}\right\}$ **108.** 33 years

---

CHAPTER 10 TEST

If $f(x) = x$, $g(x) = x - 7$, and $h(x) = x^2 - 6x + 5$, find the following.

1. $(f \circ h)(0)$ 5

2. $(g \circ f)(x)$ $x - 7$

3. $(g \circ h)(x)$ $x^2 - 6x - 2$

On the same set of axes, graph the given one-to-one function and its inverse.

4. $f(x) = 7x - 14$ See Appendix D.

Determine whether the given graph is the graph of a one-to-one function.

5.

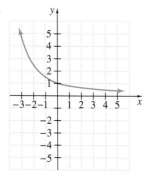

one-to-one

6.

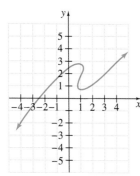

not one-to-one

Determine whether each function is one-to-one. If it is one-to-one, find an equation or a set of ordered pairs that defines the inverse function of the given function.

7. $y = 6 - 2x$ one-to-one; $f^{-1}(x) = \dfrac{-x + 6}{2}$

8. $f = \{(0, 0), (2, 3), (-1, 5)\}$

9.

WORD (INPUT)	DOG	CAT	HOUSE	DESK	CIRCLE
FIRST LETTER OF WORD (OUTPUT)	d	c	h	d	c

not one-to-one

Use the properties of logarithms to write each expression as a single logarithm.

10. $\log_3 6 + \log_3 4$ $\log_3 24$

11. $\log_5 x + 3 \log_5 x - \log_5 (x + 1)$ $\log_5\left(\dfrac{x^4}{x + 1}\right)$

12. Write the expression $\log_6 \dfrac{2x}{y^3}$ as the sum or difference of multiples of logarithms.

13. If $\log_b 3 = 0.79$ and $\log_b 5 = 1.16$, find the value of $\log_b\left(\dfrac{3}{25}\right)$. -1.53

14. Approximate $\log_7 8$ to four decimal places.   1.0686

15. Solve $8^{x-1} = \dfrac{1}{64}$ for x. Give an exact solution. $\{-1\}$

16. Solve $3^{2x+5} = 4$ for x. Give an exact solution, and also approximate the solution to four decimal places. $\left\{\dfrac{1}{2}\left(\dfrac{\log 4}{\log 3} - 5\right)\right\}; \{-1.8691\}$

Solve each logarithmic equation for x. Give an exact solution.

17. $\log_3 x = -2$ $\left\{\dfrac{1}{9}\right\}$

18. $\ln \sqrt{e} = x$ $\left\{\dfrac{1}{2}\right\}$

19. $\log_8 (3x - 2) = 2$ $\{22\}$

20. $\log_5 x + \log_5 3 = 2$ $\left\{\dfrac{25}{3}\right\}$

21. $\log_4 (x + 1) - \log_4 (x - 2) = 3$ $\left\{\dfrac{43}{21}\right\}$

22. Solve $\ln (3x + 7) = 1.31$ accurate to four decimal places. $\{-1.0979\}$

23. Graph $y = \left(\dfrac{1}{2}\right)^x + 1$.   See Appendix D.

24. Graph the functions $y = 3^x$ and $y = \log_3 x$ on the same coordinate system.   See Appendix D.

Use the formula $A = P\left(1 + \dfrac{r}{n}\right)^{nt}$ to solve Exercises 25 and 26.   **25.** $\$5234.58$

25. Find the amount in the account if $4000 is invested for 3 years at 9% interest compounded monthly.

26. Find how long it will take $2000 to grow to $3000 if the money is invested at 7% interest compounded semiannually. Round to the nearest whole.   6 years

Use the population growth formula $y = y_0 e^{kt}$ to solve Exercises 27 and 28.

27. The prairie dog population of the Grand Rapids area now stands at 57,000 animals. If the population is growing at a rate of 2.6% annually, find how many prairie dogs there will be in that area 5 years from now.   64,913 prairie dogs

28. In an attempt to save an endangered species of wood duck, naturalists would like to increase the wood duck population from 400 to 1000 ducks. If the annual population growth rate is 6.2%, find how long it will take the naturalists to reach their goal. Round to the nearest whole year.   15 years

8. one-to-one; $f^{-1} = \{(0, 0), (3, 2), (5, -1)\}$ **12.** $\log_6 2 + \log_6 x - 3 \log_6 y$

29. The formula $\log (1 + k) = \dfrac{0.3}{D}$ relates the doubling time D, in days, and the growth rate k for a population of mice. Find the rate at which the population is increasing if the doubling time is 56 days. Round to the nearest tenth of a percent.   1.2%

30. Use a graphing calculator to approximate the solution of

$$e^{0.2x} = e^{-0.4x} + 2$$

to two decimal places.   {3.95}; See Appendix D.

CHAPTER 10   CUMULATIVE REVIEW

1. Write each of the following as an algebraic expression.

   a. Two numbers have a sum of 20. If one number is x, represent the other number as an expression in x. $20 - x$

   b. The older sister is 8 years older than her younger sister. If the age of the younger sister is x, represent the age of the older sister as an expression in x. $x + 8$

   c. Two angles are complementary if the sum of their measures is 90°. If the measure of one angle is x degrees, represent the measure of the other angle as an expression in x. $90 - x$

   d. If x is the first of two consecutive integers, represent the second integer as an expression in x. $x + 1$; Sec. 1.4, Ex. 5

2. Solve for x: $\dfrac{2}{5}(x - 6) \geq x - 1$.

3. Use the graph of $y = 1500 + \dfrac{1}{10}x$ to answer the questions that follow.   Sec. 3.1, Ex. 3

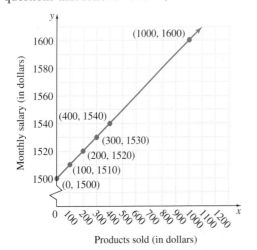

Products sold (in dollars)

2. $\left(-\infty, -\dfrac{7}{3}\right]$; Sec. 2.4, Ex. 5

   a. If a salesperson has $800 of products sold for a particular month, what is the salary for that month? $1580

   b. If the salesperson wants to make more than $1600 per month, what must be the total amount of products sold?   greater than $1000

4. Graph $x = 2$.   See Appendix D; Sec. 3.3, Ex. 6

5. Use the elimination method to solve the following system:

$$\begin{cases} -5x - 3y = 9 \\ 10x + 6y = -18 \end{cases}$$ $\{(x, y)|-5x - 3y = 9\}$; Sec. 4.1, Ex. 7

6. Solve the system:

$$\begin{cases} 2x + 4y = 1 & (1) \\ 4x - 4z = -1 & (2) \\ y - 4z = -3 & (3) \end{cases}$$ $\left\{\left(\dfrac{1}{2}, 0, \dfrac{3}{4}\right)\right\}$; Sec. 4.2, Ex. 3

7. Use the quotient rule to simplify.   Sec. 5.1, Ex. 4

   a. $\dfrac{x^7}{x^4}$ x^3

   b. $\dfrac{5^8}{5^2}$ 5^6

   c. $\dfrac{20x^6}{4x^5}$ $5x$

   d. $\dfrac{12y^{10}z^7}{14y^8z^7}$ $\dfrac{6y^2}{7}$

8. Add $11x^3 - 12x^2 + x - 3$ and $x^3 - 10x + 5$ vertically. $12x^3 - 12x^2 - 9x + 2$; Sec. 5.3, Ex. 8

9. Multiply: Sec. 5.4, Ex. 1

   a. $(2x^3)(5x^6)$ $10x^9$

   b. $(7y^4z^4)(-xy^{11}z^5)$ $-7xy^{15}z^9$

10. Find the product of $(4x^2 + 7)$ and $(x^2 + 2x + 8)$.

11. Solve $2x^2 + 9x - 5 = 0$. $\left\{-5, \dfrac{1}{2}\right\}$; Sec. 5.8, Ex. 2

12. Graph $f(x) = x^3 - 4x$. Find any intercepts.

13. Solve $\dfrac{3}{x} - \dfrac{x + 21}{3x} = \dfrac{5}{3}$. $\{-2\}$; Sec. 6.7, Ex. 2

14. Write with radical notation. Then simplify if possible.   Sec. 7.2, Ex. 2

10. $4x^4 + 8x^3 + 39x^2 + 14x + 56$; Sec. 5.4, Ex. 4   **12.** See Appendix D; Sec. 5.9, Ex. 5

a. $4^{3/2}$ 8

b. $-16^{3/4}$ -8

c. $(-27)^{2/3}$ 9

d. $\left(\dfrac{1}{9}\right)^{3/2}$ $\dfrac{1}{27}$

e. $(4x-1)^{3/5}$ $\sqrt[5]{(4x-1)^3}$

15. Multiply the complex numbers. Write the product in the form $a + bi$.   *Sec. 7.7, Ex. 3*

a. $(2-5i)(4+i)$ $13-18i$

b. $(2-i)^2$ $3-4i$

c. $(7+3i)(7-3i)$ 58

16. Solve $(x+2)(x-1)(x-5) \le 0$.

17. Sketch the graph of $f(x) = -2x^2$.

18. Graph $x^2 + y^2 = 4$.   See Appendix D; *Sec. 9.1, Ex. 7*

19. If $f(x) = x^2$ and $g(x) = x + 3$, find the following.

a. $(f \circ g)(2)$ and $(g \circ f)(2)$

b. $(f \circ g)(x)$ and $(g \circ f)(x)$

20. Solve each equation for x.   *Sec. 10.2, Ex. 3*

a. $2^x = 16$ $\{4\}$

b. $9^x = 27$ $\left\{\dfrac{3}{2}\right\}$

c. $4^{x+3} = 8^x$ $\{6\}$

21. If $\log_b 2 = 0.43$ and $\log_b 3 = 0.68$, use the properties of logarithms to evaluate   *Sec. 10.4, Ex. 5*

a. $\log_b 6$ 1.11

b. $\log_b 9$ 1.36

c. $\log_b \sqrt{2}$ 0.215

22. Solve the equation $\ln 3x = 5$ for x. Give an exact solution and then approximate the solution to four decimal places.

23. Solve $\log_2 x + \log_2 (x-1) = 1$ for x.
$\{2\}$; *Sec. 10.6, Ex. 3*

16. $(-\infty, -2] \cup [1, 5]$; *Sec. 8.4, Ex. 3*   **17.** See Appendix D; *Sec. 8.5, Ex. 7*   **19a.** $(f \circ g)(2) = 25$; $(g \circ f)(2) = 7$

19b. $(f \circ g)(x) = x^2 + 6x + 9$; $(g \circ f)(x) = x^2 + 3$; *Sec. 10.1, Ex. 1*   **22.** $\dfrac{e^5}{3} \approx 49.4711$; *Sec. 10.5, Ex. 7*

SEQUENCES, SERIES, AND THE BINOMIAL THEOREM

MODELING COLLEGE TUITION

Annual college tuition has steadily increased over the past few decades. During the 1969–1970 academic year, the average annual tuition was $427 at a public four-year university and $1089 at a private four-year university. By the 1993–1994 academic year, the average annual tuitions had increased to $2543 and $10,994, respectively. (*Source*: National Center for Education Statistics) Over the past several years, average annual tuition has been increasing at a rate greater than the annual rate of inflation.

IN THE CHAPTER GROUP ACTIVITY ON PAGE 652, YOU WILL HAVE THE OPPORTUNITY TO MODEL AND INVESTIGATE THE TREND IN INCREASING TUITION AT PUBLIC AND PRIVATE UNIVERSITIES.

H aving explored in some depth the concept of function, we turn now in this final chapter to *sequences*. In one sense, a sequence is simply an ordered list of numbers. In another sense, a sequence is itself a function. Phenomena modeled by such functions are everywhere around us. The starting place for all mathematics is the sequence of natural numbers: 1, 2, 3, 4, and so on.

Sequences lead us to *series*, which are a sum of ordered numbers. Through series we gain new insight, for example, about the expansion of a binomial $(a + b)^n$, the concluding topic of this book.

11.1 | SEQUENCES

TAPE IA 11.1

O B J E C T I V E S

1 Write the terms of a sequence given its general term.
2 Find the general term of a sequence.
3 Solve applications that involve sequences.

Suppose that a town's present population of 100,000 is growing by 5% each year. After the first year, the town's population will be

$$100,000 + 0.05(100,000) = 105,000$$

After the second year, the town's population will be

$$105,000 + 0.05(105,000) = 110,250$$

After the third year, the town's population will be

$$110,250 + 0.05(110,250) \approx 115,763$$

If we continue to calculate, the town's yearly population can be written as the **infinite sequence** of numbers

$$105,000, 110,250, 115,763, \ldots$$

If we decide to stop calculating after a certain year (say, the fourth year), we obtain the **finite sequence**

$$105,000, 110,250, 115,763, 121,551$$

SEQUENCES

An infinite sequence is a function whose domain is the set of natural numbers $\{1, 2, 3, 4, \ldots\}$.

A finite sequence is a function whose domain is the set of natural numbers $\{1, 2, 3, 4, \ldots, n\}$, where n is some natural number.

Given the sequence 2, 4, 8, 16, . . . , we say that each number is a **term** of the sequence. Because a sequence is a function, we could describe it by writing $f(n) = 2^n$, where n is a natural number. Instead, we use the notation

$$a_n = 2^n$$

Some function values are

$a_1 = 2^1 = 2$	First term of sequence.
$a_2 = 2^2 = 4$	Second term.
$a_3 = 2^3 = 8$	Third term.
$a_4 = 2^4 = 16$	Fourth term.
$a_{10} = 2^{10} = 1024$	Tenth term.

The nth term of the sequence a_n is called the **general term.**

EXAMPLE 1   Write the first five terms of the sequence whose general term is given by

$$a_n = n^2 - 1$$

Solution: Evaluate a_n, where n is 1, 2, 3, 4, and 5.

$a_n = n^2 - 1$	
$a_1 = 1^2 - 1 = 0$	Replace n with 1
$a_2 = 2^2 - 1 = 3$	Replace n with 2.
$a_3 = 3^2 - 1 = 8$	Replace n with 3.
$a_4 = 4^2 - 1 = 15$	Replace n with 4.
$a_5 = 5^2 - 1 = 24$	Replace n with 5.

Thus, the first five terms of the sequence $a_n = n^2 - 1$ are 0, 3, 8, 15, and 24.

EXAMPLE 2   If the general term of a sequence is given by $a_n = \dfrac{(-1)^n}{3n}$, find

a. the first term of the sequence

b. a_8

c. the one-hundredth term of the sequence

d. a_{15}

Solution: **a.** $a_1 = \dfrac{(-1)^1}{3(1)} = -\dfrac{1}{3}$ Replace n with 1.

b. $a_8 = \dfrac{(-1)^8}{3(8)} = \dfrac{1}{24}$ Replace n with 8.

c. $a_{100} = \dfrac{(-1)^{100}}{3(100)} = \dfrac{1}{300}$ Replace n with 100.

d. $a_{15} = \dfrac{(-1)^{15}}{3(15)} = -\dfrac{1}{45}$ Replace n with 15.

---

2 Suppose we know the first few terms of a sequence and want to find a general term that fits the pattern of the first few terms.

EXAMPLE 3   Find a general term a_n of the sequence whose first few terms are given.

a. 1, 4, 9, 16, . . .

b. $\dfrac{1}{1}, \dfrac{1}{2}, \dfrac{1}{3}, \dfrac{1}{4}, \dfrac{1}{5}, \ldots$

c. $-3, -6, -9, -12, \ldots$

d. $\dfrac{1}{2}, \dfrac{1}{4}, \dfrac{1}{8}, \dfrac{1}{16}, \ldots$

Solution:

a. These numbers are the squares of the first four natural numbers, so a general term might be $a_n = n^2$.

b. These numbers are the reciprocals of the first five natural numbers, so a general term might be $a_n = \dfrac{1}{n}$.

c. These numbers are the product of -3 and the first four natural numbers, so a general term might be $a_n = -3n$.

d. Notice that the denominators double each time.

$$\dfrac{1}{2}, \quad \dfrac{1}{2 \cdot 2}, \quad \dfrac{1}{2(2 \cdot 2)}, \quad \dfrac{1}{2(2 \cdot 2 \cdot 2)}$$

or $$\dfrac{1}{2}, \quad \dfrac{1}{2^2}, \quad \dfrac{1}{2^3}, \quad \dfrac{1}{2^4}$$

We might then suppose that the general term is $a_n = \dfrac{1}{2^n}$.

---

3 Sequences model many phenomena of the physical world, as illustrated by the following example.

EXAMPLE 4   The amount of weight, in pounds, a puppy gains in each month of its first year is modeled by a sequence whose general term is $a_n = n + 4$, where n is the number of the month. Write the first five terms of the sequence, and find how much weight the puppy should gain in its fifth month.

Solution: Evaluate $a_n = n + 4$ when n is 1, 2, 3, 4, and 5:

$$a_1 = 1 + 4 = 5$$
$$a_2 = 2 + 4 = 6$$
$$a_3 = 3 + 4 = 7$$
$$a_4 = 4 + 4 = 8$$
$$a_5 = 5 + 4 = 9$$

The puppy should gain 9 pounds in its fifth month.

3. $-1, 1, -1, 1, -1$ **4.** $-2, 4, -8, 16, -32$ **5.** $\dfrac{1}{4}, \dfrac{1}{5}, \dfrac{1}{6}, \dfrac{1}{7}, \dfrac{1}{8}$ **8.** $-6, -12, -18, -24, -30$

9. $-1, -4, -9, -16, -25$ **10.** $3, 6, 11, 18, 27$ **13.** $7, 9, 11, 13, 15$ **14.** $-2, -5, -8, -11, -14$

15. $-1, 4, -9, 16, -25$ **16.** $0, -1, 2, -3, 4$

EXERCISE SET 11.1

Write the first five terms of each sequence whose general term is given. See Example 1.

1. $a_n = n + 4$ 5, 6, 7, 8, 9   **2.** $a_n = 5 - n$ 4, 3, 2, 1, 0

3. $a_n = (-1)^n$ **4.** $a_n = (-2)^n$

5. $a_n = \dfrac{1}{n + 3}$ **6.** $a_n = \dfrac{1}{7 - n}$ $\dfrac{1}{6}, \dfrac{1}{5}, \dfrac{1}{4}, \dfrac{1}{3}, \dfrac{1}{2}$

7. $a_n = 2n$ 2, 4, 6, 8, 10   **8.** $a_n = -6n$

9. $a_n = -n^2$ **10.** $a_n = n^2 + 2$

11. $a_n = 2^n$ 2, 4, 8, 16, 32   **12.** $a_n = 3^{n-2}$ $\dfrac{1}{3}$, 1, 3, 9, 27

13. $a_n = 2n + 5$ **14.** $a_n = 1 - 3n$

15. $a^n = (-1)^n n^2$ **16.** $a_n = (-1)^{n+1}(n - 1)$

Find the indicated term for each sequence whose general term is given. See Example 2.

17. $a_n = 3n^2$; a_5 75   **18.** $a_n = -n^2$; a_{15} -225

19. $a_n = 6n - 2$; a_{20} 118   **20.** $a_n = 100 - 7n$; a_{50} -250

21. $a_n = \dfrac{n + 3}{n}$; a_{15} $\dfrac{6}{5}$ **22.** $a_n = \dfrac{n}{n + 4}$; a_{24} $\dfrac{6}{7}$

23. $a_n = (-3)^n$; a_6 729   **24.** $a_n = 5^{n+1}$; a_3 625

25. $a_n = \dfrac{n - 2}{n + 1}$; a_6 $\dfrac{4}{7}$ **26.** $a_n = \dfrac{n + 3}{n + 4}$; a_8 $\dfrac{11}{12}$

27. $a_n = \dfrac{(-1)^n}{n}$; a_8 $\dfrac{1}{8}$ **28.** $a_n = \dfrac{(-1)^n}{2n}$; a_{100} $\dfrac{1}{200}$

29. $a_n = -n^2 + 5$; a_{10} -95 **30.** $a_n = 8 - n^2$; a_{20} -392

31. $a_n = \dfrac{(-1)^n}{n + 6}$; a_{19} $-\dfrac{1}{25}$ **32.** $a_n = \dfrac{n - 4}{(-2)^n}$; a_6 $\dfrac{1}{32}$

Find a general term a_n for each sequence whose first four terms are given. See Example 3.

33. 3, 7, 11, 15 $a_n = 4n - 1$ **34.** 2, 7, 12, 17

35. $-2, -4, -8, -16$ **36.** $-4, 16, -64, 256$

37. $\dfrac{1}{3}, \dfrac{1}{9}, \dfrac{1}{27}, \dfrac{1}{81}$ $a_n = \dfrac{1}{3^n}$ **38.** $\dfrac{2}{5}, \dfrac{2}{25}, \dfrac{2}{125}, \dfrac{2}{625}$ $a_n = \dfrac{2}{5^n}$

Solve. See Example 4.

39. The distance, in feet, that a Thermos dropped from a cliff falls in each consecutive second is modeled by a sequence whose general term is $a_n = 32n - 16$, where n is the number of seconds. Find the distance the Thermos falls in the second, third, and fourth seconds.   48 ft, 80 ft, and 112 ft

40. The population size of a culture of bacteria triples every hour such that its size is modeled by the sequence $a_n = 50(3)^{n-1}$, where n is the number of the hour just beginning. Find the size of the culture at the beginning of the fourth hour and the size of the culture originally.   1350 bacteria; 50 bacteria

41. Mrs. Laser agrees to give her son Mark an allowance of $0.10 on the first day of his 14-day vacation, $0.20 on the second day, $0.40 on the third day, and so on. Write an equation of a sequence whose terms correspond to Mark's allowance. Find the allowance Mark will receive on the last day of his vacation. $a_n = 0.10(2)^{n-1}$; $819.20

42. A small theater has 10 rows with 12 seats in the first row, 15 seats in the second row, 18 seats in the third row, and so on. Write an equation of a se-

34. $a_n = 2 + 5(n - 1)$ **35.** $a_n = -2^n$ **36.** $a_n = (-4)^n$

43. 2400 cases; 75 cases   **44.** 2700, 2850, 3000, 3150, 3300

quence whose terms correspond to the seats in each row. Find the number of seats in the eighth row. $a_n = 12 + 3(n - 1)$; 33 seats

43. The number of cases of a new infectious disease is doubling every year such that the number of cases is modeled by a sequence whose general term is $a_n = 75(2)^{n-1}$, where n is the number of the year just beginning. Find how many cases there will be at the beginning of the sixth year. Find how many cases there were at the beginning of the first year.

44. A new college had an initial enrollment of 2700 students in 1995, and each year the enrollment increases by 150 students. Find the enrollment for each of 5 years, beginning with 1995.

45. An endangered species of sparrow had an estimated population of 800 in 1996, and scientists predict that its population will decrease by half each year. Estimate the population in 2000. Estimate the year the sparrow will be extinct.

46. A **Fibonacci sequence** is a special type of sequence in which the first two terms are 1 and each term thereafter is the sum of the two previous terms: 1, 1, 2, 3, 5, 8, ... Many plants and animals seem to grow according to a Fibonacci sequence, including pine cones, pineapple scales, nautilus shells, and

45. 50 sparrows in 2000; extinct in 2006 or 2007 certain flowers. Write the first 15 terms of the Fibonacci sequence.

Find the first five terms of each sequence. Round each term after the first to four decimal places.

47. $a_n = \dfrac{1}{\sqrt{n}}$ **48.** $a_n = \dfrac{\sqrt{n}}{\sqrt{n} + 1}$

49. $a_n = \left(1 + \dfrac{1}{n}\right)^n$ **50.** $a_n = \left(1 + \dfrac{0.05}{n}\right)^n$

Review Exercises

Sketch the graph of each quadratic function. See Section 8.5. For Exercises 51–54, see Appendix D.

51. $f(x) = (x - 1)^2 + 3$ **52.** $f(x) = (x - 2)^2 + 1$

53. $f(x) = 2(x + 4)^2 + 2$ **54.** $f(x) = 3(x - 3)^2 + 4$

Find the distance between each pair of points. See Section 9.1.

55. $(-4, -1)$ and $(-7, -3)$ $\sqrt{13}$ units

56. $(-2, -1)$ and $(-1, 5)$ $\sqrt{37}$ units

57. $(2, -7)$ and $(-3, -3)$ $\sqrt{41}$ units

58. $(10, -14)$ and $(5, -11)$ $\sqrt{34}$ units

46. 1, 1, 2, 3, 5, 8, 13, 21, 34, 55, 89, 144, 233, 377, 610   **47.** 1, 0.7071, 0.5774, 0.5, 0.4472
48. 0.5, 0.5858, 0.6340, 0.6667, 0.6910   **49.** 2, 2.25, 2.3704, 2.4414, 2.4883
50. 1.05, 1.0506, 1.0508, 1.0509, 1.0510

11.2   ARITHMETIC AND GEOMETRIC SEQUENCES

O B J E C T I V E S

  1 Identify arithmetic sequences and their common differences.

  2 Identify geometric sequences and their common ratios.

TAPE IA 11.2

1   Find the first four terms of the sequence whose general term is

$$a_n = 5 + (n - 1)3$$

$$a_1 = 5 + (1 - 1)3 = 5 \qquad \text{Replace } n \text{ with 1.}$$
$$a_2 = 5 + (2 - 1)3 = 8 \qquad \text{Replace } n \text{ with 2.}$$
$$a_3 = 5 + (3 - 1)3 = 11 \qquad \text{Replace } n \text{ with 3.}$$
$$a_4 = 5 + (4 - 1)3 = 14 \qquad \text{Replace } n \text{ with 4.}$$

The first four terms are 5, 8, 11, and 14. Notice that the difference of any two successive terms is 3:

$$8 - 5 = 3$$
$$11 - 8 = 3$$
$$14 - 11 = 3$$
$$\vdots$$
$$a_n - a_{n-1} = 3$$
$$\uparrow \qquad \uparrow$$

nth   previous
term   term

Because the difference of any two successive terms is a constant, we call the sequence an **arithmetic sequence,** or an **arithmetic progression.** The constant difference d in successive terms is called the **common difference.** In this example, d is 3.

ARITHMETIC SEQUENCE AND COMMON DIFFERENCE

An **arithmetic sequence** is a sequence in which each term (after the first) differs from the preceding term by a constant amount d. The constant d is called the **common difference** of the sequence.

The sequence 2, 6, 10, 14, 18, . . . is an arithmetic sequence. Its common difference is 4. Given the first term, a_1, and the common difference, d, of an arithmetic sequence, we can find any term of the sequence.

EXAMPLE 1 Write the first five terms of the arithmetic sequence whose first term is 7 and whose common difference is 2.

Solution:

$$a_1 = 7$$
$$a_2 = 7 + 2 = 9$$
$$a_3 = 9 + 2 = 11$$
$$a_4 = 11 + 2 = 13$$
$$a_5 = 13 + 2 = 15$$

The first five terms are 7, 9, 11, 13, 15.

Notice the general pattern of the terms in Example 1:

$$a_1 = 7$$
$$a_2 = 7 + 2 = 9 \quad \text{or} \quad a_2 = a_1 + d$$
$$a_3 = 9 + 2 = 11 \quad \text{or} \quad a_3 = a_2 + d = (a_1 + d) + d = a_1 + 2d$$
$$a_4 = 11 + 2 = 13 \quad \text{or} \quad a_4 = a_3 + d = (a_1 + 2d) + d = a_1 + 3d$$
$$a_5 = 13 + 2 = 15 \quad \text{or} \quad a_5 = a_4 + d = (a_1 + 3d) + d = a_1 + 4d$$

$\hookrightarrow$ (subscript $- 1$) is multiplier $\longrightarrow\uparrow$

The pattern on the right suggests that the general term a_n of an arithmetic sequence is given by

$$a_n = a_1 + (n - 1)d$$

GENERAL TERM OF AN ARITHMETIC SEQUENCE

The general term a_n of an arithmetic sequence is given by

$$a_n = a_1 + (n - 1)d$$

where a_1 is the first term and d is the common difference.

EXAMPLE 2   Consider the arithmetic sequence whose first term is 3 and common difference is -5.

a. Write an expression for the general term a_n.

b. Find the twentieth term of this sequence.

Solution: **a.** Since this is an arithmetic sequence, the general term a_n is given by $a_n = a_1 + (n - 1)d$. Here, $a_1 = 3$ and $d = -5$, so

$$a_n = 3 + (n - 1)(-5) \qquad \text{Let } a_1 = 3 \text{ and } d = -5.$$
$$= 3 - 5n + 5 \qquad \text{Multiply.}$$
$$= 8 - 5n \qquad \text{Simplify.}$$

b. $a_n = 8 - 5n$

$$a_{20} = 8 - 5 \cdot 20 \qquad \text{Let } n = 20.$$
$$= 8 - 100 = -92$$

EXAMPLE 3   Find the eleventh term of the arithmetic sequence whose first three terms are 2, 9, and 16.

Solution: Since the sequence is arithmetic, the eleventh term is

$$a_{11} = a_1 + (11 - 1)d = a_1 + 10d$$

We know a_1 is the first term of the sequence, so $a_1 = 2$. Also, d is the constant difference of terms, so $d = a_2 - a_1 = 9 - 2 = 7$. Thus,

$$a_{11} = a_1 + 10d$$
$$= 2 + 10 \cdot 7 \qquad \text{Let } a_1 = 2 \text{ and } d = 7.$$
$$= 72$$

EXAMPLE 4   If the third term of an arithmetic progression is 12 and the eighth term is 27, find the fifth term.

Solution: We need to find a_1 and d to write the general term, which then enables us to find a_5, the fifth term. The given facts about terms a_3 and a_8 lead to a system of linear equations:

$$\begin{cases} a_3 = a_1 + (3-1)d \\ a_8 = a_1 + (8-1)d \end{cases} \quad \text{or} \quad \begin{cases} 12 = a_1 + 2d \\ 27 = a_1 + 7d \end{cases}$$

Next, we solve the system $\begin{cases} 12 = a_1 + 2d \\ 27 = a_1 + 7d \end{cases}$ by elimination. Multiply both sides of the second equation by -1 so that

$$\begin{cases} 12 = a_1 + 2d \\ -1(27) = -1(a_1 + 7d) \end{cases} \quad \begin{matrix} \text{simplifies} \\ \text{to} \end{matrix} \quad \begin{cases} 12 = a_1 + 2d \\ -27 = -a_1 - 7d \end{cases}$$

$$\overline{}$$
$$-15 = \quad -5d \quad \text{Add the equations.}$$
$$3 = d \quad \text{Divide both sides by } -5.$$

To find a_1, let $d = 3$ in $12 = a_1 + 2d$. Then

$$12 = a_1 + 2(3)$$
$$12 = a_1 + 6$$
$$6 = a_1$$

Thus, $a_1 = 6$ and $d = 3$, so

$$a_n = 6 + (n-1)(3)$$
$$= 6 + 3n - 3$$
$$= 3 + 3n$$

and

$$a_5 = 3 + 3 \cdot 5 = 18$$

EXAMPLE 5   Donna has an offer for a job starting at $20,000 per year and guaranteeing her a raise of $800 per year for the next 5 years. Write the general term for the arithmetic sequence that models Donna's potential annual salaries, and find her salary for the fourth year.

Solution: The first term $a_1 = 20{,}000$, and $d = 800$. So

$$a_n = 20{,}000 + (n - 1)(800) = 19{,}200 + 800n$$
$$a_4 = 19{,}200 + 800 \cdot 4 = 22{,}400$$

Her salary for the fourth year will be $22,400. ▬▬▬

We now investigate a **geometric sequence,** also called a **geometric progression.** In the sequence 5, 15, 45, 135, . . . , each term after the first is the *product* of 3 and the preceding term. This pattern of multiplying by a constant to get the next term defines a geometric sequence. The constant is called the **common ratio** because it is the ratio of any term (after the first) to its preceding term.

$$\frac{15}{5} = 3$$

$$\frac{45}{15} = 3$$

$$\frac{135}{45} = 3$$

$$\vdots$$

$$\begin{matrix} n\text{th term} \to \\ \text{previous term} \to \end{matrix} \quad \frac{a_n}{a_{n-1}} = 3$$

GEOMETRIC SEQUENCE AND COMMON RATIO

A **geometric sequence** is a sequence in which each term (after the first) is obtained by multiplying the preceding term by a constant r. The constant r is called the **common ratio** of the sequence.

The sequence $12, 6, 3, \frac{3}{2}, \ldots$ is geometric since each term after the first is the product of the previous term and $\frac{1}{2}$.

EXAMPLE 6 Write the first five terms of a geometric sequence whose first term is 7 and whose common ratio is 2.

Solution:

$$a_1 = 7$$
$$a_2 = 7(2) = 14$$
$$a_3 = 14(2) = 28$$
$$a_4 = 28(2) = 56$$
$$a_5 = 56(2) = 112$$

The first five terms are 7, 14, 28, 56, and 112. ▬▬▬

Notice the general pattern of the terms in Example 6:

$a_1 = 7$

$a_2 = 7(2) = 14$ or $a_2 = a_1(r)$

$a_3 = 14(2) = 28$ or $a_3 = a_2(r) = (a_1 \cdot r) \cdot r = a_1 r^2$

$a_4 = 28(2) = 56$ or $a_4 = a_3(r) = (a_1 \cdot r^2) \cdot r = a_1 r^3$

$a_5 = 56(2) = 112$ or $a_5 = a_4(r) = (a_1 \cdot r^3) \cdot r = a_1 r^4$ ⟵

$\quad\quad\quad\quad\quad\quad\quad\quad \longrightarrow$ (subscript $-$ 1) is power ⟶

The pattern on the right suggests that the general term of a geometric sequence is given by $a_n = a_1 r^{n-1}$.

GENERAL TERM OF A GEOMETRIC SEQUENCE

The general term a_n of a geometric sequence is given by

$$a_n = a_1 r^{n-1}$$

where a_1 is the first term and r is the common ratio.

EXAMPLE 7   Find the eighth term of the geometric sequence whose first term is 12 and whose common ratio is $\dfrac{1}{2}$.

Solution: Since this is a geometric sequence, the general term a_n is given by

$$a_n = a_1 r^{n-1}$$

Here $a_1 = 12$ and $r = \dfrac{1}{2}$, so $a_n = 12\left(\dfrac{1}{2}\right)^{n-1}$. Evaluate a_n for $n = 8$:

$$a_8 = 12\left(\frac{1}{2}\right)^{8-1} = 12\left(\frac{1}{2}\right)^7 = 12\left(\frac{1}{128}\right) = \frac{3}{32}$$

EXAMPLE 8   Find the fifth term of the geometric sequence whose first three terms are 2, -6, and 18.

Solution: Since the sequence is geometric and $a_1 = 2$, the fifth term must be $a_1 r^{5-1}$, or $2r^4$. We know that r is the common ratio of terms, so r must be $\dfrac{-6}{2}$, or -3. Thus,

$$a_5 = 2r^4$$

$$a_5 = 2(-3)^4 = 162$$

EXAMPLE 9   If the second term of a geometric sequence is $\dfrac{5}{4}$ and the third term is $\dfrac{5}{16}$, find the first term and the common ratio.

Solution: Notice that $\dfrac{5}{16} \div \dfrac{5}{4} = \dfrac{1}{4}$, so $r = \dfrac{1}{4}$. Then

$$a_2 = a_1 \left(\frac{1}{4}\right)^{2-1}$$

$$\frac{5}{4} = a_1 \left(\frac{1}{4}\right)^{1}, \text{ or } a_1 = 5 \qquad \text{Replace } a_2 \text{ with } \frac{5}{4}.$$

The first term is 5.

EXAMPLE 10   The population size of a bacterial culture growing under controlled conditions is doubling each day. Predict how large the culture will be at the beginning of day 7 if it measures 10 units at the beginning of day 1.

Solution: Since the culture doubles in size each day, the population sizes are modeled by a geometric sequence. Here $a_1 = 10$ and $r = 2$. Thus,

$$a_n = a_1 r^{n-1} = 10(2)^{n-1} \quad \text{and} \quad a_7 = 10(2)^{7-1} = 640$$

The bacterial culture should measure 640 units at the beginning of day 7.

Exercise Set 11.2

Write the first five terms of the arithmetic or geometric sequence whose first term, a_1, and common difference, d, or common ratio, r, are given. See Examples 1 and 6.

1. $a_1 = 4; d = 2$ **2.** $a_1 = 3; d = 10$

3. $a_1 = 6; d = -2$ **4.** $a_1 = -20; d = 3$

5. $a_1 = 1; r = 3$ **6.** $a_1 = -2; r = 2$

7. $a_1 = 48; r = \dfrac{1}{2}$ **8.** $a_1 = 1; r = \dfrac{1}{3}$

Find the indicated term of each sequence. See Examples 2 and 7.

9. The eighth term of the arithmetic sequence whose first term is 12 and whose common difference is 3. 33

10. The twelfth term of the arithmetic sequence whose first term is 32 and whose common difference is −4.   −12

11. The fourth term of the geometric sequence whose first term is 7 and whose common ratio is −5. −875

12. The fifth term of the geometric sequence whose first term is 3 and whose common ratio is 3.   243

13. The fifteenth term of the arithmetic sequence whose first term is −4 and whose common difference is −4.   −60

14. The sixth term of the geometric sequence whose first term is 5 and whose common ratio is −4. −5120

Find the indicated term of each sequence. See Examples 3 and 8.

15. The ninth term of the arithmetic sequence 0, 12, 24, ...   96

16. The thirteenth term of the arithmetic sequence −3, 0, 3, ...   33

1. 4, 6, 8, 10, 12   **2.** 3, 13, 23, 33, 43   **3.** 6, 4, 2, 0, −2   **4.** −20, −17, −14, −11, −8   **5.** 1, 3, 9, 27, 81

6. −2, −4, −8, −16, −32   **7.** 48, 24, 12, 6, 3   **8.** $1, \dfrac{1}{3}, \dfrac{1}{9}, \dfrac{1}{27}, \dfrac{1}{81}$

17. The twenty-fifth term of the arithmetic sequence 20, 18, 16, ...   −28

18. The ninth term of the geometric sequence 5, 10, 20, ...   1280

19. The fifth term of the geometric sequence 2, −10, 50, ...   1250

20. The sixth term of the geometric sequence $\frac{1}{2}$, $\frac{3}{2}$, $\frac{9}{2}$, ... $\frac{243}{2}$

Find the indicated term of each sequence. See Examples 4 and 9.

21. The eighth term of the arithmetic sequence whose fourth term is 19 and whose fifteenth term is 52.   31

22. If the second term of an arithmetic sequence is 6 and the tenth term is 30, find the twenty-fifth term.   75

23. If the second term of an arithmetic progression is −1 and the fourth term is 5, find the ninth term.   20

24. If the second term of a geometric progression is 15 and the third term is 3, find a_1 and r. $a_1 = 75; r = \frac{1}{5}$

25. If the second term of a geometric progression is $-\frac{4}{3}$ and the third term is $\frac{8}{3}$, find a_1 and r. $a_1 = \frac{2}{3}; r = -2$

26. If the third term of a geometric sequence is 4 and the fourth term is −12, find a_1 and r. $a_1 = \frac{4}{9}; r = -3$

27. Explain why 14, 10, 6 may be the first three terms of an arithmetic sequence when it appears we are subtracting instead of adding to get the next term.

28. Explain why 80, 20, 5 may be the first three terms of a geometric sequence when it appears we are dividing instead of multiplying to get the next term.

Given are the first three terms of a sequence that is either arithmetic or geometric. If the sequence is arithmetic, find a_1 and d. If a sequence is geometric, find a_1 and r.

29. 2, 4, 6 $a_1 = 2; d = 2$ **30.** 8, 16, 24 $a_1 = 8; d = 8$

31. 5, 10, 20 $a_1 = 5; r = 2$

32. 2, 6, 18 $a_1 = 2; r = 3$

33. $\frac{1}{2}, \frac{1}{10}, \frac{1}{50}$ $a_1 = \frac{1}{2}; r = \frac{1}{5}$

34. $\frac{2}{3}, \frac{4}{3}, 2$ $a_1 = \frac{2}{3}; d = \frac{2}{3}$

35. x, $5x$, $25x$ $a_1 = x; r = 5$

36. y, $-3y$, $9y$ $a_1 = y; r = -3$

37. p, $p + 4$, $p + 8$ $a_1 = p; d = 4$

38. t, $t - 1$, $t - 2$ $a_1 = t; d = -1$

Find the indicated term of each sequence.

39. The twenty-first term of the arithmetic sequence whose first term is 14 and whose common difference is $\frac{1}{4}$.   19

40. The fifth term of the geometric sequence whose first term is 8 and whose common ratio is −3.   648

41. The fourth term of the geometric sequence whose first term is 3 and whose common ratio is $-\frac{2}{3}$. $-\frac{8}{9}$

42. The fourth term of the arithmetic sequence whose first term is 9 and whose common difference is 5. 24

43. The fifteenth term of the arithmetic sequence $\frac{3}{2}$, 2, $\frac{5}{2}$, ... $\frac{17}{2}$

44. The eleventh term of the arithmetic sequence 2, $\frac{5}{3}$, $\frac{4}{3}$, ... $-\frac{4}{3}$

45. The sixth term of the geometric sequence 24, 8, $\frac{8}{3}$, ... $\frac{8}{81}$

46. The eighteenth term of the arithmetic sequence 5, 2, −1, ...   −46

47. If the third term of an arithmetic progression is 2 and the seventeenth term is −40, find the tenth term.   −19

48. If the third term of a geometric sequence is −28 and the fourth term is −56, find a_1 and r.
$a_1 = -7; r = 2$

Solve. See Examples 5 and 10.

49. An auditorium has 54 seats in the first row, 58 seats in the second row, 62 seats in the third row, and so on. Find the general term of this arithmetic sequence and the number of seats in the twentieth row. $a_n = 4n + 50$; 130 seats

50. A triangular display of cans in a grocery store has 20 cans in the first row, 17 cans in the next row, and so on, in an arithmetic sequence. Find the general term and the number of cans in the fifth row. Find how many rows there are in the display and how many cans are in the top row.

51. The initial size of a virus culture is 6 units, and it triples its size every day. Find the general term of the geometric sequence that models the culture's size. $a_n = 6(3)^{n-1} = 2(3)^n$

52. A real-estate investment broker predicts that a certain property will increase in value 15% each year. Thus, the yearly property values can be modeled by a geometric sequence whose common ratio r is 1.15. If the initial property value was $500,000, write the first four terms of the sequence and predict the value at the end of the third year.

50. $a_n = 20 + (n - 1)(-3)$; or $a_n = 23 - 3n$; 8 cans in the fifth row; 7 rows in the display; top row, 2 cans
52. $500,000, $575,000, $661, 250, $760,437.50

53. A rubber ball is dropped from a height of 486 feet, and it continues to bounce one-third the height from which it last fell. Write out the first five terms of this geometric sequence and find the general term. Find how many bounces it takes for the ball to rebound less than 1 foot.

54. On the first swing, the length of the arc through which a pendulum swings is 50 inches. The length of each successive swing is 80% of the preceding swing. Determine whether this sequence is arithmetic or geometric. Find the length of the fourth swing.   geometric; 25.6 in.

55. Jose takes a job that offers a monthly starting salary of $2000 and guarantees him a monthly raise of $125 during his first year of training. Find the general term of this arithmetic sequence and his monthly salary at the end of his training.

56. At the beginning of Claudia Schaffer's exercise program, she rides 15 minutes on the Lifecycle. Each week she increases her riding time by 5 minutes. Write the general term of this arithmetic sequence, and find her riding time after 7 weeks. Find how many weeks it takes her to reach a riding time of 1 hour.

57. If a radioactive element has a half-life of 3 hours, then x grams of the element dwindles to $\dfrac{x}{2}$ grams after 3 hours. If a nuclear reactor has 400 grams of that radioactive element, find the amount of radioactive material after 12 hours.   25 grams

Write the first four terms of the arithmetic or geometric sequence whose first term, a_1, and common difference, d, or common ratio, r, are given.

58. $a_1 = \$3720, d = -\$268.50$

59. $a_1 = \$11{,}782.40, r = 0.5$

60. $a_1 = 26.8, r = 2.5$ 26.8, 67, 167.5, 418.75

61. $a_1 = 19.652; d = -0.034$

62. Describe a situation in your life that can be modeled by a geometric sequence. Write an equation for the sequence.   Answers will vary.

63. Describe a situation in your life that can be modeled by an arithmetic sequence. Write an equation for the sequence.   Answers will vary.

Review Exercises

Evaluate. See Section 1.4.

64. $5(1) + 5(2) + 5(3) + 5(4)$ 50

65. $\dfrac{1}{3(1)} + \dfrac{1}{3(2)} + \dfrac{1}{3(3)}$ $\dfrac{11}{18}$

66. $2(2 - 4) + 3(3 - 4) + 4(4 - 4)$ -7

67. $3^0 + 3^1 + 3^2 + 3^3$ 40

68. $\dfrac{1}{4(1)} + \dfrac{1}{4(2)} + \dfrac{1}{4(3)}$ $\dfrac{11}{24}$

69. $\dfrac{8 - 1}{8 + 1} + \dfrac{8 - 2}{8 + 2} + \dfrac{8 - 3}{8 + 3}$ $\dfrac{907}{495}$

53. 486, 162, 54, 18, 6; $a_n = \dfrac{486}{3^{n-1}}$; 7 bounces   **55.** $a_n = 125n + 1875$; $3375

56. $a_n = 15 + (n - 1)(5)$; or $a_n = 10 + 5n$; 45 min; 10 weeks   **58.** $3720, $3451.50, $3183, $2914.50

59. $11,782.40, $5891.20, $2945.60, $1472.80   **61.** 19.652, 19.618, 19.584, 19.55

11.3 | SERIES

TAPE IA 11.3

O B J E C T I V E S

1 Identify finite and infinite series.

2 Use summation notation.

3 Find partial sums.

1 A person who conscientiously saves money by saving first $100 and then saving $10 more each month than he saved the preceding month is saving money according to the arithmetic sequence

$$a_n = 100 + 10(n - 1)$$

Following this sequence, he can predict how much money he should save for any particular month. But if he also wants to know how much money *in total* he has saved, say, by the fifth month, he must find the *sum* of the first five terms of the sequence

$$100 + \underbrace{100 + 10}_{} + \underbrace{100 + 20}_{} + \underbrace{100 + 30}_{} + \underbrace{100 + 40}_{}$$
$$a_1 \qquad a_2 \qquad a_3 \qquad a_4 \qquad a_5$$

A sum of the terms of a sequence is called a **series** (the plural is also "series"). As our example here suggests, series are frequently used to model financial and natural phenomena.

A series is a **finite series** if it is the sum of only the first k terms of the sequence, for some natural number k. A series is an **infinite series** if it is the sum of all the terms of the sequence. For example,

SEQUENCE	SERIES	
$5, 9, 13$	$5 + 9 + 13$	Finite, k is 3.
$5, 9, 13, \ldots$	$5 + 9 + 13 + \cdots$	Infinite.
$4, -2, 1, -\dfrac{1}{2}, \dfrac{1}{4}$	$4 + (-2) + 1 + \left(-\dfrac{1}{2}\right) + \left(\dfrac{1}{4}\right)$	Finite, k is 5.
$4, -2, 1, \ldots$	$4 + (-2) + 1 + \cdots$	Infinite.
$3, 6, \ldots, 99$	$3 + 6 + \cdots + 99$	Finite, k is 33.

2 A shorthand notation for denoting a series when the general term of the sequence is known is called **summation notation.** The Greek uppercase letter **sigma,** Σ, is used to mean "sum." The expression $\displaystyle\sum_{n=1}^{5}(3n + 1)$ is read "the sum of $3n + 1$ as n goes from 1 to 5"; this expression means the sum of the first five terms of the sequence whose general term is $a_n = 3n + 1$. Often, the variable i is used instead of n in summation notation: $\displaystyle\sum_{i=1}^{5}(3i + 1)$. Whether we use n, i, k, or some other variable, the variable is called the **index of summation.** The notation $i = 1$ below the symbol Σ indicates the beginning value of i, and the number 5 above the symbol Σ indicates the ending value of i. Thus, the terms of the sequence are found by successively replacing i with the natural numbers 1, 2, 3, 4, 5. To find the sum, we write out the terms and then add:

$$\sum_{i=1}^{5}(3i + 1) = (3 \cdot 1 + 1) + (3 \cdot 2 + 1) + (3 \cdot 3 + 1)$$
$$+ (3 \cdot 4 + 1) + (3 \cdot 5 + 1)$$
$$= 4 + 7 + 10 + 13 + 16 = 50$$

EXAMPLE 1   Evaluate.

a. $\displaystyle\sum_{i=0}^{6} \frac{i - 2}{2}$

b. $\displaystyle\sum_{i=3}^{5} 2^i$

Solution:

a. $\displaystyle\sum_{i=0}^{6} \frac{i-2}{2} = \frac{0-2}{2} + \frac{1-2}{2} + \frac{2-2}{2} + \frac{3-2}{2} + \frac{4-2}{2} + \frac{5-2}{2} + \frac{6-2}{2}$

$\phantom{\displaystyle\sum_{i=0}^{6} \frac{i-2}{2}} = (-1) + \left(-\frac{1}{2}\right) + 0 + \frac{1}{2} + 1 + \frac{3}{2} + 2$

$\phantom{\displaystyle\sum_{i=0}^{6} \frac{i-2}{2}} = \frac{7}{2}, \text{ or } 3\frac{1}{2}$

b. $\displaystyle\sum_{i=3}^{5} 2^i = 2^3 + 2^4 + 2^5$

$\phantom{\displaystyle\sum_{i=3}^{5} 2^i} = 8 + 16 + 32$

$\phantom{\displaystyle\sum_{i=3}^{5} 2^i} = 56$

▬▬▬

EXAMPLE 2   Write each series with summation notation.

a. $3 + 6 + 9 + 12 + 15$ **b.** $\dfrac{1}{2} + \dfrac{1}{4} + \dfrac{1}{8} + \dfrac{1}{16}$

Solution: **a.** Since the *difference* of each term and the preceding term is 3, the terms correspond to the first five terms of the arithmetic sequence $a_n = a_1 + (n-1)d$ with $a_1 = 3$ and $d = 3$. So $a_n = 3 + (n-1)3$. Thus, in summation notation,

$$3 + 6 + 9 + 12 + 15 = \sum_{i=1}^{5} 3 + (i-1)3$$

b. Since each term is the *product* of the preceding term and $\dfrac{1}{2}$, these terms correspond to the first four terms of the geometric sequence $a_n = a_1 r^{n-1}$. Here $a_1 = \dfrac{1}{2}$ and $r = \dfrac{1}{2}$, so $a_n = \left(\dfrac{1}{2}\right)\left(\dfrac{1}{2}\right)^{n-1} = \left(\dfrac{1}{2}\right)^{1+(n-1)} = \left(\dfrac{1}{2}\right)^{n}$. In summation notation,

$$\frac{1}{2} + \frac{1}{4} + \frac{1}{8} + \frac{1}{16} = \sum_{i=1}^{4} \left(\frac{1}{2}\right)^{i}$$

▬▬▬

3   The sum of the first n terms of a sequence is a finite series known as a **partial sum,** S_n. Thus, for the sequence $a_1, a_2, \ldots, a_n$, the first three partial sums are

$$S_1 = a_1$$
$$S_2 = a_1 + a_2$$
$$S_3 = a_1 + a_2 + a_3$$

In general, S_n is the sum of the first n terms of a sequence:

$$S_n = \sum_{i=1}^{n} a_n$$

EXAMPLE 3   Find the sum of the first three terms of the sequence whose general term is
$a_n = \dfrac{n+3}{2n}$.

Solution: $S_3 = \displaystyle\sum_{i=1}^{3} \dfrac{i+3}{2i} = \dfrac{1+3}{2\cdot 1} + \dfrac{2+3}{2\cdot 2} + \dfrac{3+3}{2\cdot 3}$

$= 2 + \dfrac{5}{4} + 1 = 4\dfrac{1}{4}.$

The next example illustrates how these sums model real-life phenomena.

EXAMPLE 4   The number of baby gorillas born at the San Diego Zoo is a sequence defined by $a_n = n(n-1)$, where n is the number of years the zoo has owned gorillas. Find the *total* number of baby gorillas born in the *first 4 years*.

Solution: To solve, find the sum

$$S_4 = \sum_{i=1}^{4} i(i-1)$$

$$= 1(1-1) + 2(2-1) + 3(3-1) + 4(4-1)$$
$$= 0 + 2 + 6 + 12 = 20$$

There were 20 gorillas born in the first 4 years.

EXERCISE SET 11.3

Evaluate. See Example 1.

1. $\displaystyle\sum_{i=1}^{4} (i-3)$ −2

2. $\displaystyle\sum_{i=1}^{5} (i+6)$ 45

3. $\displaystyle\sum_{i=4}^{7} (2i+4)$ 60

4. $\displaystyle\sum_{i=2}^{3} (5i-1)$ 23

5. $\displaystyle\sum_{i=2}^{4} (i^2-3)$ 20

6. $\displaystyle\sum_{i=3}^{5} i^3$ 216

7. $\displaystyle\sum_{i=1}^{3} \left(\dfrac{1}{i+5}\right)$ $\dfrac{73}{168}$

8. $\displaystyle\sum_{i=2}^{4} \left(\dfrac{2}{i+3}\right)$ $\dfrac{107}{105}$

9. $\displaystyle\sum_{i=1}^{3} \dfrac{1}{6i}$ $\dfrac{11}{36}$

10. $\displaystyle\sum_{i=1}^{3} \dfrac{1}{3i}$ $\dfrac{11}{18}$

11. $\displaystyle\sum_{i=2}^{6} 3i$ 60

12. $\displaystyle\sum_{i=3}^{6} -4i$ −72

13. $\displaystyle\sum_{i=3}^{5} i(i+2)$ 74

14. $\displaystyle\sum_{i=2}^{4} i(i-3)$ 2

15. $\displaystyle\sum_{i=1}^{5} 2^i$ 62

16. $\displaystyle\sum_{i=1}^{4} 3^{i-1}$ 40

17. $\displaystyle\sum_{i=1}^{4} \dfrac{4i}{i+3}$ $\dfrac{241}{35}$

18. $\displaystyle\sum_{i=2}^{5} \dfrac{6-i}{6+i}$ $\dfrac{371}{330}$

Write each series with summation notation. See Example 2.

19. $1 + 3 + 5 + 7 + 9$

20. $4 + 7 + 10 + 13$

21. $4 + 12 + 36 + 108$

22. $5 + 10 + 20 + 40 + 80 + 160$

23. $12 + 9 + 6 + 3 + 0 + (-3)$

24. $5 + 1 + (-3) + (-7)$

19. $\displaystyle\sum_{i=1}^{5} (2i-1)$ **20.** $\displaystyle\sum_{i=1}^{4} (3i+1)$ **21.** $\displaystyle\sum_{i=1}^{4} (3)^{i-1}$ **22.** $\displaystyle\sum_{i=1}^{6} 5(2)^{i-1}$ **23.** $\displaystyle\sum_{i=1}^{6} (-3i+15)$ **24.** $\displaystyle\sum_{i=1}^{4} (-4i+9)$

25. $12 + 4 + \dfrac{4}{3} + \dfrac{4}{9}$ $\displaystyle\sum_{i=1}^{4} \dfrac{4}{3^{i-2}}$

26. $80 + 20 + 5 + \dfrac{5}{4} + \dfrac{5}{16}$ $\displaystyle\sum_{i=1}^{5} \dfrac{5}{4^{i-3}}$

27. $1 + 4 + 9 + 16 + 25 + 36 + 49$ $\displaystyle\sum_{i=1}^{7} i^2$

28. $1 + (-4) + 9 + (-16)$ $\displaystyle\sum_{i=1}^{4} (-1)^{i-1} i^2$

Find each partial sum. See Example 3.

29. Find the sum of the first two terms of the sequence whose general term is $a_n = (n + 2)(n - 5)$.   −24

30. Find the sum of the first six terms of the sequence whose general term is $a_n = (-1)^n$.   0

31. Find the sum of the first two terms of the sequence whose general term is $a_n = n(n - 6)$.   −13

32. Find the sum of the first seven terms of the sequence whose general term is $a_n = (-1)^{n-1}$.   1

33. Find the sum of the first four terms of the sequence whose general term is $a_n = (n + 3)(n + 1)$.   82

34. Find the sum of the first five terms of the sequence whose general term is $a_n = \dfrac{(-1)^n}{2n}$. $-\dfrac{47}{120}$

35. Find the sum of the first four terms of the sequence whose general term is $a_n = -2n$.   −20

36. Find the sum of the first five terms of the sequence whose general term is $a_n = (n - 1)^2$.   30

37. Find the sum of the first three terms of the sequence whose general term is $a_n = -\dfrac{n}{3}$.   −2

38. Find the sum of the first three terms of the sequence whose general term is $a_n = (n + 4)^2$.   110

Solve. See Example 4.

39. A gardener is making a triangular planting with 1 tree in the first row, 2 trees in the second row, 3 trees in the third row, and so on for 10 rows. Write the sequence that describes the number of trees in each row. Find the total number of trees planted.
1, 2, 3, . . . , 10; 55 trees

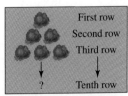

First row
Second row
Third row
?   Tenth row

40. Some surfers at the beach form a human pyramid with 2 surfers in the top row, 3 surfers in the second row, 4 surfers in the third row, and so on. If there are 6 rows in the pyramid, write the sequence that describes the number of surfers in each row of the pyramid. Find the total number of surfers.   2, 3, 4, 5, 6, 7; 27 surfers

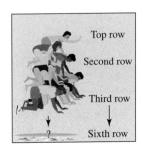

Top row
Second row
Third row
?   Sixth row

41. A culture of fungus starts with 6 units and doubles every day. Write the general term of the sequence that describes the growth of this fungus. Find the number of fungus units there will be at the beginning of the fifth day. $a_n = 6(2)^{n-1}$; 96 units

42. A bacterial colony begins with 100 bacteria and doubles every 6 hours. Write the general term of the sequence describing the growth of the bacteria. Find the number of bacteria there will be after 24 hours.

43. A bacterial colony begins with 50 bacteria and doubles every 12 hours. Write the sequence that describes the growth of the bacteria. Find the number of bacteria there will be after 48 hours.

44. The number of otters born each year in a new aquarium forms a sequence whose general term is $a_n = (n - 1)(n + 3)$. Find the number of otters born in the third year, and find the total number of otters born in the first three years.

45. The number of opossums killed each month on a new highway describes the sequence whose general term is $a_n = (n + 1)(n + 2)$, where n is the number of the months. Find the number of opossums killed in the fourth month, and find the total number killed in the first four months.

46. In 1993 the population of an endangered fish was estimated by environmentalists to be decreasing each year. The size of the population in a given year is $24 - 4n$ thousand fish fewer than the previous year. Find the decrease in population in 1995,

42. $a_n = 100(2)^n$; n represents the number of 6-hr. periods; 1600 bacteria   **43.** $a_n = 50(2)^n$; n represents the number of 12-hr. periods; 800 bacteria   **44.** 12 otters; 17 otters   **45.** 30 opossums; 68 opossums

if year 1 is 1993. Find how many total fish died from 1993 through 1995.   12,000 fish; 48,000 fish

47. The amount of decay in pounds of a radioactive isotope each year is given by the sequence whose general term is $a_n = 100(0.5)^n$, where n is the number of the year. Find the amount of decay in the fourth year, and find the total amount of decay in the first four years.   6.25 lb.; 93.75 lb.

48. Susan has a choice between two job offers. Job A has an annual starting salary of $20,000 with guaranteed annual raises of $1200 for the next four years, whereas job B has an annual starting salary of $18,000 with guaranteed annual raises of $2500 for the next four years. Compare the fifth partial sums for each sequence to determine which job would pay Susan more money over the next 5 years.

49. A pendulum swings a length of 40 inches on its first swing. Each successive swing is $\frac{4}{5}$ of the preceding swing. Find the length of the fifth swing and the total length swung during the first five swings. (Round to the nearest tenth of an inch.)   16.4 in.; 134.5 in.

50. Explain the difference between a sequence and a series.

51. a. Write the sum $\sum\limits_{i=1}^{7} i + i^2$ without summation notation. $2 + 6 + 12 + 20 + 30 + 42 + 56$

b. Write the sum $\sum\limits_{i=1}^{7} i + \sum\limits_{i=1}^{7} i^2$ without summation notation.

c. Compare the results of parts (a) and (b).

d. Do you think the following is true or false? Explain your answer.

$$\sum_{i=1}^{n} (a_n + b_n) = \sum_{i=1}^{n} a_n + \sum_{i=1}^{n} b_n \quad \text{true}$$

48. Job B pays $3000 more over the next 5 yrs.
51b. $1 + 2 + 3 + 4 + 5 + 6 + 7 + 1 + 4 + 9 + 16 + 25 + 36 + 49$

52. a. Write the sum $\sum\limits_{i=1}^{6} 5i^3$ without summation notation. $5 + 40 + 135 + 320 + 625 + 1080$

b. Write the expression $5 \cdot \sum\limits_{i=1}^{6} i^3$ without summation notation. $5(1 + 8 + 27 + 64 + 125 + 216)$

c. Compare the results of parts (a) and (b).

d. Do you think the following is true or false? Explain your answer.   true

$$\sum_{i=1}^{n} c \cdot a_n = c \cdot \sum_{i=1}^{n} a_n \quad \text{where } c \text{ is a constant}$$

Review Exercises

Evaluate. See Section 1.4.

53. $\dfrac{5}{1 - \dfrac{1}{2}}$ 10

54. $\dfrac{-3}{1 - \dfrac{1}{7}}$ $-\dfrac{7}{2}$

55. $\dfrac{\dfrac{1}{3}}{1 - \dfrac{1}{10}}$ $\dfrac{10}{27}$

56. $\dfrac{\dfrac{6}{11}}{1 - \dfrac{1}{10}}$ $\dfrac{20}{33}$

57. $\dfrac{3(1 - 2^4)}{1 - 2}$ 45

58. $\dfrac{2(1 - 5^3)}{1 - 5}$ 62

59. $\dfrac{10}{2}(3 + 15)$ 90

60. $\dfrac{12}{2}(2 + 19)$ 126

TAPE IA 11.4

11.4 | PARTIAL SUMS OF ARITHMETIC AND GEOMETRIC SEQUENCES

O B J E C T I V E S

1 Find the partial sum of an arithmetic sequence.

2 Find the partial sum of a geometric sequence.

3 Find the infinite series of a geometric sequence.

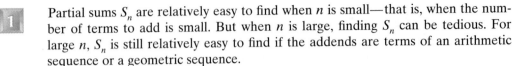

Partial sums S_n are relatively easy to find when n is small—that is, when the number of terms to add is small. But when n is large, finding S_n can be tedious. For large n, S_n is still relatively easy to find if the addends are terms of an arithmetic sequence or a geometric sequence.

For an arithmetic sequence, $a_n = a_1 + (n-1)d$ for some first term a_1 and some common difference d. So S_n, the sum of the first n terms, is

$$S_n = a_1 + (a_1 + d) + (a_1 + 2d) + \cdots + (a_1 + (n-1)d)$$

We might also find S_n by "working backward" from the nth term a_n, finding the preceding term a_{n-1}, by subtracting d each time:

$$S_n = a_n + (a_n - d) + (a_n - 2d) + \cdots + (a_n - (n-1)d)$$

Now add the left sides of these two equations, and add the right sides:

$$2S_n = (a_1 + a_n) + (a_1 + a_n) + (a_1 + a_n) + \cdots + (a_1 + a_n)$$

The d terms subtract out, leaving n sums of the first term, a_1, and last term, a_n. Thus, we write

$$2S_n = n(a_1 + a_n)$$

or

$$S_n = \frac{n}{2}(a_1 + a_n)$$

Partial Sum S_n of an Arithmetic Sequence

The partial sum S_n of the first n terms of an arithmetic sequence is given by

$$S_n = \frac{n}{2}(a_1 + a_n)$$

where a_1 is the first term of the sequence and a_n is the nth term.

EXAMPLE 1   Use the partial sum formula to find the sum of the first six terms of the arithmetic sequence 2, 5, 8, 11, 14, 17, ...

Solution: Use the formula for S_n of an arithmetic sequence, replacing n with 6, a_1 with 2, and a_n with 17.

$$S_n = \frac{n}{2}(a_1 + a_n) = \frac{6}{2}(2 + 17) = 3(19) = 57$$

EXAMPLE 2   Find the sum of the first 30 positive integers.

Solution: Because $1, 2, 3, \ldots, 30$ is an arithmetic sequence, use the formula for S_n with $n = 30$, $a_1 = 1$, and $a_n = 30$. Thus,

$$S_n = \frac{n}{2}(a_1 + a_n) = \frac{30}{2}(1 + 30) = 15(31) = 465$$

EXAMPLE 3   Rolls of carpet are stacked in 20 rows with 3 rolls in the top row, 4 rolls in the next row, and so on, forming an arithmetic sequence. Find the total number of carpet rolls if there are 22 rolls in the bottom row.

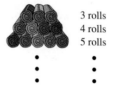

3 rolls
4 rolls
5 rolls

Solution: The list $3, 4, 5, \ldots, 22$ is the first 20 terms of an arithmetic sequence. Use the formula for S_n with $a_1 = 3$, $a_n = 22$, and $n = 20$ terms. Thus,

$$S_{20} = \frac{20}{2}(3 + 22) = 10(25) = 250$$

There are a total of 250 rolls of carpet in the display.

We can also derive a formula for the partial sum S_n of the first n terms of a geometric series. If $a_n = a_1 r^{n-1}$, then

$$S_n = a_1 + a_1 r + a_1 r^2 + \cdots + a_1 r^{n-1}$$

$$\uparrow \quad \uparrow \quad \uparrow \qquad\qquad \uparrow$$

1st   2nd   3rd nth
term   term   term   term

Multiply each side of the equation by $-r$:

$$-rS_n = -a_1 r - a_1 r^2 - a_1 r^3 - \cdots - a_1 r^n$$

Add the two equations:

$$S_n - rS_n = a_1 + (a_1 r - a_1 r) + (a_1 r^2 - a_1 r^2) + (a_1 r^3 - a_1 r^3) + \cdots - a_1 r^n$$
$$S_n - rS_n = a_1 - a_1 r^n$$

Now factor each side:

$$S_n(1 - r) = a_1(1 - r^n)$$

Solve for S_n by dividing both sides by $1 - r$. Thus,

$$S_n = \frac{a_1(1 - r^n)}{1 - r}$$

as long as r is not 1.

PARTIAL SUM S_n OF A GEOMETRIC SEQUENCE

The partial sum S_n of the first n terms of a geometric sequence is given by

$$S_n = \frac{a_1(1 - r^n)}{1 - r}$$

where a_1 is the first term of the sequence, r is the common ratio, and $r \neq 1$.

EXAMPLE 4   Find the sum of the first six terms of the geometric sequence 5, 10, 20, 40, 80, 160.

Solution: Use the formula for the partial sum S_n of the terms of a geometric sequence. Here, $n = 6$, the first term $a_1 = 5$, and the common ratio $r = 2$:

$$S_n = \frac{a_1(1 - r^n)}{1 - r}$$

$$S_6 = \frac{5(1 - 2^6)}{1 - 2} = \frac{5(-63)}{-1} = 315$$

EXAMPLE 5   A grant from an alumnus to a university specified that the university was to receive \$800,000 during the first year and 75% of the preceding year's donation during each of the following five years. Find the total amount donated during the 6 years.

Solution: The donations are modeled by the first six terms of a geometric sequence. Evaluate S_n when $n = 6$, $a_1 = 800,000$, and $r = 0.75$.

$$S_6 = \frac{800,000[1 - (0.75)^6]}{1 - 0.75}$$

$$= \$2,630,468.75$$

The total amount donated during the 6 years is \$2,630,468.75.

3 Is it possible to find the sum of all the terms of an infinite sequence? Examine the partial sums of the geometric sequence $\frac{1}{2}, \frac{1}{4}, \frac{1}{8}, \ldots$.

$$S_1 = \frac{1}{2}$$

$$S_2 = \frac{1}{2} + \frac{1}{4} = \frac{3}{4}$$

$$S_3 = \frac{1}{2} + \frac{1}{4} + \frac{1}{8} = \frac{7}{8}$$

$$S_4 = \frac{1}{2} + \frac{1}{4} + \frac{1}{8} + \frac{1}{16} = \frac{15}{16}$$

$$S_5 = \frac{1}{2} + \frac{1}{4} + \frac{1}{8} + \frac{1}{16} + \frac{1}{32} = \frac{31}{32}$$

$$\vdots$$

$$S_{10} = \frac{1}{2} + \frac{1}{4} + \frac{1}{8} + \cdots + \frac{1}{2^{10}} = \frac{1023}{1024}$$

Even though each partial sum is larger than the preceding partial sum, we see that each partial sum is closer to 1 than the preceding partial sum. If n gets larger and larger, then S_n gets closer and closer to 1. We say that 1 is the **limit** of S_n and also that 1 is the sum of the terms of this infinite sequence. In general, if $|r| < 1$, the following formula gives the sum of the terms of an infinite geometric sequence.

SUM OF THE TERMS OF AN INFINITE GEOMETRIC SEQUENCE

The sum S_∞ of the terms of an infinite geometric sequence is given by

$$S_\infty = \frac{a_1}{1 - r}$$

where a_1 is the first term of the sequence, r is the common ratio, and $|r| < 1$. If $|r| \geq 1$, S_∞ does not exist.

What happens for other values of r? For example, in the following geometric sequence, $r = 3$:

6, 18, 54, 162, . . .

Here, as n increases, the sum S_n increases also. This time, though, S_n does not get closer and closer to a fixed number but instead increases without bound.

EXAMPLE 6 Find the sum of the terms of the geometric sequence $2, \dfrac{2}{3}, \dfrac{2}{9}, \dfrac{2}{27}, \ldots$

Solution: For this geometric sequence, $r = \dfrac{1}{3}$. Since $|r| < 1$, we may use the formula for S_∞ of a geometric sequence with $a_1 = 2$ and $r = \dfrac{1}{3}$:

$$S_\infty = \frac{a_1}{1-r} = \frac{2}{1-\dfrac{1}{3}} = \frac{2}{\dfrac{2}{3}} = 3$$

The formula for the sum of the terms of an infinite geometric sequence can be used to write a repeating decimal as a fraction. For example,

$$0.33\overline{3} = \frac{3}{10} + \frac{3}{100} + \frac{3}{1000} + \cdots$$

This sum is the sum of the terms of an infinite geometric sequence whose first term $a_1 = \dfrac{3}{10}$ and whose common ratio $r = \dfrac{1}{10}$. Using the formula for S_∞,

$$S_\infty = \frac{a_1}{1-r} = \frac{\dfrac{3}{10}}{1-\dfrac{1}{10}} = \frac{1}{3}$$

So $0.33\overline{3} = \dfrac{1}{3}$.

EXAMPLE 7 On its first pass, a pendulum swings through an arc whose length is 24 inches. On each pass thereafter, the arc length is 75% of the arc length on the preceding pass. Find the total distance the pendulum travels before it comes to rest.

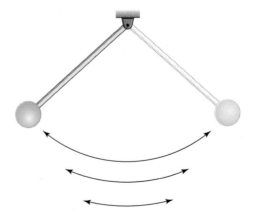

Solution: We must find the sum of the terms of an infinite geometric sequence whose first term, a_1, is 24 and whose common ratio, r, is 0.75. Since $|r| < 1$, we may use the formula for S_∞:

$$S_\infty = \frac{a_1}{1 - r} = \frac{24}{1 - 0.75} = \frac{24}{0.25} = 96$$

The pendulum travels a total distance of 96 inches before it comes to rest.

Exercise Set 11.4

Use the partial sum formula to find the partial sum of the given arithmetic or geometric sequence. See Examples 1 and 4.

1. Find the sum of the first six terms of the arithmetic sequence 1, 3, 5, 7, ... **36**

2. Find the sum of the first seven terms of the arithmetic sequence $-7, -11, -15, \ldots$. **−133**

3. Find the sum of the first five terms of the geometric sequence 4, 12, 36, ... **484**

4. Find the sum of the first eight terms of the geometric sequence $-1, 2, -4, \ldots$. **85**

5. Find the sum of the first six terms of the arithmetic sequence 3, 6, 9, ... **63**

6. Find the sum of the first four terms of the arithmetic sequence $-4, -8, -12, \ldots$. **−40**

7. Find the sum of the first four terms of the geometric sequence $2, \frac{2}{5}, \frac{2}{25}, \ldots$. **2.496**

8. Find the sum of the first five terms of the geometric sequence $\frac{1}{3}, -\frac{2}{3}, \frac{4}{3}, \ldots$. $\frac{11}{3}$

Solve. See Example 2.

9. Find the sum of the first ten positive integers.   **55**

10. Find the sum of the first eight negative integers. **−36**

11. Find the sum of the first four positive odd integers. **16**

12. Find the sum of the first five negative odd integers. **−25**

Find the sum of the terms of each infinite geometric sequence. See Example 6.

13. 12, 6, 3, ...   **24**

14. 45, 15, 5, ...   **67.5**

15. $\frac{1}{10}, \frac{1}{100}, \frac{1}{1000}, \cdots$ $\frac{1}{9}$

16. $\frac{3}{5}, \frac{3}{20}, \frac{3}{80}, \cdots$ $\frac{4}{5}$

17. $-10, -5, -\frac{5}{2}, \ldots$ **−20**

18. $-16, -4, -1, \ldots$ $-\frac{64}{3}$

19. $2, -\frac{1}{4}, \frac{1}{32}, \ldots$ $\frac{16}{9}$

20. $-3, \frac{3}{5}, -\frac{3}{25}, \ldots$ $-\frac{5}{2}$

21. $\frac{2}{3}, -\frac{1}{3}, \frac{1}{6}, \cdots$ $\frac{4}{9}$

22. $6, -4, \frac{8}{3}, \ldots$ $\frac{18}{5}$

Solve.

23. Find the sum of the first ten terms of the sequence $-4, 1, 6, \ldots$.   **185**

24. Find the sum of the first twelve terms of the sequence $-3, -13, -23, \ldots$.   **−696**

25. Find the sum of the first seven terms of the sequence $3, \frac{3}{2}, \frac{3}{4}, \ldots$. $\frac{381}{64}$, or 5.95

26. Find the sum of the first five terms of the sequence $-2, -6, -18, \ldots$.   **−242**

27. Find the sum of the first five terms of the sequence $-12, 6, -3, \ldots$. $-\frac{33}{4}$, or −8.25

28. Find the sum of the first four terms of the sequence $-\frac{1}{4}, -\frac{3}{4}, -\frac{9}{4}, \ldots$.   **−10**

29. Find the sum of the first twenty terms of the sequence $\frac{1}{2}, \frac{1}{4}, 0, \ldots$. $-\frac{75}{2}$

30. Find the sum of the first fifteen terms of the sequence $-5, -9, -13, \ldots$.   **−495**

31. If a_1 is 8 and r is $-\frac{2}{3}$, find S_3. $\frac{56}{9}$

32. If a_1 is 10 and d is $-\frac{1}{2}$, find S_{18}. $\frac{207}{2}$

Solve. See Example 3.

33. Modern Car Company has come out with a new car model. Market analysts predict that 4000 cars will be sold in the first month and that sales will drop by 50 cars per month after that during the first year. Write out the first five terms of the se-

33. 4000, 3950, 3900, 3850, 3800; 3450 cars; 44,700 cars quence, and find the number of sold cars predicted for the twelfth month. Find the total predicted number of sold cars for the first year.

34. A company that sends faxes charges $3 for the first page sent and $0.10 less than the preceding page for each additional page sent. The cost per page forms an arithmetic sequence. Write the first five terms of this sequence, and use a partial sum to find the cost of sending a nine-page document.

35. Sal has two job offers: Firm *A* starts at $22,000 per year and guarantees raises of $1000 per year, whereas Firm *B* starts at $20,000 and guarantees raises of $1200 per year. Over a 10-year period, determine the more profitable offer.

36. The game of pool uses 15 balls numbered 1 to 15. In the variety called rotation, a player who sinks a ball receives as many points as the number on the ball. Use an arithmetic series to find the score of a player who sinks all 15 balls.   120 points

Solve. See Example 5.

37. A woman made $30,000 during the first year she owned her business and made an additional 10% over the previous year in each subsequent year. Find how much she made during her fourth year of business. Find her total earnings during the first four years. $39,930; $139,230

38. In free fall, a parachutist falls 16 feet during the first second, 48 feet during the second second, 80 feet during the third second, and so on. Find how far she falls during the eighth second. Find the total distance she falls during the first 8 seconds. 240 ft; 1024 ft

39. A trainee in a computer company takes 0.9 times as long to assemble each computer as he took to assemble the preceding computer. If it took him 30 minutes to assemble the first computer, find how long it takes him to assemble the fifth computer. Find the total time he takes to assemble the first five computers (round to the nearest minute).

40. On a gambling trip to Reno, Carol doubled her bet each time she lost. If her first losing bet was $5 and she lost six consecutive bets, find how much she lost on the sixth bet. Find the total amount lost on these six bets. $160; $315

Solve. See Example 7.

41. A ball is dropped from a height of 20 feet and repeatedly rebounds to a height that is $\frac{4}{5}$ of its previ-

34. 3, 2.90, 2.80, 2.70, 2.60; $23.40
ous height. Find the total distance the ball covers before it comes to rest.   180 ft

42. A rotating flywheel coming to rest makes 300 revolutions in the first minute and in each minute thereafter makes $\frac{2}{5}$ as many revolutions as in the preceding minute. Find how many revolutions the wheel makes before it comes to rest.

500 revolutions

Solve.   **43.** Player *A*, 45 points; Player *B*, 75 points

43. In the pool game of rotation, player *A* sinks balls numbered 1 to 9, and player *B* sinks the rest of the balls. Use arithmetic series to find each player's score (see Exercise 36).

44. A godfather deposited $250 in a savings account on the day his godchild was born. On each subsequent birthday he deposited $50 more than he deposited the previous year. Find how much money he deposited on his godchild's twenty-first birthday. Find the total amount deposited over the 21 years. $1300; $17,050

45. During the holiday rush a business can rent a computer system for $200 the first day, with the rental fee decreasing $5 for each additional day. Find the fee paid for 20 days during the holiday rush. $3050

46. The spraying of a field with insecticide killed 6400 weevils the first day, 1600 the second day, 400 the third day, and so on. Find the total number of weevils killed during the first 5 days.   8525 weevils

47. A college student humorously asks his parents to charge him room and board according to this geometric sequence: $0.01 for the first day of the month, $0.02 for the second day, $0.04 for the third day, and so on. Find the total room and board he would pay for 30 days. $10,737,418.23

48. Following its television advertising campaign, a bank attracted 80 new customers the first day, 120 the second day, 160 the third day, and so on, in an arithmetic sequence. Find how many new customers were attracted during the first 5 days following its television campaign.   800 customers

49. Write $0.88\overline{8}$ as an infinite geometric series and use the formula for S_∞ to write it as a rational number.

50. Write $0.54\overline{54}$ as an infinite geometric series and use the formula S_∞ to write it as a rational number.

35. Firm *A* (Firm *A*, $265,000; Firm *B*, $254,000)   **39.** 20 min.; 123 min.   **49.** $\dfrac{8}{10} + \dfrac{8}{100} + \dfrac{8}{1000} + \cdots ; \dfrac{8}{9}$

50. $\dfrac{54}{100} + \dfrac{54}{10,000} + \dfrac{54}{1,000,000} + \cdots ; \dfrac{6}{11}$

51. Explain whether the sequence 5, 5, 5, . . . is arithmetic, geometric, neither, or both.

52. Describe a situation in everyday life that can be modeled by an infinite geometric series.

Review Exercises

Evaluate. See Section 1.4.

53. $6 \cdot 5 \cdot 4 \cdot 3 \cdot 2 \cdot 1$ 720

54. $8 \cdot 7 \cdot 6 \cdot 5 \cdot 4 \cdot 3 \cdot 2 \cdot 1$ 40,320

55. $\dfrac{3 \cdot 2 \cdot 1}{2 \cdot 1}$ 3

56. $\dfrac{5 \cdot 4 \cdot 3 \cdot 2 \cdot 1}{3 \cdot 2 \cdot 1}$ 20

Multiply. See Section 5.4.

57. $(x + 5)^2$

58. $(x - 2)^2$ $x^2 - 4x + 4$

59. $(2x - 1)^3$

60. $(3x + 2)^3$

57. $x^2 + 10x + 25$ **59.** $8x^3 - 12x^2 + 6x - 1$ **60.** $27x^3 + 54x^2 + 36x + 8$

11.5 | THE BINOMIAL THEOREM

TAPE IA 11.5

O B J E C T I V E S

1. Use Pascal's triangle to expand binomials.
2. Evaluate factorials.
3. Use the binomial theorem to expand binomials.
4. Find the nth term in the expansion of a binomial raised to a positive power.

In this section, we learn how to **expand** binomials of the form $(a + b)^n$ easily. Expanding a binomial such as $(a + b)^n$ means to write the factored form as a sum. First, we review the patterns in the expansions of $(a + b)^n$.

$$(a + b)^0 = 1 \qquad \text{1 term}$$
$$(a + b)^1 = a + b \qquad \text{2 terms}$$
$$(a + b)^2 = a^2 + 2ab + b^2 \qquad \text{3 terms}$$
$$(a + b)^3 = a^3 + 3a^2b + 3ab^2 + b^3 \qquad \text{4 terms}$$
$$(a + b)^4 = a^4 + 4a^3b + 6a^2b^2 + 4ab^3 + b^4 \qquad \text{5 terms}$$
$$(a + b)^5 = a^5 + 5a^4b + 10a^3b^2 + 10a^2b^3 + 5ab^4 + b^5 \qquad \text{6 terms}$$

Notice the following patterns:

1. The expansion of $(a + b)^n$ contains $n + 1$ terms. For example, for $(a + b)^3$, $n = 3$, and the expansion contains $3 + 1$ terms, or 4 terms.

2. The first term of the expansion of $(a + b)^n$ is a^n, and the last term is b^n.

3. The powers of a decrease by 1 for each term, whereas the powers of b increase by 1 for each term.

4. For each term of the expansion of $(a + b)^n$, the sum of the exponents of a and b is n. (For example, the sum of the exponents of $5a^4b$ is $4 + 1$, or 5, and the sum of the exponents of $10a^3b^2$ is $3 + 2$, or 5.)

1 There are patterns in the coefficients of the terms as well. Written in a triangular array, the coefficients are called **Pascal's triangle.**

$(a + b)^0$:						1						Row 1
$(a + b)^1$:					1		1					Row 2
$(a + b)^2$:				1		2		1				Row 3
$(a + b)^3$:			1		3		3		1			Row 4
$(a + b)^4$:		1		4		6		4		1		Row 5
$(a + b)^5$:	1		5		10		10		5		1	Row 6

Each row in Pascal's triangle begins and ends with 1. Any other number in a row is the sum of the two closest numbers above it. Using this pattern, we can write the next row, the seventh row, by first writing the number 1. Then we can add the consecutive numbers in row 6 and write each sum "between and below" the pair. We complete the row by writing a 1.

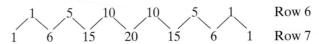

1		5		10		10		5		1	Row 6
1	6		15		20		15		6	1	Row 7

We can use Pascal's triangle and the patterns noted to expand $(a + b)^n$ without actually multiplying any terms.

EXAMPLE 1   Expand $(a + b)^6$.

Solution: Using the seventh row of Pascal's triangle as the coefficients and following the patterns noted, $(a + b)^6$ can be expanded as

$$a^6 + 6a^5b + 15a^4b^2 + 20a^3b^3 + 15a^2b^4 + 6ab^5 + b^6$$

2 For large n, the use of Pascal's triangle to find coefficients for $(a + b)^n$ can be tedious. An alternative method for determining these coefficients is based on the concept of a **factorial.**

The **factorial of n,** written $n!$ (read "n factorial"), is the product of the first n consecutive natural numbers.

FACTORIAL OF n: $n!$

If n is a natural number, then $n! = n(n - 1)(n - 2)(n - 3) \cdot \cdots \cdot 3 \cdot 2 \cdot 1$.
The factorial of 0, written 0!, is defined to be 1.

For example, $3! = 3 \cdot 2 \cdot 1 = 6$, $5! = 5 \cdot 4 \cdot 3 \cdot 2 \cdot 1 = 120$, and $0! = 1$.

EXAMPLE 2   Evaluate each expression.

a. $\dfrac{5!}{6!}$

b. $\dfrac{10!}{7!3!}$

c. $\dfrac{3!}{2!1!}$

d. $\dfrac{7!}{7!0!}$

Solution: **a.** $\dfrac{5!}{6!} = \dfrac{5 \cdot 4 \cdot 3 \cdot 2 \cdot 1}{6 \cdot 5 \cdot 4 \cdot 3 \cdot 2 \cdot 1} = \dfrac{1}{6}$

b. $\dfrac{10!}{7!3!} = \dfrac{10 \cdot 9 \cdot 8 \cdot 7!}{7! \cdot 3 \cdot 2 \cdot 1} = \dfrac{10 \cdot 9 \cdot 8}{3 \cdot 2 \cdot 1} = 10 \cdot 3 \cdot 4 = 120$

c. $\dfrac{3!}{2!1!} = \dfrac{3 \cdot 2 \cdot 1}{2 \cdot 1 \cdot 1} = 3$

d. $\dfrac{7!}{7!0!} = \dfrac{7!}{7! \cdot 1} = 1$

R E M I N D E R   We can use a calculator with a factorial key to evaluate a factorial. A calculator uses scientific notation for large results.

---

3   It can be proved, although we won't do so here, that the coefficients of terms in the expansion of $(a + b)^n$ can be expressed in terms of factorials. Following patterns 1 through 4 given earlier and using the factorial expressions of the coefficients, we have what is known as the **binomial theorem.**

BINOMIAL THEOREM

If n is a positive integer, then

$$(a + b)^n = a^n + \frac{n}{1!} a^{n-1}b^1 + \frac{n(n - 1)}{2!} a^{n-2}b^2$$

$$+ \frac{n(n - 1)(n - 2)}{3!} a^{n-3}b^3 + \cdots + b^n$$

We call the formula for $(a + b)^n$ given by the binomial theorem the **binomial formula.**

EXAMPLE 3   Use the binomial theorem to expand $(x + y)^{10}$.

Solution: Let $a = x$, $b = y$, and $n = 10$ in the binomial formula:

$$(x + y)^{10} = x^{10} + \frac{10}{1!}x^9y + \frac{10 \cdot 9}{2!}x^8y^2 + \frac{10 \cdot 9 \cdot 8}{3!}x^7y^3 + \frac{10 \cdot 9 \cdot 8 \cdot 7}{4!}x^6y^4$$

$$+ \frac{10 \cdot 9 \cdot 8 \cdot 7 \cdot 6}{5!}x^5y^5 + \frac{10 \cdot 9 \cdot 8 \cdot 7 \cdot 6 \cdot 5}{6!}x^4y^6$$

$$+ \frac{10 \cdot 9 \cdot 8 \cdot 7 \cdot 6 \cdot 5 \cdot 4}{7!}x^3y^7$$

$$+ \frac{10 \cdot 9 \cdot 8 \cdot 7 \cdot 6 \cdot 5 \cdot 4 \cdot 3}{8!}x^2y^8$$

$$+ \frac{10 \cdot 9 \cdot 8 \cdot 7 \cdot 6 \cdot 5 \cdot 4 \cdot 3 \cdot 2}{9!}xy^9 + y^{10}$$

$$= x^{10} + 10x^9y + 45x^8y^2 + 120x^7y^3 + 210x^6y^4 + 252x^5y^5 + 210x^4y^6$$

$$+ 120x^3y^7 + 45x^2y^8 + 10xy^9 + y^{10}$$

EXAMPLE 4   Use the binomial theorem to expand $(x + 2y)^5$.

Solution: Let $a = x$ and $b = 2y$ in the binomial formula:

$$(x + 2y)^5 = x^5 + \frac{5}{1!}x^4(2y) + \frac{5 \cdot 4}{2!}x^3(2y)^2 + \frac{5 \cdot 4 \cdot 3}{3!}x^2(2y)^3$$

$$+ \frac{5 \cdot 4 \cdot 3 \cdot 2}{4!}x(2y)^4 + (2y)^5$$

$$= x^5 + 10x^4y + 40x^3y^2 + 80x^2y^3 + 80xy^4 + 32y^5$$

EXAMPLE 5   Use the binomial theorem to expand $(3m - n)^4$.

Solution: Let $a = 3m$ and $b = -n$ in the binomial formula:

$$(3m - n)^4 = (3m)^4 + \frac{4}{1!}(3m)^3(-n) + \frac{4 \cdot 3}{2!}(3m)^2(-n)^2$$

$$+ \frac{4 \cdot 3 \cdot 2}{3!}(3m)(-n)^3 + (-n)^4$$

$$= 81m^4 - 108m^3n + 54m^2n^2 - 12mn^3 + n^4$$

4   Sometimes it is convenient to find a specific term of a binomial expansion without writing out the entire expansion. By studying the expansion of binomials, a pattern forms for each term. This pattern is most easily stated for the $(r + 1)$st term.

> **$(r + 1)$st Term in a Binomial Expansion**
>
> The $(r + 1)$st term of the expansion of $(a + b)^n$ is $\dfrac{n!}{r!(n - r)!}a^{n-r}b^r$.

EXAMPLE 6   Find the eighth term in the expansion of $(2x - y)^{10}$.

Solution: Use the formula, with $n = 10$, $a = 2x$, $b = -y$, and $r + 1 = 8$. Notice that, since $r + 1 = 8$, $r = 7$:

$$\frac{n!}{r!(n - r)!}a^{n-r}b^r = \frac{10!}{7!3!}(2x)^3(-y)^7$$

$$= 120(8x^3)(-y)^7$$

$$= -960x^3y^7$$

1. $m^3 + 3m^2n + 3mn^2 + n^3$
2. $x^4 + 4x^3y + 6x^2y^2 + 4xy^3 + y^4$
3. $c^5 + 5c^4d + 10c^3d^2 + 10c^2d^3 + 5cd^4 + d^5$ 4. $a^6 + 6a^5b + 15a^4b^2 + 20a^3b^3 + 15a^2b^4 + 6ab^5 + b^6$
5. $y^5 - 5y^4x + 10y^3x^2 - 10y^2x^3 + 5yx^4 - x^5$ 6. $q^7 - 7q^6r + 21q^5r^2 - 35q^4r^3 + 35q^3r^4 - 21q^2r^5 + 7qr^6 - r^7$
8. 1 8 28 56 70 56 28 8 1   17. $a^7 + 7a^6b + 21a^5b^2 + 35a^4b^3 + 35a^3b^4 + 21a^2b^5 + 7ab^6 + b^7$
18. $x^8 + 8x^7y + 28x^6y^2 + 56x^5y^3 + 70x^4y^4 + 56x^3y^5 + 28x^2y^6 + 8xy^7 + y^8$
19. $a^5 + 10a^4b + 40a^3b^2 + 80a^2b^3 + 80ab^4 + 32b^5$

Exercise Set 11.5

20. $x^6 + 18x^5y + 135x^4y^2 + 540x^3y^3 + 1215x^2y^4 + 1458xy^5 + 729y^6$
21. $q^9 + 9q^8r + 36q^7r^2 + 84q^6r^3 + 126q^5r^4 + 126q^4r^5 + 84q^3r^6 + 36q^2r^7 + 9qr^8 + r^9$

Use Pascal's triangle to expand the binomial. See Example 1.

1. $(m + n)^3$ 2. $(x + y)^4$ 3. $(c + d)^5$
4. $(a + b)^6$ 5. $(y - x)^5$ 6. $(q - r)^7$

7. Explain how to generate a row of Pascal's triangle.
8. Write the ninth row of Pascal's triangle.

Evaluate each expression. See Example 2.

9. $\dfrac{8!}{7!}$ 8   10. $\dfrac{6!}{0!}$ 720   11. $\dfrac{7!}{5!}$ 42   12. $\dfrac{8!}{5!}$ 336

13. $\dfrac{10!}{7!2!}$ 360 14. $\dfrac{9!}{5!3!}$ 504 15. $\dfrac{8!}{6!0!}$ 56 16. $\dfrac{10!}{4!6!}$ 210

Use the binomial formula to expand each binomial. See Examples 3 through 5.

17. $(a + b)^7$ 18. $(x + y)^8$ 19. $(a + 2b)^5$
20. $(x + 3y)^6$ 21. $(q + r)^9$ 22. $(b + c)^6$
23. $(4a + b)^5$ 24. $(3m + n)^4$ 25. $(5a - 2b)^4$
26. $(m - 4)^6$ 27. $(2a + 3b)^3$ 28. $(4 - 3x)^5$
29. $(x + 2)^5$ 30. $(3 + 2a)^4$

Find the indicated term. See Example 6.

31. The fifth term of the expansion of $(c - d)^5$ $5cd^4$
32. The fourth term of the expansion of $(x - y)^6$
33. The eighth term of the expansion of $(2c + d)^7$ d^7
34. The tenth term of the expansion of $(5x - y)^9$ $-y^9$
35. The fourth term of the expansion of $(2r - s)^5$
36. The first term of the expansion of $(3q - 7r)^6$ $729q^6$
37. The third term of the expansion of $(x + y)^4$ $6x^2y^2$
38. The fourth term of the expansion of $(a + b)^8$ $56a^5b^3$
39. The second term of the expansion of $(a + 3b)^{10}$
40. The third term of the expansion of $(m + 5n)^7$
$525m^5n^2$

Review Exercises
For Exercises 41–46, see Appendix D.
Sketch the graph of each function. Decide whether each function is one-to-one. See Sections 8.5 and 10.1.

41. $f(x) = |x|$ 42. $g(x) = 3(x - 1)^2$
43. $H(x) = 2x + 3$ 44. $F(x) = -2$
45. $f(x) = x^2 + 3$ 46. $h(x) = -(x + 1)^2 - 4$

22. $b^6 + 6b^5c + 15b^4c^2 + 20b^3c^3 + 15b^2c^4 + 6bc^5 + c^6$
23. $1024a^5 + 1280a^4b + 640a^3b^2 + 160a^2b^3 + 20ab^4 + b^5$
24. $81m^4 + 108m^3n + 54m^2n^2 + 12mn^3 + n^4$ 25. $625a^4 - 1000a^3b + 600a^2b^2 - 160ab^3 + 16b^4$
26. $m^6 - 24m^5 + 240m^4 - 1280m^3 + 3840m^2 - 6144m + 4096$ 27. $8a^3 + 36a^2b + 54ab^2 + 27b^3$
28. $1024 - 3840x + 5760x^2 - 4320x^3 + 1620x^4 - 243x^5$ 29. $x^5 + 10x^4 + 40x^3 + 80x^2 + 80x + 32$
30. $81 + 216a + 216a^2 + 96a^3 + 16a^4$ 32. $-20x^3y^3$ 35. $-40r^2s^3$ 39. $30a^9b$

GROUP ACTIVITY

MODELING COLLEGE TUITION

MATERIALS:
• Newspapers, news magazines.

Annual tuitions at public and private universities alike are increasing at a rate of 6% per year. During the 1995–1996 academic year, the average tuition was $2860 at a public four-year university and $12,432 at a private four-year university. (*Source: USA Today*, February 20, 1996)

1. Find the general term of the sequence that describes the pattern of average annual tuition for public universities. Find the general term of the sequence that describes the pattern of average annual tuition for private universities. In each case let $n = 1$ represent the 1995–1996 academic year.

2. Assuming that the rate of tuition increase remains the same, find the annual tuition for the 1999–2000 academic year for each type of university.

3. Use partial sums to find the average cost of a four-year college education at both a public university and a private university for a student starting college in the 1997–1998 academic year.

4. Find the year in which tuition at a public university will reach $4300 per year.

5. Use newspapers and/or news magazines to find a situation that can be modeled by a sequence. Briefly describe the situation and write several reasonable questions about the situation. Exchange your problem with another member of your group to solve. Compare your answers.

See Appendix D for Group Activity answers and suggestions.

CHAPTER 11   HIGHLIGHTS

DEFINITIONS AND CONCEPTS	EXAMPLES
SECTION 11.1 SEQUENCES	
An **infinite sequence** is a function whose domain is the set of natural numbers $\{1, 2, 3, 4, \ldots\}$.	Infinite sequence: $2, 4, 6, 8, 10, \ldots$
A **finite sequence** is a function whose domain is the set of natural numbers $\{1, 2, 3, 4, \ldots, n\}$, where n is some natural number.	Finite sequence: $1, -2, 3, -4, 5, -6$
	(continued)

DEFINITIONS AND CONCEPTS	EXAMPLES

SECTION 11.1   SEQUENCES

The notation a_n, where n is a natural number, is used to denote a sequence.

Write the first four terms of the sequence whose general term is $a_n = n^2 + 1$:

$$a_1 = 1^2 + 1 = 2$$
$$a_2 = 2^2 + 1 = 5$$
$$a_3 = 3^2 + 1 = 10$$
$$a_4 = 4^2 + 1 = 17$$

SECTION 11.2   ARITHMETIC AND GEOMETRIC SEQUENCES

An **arithmetic sequence** is a sequence in which each term differs from the preceding term by a constant amount d, called the **common difference.**

Arithmetic sequence:

$$5, 8, 11, 14, 17, 20, \ldots$$

Here, $a_1 = 5$ and $d = 3$.

The **general term** a_n of an arithmetic sequence is given by

$$a_n = a_1 + (n - 1)d$$

where a_1 is the first term and d is the common difference.

The general term is

$$a_n = a_1 + (n - 1)d \text{ or}$$
$$a_n = 5 + (n - 1)3$$

A **geometric sequence** is a sequence in which each term is obtained by multiplying the preceding term by a constant r, called the **common ratio.**

Geometric sequence:

$$12, -6, 3, -\frac{3}{2}, \ldots$$

Here $a_1 = 12$ and $r = -\frac{1}{2}$.

The **general term** a_n of a geometric sequence is given by

$$a_n = a_1 r^{n-1}$$

where a_1 is the first term and r is the common ratio.

The general term is

$$a_n = a_1 r^{n-1} \text{ or}$$
$$a_n = 12\left(-\frac{1}{2}\right)^{n-1}$$

SECTION 11.3   SERIES

A sum of the terms of a sequence is called a **series.**

A shorthand notation for denoting a series is called **summation notation:**

$$\underset{i=1}{\overset{4}{\sum}}$$

index of summation → $\;$ → Greek letter sigma used to mean sum

SEQUENCE	SERIES	
3, 7, 11, 15	3 + 7 + 11 + 15	finite
3, 7, 11, 15, . . .	3 + 7 + 11 + 15 + $\cdots$	infinite

$$\sum_{i=1}^{4} 3^i = 3^1 + 3^2 + 3^3 + 3^4$$
$$= 3 + 9 + 27 + 81$$
$$= 120$$

DEFINITIONS AND CONCEPTS	EXAMPLES

SECTION 11.4 PARTIAL SUMS OF ARITHMETIC AND GEOMETRIC SEQUENCES

Partial sum, S_n, of the first n terms of an arithmetic sequence:

$$S_n = \frac{n}{2}(a_1 + a_n)$$

where a_1 is the first term and a_n is the nth term.

The sum of the first five terms of the arithmetic sequence

$$12, 24, 36, 48, 60, \ldots \text{ is}$$

$$S_n = \frac{5}{2}(12 + 60) = 180$$

Partial sum, S_n, of the first n terms of a geometric sequence:

$$S_n = \frac{a_1(1 - r^n)}{1 - r}$$

where a_1 is the first term, r is the common ratio, and $r \neq 1$.

The sum of the first five terms of the geometric sequence

$$15, 30, 60, 120, 240, \ldots \text{ is}$$

$$S_5 = \frac{15(1 - 2^5)}{1 - 2} = 465$$

Sum of the terms of an infinite geometric sequence:

$$S_\infty = \frac{a_1}{1 - r}$$

where a_1 is the first term, r is the common ratio, and $|r| < 1$. (If $|r| \geq 1$, S_∞ does not exist.)

The sum of the terms of the infinite geometric sequence

$$1, \frac{1}{3}, \frac{1}{9}, \frac{1}{27}, \ldots \text{ is}$$

$$S_\infty = \frac{1}{1 - \frac{1}{3}} = \frac{3}{2}$$

SECTION 11.5 THE BINOMIAL THEOREM

The **factorial of n**, written $n!$, is the product of the first n consecutive natural numbers.

Binomial theorem:

If n is a positive integer, then

$$(a + b)^n = a^n + \frac{n}{1!}a^{n-1}b^1 + \frac{n(n-1)}{2!}a^{n-2}b^2$$

$$+ \frac{n(n-1)(n-2)}{3!}a^{n-3}b^3 + \cdots + b^n$$

$$5! = 5 \cdot 4 \cdot 3 \cdot 2 \cdot 1 = 120$$

Expand $(3x + y)^4$.

$$(3x + y)^4 = (3x)^4 + \frac{4}{1!}(3x)^3(y)^1$$

$$+ \frac{4 \cdot 3}{2!}(3x)^2(y)^2 + \frac{4 \cdot 3 \cdot 2}{3!}(3x)^1y^3 + y^4$$

$$= 81x^4 + 108x^3y + 54x^2y^2 + 12xy^3 + y^4$$

CHAPTER 11 REVIEW

(11.1) *Find the indicated term(s) of the given sequence.*

1. The first five terms of the sequence $a_n = -3n^2$.

2. The first five terms of the sequence $a_n = n^2 + 2n$.

3. The one-hundredth term of the sequence $a_n = \frac{(-1)^n}{100} \cdot \frac{1}{100}$

4. The fiftieth term of the sequence $a_n = \frac{2n}{(-1)^2}$. 100

1. $-3, -12, -27, -48, -75$ **2.** $3, 8, 15, 24, 35$

5. $a_n = \dfrac{1}{6n}$

5. The general term a_n of the sequence $\dfrac{1}{6}, \dfrac{1}{12}, \dfrac{1}{18}, \dots$.

6. The general term a_n of the sequence $-1, 4, -9, 16, \dots$.
$$a_n = (-1)^n n^2$$

Solve the following applications.

7. The distance in feet that an olive falling from rest in a vacuum will travel during each second is given by an arithmetic sequence whose general term is $a_n = 32n - 16$, where n is the number of the second. Find the distance the olive will fall during the fifth, sixth, and seventh seconds.   144 ft, 176 ft, 208 ft

8. A culture of yeast doubles every day in a geometric progression whose general term is $a_n = 100(2)^{n-1}$, where n is the number of the day just beginning. Find how many days it takes the yeast culture to measure at least 10,000. Find the original measure of the yeast culture.   8 days; 100

9. The Centers for Disease Control and Prevention (CDC) reported that a new type of virus infected approximately 450 people during 1995, the year it was first discovered. The CDC predicts that during the next decade the virus will infect three times as many people each year as the year before. Write out the first five terms of this geometric sequence, and predict the number of infected people there will be in 1999.

10. The first row of an amphitheater contains 50 seats, and each row thereafter contains 8 additional seats. Write the first ten terms of this arithmetic progression, and find the number of seats in the tenth row.

(11.2)   **11.** $-2, -\dfrac{4}{3}, -\dfrac{8}{9}, -\dfrac{16}{27}, -\dfrac{32}{81}$

11. Find the first five terms of the geometric sequence whose first term is -2 and whose common ratio is $\dfrac{2}{3}$.

12. Find the first five terms for the arithmetic sequence whose first term is 12 and whose common difference is -1.5.   12, 10.5, 9, 7.5, 6

13. Find the thirtieth term of the arithmetic sequence whose first term is -5 and whose common difference is 4.   111

14. Find the eleventh term of the arithmetic sequence whose first term is 2 and whose common difference is $\dfrac{3}{4}$. $\dfrac{19}{2}$

15. Find the twentieth term of the arithmetic sequence whose first three terms are 12, 7, and 2. -83

16. Find the sixth term of the geometric sequence whose first three terms are 4, 6, and 9. $\dfrac{243}{8}$

17. If the fourth term of an arithmetic sequence is 18 and the twentieth term is 98, find the first term and the common difference. $a_1 = 3; d = 5$

18. If the third term of a geometric sequence is -48 and the fourth term is 192, find the first term and the common ratio. $a_1 = -3; r = -4$

19. Find the general term of the sequence $\dfrac{3}{10}, \dfrac{3}{100}, \dfrac{3}{1000}, \dots$.

20. Find a general term that satisfies the terms shown for the sequence 50, 58, 66, $\dots$. $a_n = 50 + (n-1)(8)$

Determine whether each of the following sequences is arithmetic, geometric, or neither. If a sequence is arithmetic, find a_1 and d. If a sequence is geometric, find a_1 and r.

21. $\dfrac{8}{3}, 4, 6, \dots$ $a_1 = \dfrac{8}{3}, r = \dfrac{3}{2}$ **22.** $-10.5, -6.1, -1.7$

23. $7x, -14x, 28x$ **24.** $3x^2, 9x^4, 81x^8, \dots$ neither
$a_1 = 7x, r = -2$

Solve the following applications.

25. To test the bounce of a racquetball, the ball is dropped from a height of 8 feet. The ball is judged "good" if it rebounds at least 75% of its previous height with each bounce. Write out the first six terms of this geometric sequence (round to the nearest tenth). Determine if a ball is "good" that rebounds to a height of 2.5 feet after the fifth bounce.

26. A display of oil cans in an auto parts store has 25 cans in the bottom row, 21 cans in the next row, and so on, in an arithmetic progression. Find the general term and the number of cans in the top row.

27. Suppose that you save $1 the first day of a month, $2 the second day, $4 the third day, continuing to double your savings each day. Write the general term of this geometric sequence and find the amount you will save on the tenth day. Estimate the amount you will save on the thirtieth day of the month, and check your estimate with a calculator.

28. On the first swing, the length of an arc through which a pendulum swings is 30 inches. The length of the arc for each successive swing is 70% of the preceding swing. Find the length of the arc for the fifth swing.   7.203 in.

29. Rosa takes a job that has a monthly starting salary of $900 and guarantees her a monthly raise of $150 during her 6-month training period. Find the general term of this sequence and her salary at the end of her training. $a_n = 150n + 750;$ $1650/month

9. 450, 1350, 4050, 12,150, 36,450; 36,450 infected people in 1999

10. 50, 58, 66, 74, 82, 90, 98, 106, 114, 122; 122 seats   **19.** $a_n = \dfrac{3}{10^n}$ **22.** $a_1 = -10.5, d = 4.4$

25. 8, 6, 4.5, 3.4, 2.5, 1.9; good   **26.** $a_n = 25 + (n-1)(-4);$ 1 can   **27.** $a_n = 2^{n-1},$ $512, $536,870,912

30. A sheet of paper is $\frac{1}{512}$-inch thick. By folding the sheet in half, the total thickness will be $\frac{1}{256}$ inch: A second fold produces a total thickness of $\frac{1}{128}$ inch. Estimate the thickness of the stack after 15 folds, and then check your estimate with a calculator. 64 in.

(11.3) *Write out the terms and find the sum for each of the following.* **31.** $1 + 3 + 5 + 7 + 9 = 25$

31. $\displaystyle\sum_{i=1}^{5} 2i - 1$

32. $\displaystyle\sum_{i=1}^{5} i(i + 2)$

33. $\displaystyle\sum_{i=2}^{4} \frac{(-1)^i}{2i}$

34. $\displaystyle\sum_{i=3}^{5} 5(-1)^{i-1}$ $5 - 5 + 5 = 5$

Find the partial sum of the given sequence.

35. S_4 of the sequence $a_n = (n - 3)(n + 2)$ -4

36. S_6 of the sequence $a_n = n^2$ 91

37. S_5 of the sequence $a_n = -8 + (n - 1)3$ -10

38. S_3 of the sequence $a_n = 5(4)^{n-1}$ 105

Write the sum with Σ notation.

39. $1 + 3 + 9 + 27 + 81 + 243$ $\displaystyle\sum_{i=1}^{6} 3^{i-1}$

40. $6 + 2 + (-2) + (-6) + (-10) + (-14) + (-18)$

41. $\dfrac{1}{4} + \dfrac{1}{16} + \dfrac{1}{64} + \dfrac{1}{256}$ $\displaystyle\sum_{i=1}^{4} \frac{1}{4^i}$

42. $1 + \left(-\dfrac{3}{2}\right) + \dfrac{9}{4}$ $\displaystyle\sum_{i=1}^{3} \left(\frac{3}{-2}\right)^{i-1}$

43. $a_n = 20(2)^n$; n represents the number of *Solve.* 8-hour periods; 1280 yeast

43. A yeast colony begins with 20 yeast and doubles every 8 hours. Write the sequence that describes the growth of the yeast, and find the total yeast after 48 hours.

44. The number of cranes born each year in a new aviary forms a sequence whose general term is $a_n = n^2 + 2n - 1$. Find the number of cranes born in the fourth year and the total number of cranes born in the first four years. 23 cranes; 46 cranes

45. Harold has a choice between two job offers. Job A has an annual starting salary of \$19,500 with guaranteed annual raises of \$1100 for the next four years, whereas job B has an annual starting salary of \$21,000 with guaranteed annual raises of \$700 for the next four years. Compare the salaries for the fifth year under each job offer.

46. A sample of radioactive waste is decaying such that the amount decaying in kilograms during year n is $a_n = 200(0.5)^n$. Find the amount of decay in the third year, and the total amount of decay in the first three years. 25 kg; 175 kg

(11.4) *Find the partial sum of the given sequence.*

47. The sixth partial sum of the sequence $15, 19, 23, \ldots$.

48. The ninth partial sum of the sequence $5, -10, 20, \ldots$.

49. The sum of the first 30 odd positive integers. 900

50. The sum of the first 20 positive multiples of 7. 1470

51. The sum of the first 20 terms of the sequence $8, 5, 2, \ldots$. -410

52. The sum of the first eight terms of the sequence $\dfrac{3}{4}, \dfrac{9}{4}, \dfrac{27}{4}, \ldots$. 2460

53. S_4 if $a_1 = 6$ and $r = 5$. 936

54. S_{100} if $a_1 = -3$ and $d = -6$. $-30{,}000$

Find the sum of each infinite geometric sequence.

55. $5, \dfrac{5}{2}, \dfrac{5}{4}, \ldots$ 10

56. $18, -2, \dfrac{2}{9}, \ldots$ $\dfrac{81}{5}$

57. $-20, -4, -\dfrac{4}{5}, \ldots$ -25

58. $0.2, 0.02, 0.002, \ldots$ $\dfrac{2}{9}$

Solve. **59.** \$30,418; \$99,868

59. A frozen-yogurt store owner cleared \$20,000 the first year he owned his business and made an additional 15% over the previous year in each subsequent year. Find how much he made during his fourth year of business. Find his total earnings during the first 4 years (round to the nearest dollar).

60. On his first morning in a television assembly factory, a trainee takes 0.8 times as long to assemble each television as he took to assemble the one before. If it took him 40 minutes to assemble the first television, find how long it takes him to assemble the fourth television. Find the total time he takes to assemble the first four televisions (round to the nearest minute). 20 min.; 118 min.

61. During the harvest season a farmer can rent a combine machine for \$100 the first day, with the rental fee decreasing \$7 for each additional day. Find how much the farmer pays for the rental on the seventh day. Find how much total rent the farmer pays for 7 days. \$58; \$553

32. $3 + 8 + 15 + 24 + 35 = 85$ **33.** $\dfrac{1}{4} - \dfrac{1}{6} + \dfrac{1}{8} = \dfrac{5}{24}$ **40.** $\displaystyle\sum_{i=1}^{7} 6 + (i - 1)(-4)$

45. Job A, \$23,900; Job B, \$23,800 **47.** 150 **48.** 855

63. 2696 mosquitoes   **67.** $x^5 + 5x^4z + 10x^3z^2 + 10x^2z^3 + 5xz^4 + z^5$

62. A rubber ball is dropped from a height of 15 feet and rebounds 80% of its previous height after each bounce. Find the total distance the ball travels before it comes to rest.   135 ft

63. After a pond was sprayed once with insecticide, 1800 mosquitoes were killed the first day, 600 the second day, 200 the third day, and so on. Find the total number of mosquitoes killed during the first 6 days after the spraying (round to the nearest unit).

64. See Exercise 63. Find the day on which the insecticide is no longer effective, and find the total number of mosquitoes killed (round to the nearest mosquito).   eighth day; 2700 mosquitoes

65. Use the formula S_∞ to write $0.5\overline{5}$ as a fraction. $\dfrac{5}{9}$

66. A movie theater has 27 seats in the first row, 30 seats in the second row, 33 seats in the third row,

and so on. Find the total number of seats in the theater if there are 20 rows.   1110 seats

(11.5)   Use Pascal's triangle to expand the binomial.

67. $(x + z)^5$ **68.** $(y - r)^6$

69. $(2x + y)^4$ $16x^4 + 32x^3y + 24x^2y^2 + 8xy^3 + y^4$

70. $(3y - z)^4$ $81y^4 - 108y^3z + 54y^2z^2 - 12yz^3 + z^4$

Use the binomial formula to expand the following.

71. $(b + c)^8$ **72.** $(x - w)^7$

73. $(4m - n)^4$ **74.** $(p - 2r)^5$

Find the indicated term.   **75.** $35a^4b^3$ **76.** $1024z^{10}$

75. The fourth term of the expansion of $(a + b)^7$.

76. The eleventh term of the expansion of $(y + 2z)^{10}$.

68. $y^6 - 6y^5r + 15y^4r^2 - 20y^3r^3 + 15y^2r^4 - 6yr^5 + r^6$

71. $b^8 + 8b^7c + 28b^6c^2 + 56b^5c^3 + 70b^4c^4 + 56b^3c^5 + 28b^2c^6 + 8bc^7 + c^8$

72. $x^7 - 7x^6w + 21x^5w^2 - 35x^4w^3 + 35x^3w^4 - 21x^2w^5 + 7xw^6 - w^7$

73. $256m^4 - 256m^3n + 96m^2n^2 - 16mn^3 + n^4$

74. $p^5 - 10p^4r + 40p^3r^2 - 80p^2r^3 + 80pr^4 - 32r^5$

CHAPTER 11 TEST

Find the indicated term(s) of the given sequence.

1. The first five terms of the sequence $a_n = \dfrac{(-1)^n}{n + 4}$.

2. The first five terms of the sequence $a_n = \dfrac{3}{(-1)^n}$. $-3, 3, -3, 3, -3$

3. The eightieth term of the sequence $a_n = 10 + 3(n - 1)$.   247

4. The two-hundredth term of the sequence $a_n = (n + 1)(n - 1)(-1)^n$.   39,999

5. The general term of the sequence $\dfrac{2}{5}, \dfrac{2}{25}, \dfrac{2}{125}, \ldots$ $a_n = \dfrac{2}{5}\left(\dfrac{1}{5}\right)^{n-1}$

6. The general term of the sequence $-9, 18, -27, 36, \ldots$ $(-1)^n 9n$

Find the partial sum of the given sequence.

7. S_5 of the sequence $a_n = 5(2)^{n-1}$ 155

8. S_{30} of the sequence $a_n = 18 + (n - 1)(-2)$ -330

9. S_∞ of the sequence $a_1 = 24$ and $r = \dfrac{1}{6}$ $\dfrac{144}{5}$

10. S_∞ of the sequence $\dfrac{3}{2}, -\dfrac{3}{4}, \dfrac{3}{8}, \ldots$ 1

11. $\displaystyle\sum_{i=1}^{4} i(i - 2)$ 10   **12.** $\displaystyle\sum_{i=2}^{4} 5(2)^i(-1)^{i-1}$ -60

1. $-\dfrac{1}{5}, \dfrac{1}{6}, -\dfrac{1}{7}, \dfrac{1}{8}, -\dfrac{1}{9}$ **13.** $a^6 - 6a^5b + 15a^4b^2 - 20a^3b^3 + 15a^2b^4 - 6ab^5 + b^6$

Expand the binomial by using Pascal's triangle.

13. $(a - b)^6$ **14.** $(2x + y)^5$

14. $32x^5 + 80x^4y + 80x^3y^2 + 40x^2y^3 + 10xy^4 + y^5$

Expand the binomial by using the binomial formula.

15. $(y + z)^8$ **16.** $(2p + r)^7$

Solve the following applications.

17. The population of a small town is growing yearly according to the sequence defined by $a_n = 250 + 75(n - 1)$, where n is the number of the year just beginning. Predict the population at the beginning of the tenth year. Find the town's initial population. 925 people; 250 people initially

18. A gardener is making a triangular planting with one shrub in the first row, three shrubs in the second row, five shrubs in the third row, and so on, for eight rows. Write the finite series of this sequence, and find the total number of shrubs planted.

19. A pendulum swings through an arc of length 80 centimeters on its first swing. On each successive swing, the length of the arc is $\dfrac{3}{4}$ the length of the arc on the preceding swing. Find the length of the arc on the fourth swing, and find the total arc lengths for the first four swings. 33.75 cm, 218.75 cm

15. $y^8 + 8y^7z + 28y^6z^2 + 56y^5z^3 + 70y^4z^4 + 56y^3z^5 + 28y^2z^6 + 8yz^7 + z^8$

16. $128p^7 + 448p^6r + 672p^5r^2 + 560p^4r^3 + 280p^3r^4 + 84p^2r^5 + 14pr^6 + r^7$

18. $1 + 3 + 5 + 7 + 9 + 11 + 13 + 15$; 64 shrubs

20. See Exercise 19. Find the total arc lengths before the pendulum comes to rest.   320 cm

21. A parachutist in free fall falls 16 feet during the first second, 48 feet during the second second, 80 feet

during the third second, and so on. Find how far he falls during the tenth second. Find the total distance he falls during the first 10 seconds.   304 ft; 1600 ft

22. Use the formula S_∞ to write $0.42\overline{42}$ as a fraction. $\frac{14}{33}$

CHAPTER 11   CUMULATIVE REVIEW

1. Solve for c: $25 - 3.5c = 6$. $\left[\frac{38}{7}\right]$; Sec. 2.1, Ex. 2

2. Solve $|x| \leq 3$. $[-3, 3]$; Sec. 2.7, Ex. 1

3. Which of the following relations are also functions?

a. $\{(-2, 5), (2, 7), (-3, 5), (9, 9)\}$ function

b.   not a function

    (6, 6)
    (1, 5)
    (0, 3)
    (-3, 2)
    (0, -2)

c.   Input: Correspondence: Output:

People in a   Each person's   The set of
certain city   age   nonnegative
function; Sec. 3.2, Ex. 2   integers

4. Write a function that describes the line that contains the point $(4, 4)$ and is perpendicular to the line $2x + 3y = -6$.

5. A small manufacturing company manufactures and sells compact disc storage units. The revenue equation for these units is

$$y = 50x$$

where x is the number of units sold and y is the revenue, or income, for selling x units. The cost equation for these units is

$$y = 30x + 10,000$$

500 units to break even; Sec. 4.1, Ex. 8

where x is the number of units sold and y is the total cost of manufacturing x units. Use these equations to find the number of units to be sold for the company to break even.

6. The first number is 4 less than a second number.

8. $12z^5 + 3z^4 - 13z^3 - 11z$; Sec. 5.3, Ex. 9

10. $\dfrac{7x^2 - 9x - 13}{(2x + 1)(x - 5)(3x - 2)}$; Sec. 6.3, Ex. 5

11. $3x^2 + 2x + 3 + \dfrac{-6x + 9}{x^2 - 1}$; Sec. 6.5, Ex. 6

4. $f(x) = \dfrac{3}{2}x - 2$; Sec. 3.5, Ex. 8

6. The numbers are 7 and 11; Sec. 4.3, Ex. 1

Four times the first number is 6 more than twice the second. Find the numbers.

7. Write the following with only positive exponents. Simplify if possible.   Sec. 5.1, Ex. 5

a. 5^{-2} $\dfrac{1}{25}$ **b.** $2x^{-3}$ $\dfrac{2}{x^3}$ **c.** $(3x)^{-1}$ $\dfrac{1}{3x}$ **d.** $\dfrac{m^5}{m^{15}}$ $\dfrac{1}{m^{10}}$

e. $\dfrac{3^3}{3^6}$ $\dfrac{1}{27}$ **f.** $2^{-1} + 3^{-2}$ $\dfrac{11}{18}$ **g.** $\dfrac{1}{t^{-5}}$ t^5

8. Subtract: $(12z^5 - 12z^3 + z) - (-3z^4 + z^3 + 12z)$.

9. Factor $y^3 - 64$. $(y - 4)(y^2 + 4y + 16)$; Sec. 5.7, Ex. 7

10. Add:

$$\frac{2x - 1}{2x^2 - 9x - 5} + \frac{x + 3}{6x^2 - x - 2}$$

11. Divide $2x^3 + 3x^4 - 8x + 6$ by $x^2 - 1$.

12. Use the quotient rule to simplify.   Sec. 7.3, Ex. 2

a. $\sqrt{\dfrac{25}{49}}$ $\dfrac{5}{7}$ **b.** $\sqrt[3]{\dfrac{8}{27}}$ $\dfrac{2}{3}$

c. $\sqrt{\dfrac{x}{9}}$ $\dfrac{\sqrt{x}}{3}$ **d.** $\sqrt[4]{\dfrac{3}{16x^4}}$ $\dfrac{\sqrt[4]{3}}{2y}$

13. Solve $\sqrt{-10x - 1} + 3x = 0$ for x.

14. Solve $3x^2 + 16x + 5 = 0$ for x.

15. Graph $f(x) = 3x^2 + 3x + 1$. Find the vertex and any intercepts.   See Appendix D; Sec. 8.6, Ex. 2

16. Graph $\dfrac{x^2}{9} + \dfrac{y^2}{16} = 1$. See Appendix D; Sec. 9.2, Ex. 1

17. Solve the system

$$\begin{cases} x^2 + 2y^2 = 10 \\ x^2 - y^2 = 1 \end{cases}$$

$\{(2, \sqrt{3}), (-2, \sqrt{3}), (2, -\sqrt{3}), (-2, -\sqrt{3})\}$; Sec. 9.3, Ex. 3

18. Determine whether each function described is one-to-one.

a. $f = \{(6, 2), (5, 4), (-1, 0), (7, 3)\}$ one-to-one

b. $g = \{(3, 9), (-4, 2), (-3, 9), (0, 0)\}$ not one-to-one

c. $h = \{(1, 1), (2, 2), (10, 10), (-5, -5)\}$ one-to-one

13. $\{-\frac{1}{9}, -1\}$; Sec. 7.6, Ex. 2   **14.** $\{-\frac{1}{3}, -5\}$; Sec. 8.2, Ex. 1

d.

MINERAL (INPUT)	HARDNESS ON THE MOHS SCALE (OUTPUT)
Talc	1
Gypsum	2
Diamond	10
Topaz	8
Stibnite	2

not one-to-one

e.

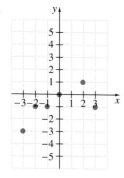

not one-to-one; Sec. 10.1, Ex. 3

19. Find the amount owed at the end of 5 years if $1600 is loaned at a rate of 9% compounded monthly. $2505.09; *Sec. 10.2, Ex. 5*

20. Solve each equation for x.

a. $\log_4 \frac{1}{4} = x$ $\{-1\}$ **b.** $\log_5 x = 3$ $\{125\}$

c. $\log_x 25 = 2$ $\{5\}$ **d.** $\log_3 1 = x$ $\{0\}$

e. $\log_b 1 = x$ $\{0\}$; *Sec. 10.3, Ex. 2*

21. Solve $\log (x + 2) - \log x = 2$ for x.

22. If the general term of a sequence is given by $a_n = \frac{(-1)^n}{3n}$, find:

a. the first term of the sequence $-\dfrac{1}{3}$

b. a_8 $\dfrac{1}{24}$

c. the one-hundredth term of the sequence $\dfrac{1}{300}$

d. a_{15} $-\dfrac{1}{45}$; *Sec. 11.1, Ex. 2*

23. Donna has an offer for a job starting at $20,000 per year and guaranteeing her a raise of $800 per year for the next 5 years. Write the general term for the arithmetic sequence that models Donna's potential annual salaries, and find her salary for the fourth year.

24. Use the binomial theorem to expand $(x + 2y)^5$.

21. $\{\frac{2}{99}\}$; *Sec. 10.6, Ex. 4* **23.** $a_n = 20,000 + (n - 1)(800)$, $22,400$; *Sec. 11.2, Ex. 5*

24. $x^5 + 10x^4y + 40x^3y^2 + 80x^2y^3 + 80xy^4 + 32y^5$; *Sec. 11.5, Ex. 4*

REVIEW OF ANGLES, LINES, AND SPECIAL TRIANGLES

The word **geometry** is formed from the Greek words, **geo,** meaning earth, and **metron,** meaning measure. Geometry literally means to measure the earth.

This section contains a review of some basic geometric ideas. It will be assumed that fundamental ideas of geometry such as point, line, ray, and angle are known. In this appendix, the notation $\angle 1$ is read "angle 1" and the notation $m\angle 1$ is read "the measure of angle 1."

We first review types of angles.

ANGLES

A **right angle** is an angle whose measure is 90°. A right angle can be indicated by a square drawn at the vertex of the angle, as shown below.

An angle whose measure is more than 0° but less than 90° is called an **acute angle**.

An angle whose measure is greater than 90° but less than 180° is called an **obtuse angle**.

An angle whose measure is 180° is called a **straight angle**.

Two angles are said to be **complementary** if the sum of their measures is 90°. Each angle is called the **complement** of the other.

Two angles are said to be **supplementary** if the sum of their measures is 180°. Each angle is called the **supplement** of the other.

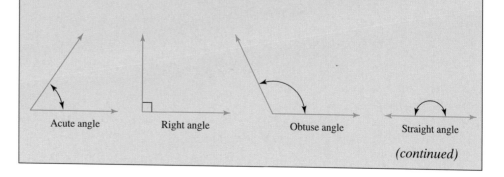

Acute angle   Right angle   Obtuse angle   Straight angle

(continued)

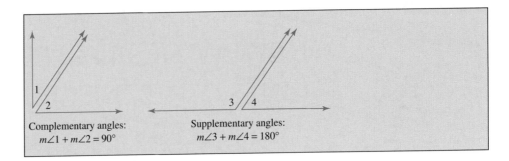

Complementary angles:
$m\angle 1 + m\angle 2 = 90°$

Supplementary angles:
$m\angle 3 + m\angle 4 = 180°$

EXAMPLE 1 If an angle measures 28°, find its complement.

Solution: Two angles are complementary if the sum of their measures is 90°. The complement of a 28° angle is an angle whose measure is $90° - 28° = 62°$. To check, notice that $28° + 62° = 90°$.

Plane is an undefined term that we will describe. A plane can be thought of as a flat surface with infinite length and width, but no thickness. A plane is two dimensional. The arrows in the following diagram indicate that a plane extends indefinitely and has no boundaries.

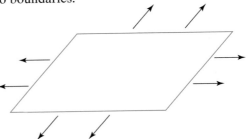

Figures that lie on a plane are called **plane figures.** (See the description of common plane figures in Appendix B.) Lines that lie in the same plane are called **coplanar.**

LINES

Two lines are **parallel** if they lie in the same plane but never meet.

Intersecting lines meet or cross in one point.

Two lines that form right angles when they intersect are said to be **perpendicular.**

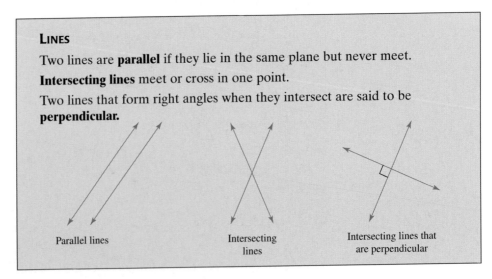

Parallel lines

Intersecting lines

Intersecting lines that are perpendicular

Two intersecting lines form **vertical angles.** Angles 1 and 3 are vertical angles. Also angles 2 and 4 are vertical angles. It can be shown that **vertical angles have equal measures.**

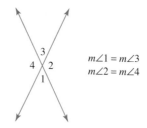

$$m\angle 1 = m\angle 3$$
$$m\angle 2 = m\angle 4$$

Adjacent angles have the same vertex and share a side. Angles 1 and 2 are adjacent angles. Other pairs of adjacent angles are angles 2 and 3, angles 3 and 4, and angles 4 and 1.

A **transversal** is a line that intersects two or more lines in the same plane. Line l is a transversal that intersects lines m and n. The eight angles formed are numbered and certain pairs of these angles are given special names.

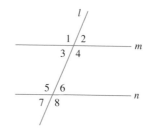

Corresponding angles: $\angle 1$ and $\angle 5$, $\angle 3$ and $\angle 7$, $\angle 2$ and $\angle 6$, and $\angle 4$ and $\angle 8$.

Exterior angles: $\angle 1$, $\angle 2$, $\angle 7$, and $\angle 8$.

Interior angles: $\angle 3$, $\angle 4$, $\angle 5$, and $\angle 6$.

Alternate interior angles: $\angle 3$ and $\angle 6$, $\angle 4$ and $\angle 5$.

These angles and parallel lines are related in the following manner.

PARALLEL LINES CUT BY A TRANSVERSAL

1. If two parallel lines are cut by a transversal, then
   a. **corresponding angles are equal** and
   b. **alternate interior angles are equal.**

2. If corresponding angles formed by two lines and a transversal are equal, then the lines are parallel.

3. If alternate interior angles formed by two lines and a transversal are equal, then the lines are parallel.

EXAMPLE 2   Given that lines m and n are parallel and that the measure of angle 1 is $100°$, find the measures of angles 2, 3, and 4.

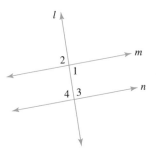

Solution: $m\angle 2 = 100°,$ since angles 1 and 2 are vertical angles.

$m\angle 4 = 100°,$ since angles 1 and 4 are alternate interior angles.

$m\angle 3 = 180° - 100° = 80°,$ since angles 4 and 3 are supplementary angles.

A **polygon** is the union of three or more coplanar line segments that intersect each other only at each end point, with each end point shared by exactly two segments.

A **triangle** is a polygon with three sides. The sum of the measures of the three angles of a triangle is $180°$. In the following figure, $m\angle 1 + m\angle 2 + m\angle 3 = 180°$.

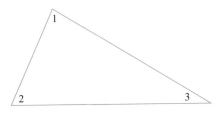

EXAMPLE 3   Find the measure of the third angle of the triangle shown.

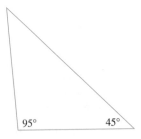

Solution: The sum of the measures of the angles of a triangle is $180°$. Since one angle measures $45°$ and the other angle measures $95°$, the third angle measures $180° - 45° - 95° = 40°$.

Two triangles are **congruent** if they have the same size and the same shape. In congruent triangles, the measures of corresponding angles are equal and the lengths of corresponding sides are equal. The following triangles are congruent.

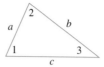 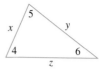

Corresponding angles are equal: $m\angle 1 = m\angle 4$, $m\angle 2 = m\angle 5$, and $m\angle 3 = m\angle 6$. Also, lengths of corresponding sides are equal: $a = x$, $b = y$, and $c = z$.

Any one of the following may be used to determine whether two triangles are congruent.

CONGRUENT TRIANGLES

1. If the measures of two angles of a triangle equal the measures of two angles of another triangle and the lengths of the sides between each pair of angles are equal, the triangles are congruent.

$m\angle 1 = m\angle 3$
$m\angle 2 = m\angle 4$
and
$a = x$

2. If the lengths of the three sides of a triangle equal the lengths of corresponding sides of another triangle, the triangles are congruent.

$a = x$
$b = y$
and
$c = z$

3. If the lengths of two sides of a triangle equal the lengths of corresponding sides of another triangle, and the measures of the angles between each pair of sides are equal, the triangles are congruent.

$a = x$
$b = y$
and
$m\angle 1 = m\angle 2$

Two triangles are **similar** if they have the same shape. In similar triangles, the measures of corresponding angles are equal and corresponding sides are in proportion. The following triangles are similar. (All similar triangles drawn in this appendix will be oriented the same.)

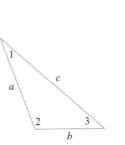
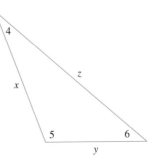

Corresponding angles are equal: $m\angle1 = m\angle4, m\angle2 = m\angle5,$ and $m\angle3 = m\angle6.$

Also, corresponding sides are proportional: $\dfrac{a}{x} = \dfrac{b}{y} = \dfrac{c}{z}.$

Any one of the following may be used to determine whether two triangles are similar.

SIMILAR TRIANGLES

1. If the measures of two angles of a triangle equal the measures of two angles of another triangle, the triangles are similar.

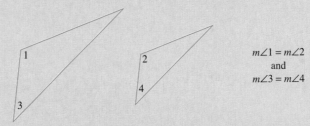

$$m\angle1 = m\angle2$$
and
$$m\angle3 = m\angle4$$

2. If three sides of one triangle are proportional to three sides of another triangle, the triangles are similar.

$$\frac{a}{x} = \frac{b}{y} = \frac{c}{z}$$

3. If two sides of a triangle are proportional to two sides of another triangle and the measures of the included angles are equal, the triangles are similar.

$$m\angle1 = m\angle2$$
and
$$\frac{a}{x} = \frac{b}{y}$$

EXAMPLE 4   Given that the following triangles are similar, find the missing length x.

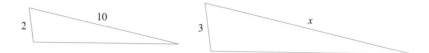

Solution: Since the triangles are similar, corresponding sides are in proportion. Thus, $\dfrac{2}{3} = \dfrac{10}{x}$. To solve this equation for x, we multiply both sides by the LCD $3x$.

$$3x\left(\frac{2}{3}\right) = 3x\left(\frac{10}{x}\right)$$
$$2x = 30$$
$$x = 15$$

The missing length is 15 units.

A **right triangle** contains a right angle. The side opposite the right angle is called the **hypotenuse,** and the other two sides are called the **legs.** The **Pythagorean theorem** gives a formula that relates the lengths of the three sides of a right triangle.

THE PYTHAGOREAN THEOREM

If a and b are the lengths of the legs of a right triangle, and c is the length of the hypotenuse, then $a^2 + b^2 = c^2$.

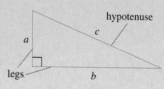

EXAMPLE 5   Find the length of the hypotenuse of a right triangle whose legs have lengths of 3 centimeters and 4 centimeters.

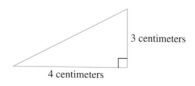

Solution: Because we have a right triangle, we use the Pythagorean theorem. The legs are 3 centimeters and 4 centimeters, so let $a = 3$ and $b = 4$ in the formula.

$$a^2 + b^2 = c^2$$
$$3^2 + 4^2 = c^2$$
$$9 + 16 = c^2$$
$$25 = c^2$$

Since c represents a length, we assume that c is positive. Thus, if c^2 is 25, c must be 5. The hypotenuse has a length of 5 centimeters.

APPENDIX A EXERCISE SET

Find the complement of each angle. See Example 1.

1. 19°   71°   **2.** 65°   25°   **3.** 70.8°   19.2°

4. $45\frac{2}{3}°$ $44\frac{1}{3}°$ **5.** $11\frac{1}{4}°$ $78\frac{3}{4}°$ **6.** 19.6°   70.4°

Find the supplement of each angle.

7. 150°   30°   **8.** 90°   90°   **9.** 30.2°   149.8°

10. 81.9°   98.1°   **11.** $79\frac{1}{2}°$ $100\frac{1}{2}°$ **12.** $165\frac{8}{9}°$ $14\frac{1}{9}°$

13. If lines m and n are parallel, find the measures of angles 1 through 7. See Example 2.

$$m\angle 1 = m\angle 5 = m\angle 7 = 110°$$
$$m\angle 2 = m\angle 3 = m\angle 4 = m\angle 6 = 70°$$

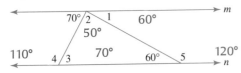

14. If lines m and n are parallel, find the measures of angles 1 through 5. See Example 2.

In each of the following, the measures of two angles of a triangle are given. Find the measure of the third angle. See Example 3.

15. 11°, 79°   90°   **16.** 8°, 102°   70°   **17.** 25°, 65°   90°

18. 44°, 19°   117°   **19.** 30°, 60°   90°   **20.** 67°, 23°   90°

In each of the following, the measure of one angle of a right triangle is given. Find the measures of the other two angles.

21. 45°   45°, 90°   **22.** 60°   30°, 90°   **23.** 17°   73°, 90°

24. 30°   60°, 90°   **25.** $39\frac{3}{4}°$ $50\frac{1}{4}°$, 90°   **26.** 72.6°   17.4°, 90°

Given that each of the following pairs of triangles is similar, find the missing lengths. See Example 4.

27.

$x = 6$

28.
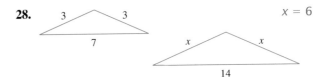
$x = 6$

29.
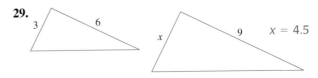
$x = 4.5$

30.
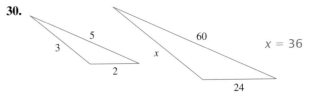
$x = 36$

Use the Pythagorean Theorem to find the missing lengths in the right triangles. See Example 5.

31.

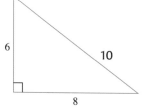

32.

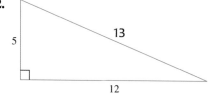

33.

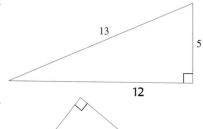

34.

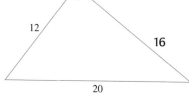

B | REVIEW OF GEOMETRIC FIGURES

PLANE FIGURES HAVE LENGTH AND WIDTH BUT NO THICKNESS OR DEPTH.		
NAME	**DESCRIPTION**	**FIGURE**
POLYGON	Union of three or more coplanar line segments that intersect with each other only at each end point, with each end point shared by two segments.	
TRIANGLE	Polygon with three sides (sum of measures of three angles is 180°).	
SCALENE TRIANGLE	Triangle with no sides of equal length.	
ISOSCELES TRIANGLE	Triangle with two sides of equal length.	
EQUILATERAL TRIANGLE	Triangle with all sides of equal length.	
RIGHT TRIANGLE	Triangle that contains a right angle.	leg, hypotenuse, leg
QUADRILATERAL	Polygon with four sides (sum of measures of four angles is 360°).	

PLANE FIGURES HAVE LENGTH AND WIDTH BUT NO THICKNESS OR DEPTH.		
NAME	**DESCRIPTION**	**FIGURE**
TRAPEZOID	Quadrilateral with exactly one pair of opposite sides parallel.	
ISOSCELES TRAPEZOID	Trapezoid with legs of equal length.	
PARALLELOGRAM	Quadrilateral with both pair of opposite sides parallel and equal in length.	
RHOMBUS	Parallelogram with all sides of equal length.	
RECTANGLE	Parallelogram with four right angles.	
SQUARE	Rectangle with all sides of equal length.	
CIRCLE	All points in a plane the same distance from a fixed point called the **center.**	

SOLID FIGURES HAVE LENGTH, WIDTH, AND HEIGHT OR DEPTH.		
NAME	**DESCRIPTION**	**FIGURE**
RECTANGULAR SOLID	A solid with six sides, all of which are rectangles.	
CUBE	A rectangular solid whose six sides are squares.	
SPHERE	All points the same distance from a fixed point, called the **center.**	radius / center
RIGHT CIRCULAR CYLINDER	A cylinder consisting of two circular bases that are perpendicular to its altitude.	
RIGHT CIRCULAR CONE	A cone with a circular base that is perpendicular to its altitude.	

C

AN INTRODUCTION TO USING A GRAPHING UTILITY

VIEWING WINDOW AND INTERPRETING WINDOW SETTINGS

In this appendix, we will use the term **graphing utility** to mean a graphing calculator or a computer software graphing package. All graphing utilities graph equations by plotting points on a screen. While plotting several points can be slow and sometimes tedious for us, a graphing utility can quickly and accurately plot hundreds of points. How does a graphing utility show plotted points? A computer or calculator screen is made up of a grid of small rectangular areas called **pixels.** If a pixel contains a point to be plotted, the pixel is turned "on"; otherwise, the pixel remains "off." The graph of an equation is then a collection of pixels turned "on." The graph of $y = 3x + 1$ from a graphing calculator is shown in Figure A-1. Notice the irregular shape of the line caused by the rectangular pixels.

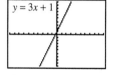

Figure A-1

The portion of the coordinate plane shown on the screen in Figure A-1 is called the **viewing window** or the **viewing rectangle.** Notice the x-axis and the y-axis on the graph. While tick marks are shown on the axes, they are not labeled. This means that from this screen alone, we do not know how many units each tick mark represents. To see what each tick mark represents and the minimum and maximum values on the axes, check the *window setting* of the graphing utility. It defines the viewing window. The window of the graph of $y = 3x + 1$ shown in Figure A-1 has the following setting (Figure A-2):

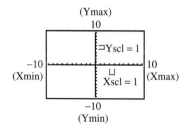

Figure A-2

$X\text{min} = -10$	The minimum x-value is -10.
$X\text{max} = 10$	The maximum x-value is 10.
$X\text{scl} = 1$	The x-axis scale is 1 unit per tick mark.
$Y\text{min} = -10$	The minimum y-value is -10.
$Y\text{max} = 10$	The maximum y-value is 10.
$Y\text{scl} = 1$	The y-axis scale is 1 unit per tick mark.

By knowing the scale, we can find the minimum and the maximum values on the axes simply by counting tick marks. For example, if both the $X\text{scl}$ (x-axis scale) and the $Y\text{scl}$ are 1 unit per tick mark on the graph in Figure A-3, we can count the tick marks and find that the minimum x-value is -10 and the maximum x-value is 10. Also, the minimum y-value is -10 and the maximum y-value is 10. If the $X\text{scl}$ (x-axis scale) changes to 2 units per tick mark (shown in Figure A-4), by counting tick marks, we see that the minimum x-value is now -20 and the maximum x-value is now 20.

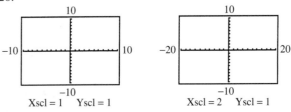

Figure A-3   Figure A-4

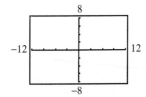

Figure A-5

It is also true that if we know the Xmin and the Xmax values, we can calculate the Xscl by the displayed axes. For example, the Xscl of the graph in Figure A-5 must be 3 units per tick mark for the maximum and minimum x-values to be as shown. Also, the Yscl of that graph must be 2 units per tick mark for the maximum and minimum y-values to be as shown.

We will call the viewing window in Figure A-3 a *standard* viewing window or rectangle. Although a standard viewing window is sufficient for much of this text, special care must be taken to ensure that all key features of a graph are shown. Figures A-6, A-7, and A-8 show the graph of $y = x^2 + 11x - 1$ on three different viewing windows. Note that certain viewing windows for this equation are misleading.

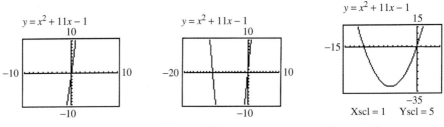

Figure A-6   Figure A-7   Figure A-8

How do we ensure that all distinguishing features of the graph of an equation are shown? It helps to know about the equation that is being graphed. For example, the equation $y = x^2 + 11x - 1$ is not a linear equation, and its graph is not a line. This equation is a quadratic equation, and therefore its graph is a parabola. By knowing this information, we know that the graph shown in Figure A-6, although correct, is misleading. Of the three viewing rectangles shown, the graph in Figure A-8 is best because it shows more of the distinguishing features of the parabola. Properties of equations needed for graphing will be studied in this text.

VIEWING WINDOW AND INTERPRETING WINDOW SETTINGS EXERCISE SET

In Exercises 1–4, determine whether all ordered pairs listed will lie within a standard viewing rectangle.

1. $(-9, 0), (5, 8), (1, -8)$ yes

2. $(4, 7), (0, 0), (-8, 9)$ yes

3. $(-11, 0), (2, 2), (7, -5)$ no

4. $(3, 5), (-3, -5), (15, 0)$ no

In Exercises 5–10, choose an Xmin, Xmax, Ymin, and Ymax so that all ordered pairs listed will lie within the viewing rectangle.

For Exercises 5–10, answers will vary.

5. $(-90, 0), (55, 80), (0, -80)$

6. $(4, 70), (20, 20), (-18, 90)$

7. $(-11, 0), (2, 2), (7, -5)$

8. $(3, 5), (-3, -5), (15, 0)$

9. $(200, 200), (50, -50), (70, -50)$

10. $(40, 800), (-30, 500), (15, 0)$

Write the window setting for each viewing window shown. Use the following format:

$$\begin{array}{ll} X\min = & Y\min = \\ X\max = & Y\max = \\ X\text{scl} = & Y\text{scl} = \end{array}$$

11.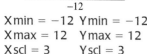

$\begin{array}{ll} X\min = -12 & Y\min = -12 \\ X\max = 12 & Y\max = 12 \\ X\text{scl} = 3 & Y\text{scl} = 3 \end{array}$

12.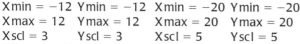

$\begin{array}{ll} X\min = -20 & Y\min = -20 \\ X\max = 20 & Y\max = 20 \\ X\text{scl} = 5 & Y\text{scl} = 5 \end{array}$

13.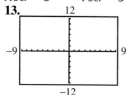

$\begin{array}{ll} X\min = -9 & Y\min = -12 \\ X\max = 9 & Y\max = 12 \\ X\text{scl} = 1 & Y\text{scl} = 2 \end{array}$

14.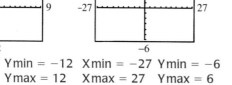

$\begin{array}{ll} X\min = -27 & Y\min = -6 \\ X\max = 27 & Y\max = 6 \\ X\text{scl} = 3 & Y\text{scl} = 1 \end{array}$

15.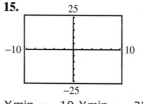

$\begin{array}{ll} X\min = -10 & Y\min = -25 \\ X\max = 10 & Y\max = 25 \\ X\text{scl} = 2 & Y\text{scl} = 5 \end{array}$

16.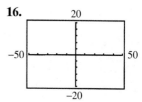

$\begin{array}{ll} X\min = -50 & Y\min = -20 \\ X\max = 50 & Y\max = 20 \\ X\text{scl} = 10 & Y\text{scl} = 4 \end{array}$

17.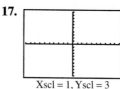

Xscl = 1, Yscl = 3

$\begin{array}{ll} X\min = -10 & Y\min = -30 \\ X\max = 10 & Y\max = 30 \\ X\text{scl} = 1 & Y\text{scl} = 3 \end{array}$

18.

Xscl = 10, Yscl = 2

$\begin{array}{ll} X\min = -100 & Y\min = -20 \\ X\max = 100 & Y\max = 20 \\ X\text{scl} = 10 & Y\text{scl} = 2 \end{array}$

19.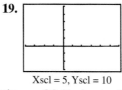

Xscl = 5, Yscl = 10

$\begin{array}{ll} X\min = -20 & Y\min = -30 \\ X\max = 30 & Y\max = 50 \\ X\text{scl} = 5 & Y\text{scl} = 10 \end{array}$

20.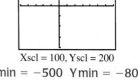

Xscl = 100, Yscl = 200

$\begin{array}{ll} X\min = -500 & Y\min = -800 \\ X\max = 700 & Y\max = 400 \\ X\text{scl} = 100 & Y\text{scl} = 200 \end{array}$

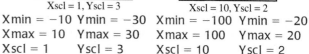

GRAPHING EQUATIONS AND SQUARE VIEWING WINDOW

In general, the following steps may be used to graph an equation on a standard viewing window.

> **TO GRAPH AN EQUATION IN *X* AND *Y* WITH A GRAPHING UTILITY ON A STANDARD VIEWING WINDOW**
>
> *Step 1.* Solve the equation for *y*.
>
> *Step 2.* Use your graphing utility and enter the equation in the form
> Y = *expression involving x*
>
> *Step 3.* Activate the graphing utility.

Special care must be taken when entering the *expression involving x* in *step 2*. You must be sure that the graphing utility you are using interprets the expression as you want it to. For example, let's graph $3y = 4x$. To do so,

Step 1.   Solve the equation for *y*.

$$3y = 4x$$

$$\frac{3y}{3} = \frac{4x}{3}$$

$$y = \frac{4}{3}x$$

Step 2.   Using your graphing utility, enter the expression $\frac{4}{3}x$ after the Y = prompt. In order for your graphing utility to correctly interpret the expression, you may need to enter $(4/3)x$ or $(4 \div 3)x$.

Step 3.   Activate the graphing utility. The graph should appear as in Figure A-9.

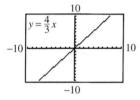

Figure A-9

Distinguishing features of the graph of a line include showing all the intercepts of the line. For example, the window of the graph of the line in Figure A-10 does not show both intercepts of the line, but the window of the graph of the same line in Figure A-11 does show both intercepts. Notice the notation below each graph. This is a shorthand notation of the range setting of the graph. This notation means [Xmin, Xmax] by [Ymin, Ymax].

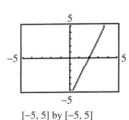

[−5, 5] by [−5, 5]

Figure A-10

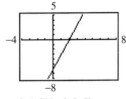

[−4, 8] by [−8, 5]

Figure A-11

On a standard viewing window, the tick marks on the *y*-axis are closer than the tick marks on the *x*-axis. This happens because the viewing window is a rectangle, and so 10 equally spaced tick marks on the positive *y*-axis will be closer together than 10 equally spaced tick marks on the positive *x*-axis. This causes the appearance of graphs to be distorted.

For example, notice the different appearances of the same line graphed using different viewing windows. The line in Figure A-12 is distorted because the tick marks along the *x*-axis are farther apart than the tick marks along the *y*-axis. The graph of the same line in Figure A-13 is not distorted because the viewing rectangle has been selected so that there is equal spacing between tick marks on both axes.

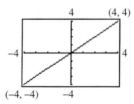

Figure A-12

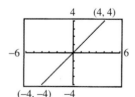

Figure A-13

We say that the line in Figure A-13 is graphed on a *square* setting. Some graphing utilities have a built-in program that, if activated, will automatically provide a square setting. A square setting is especially helpful when we are graphing perpendicular lines, circles, or when a true geometric perspective is desired. Some examples of square screens are shown in Figures A-14 and A-15.

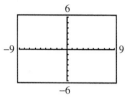

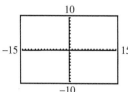

Figure A-14   Figure A-15

Other features of a graphing utility such as Trace, Zoom, Intersect, and Table are discussed in appropriate Graphing Calculator Explorations in this text.

GRAPHING EQUATIONS AND SQUARE VIEWING WINDOW EXERCISE SET

Graph each linear equation in two variables, using the two different range settings given. Determine which setting shows all intercepts of a line.

1. $y = 2x + 12$

   Setting A: $[-10, 10]$ by $[-10, 10]$

   Setting B: $[-10, 10]$ by $[-10, 15]$ Setting B

2. $y = -3x + 25$

   Setting A: $[-5, 5]$ by $[-30, 10]$

   Setting B: $[-10, 10]$ by $[-10, 30]$ Setting B

3. $y = -x - 41$

   Setting A: $[-50, 10]$ by $[-10, 10]$

   Setting B: $[-50, 10]$ by $[-50, 15]$ Setting B

4. $y = 6x - 18$

   Setting A: $[-10, 10]$ by $[-20, 10]$

   Setting B: $[-10, 10]$ by $[-10, 10]$ Setting A

5. $y = \frac{1}{2}x - 15$

   Setting A: $[-10, 10]$ by $[-20, 10]$

   Setting B: $[-10, 35]$ by $[-20, 15]$ Setting B

6. $y = -\frac{2}{3}x - \frac{29}{3}$

   Setting A: $[-10, 10]$ by $[-10, 10]$

   Setting B: $[-15, 5]$ by $[-15, 5]$ Setting B

The graph of each equation is a line. Use a graphing utility and a standard viewing window to graph each equation.   For Exercises 7–22, see Appendix D.

7. $3x = 5y$ **8.** $7y = -3x$ **9.** $9x - 5y = 30$

10. $4x + 6y = 20$ **11.** $y = -7$ **12.** $y = 2$

13. $x + 10y = -5$ **14.** $x - 5y = 9$

Graph the following equations using the square setting given. Some keystrokes that may be helpful are given.

15. $y = \sqrt{x}$ $[-12, 12]$ by $[-8, 8]$

   Suggested keystrokes: $\sqrt{}$ x

16. $y = \sqrt{2x}$ $[-12, 12]$ by $[-8, 8]$

   Suggested keystrokes: $\sqrt{}$ (2x)

17. $y = x^2 + 2x + 1$ $[-15, 15]$ by $[-10, 10]$

   Suggested keystrokes: x ^ 2 + 2 x + 1

18. $y = x^2 - 5$ $[-15, 15]$ by $[-10, 10]$

   Suggested keystrokes: x ^ 2 - 5

19. $y = |x|$ $[-9, 9]$ by $[-6, 6]$

   Suggested keystrokes: ABS x

20. $y = |x - 2|$ $[-9, 9]$ by $[-6, 6]$

   Suggested keystrokes: ABS (x − 2)

Graph the line on a single set of axes. Use a standard viewing window; then, if necessary, change the viewing window so that all intercepts of the line show.

21. $x + 2y = 30$ **22.** $1.5x - 3.7y = 40.3$

D

INSTRUCTOR'S ANSWERS

- Answers to Group Activities
- Answers to Exercises
  Requiring Graphical Solutions
- Notes to the Instructor

ANSWERS TO GROUP ACTIVITIES

CHAPTER 1

Group Activity (page 33)

Suggestions:

* For Question 2, consider providing your students with the beginnings of a bar chart with preset axes such as the one below.

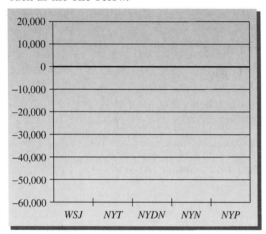

* Consider showing your students how to make a bar graph by means of a graphing calculator or computer software (such as a spreadsheet program).
* Consider asking each group to write a brief report about the changes in the newspaper publishing industry (as a business analyst might do) to hand in.
* Encourage students who finish early to help others in their group. Encourage group members to check each others' work as time permits.

Answers

1. Yes, the *New York Post,* 10,887
2. Greatest change: *Newsday*

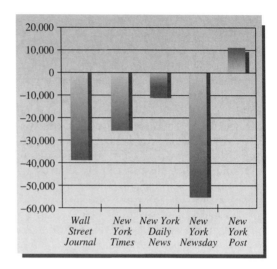

3. 1993: 4,866,319; 1994: 4,747,245; decreased by 119,074
4. Answers will vary. Some factors contributing to the decrease in New York newspaper circulation include prevalence of and competition with television news programs such as CNN, increased use of Internet news services, difficulties with labor, and increased costs of production resulting in higher newsstand prices.
5. NYC circulation per person: 0.648; Metro NYC circulation per person: 0.292. More meaningful: Metro NYC circulation per person, because it is more realistic to consider that these New York newspapers are sold outside the city limits. These figures can't fully describe the circulation situation, because many of these newspapers (especially the *Wall Street Journal* and *New York Times*) are also sold across the country .
6. Total daily amount: $2,930,223.50
   Total spent annually:
   260 × $2,930,223.50 = $761,858,110 (According to

a 1994 calendar, there were actually 260 weekdays in 1994; alternatively, your students may figure 261 weekdays by subtracting 52 Saturdays and 52 Sundays from 365 and get $764,788,333.50.)

CHAPTER 2

Group Activity (page 96)

Suggestions:

- For Question 1, encourage students to double-check their calculations. For example, two students could make the computations simultaneously and then compare their results.
- Should you choose to assign Question 4, you might have available the necessary advertisements for each group in case it proves difficult for each group to bring their own. To save time in class, you might also consider collecting information on investment opportunities in advance and either handing out this information to each group or displaying the information on an overhead projector. Alternatively, you might consider assigning Question 4 as a project, allowing sufficient time for completion.

Answers

1.

BANK	1 YEAR	2 YEARS	10 YEARS
A	$1067.60	$1139.78	$1923.56
B	$1067.84	$1140.29	$1927.83
C	$1067.00	$1138.49	$1912.69
D	$5356.13	$5737.62	$9948.94

2. Answers will vary. Take note: If you open the CD for Option D, you will have no money left to open your checking and savings accounts. With that in mind, Option B would be best, because it earns more interest than do options A or C. Other options to consider include items such as penalties for early withdrawal, fees, etc.

3. Option A: APY is 6.76%; Option B: APY is 6.84%; Option C: APY is 6.7%; Option D: APY is 7.12%

4. Answers will vary depending on options considered.

CHAPTER 3

Group Activity (page 173)

Suggestions:

- Consider pointing out to students that it is possible that each group member could have a different idea about which line fits best through the points on the plot. Suggest that if group members have difficulty agreeing on a single line, they could find several different lines and compare their results.
- If graphing calculators are not available for Question 6, consider asking students to find the equation of the best-fitting line by using the linear curve-fitting features of a computer spreadsheet program.
- Optional Question 6 could be assigned as a separate project, allowing sufficient time for completion.

Answers

1.

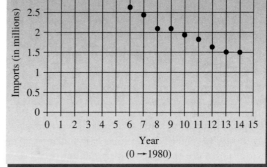

2. Answers will vary.

3. Answers will vary depending on the line drawn. It should, however, be somewhat close to $f(x) = 3.34 - 0.14x$.

4. Values of $f(19)$ will vary. Interpretation: The value of $f(19)$ represents the Japanese automobile imports (in millions) for the year 1999.

5. Answers will vary.

6. $y = 3.34 - 0.14x$; $f(19) = 0.68$. The total Japanese automobile imports for 1999 is estimated to be 0.68 million, or 680,000.

CHAPTER 4

Group Activity (page 229)

Suggestions:

- If you are using graphing calculators in your course, consider groups of five for this activity so that each group member can solve the system of equations in a different way in Question 3. If you do not plan to assign optional part (e) in Question 3, consider groups of four or less.

- Alternatively, divide the entire class into five (or as many groups as the number of parts you plan to assign) different groups, and assign each group to work on one of the five solution methods. After a few minutes, as a class discuss the solution of the system of equations. Differences among solution methods can be discussed as a class.

- Note: Students do not need a background in angle measurement to complete this activity.

Answers

1.

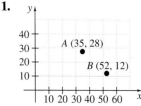

2. Station A: $y = -1.732x + 88.62$; Station B: $y = 0.577x - 18.004$
3. $(46.18, 8.64)$

CHAPTER 5

Group Activity (page 320)

Suggestions:

- Suggest to your students that they work together to fill in the table. Each pair of students in the group might be responsible for filling in five to ten rows or one or two columns of the table. Alternatively, everyone could be responsible for filling in the table (a few rows at a time) and then compare their results with another student(s), discuss any discrepancies, and revise the results as needed.

- Consider reminding your students that a graphing calculator capable of producing tables could help.

- As a first step, students might try to model the frame with the largest interior area with string or sketches on paper by trial and error.

Answers

1. $2x + 2y = 50$
2. Largest interior area: 110.25 square inches; exterior dimensions: 12.5 in. $\times$ 12.5 in.

		FRAME'S INTERIOR DIMENSIONS		
x	***y***	**INTERIOR WIDTH**	**INTERIOR HEIGHT**	***A(x)*** **INTERIOR AREA**
2.0	23.0	0.0	21.0	0.00
2.5	22.5	0.5	20.5	10.25
3.0	22.0	1.0	20.0	20.00
3.5	21.5	1.5	19.5	29.25
4.0	21.0	2.0	19.0	38.00
4.5	20.5	2.5	18.5	46.25
5.0	20.0	3.0	18.0	54.00
5.5	19.5	3.5	17.5	61.25
6.0	19.0	4.0	17.0	68.00
6.5	18.5	4.5	16.5	74.25
7.0	18.0	5.0	16.0	80.00
7.5	17.5	5.5	15.5	85.25
8.0	17.0	6.0	15.0	90.00
8.5	16.5	6.5	14.5	94.25
9.0	16.0	7.0	14.0	98.00
9.5	15.5	7.5	13.5	101.25
10.0	15.0	8.0	13.0	104.00
10.5	14.5	8.5	12.5	106.25
11.0	14.0	9.0	12.0	108.00
11.5	13.5	9.5	11.5	109.25
12.0	13.0	10.0	11.0	110.00
12.5	12.5	10.5	10.5	110.25
13.0	12.0	11.0	10.0	110.00
13.5	11.5	11.5	9.5	109.25
14.0	11.0	12.0	9.0	108.00
14.5	10.5	12.5	8.5	106.25
15.0	10.0	13.0	8.0	104.00

3. The interior area of the frame is given by the function $A(x) = (x - 2)(23 - x) = -x^2 + 25x - 46$.

4. The point representing the dimensions that yield the maximum interior area is the highest point on the graph.

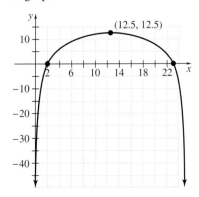

3.

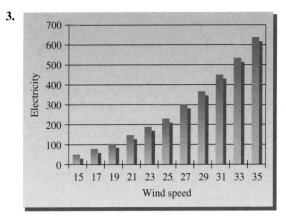

4. No

5. 18.75 watts per hour; $\frac{1}{8}c$

CHAPTER 6

Group Activity (page 399)

Suggestions:

- While this is not necessary, you might consider using a graphing calculator with the capability to make tables in conjunction with this activity.

Answers

1. Let y represent the amount of electricity generated per hour, and let x represent the speed of the wind.

$$y = 0.015x^3$$

2.

WIND SPEED (MILES PER HOUR)	ELECTRICITY (WATTS PER HOUR)
15	50.625
17	73.695
19	102.885
21	138.915
23	182.505
25	234.375
27	295.245
29	365.835
31	446.865
33	539.055
35	643.125

Approximately $29.88 \approx 30$ mph

CHAPTER 7

Group Activity (page 459)

Suggestions:

- For the pendulum weight, you could use a metal washer, spool, or fishing sinker. It would be best to use string that does not stretch, such as fishing line.

- Accurately measuring the length of the pendulum is important in this activity. You may wish to demonstrate this before students start the activity on their own. If the pendulum is to be 30 cm long, this means that the distance from the center of the weight to the point at which a student grasps the string is to be 30 cm long. Remind students that if they measure and cut 30 cm of string first and then tie it to the weight, some of the length of the string is lost due to the tying.

- Remind students that to get a more accurate measurement of the time it takes the pendulum to complete one swing, find an average time over several swings. That way any error involved in starting and stopping the stopwatch is spread out over several swings instead of included in only one swing.

- As an extra activity, you might consider asking your students to investigate the effects of measuring the length of the pendulum to the top of the pendulum weight versus the bottom of the pendulum weight.

Answers

1. Answers will vary depending on measurements.
2. Answers will vary depending on measurements.
3. The greater the pendulum's length, the greater the period.
4. Answers will vary. The differences will most likely be due to error in measurement or timing.

CHAPTER 8

Group Activity (page 516)

Suggestions:

- This activity will give your students practice in matching descriptions of real-life situations involving freely falling objects to the equations that model them and to the graphs that represent them. Note that there are more equations and graphs from which to choose than there are descriptions of situations. You might consider allowing the groups to use a graphing calculator as a tool to verify their answers by matching up equations to graphs. For instance, one member of the group could have the responsibility of checking graphs with the graphing calculator.

- Another way to use this activity is as a "partner quiz." Alternatively, a shortened individual quiz similar to this activity could be given to the class the following day, and each individual could receive his or her own score. If every member of the original group scores above a set minimum percent on the quiz, the group can be declared a success and receive some amount of extra credit points.

Answers

1. Equation: (g); graph: (vi); approximately 1.52 seconds
2. Equation: (f); graph: (i); approximately 37.93 seconds
3. Equation: (d); graph: (iii); approximately 10.20 seconds
4. Equation: (a); graph: (v); approximately 4.52 seconds
5. Equation: (c); graph: (vii); approximately 11.92 seconds
6. Equation: (e); graph: (iv); approximately 3.68 seconds

CHAPTER 9

Group Activity (page 557)

Suggestions:

- This activity can be used before or after a formal discussion of ellipses. Point out to your students that the length of the string *between the two thumbtacks* should be nine inches. One way to accomplish this is to cut a piece of string that is longer than nine inches and tie two knots that are nine inches apart. Then the thumbtacks can be pushed through the knots into the cardboard.

- Alternatively, students can use a tape measure or ruler to measure the exact length of string between the two tacks once the string has been attached to the cardboard with the tacks.

- Encourage students to see how the definition of an ellipse relates to their drawing.

- Optional Question 8 is an exploration in geometry. Consider demonstrating how to draw tangent lines accurately with a straight edge before students start. This activity also works for a circle. You might wish to ask students to make a conjecture about what will happen if a circle is used and then verify their conjecture.

Answers

1. Answers will vary.
2. Answers will vary.
3. Answers will vary.
4. Answers will vary.
5. Answers will vary.
6. Answers will vary.
7. Answers will vary.
8. Students should observe/conjecture that the three lines connecting opposite points of intersection meet in a single point.

CHAPTER 10

Group Activity (page 610)

Suggestions:

- Keep in mind that the more temperature readings your students can take, the easier it will be for them to discern the shape of the graph of these data.

- In Method 1, finer degree markings on the thermometer will result in more accurate temperature readings, which in turn result in data that are more easily recognized as data from an exponential model (as opposed to a linear or quadratic model) when they are graphed. If degree markings on the thermometer aren't very fine, use longer time intervals between temperature readings. If markings are somewhat fine, you can use shorter time intervals between readings. In any case, encourage your students to estimate temperatures that fall between markings rather than automatically rounding to the nearest marking.

- For Method 2, collect liquid temperature readings at 30-second or 60-second intervals. You might also consider using heated or cooled aluminum foil instead of water. Aluminum foil heats and cools more quickly than water and is well suited to the method of data collection used by the CBL. For aluminum foil, you can take temperature readings every 5 to 10 seconds. (Note: Aluminum foil is not suited to the manual data collection used in Method 1, because the temperature changes too quickly for accurate measurements to be made manually.)

Method 2 Setup and Description

Enter the HEAT program (see the following) into a TI-82, TI-83, or TI-85 graphing calculator. This program collects 36 data points and plots them in real time. L_3 contains time data in seconds, and L_4 contains temperature data in degrees Celsius. If you enter 1 when prompted for the time between points, data are collected every second for 36 seconds. If you enter 60 for the time between points, data are collected once each minute for 36 minutes.

To connect the equipment:

1. Connect the CBL unit to the TI-82, TI-83, or TI-85 graphing calculator with the unit-to-unit cable by means of the input/output ports located on the bottom edge of each unit. Press the cable ends in firmly.
2. Connect the temperature probe to Channel 1 (CH1) on the top edge of the CBL unit.
3. Turn on the CBL unit and the calculator. The CBL system is now ready to receive commands from the calculator.

TI-82 OR TI-83 PROGRAM

```
PROGRAM:HEAT82
:PlotsOff
:Func
:FnOff
:AxesOn
:ClrDraw
:ClrList L3,L4
:-10→Ymin
:90→Ymax
:10→Yscl
:ClrHome
:{1,0}→L1
:Send(L1)
:{1,1,1}→L1
:Send(L1)
:36→dim L3
:36→dim L4
:Disp "HOW MUCH TIME"
:Disp "BETWEEN POINTS"
:Disp "IN SECONDS?"
:Input T
:-2*T→Xmin
:36*T→Xmax
:T→Xscl
:seq(K,K,T,36*T,T)→L3
:ClrHome
:Disp "PRESS ENTER"
:Disp "TO START"
:Pause
:ClrHome
:{3,T,-1,0}→L1
:Send(L1)
:For(K,1,36,1)
:Get(L4(K))
:Pt-On(L3(K),L4(K))
:End
:ClrHome
:Plot1(Scatter,L3,L4,·)
:DispGraph
:Stop
```

TI-85 PROGRAM

```
PROGRAM:HEAT85
:Func
:FnOff
:AxesOn
:ClDrw
:1→dimL L3:1→dimL L4
:-10→yMin
:90→yMax
:10→yScl
:ClLCD
:{1,0}→L1
:Outpt("CBLSEND",L1)
:{1,1,1}→L1
:Outpt("CBLSEND",L1)
:36→dimL L3
:36→dimL L4
:Disp "HOW MUCH TIME"
:Disp "BETWEEN POINTS"
:Disp "IN SECONDS"
:Input T
:-2*T→xMin
:36*T→xMax
:T→xScl
:seq(K,K,T,36*T,T)→L3
:ClLCD
:Disp "PRESS ENTER"
:Disp "TO START"
:Pause
:ClLCD
:{3,T,-1,0}→L1
:Outpt("CBLSEND",L1)
:For(K,1,36,1)
:Input "CBLGET",L4(K)
:PtOn(L3(K),L4(K))
:End
:ClLCD
:Scatter L3,L4
:DispG
:Stop
```

More Suggestions:

- For Optional Question 3, you might want to point out to your students that if they collected the temperature data by the CBL method, they can subtract the value of c from each observation of T with the following method. Go to STAT EDIT and place the cursor on L_5. Press CLEAR ENTER to clear any existing data. Place the cursor back on L_5 and enter $L_4 - c$ (where c is the measured room temperature). This subtracts the room temperature from each value in L_4 and places the new value in L_5. Redraw the graph by creating a scatter plot of L_3 and L_5.

- Remind your students that once they have fit an appropriate exponential model to the data from which the room temperature has been subtracted, they must add the term $+ c$ to the model to get the model of the original temperature data.

- Additional project: Consider asking students to investigate whether different types of containers allow the liquid to heat up or cool down faster.

Answers

1. For hot liquid cooling down, model (b); for cold liquid warming up, model (c)
2. The constant c represents the temperature of the room.
3. Answers will vary depending on the data.

CHAPTER 11

Group Activity (page 652)

Suggestions:

- To save time in class, you might supply copies of pertinent magazine or newspaper articles to each group rather than asking students to find and bring in their own. You might also consider collecting information yourself in advance on situations that could be modeled by a sequence and either handing out this information to each group or displaying the information on an overhead projector.

- Alternatively, consider assigning Question 5 as a project, allowing sufficient time for completion.

Answers

1. Public: $a_n = 2860(1.06)^{n-1}$; private: $a_n = 12,432(1.06)^{n-1}$
2. Public: \$3610.68; private: \$15,695.11
3. Public: \$14,057.81; private: \$61,107.24
4. Academic year 2002–2003
5. Answers will vary.

ANSWERS TO EXERCISES REQUIRING GRAPHICAL SOLUTIONS

Exercise Set 3.1

1. quadrant I

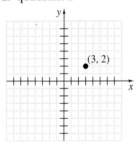
(3, 2)

2. quadrant IV

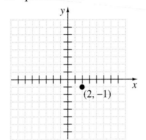
(2, −1)

3. quadrant II

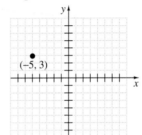
(−5, 3)

4. quadrant III

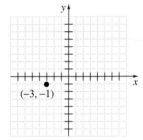
(−3, −1)

5. quadrant IV

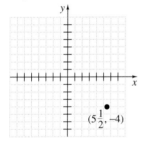
$(5\frac{1}{2}, -4)$

6. quadrant II

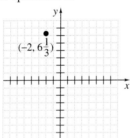
$(-2, 6\frac{1}{3})$

7. *y*-axis

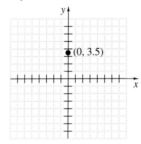
(0, 3.5)

8. quadrant II

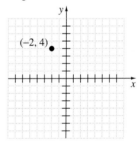
(−2, 4)

9. quadrant III

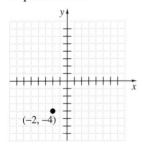
(−2, −4)

10. *x*-axis

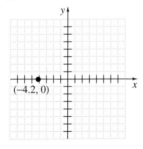
(−4.2, 0)

53.

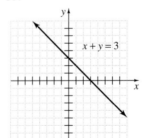
$x + y = 3$

54.

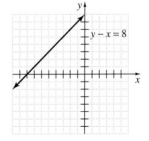
$y − x = 8$

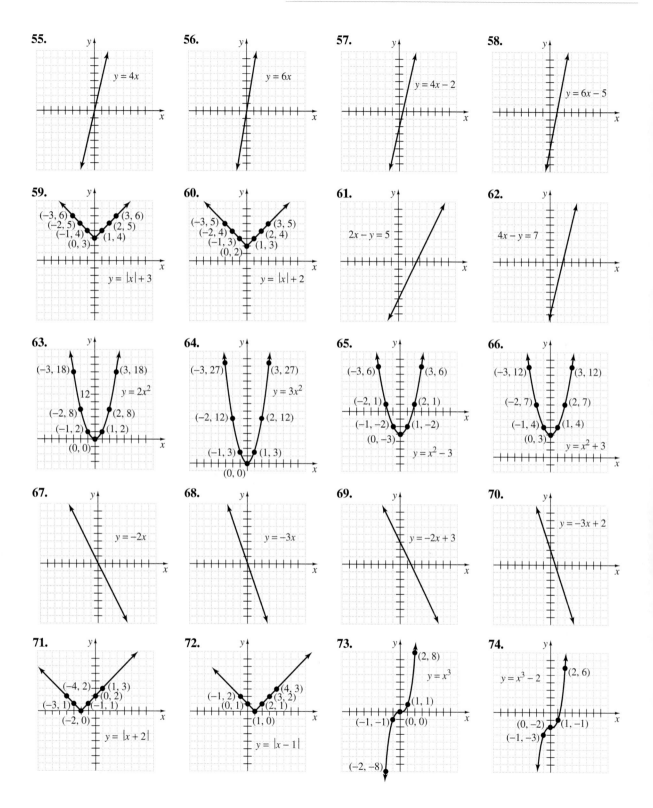

55. $y = 4x$

56. $y = 6x$

57. $y = 4x - 2$

58. $y = 6x - 5$

59. $y = |x| + 3$
(-3, 6) (3, 6)
(-2, 5) (2, 5)
(-1, 4) (1, 4)
(0, 3)

60. $y = |x| + 2$
(-3, 5) (3, 5)
(-2, 4) (2, 4)
(-1, 3) (1, 3)
(0, 2)

61. $2x - y = 5$

62. $4x - y = 7$

63. $y = 2x^2$
(-3, 18) (3, 18)
12
(-2, 8) (2, 8)
(-1, 2) (1, 2)
(0, 0)

64. $y = 3x^2$
(-3, 27) (3, 27)
(-2, 12) (2, 12)
(-1, 3) (1, 3)
(0, 0)

65. $y = x^2 - 3$
(-3, 6) (3, 6)
(-2, 1) (2, 1)
(-1, -2) (1, -2)
(0, -3)

66. $y = x^2 + 3$
(-3, 12) (3, 12)
(-2, 7) (2, 7)
(-1, 4) (1, 4)
(0, 3)

67. $y = -2x$

68. $y = -3x$

69. $y = -2x + 3$

70. $y = -3x + 2$

71. $y = |x + 2|$
(-4, 2) (1, 3)
(0, 2)
(-3, 1) (-1, 1)
(-2, 0)

72. $y = |x - 1|$
(-1, 2) (4, 3)
(0, 1) (3, 2)
(2, 1)
(1, 0)

73. $y = x^3$
(2, 8)
(1, 1)
(-1, -1) (0, 0)
(-2, -8)

74. $y = x^3 - 2$
(2, 6)
(0, -2) (1, -1)
(-1, -3)

75.

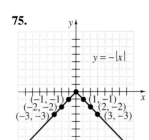

76.

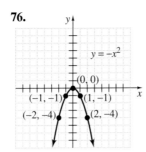

77.

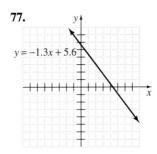

78.

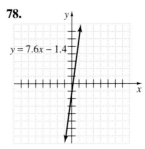

79.

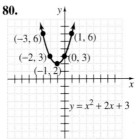

80.

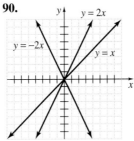

81. (a)

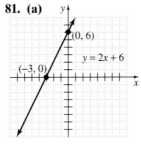

82. (a)

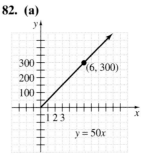

89.

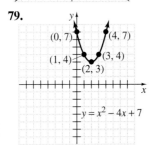

90.

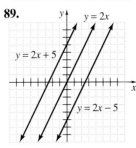

96.

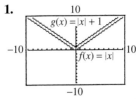

97.

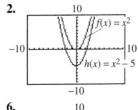

98.

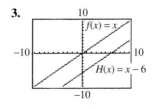

99.

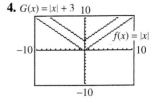

Graphing Calculator Explorations 3.2

1.

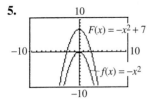

2.

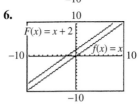

3.

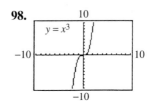

4. $G(x) = |x| + 3$

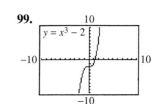

5.

6.

Exercise Set 3.2

85. $(0, 5), (-5, 0), (1, 6)$ **86.** $\left(0, \dfrac{10}{3}\right), (5, 0), (2, 2)$ **87.** $(0, 2), \left(\dfrac{8}{7}, 0\right), \left(\dfrac{12}{7}, -1\right)$ **88.** $(0, -3), (15, 0), \left(-2, -\dfrac{17}{5}\right)$

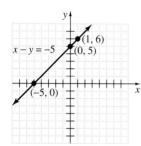 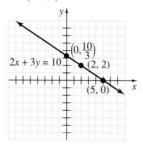 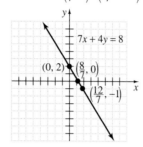 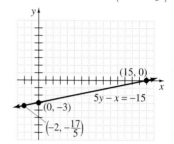

89. $(0, 0), (0, 0), (-1, -6)$ **90.** $(0, 0), (0, 0), (-2, 4)$

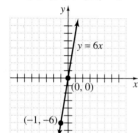 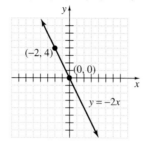

Graphing Calculator Explorations 3.3

1. $y = \dfrac{x}{3.5}$

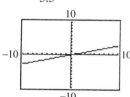

2. $y = -\dfrac{x}{2.7}$

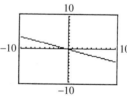

3. $y = -\dfrac{5.78}{2.31}x + \dfrac{10.98}{2.31}$

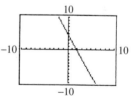

4. $y = \dfrac{7.22}{3.89}x + \dfrac{12.57}{3.89}$

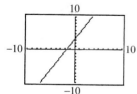

5. $y = |x| + 3.78$

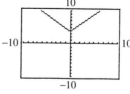

6. $y = \dfrac{5}{3}x^2 + 2x - \dfrac{4}{3}$

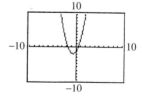

7. $y = 5.6x^2 + 7.7x + 1.5$

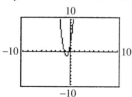

8. $y = -2.6|x| - 3.2$

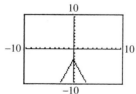

Exercise Set 3.3

1.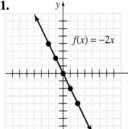
$f(x) = -2x$

2.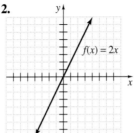
$f(x) = 2x$

3.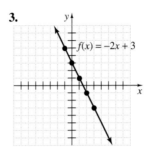
$f(x) = -2x + 3$

4.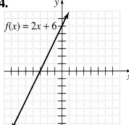
$f(x) = 2x + 6$

5.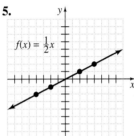
$f(x) = \frac{1}{2}x$

6.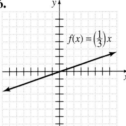
$f(x) = \left(\frac{1}{3}\right)x$

7.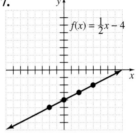
$f(x) = \frac{1}{2}x - 4$

8.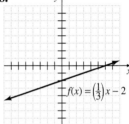
$f(x) = \left(\frac{1}{3}\right)x - 2$

13.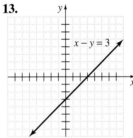
$x - y = 3$

14.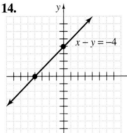
$x - y = -4$

15.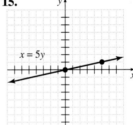
$x = 5y$

16.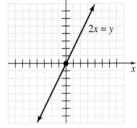
$2x = y$

17.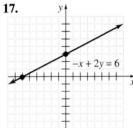
$-x + 2y = 6$

18.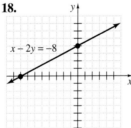
$x - 2y = -8$

19.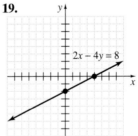
$2x - 4y = 8$

20.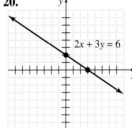
$2x + 3y = 6$

23.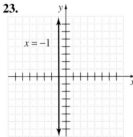
$x = -1$

24.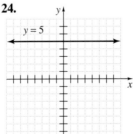
$y = 5$

25.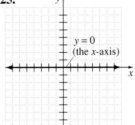
$y = 0$
(the x-axis)

26.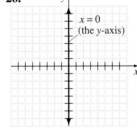
$x = 0$
(the y-axis)

27.
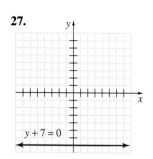
$y + 7 = 0$

28.
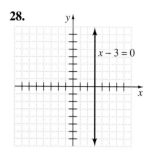
$x - 3 = 0$

35.
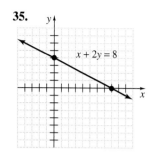
$x + 2y = 8$

36.
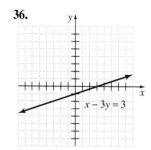
$x - 3y = 3$

37.
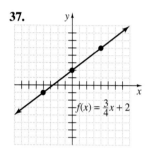
$f(x) = \frac{3}{4}x + 2$

38.
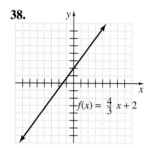
$f(x) = \frac{4}{3}x + 2$

39.
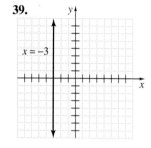
$x = -3$

40.
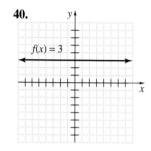
$f(x) = 3$

41.
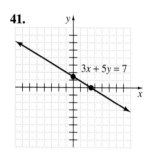
$3x + 5y = 7$

42.
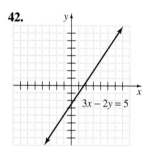
$3x - 2y = 5$

43.
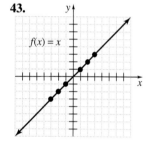
$f(x) = x$

44.
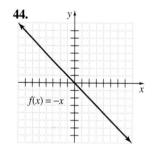
$f(x) = -x$

45.
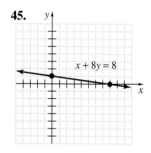
$x + 8y = 8$

46.
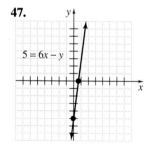
$x - 3y = 9$

47.
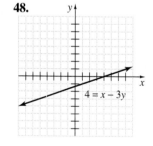
$5 = 6x - y$

48.
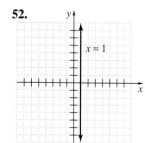
$4 = x - 3y$

49.
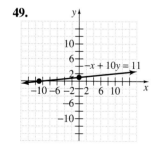
$-x + 10y = 11$

50.
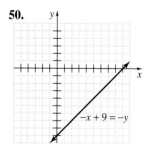
$-x + 9 = -y$

51.
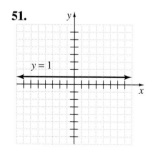
$y = 1$

52.
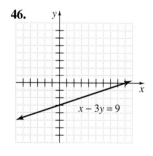
$x = 1$

53.

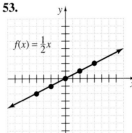

54.

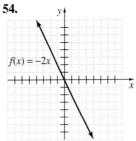

55.

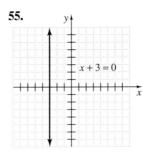

56.

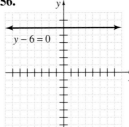

57.

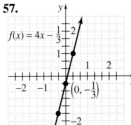

58.

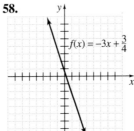

59.

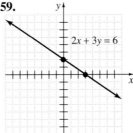

60.

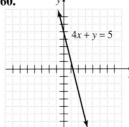

63. (b)

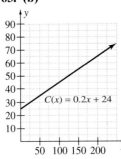

64. (b)

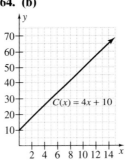

67.

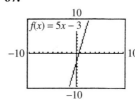

68.

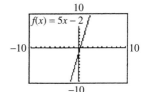

69.

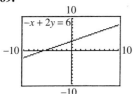

70.

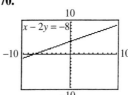

Graphing Calculator Explorations 3.4

1. 18.4

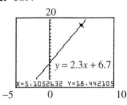

2. 11.5

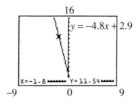

3. -1.5

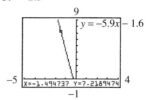

4. 10.5

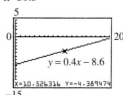

5. 14.0; 4.2, −9.4

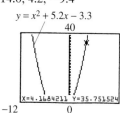

$y = x^2 + 5.2x - 3.3$

X=4.1684211 Y=35.751524

6. 23.1; 2.7, −1.5

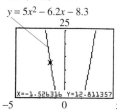

$y = 5x^2 - 6.2x - 8.3$

X=−1.526316 Y=12.811357

Exercise Set 3.4

85.

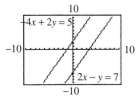

86.

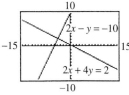

87. (a)

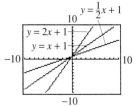

87. (b)

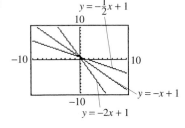

Exercise Set 3.5

7.

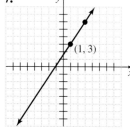

8.

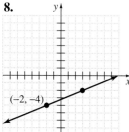

9.

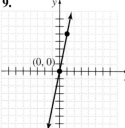

10.

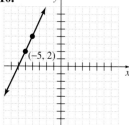

11.

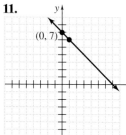

12.

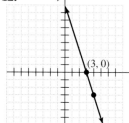

83.

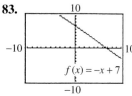

84.

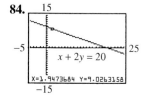

85.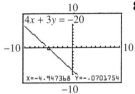

86.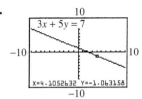

Exercise Set 3.6

1.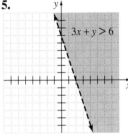

2.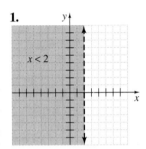

3.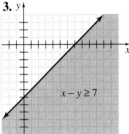

4.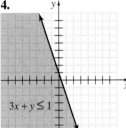

5.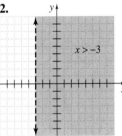

6.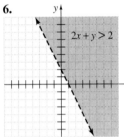

7.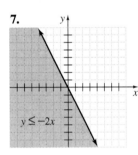

8.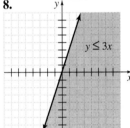

9.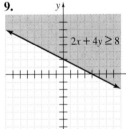

10.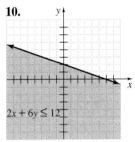

11.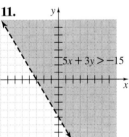

12.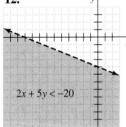

15.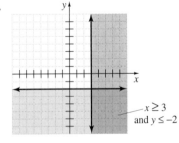

16.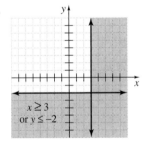

17.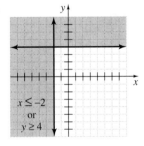

18. $x \le -2$ and $y \ge 4$
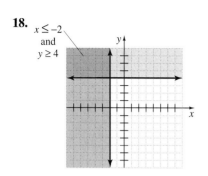

19. $x - y < 3$ and $x > 4$
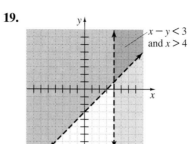

20. $2x > y$ and $y > x + 2$
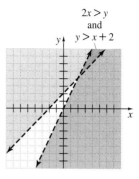

21. $x + y \le 3$ or $x - y \ge 5$
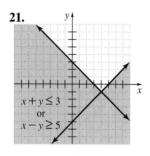

22. $x - y \le 3$ or $x + y > -1$
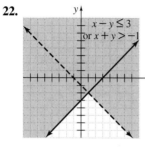

23. $y \ge -2$
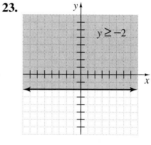

24. $y \le 4$
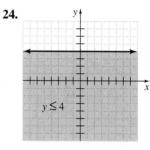

25. $x - 6y < 12$
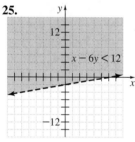

26. $x - 4y < 8$
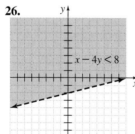

27. $x > 5$
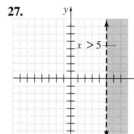

28. $y \ge -2$
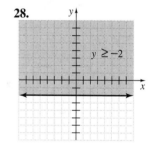

29. $-2x + y \le 4$
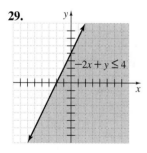

30. $-3x + y \le 9$
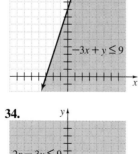

31. $x - 3y < 0$
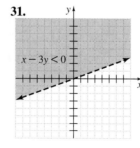

32. $x + 2y > 0$
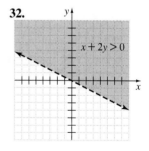

33. $3x - 2y \le 12$
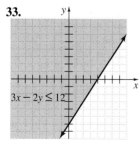

34. $2x - 3y \le 9$
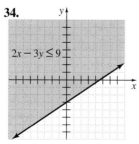

35. $x - y \ge 2$ or $y < 5$
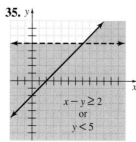

36. $x - y < 3$ or $x > 4$
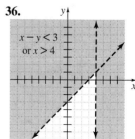

37.

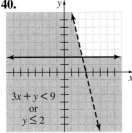

$x + y \leq 1$
and $y \leq -1$

38.

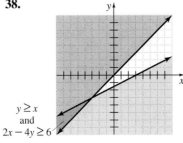

$y \geq x$
and
$2x - 4y \geq 6$

39.

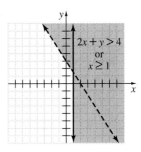

$2x + y > 4$
or
$x \geq 1$

40.

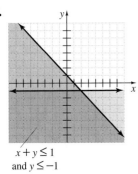

$3x + y < 9$
or
$y \leq 2$

41.

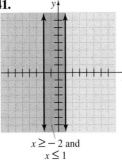

$x \geq -2$ and
$x \leq 1$

42.

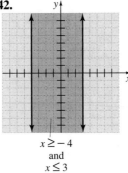

$x \geq -4$
and
$x \leq 3$

43.

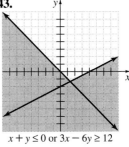

$x + y \leq 0$ or $3x - 6y \geq 12$

44.

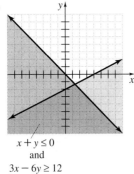

$x + y \leq 0$
and
$3x - 6y \geq 12$

45.

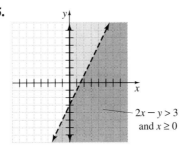

$2x - y > 3$
and $x \geq 0$

46.

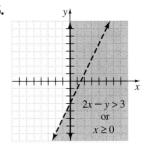

$2x - y > 3$
or
$x \geq 0$

59.

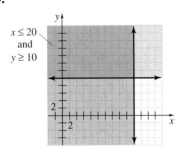

$x \leq 20$
and
$y \geq 10$

60.

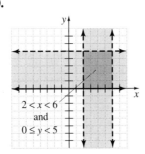

$2 < x < 6$
and
$0 \leq y < 5$

61.

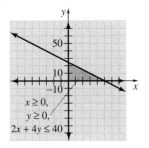

$x \geq 0,$
$y \geq 0,$
$2x + 4y \leq 40$

Chapter 3 Review

1.

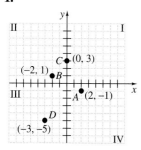

2.

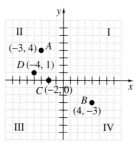

9. $(-3, 0), (1, 3), (9, 9)$

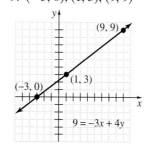

10. $(7, -14), (-7, 14), (0, 0)$

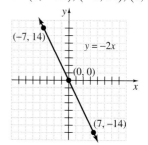

11. $(0, 0), (10, 5), (-10, 5)$

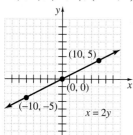

18.

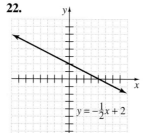

19.

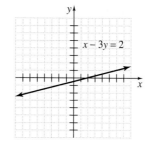

20.

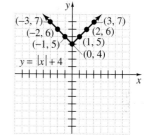

21.

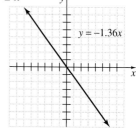

22.

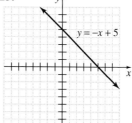

23.

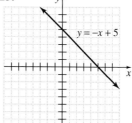

24.

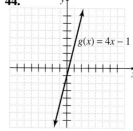

25.

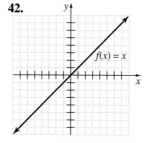

42.

43.

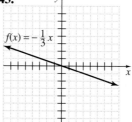

44.

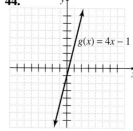

49.

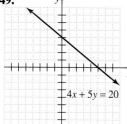

50.

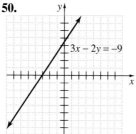

51.

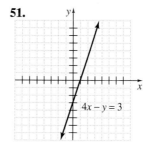

52.

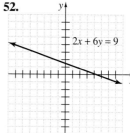

53.

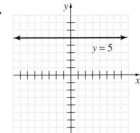

54.

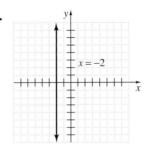

55. (b)

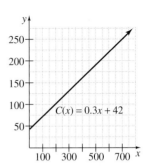

86.

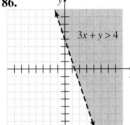

87.

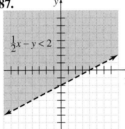

88.

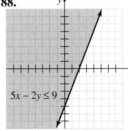

89.

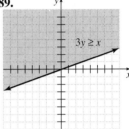

90.

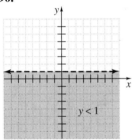

91.

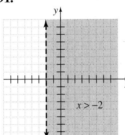

92.

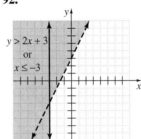

93.

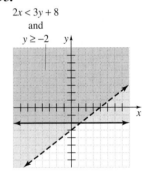

Chapter 3 Test

1.

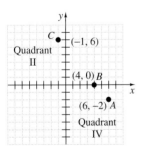

3.

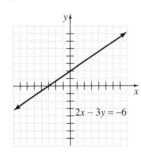

4.

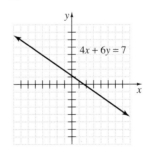

5.

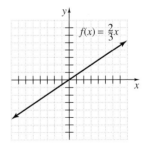

6.

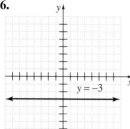

9.

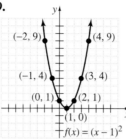

10.

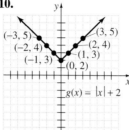

20.

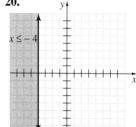

21.

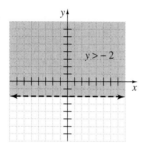

22.

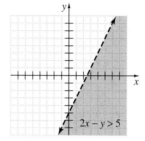

23.

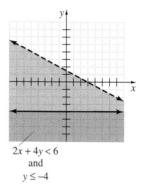

Chapter 3 Cumulative Review

10. *(Sec 3.1, Ex. 1)*
   (a) quadrant IV
   (b) on y-axis
   (c) quadrant II
   (d) x-axis
   (e) quadrant III

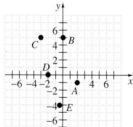

11. *(Sec. 3.1, Ex. 7)*

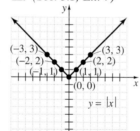

14. *(Sec. 3.3, Ex. 4)*

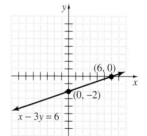

15. *(Sec. 3.3, Ex. 7)*

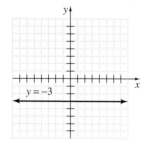

20. *(Sec. 3.6, Ex. 2)*

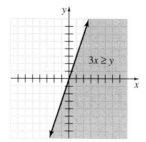

21. *(Sec. 3.6, Ex. 4)*

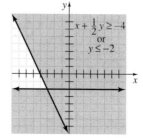

CHAPTER 4

Systems of Equations

Graphing Calculator Explorations 4.1

1. $\{(2.11, 0.17)\}$

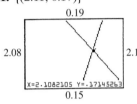

2. $\{(-1.12, -5.02)\}$

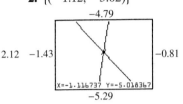

3. $\{(0.57, -1.97)\}$

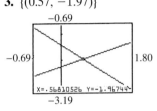

4. $\{(-1.38, 1.35)\}$

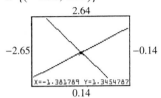

Exercise Set 4.1

1. $\{(2, -1)\}$

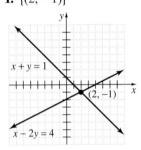

2. $\{(5, 2)\}$

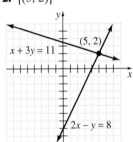

3. $\{(1, 2)\}$

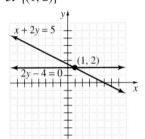

4. $\{(2, 2)\}$

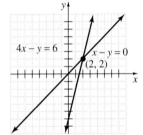

5. $\{\ \}$

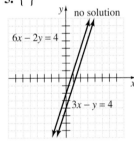

6. $\{\ \}$

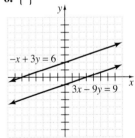

Exercise Set 4.5

55.

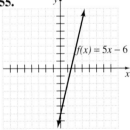

56.

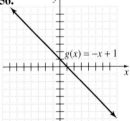

57.

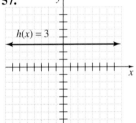

58.

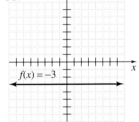

Chapter 4 Review

1. $\{(-3, 1)\}$

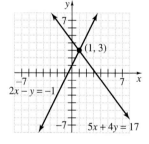

2. $\left\{\left(0, \frac{2}{3}\right)\right\}$

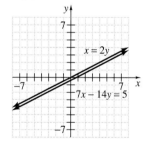

3. $\{\ \}$

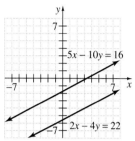

4. $\{(x, y) \mid 3x - 6y = 12)\}$

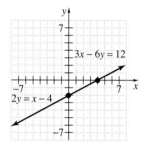

5. $\left\{\left(3, \frac{8}{3}\right)\right\}$

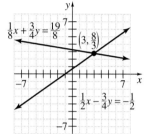

Chapter 4 Test

3. $\{(1, 3)\}$

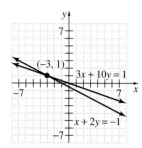

4. $\{\ \}$

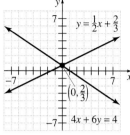

Chapter 4 Cumulative Review

8. *(Sec. 3.1, Ex. 4)*

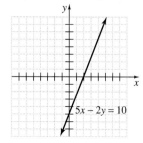

10. *(Sec. 3.3, Ex. 5)*

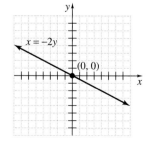

14. *(Sec. 3.6, Ex. 1)*

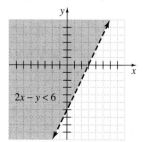

CHAPTER 5

Exponents, Polynomials, and Polynomial Functions

Graphing Calculator Explorations 5.3

1. $x^3 - 4x^2 + 7x - 8$

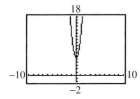

2. $-15x^3 + 3x^2 + 3x + 2$

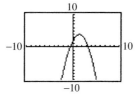

3. $-2.1x^2 - 3.2x - 1.7$

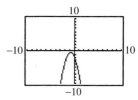

4. $-7.9x^2 + 20.3x - 7.8$

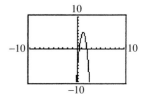

5. $7.69x^2 - 1.26x + 5.3$

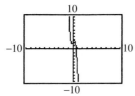

6. $-0.98x^2 + 2.8x + 1.86$

Graphing Calculator Explorations 5.4

1. $x^2 - 16$

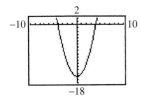

2. $x^2 + 6x + 9$

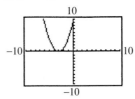

3. $9x^2 - 42x + 49$

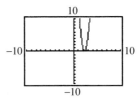

4. $25x^2 - 20x + 4$

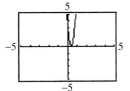

5. $5x^3 - 14x^2 - 13x - 2$

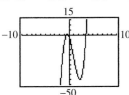

6. $14x^3 + 29x^2 - 23x - 20$

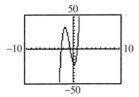

Exercise Set 5.6

91. $x^2(x + 5)(x + 1)$

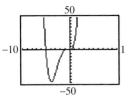

92. $x(x + 2)(x + 4)$

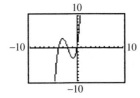

93. $3x(5x - 1)(2x + 1)$

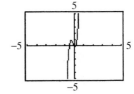

94. $-2x^2(3x + 1)(x - 2)$

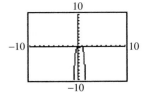

Graphing Calculator Explorations 5.8

1. $\{-3.562, 0.562\}$

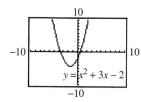

$y = x^2 + 3x - 2$

2. $\{0.161, 1.239\}$

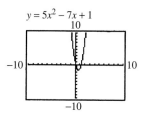

$y = 5x^2 - 7x + 1$

3. $\{-0.874, 2.787\}$

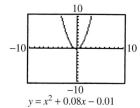

$y = 2.3x^2 - 4.4x - 5.6$

4. $\{-30.655, -0.342\}$

$y = 0.2x^2 + 6.2x + 2.1$

5. $\{-0.465, 1.910\}$

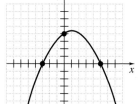

$y = 0.09x^2 - 0.13x - 0.08$

6. $\{-0.148, 0.068\}$

$y = x^2 + 0.08x - 0.01$

Exercise Set 5.8

96. Answers may vary. For example:

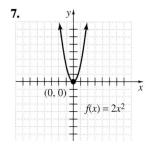

97. Answers may vary. For example:

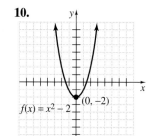

Exercise Set 5.9

7.

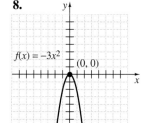

$(0, 0)$

$f(x) = 2x^2$

8.

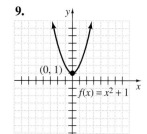

$f(x) = -3x^2$ $(0, 0)$

9.

$(0, 1)$

$f(x) = x^2 + 1$

10.

$f(x) = x^2 - 2$ $(0, -2)$

11.

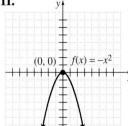

$f(x) = -x^2$ with vertex $(0, 0)$

12.

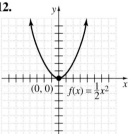

$f(x) = \frac{1}{2}x^2$ with vertex $(0, 0)$

24.

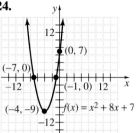

$f(x) = x^2 + 8x + 7$ with points $(-7, 0)$, $(-1, 0)$, $(0, 7)$, $(-4, -9)$; 12, -12

25.

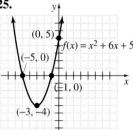

$f(x) = x^2 + 6x + 5$ with points $(0, 5)$, $(-5, 0)$, $(-1, 0)$, $(-3, -4)$

26.

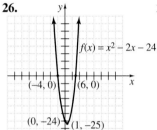

$f(x) = x^2 - 2x - 24$ with points $(-4, 0)$, $(6, 0)$, $(0, -24)$, $(1, -25)$

27.

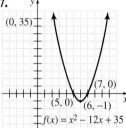

$f(x) = x^2 - 12x + 35$ with points $(0, 35)$, $(5, 0)$, $(7, 0)$, $(6, -1)$

28.

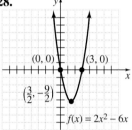

$f(x) = 2x^2 - 6x$ with points $(0, 0)$, $(3, 0)$, $\left(\frac{3}{2}, -\frac{9}{2}\right)$

29.

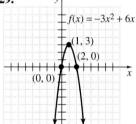

$f(x) = -3x^2 + 6x$ with points $(1, 3)$, $(2, 0)$, $(0, 0)$

30.

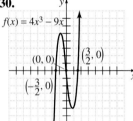

$f(x) = 4x^3 - 9x$ with points $(0, 0)$, $\left(\frac{3}{2}, 0\right)$, $\left(-\frac{3}{2}, 0\right)$

31.

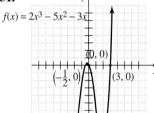

$f(x) = 2x^3 - 5x^2 - 3x$ with points $(0, 0)$, $\left(-\frac{1}{2}, 0\right)$, $(3, 0)$

32.

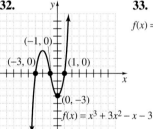

$f(x) = x^3 + 3x^2 - x - 3$ with points $(-1, 0)$, $(-3, 0)$, $(1, 0)$, $(0, -3)$

33.

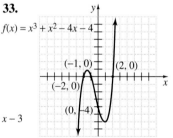

$f(x) = x^3 + x^2 - 4x - 4$ with points $(-1, 0)$, $(2, 0)$, $(-2, 0)$, $(0, -4)$

36.

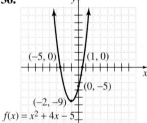

$f(x) = x^2 + 4x - 5$ with points $(-5, 0)$, $(1, 0)$, $(0, -5)$, $(-2, -9)$

37.

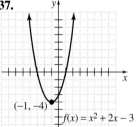

$f(x) = x^2 + 2x - 3$ with point $(-1, -4)$

38.

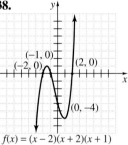

$f(x) = (x - 2)(x + 2)(x + 1)$ with points $(-1, 0)$, $(-2, 0)$, $(2, 0)$, $(0, -4)$

39.

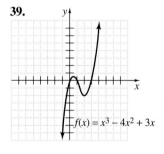

$f(x) = x^3 - 4x^2 + 3x$

40.

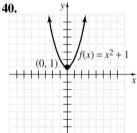

$f(x) = x^2 + 1$ with point $(0, 1)$

41.

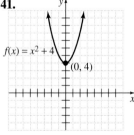

$f(x) = x^2 + 4$ with point $(0, 4)$

42.

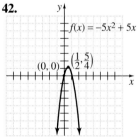

$f(x) = -5x^2 + 5x$ with points $(0, 0)$, $\left(\frac{1}{2}, \frac{5}{4}\right)$

43.

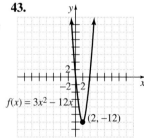

$f(x) = 3x^2 - 12x$ with points 2, -2, 2; $(2, -12)$

44.
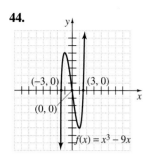
$(-3, 0)$ $(3, 0)$
$(0, 0)$
$f(x) = x^3 - 9x$

45.
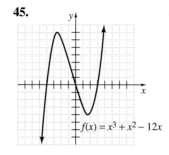
$f(x) = x^3 + x^2 - 12x$

46.
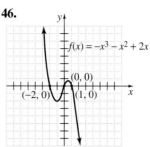
$f(x) = -x^3 - x^2 + 2x$
$(0, 0)$
$(-2, 0)$ $(1, 0)$

47.
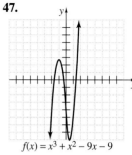
$f(x) = x^3 + x^2 - 9x - 9$

48.
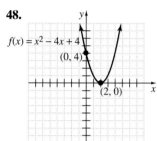
$f(x) = x^2 - 4x + 4$
$(0, 4)$
$(2, 0)$

49.
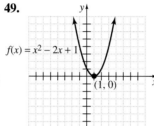
$f(x) = x^2 - 2x + 1$
$(1, 0)$

50.
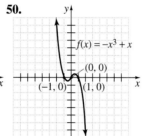
$f(x) = -x^3 + x$
$(0, 0)$
$(-1, 0)$ $(1, 0)$

51.
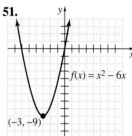
$f(x) = x^2 - 6x$
$(-3, -9)$

52.
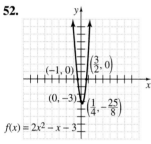
$(-1, 0)$ $\left(\frac{3}{2}, 0\right)$
$(0, -3)$ $\left(\frac{1}{4}, -\frac{25}{8}\right)$
$f(x) = 2x^2 - x - 3$

53.
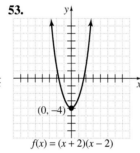
$(0, -4)$
$f(x) = (x + 2)(x - 2)$

54.
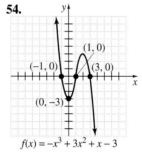
$(1, 0)$
$(-1, 0)$ $(3, 0)$
$(0, -3)$
$f(x) = -x^3 + 3x^2 + x - 3$

55.
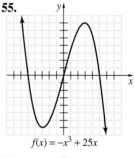
$f(x) = -x^3 + 25x$

56.
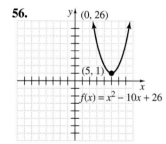
$(0, 26)$
$(5, 1)$
$f(x) = x^2 - 10x + 26$

57.
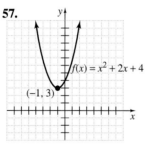
$f(x) = x^2 + 2x + 4$
$(-1, 3)$

58.
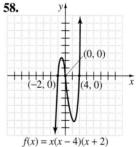
$(0, 0)$
$(-2, 0)$ $(4, 0)$
$f(x) = x(x - 4)(x + 2)$

59.
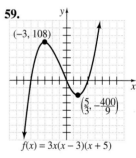
$(-3, 108)$
$\left(\frac{5}{3}, -\frac{400}{9}\right)$
$f(x) = 3x(x - 3)(x + 5)$

60.
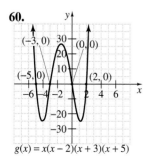
$(-3, 0)$ 30 $(0, 0)$
20
10
$(-5, 0)$
$(2, 0)$
-6 -4 -2 2   4   6
-20
-30
$g(x) = x(x - 2)(x + 3)(x + 5)$

61.
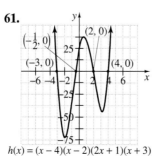
$\left(-\frac{1}{2}, 0\right)$ $(2, 0)$
25
$(-3, 0)$ $(4, 0)$
-6 -4 -2 2   4   6
-25
-50
-75
$h(x) = (x - 4)(x - 2)(2x + 1)(x + 3)$

62.
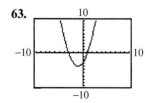

63.
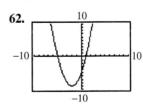

64.

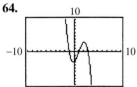

65.

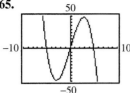

66.

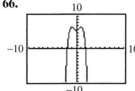

67.

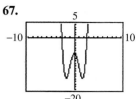

Chapter 5 Review

140.

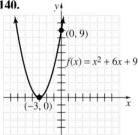

141.

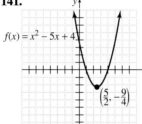

$f(x) = x^2 - 5x + 4$

142.

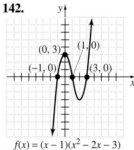

$f(x) = (x - 1)(x^2 - 2x - 3)$

143.

$f(x) = (x + 3)(x^2 - 4x + 3)$

144.

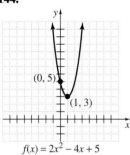

$f(x) = 2x^2 - 4x + 5$

145.

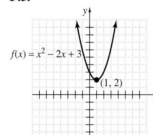

$f(x) = x^2 - 2x + 3$

146.

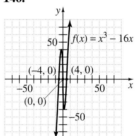

$f(x) = x^3 - 16x$

147.

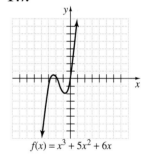

$f(x) = x^3 + 5x^2 + 6x$

Chapter 5 Test

29.

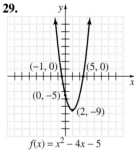

$f(x) = x^2 - 4x - 5$

30.

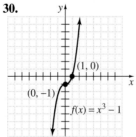

$f(x) = x^3 - 1$

Chapter 5 Cumulative Review

6. *(Sec. 3.1, Ex. 1)*
   (a) quadrant IV
   (b) y-axis
   (c) quadrant II
   (d) x-axis
   (e) quadrant III
   (f) quadrant I

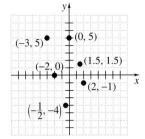

8. *(Sec. 3.3, Ex. 4)*

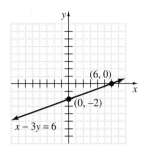

10. *(Sec. 3.6, Ex. 3)* $x \geq 1$ and $y \geq 2x - 1$

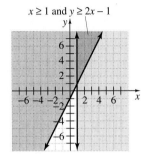

23. *(Sec. 5.9, Ex. 3)*

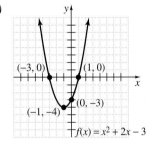

CHAPTER 6

Rational Expressions

Graphing Calculator Explorations 6.1

1. $\{x \mid x$ is a real number and $x \neq -2, x \neq 2\}$

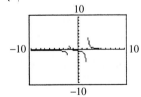

2. $\{x \mid x$ is a real number and $x \neq -3, x \neq 3\}$

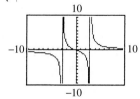

3. $\left\{x \mid x$ is a real number and $x \neq -4, x \neq \dfrac{1}{2}\right\}$

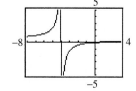

4. $\left\{x \mid x$ is a real number and $x \neq -\dfrac{1}{4}, x \neq 5\right\}$

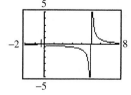

Exercise Set 6.1

75. $0, \dfrac{20}{9}, \dfrac{60}{7}, 20, \dfrac{140}{3}, 180, 350, 1980$

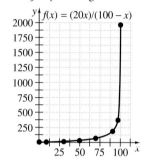

76. $4, 2, 1, \dfrac{1}{2}, \dfrac{1}{4}; -\dfrac{1}{4}, -\dfrac{1}{2}, -1, -2, -4$

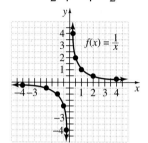

Exercise Set 6.7

72.

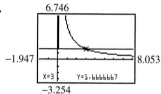

73.

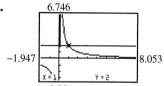

74.

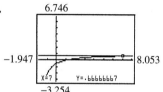

75.

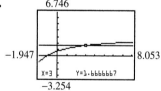

Exercise Set 6.9

47. $4, 2, 1, \dfrac{1}{2}, \dfrac{1}{4}$

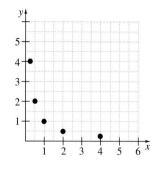

48. $12, 6, 3, \dfrac{3}{2}, \dfrac{3}{4}$

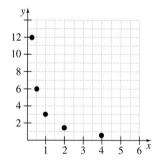

49. $20, 10, 5, \dfrac{5}{2}, \dfrac{5}{4}$

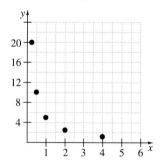

50. $2, 1, \dfrac{1}{2}, \dfrac{1}{4}, \dfrac{1}{8}$

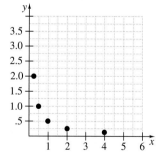

59.

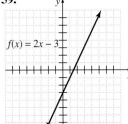

$f(x) = 2x - 3$

60.

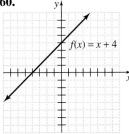

$f(x) = x + 4$

61.

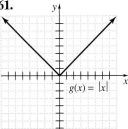

$g(x) = |x|$

62.

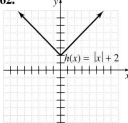

$h(x) = |x| + 2$

63.

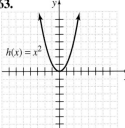

$h(x) = x^2$

64.

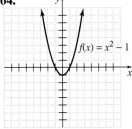

$f(x) = x^2 - 1$

Chapter 6 Cumulative Review

4. *(Sec. 3.1, Ex. 4)*

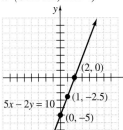

$(2, 0)$
$5x - 2y = 10$
$(1, -2.5)$
$(0, -5)$

6. *(Sec. 3.3, Ex. 7)*

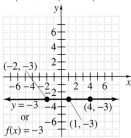

$(-2, -3)$
$y = -3$
or
$f(x) = -3$
$(4, -3)$
$(1, -3)$

CHAPTER 7

Rational Exponents, Radicals, and Complex Numbers

Exercise Set 7.1

83. $[0, \infty)$

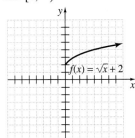

$f(x) = \sqrt{x} + 2$

84. $[0, \infty)$

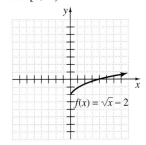

$f(x) = \sqrt{x} - 2$

85. domain: $[3, \infty)$; $(3, 0)$, $(4, 1)$, $(7, 2)$, $(12, 3)$

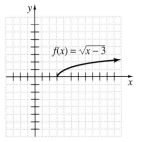

$f(x) = \sqrt{x - 3}$

86. domain: $[-1, \infty)$; $(-1, 0)$, $(0, 1)$, $(3, 2)$, $(8, 3)$

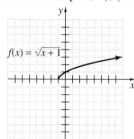

87. $(-\infty, \infty)$

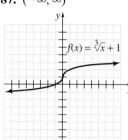

88. $(-\infty, \infty)$

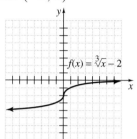

89. domain: $(-\infty, \infty)$; $(1, 0)$, $(2, 1)$, $(0, -1)$, $(9, 2)$, $(-7, -2)$

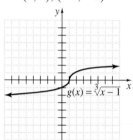

90. domain $(-\infty, \infty)$; $(-1, 0)$, $(0, 1)$, $(-2, -1)$, $(7, 2)$, $(-9, -2)$

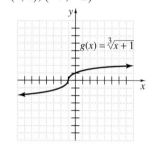

91.
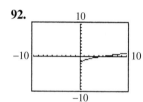

92.

93.

94.

95.
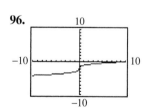

96.

97.
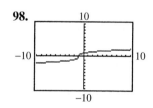

98.
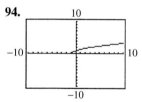

Graphing Calculator Explorations 7.6

1. {3.19}

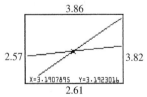

2. {1.55}

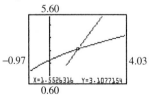

3. { }

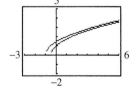

4. {0.34}

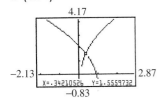

5. {3.23}

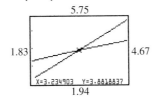

6. {−5.44, 1.63}

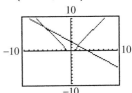

Chapter 7 Review

24. $[0, \infty)$

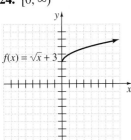

25. $(-\infty, \infty)$

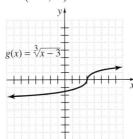

Chapter 7 Test

32. domain: $[-2, \infty)$; $(-2, 0), (-1, 1), (2, 2), (7, 3)$

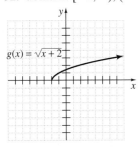

Chapter 7 Cumulative Review

4. *(Sec. 3.3, Ex. 1)* The graph of $g(x) = 2x + 1$ is the graph of $f(x) = 2x$ shifted upward 1 unit.

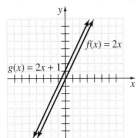

CHAPTER 8

Quadratic Equations and Functions

Graphing Calculator Explorations 8.1

1. $\{-1.27, 6.27\}$

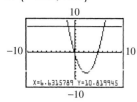

2. $\{-3.45, 1.45\}$

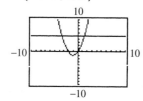

3. $\{-1.10, 0.90\}$

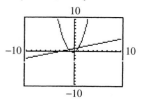

4. $\{-1.54, 1.94\}$

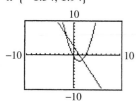

5. $\{\ \}$

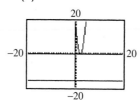

Exercise Set 8.2

73.

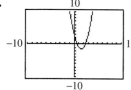

74.

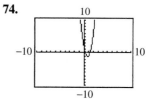

Exercise Set 8.3

65.

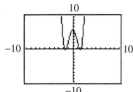

66.

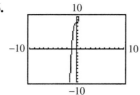

Exercise Set 8.4

61.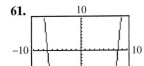

62.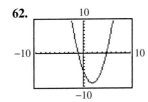

63.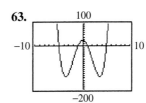

64.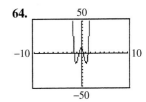

65.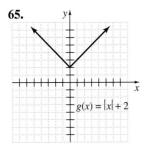

66.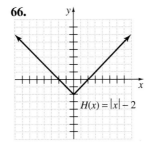

67.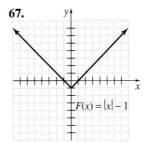

68.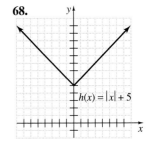

69.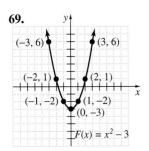

70.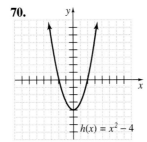

71.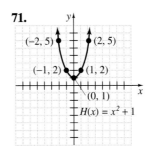

72.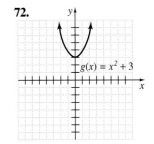

Graphing Calculator Explorations 8.5

1.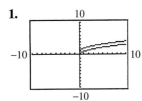

2.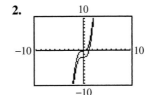

3.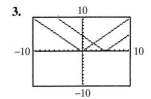

4.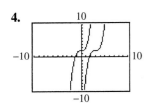

5.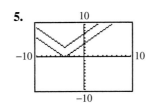

6.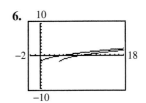

Exercise Set 8.5

1.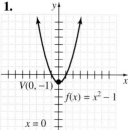

2.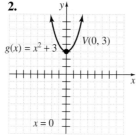

3.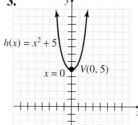

4.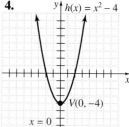

5.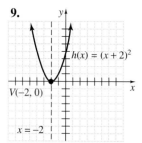

6.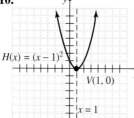

7.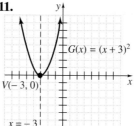

8.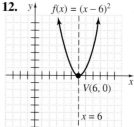

9.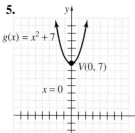

10.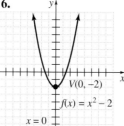

11.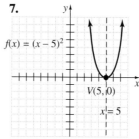

12.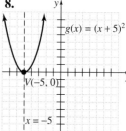

13.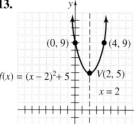

14.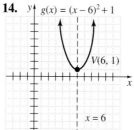

15.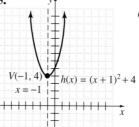

16.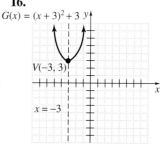

17.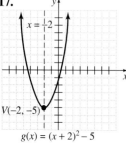

18.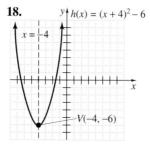

19.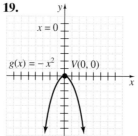

20.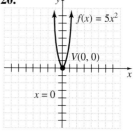

21.

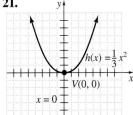

$h(x) = \frac{1}{3}x^2$

$V(0, 0)$

$x = 0$

22.

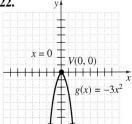

$x = 0$

$V(0, 0)$

$g(x) = -3x^2$

23.

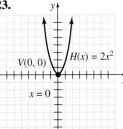

$V(0, 0)$

$H(x) = 2x^2$

$x = 0$

24.

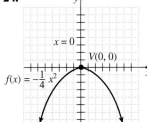

$x = 0$

$V(0, 0)$

$f(x) = -\frac{1}{4}x^2$

25.

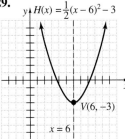

$f(x) = 2(x - 1)^2 + 3$

$V(1, 3)$

$x = 1$

26.

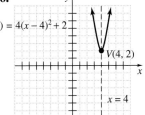

$g(x) = 4(x - 4)^2 + 2$

$V(4, 2)$

$x = 4$

27.

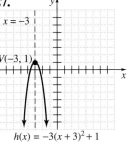

$x = -3$

$V(-3, 1)$

$h(x) = -3(x + 3)^2 + 1$

28.

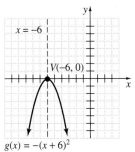

$x = 2$

$f(x) = -(x - 2)^2 - 6$

$V(2, -6)$

29.

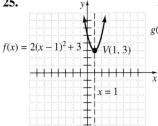

$H(x) = \frac{1}{2}(x - 6)^2 - 3$

$V(6, -3)$

$x = 6$

30.

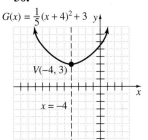

$G(x) = \frac{1}{5}(x + 4)^2 + 3$

$V(-4, 3)$

$x = -4$

31.

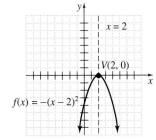

$x = 2$

$V(2, 0)$

$f(x) = -(x - 2)^2$

32.

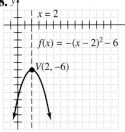

$x = -6$

$V(-6, 0)$

$g(x) = -(x + 6)^2$

33.

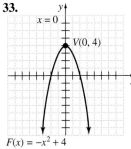

$x = 0$

$V(0, 4)$

$F(x) = -x^2 + 4$

34.

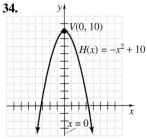

$V(0, 10)$

$H(x) = -x^2 + 10$

$x = 0$

35.

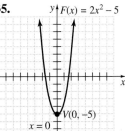

$F(x) = 2x^2 - 5$

$V(0, -5)$

$x = 0$

36.

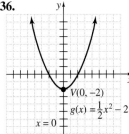

$V(0, -2)$

$g(x) = \frac{1}{2}x^2 - 2$

$x = 0$

37.

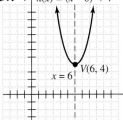

$h(x) = (x - 6)^2 + 4$

$V(6, 4)$

$x = 6$

38.

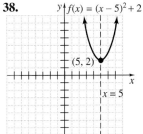

$f(x) = (x - 5)^2 + 2$

$(5, 2)$

$x = 5$

39.

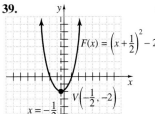

$F(x) = \left(x + \frac{1}{2}\right)^2 - 2$

$V\left(-\frac{1}{2}, -2\right)$

$x = -\frac{1}{2}$

40.

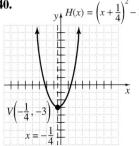

$H(x) = \left(x + \frac{1}{4}\right)^2 - 3$

$V\left(-\frac{1}{4}, -3\right)$

$x = -\frac{1}{4}$

41.
$F(x) = \frac{3}{2}(x+7)^2 + 1$

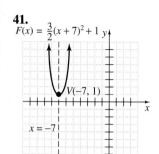

$V(-7, 1)$
$x = -7$

42.

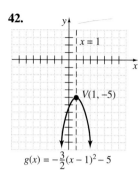

$x = 1$
$V(1, -5)$
$g(x) = -\frac{3}{2}(x-1)^2 - 5$

43.
$f(x) = \frac{1}{4}x^2 - 9$

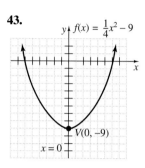

$V(0, -9)$
$x = 0$

44.

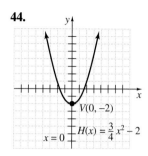

$V(0, -2)$
$H(x) = \frac{3}{4}x^2 - 2$
$x = 0$

45.

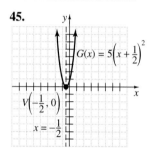

$G(x) = 5\left(x + \frac{1}{2}\right)^2$
$V\left(-\frac{1}{2}, 0\right)$
$x = -\frac{1}{2}$

46.

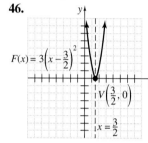

$F(x) = 3\left(x - \frac{3}{2}\right)^2$
$V\left(\frac{3}{2}, 0\right)$
$x = \frac{3}{2}$

47.

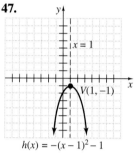

$x = 1$
$V(1, -1)$
$h(x) = -(x-1)^2 - 1$

48.

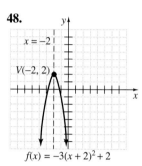

$x = -2$
$V(-2, 2)$
$f(x) = -3(x+2)^2 + 2$

49.
$g(x) = \sqrt{3}(x+5)^2 + \frac{3}{4}$

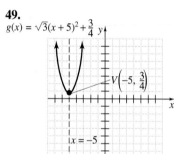

$V\left(-5, \frac{3}{4}\right)$
$x = -5$

50.
$G(x) = \sqrt{5}(x-7)^2 - \frac{1}{2}$

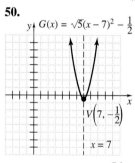

$V\left(7, -\frac{1}{2}\right)$
$x = 7$

51.
$h(x) = 10(x+4)^2 - 6$

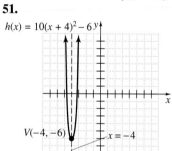

$V(-4, -6)$
$x = -4$

52.

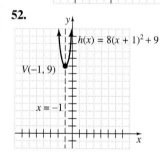

$h(x) = 8(x+1)^2 + 9$
$V(-1, 9)$
$x = -1$

53.

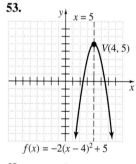

$x = 5$
$V(4, 5)$
$f(x) = -2(x-4)^2 + 5$

54.

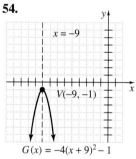

$x = -9$
$V(-9, -1)$
$G(x) = -4(x+9)^2 - 1$

59.

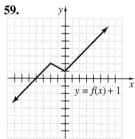

$y = f(x) + 1$

60.

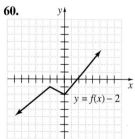

$y = f(x) - 2$

61.

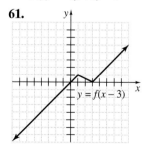

$y = f(x - 3)$

62.

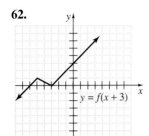

$y = f(x + 3)$

63.

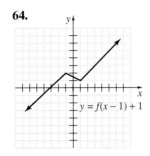

$y = f(x + 2) + 2$

64.

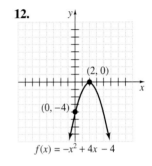

$y = f(x - 1) + 1$

Exercise Set 8.6

9.

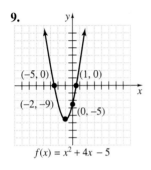

$(-5, 0)$ $(1, 0)$ $(-2, -9)$ $(0, -5)$

$f(x) = x^2 + 4x - 5$

10.

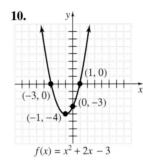

$(-3, 0)$ $(1, 0)$ $(0, -3)$ $(-1, -4)$

$f(x) = x^2 + 2x - 3$

11.

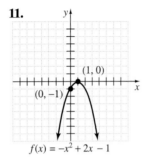

$(1, 0)$ $(0, -1)$

$f(x) = -x^2 + 2x - 1$

12.

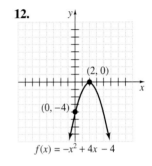

$(2, 0)$ $(0, -4)$

$f(x) = -x^2 + 4x - 4$

13.

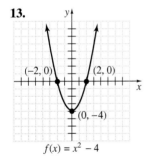

$(-2, 0)$ $(2, 0)$ $(0, -4)$

$f(x) = x^2 - 4$

14.

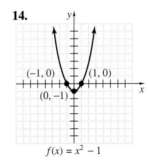

$(-1, 0)$ $(1, 0)$ $(0, -1)$

$f(x) = x^2 - 1$

15.

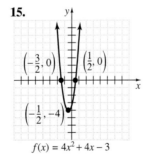

$\left(-\frac{3}{2}, 0\right)$ $\left(\frac{1}{2}, 0\right)$ $\left(-\frac{1}{2}, -4\right)$

$f(x) = 4x^2 + 4x - 3$

16.

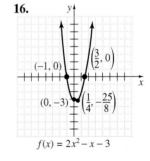

$(-1, 0)$ $\left(\frac{3}{2}, 0\right)$ $(0, -3)$ $\left(\frac{1}{4}, -\frac{25}{8}\right)$

$f(x) = 2x^2 - x - 3$

17.

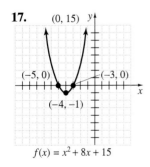

$(0, 15)$ $(-5, 0)$ $(-3, 0)$ $(-4, -1)$

$f(x) = x^2 + 8x + 15$

18.

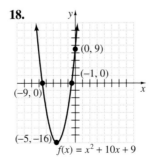

$(0, 9)$ $(-1, 0)$ $(-9, 0)$ $(-5, -16)$

$f(x) = x^2 + 10x + 9$

19.

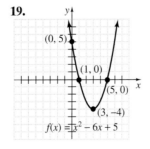

$(0, 5)$ $(1, 0)$ $(5, 0)$ $(3, -4)$

$f(x) = x^2 - 6x + 5$

20.

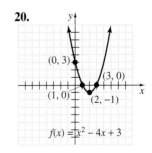

$(0, 3)$ $(3, 0)$ $(1, 0)$ $(2, -1)$

$f(x) = x^2 - 4x + 3$

21.

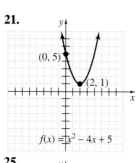

(0, 5) (2, 1)

$f(x) = x^2 - 4x + 5$

22.

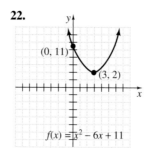

(0, 11) (3, 2)

$f(x) = x^2 - 6x + 11$

23.

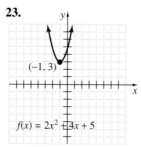

(-1, 3)

$f(x) = 2x^2 + 4x + 5$

24.

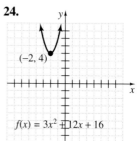

(-2, 4)

$f(x) = 3x^2 + 12x + 16$

25.

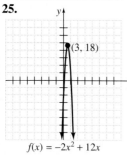

(3, 18)

$f(x) = -2x^2 + 12x$

26.

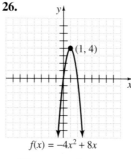

(1, 4)

$f(x) = -4x^2 + 8x$

27.

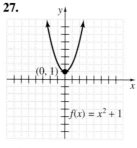

(0, 1)

$f(x) = x^2 + 1$

28.

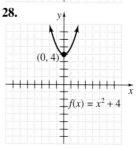

(0, 4)

$f(x) = x^2 + 4$

29.

$f(x) = x^2 - 2x - 15$

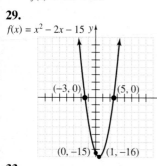

(-3, 0) (5, 0)

(0, -15) (1, -16)

30.

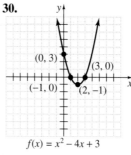

(0, 3) (3, 0)

(-1, 0) (2, -1)

$f(x) = x^2 - 4x + 3$

31.

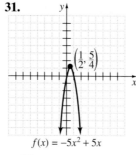

$\left(\frac{1}{2}, \frac{5}{4}\right)$

$f(x) = -5x^2 + 5x$

32.

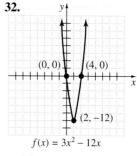

(0, 0) (4, 0)

(2, -12)

$f(x) = 3x^2 - 12x$

33.

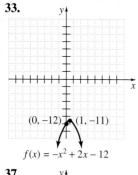

(0, -12) (1, -11)

$f(x) = -x^2 + 2x - 12$

34.

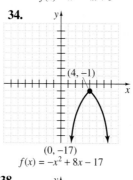

(4, -1)

(0, -17)

$f(x) = -x^2 + 8x - 17$

35.

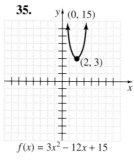

(0, 15)

(2, 3)

$f(x) = 3x^2 - 12x + 15$

36.

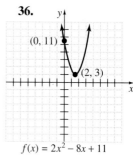

(0, 11)

(2, 3)

$f(x) = 2x^2 - 8x + 11$

37.

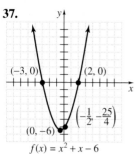

(-3, 0) (2, 0)

$\left(-\frac{1}{2}, -\frac{25}{4}\right)$

(0, -6)

$f(x) = x^2 + x - 6$

38.

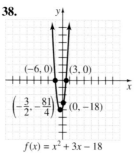

(-6, 0) (3, 0)

$\left(-\frac{3}{2}, -\frac{81}{4}\right)$ (0, -18)

$f(x) = x^2 + 3x - 18$

39.

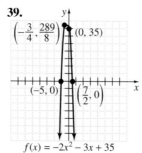

$\left(-\frac{3}{4}, \frac{289}{8}\right)$ (0, 35)

(-5, 0) $\left(\frac{7}{2}, 0\right)$

$f(x) = -2x^2 - 3x + 35$

40.

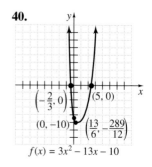

$\left(-\frac{2}{3}, 0\right)$ (5, 0)

(0, -10) $\left(\frac{13}{6}, -\frac{289}{12}\right)$

$f(x) = 3x^2 - 13x - 10$

47.

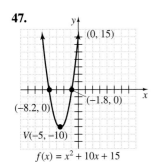

$f(x) = x^2 + 10x + 15$

48.

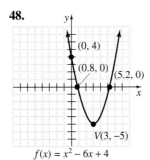

$f(x) = x^2 - 6x + 4$

49.

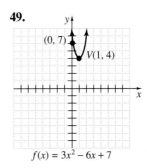

$f(x) = 3x^2 - 6x + 7$

50.

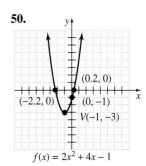

$f(x) = 2x^2 + 4x - 1$

51.

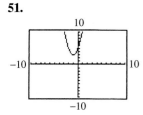

52.

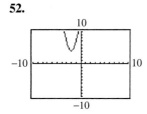

53.

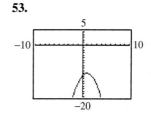

54.

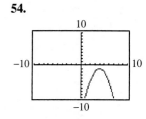

59.

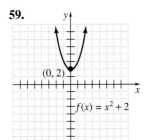

$f(x) = x^2 + 2$

60.

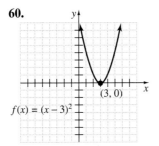

$f(x) = (x - 3)^2$

61.

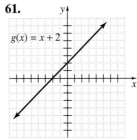

$g(x) = x + 2$

62.

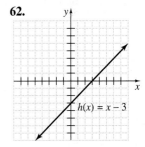

$h(x) = x - 3$

63.

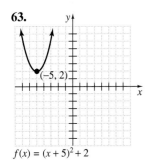

$f(x) = (x + 5)^2 + 2$

64.

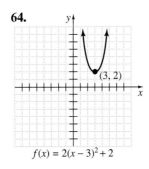

$f(x) = 2(x - 3)^2 + 2$

65.

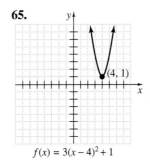

$f(x) = 3(x - 4)^2 + 1$

66.

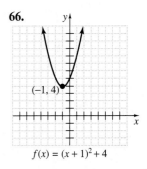

$f(x) = (x + 1)^2 + 4$

67.

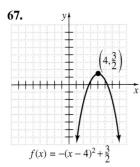

$f(x) = -(x - 4)^2 + \frac{3}{2}$

68.

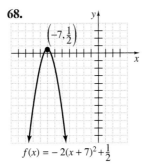

$f(x) = -2(x + 7)^2 + \frac{1}{2}$

Chapter 8 Review

53.

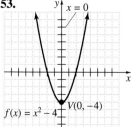

$f(x) = x^2 - 4$; $x = 0$; $V(0, -4)$

54.

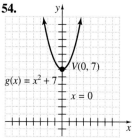

$g(x) = x^2 + 7$; $V(0, 7)$; $x = 0$

55.

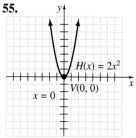

$H(x) = 2x^2$; $V(0, 0)$; $x = 0$

56.

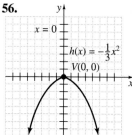

$x = 0$; $h(x) = -\frac{1}{3}x^2$; $V(0, 0)$

57.

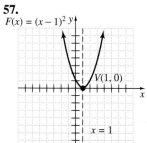

$F(x) = (x - 1)^2$; $V(1, 0)$; $x = 1$

58.

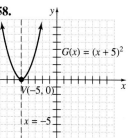

$G(x) = (x + 5)^2$; $V(-5, 0)$; $x = -5$

59.

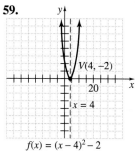

$V(4, -2)$; $x = 4$; $f(x) = (x - 4)^2 - 2$

60.

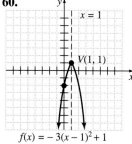

$x = 1$; $V(1, 1)$; $f(x) = -3(x - 1)^2 + 1$

61.

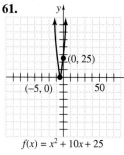

$(0, 25)$; $(-5, 0)$; $f(x) = x^2 + 10x + 25$

62.

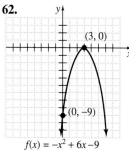

$(3, 0)$; $(0, -9)$; $f(x) = -x^2 + 6x - 9$

63.

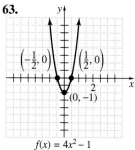

$\left(-\frac{1}{2}, 0\right)$; $\left(\frac{1}{2}, 0\right)$; $(0, -1)$; $f(x) = 4x^2 - 1$

64.

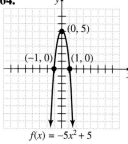

$(0, 5)$; $(-1, 0)$; $(1, 0)$; $f(x) = -5x^2 + 5$

65.

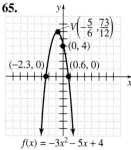

$V\left(-\frac{5}{6}, \frac{73}{12}\right)$; $(0, 4)$; $(-2.3, 0)$; $(0.6, 0)$; $f(x) = -3x^2 - 5x + 4$

Chapter 8 Test

17.
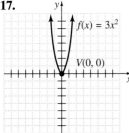
$f(x) = 3x^2$
$V(0, 0)$

18.
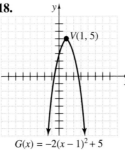
$V(1, 5)$
$G(x) = -2(x - 1)^2 + 5$

19.
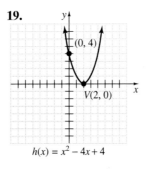
$(0, 4)$
$V(2, 0)$
$h(x) = x^2 - 4x + 4$

20.
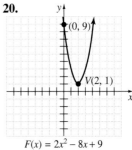
$(0, 9)$
$V(2, 1)$
$F(x) = 2x^2 - 8x + 9$

Chapter 8 Cumulative Review

7. *(Sec. 5.9, Ex. 4)*
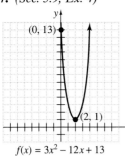
$(0, 13)$
$(2, 1)$
$f(x) = 3x^2 - 12x + 13$

23. *(Sec. 8.5, Ex. 5)*
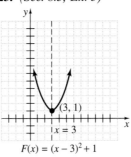
$(3, 1)$
$x = 3$
$F(x) = (x - 3)^2 + 1$

CHAPTER 9

Conic Sections

Graphing Calculator Explorations 9.1

1.
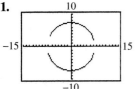

2.
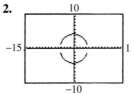

3.
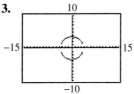

4.
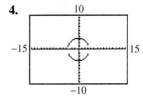

Exercise Set 9.1

1.
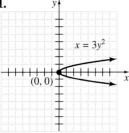
$x = 3y^2$
$(0, 0)$

2.
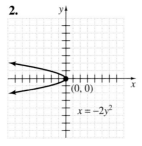
$(0, 0)$
$x = -2y^2$

3.
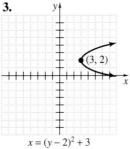
$(3, 2)$
$x = (y - 2)^2 + 3$

4.
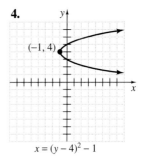
$(-1, 4)$
$x = (y - 4)^2 - 1$

5.

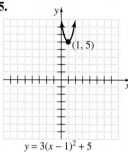

(1, 5)

$$y = 3(x - 1)^2 + 5$$

6.

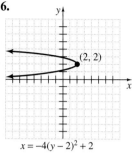

(2, 2)

$$x = -4(y - 2)^2 + 2$$

7.

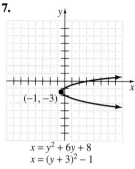

(-1, -3)

$$x = y^2 + 6y + 8$$
$$x = (y + 3)^2 - 1$$

8.

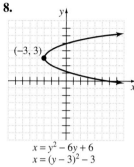

(-3, 3)

$$x = y^2 - 6y + 6$$
$$x = (y - 3)^2 - 3$$

9.

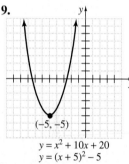

(-5, -5)

$$y = x^2 + 10x + 20$$
$$y = (x + 5)^2 - 5$$

10.

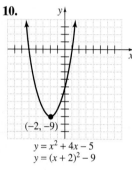

(-2, -9)

$$y = x^2 + 4x - 5$$
$$y = (x + 2)^2 - 9$$

11.

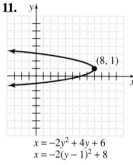

(8, 1)

$$x = -2y^2 + 4y + 6$$
$$x = -2(y - 1)^2 + 8$$

12.

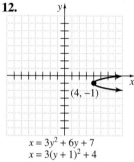

(4, -1)

$$x = 3y^2 + 6y + 7$$
$$x = 3(y + 1)^2 + 4$$

37.

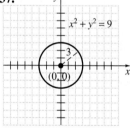

$x^2 + y^2 = 9$

3

(0, 0)

38.

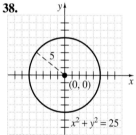

5

(0, 0)

$$x^2 + y^2 = 25$$

39.

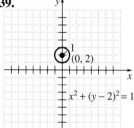

1

(0, 2)

$$x^2 + (y - 2)^2 = 1$$

40.

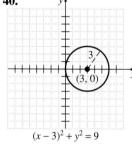

3

(3, 0)

$$(x - 3)^2 + y^2 = 9$$

41.

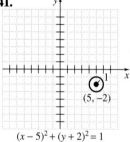

1

(5, -2)

$$(x - 5)^2 + (y + 2)^2 = 1$$

42.

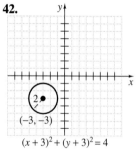

2

(-3, -3)

$$(x + 3)^2 + (y + 3)^2 = 4$$

43.

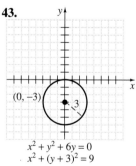

(0, -3)

3

$$x^2 + y^2 + 6y = 0$$
$$x^2 + (y + 3)^2 = 9$$

44.

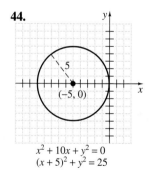

5

(-5, 0)

$$x^2 + 10x + y^2 = 0$$
$$(x + 5)^2 + y^2 = 25$$

45.

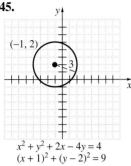

$x^2 + y^2 + 2x - 4y = 4$
$(x + 1)^2 + (y - 2)^2 = 9$

46.

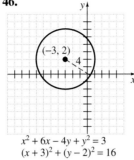

$x^2 + 6x - 4y + y^2 = 3$
$(x + 3)^2 + (y - 2)^2 = 16$

54.

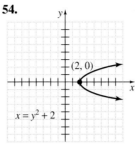

$x = y^2 + 2$

55.

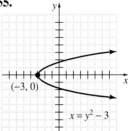

$x = y^2 - 3$

56.

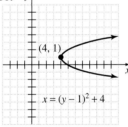

$y = (x + 3)^2 + 3$

57.

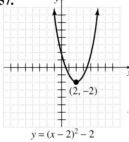

$y = (x - 2)^2 - 2$

58.

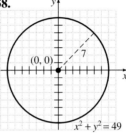

$x^2 + y^2 = 49$

59.

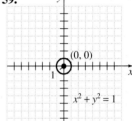

$x^2 + y^2 = 1$

60.

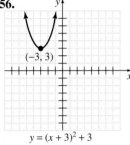

$x = (y - 1)^2 + 4$

61.

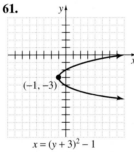

$x = (y + 3)^2 - 1$

62.

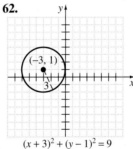

$(x + 3)^2 + (y - 1)^2 = 9$

63.

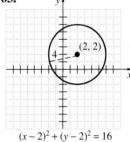

$(x - 2)^2 + (y - 2)^2 = 16$

64.

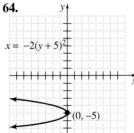

$x = -2(y + 5)^2$

65.

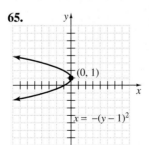

$x = -(y - 1)^2$

66.

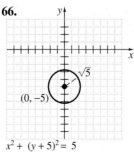

$x^2 + (y + 5)^2 = 5$

67.

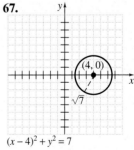

$(x - 4)^2 + y^2 = 7$

68.

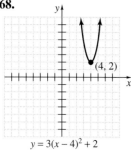

$y = 3(x - 4)^2 + 2$

69.

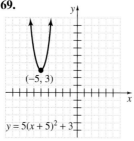

$y = 5(x + 5)^2 + 3$

70.

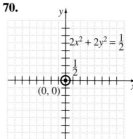

$2x^2 + 2y^2 = \frac{1}{2}$

71.

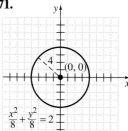

$\dfrac{x^2}{8} + \dfrac{y^2}{8} = 2$

72.

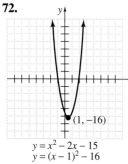

$y = x^2 - 2x - 15$
$y = (x - 1)^2 - 16$

73.

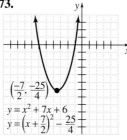

$y = x^2 + 7x + 6$
$y = \left(x + \dfrac{7}{2}\right)^2 - \dfrac{25}{4}$

74.

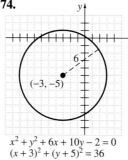

$x^2 + y^2 + 6x + 10y - 2 = 0$
$(x + 3)^2 + (y + 5)^2 = 36$

75.

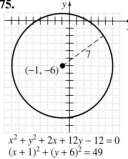

$x^2 + y^2 + 2x + 12y - 12 = 0$
$(x + 1)^2 + (y + 6)^2 = 49$

76.

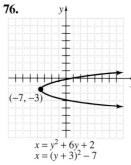

$x = y^2 + 6y + 2$
$x = (y + 3)^2 - 7$

77.

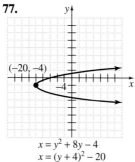

$x = y^2 + 8y - 4$
$x = (y + 4)^2 - 20$

78.

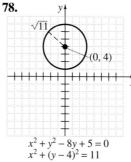

$x^2 + y^2 - 8y + 5 = 0$
$x^2 + (y - 4)^2 = 11$

79.

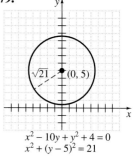

$x^2 - 10y + y^2 + 4 = 0$
$x^2 + (y - 5)^2 = 21$

80.

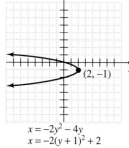

$x = -2y^2 - 4y$
$x = -2(y + 1)^2 + 2$

81.

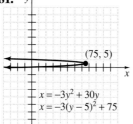

$x = -3y^2 + 30y$
$x = -3(y - 5)^2 + 75$

82.

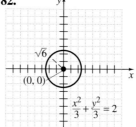

$\dfrac{x^2}{3} + \dfrac{y^2}{3} = 2$

83.

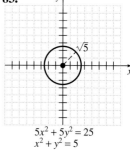

$5x^2 + 5y^2 = 25$
$x^2 + y^2 = 5$

84.
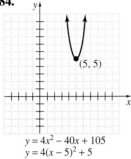
$y = 4x^2 - 40x + 105$
$y = 4(x - 5)^2 + 5$

85.
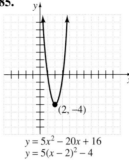
$y = 5x^2 - 20x + 16$
$y = 5(x - 2)^2 - 4$

90.
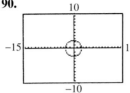

91.
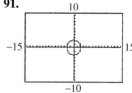

92.
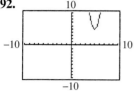

93.
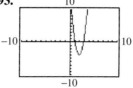

94.
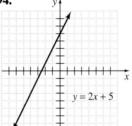
$y = 2x + 5$

95.
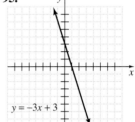
$y = -3x + 3$

96.
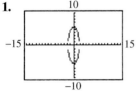
$y = 3$

97.
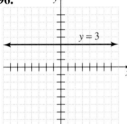
$x = -2$

Graphing Calculator Explorations 9.2

1.
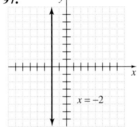

2.
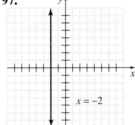

3.
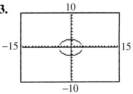

4.
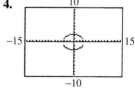

Exercise Set 9.2

1.
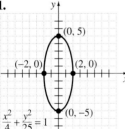
$(0, 5)$
$(-2, 0)$ $(2, 0)$
$(0, -5)$
$\dfrac{x^2}{4} + \dfrac{y^2}{25} = 1$

2.
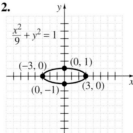
$\dfrac{x^2}{9} + y^2 = 1$
$(-3, 0)$ $(0, 1)$
$(0, -1)$ $(3, 0)$

3.
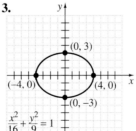
$(0, 3)$
$(-4, 0)$ $(4, 0)$
$(0, -3)$
$\dfrac{x^2}{16} + \dfrac{y^2}{9} = 1$

4.
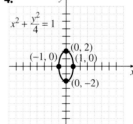
$x^2 + \dfrac{y^2}{4} = 1$
$(0, 2)$
$(-1, 0)$ $(1, 0)$
$(0, -2)$

5.

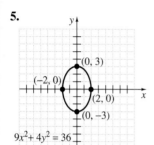

$9x^2 + 4y^2 = 36$

6.

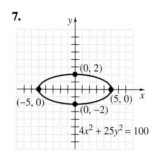

$x^2 + 4y^2 = 16$

7.

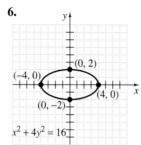

$4x^2 + 25y^2 = 100$

8.

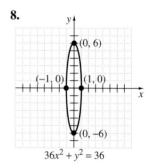

$36x^2 + y^2 = 36$

9.

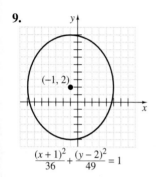

$\dfrac{(x+1)^2}{36} + \dfrac{(y-2)^2}{49} = 1$

10.

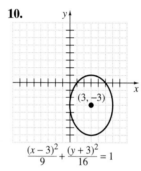

$\dfrac{(x-3)^2}{9} + \dfrac{(y+3)^2}{16} = 1$

11.

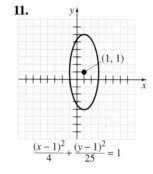

$\dfrac{(x-1)^2}{4} + \dfrac{(y-1)^2}{25} = 1$

12.

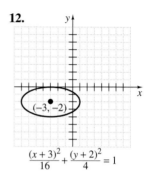

$\dfrac{(x+3)^2}{16} + \dfrac{(y+2)^2}{4} = 1$

13.

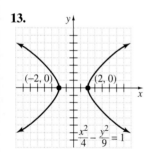

$\dfrac{x^2}{4} - \dfrac{y^2}{9} = 1$

14.

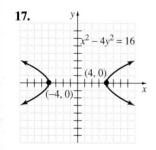

$\dfrac{x^2}{36} - \dfrac{y^2}{36} = 1$

15.

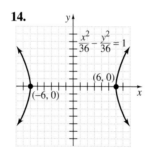

$\dfrac{y^2}{25} - \dfrac{x^2}{16} = 1$

16.

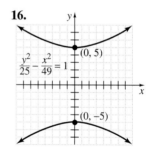

$\dfrac{y^2}{25} - \dfrac{x^2}{49} = 1$

17.

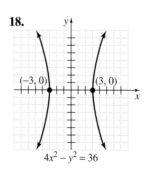

$x^2 - 4y^2 = 16$

18.

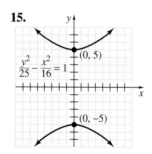

$4x^2 - y^2 = 36$

19.

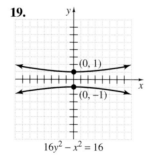

$16y^2 - x^2 = 16$

20.

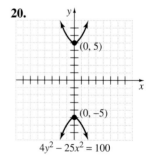

$4y^2 - 25x^2 = 100$

22. circle

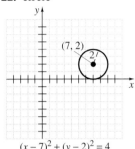

(7, 2)

$(x - 7)^2 + (y - 2)^2 = 4$

23. parabola

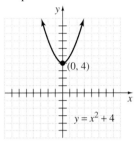

(0, 4)

$y = x^2 + 4$

24. parabola

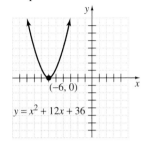

(−6, 0)

$y = x^2 + 12x + 36$

25. ellipse

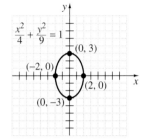

$\dfrac{x^2}{4} + \dfrac{y^2}{9} = 1$

(0, 3)

(−2, 0)

(2, 0)

(0, −3)

26. hyperbola

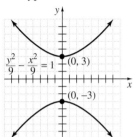

$\dfrac{y^2}{9} - \dfrac{x^2}{9} = 1$

(0, 3)

(0, −3)

27. hyperbola

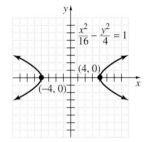

$\dfrac{x^2}{16} - \dfrac{y^2}{4} = 1$

(4, 0)

(−4, 0)

28. ellipse

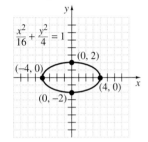

$\dfrac{x^2}{16} + \dfrac{y^2}{4} = 1$

(0, 2)

(−4, 0)

(4, 0)

(0, −2)

29. circle

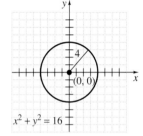

4

(0, 0)

$x^2 + y^2 = 16$

30. parabola

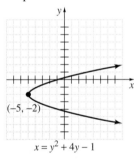

(−5, −2)

$x = y^2 + 4y - 1$

31. parabola

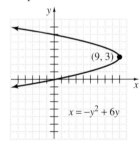

(9, 3)

$x = -y^2 + 6y$

32. hyperbola

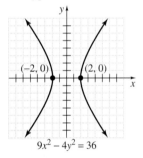

(−2, 0)

(2, 0)

$9x^2 - 4y^2 = 36$

33. ellipse

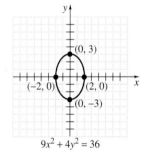

(0, 3)

(−2, 0)

(2, 0)

(0, −3)

$9x^2 + 4y^2 = 36$

34. ellipse

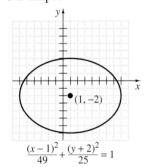

(1, −2)

$\dfrac{(x - 1)^2}{49} + \dfrac{(y + 2)^2}{25} = 1$

35. hyperbola

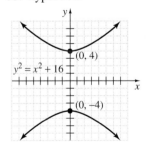

$y^2 = x^2 + 16$

(0, 4)

(0, −4)

36. circle

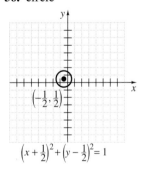

$\left(-\dfrac{1}{2}, \dfrac{1}{2}\right)$

$\left(x + \dfrac{1}{2}\right)^2 + \left(y - \dfrac{1}{2}\right)^2 = 1$

37. parabola

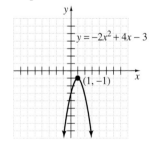

$y = -2x^2 + 4x - 3$

(1, −1)

40.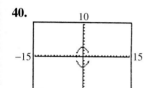

41.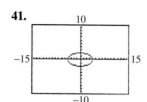

50.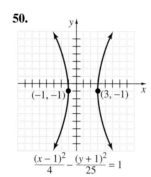

$$\frac{(x-1)^2}{4} - \frac{(y+1)^2}{25} = 1$$

51.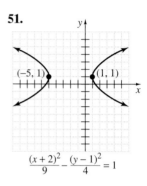

$$\frac{(x+2)^2}{9} - \frac{(y-1)^2}{4} = 1$$

52.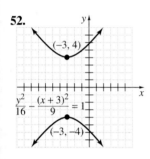

$$\frac{y^2}{16} - \frac{(x+3)^2}{9} = 1$$

53.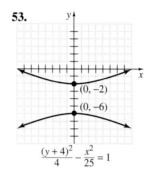

$$\frac{(y+4)^2}{4} - \frac{x^2}{25} = 1$$

54.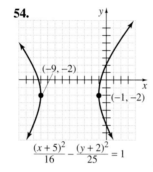

$$\frac{(x+5)^2}{16} - \frac{(y+2)^2}{25} = 1$$

55.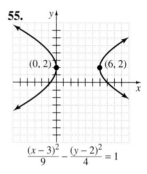

$$\frac{(x-3)^2}{9} - \frac{(y-2)^2}{4} = 1$$

Exercise Set 9.3

39.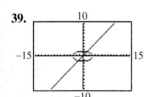

40.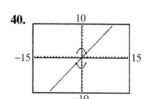

41.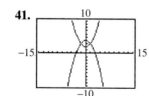

42.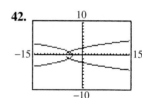

43.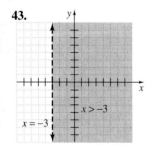

44.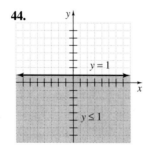

45.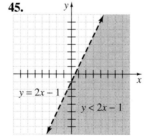

46.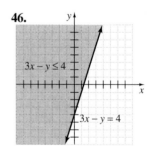

Exercise Set 9.4

1.
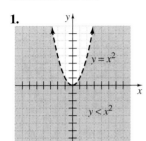
$y = x^2$
$y < x^2$

2.
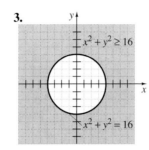
$y = -x^2$ $y < -x^2$

3.
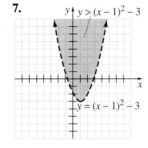
$x^2 + y^2 \geq 16$
$x^2 + y^2 = 16$

4.
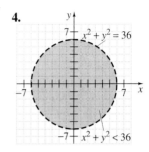
$x^2 + y^2 = 36$
$x^2 + y^2 < 36$

5.
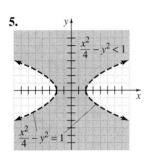
$\dfrac{x^2}{4} - y^2 < 1$
$\dfrac{x^2}{4} - y^2 = 1$

6.
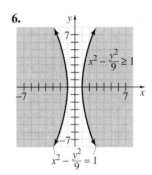
$x^2 - \dfrac{y^2}{9} \geq 1$
$x^2 - \dfrac{y^2}{9} = 1$

7.
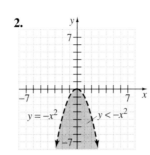
$y > (x - 1)^2 - 3$
$y = (x - 1)^2 - 3$

8.
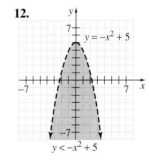
$y > (x + 3)^2 + 2$
$y = (x + 3)^2 + 2$

9.
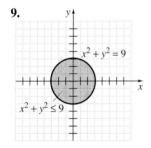
$x^2 + y^2 = 9$
$x^2 + y^2 \leq 9$

10.
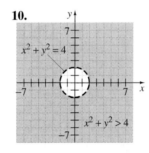
$x^2 + y^2 = 4$
$x^2 + y^2 > 4$

11.
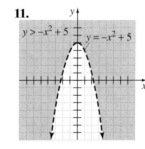
$y > -x^2 + 5$ $y = -x^2 + 5$

12.
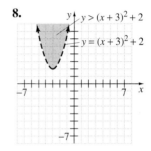
$y = -x^2 + 5$
$y < -x^2 + 5$

13.
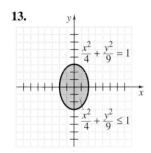
$\dfrac{x^2}{4} + \dfrac{y^2}{9} = 1$
$\dfrac{x^2}{4} + \dfrac{y^2}{9} \leq 1$

14.
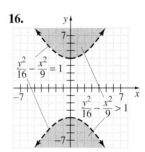
$\dfrac{x^2}{25} + \dfrac{y^2}{4} = 1$
$\dfrac{x^2}{25} + \dfrac{y^2}{4} \geq 1$

15.
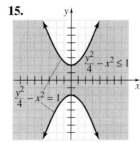
$\dfrac{y^2}{4} - x^2 \leq 1$
$\dfrac{y^2}{4} - x^2 = 1$

16.
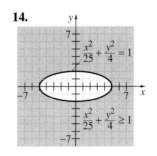
$\dfrac{y^2}{16} - \dfrac{x^2}{9} = 1$
$\dfrac{y^2}{16} - \dfrac{x^2}{9} > 1$

17.

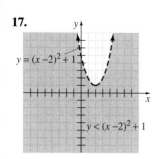

$y = (x-2)^2 + 1$

$y < (x-2)^2 + 1$

18.

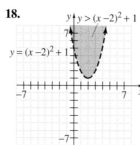

$y > (x-2)^2 + 1$

$y = (x-2)^2 + 1$

19.

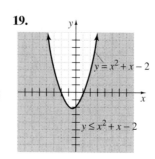

$y = x^2 + x - 2$

$y \leq x^2 + x - 2$

20.

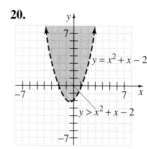

$y = x^2 + x - 2$

$y > x^2 + x - 2$

23.

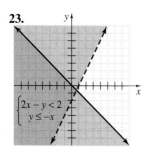

$\begin{cases} 2x - y < 2 \\ y \leq -x \end{cases}$

24.

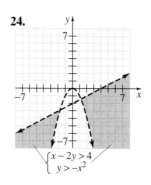

$\begin{cases} x - 2y > 4 \\ y > -x^2 \end{cases}$

25.

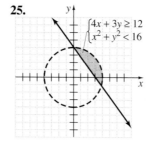

$\begin{cases} 4x + 3y \geq 12 \\ x^2 + y^2 < 16 \end{cases}$

26.

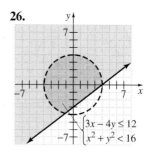

$\begin{cases} 3x - 4y \leq 12 \\ x^2 + y^2 < 16 \end{cases}$

27.

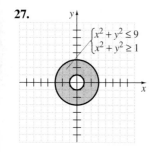

$\begin{cases} x^2 + y^2 \leq 9 \\ x^2 + y^2 \geq 1 \end{cases}$

28.

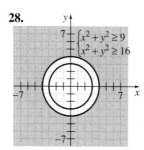

$\begin{cases} x^2 + y^2 \geq 9 \\ x^2 + y^2 \geq 16 \end{cases}$

29.

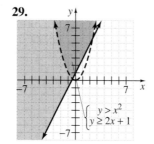

$\begin{cases} y > x^2 \\ y \geq 2x + 1 \end{cases}$

30.

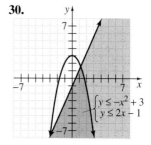

$\begin{cases} y \leq -x^2 + 3 \\ y \leq 2x - 1 \end{cases}$

31.

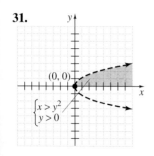

$(0, 0)$

$\begin{cases} x > y^2 \\ y > 0 \end{cases}$

32.

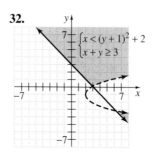

$\begin{cases} x < (y + 1)^2 + 2 \\ x + y \geq 3 \end{cases}$

33.

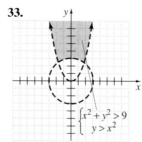

$\begin{cases} x^2 + y^2 > 9 \\ y > x^2 \end{cases}$

34.

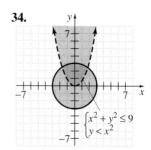

$\begin{cases} x^2 + y^2 \leq 9 \\ y < x^2 \end{cases}$

35.

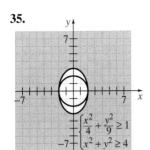

$$\begin{cases} \dfrac{x^2}{4} + \dfrac{y^2}{9} \geq 1 \\ x^2 + y^2 \geq 4 \end{cases}$$

36.

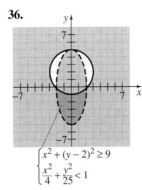

$$\begin{cases} x^2 + (y-2)^2 \geq 9 \\ \dfrac{x^2}{4} + \dfrac{y^2}{25} < 1 \end{cases}$$

37.

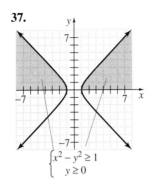

$$\begin{cases} x^2 - y^2 \geq 1 \\ y \geq 0 \end{cases}$$

38.

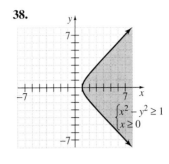

$$\begin{cases} x^2 - y^2 \geq 1 \\ x \geq 0 \end{cases}$$

39.

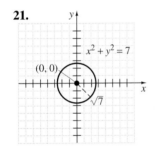

$$\begin{cases} x + y \geq 1 \\ 2x + 3y < 1 \\ x > -3 \end{cases}$$

40.

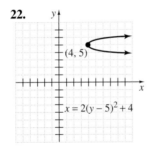

$$\begin{cases} x - y < -1 \\ 4x - 3y > 0 \\ y > 0 \end{cases}$$

41.

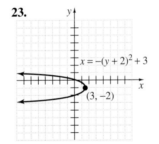

$$\begin{cases} x^2 - y^2 < 1 \\ \dfrac{x^2}{16} + y^2 \leq 1 \\ x \geq -2 \end{cases}$$

42.

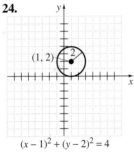

$$\begin{cases} x^2 - y^2 \geq 1 \\ \dfrac{x^2}{16} + \dfrac{y^2}{4} \leq 1 \\ y \geq 1 \end{cases}$$

Chapter 9 Review

21.

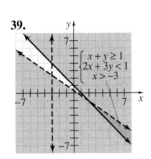

$x^2 + y^2 = 7$
$(0, 0)$
$\sqrt{7}$

22.

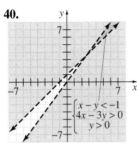

$(4, 5)$
$x = 2(y - 5)^2 + 4$

23.

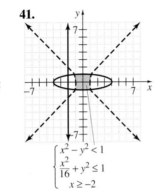

$x = -(y + 2)^2 + 3$
$(3, -2)$

24.

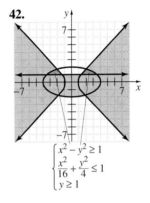

$(1, 2)$ 2
$(x - 1)^2 + (y - 2)^2 = 4$

25.

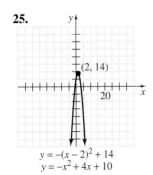

$(2, 14)$
20
$y = -(x - 2)^2 + 14$
$y = -x^2 + 4x + 10$

26.

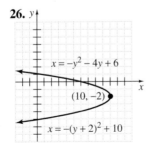

$x = -y^2 - 4y + 6$
$(10, -2)$
$x = -(y + 2)^2 + 10$

27.

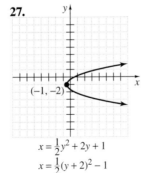

$(-1, -2)$
$x = \dfrac{1}{2}y^2 + 2y + 1$
$x = \dfrac{1}{2}(y + 2)^2 - 1$

28.

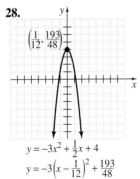

$\left(\dfrac{1}{12}, \dfrac{193}{48}\right)$
$y = -3x^2 + \dfrac{1}{2}x + 4$
$y = -3\left(x - \dfrac{1}{12}\right)^2 + \dfrac{193}{48}$

29.

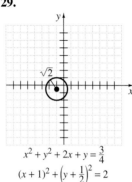

$$x^2 + y^2 + 2x + y = \frac{3}{4}$$
$$(x+1)^2 + \left(y + \frac{1}{2}\right)^2 = 2$$

30.

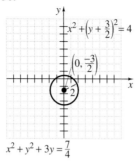

$$x^2 + y^2 + 3y = \frac{7}{4}$$

31.

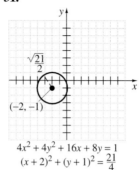

$$4x^2 + 4y^2 + 16x + 8y = 1$$
$$(x+2)^2 + (y+1)^2 = \frac{21}{4}$$

32.

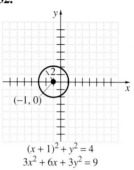

$$(x+1)^2 + y^2 = 4$$
$$3x^2 + 6x + 3y^2 = 9$$

33.

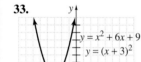

34.

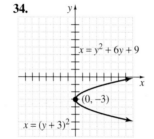

36.

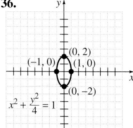

37.

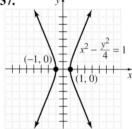

38.

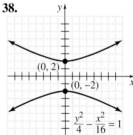

39.

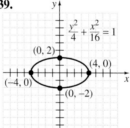

40.

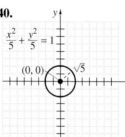

41.

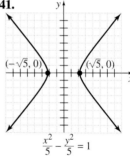

42.

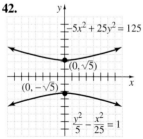

43.

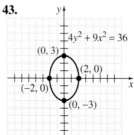

44.

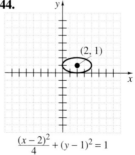

$$\frac{(x-2)^2}{4} + (y-1)^2 = 1$$

45.

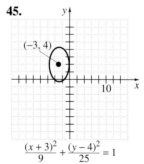

$$\frac{(x+3)^2}{9} + \frac{(y-4)^2}{25} = 1$$

46.

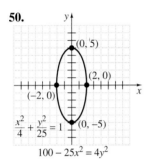

$(-1, 0)$ $(1, 0)$

$x^2 - y^2 = 1$

47.

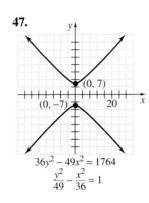

$(0, 7)$

$(0, -7)$ 20

$36y^2 - 49x^2 = 1764$

$$\frac{y^2}{49} - \frac{x^2}{36} = 1$$

48.

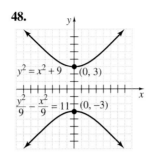

$y^2 = x^2 + 9$ $(0, 3)$

$\dfrac{y^2}{9} - \dfrac{x^2}{9} = 11$ $(0, -3)$

49.

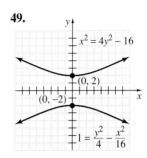

$x^2 = 4y^2 - 16$

$(0, 2)$

$(0, -2)$

$1 = \dfrac{y^2}{4} - \dfrac{x^2}{16}$

50.

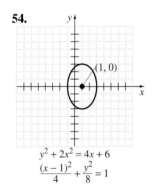

$(0, 5)$

$(2, 0)$

$(-2, 0)$

$\dfrac{x^2}{4} + \dfrac{y^2}{25} = 1$ $(0, -5)$

$100 - 25x^2 = 4y^2$

51.

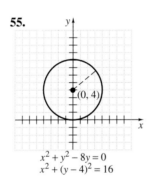

$(-2, 2)$

10

$y = x^2 - 4x + 6$

52.

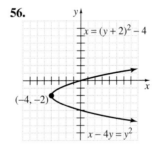

$(0, \sqrt{6})$

$y^2 = x^2 + 6$ $(0, -\sqrt{6})$

$$\frac{y^2}{6} - \frac{x^2}{6} = 1$$

53.

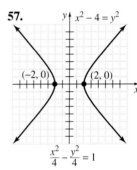

$\sqrt{10}$

$(2, 0)$

$y^2 + x^2 = 4x + 6$

$(x - 2)^2 + y^2 = 10$

54.

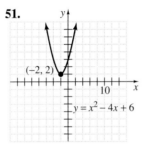

$(1, 0)$

$y^2 + 2x^2 = 4x + 6$

$$\frac{(x-1)^2}{4} + \frac{y^2}{8} = 1$$

55.

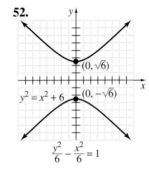

$(0, 4)$

$x^2 + y^2 - 8y = 0$

$x^2 + (y - 4)^2 = 16$

56.

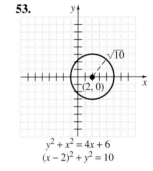

$x = (y + 2)^2 - 4$

$(-4, -2)$

$x - 4y = y^2$

57.

$x^2 - 4 = y^2$

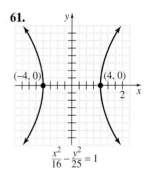

$(-2, 0)$ $(2, 0)$

$$\frac{x^2}{4} - \frac{y^2}{4} = 1$$

58.

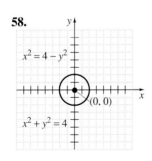

$x^2 = 4 - y^2$

$(0, 0)$

$x^2 + y^2 = 4$

59.

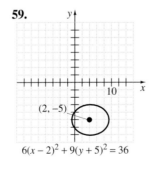

10

$(2, -5)$

$6(x - 2)^2 + 9(y + 5)^2 = 36$

60.

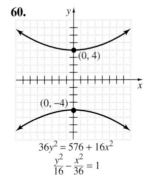

$(0, 4)$

$(0, -4)$

$36y^2 = 576 + 16x^2$

$$\frac{y^2}{16} - \frac{x^2}{36} = 1$$

61.

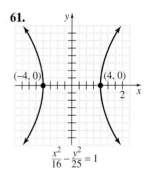

$(-4, 0)$ $(4, 0)$

2

$$\frac{x^2}{16} - \frac{y^2}{25} = 1$$

62.

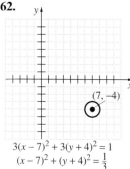

$$3(x-7)^2 + 3(y+4)^2 = 1$$
$$(x-7)^2 + (y+4)^2 = \tfrac{1}{3}$$

63.

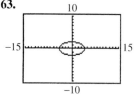

64.

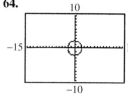

65.

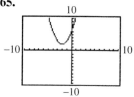

66.

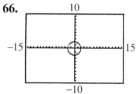

79.

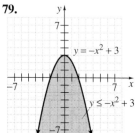

80.

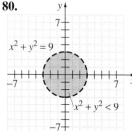

81.

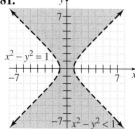

82.

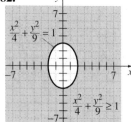

83.
$$\begin{cases} 2x \le 4 \\ x+y \ge 1 \end{cases}$$

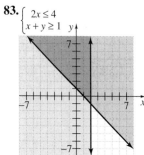

84.

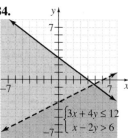

$$\begin{cases} 3x+4y \le 12 \\ x-2y > 6 \end{cases}$$

85.
$$\begin{cases} y > x^2 \\ y+x \ge 3 \end{cases}$$

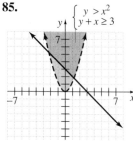

86.

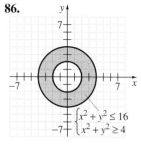

$$\begin{cases} x^2+y^2 \le 16 \\ x^2+y^2 \ge 4 \end{cases}$$

87.

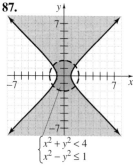

$$\begin{cases} x^2+y^2 < 4 \\ x^2-y^2 \le 1 \end{cases}$$

88.

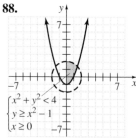

$$\begin{cases} x^2+y^2 < 4 \\ y \ge x^2-1 \\ x \ge 0 \end{cases}$$

Chapter 9 Test

5.

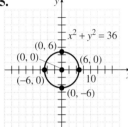

$x^2 + y^2 = 36$

6.

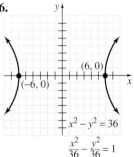

$x^2 - y^2 = 36$

$\dfrac{x^2}{36} - \dfrac{y^2}{36} = 1$

7.

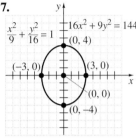

$\dfrac{x^2}{9} + \dfrac{y^2}{16} = 1$

$16x^2 + 9y^2 = 144$

8.

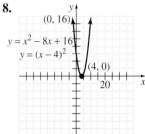

$y = x^2 - 8x + 16$

$y = (x - 4)^2$

9.

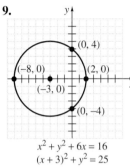

$x^2 + y^2 + 6x = 16$

$(x + 3)^2 + y^2 = 25$

10.

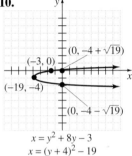

$x = y^2 + 8y - 3$

$x = (y + 4)^2 - 19$

11.

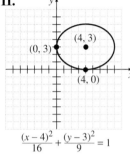

$\dfrac{(x - 4)^2}{16} + \dfrac{(y - 3)^2}{9} = 1$

12.

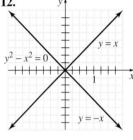

$y^2 - x^2 = 0$

$y = x$

$y = -x$

17.

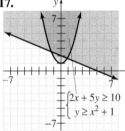

$\begin{cases} 2x + 5y \geq 10 \\ y \geq x^2 + 1 \end{cases}$

18.

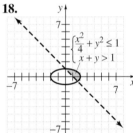

$\begin{cases} \dfrac{x^2}{4} + y^2 \leq 1 \\ x + y > 1 \end{cases}$

19.

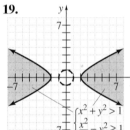

$\begin{cases} x^2 + y^2 > 1 \\ \dfrac{x^2}{4} - y^2 \geq 1 \end{cases}$

20.

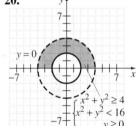

$y = 0$

$\begin{cases} x^2 + y^2 \geq 4 \\ x^2 + y^2 < 16 \\ y \geq 0 \end{cases}$

Chapter 9 Cumulative Review

6. *(Sec. 3.1, Ex. 7)*

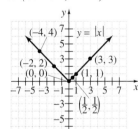

$y = |x|$

21. *(Sec. 8.5, Ex. 2)*

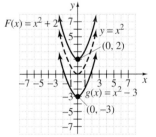

$F(x) = x^2 + 2$

$y = x^2$

$g(x) = x^2 - 3$

CHAPTER 10

Exponential and Logarithmic Functions

Exercise Set 10.1

43.

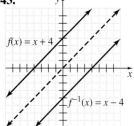

$f(x) = x + 4$

$f^{-1}(x) = x - 4$

44.

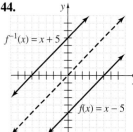

$f^{-1}(x) = x + 5$

$f(x) = x - 5$

45.

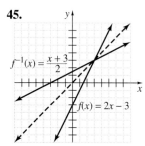

$f^{-1}(x) = \dfrac{x+3}{2}$

$f(x) = 2x - 3$

46.

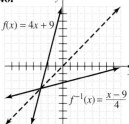

$f(x) = 4x + 9$

$f^{-1}(x) = \dfrac{x-9}{4}$

47.

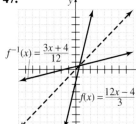

$f^{-1}(x) = \dfrac{3x+4}{12}$

$f(x) = \dfrac{12x-4}{3}$

48.

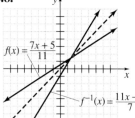

$f(x) = \dfrac{7x+5}{11}$

$f^{-1}(x) = \dfrac{11x-5}{7}$

49.

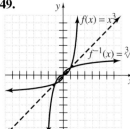

$f(x) = x^3$

$f^{-1}(x) = \sqrt[3]{x}$

50.

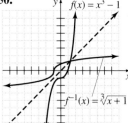

$f(x) = x^3 - 1$

$f^{-1}(x) = \sqrt[3]{x+1}$

51.

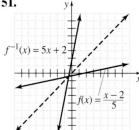

$f^{-1}(x) = 5x + 2$

$f(x) = \dfrac{x-2}{5}$

52.

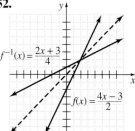

$f^{-1}(x) = \dfrac{2x+3}{4}$

$f(x) = \dfrac{4x-3}{2}$

53.

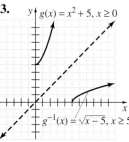

$g(x) = x^2 + 5,\ x \geq 0$

$g^{-1}(x) = \sqrt{x-5},\ x \geq 5$

54.

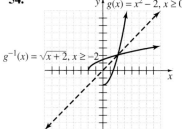

$g(x) = x^2 - 2,\ x \geq 0$

$g^{-1}(x) = \sqrt{x+2},\ x \geq -2$

59. (c)

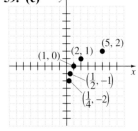

$(1, 0)$ $(2, 1)$ $(5, 2)$ $\left(\frac{1}{2}, -1\right)$ $\left(\frac{1}{4}, -2\right)$

59. (d)

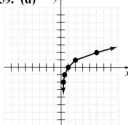

60. (c)

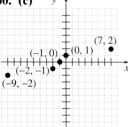

$(-1, 0)$ $(0, 1)$ $(7, 2)$ $(-2, -1)$ $(-9, -2)$

60. (d)

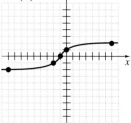

61. 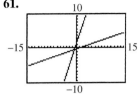   **62.** 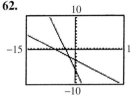   **63.** 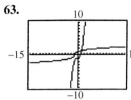   **64.**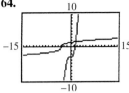

Graphing Calculator Explorations 10.2

1. 81.98%   **2.** 37.04%   **3.** 22.54%   **4.** 8.35%

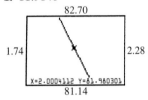   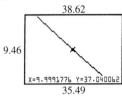   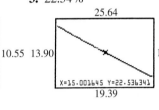   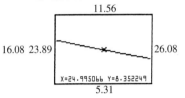

Exercise Set 10.2

1. 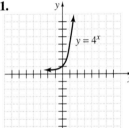   **2.** 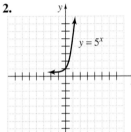   **3.** 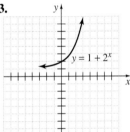   **4.**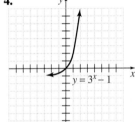

5. 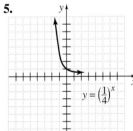   **6.** 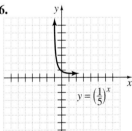   **7.** 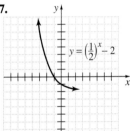   **8.**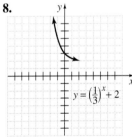

9. 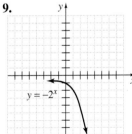   **10.** 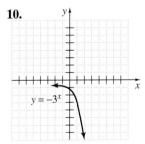   **11.** 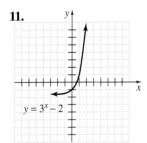   **12.**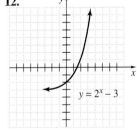

13.
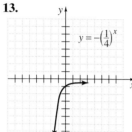
$y = -\left(\frac{1}{4}\right)^x$

14.
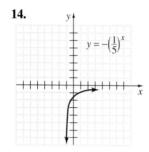
$y = -\left(\frac{1}{5}\right)^x$

15.
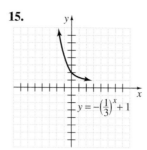
$y = -\left(\frac{1}{3}\right)^x + 1$

16.
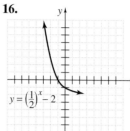
$y = \left(\frac{1}{2}\right)^x - 2$

51.
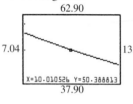

52.
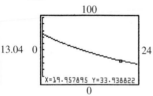

53. 18.62 lb
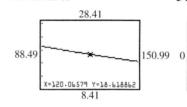

54.
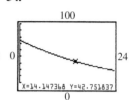

55. 50.41 grams
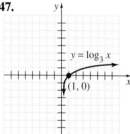

56.
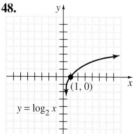

Exercise Set 10.3

47.
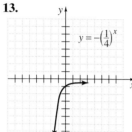
$y = \log_3 x$
$(1, 0)$

48.
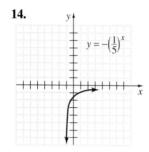
$(1, 0)$
$y = \log_2 x$

49.
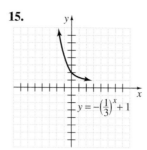
$(1, 0)$
$y = \log_{1/4} x$

50.
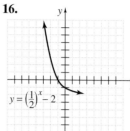
$(1, 0)$
$f(x) = \log_{1/2} x$

51.
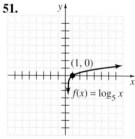
$(1, 0)$
$f(x) = \log_5 x$

52.
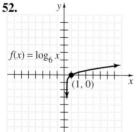
$f(x) = \log_6 x$
$(1, 0)$

53.
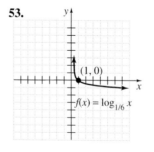
$(1, 0)$
$f(x) = \log_{1/6} x$

54.
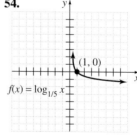
$(1, 0)$
$f(x) = \log_{1/5} x$

55.
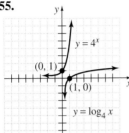

56.
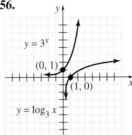

57.
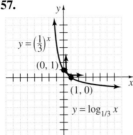

58.
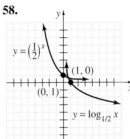

Exercise Set 10.4

61.
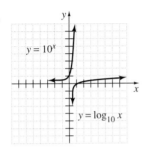

Exercise Set 10.5

69.
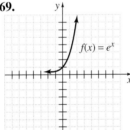

70.
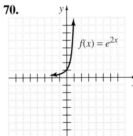

71.
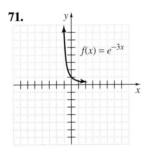

72.
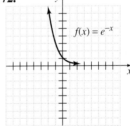

73.
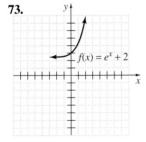

74.
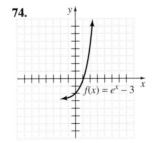

75.
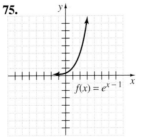

76.
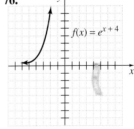

77.

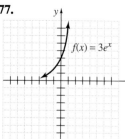

$f(x) = 3e^x$

78.

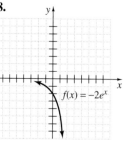

$f(x) = -2e^x$

79.

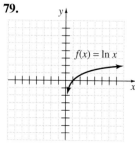

$f(x) = \ln x$

80.

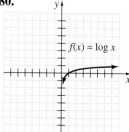

$f(x) = \log x$

81.

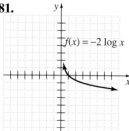

$f(x) = -2 \log x$

82.

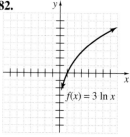

$f(x) = 3 \ln x$

83.

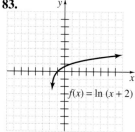

$f(x) = \ln (x + 2)$

84.

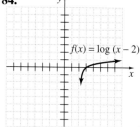

$f(x) = \log (x - 2)$

85.

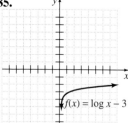

$f(x) = \log x - 3$

86.

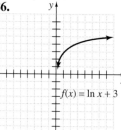

$f(x) = \ln x + 3$

87.

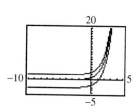

88.

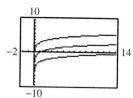

Graphing Calculator Explorations 10.6

1. 3.67 years or 3 years and 8 months

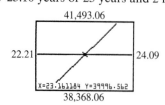

2. 15.40 years or 15 years and 5 months

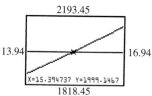

3. 23.16 years or 23 years and 2 months

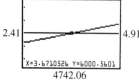

4. 11.87 years or 11 years and 10 months

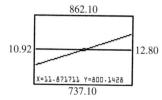

Exercise Set 10.6

61.
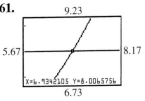
X=6.9342105 Y=8.0065756

62.
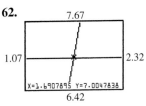
X=1.6907895 Y=7.0047838

63.
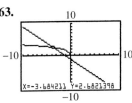
X=-3.684211 Y=2.6821398

64.
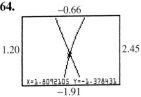
X=1.8092105 Y=-1.378431

65.
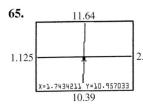
X=1.7434211 Y=10.957033

66.
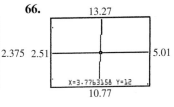
X=3.7763158 Y=12

67.
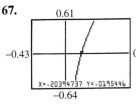
X=.20394737 Y=.0195446

68.
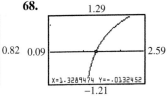
X=1.3289474 Y=-.0132452

Chapter 10 Review

22.
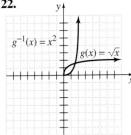

23.
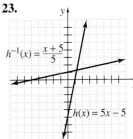

24.
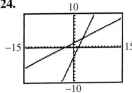

31.
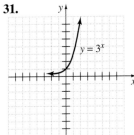

32.
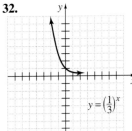

33.
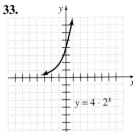

34.
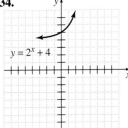

37.
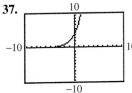

56.
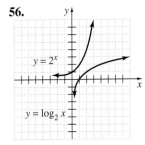

57.
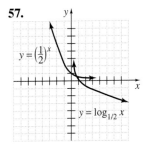

112.
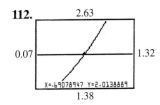
X=.69078947 Y=2.0138889

113. {2.82}

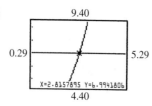

Chapter 10 Test

4.

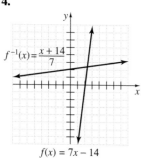

$f^{-1}(x) = \dfrac{x + 14}{7}$

$f(x) = 7x - 14$

23.

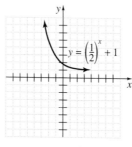

$y = \left(\dfrac{1}{2}\right)^x + 1$

24.

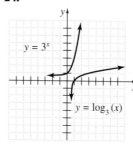

$y = 3^x$

$y = \log_3(x)$

30. {3.95}

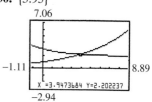

Chapter 10 Cumulative Review

4. *(Sec. 3.3, Ex. 6)*

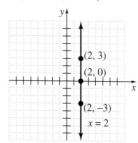

(2, 3)
(2, 0)
(2, −3)
$x = 2$

12. *(Sec. 5.9, Ex. 5)*

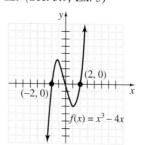

(2, 0)
(−2, 0)
$f(x) = x^3 - 4x$

17. *(Sec. 8.5, Ex. 7)*

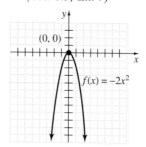

(0, 0)
$f(x) = -2x^2$

18. *(Sec. 9.1, Ex. 7)*

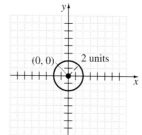

(0, 0)
2 units

CHAPTER 11

Sequences, Series, and the Binomial Theorem

Exercise Set 11.1

51.

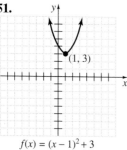

(1, 3)

$f(x) = (x - 1)^2 + 3$

52.

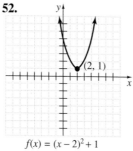

(2, 1)

$f(x) = (x - 2)^2 + 1$

53.

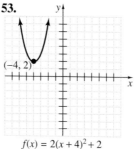

(−4, 2)

$f(x) = 2(x + 4)^2 + 2$

54.

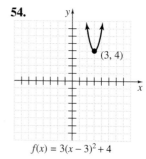

(3, 4)

$f(x) = 3(x - 3)^2 + 4$

Exercise Set 11.5

41. not one-to-one

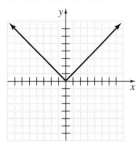

42. not one-to-one

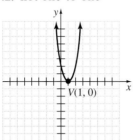

43. one-to-one

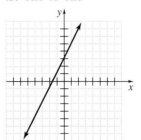

44. not one-to-one

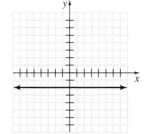

45. not one-to-one

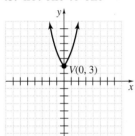

46. not one-to-one

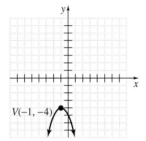

Chapter 11 Cumulative Review

15. *(Sec. 8.6, Ex. 2)*

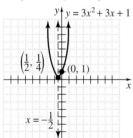

16. *(Sec. 9.2, Ex. 1)*

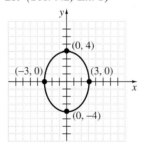

APPENDIX C

An Introduction to Using a Graphing Utility

Graphing Equations and Square Viewing Window Exercise Set

7.

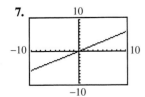

8.

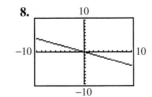

9.

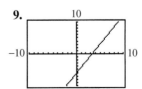

10.

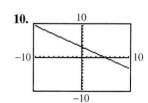

11.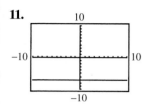

12.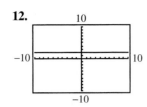

13.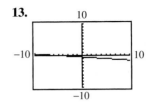

14.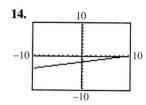

15.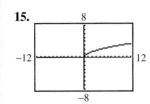

16.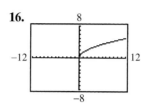

17.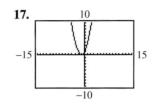

18.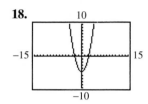

19.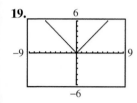

20.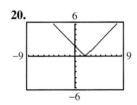

21.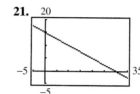

22.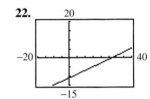

Notes to the Instructor

An Overview

Research on how students learn and regular advances in technology continue to affect how and what instructors teach. Many of these changes are reflected in the AMATYC Crossroads Guidelines and the NCTM Standards. The material that follows is intended to give a quick overview of several important related topics together with some suggestions to try with your students. Wherever possible, we have drawn on the experiences of and suggestions from instructors who have used these approaches successfully. The topics included in these Notes to the Instructor are:

- Interactive and Cooperative Learning
- Interpreting Graphs and Data
- Alternative Assessment
- Using Technology
- Helping Students Succeed

If you have suggestions for other topics or ideas that should be included, or you want to share your own successful approaches or instructional strategies, please call 1-800-435-3499, ext. 7404, and ask for a developmental mathematics editor, or e-mail ann_marie_jones@prenhall.com.

Interactive and Cooperative Learning

Many students absorb concepts well when working and learning together in small groups. For this reason you may wish to consider incorporating some interactive and cooperative learning strategies in your classroom. Types of interactive and cooperative learning activities are varied. They include discussion, skill practice and review, concept investigation, concept synthesis, and concept extension. *Intermediate Algebra*, Second Edition, includes end-of-section exercises, many of which are appropriate for group work. In addition, chapter-ending Group Activities combine several of the aspects listed above. No matter what type of activity you choose to use, here are some general points to keep in mind:

1. Clearly explain the goals of the cooperative learning activity to students. Encourage active participation by all members of the group.

2. As appropriate, describe how teams will be rewarded for the success of their group.

3. Emphasize that students should share equally in the responsibility for completing the tasks involved in the activity.

4. Stress that each member of the team is accountable for the other team members' understanding of concepts involved in the activity.

5. Highlight how the gains in competence of each individual will contribute to the success of the team.

One way to implement cooperative learning strategies is to structure the groups yourself rather than allowing groups to be self-selecting. Before assigning a cooperative learning activity, decide on the optimal group size for the activity and who should be in each group. Consider the following strategies:

■ Try to establish groups with an equal likelihood of success.

■ Predetermine groups with similar female-to-male ratios, ethnic make-up, and low-to-high-achiever ratios.

■ Groups can be formed with students' majors as a factor, with each group having a business major, nursing major, engineering major, and so on.

Methods for random assignment include the following:

■ Groups can be formed by counting off. For instance, if the class size is 32 and groups of size four are desired, then assign each of the first eight students numbers from 1 to 8, and repeat assigning numbers 1 to 8 in this fashion to the remaining students. Then all the students assigned the number 1 form a group, all the 2's form a group, all the 3's form a group, and so on.

■ Groups can be formed by using an appropriate subset of a deck of cards: all the aces in one group, kings in another, and so on.

■ Groups can be created according to the alphabetical order of first or last names.

■ Partners can be pointed out based on proximity—you two, you two, ...

■ Larger groups can be quickly selected based on location in the classroom—for instance, by rows or by proximity to each corner of the room.

Your role in interactive and cooperative learning can also include physical setup for the activity, such as rearranging the room, if necessary. Be sure to either provide any necessary special materials or ask students in advance to bring their own. You should be available for assistance with any portion of the activity, but remind students that their primary resource for aid should be the group itself. You may also suggest guidelines for working together and help create a positive atmosphere for learning. These may include the following.

GUIDELINES FOR WORKING TOGETHER

1. Agree on what your group must do and how you will get it done.

2. Be courteous and listen carefully to other group members.

3. Create an atmosphere that allows group members to be comfortable in asking for help when needed.

4. Remember that not everyone works at the same pace. Be patient. Offer your help. Receive help graciously.

5. Keep in mind that all contributions made by group members are valuable.

6. Ask for your instructor's help only when you have exhausted all other sources of assistance.

7. If you finish early, double-check your work. As appropriate, reflect on your work. Can you make a general rule about the solution or describe a real-world use for what you learned?

Some classroom strategies that have been used successfully during group activities also include the following:

■ Circulate around the room from group to group to monitor progress, steer students in the right direction, and give encouragement.

■ Consider giving each group a blank acetate/transparency and overhead pens at the start of the group activity. Each group can write its results on the transparency. You or your students can then share some of these with the class on an overhead projector.

■ If the group is working with a calculator, have everyone discuss and agree which keys to push and why. During the course, have students of all abilities take turns being in charge of using the calculator.

- Consider selecting some group activities with the goal of previewing and exploring a new topic, while you select other group activities to reinforce what has already been discussed.

- "Partner Quizzes" have been used successfully to try to lower test-taking anxiety and to encourage cooperation. (See Alternative Assessment for more information.)

- Consider assigning roles for members of a group of three or four, such as recorder, facilitator, checker, and reporter.

Individual accountability can be key to the success of an interactive or cooperative learning activity. To guard against "freeloading," you might consider asking each student in the class (or one randomly selected student from each group) to summarize briefly the findings of his or her group or the investigation process used by the group in a sentence or two. You might also consider judging the success of the group based on each individual's improvement. For instance, suppose you give a quiz on factoring and then incorporate a cooperative learning activity for practicing factoring skills. A follow-up quiz on factoring can be used to assess each individual's understanding of factoring. If each member of a group is able to achieve a predetermined goal, such as improving his or her score on the follow-up quiz as compared to the original quiz, then reward the entire group.

You might consider asking your students to report the results of an extensive group activity in the form of a lab report. A lab report might consist of three parts: (1) an introduction in which the title of the lab or group activity is given, the goals of the activity are explained, the necessary materials are listed, and the group members are recorded; (2) an organized description of the process the group used to accomplish the goals of the group activity, including any research performed, tables and/or graphs of data collected, and mathematical work completed; and (3) a summary of the group's findings and/or conclusions and, if appropriate, generalizations or conjectures to which the group was led during the

course of the activity. If you choose to use this lab reporting method, be sure to emphasize that work done to complete the lab must be done cooperatively. You might divide the activity into sections for which each group member is responsible. If so, the lab report should indicate who played which role and/or who wrote which sections of the lab report. Remind students that they are all responsible for the results given in their group's lab report. Suggest that after each student has made his or her written contribution to the report, the group should review all the contributions together to increase their understanding of the activity.

INTERPRETING GRAPHS AND DATA

Both the NCTM Standards and Addenda and the AMATYC Crossroads Guidelines encourage the inclusion of graphical and data interpretation in mathematics courses. This policy reflects changing needs in the workplace. Consider the following facts, which show a trend toward an increased need to analyze data and information.

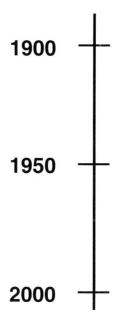

"During the early 1900's, 85% of our workers were in agriculture. Now agriculture involves less than 3% of the workforce."

"In 1950, 73% of U.S. employees worked in production or manufacturing. Now less than 15% do so."

"The Department of Labor estimates that by the year 2000 at least 44% of all workers will be in data services, for example, gathering, processing, retrieving, or analyzing information."

(*Source: The Employee Handbook of New Work Habits for a Radically Changing World,* by Price Pritchett. Used with full permission of Pritchett & Associates. All rights reserved.)

Intermediate Algebra, Second Edition, provides ample opportunity for several types of interpretation throughout the text. Circle graphs, line graphs, and bar graphs are used often to convey statistical information and are accompanied by questions that reinforce understanding.

Consider bringing into your classroom (or having your students bring in) other examples of current real data from newspapers, magazines, or the Internet. Note that one supplement available with this text is *The New York Times/Themes of the Times,* which is created new each year. Another supplement is an *Internet Guide.* In addition, a great source of real data is your students. Consider having students list information such as age, height, shoe size, number of hours worked at a job during the last week, and birth dates. This information could be used to develop a range of graphs and charts in which they have a personal interest.

There are many key ideas that could be discussed when graphs and data are created or analyzed. Many such ideas will be specific to the particular graph and the information it conveys. However, some more general ideas you may choose to keep in mind include the following.

READING AND USING GRAPHS

- To show trends over time, a line graph is often useful. The horizontal axis generally indicates the period of time. Look for relationships, if any, between the two quantities. (In a line graph, a break in the axes indicates that some numbers have been omitted. This break is usually shown by a jagged line near the origin.)

- To compare total amounts, such as results from a quantitative survey, a horizontal or vertical bar graph can be a good way to represent the information. One of the axes could be labeled with word phrases rather than with numbers.

- A circle graph (pie graph or pie chart) is shown in this text to display information visually by using a circle. The circle is divided into wedges called sectors. Often the sectors are labeled with percents.

Encourage your students to study the graphs in this text and to become acquainted with the power of communicating complex data graphically.

ALTERNATIVE ASSESSMENT

Alternative assessment provides a means for evaluating students' progress and/or understanding with tools other than the traditional testing methods. When used with traditional assessment methods, alternative methods can help give a truer picture of a student's comprehension of mathematical concepts. For instance, a student who experiences test anxiety may perform poorly on quizzes or exams but still fully understand the concepts being tested. An alternative assessment method should uncover that understanding. Several common forms of alternative assessment include topic interviews, self-evaluation, partner quizzes, homework notebooks, journals, portfolios, conceptual and writing exercises, research projects, and demonstrations.

Topic Interviews In this type of assessment, you schedule a short, one-on-one "interview" with each student to discuss particular mathematics concepts informally. For example, you and the student could discuss the graph of a particular linear equation, including whether the slope of the line is positive or negative, approximately where the x- and y-intercepts are located, a possible situation that the graph might describe, and so on.

Self-Evaluation You ask several questions to which students must respond confidentially in writing. These questions could allow students to give their own summaries of the day's class, list concepts they do not fully understand, and suggest ways in which material could be covered more clearly. Students could be asked to rate their own understanding of a topic and supply an explanation or demonstration of why they chose their ratings. Students could also be asked to contribute a quiz or exam question, with the accompanying answer, that they think does a fair job of gauging understanding. As an alternative, students could be asked only one short question. This can quickly provide you with feedback.

Partner Quizzes This type of assessment may take several forms. In one form, partners are asked to write a short quiz that they think adequately and fairly tests their knowledge from a particular section of material. Partners then exchange their quizzes, take their partners' quiz, and then grade the quiz they wrote. This activity requires students to write easily solvable quiz questions by working backward and requires them to be the "answer authority" when grading the quizzes they wrote. Analyzing the logic another student used in solving a problem can be quite educational to the quiz creator/grader. Another form of partner quizzes asks two students to work together to take a quiz you provide. Students then help each other understand the concepts required to answer the quiz questions correctly. These may or may not be open-book partner quizzes.

Homework Notebooks Students are asked to do homework assignments regularly in a notebook. Occasionally, you could check a few selected exercises for effort and correctness. Evaluation could include awarding some credit for effort (tried the exercise), partial credit for correct process, and credit for correct results. You may choose to award some credit or bonus points for overall completeness and neatness of homework assignments. Using notebooks provides feedback and can encourage students to attend class and try the exercises.

Ways to organize a notebook that you might suggest to students include writing the text section number or title at the top of each page. Another suggestion can be to leave room for class notes to be written beside the attempted exercises. That is, if a homework exercise is discussed during the next class, a student can make notes or modifications to his or her solution, if needed. When reviewing for a test, the notes can help students study more effectively. (One way for students to plan for notes is to draw a vertical line on each notebook page to divide the page into two unequal parts (columns). The

smaller part can represent 1/3 page and the larger part 2/3 of a page. Exercises can be written in the larger area, and any notes can be added in the smaller column.)

Journals Students are asked to keep an ongoing notebook or journal of thoughts about mathematics. Journal writing assignments might include such items as listing the main topics of the day's class, describing the concept that was easiest or hardest to understand, writing a "note" to a friend explaining how to solve a particular problem, listing questions that remain unanswered or writing a question that the student would like to ask the text's author about the topic, describing a real-life situation in which the concepts learned would be useful, creating a "crib sheet" of formulas that could be used on the next test, summarizing certain properties (such as the properties of inequalities or the properties of exponents) and giving an example of each one, or giving advice to students who will be taking this course during the next term.

Portfolios This type of assessment allows students to collect examples of their mathematical work, just as an artist would assemble an art portfolio. Portfolio assignments might include such items as selecting what the student thinks is the best example of his or her work during a week, creating an "original piece" such as an application problem written by the student, choosing a particularly well-drawn graph, picking an interesting piece of graphing technology output, adding examples of effective problem solving or student-corrected homework, or collecting newspaper/magazine articles or comic strips that have a connection to math. You might consider asking the students to supply written explanations of why they chose each piece to add to their portfolios.

Conceptual and Writing Exercises This text contains conceptual and writing exercises. These exercises often ask students to use two or more concepts together. Some require students to stop, think, and explain in their own words the concepts used in the exercises just completed. These exercises can be used alone or as part of a portfolio to assess understanding.

Research Projects Projects that require library research or the collection of data outside the classroom provide opportunities for students to synthesize what they have been learning. Projects may be assigned to groups or individuals. Some of the chapter-ending Group Activities in this text incorporate aspects of independent research.

Demonstrations There are a variety of ways to gauge students' understanding through demonstrations that could be made either to the entire class or to a small group. Students might be asked to report the results of a cooperative learning activity or the findings of a research project to the class, prepare an example or the solution to an exercise that will be demonstrated on the blackboard or overhead during the next class, learn a new graphing calculator function and teach its use to the class, or illustrate a mathematical concept with the use of manipulatives.

Face-to-face meetings or written submissions can address all these methods of alternative assessment. However, if e-mail is available to you and your students, it can offer another way for them to submit written work and ideas to you and for you to respond.

USING TECHNOLOGY

The use of a graphing calculator with *Intermediate Algebra,* Second Edition, is optional. Graphing Calculator Explorations and Scientific Calculator Explorations give instruction, at point of use, on how to use a calculator as a problem-solving tool in conjunction with the material of the section and allow students to practice particular calculator skills on a short set of exercises. Before the course starts,

decide on the role you wish technology to play in your course. Clearly explain your policies on the use of scientific and/or graphing calculators with homework, group work, quizzes, and tests.

You may also want to give students guidance on *when* to use calculators. Scientific calculators, graphing calculators, and computers are tools that can be used to help produce desired results more efficiently than can other alternatives.

For example, it is quicker to calculate 5^2 by using mental math than by using a calculator. However, whereas finding 219^3 by pencil and paper eventually yields an exact answer, a calculator can quickly and accurately produce an exact answer. It would be wise to turn to the calculator in this case to save time. Also, simple calculations or the determination of simple function values may be checked with the aid of a calculator. The calculator can be used to help recognize a pattern that reinforces the understanding of a rule or concept. Whether a scientific or graphing calculator is available, the process of discovering or recognizing patterns or trends can be enhanced with this technology. For example, with a graphing calculator the equations $y = x^2, y = (x + 2)^2, y = x^2 + 2,$ and $y = 2x^2$ can all be quickly graphed and compared.

A scientific calculator or graphing calculator can be used to help check results. For instance, if the solution to a linear equation in one variable is found to be 2, a calculator might be used to substitute 2 into the original equation. Or, on a graphing calculator, the equation could be solved graphically.

You might pose a series of questions to your students to check their understanding of the appropriate uses of technology. Consider one example. For the quadratic equation $f(x) = x^2 - 5x + 1,$

(a) would you use paper and pencil or a calculator to find $f(0)$? $f(1)$? $f(8)$? $f(99)$?
(b) how could you mentally approximate $f(99)$? How could you use this approximation to check an exact answer that you found by using a calculator? [Hint: Find $f(100)$ or even $(100)^2$.]
(c) is the calculator's answer to $f(99)$ likely to be an exact answer or an approximation? Is

the calculator's answer to $f(999,999)$ likely to be an exact answer or an approximation?

Demonstration Graphing technology provides a means for exploring mathematical concepts. You might consider using a graphing calculator as a demonstration tool. It is generally easier to show the effect of different values of b in the equation $y = mx + b$ with a graphing calculator or computer graphing tool than to graph the equations of many lines by hand. Similarly, graphical solutions to systems of two linear equations in two variables may be discovered by demonstrating graphs with a graphing utility. To involve students in the demonstration, you may occasionally want to have a student hold the calculator and press the keys under your direction as you circulate around the room and lead a discussion of the concept being illustrated.

Using Graphing Calculators During Class If students in your classroom have a mix of graphing calculator models, you might consider physically grouping students with the same models together so that they may answer one another's questions about the details of using a certain model. Also, encourage your students to refer to the owner's manuals that came with their calculators for help.

Rules for Rounding When students use a calculator, they often need to know how to round an answer to present it in the decimal form. Advise students that in this text, numbers are rounded *up* if the digit to the right of the given place value is a 5 or greater and rounded *down* if the digit to the right is a 4 or less. For example, 82.6259 rounded to two decimal places is rounded *up* to 82.63; 82.6259 rounded to the nearest tenth is rounded *down* to 82.6.

You may also want to inform students that many calculators carry internally more digits than are shown on the display. Encourage students not to round during intermediate steps of a calculation. Each rounding during an intermediate step of a calculation may increase the error of the final calculation.

HELPING STUDENTS SUCCEED

Convey a positive attitude, and encourage your students to have a positive attitude. Also, consider taking the first step to help your students know how to find help, if needed. In addition to your office location and office hours, provide students with your office phone number, e-mail as appropriate, and where to obtain help if you are not available. Consider encouraging students to exchange names and phone numbers with at least one other person in the class. This contact from class can be someone with whom a student can discuss the algebra concepts and exercises, so both students improve their understanding. Depending on the size of your class, spending a little time on student introductions, why they are taking the course, and what their major and interests are may assist them in finding study partners.

Resources Consider letting students know what other resources are available, such as a resource center or library with tutorial software, videotapes, or tutors. At the bookstore, print supplements such as a student solutions manual or study guide may also be available. Encourage students to seek help as soon as they have questions. There are many resources they can use to their advantage.

Study Strategies Having good study skills and strategies can be very helpful for students' success at the college level. Study skills particularly appropriate for mathematics are described in this text's Student Solutions Manual and Study Guide to help students. Referring students to these lists may prove useful. Topics covered include suggestions on preparing for class, attending class, taking notes, allocating study time, doing homework, checking exercises, preparing for tests, and taking tests.

In addition, you may choose to elaborate on time management. Having students prepare bar charts or lists on how their time was spent on one or several days can be helpful. Students could allocate time into categories such as sleeping, eating, showering and dressing, traveling, attending classes, studying, working at a job, watching television, exerising, performing household tasks, spending time with friends, and other. Examining the charts can help students understand some of the conflicting demands on their time and some trade-offs they may have to make. Encourage students to write down on something they will consult often (such as a calendar) how much time they plan to spend studying algebra. This will help them follow through on their plan.

Some students experience difficulty with math classes not because the material is difficult for them but as a result of the way the information is presented. Two possible areas of difficulty include differences in learning styles and proficiency in language.

Learning Styles Society is becoming more and more visually oriented. It shouldn't be surprising that more and more students are, to some degree, visual learners. *Intermediate Algebra,* Second Edition, presents a wide variety of tools for learning that students of many different learning styles should find useful. For instance, the steps for problem solving taught in *Intermediate Algebra* include "illustrate the problem" as a step in the problem-solving process. Other such tools include clear, worked-out examples, visual emphasis of important concepts and definitions, use of color in the explanations, and numerous graphs, including bar graphs and circle graphs, as well as diagrams. Kinesthetic learners benefit from "hands-on" activities and can benefit from working in groups. Real-world examples can make it easier for these students to relate to the underlying mathematics, as it becomes something real that they may have experienced.

Language Issues For students whose first language is not English, mathematics may become difficult due only to the lack of understanding of the language of instruction. For these students, care must be used when you introduce new mathematical terms; write out definitions and clearly label each with its term. Try to write down any terms or statements that require emphasis. It may be mutually beneficial to pair a student whose grasp of the English language is good but whose math skills are weak with a student for whom English is a second language but who is proficient in mathematics. A picture is worth a thousand words; drawing a diagram when you explain a concept will be extremely beneficial to any student who is struggling with the English explanation of the concept.

This is just a brief review of some key suggestions that have worked for other instructors around the United States. We hope that you will find some of the ideas useful and that we have stimulated you to think of new approaches appropriate to your students.

INDEX

Photo Credits

p. xxiv (top) Jim Williamson **(center)** Laima Druskis

CHAPTER 1 CO Larry Mulvehill/Rainbow

CHAPTER 2 CO Jeff Greenberg/PhotoEdit **p. 59** Howard Bluestein/Photo Researchers, Inc. **p. 64** Richard Pasley/Stock Boston **p. 67** Photo Researchers, Inc.

CHAPTER 3 CO Mathew Neal McVay/Tony Stone Images

CHAPTER 4 CO, p. 229 Tom Ives/The Stock Market

CHAPTER 5 CO Myrleen Ferguson/PhotoEdit **p. 250** Chris Butler/Science Photo Library/Photo Researchers, Inc.

CHAPTER 6 CO Jim Corwin/Photo Researchers, Inc.

CHAPTER 7 CO, p. 459 Richard Megna/Fundamental Photographs **p. 451** Steve Gottlieb/FPG International

CHAPTER 8 CO Ken Fisher/Tony Stone Images

CHAPTER 9 CO Bob Daemmrich Photos/The Image Works

CHAPTER 10 CO Nino Mascardi/The Image Bank

CHAPTER 11 CO Chuck Savage/The Stock Market

COMMON GRAPHS AND MODELS

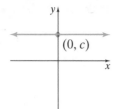

Horizontal Line;
Zero Slope
$y = c$

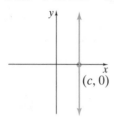

Vertical Line;
Undefined Slope
$x = c$

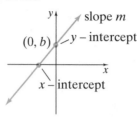

Linear Function;
Positive Slope
$y = mx + b; m > 0$

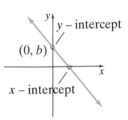

Linear Function;
Negative Slope
$y = mx + b; m < 0$

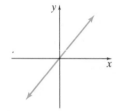

$y = x$

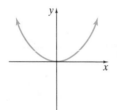

$y = x^2$

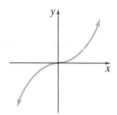

$y = x^3$

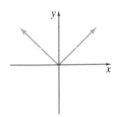

$y = |x|$

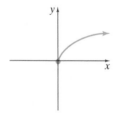

$y = \sqrt{x}; x \geq 0$

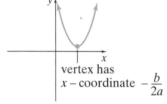

Quadratic Function
$y = ax^2 + bx + c; a \neq 0$
Parabola opens upward if $a > 0$
Parabola opens downward if $a < 0$

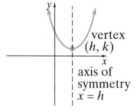

Quadratic Function
$y = a(x - h)^2 + k; a \neq 0$
Parabola opens upward if $a > 0$
Parabola opens downward if $a < 0$

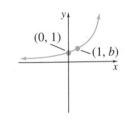

Exponential Function
$y = b^x$ for $b > 1$

SYSTEMS OF LINEAR EQUATIONS

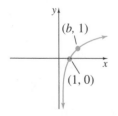

Logarithmic Function
$y = \log_b x$ for $b > 1$

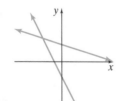

Independent and
consistent; one solution

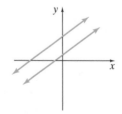

Independent and
inconsistent; no solution

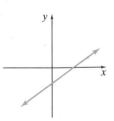

Dependent and consistent
infinitely many solutions